33=UAKN1013

SELECTED WORKS OF LOUIS NÉEL

Corrigendum to
SELECTED WORKS
OF
LOUIS NÉEL

Translation from the French authorized by the
CENTRE NATIONAL DE LA RECHERCHE SCIENTIFIQUE

Editor and Translation Coordinator

NICHOLAS KURTI

Department of Engineering Science
University of Oxford, U.K.

GORDON AND BREACH SCIENCE PUBLISHERS
New York London Paris Montreux Tokyo Melbourne

LOUIS NÉEL - CHRONOLOGY

Né à Lyon (3e), le 22 Novembre 1904

1924	Elève de l'Ecole Normale Supérieure
1928	Agrégé des Science Physiques
1928	Assistant à la Faculté des Sciences de Strasbourg
1932	Docteur ès Sciences
1934	Maître de Conférence à la Faculté de Science de Strasbourg
1937	Professeur à la Faculté de Science de Strasbourg
1945	Professeur à la Faculté de Science de Grenoble
1945	Directeur de Laboratoire d'Electrostatique et de Physique du Métal (CNRS)
1947	Membre du Comité Consultatif des Universités
1948	Médaille André Blondel
1952	Prix Holweck
1952	Conseiller scientifique de la Marine Nationale
1952	Membre du Directoire et du Conseil d'Administration du CNRS
1952	Membre de l'Académie des Sciences
1954	Directeur de l'Institut Polytechnique de Grenoble
1956	Directeur du Centre d'Etudes Nucléaires de Grenoble
1957	Président de la Société Française de Physique
1960	Représentant de la France au Comité Scientifique de l'OTAN
1963	Prix des Trois Physiciens
1963	Président de l'Union Internationale de Physique Pure et Appliquée
1965	Médaille d'Or du CNRS
1970	Prix Nobel de Physique
1971	Directeur du Laboratoire de Magnétisme de Grenoble (CNRS)
1971	Président de l'Institut National Polytechnique de Grenoble
1971	Grande Médaille d'Or de l'Electronique

Docteur "honoris causa" des Universités de Graz, Nottingham, Oxford, Louvain, Newcastle, Coimbra, Sherbrooke, Iassy, Madrid.

Membre Etranger des Académies des Sciences de l'URSS, de l'Académie Royale Néerlandaise, de l'Académie des "Naturforscher Leopoldina", de l'Académie de la République Populaire Roumaine, de la Royal Society (London), de l'American Academy of Arts and Sciences, de l'Académie Polonaise des Sciences.

Grand-Croix de la Légion d'Honneur.
Grand-Croix de l'Ordre National du Mérite.
Croix de Guerre avec Palmes.

PREFACE TO THE ENGLISH EDITION

I have the greatest admiration for Néel's scientific work and I am glad to supplement, on the occasion of this English edition of Néel's papers, Professor Chabbal's Preface. Néel added two new varieties of magnetism, antiferro- and ferrimagnetism, to the three previously existing types, namely dia-, para-, and ferromagnetism, and he did so with the elegant simplicity and truly physical intuition that is the hallmark of his work. It is comforting to encounter such qualities in these days of giant accelerators and computers.

Quite a few years ago I wrote an article entitled 'Nonmathematical Theoretical Physics' on the general theme that most of the important discoveries in theoretical physics were the result of the use of a new and imaginative, but essentially simple, concept that had eluded others. As an example, I told the story of what happened when I was writing my book Electric and Magnetic Susceptibilities in the early 1930s. It was well known at that time that a positive exchange integral can produce ferromagnetism, so, in my book, I correctly emphasized the fact that a negative exchange lowers the susceptibility. However, it had not occurred to me that there existed a critical temperature below which a negative exchange integral can produce antiferromagnetic ordering; I therefore wrote in my later review article that it was left to Néel to have the perspicacity to propose antiferromagnetic ordering. I reproach myself to this day for not having thought of it myself, though it is some consolation that the idea eluded all the specialists in ferromagnetism. I might also mention that the concept of antiferromagnetic ordering could have been introduced as early as 1907 by means of the negative Weiss molecular field constant.

In my review article I also emphasized the fact that many of the great theoretical discoveries were the result of supreme physical insight rather than of rigid mathematical formalism. To understand Pauli's exclusion principle no knowledge of arithmetic is necessary, an ability to count is sufficient. It is true that Néel's studies of the saturation magnetization of ferrites and other ferrimagnetic materials also required an ability to add and substract. However his success was not simply due to his knowledge of arithmetics, but to his having had the imagination and the courage to propose, well before it could be proved experimentally, that the ferrites have inverse spinel structure.

Néel's work helps to maintain and, indeed, to enhance the distinguished position in the field of magnetism occupied by France for many years, as exemplified by names like Ampère, Pierre Curie, Langevin, Weiss, Jean Becquerel, Brillouin, Kastler and Néel. The name of the Place de l'Etoile may be changed, but 'Curie-point' and 'Néel-point' will forever belong in the terminology of physics.

J.H. van Vleck

Editor's Note

At the end of August 1980, when plans for this book were taking shape, I wrote to Professor J.H. van Vleck, a good friend and colleague of Professor Néel's, to ask him whether he would be willing to write an additional preface to the English edition of Néel's scientific work. In his reply, dated September 20, 1980, van Vleck said that at the age of 81 he was trying to keep all new commitments to a minimum but suggested that I could use as the basis of a preface the tribute he sent on the occasion of Néel's Nobel prize (see *Bull. Soc. Française de Physique*, April–May 1971).

Professor van Vleck died on 27 October 1980 and did not see the draft of this preface. But it was clearly his wish to be associated with this volume which, it is hoped, will be regarded by the English speaking physics community as a modest but grateful and warm tribute to Professor Néel.

PREFACE TO THE FRENCH EDITION

The broad outlines of Louis Néel's scientific work are generally well known, not only to physicists but also to the public thanks to the awarding of the 1970 Nobel Prize in Physics to this brilliant physicist. The publication by the CNRS of this collection of Louis Néel's works makes it possible to gain a more intimate knowledge of the development of his scientific ideas. The reader can thus grasp an entire cross-section of scientific research and experience the writing of an important page of science.

All the characteristics of a great scientist's development emerge from this collection. First, the urge to understand the innermost behaviour of matter, rejecting all dogmatism and perceiving the phenomena as they should be on the microscopic scale; hence, the theory of antiferromagnetism. Then the increasing concern about sticking to physical reality and the experimental search for materials with new properties; here we have the ferrimagnetic substances. And, finally, the translation of these ideas to the world outside physics, leading to applications in many fields. Thus, Louis Néel made important contributions to the interpretation of the magnetic memory of rocks, which ultimately provided a key to the phenomenon of continental drift. He also devised, at the beginning of World War II, a method of degaussing ships, which was used successfully to protect the French fleet against magnetic mines. This sketch, brief as it is, of an impressive list of achievements may at least bring out the various facets of Louis Néel's scientific influence.

I do not wish to anticipate the unfolding of the progress of Louis Néel's thoughts; however, at his request, I should like to recall the various roles of the CNRS in the blossoming of his remarkable career. As early as 1931, when he was a young research worker in Pierre Weiss's laboratory in Strasbourg, Louis Néel was awarded a research bursary by the Caisse Nationale des Sciences, the precursor of the CNRS. Such grants were made to encourage and facilitate the scientific work of young research workers in important laboratories.

When Néel settled in Grenoble after the 1940 armistice, the CNRS continued to support him in his endeavour to start new experiments. In 1945, the CNRS decided to put this support on an institutional basis by creating a laboratory of electrostatics and metal physics and making him at once its Director. Fifteen years later in view of the volume and the diversity of the laboratory's researches the CNRS decided to create, starting from this initial laboratory, three new Laboratories, one of them being the Laboratory for Magnetism under Néel's direction.

The way the CNRS acts in developing scientific research can well be exemplified by the successive stages of a career. It all begins with finding a young scientific talent whom the CNRS provides with the means of developing his own research, possibly by appointing him to a full-time research post and placing him with a group renowned for the quality of its scientific output. As the work initiated by the research worker achieves international influence, the CNRS makes it support institutional by creating first a research team and then, possibly, a laboratory. Finally, when the research requirements demand it, the CNRS undertakes the introduction of big scientific equipment. Thus, in Grenoble, with its host of solid state physics laboratories, the CNRS established a national centre for intense magnetic fields and participated in setting up the high neutron flux reactor of the Laue-Langevin Institute.

Such a development of scientific research does not come about spontaneously, it is the result of protracted perseverance maintained despite the often conflicting trends set by the CNRS which relies on the best scientists to define a policy and to realize it. Here Néel showed organizational talents fully up to the level of his scientific achievements.

By accepting heavier and heavier responsibilities for science policy and administration, Louis Néel has built up the scientific complex of Grenoble, indisputably one of the top centres of scientific research in France. In 1939 he became *chargé de mission* to advise the management of the CNRS. From 1949 to 1970 he was a member of the national committee for scientific research and of the *Directoire*, later the Council of Administration of CNRS.

I hope that I have clearly demonstrated, in accordance with Louis Néel's wishes, the symbiosis of his work and the actions of the CNRS. The development of the function of research is a complex process that is achieved thanks to the researchers, the administrators, and the *batisseurs*. Louis Néel managed to fulfil these functions remarkably well in each case, and I now leave it to the reader to rediscover the most distinctive aspects of the work of a great French scientist.

Robert Chabbal
Directeur Général du CNRS
(Translation by N. Kurti)

TRANSLATORS

J.M.D. Coey, Ph.D., F.Inst.P.
Professor, Department of Pure and Applied Physics
Trinity College
University of Dublin
Ireland

E. Geissler, D.Phil.
Laboratoire de Spectrométrie Physique
Université Joseph Fourier
38402 Grenoble, France

K. Hoselitz, Ph.D., F.Inst.P., F.I.E.E.
Formerly Director Mullard (now Philips) Research Laboratories
Redhill, England
Address:
4 Rochester Gardens
Croydon
Surrey CR0 5NN, England

P.D. Johnston, D. Phil., M.Inst.P.
Formerly with OECD Nuclear Energy Agency, Paris
Now with the European Commission, Brussels
Address:
4 Wrights Walk
Mortlake
London SW14, England

G.A. Jones, Ph.D., F.Inst.P.
Lecturer, Department of Pure and Applied Physics
University of Salford
Salford M5 4WT, England

R.J. Nicholas, D.Phil., M.Inst.P.
Lecturer in Physics
Clarendon Laboratory
Oxford OX1 3PU, England

P.W. Readman, Ph.D.
School of Cosmic Physics, Dublin Institute for Advanced Studies
Dublin, Ireland

P. Rhodes, Ph.D., M.Inst.P.
Senior Lecturer, Department of Physics
University of Leeds
Leeds, England

R.H. Wade, Ph.D., F.Inst.P.
Departement de Recherche Fondamentale
Centre d'Etudes Nucléaires de Grenoble
38041 Grenoble Cedex, France

CONTENTS

Preface to the English Edition

Preface to the French Edition

Translators

Louis Néel - Chronology

I	General theory of magnetism A7, A17, A40, A22, A35, C5, C29. (Translator: P. Rhodes)	1
II	Ferrimagnetism A61 (Translator P Johnston) A70 A72 A77 A94 C42 (Translator: K. Hoselitz)	61
III	Antiferromagnetism A29, A112, A113, A114, A115 (Translator: J.M.D. Coey)	103
IV	Magnetic interactions and their variations A18, A23, A33, A30 (Translator: P. Johnston)	123
V	Approach to saturation in ferromagnets A62, A63, A93 (Translator: G.A. Jones)	159
VI	Elementary domains; phases and modes A52, C29, A58, A127, A85 (Translator: R.H. Wade)	179
VII	Bloch and Néel walls A50, A97, A118 (Translator: R.H. Wade)	219
VIII	Weak fields and anhysteretic magnetization A45, A44, A47, A48, A144 (Translator: G.A. Jones)	239
IX	Coercive force; magnets B1, A51, A57, A59, A60, A91 (Translator: K. Hoselitz)	271
X	Random fields; reptation and tilting A101, A102, A106, A107, A109, A108 (Translator: E. Geissler)	313
XI	Magnetic after-effects A73, A78, A84 (Translator: P.W. Readman)	341
XII	Directional order; irradiation A92, A119 (Translator: E. Geissler)	385
XIII	Rocks and baked clays A69 (Translator: J. M. Coey), C27 (originally published in English)	405
XIV	Surface problems A121, A122 (Translator: R.J. Nicholas)	457
XV	Antiferromagnetic hysteresis A142 (Translator: R.J. Nicholas)	467
XVI	Apparatus and Techniques A11, A31, A39 (Translator: R.J. Nicholas)	489
XVII	Experimental work A7, A9, A20, A36 (Translator: R.J. Nicholas)	497
	Index	519

LIST OF LOUIS NÉEL'S SCIENTIFIC PUBLICATIONS

Out of a total of just over 200 papers 65 were selected as representing the most interesting and most important aspects of Néel's work. These papers are indicated by their page-numbers in this volume.

In several cases the original abstracts which precede the papers are supplemented by Notes or Remarks or Summaries, usually followed by a date in brackets. They were specially written for the French Edition of this volume.

A. ORIGINAL ARTICLES
B. PATENTS
C. PUBLICATIONS OF A GENERAL CHARACTER RELEVANT TO MAGNETISM
D. MISCELLANEOUS PUBLICATIONS (up to 1970)

LISTE DES PUBLICATIONS SCIENTIFIQUES
DE LOUIS NÉEL

A – ARTICLES ORIGINAUX

A 1 – Action du champ magnétique sur quelques raies d'étincelle du spectre du chlore, Diplôme d'Etudes supérieures, Paris, 1927 (Les Presses Modernes, Paris, 1927).
Comm. Soc. Fr. Phys., n° 272, 1929, p. 27 S.

A 2 – Le calcul statistique de l'aimantation rémanente.
Comm. Soc. Fr. Phys., Bull. n° 275, 1929, p. 55 S et *J. Phys. Rad.*, 10, 1929, p. 263–266.

A 3 – Le champ moléculaire et les deux points de Curie dans les substances ferromagnétiques.
Comm. Soc. Fr. Phys., Bull. n° 295, 1930, p. 113 S.

A 4 – Le champ moléculaire dans les alliages.
Comm. Soc. Fr. Phys., Bull. n° 300, 1930, p. 159 S.

A 5 – Propriétés magnétiques du fer au-dessus du point de Curie.
C.R. Ac. Sc., 193, 1931, p. 1325–1326.

A 6 – Susceptibilité magnétique du fer à quelques degrés au-dessus du point de Curie.
C.R. Ac. Sc. 194, 1932, p. 263–265.

A 7 – Influence des fluctuations du champ moléculaire sur les propriétés magnétiques des corps. **3**
Ann. de Phys., 17, 1932, p. 5–105.

A 8 – L'aimantation du fer immédiatement au-dessus du point de Curie.
Comm. Soc. Fr. Phys., Bull. n° 317, 1932, p. 14 S.

A 9 – Propriétés magnétiques du manganèse et du chrome en solution solide étendue. **501**
J. Phys. Rad., 3, 1932, p. 160–171.

A 10 – Susceptibilité magnétique de la vapeur de soufre.
C.R. Ac. Sc., 194, 1932, p. 2035–2037.

A 11 – Potentiomètre de haute précision pour la mesure des forces électromotrices **491**
thermoélectriques.
Comm. Soc. Fr. Phys., Bull. n° 334, 1933, p. 20–21 S.

A 12 – Calcul des constantes de la loi de Curie-Weiss généralisée.
Comm. Soc. Fr. Phys., Bull. n° 336, 1933, p. 43–44 S.

A 13 – Nouvelle forme donnée au pendule de translation pour la mesure des susceptibilités magnétiques. Emploi de cellules photoélectriques pour le repérage du zéro.
Comm. Soc. Fr. Phys., Bull. n° 342, 1933, p. 118 S.

A 14 – Calcul de la susceptibilité du nickel au voisinage du point de Curie.
C.R. Ac. Sc., 197, 1933, p. 1195–1197.

A 15 – Les fluctuations du champ moléculaire et l'équation d'état magnétique.
    C.R. Ac. Sc., 197, 1933, p. 1310–1312.

A 16 – Sur l'équation d'état magnétique du nickel.
    Comm. Soc. Fr. Phys., Bull. n° 349, 1933, p. 13 S.

A 17 – L'équation d'état et le porteur élémentaire de magnétisme du nickel. **26**
    J. Phys. Rad., 5, 1934, p. 104–120.

A 18 – Sur l'interprétation des propriétés magnétique des alliages. **125**
    C.R. Ac. Sc., 198, 1934, p. 1311–1313.

A 19 – Le paramagnétisme des solutions solides. Le cas des nickel-cobalt.
    Comm. Soc. Fr. Phys., Bull. n° 335, 1934, p. 87–88 S.

A 20 – Propriétés magnétiques du nickel pur à proximité du point de Curie. **509**
    J. Phys. Rad., 6, 1935, p. 27–34.

A 21 – Champ moléculaire et distances interatomiques.
    Comm. Soc. Fr. Phys., Bull. n° 374, 1935, p. 93–95 S.

A 22 – Nombre des électrons qui contribuent au paramagnétisme du nickel. **44**
    C.R. Ac. Sc., 201, 1935, p. 135–137.

A 23 – Propriétés magnétiques de l'état métallique et énergie d'interaction entre atomes **127**
    magnétiques.
    Ann. de Phys., 5, 1936, p. 232–279.

A 24 – Les anomalies de dilatation du nickel et du fer.
    Comm. Soc. Fr. Phys., Bull. n° 385, 1936, p. 45–46 S.

A 25 – Théorie des anomalies de volume des substances ferromagnétiques.
    C.R. Ac. Sc., 202, 1936, p. 742–744.

A 26 – Influence de la variation thermique du champ moléculaire sur la constante de Curie.
    C.R. Ac. Sc., 202, 1936, p. 1038–1040.

A 27 – Essai d'interprétation du moment à saturation des ferromagnétiques.
    C.R. Ac. Sc., 202, 1936, p. 1269–1271.

A 28 – La variation thermique du champ moléculaire.
    Comm. Soc. Fr. Phys., Bull. n° 387, 1936, p. 70 S.

A 29 – Théorie du paramagnétisme constant; application au manganèse. **105**
    C.R. Ac. Sc., 203, p. 304–306.

A 30 – Etudes sur le moment et le champ moléculaire des ferromagnétiques. **138**
    Ann. de Phys, 8, 1937, p. 237–308.

A 31 – Joulemètre (avec B. Persoz). **493**
    Comm. Soc. Fr. Phys., Bull. n° 409, 1937, p. 121–122 S.

A 32 – Points de Curie des métaux du groupe yttrique.
    Comm. Soc. Fr. Phys., Bull. n° 416, 1938, p. 50–51 S.

A 33 – Interprétation du point de Curie paramagnétique des éléments du groupe des **136**
    terres rares.
    C.R. Ac. Sc., 206, 1938, p. 49–51.

A 34 – Remarques sur la chaleur spécifique des ferromagnétiques.
    Comm. Soc. Fr. Phys., Bull. n° 421, 1938, p. 104–105 S.

A 35 – Paramagnétisme d'électrons dans une bande rectangulaire. **46**
    C.R. Ac. Sc., 206, 1938, p. 471–473.

A 36 – Application au nickel d'une nouvelle méthode de mesure des chaleurs spécifiques vraies. **517**
    C.R. Ac. Sc., 207, 1938, p. 1384–1385.

A 37 – Chaleur spécifique et fluctuations du champ moléculaire.
      C.R. Ac. Sc., 208, 1939, p. 177–179.

A 38 – Beziehungen zwischen den Curie-punkten der Metalle der Yttererden.
      Z. für Elektrochemie, 45, 1939, p. 378–379.

A 39 – Nouvelle méthode de mesure des chaleurs spécifiques vraies à haute température (avec B. Persoz). **495**
      C.R. Ac. Sc., 208, 1939, p. 642–643.

A 40 – Remarques sur les propriétés magnétiques d'un gaz obéissant à la statistique de Bose-Einstein. **43**
      Comm. Soc. Fr. Phys., Bull. n° 434, 1939, p. 95–96 S.

A 41 – Neutralisation.
      Centre de Recherches de la Marine, Publ. n° 606 (66 p.), Alger, 6 Aout 1940.

A 42 – La forme des noyaux d'inversion dans les grandes discontinuités Barkausen.
      Cah. Phys., n°4, 1941, p. 57–61.

A 43 – Théorie des lois de Rayleigh, du champ coercitif et de l'aimantation idéale.
      Comm. Soc. Fr. Phys., Cah, Phys., n° 8, 1942, p. 65.

A 44 – Les lois de Lord Rayleigh pour certains aciers à aimants (avec G. Glinksi). **259**
      Cah. Phys., n° 11, 1942, p. 73–74.

A 45 – Théorie des lois d'aimantation de Lord Rayleigh. **241**
      1ère partie : Les déplacements d'une paroi isolée.
          Cah. Phys., n° 12, 1942, p. 1–20.
      2ème partie : Multiples domaines et champ coercitif.
          Cah. Phys., n° 13, 1943, p. 18–30.

A 46 – Nouvelles données expérimentales sur l'aimantation dans les champs faibles des substances dures. Comparaison avec la théorie.
      Cah. Phys., n° 14, 1943, p. 69.

A 47 – Théorie de l'influence du champ démagnétisant sur l'aimantation anhystérétique. **260**
      Cah. Phys., n° 17, 1943, p. 47–50.

A 48 – Aimantation anhystérétique et champ démagnétisant : l'expérience et la théorie (avec R. Forrer, N. Janet et R. Baffie). **263**
      Cah. Phys., n° 17, 1943, p. 51–56.

A 49 – Domaines élémentaires et diagrammes de poudres d'un monocristal de fer.
      J. Phys. Rad., 4, 1943, p. 31 S et p. 319.

A 50 – Quelques propriétés des parois des domaines élémentaires ferromagnétiques. **221**
      Cah Phys., n° 25, 1944, p. 1–20.

A 51 – Effet des cavités et des inclusions sur le champ coercitif. **274**
      Cah, Phys., n° 25, 1944, p. 21–44.

A 52 – Les lois de l'aimantation et de la subdivision en domaines élémentaires d'un monocristal de fer. **181**
      1ère partie : Différents modes d'aimantation d'un monocristal.
          J. Phys. Rad., 5, 1944, p. 241–251.
      2ème partie : Forme et orientation des domaines élémentaires.
      3ème partie : Structure secondaire superficielle et épaisseur des feuillets.
          J. Phys. Rad., 5, 1944, p. 265–279.

A 53 – Loi d'approche à la saturation d'un ferromagnétique à aimantation spontanée irrégulière.
      C.R. Ac. Sc., 220, 1945, p. 738–740.

A 54 – Loi d'approche à la saturation et forces magnétocristallines dans les ferromagnétiques cubiques monocristallins.
      C.R. Ac. Sc., 220, 1945, p. 814–815.

A 55 – Le calcul du champ coercitif d'après les théories de Becker et de Kersten.
　　　　C.R. Ac. Sc., *223*, 1946, p. 141–142.

A 56 – Une nouvelle théorie générale du champ coercitif.
　　　　C.R. Ac. Sc., *223*, 1946, p. 198.

A 57 – Bases d'une nouvelle théorie générale du champ coercitif. **286**
　　　　Ann. Univ. Grenoble, *22*, 1946, p. 299–343.

A 58 – Propriétés d'un ferromagnétique cubique en grains fins. **206**
　　　　C.R. Ac. Sc., *224*, 1947, p. 1488–1490.

A 59 – Le champ coercitif d'une poudre ferromagnétique à grains anisotropes. **304**
　　　　C.R. Ac. Sc., *224*, 1947, p. 1550–1551.

A 60 – Théorie de l'anisotropie de certains aciers à aimants traités à chaud dans un champ **306**
　　　　magnétique.
　　　　C.R. Ac. Sc., *225*, 1947, p. 109–111.

A 61 – Propriétés magnétiques des ferrites. Ferrimagnétisme et Antiferromagnétisme. **63**
　　　　Ann. de Phys., *3*, 1948, p. 137–198.

A 62 – La loi d'approche en a : H et une nouvelle théorie de la dureté magnétique. **161**
　　　　J. Phys. Rad., *9*, 1948, p. 184–192.

A 63 – Relation entre la constante d'anisotropie et la loi d'approche à la saturation des **170**
　　　　ferromagnétiques.
　　　　J. Phys. Rad., *9*, 1948, p. 193–199.

A 64 – Les propriétés magnétiques du sesquioxyde de fer rhomboèdrique.
　　　　C.R. Ac. Sc., *228*, 1949, p. 64–66.

A 65 – Influence des fluctuations thermiques sur l'aimantation de grains ferromagnétiques très fins.
　　　　C.R. Ac. Sc., *228*, 1949, p. 664–668.

A 66 – Influences des fluctuations thermiques sur l'aimantation des substances ferromagnétiques
　　　　massives.
　　　　C.R. Ac. Sc., *228*, 1949, p. 1210–1212.

A 67 – Nouvelle théorie du champ coercitif.
　　　　Physica, *15*, 1949, p. 225–234.

A 68 – Essai d'interprétation des propriétés du sesquioxyde de fer rhomboèdrique.
　　　　Ann. de Phys., *4*, 1949, p. 249–268.

A 69 – Théorie du trainage magnétique des ferromagnétiques en grains fins avec **407**
　　　　application aux terres cuites.
　　　　Ann. Géophys., *5*, 1949, p. 99–136.

A 70 – Aimantation à saturation de certains ferrites. **93**
　　　　C.R. Ac. Sc., *230*, 1950, p. 190–192.

A 71 – Les coefficients de champ moléculaire des ferrites mixtes de nickel et de zinc (avec P. Brochet).
　　　　C.R. Ac. Sc., *230*, 1950, p. 280–282.

A 72 – Aimantation à saturation des ferrites mixtes de nickel et de zinc. **95**
　　　　C.R. Ac. Sc., *230*, 1950, p. 375–377.

A 73 – Théorie du trainage magnétique des substances massives dans le domaine de Rayleigh. **343**
　　　　J. Phys. Rad., *11*, 1950, p. 49–61.

A 74 – Preuves expérimentales du ferrimagnétisme et de l'antiferromagnétisme.
　　　　Ann. Inst. Fourier (Grenoble), 1, 1949, p. 163–183.

A 75 – Sur le ferromagnétisme des ferrites ou ferrimagnétisme.
　　　　Physica, *16*, 1950, p. 350–351.

A 76 – Magnetische Eigenschaften der Ferrite und der Ferrimagnetismus.
　　　　Z. für Anorg. Chemie, *262*, 1950, p. 175–184.

A 77 – Effet de la dilatation thermique sur la valeur de la constante de Curie des ferrites. **97**
　　　　J. Phys. Rad., *12*, 1951, p. 258–259.

A 78 – Le trainage magnétique. **356**
J. Phys. Rad., *12*, 1951, p. 339–351.

A 79 – Le signe de l'aimantation thermorémanente des roches.
J. Phys. Rad., *12*, 1951, p. 11 S.

A 80 – L'inversion de l'aimantation permanente des roches.
Ann. Géophys., *7*, 1951, p. 90–102.

A 81 – Quelques résultats nouveaux sur l'aimantation des laves en sens inverse du champ magnétique terrestre.
J. Phys. Rad., *12*, 1952, p. 953.

A 82 – Etude thermomagnétique d'un monocristal de Fe_2O_3 alpha (avec R. Pauthenet).
C.R. Ac. Sc., *234*, 1952, p. 2172–2174.

A 83 – Confirmation expérimentale d'un mécanisme d'inversion de l'aimantation thermorémanente.
C.R. Ac. Sc., *234*, 1952, p. 1991–1993.

A 84 – Théorie du trainage magnétique de diffusion. **369**
J. Phys. Rad., *13*, 1952, p. 249–263.

A 85 – Influence de la subdivision en domaines élémentaires sur la perméabilité en haute **211**
fréquence des corps ferromagnétiques conducteurs.
Ann. Inst. Fourier (Grenoble), *3*, 1951, p. 301.

A 86 – Les propriétés magnétiques de certains bisulfures (avec R. Benoit).
C.R. Ac. Sc., *237*, 1953, p. 444–447.

A 87 – Les processus d'aimantation des corps ferromagnétiques en très haute fréquence.
Ann. Inst. Polytechn. Grenoble, *2*, 1953, p. 5.

A 88 – L'anisotropie superficielle des substances ferromagnétiques.
C.R. Ac. Sc., *237*, 1953, p. 1468–1470.

A 89 – Les surstructures d'orientation.
C.R. Ac. Sc., *237*, 1953, p. 1613–1616.

A 90 – Surstructures d'orientation dues aux déformations mécaniques.
C.R. Ac. Sc., *238*, 1954, p. 305–308.

A 91 – Remarques sur les propriétés magnétiques des substances dures. **308**
Appl. Sci. Res., B, *4*, 1954, p. 13–24.

A 92 – Anisotropie superficielle et surstructures d'orientation magnétique. **387**
J. Phys. Rad., *15*, 1954, p. 225–239.

A 93 – L'approche à la saturation de la magnétostriction. **177**
J. Phys. Rad., *15*, 1954, p. 376–378.

A 94 – Sur l'interprétation des propriétés magnétiques des ferrites de terres rares. **99**
C.R. Ac. Sc., *239*, 1954, p. 8–11.

A 95 – La loi en $T^{3/2}$ de l'approche à la saturation des substances ferromagnétiques.
J. Phys. Rad., *15*, 1954, p. 74–75 S.

A 96 – Le problème du permalloy.
J. Phys. Rad., *15*, 1954, p. 92–93 S.

A 97 – Energie des parois de Bloch dans les couches minces. **232**
C.R. Ac. Sc., *241*, 1955, p. 533–538.

A 98 – Remarques sur la théorie des propriétés magnétiques des couches minces et des grains fins.
J. Phys. Rad., *17*, 1956, p. 250–255.

A 99 – Métamagnétisme et propriétés magnétiques de $MnAu_2$.
C.R. Ac. Sc., *242*, 1956, p. 1549–1554.

A 100 – Interprétation des propriétés magnétiques du dysprosium et de l'erbium.
  C.R. Ac. Sc., 242, 1956, p. 1824–1828.

A 101 – Action de champs magnétiques successifs de caractère aléatoire sur l'aimantation des substances ferromagnétiques. **315**
  C.R. Ac. Sc., 244, 1957, p. 2441–2446.

A 102 – Essai d'interprétation de la reptation des cycles d'hystérésis. **318**
  C.R. Ac. Sc., 244, 1957, p. 2668–2674.

A 103 – Interprétation des propriétés magnétiques du dysprosium et de l'erbium.
  J. Phys. Rad., 18, 1957, p. 32 S.

A 104 – Les métamagnétiques ou substances antiferromagnétiques à champ seuil (en russe).
  Isvest. Akad. Nauk. SSSR, Moscou, 21, 1957, p. 890–903.

A 105 – Les métamagnétiques ou substances antiferromagnétiques à champ seuil.
  Nuovo Cimento, 6, 1957, Suppl. p. 942–960.

A 106 – Sur les effects d'un couplage entre grains ferromagnétiques doués d'hystérésis. **322**
  C.R. Ac. Sc., 246, 1958, p. 2313–2319.

A 107 – Couplage entre domaines élémentaires ferromagnétiques : effect de bascule. **326**
  C.R. Ac. Sc., 246, 1958, p. 2963–2968.

A 108 – Action combinée des champs aléatoires de reptation et de fluctuations thermiques. **336**
  C.R. Ac. Sc., 248, 1959, p. 2676–2681.

A 109 – Sur les effets des interactions entre les domaines élémentaires : bascule et reptation. **329**
  J. Phys. Rad., 20, 1959, p. 215–221.

A 110 – Directional order and diffusion after-effect.
  J. Appl. Phys., Suppl., 30, 1959, p. 3–8 S.

A 111 – On the laws of magnetization of ferromagnetic single crystals and polycrystals. Application to uniaxial compounds (avec R. Pauthenet, G. Rimet et V.S. Giron).
  J. Appl. Phys., 31, 1960, p. 27–29 S.

A 112 – Superparamagnétisme de grains très fins antiferromagnétiques. **107**
  C.R. Ac. Sc., 252, 1961, p. 4075–4080.

A 113 – Superposition de l'antiferromagnétisme et du superparamagnétisme dans un grain très fin. **111**
  C.R. Ac. Sc., 253, 1961, p. 9–12.

A 114 – Superantiferromagnétisme dans les grains très fins. **114**
  C.R. Ac. Sc., 253, 1961, p. 203–208.

A 115 – Sur le calcul de la susceptibilité additionnelle superantiferromagnétique des grains fins et sa variation thermique. **118**
  C.R. Ac. Sc., 253, 1961, p. 1286–1291.

A 116 – Influence des couplages magnétocristallins sur le superantiferromagnétisme des grains fins.
  C.R. Ac. Sc., 254, 1962, p. 598–602.

A 117 – Remarques sur les interactions magnétiques.
  J. Phys. Rad., 23, 1962, p. 449–452.

A 118 – Nouvelle méthode de mesure de l'énergie des parois de Bloch. **234**
  C.R. Ac. Sc., 254, 1962, p. 2891–2896.

A 119 – Etablissement d'une structure ordonnée FeNi par irradiation aux neutrons (avec J. Paulevé, D. Dautreppe, et J. Laugier). **403**
  C.R. Ac. Sc., 254, 1962, p. 965–968.

A 120 – Propriétés magnétiques des grains fins antiferromagnétiques, superparamagnétisme et superantiferromagnétisme.
  J. Phys. Soc. Japan, 17, Suppl. B 1, 1962, p. 676–685.

A 121 – Sur un problème de magnétostatique relatif à des couches minces ferromagnétiques. **459**
C.R. Ac. Sc., 255, 1962, p. 1545–1550.

A 122 – Sur un nouveau mode de couplage entre les aimantations de deux couches minces ferromagnétiques. **463**
C.R. Ac. Sc., 255, 1962, p. 1676–1681.

A 123 – Irradiation aux électrons de I MeV d'un alliage Fe–Ni (50–50) (avec J. Paulevé, D. Dautreppe et W. Chambon).
C.R. Ac. Sc., 255, 1962, p. 2037–2039.

A 124 – A new method to measure directly the 180° Bloch wall energy (avec P. Brissonneau et R. Aléonard).
J. Appl. Phys., 34, 4, 1963, p. 1321.

A 125 – Défauts ponctuels dans les solides ferromagnétiques et ordre directionnel.
J. Phys. Rad., 24, 1963, p. 513–516.

A 126 – Le paramagnétisme des alliages fer-cobalt (avec Y. Barnier et R. Pauthenet).
C.R. Ac. Sc., 256, 1963, p. 5011–5015.

A 127 – Energie magnétocristalline d'un monocristal subdivisé en cristallites quadratiques. **208**
C.R. Ac. Sc., 257, 1963, p. 2917–2921.

A 128 – Sur certaines parois à divergence de l'aimantation non nulle.
C.R. Ac. Sc., 257, 1963, p. 4092.

A 129 – Contribution à l'étude du paramagnétisme des alliages fer-cobalt (avec Y. Barnier et R. Pauthenet).
Cobalt (Bruxelles) 21, 1963, p. 153–160.

A 130 – Interactions magnétiques entre deux couches minces ferromagnétiques séparées par une couche de chrome ou de palladium d'épaisseur inférieure à 300 A (avec J.C. Bruyère, O. Massenet et R. Montmory).
C.R. Ac. Sc., 258, 1964, p. 1423.

A 131 – Magnetic properties of multilayers films of FeNi–Mn–FeNiCo and of FeNi–Mn (avec O. Massenet, R. Montmory, D. Paccard et C. Schnaider).
Intermag, 1964, I.E.E.E. Trans. on Magn., I, 1965, p. 63–65.

A 132 – A coupling phenomenon between the magnetization of two thin ferromagnetic films separated by a thin metallic film; application to magnetic memories (avec J. Bruyère, O. Massenet et R. Montmory).
Intermag, 1964, I.E.E.E. Trans.on Magn., I., 1965, p. 10–12.

A 134 – Sur un nouveau phénomène de couplage entre couches minces ferromagnétiques séparées par un matériau non ferromagnétique (avec J. Bruyère, O. Massenet et R. Montmory).
C.R. Ac. Sc., 258, 1964, p. 841–844.

A 135 – Magnetic properties of an iron-nickel single crystal ordered by neutron bombardment (avec J. Paulevé, R. Pauthenet, J. Laugier et D. Dautreppe).
J. Appl. Phys., 35, 3 (part 2), 1964, p. 873–876.

A 136 – Coupling effect between the magnetization of two thin layers separated by a thin non magnetic metallic layer (avec R. Montmory, J. Bruyère, G. Clerc, O. Massenet, D. Paccard et A. Yelon).
10th Conf. Magnetism J. Appl. Phys., 1965, 36, p. 944–945.

A 137 – Sur les variations irréversibles d'aimantation provoquées par les contraintes élastiques.
Inédit, 1964.

A 138 – Coupling phenomenon observed in multilayer films composed of two ferromagnetic layers separated by a non magnetic intermediate layer (avec R. Montmory, J. Bruyère, J. Devenyi et O. Massenet).
Nottingham Conf., 3, 1964, L. 8–I–I.

A 139 – Indirectly coupled films (avec J. Bruyère, G. Clerc, O. Massenet, D. Paccard, R. Montmory, J. Valin et A. Yelon).
I.E.E.E. Trans. on Magn., Vol. Mag–I, n° 3, 1965, p. 174–180.

A 140 – Etude par microscopie électronique de parois couplées en défocalisation variable (avec E. Biraguet, J. Devenyi, G. Clerc, O. Massenet et A. Yelon).
    C.R. Ac. Sc., 263, 1966, p. 166.

A 141 – Sur les effects de l'hystérésis antiferromagnétique.
    C.R. Ac. Sc., 264, 1967, p. 1002–1006.

A 142 – Etude théorique du couplage ferro-antiferromagnétique dans les couches minces. **469**
    Ann. de Phys., 1967, *2,* p. 61–80.

A 143 – Parois dans les films minces.
    *J. Phys. Rad.,*Suppl. *29,* 1968, p. C2-87 à C2-94.

A 144 – Sur l'augmentation de la susceptibilité apparente d'un corps ferromagnétique **267**
produite par la superposition d'un champ alternatif.
    C.R. Ac. Sc., 268, B. 1969, p. 725–729.

B – BREVETS D'INVENTION

B 1 – Perfectionnement à la fabrication des aimants permanents. **273**
déposé à Chambéry, le 7 Avril 1942, n° PV 323. Publié le 20 Avril 1951 (n° 979043).

B 2 – Perméamètre à variation de réluctance pour corps faiblement magnétiques (avec M. Toitot).
déposé le 26 Juin 1958, n° PV 717368.

B 3 – Matériau magnétique pour très haute fréquence (avec I. Epelboin).
déposé 4 Avril 1960, n° PV 823 322.

B 4 – Procédé, installation et dispositif pour l'obtention d'un corps présentant une forte anisotropie magnétique (avec W. Chambon, D. Dautreppe et J. Paulevé).
déposé le 5 Septembre 1962, n° PV 908 713.

B 5 – Structures magnétiques à films minces et application aux mémoires magnétiques (avec J. Bruyère, O. Massenet et R. Montmory).
déposé le 18 Octobre 1963, n° PV 951 108.

C – PUBLICATIONS DE CARACTERE ASSEZ GENERAL INTERESSANT LE MAGNETISME

C 1 – Tables annuelles de constantes, Paris (avec G. Foëx). Vol. IX, p. 306–324 et Vol. X, p. 239–257 (Magnétisme).

C 2 – Recherches récentes sur le magnétisme (partie théorique).
Congrès Int. Electr., Paris, 1932, Vol. II, p. 227–247.

C 3 – L'interprétation des propriétés magnétiques des métaux.
Rev. Gén. Sci., 44, 1933, p. 169–175.

C 4 – Tables annuelles de constantes, Paris.
Vol. XI (1931–1934), Sec. 24 et 25 (Ferromagnétisme et Effet Hall); Vol. XII (1935–1936), Sect. 24 (Ferromagnétisme).

C 5 – Champ moléculaire, aimantation à saturation et constantes de Curie des éléments de **48**
transition et de leurs alliages.
Le Magnétisme, tome II, Ferromagnétisme. Coll. Scient. Inst. Int. Coopération Intellectuelle, Paris 1940, p. 65–164.

C 6 – Le ferromagnétisme et l'état métallique.
J. Phys. Rad., 1, 1940, p. 242–250.

C 6 bis – Le moment magnétique dans la théorie moderne des métaux.
Cah. Phys., n° 1, 1941, p. 78–80.

C 7 – Etude de quelques problèmes actuels de ferromagnétisme.
Conf. Inst. Henri Poincaré, Paris, 1942.
    I. – Ferro- et para-magnétisme à champ moléculaire.
    II. – L'équilibre interionique.
    III. – Antiferromagnétisme.
    IV. – Interactions magnétiques et distances interatomiques.
    V. – Le fer.
    VI. – Nature des ions magnétiques.
Conf. origin. Serv. Docum., n° 102 (Arch. docum. C.N.R.S.).

C 8 – Le magnétisme en France entre 1939 et 1945.
Conf. Congrès Assoc. Fr. Avancement des Sciences, 25 Oct. 1945.

C 9 – Action des champs démagnétisants internes sur les propriétés des corps ferromagnétiques.
Soc. Fr. Phys. Comm. Paris, 6–8 Nov. 1948.

C 10 – Métaux à propriétés nouvelles obtenus par la métallurgie des poudres (avec L. Weil).
Bull. Soc. Sci. Dauphiné, 69, 1946, p. 193.

C 11 – Applications nouvelles de recherches théoriques aux aimants permanents et aux machines électrostatiques (avec L. Weil et N. Félici).
Ann. Univ. Grenoble, 22, 1946, p. 71.

C 12 – L'origine du champ coercitif dans les substances ferromagnétiques.
L'Information des Sci. Phys., 2, 1947, p. 69–73.

C 13 – Les théories modernes du magnétisme et leurs applications.
Rev. Métallurgie, 45, 1948, p. 475–480.

C 14 – The influence of porosity in ferromagnetic substances upon the coercive force and the approach to saturation.
Metal Powder Rep., 2, 1948, p. 189.

C 15 – Quelques aspects actuels de la théorie du ferromagnétisme.
Bull. Soc. Fr. Electr., 9, 1949, p. 308–318.

C 16 – Quelques aspects actuels de la théorie de Langevin-Weiss du paramagnétisme à champ moléculaire : le ferrimagnétisme.
Colloque Phénomènes Gyromagn., Paris, 1948, CNRS Ed., p. 22–28.

C 17 – Aimantation et domaines élémentaires dans les ferromagnétiques.
Bull. Union Physic., n° 398, *45*, 1951, p. 385.

C 18 – Les points de transition magnétique.
Colloque Changem. Phase, Soc. Chim. Phys., 2–6 Juin 1952, p. 301–305.

C 19 – Antiferromagnetism and ferrimagnetism.
Proc. Phys. Soc., A, 65, 1952, p. 869; même article en français dans *Ann. Inst. Polytechn. Grenoble, 2*, 1953, p. 9–19.

C 20 – Some new results on antiferromagnetism and ferrimagnetism.
Rev. Mod. Phys., 25, 1953, p. 58–63.

C 21 – La théorie du ferrimagnétisme et ses vérifications.
J. Phys. Rad., 14, 1953, p. 64 S.

C 22 – Le ferri- et l'anti-ferromagnétisme.
Proc. Int. Conf. Theoret. Phys., Kyoto and Tokyo, 1953.

C 23 – Structure cristalline et propriétés magnétiques.
Bull. Soc. Fr. Minér. Crist., 77, 1954, p. 257–274.

C 24 – Antiferromagnétisme et métamagnétisme.
10ᵉ Conseil Solvay, Rapport, Stoops, éd., Bruxelles, 1954, p. 251–281.

C 25 – Theoretical remarks on ferrimagnetism at low temperatures.
Progress in Low Temp. Physic. 1, 1955, p. 336–343.

C 26 – Ferromagnétisme, ferrimagnétisme et antiferromagnétisme.
Encycl. Fr., t. II, p. 2-58-1 à 2-58-6, Paris, 1955.

C 27 – Some theoretical aspects of rock magnetism. **428**
Phil. Mag. Suppl., 4, 1955, p. 191–243.

C 28 – Quelques problèmes de paléomagnétisme.
Bull. Soc. Sci. Dauphiné, 69, 1955, p. 4–7, n° 5.

C 29 – Le champ moléculaire de Weiss et le champ moléculaire local. **53, 203**
Colloque national Strasbourg, commémoratif de l'oeuvre de P. Weiss, 8–10 Juillet 1957, CNRS éd., 24 p.

C 30 – Reptation des cycles d'hystérésis.
ibid., p. 75–77.

C 31 – Un nouveau type de substances ferromagnétiques : les ferrites des terres rares ayant une structure du grenat (en russe avec F. Bertaut, F. Forrat et R. Pauthenet)
Isvest. Akad. Nauk, SSSR, Moscou, 21, 1957, p. 904.

C 32 – Application de la notion de susceptibilité irréversible à l'interprétation de l'hystérésis ferromagnétique.
Proc. Robert A. Welch Foundation : Atomic structure, Houston, Texas, déc. 1958, p. 167–189.

C 33 – Kopplungs Effekt in der ferromagnetischen Hysterese : Kriechen und Kippen.
Physiker Tagung Essen, 1958, Physik Verlag., Mosbach/Baden, p. 54–68.

C 34 – Quelques aspects récents des problèmes de magnétisme.
Bull. Soc. Fr. Electric., 8ᵉ Sér., I, 1960, n° 2, p. 1–7.

C 35 – Théorie des propriétés magnétiques des grains fins antiferromagnétiques : superparamagnétisme et superem ferromagnétisme.
Conf. Ecole Phys. Théor., Les Houches, Juillet 1961 (édité par C. DeWitt, CNRS, p. 413).

C 36 – Research on thermal cycles and the directional order of defects.
Symp. on Basic Science in France and USA, New York Univ. Press, 1962, p. 142.

C 37 – French contributions to the field of magnetism.
ibid., p. 129.

C 38 – Etat actuel de nos connaissances sur l'antiferromagnétisme et le ferrimagnétisme.
Symposium sur les Matériaux, Paris, 18 Mai 1961, OTAN (AGARD) et Adv. in Materials Researching the NATO Nations, Pergamon Press, 1963, p. 329.

C 39 – L'Ecole Française de Magnétisme.
La Métallurgie, 96, 1964, p. 101–112.

C 40 – Sur les bases physiques de la démagnétisation.
Colloque OTAN sur la démagnétisation, Brest, Juin 1964.

C 41 – Die magnetische Untersuchung und die Bedeutung von Punktfehlern die durch Strahlung in einem ferromagnetischen Metall erzeugt werden.
Z. für Angewandte Phys., 17, 1964, p. 113–120.

C 42 – The rare earth garnets (avec R. Pauthenet et B. Dreyfus).
Progress in Low Temp. Phys., 4, 1964, p. 344–383.

C 43 – Une nouvelle structure ordonnée créée par irradiation aux neutrons (avec D. Dautreppe, J. Paulevé, J. Laugier, R. Pauthenet et A. Marchand).
3ᵉ Conf. Int. Utilisation de l'Energie atomique à des fins pacifiques, 1964.

C 44 – L'Hystérésis ferro- et antiferromagnétique : ses lois.
Conf. Assemblée Gén. Soc. Fr. Phys., Paris, 4 Février 1967.

C 45 – Trainage magnétique de diffusion et hystérésis ferro- et antiferromagnétique.
Conf. Royal Society, London, 16 Mars 1967.

C 46 – Domaines et parois dans les corps ferromagnétiques, leur visualisation.
Bull. Soc. Fr. Minér. Crist., 91, 1968, p. 627–636.

C 47 – Die elementare Bezirke und ihre Wände in einem ferromagnetischen Kristall.
(177ᵉ Sitzung der Natur und Ingenieurwissenschaftlichen Sektion, 5 Februar 1969, Arbeitsgemeinschaft für Forschung des Landes Nordrhein-Westfalen, 43 pages).

C 48 – Magnétisme et Champ moléculaire local.
Conf. Nobel. Les prix Nobel en 1970, p. 158–177.

PUBLICATIONS DIVERSES (JUSQU'EN 1970)

D 1 – Quelques problèmes actuels de géomagnétisme.
Bull. Soc. Scient. Dauphiné, 1949, p. 61.

D 3 – Marine et Université (avec P. Bousquet)
L'Atlantique Nord, 3, 1957, p. 70.

D 4 – Notice nécrologique sur Wander Johannès de Haas.
C.R. Ac. Sc., 250, 1960, p. 3918.

D 5 – La recherche fondamentale dans les Centres de Recherches nucléaires.
Atomic Energy Rev., Vol. 4, 1966, Commemorative issue.

D 6 – Electrostatique et magnétisme.
3e Centenaire Ac. Sc., 1666–1966, p. 421, Paris, Gauthier-Villars, éd.

D 7 – L'avenir de la physique du solide.
C.R. Ac. Sc., 265, 1967, Vie académique, p. 148.

D 8 – Développement de la recherche fondamentale à Grenoble.
Rev. Fr. Elite Européenne, n° 205, déc. 1967, p. 79.

D 9 – La recherche fondamentale au Centre d'Etudes nucléaires de Grenoble.
L'Ingénieur (Rev. Union Région. Ing. Dauphiné-Savoie), *19,* 1967, n° 76, p. 15.

SELECTED WORKS
OF
LOUIS NÉEL

Translation from the French authorized by the
CENTRE NATIONAL DE LA RECHERCHE SCIENTIFIQUE

Editor and Translation Coordinator
NICHOLAS KURTI
Department of Engineering Science
University of Oxford, U.K.

GORDON AND BREACH SCIENCE PUBLISHERS
New York London Paris Montreux Tokyo Melbourne

© 1988 Gordon and Breach Science Publishers S.A. All rights reserved. Published under license by OPA Ltd. for Gordon and Breach S.A.

Gordon and Breach Science Publishers

Post Office Box 786
Cooper Station
New York, New York 10276
United States of America

Post Office Box 197
London WC2E 9PX
England

58, rue Lhomond
75005 Paris
France

Post Office Box 161
1820 Montreux 2
Switzerland

3-14-9, Okubo
Shinjuku-ku, Tokyo
Japan

Private Bag 8
Camberwell, Victoria 3124
Australia

Originally published in French in 1978 by Editions du Centre National de la Recherche Scientifique as *Oeuvres Scientifiques de Louis Néel*

Library of Congress Cataloguing in Publication Data

Néel, Louis, 1904
The selected works of Louis Néel

Translation of: Oeuvres Scientifiques de Louis Néel
Includes bibliographies

1. Magnetism - Collected works.   I Kurti, Nicholas. II Title
QC750.15.N4313 1988 538.82-12009
ISBN 2-88124-300-2

No part of this book may be reproduced by any means, electronic or mechanical, including photocopying and recording, or by any information storage or retrieval system, without permission in writing from the publishers.
Printed in Great Britain by Bell and Bain Ltd., Glasgow

CONTENTS

	Preface to the English Edition	vii
	Preface to the French Edition	viii
	Louis Néel - Chronology	x
I	General theory of magnetism A7, A17, A40, A22, A35, C5, C29. (Translator: P. Rhodes)	1
II	Ferrimagnetism A6 (Translator P Johnston A70 A72 A77 A94 C4 (Translator: K. Hoselitz)	61
III	Antiferromagnetism A29, A112, A113, A114, A115 (Translator: J.M.D. Coey)	103
IV	Magnetic interactions and their variations A18, A23, A33, A30 (Translator: P. Johnston)	123
V	Approach to saturation in ferromagnets A62, A63, A93 (Translator: G.A. Jones)	159
VI	Elementary domains; phases and modes A52, C29, A58, A127, A85 (Translator: R.D. Wade)	179
VII	Bloch and Néel walls A50, A97, A118 (Translator: R.D. Wade)	219
VIII	Weak fields and anhysteretic magnetization A45, A44, A47, A48, A144 (Translator: G.A. Jones)	239
IX	Coercive force; magnets B1, A51, A57, A59, A60, A91 (Translator: K. Hoselitz)	271
X	Random fields; reptation and tilting A101, A102, A106, A107, A109, A108 (Translator: E. Geissler)	313
XI	Magnetic after-effects A73, A78, A84 (Translator: P.W. Readman)	341
XII	Directional order; irradiation A92, A119 (Translator: E. Geissler)	385
XIII	Rocks and baked clays A6 (Translator: J.M. Coey), C2 (originally published in English)	405
XIV	Surface problems A121, A122 (Translator: R.J. Nicholas)	457
XV	Antiferromagnetic hysteresis A142 (Translator: R.J. Nicholas)	467
XVI	Apparatus and Techniques A11, A31, A39 (Translator: R.J. Nicholas)	489
XVII	Experimental work A7, A9, A20, A36 (Translator: R.J. Nicholas)	497
	Index	519

PREFACE TO THE ENGLISH EDITION

I have the greatest admiration for Néel's scientific work and I am glad to supplement, on the occasion of this English edition of Néel's papers, Professor Chabal's Preface. Néel added two new varieties of-magnetism, antiferro- and ferrimagnetism, to the three previously existing types, namely dia-, para-, and ferromagnetism, and he did so with the elegant simplicity and truly physical intuition that is the hallmark of his work. It is comforting to encounter such qualities in these days of giant accelerators and computers.

Quite a few years ago I wrote an article entitled 'Nonmathematical Theoretical Physics' on the general theme that most of the important discoveries in theoretical physics were the result of the use of a new and imaginative, but essentially simple, concept that had eluded others. As an example, I told the story of what happened when I was writing my book *Electric and Magnetic Susceptibilities* in the early 1930s. It was well known at that time that a positive exchange integral can produce ferromagnetism, so, in my book, I correctly emphasized the fact that a negative exchange integral can produce antiferromagnetic ordering; I therefore wrote in my later review article that it was left to Néel to have the perspicacity to propose antiferromagnetic ordering. I reproach myself to this day for not having thought of it myself, though it is some consolation that the idea eluded all the specialists in ferromagnetism. I might also mention that the concept of antiferromagnetic ordering could have been introduced as early as 1907 by means of the negative Weiss molecular field constant.

In my review article I also emphasized the fact that many of the great theoretical discoveries were the result of supreme physical insight rather than of rigid mathematical formalism. To understand Pauli's exclusion principle no knowledge of arithmetic is necessary, an ability to count is sufficient. It is true that Néel's studies of the saturation magnetization of ferrites and other ferrimagnetic materials also required an ability to add and subtract, however, having had the imagination and the courage to propose, well before it could be proved experimentally, that the ferrites have inverse spinel structure.

Néel's work helps to maintain and, indeed, to enhance the distinguished position in the field of magnetism occupied by France for many years, as exemplified by names like Ampère, Pierre Curie, Langevin, Weiss, Jean Becquerel, Brillouin, Kastler and Néel. The name of the Place de l'Etoile may be changed, but 'Curie-point' and 'Néel-point' will forever belong in the terminology of physics.

J.H. van Vleck

Editor's Note

At the end of August 1980, when plans for this book were taking shape, I wrote to Professor J.H. van Vleck, a good and old colleague of Professor Néel's, to ask him whether he would be willing to write an additional preface to the English edition of Néel's scientific work. In his reply, dated September 20, 1980, van Vleck said that at the age of 81 he was trying to keep all new commitments to a minimum but suggested that I could use as the basis of a preface the tribute he sent on the occasion of Néel's Nobel prize (see *Bull. Soc. Française de Physique*, April-May 1971).

Professor van Vleck died on 27 October 1980 and did not see the draft of this preface. But it was clearly his wish to be associated with this volume which, it is hoped, will be regarded by the English speaking physics community as a modest but grateful and warm tribute to Professor Néel.

PREFACE TO THE FRENCH EDITION

The broad outlines of Louis Néel's scientific work are generally well known, not only to physicists but also to the public thanks to the awarding of the 1970 Nobel Prize in Physics to this brilliant physicist. The publication by the CNRS of this collection of Louis Néel's works makes it possible to gain a more intimate knowledge of the development of his scientific ideas.The reader can thus grasp an entire cross-section of scientific research and experience the writing of an important page of science.

All the characteristics of a great scientist's development emerge from this collection. First, the urge to understand the innermost behaviour of matter, rejecting all dogmatism and perceiving the phenomena as they should be on the microscopic scale; hence, the theory of antiferromagnetism. Then the increasing concern about sticking to personal reality and the experimental search for materials with new properties; here we have the ferrimagnetic substances. And,finally, the translation of these ideas to the world outside physics, leading to applications in many fields. Thus, Louis Néel made important contributions to the interpretation of the magnetic memory of rocks, which ultimately provided a key to the phenomenon of continental drift. He also devised, at the beginning of World War II, a method of degaussing ships, which was used unsuccessfully to protect the French fleet against magnetic mines. This sketch, brief as it is, of an impressive list of achievements may at least bring out the various facets of Louis Néel's scientific influence.

I do not wish to anticipate the unfolding of the progress of Louis Néel's thoughts; however, at his request, I should like to recall the different roles of the CNRS in the blossoming of his remarkable career. As early as 1931, when he was a young research worker in Pierre Weiss's laboratory in Strasbourg, Louis Néel was awarded a research bursary by the Caisse Nationale des Sciences, the precursor of the CNRS. Such grants were made to encourage and facilitate the scientific work of young research workers in important laboratories.

When Néel settled in Grenoble after the 1940 armistice, the CNRS continued tosupport him in his endeavour to start new experiments. In 1945, the CNRS decided to put this support on an institutional basis by creating a laboratory of electrostatics and metal physics and making him its first director.

The way the CNRS acts in developing scientific research can well be exemplified by the successive stages of a career. It all begins with finding a young scientific talent whom the CNRS provides with the means of developing his own research, possibly by appointing him to a full-time research post and placing him with a group renowned for the quality of its scientific output. As the research initiated by the research worker achieves international influence, the CNRS makes its support institutional by creating first a research team and then, possibly, a laboratory. Finally, when the research requirements demand it, the CNRS undertakes the introduction of big scientific equipment. Thus, in Grenoble, in the environment of solid state physics laboratories, the CNRS established a national centre for intense magnetic fields and participated in setting up the high neutron flux reactor of the Laue-Langevin Institute.

Such a development of scientific research does not come about spontaneously, it is the result of long perseverance maintained despite the often conflicting trends set by the CNRS, which relies on the best scientists to define and realize a policy. Here Néel showed organizational talents fully up to the level of his scientific achievements.

By accepting heavier and heavier responsibilities for science policy and administration, Louis Néel built up the scientific complex of Grenoble, indisputably one of the top centres of scientific research in France. In 1939 he became *chargé de mission* to advise the management of the CNRS. From 1949 to

1970 he was a member of the national committee for scientific research and of the *Directoire*, later the Council of Administration of CNRS.

I hope that I have clearly demonstrated, in accordance with Louis Néel's wishes, the symbiosis of his work and the actions of the CNRS. The development of the function of research is a complex process that is achieved thanks to the researchers, the administrators, and the *batisseurs*. Louis Néel managed to fulfil these functions remarkably in each case, and I now leave it to the reader to rediscover the most individual aspects of the work of a great French scientist.

Robert Chabbal
Directeur Général du CNRS
(Translation by N. Kurti)

LOUIS NÉEL - CHRONOLOGY

Né à Lyon (3e), le 22 Novembre 1904

1924	Eléve de l'Ecole Normale Supérieure
1928	Agrégé des Science Physiques
1928	Assistant à la Faculté des Sciences de Strasbourg
1932	Docteur ès Sciences
1934	Mâitre de Conférence à la Faculté de Science de Strasbourg
1937	Professeur à la Faculté de Science de Strasbourg
1945	Professeur à la Faculté de Science de Grenoble
1945	Directeur de Laboratoire d'Electrostatique et de Physique du Métal (CNRS)
1947	Membre du Comité Consultatif des Universités
1948	Médaille André Blondel
1952	Prix Holweck
1952	Conseiller scientifique de la Marine Nationale
1952	Membre du Directoire et du Conseil d'Administration du CNRS
1952	Membre de l'Académie des Sciences
1954	Directeur de l'Institut Polytechnique de Grenoble
1956	Directeur du Centre d'Etudes Nucléaires de Grenoble
1957	Pràsident de la Société Française de Physique
1960	Représentant de la France au Comité Scientifique de l'OTAN
1963	Prix des Trois Physiciens
1963	Président de l'Union Internationale de Physique Pure et Appliqué
1965	Màdaille d'Or du CNRS
1970	Prix Nobel de Physique
1971	Directeur du Laboratoire de Magnétisme de Grenoble (CNRS)
1971	Pràsident de l'Institut National Polytechnique de Grenoble
1971	Grande Médaille d'Or de l'Electronique

Docteur "honori causa" des Universités des Graz, Nottingham, Oxford, Louvain, Newcastle, Iassy, Madrid.

Membre Etranger des Académies des Sciences de l'URSS, de l'Académie Royale Néerlandaise, de l'Académie des "Naturforscher Leopoldina", de l'Académie de la Republique Populaire Roumaine de la Royal Society (London), de l'American Academy of Arts and Sciences, de l'Académie Polonaise des Sciences.

Grand-Croix de la Légion d'Honneur.

Grand-Croix de l'Ordre National du Mérite.

Croix de Guerre avec Palmes.

Chapter I
GENERAL THEORY OF MAGNETISM
A7, A17, A40, A22, A35, C5, C29

A7 (1932)

INFLUENCE OF FLUCTUATIONS IN THE MOLECULAR FIELD ON THE PROPERTIES OF MAGNETIC BODIES

Summary (1977)—Pages 1 to 50 and 55 to 70 of the original memoir A–7 (Thesis by L. Néel) are reproduced here.

The magnetic coupling between the atoms of a ferromagnetic body are only effective over a short distance and can therefore not be represented by a *uniform* Weiss molecular field. It is necessary to take into account *fluctuations* of that field. To determine the effects of this a study is made of the magnetic properties of a simple model consisting of an infinite chain of atoms satisfying Lenz's law. Each atom is coupled to the p atoms which precede it in the chain and to the p atoms which follow it. One thus finds the origin of some anomalies which are not explained by the Weiss theory, such as the difference between the ferromagnetic Curie point θ_f, the temperature at which the spontaneous magnetization vanishes, and the paramagnetic Curie point θ_p deduced, in the paramagnetic region, by extrapolation of the Curie–Weiss line. In the same way, one can explain the persistence of an excess specific heat above θ_f.

It is also shown that, in solid solutions, it is necessary to use different molecular fields, depending upon the surroundings of the site at which they act. The temperature variation of the inverse susceptibility is no longer a straight line but rather a hyperbola.

Finally, a study is made of the case of negative interactions (§ 25, §§ 46–50) which tend to align antiparallel the magnetic moments of two neighbouring atoms. Using a very simple schematic model, it is shown that at low temperatures the body is paramagnetic with a susceptibility independent of temperature (constant paramagnetism). In this temperature range, and if the body has a body centred cubic lattice, it forms (§ 48) two simple cubic sub-lattices magnetized in opposite directions.

INTRODUCTION

§ 1. Theory of paramagnetism,—The theory of paramagnetism is based on the following basic hypothesis: inside a paramagnetic material there are carriers of magnetic moment. Their magnetic moment μ is fixed. One can, for example, identify the carriers as atoms.

Langevin [1] assumed that these carriers had complete freedom to rotate and that Boltzmann statistics could be applied to them. Under these conditions, the magnetization per carrier, in a magnetic field H, is given by the equation:

$$\bar{\mu} = \mu \left(\coth a - \frac{1}{a} \right) \qquad a = \frac{\mu H}{KT}. \qquad (1)$$

The carriers are then independent of each other; this is certainly not true in a dense material. Weiss [2] suggested that the interactions between carriers are equivalent to a magnetic field: the molecular field proportional to the magnetization:

$$h_m = n\bar{\mu}. \qquad (2)$$

In weak fields the law of magnetization is of the form:

$$\mu = \frac{\mu^2}{3K} \frac{H}{T - \theta} \qquad \theta = \frac{n\mu^2}{3K}$$

For $T < \theta$ the material remains spontaneously magnetized (ferromagnetism) in zero external field.

The agreement of the theory with experiment is remarkable and provides a solid confirmation of the basic hypotheses.

Also, Heisenberg's theory [3] gives a satisfying interpretation of the interactions, that is to say, of the molecular field.

§ 2.—Some unexplained facts.—Nevertheless, there is a whole category of materials the properties of which remain unexplained by the Langevin–Weiss theory. There are bodies with a paramagnetism independent of the temperature, that is to say most metals and several salts.

In the same way, in ferromagnetic metals one finds that the temperature at which the spontaneous magnetization disappears, i.e. the ferromagnetic Curie point θ_f, does not coincide with the paramagnetic

Curie point θ_p, defined by the relation (1):

$$\chi(T - \theta_p) = C.$$

Finally, interpretation of the coercive field and of hysteresis is difficult and requires additional assumptions.

§ 3.—Fluctuations of the molecular field.

A priori, is it rigorous to replace the interactions between the carriers of magnetic moment by a molecular field simply proportional to the magnetization? As a consequence of thermal agitation the interactions are subject, at a particular time, to fluctuations from one point to another in the material, and at a point they vary from one instant to another.

It is necessary to study the effect of these fluctuations on the magnetization.

Moreover, the existence of these fluctuations is independent of the interpretation of the molecular field.

§ 4.—Aim of this work.

I propose, in the following chapters to make a close study of the mechanism of the molecular field in solids and from it to show that most of the differences between experiment and the Langevin–Weiss theory disappear.

NOTATION

μ	magnetic moment of a carrier.
$\bar{\mu}$	mean magnetization per carrier.
$\mathscr{T}$	magnetization of a group of N atoms: $\mathscr{T} = N\bar{\mu}$.
$2p$	number of neighbouring carriers interacting with a given carrier.
w	potential energy of two anti-parallel carriers (the energy of two parallel carriers is zero).
K	Boltzmann's constant.
P,Q	in an alloy, fractional concentration of the two kinds of carriers.
T	absolute temperature.
U	internal energy of a group of N atoms.
χ	mass susceptibility.
H	magnetic field.
Θ	paramagnetic Curie point if there were interactions only between nearest neighbours.
θ_f	ferromagnetic Curie point.
θ_p	paramagnetic Curie point.
h_m	molecular field without fluctuations.
n	molecular field coefficient without fluctuations $h_m = n\bar{\mu}$.
μ_s	spontaneous magnetization per carrier.

---

*On the subject of two Curie points, see the paper by M. FORRER, *J. de Phys.*, 1, 1930, 49.

C	Curie constant with respect to unit mass.
C_M	atomic Curie constant.

N.B.—In the references, the first number shows the volume, the second the year and the third the page.

PART ONE

CHAPTER I

THE HYPOTHESES

§ 5.—The fundamental hypotheses

a) The carriers of magnetic moment are situated at the lattice points of a rigid space lattice. They have a constant magnetic moment μ, characteristic of the material and its crystalline state.

In this way, I shall ignore all phenomena related to magnetostriction.

The magnetic state of the material is defined when the orientation of all the carriers is given. *A priori* all the geometrical configurations are possible that can be created by orientating the carriers in all possible ways.

b) The total potential energy of any configuration is calculated by adding the relative potential energies of the carriers, taken two by two, in all possible ways.

The relative potential energy of two carriers depends, from the nature of the two carriers, on their separation and their relative orientation.

I shall not make any assumption about the origin of the relative potential energy of two carriers, but shall simply suppose that it exists, and denote it by w.

Given the potential energy of all the possible configurations, the probability of any particular configuration can be calculated from Boltzmann's theorem.

§ 6.—Difficulty of the problem.

The function w is completely unknown.

It probably varies very rapidly with the separation of two carriers. Schematically, one could say that the mutual interactions are significant when the distance between two carriers is less than a distance r, and that they are very weak when the carriers are further apart. One can thus divide the neighbours of a given carrier into two groups: those whose distance from the central carrier is less than r, and those which are more. The first group will exert a strong average influence on the central carrier, but since it contains a few atoms, fluctuations on both sides of the average will be strong. Conversely, the average influence of the second group will be weaker, but since it contains a large number of atoms, its importance in respect of fluctuations will be very much eaker.

§ 7.—**Short-range interactions.—Long range interactions.**—This division of the interactions into two categories, at first sight arbitrary, corresponds to the following fundamental fact: the long-range interactions necessarily involve a large number of carriers, so that one can replace the instantaneous interaction of the carriers by its time average: the Weiss molecular field theory is fully applicable.

On the other hand, for the short-range interactions the number of carriers involved is small, so that it is necessary to give a rigorous treatment of fluctuations.

The distinctions in the interactions which I have made are not peculiar to magnetism. Analogous problems arise in the kinetic theory of gases [4].

§ 8.—**Simplifying assumptions.**—In order to make the problem amenable to calculation, I suggest the following scheme:

c) The short range interactions will be represented by the interactions between one carrier and its nearest neighbouts. The number of neighbours thus defined is completely determined by the lattice; for a face-centred cubic system each carrier has 12 neighbours. I shall study the law of magnetization of a body which has only this type of interaction, and I shall replace the interaction with the other carriers by a molecular field in the sense of Weiss.

d) Even so, the problem is still too complicated and I shall implify it further, by supposing that each carrier can only take up two orientations with respect to an external magnetic field: parallel or anti-parallel. This is Lenz's hypothesis [5].

I propose, in Chapter II, to study the problem under these assumptions. Later, I shall discuss the extent to which these simplifying assumptions are legitimate. They are only made, in fact, on a provisional basis.

In particular, I shall emphasize the distinction between long-range and short-range interactions.

We shall see, moreover, that Lenz's hypothesis must be rejected in the low temperature region.

CHAPTER II

THE PROBLEM OF FLUCTUATIONS IN A LINEAR LATTICE

§ 9.—**Statement of the problem.**—I consider a crystalline lattice. From the point of view of statistical mechanics, I shall take account only of energies between nearest neighbours.

Two neighbouring carriers can have only two different relative orientations,*: in one, the carriers are

---

*This follows directly from assumption d). I also neglect the orientation with respect to the line joining the two carriers.

parallel, their mutual potential energy is w_1; in the other, they are anti-parallel, their mutual potential energy is w_2. I put: $w = w_2 - w_1$.

The potential energy of the system, arising from the interactions between carriers, or, more briefly, the bond energy, can be calculated when the orientation of the carriers is known.

Each of the N carriers can have two different orientations: there are thus, in total, 2^N distinct configurations.

If M is the total magnetic moment of a configuration, measured in the direction of the external magnetic field; if w is the bond energy of the same configuration added to the magnetic potential energy which arises from the moments μ being in the external magnetic field H, then under these conditions the magnetization per carrier is given by the following equation:

$$\bar{\mu} = \frac{1}{N} \frac{\sum M e^{-\frac{w}{KT}}}{\sum e^{-\frac{w}{KT}}} \qquad (3)$$

The Σ sign extends over the 2^N configurations defined above.

§ 10.—**Necessity of a simplification.**—When one attempts to carry out the calculation, for example, in the case of a face-centred cubic lattice, one is quickly brought to a halt by the complexity.

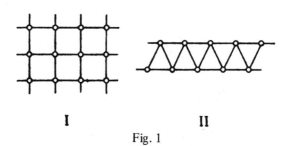

Fig. 1

It is essential to find a legitimate simplification. Now, the general form of the phenomena must depend more on the number of neighbours of each carrier than on their geometric position. In fact, it is the number of neighbours which plays an essential role in the problem of fluctuations. Intuitively, it can be seen that each time the number of neighbours is increased by one unit the relative importance of fluctuations is decreased by half. Hence one can consider as equivalent the two groups in figure 1, where each carrier has four neighbours.

Moreover, anticipating the results of the calculations, the values of the susceptibility above the paramagnetic Curie point, calculated rigorously in example II, agrees with the expansions obtained by an approximate method, which involves no assumption about the position of the neighbours, and which, consequently, is equally valid in example I as in example II. This supports this point of view.

§ 11.—Proposed method.—More precisely, I can replace the study of a three-dimensional lattice, in which each carrier is surrounded by $2p$ neighbours, by the study of a linear chain of carriers, parallel to the external field, in which I shall take account of the interaction energy of each carrier with the p carriers which precede it and the p which follow it in the chain. All these interactions will be considered as equivalent, and characterized by an energy w as in § 9.

That is the problem I propose to solve in the following sections. The enumeration of the configurations and the calculation of their potential energy is then much easier than in the case of a spatial lattice.

§ 12.—Enumeration of configurations.—In order to construct all possible configurations, I shall have recourse to the following trick: I shall partition the line into elementary groups of p consecutive carriers.

Since each carrier can take up two different orientations, there are $q = 2^p$ different elementary groups. I number the groups from 1 to q. In the following, any group will be characterized by the index i, or by the index j: i and j can take all integer values from 1 to q.

In order to reproduce all the configurations, it is necessary to form all possible arrangements between the elementary groups.

The potential energy of a configuration is the sum of the internal potential energies of the groups themselves and the potential energies of the bonds between two consecutive groups.

Thanks to the device of taking groups of p carriers, it is not necessary to consider the bond energy between two non-consecutive groups.

I shall designate the interaction between two consecutive groups by the double index ij, the first kind i, the second kind j. Any configuration is characterised, from the energy point of view, by the number of bonds of each kind which it contains.

I let

$$X = \begin{bmatrix} ik \\ n_{11} n_{12} \ldots n_{ij} \ldots n_{qq} \end{bmatrix}$$

be the number of configurations which have the following properties: they start with a group i and finish with a group k; they contain n_{11} interactions 11, n_{12} interactions 12, ... n_{ij} interactions ij ... etc.

§ 13.—Calculation of X.—To create all the possible configurations with $m+1$ bonds, I can add one bond to the configurations with m bonds. I can thus immediately write the following recurrence relation:

$$\begin{bmatrix} ik \\ n_{11} n_{12} \ldots n_{jl} \ldots \end{bmatrix} = \begin{bmatrix} i1 \\ n_{11} \ldots (n_{1k} - 1) \ldots n_{jl} \end{bmatrix}$$
$$+ \begin{bmatrix} i2 \\ n_{11} \ldots (n_{2k} - 1) \ldots n_{jl} \ldots \end{bmatrix}$$
$$+ \ldots + \begin{bmatrix} ik \\ n_{11} \ldots (n_{kk} - 1) \ldots n_{jl} \ldots \end{bmatrix}$$
$$+ \ldots + \begin{bmatrix} iq \\ n_{11} \ldots (n_{qk} - 1) \ldots n_{qq} \end{bmatrix} \quad (4)$$

by hypothesis:

$$\sum_{ij} n_{ij} = m + 1.$$

I put:

$$\sum_i n_{ij} = n_j.$$

From the recurrence relation (4), if the formula

$$X = \frac{\Pi_i n_i !}{\Pi_{ij} n_{ij} !} \quad (5)$$

is true for m bonds, it is also true for $m+1$ bonds, hence for m as large as one wishes. From the physical nature of the problem, when I let the total number of bonds tend to infinity, I can, without changing the result, arrange the first m bonds (m finite) arbitrarily, and in particular I can attribute an arbitrary potential energy to them. That is, I can assume, *a priori*, that equation (5) is exact for m bonds.

Hence, the formula which I have derived for X is independent of the indices i, k, i.e. of the nature of the first and last groups. This is also evident *a priori*.

Finally, I conclude that the probability of a complexion which contains n_{11} groups 11, n_{12} groups 12, ... etc. is:

$$W = A \frac{\Pi_i n_i !}{\Pi_{ij} n_{ij} !} \quad (5a)$$

A is a numerical coefficient which depends on q and on the total number N of carriers.

§ 14.—General equations of the problem.—According to the method used to construct the configurations, to the q relations:

$$\sum_i n_{ij} = n_j \qquad j = 1, 2 \ldots q, \quad (6)$$

it is necessary to add the q analogous relations:

$$\sum_j n_{ij} = n_i \qquad i = 1, 2 \ldots q, \qquad (7)$$

n_i, n_j are the numbers of groups type i and type j, respectively, and the first equations simply express the fact that the number of bonds which join onto a group j is equal to the number of groups of that type. In the same way equations (7) express the fact that the number of bonds which start from a group i is equal to the number of groups of that type.

Since the total number of groups is equal to N/p it is necessary to write also:

$$p \sum_i n_i = N. \qquad (8)$$

It is also necessary to specify now that the total energy of the complexion is equal to U.

Letting w_i be the energy of a group i(1), w'_{ij} the energy of a bond between two consecutive groups i and j, one has:

$$\sum_i n_i w_i + \sum_{ij} n_{ij} w'_{ij} = U. \qquad (9)$$

In the following, I shall put:

$$w_{ji} = w'_{ij} + \frac{w_i + w_j}{2}$$

the w_i will thus disappear from the calculations.

Everywhere I shall make the change of variables:

$$\frac{p n_i}{N} = x_i \qquad \frac{p n_{ij}}{N} = x_{ij}.$$

I take logarithms of the two members of the expression 5a, and as they consist of large numbers I apply Stirling's formula limited to its first two terms:

$$\log n! = n \log n - n$$

I thus obtain the following relations:

$$\frac{p}{N} \log W = \sum_i x_i \log x_i - \sum_{ij} x_{ij} \log x_{ij} + B \qquad (10)$$

$$\sum_i x_{ij} = x_j \qquad \sum_i x_{ji} = x_j \qquad j = 1, 2, \ldots q \qquad (11)$$

$$\sum_{ij} x_{ij} w_{ij} = \frac{pU}{N} \qquad (12)$$

$$\sum_i x_i = 1. \qquad (13)$$

---

*This energy w_1 contains not only the internal energy of a bond, but also the energy of the magnetic moments in the external field H.

The most probable state corresponds to the maximum of W. Then:

$$\frac{p}{N} \delta \log W$$

$$= \sum_i (1 + \log x_i) \delta x_i - \sum_{ij} (1 + \log x_{ij}) \delta x_{ij} = 0. \qquad (14)$$

Also, the δx_i and δx_{ij} are related by the following equations, obtained by differentiating the conditional equations

$$\sum_i \delta x_{ij} = \delta x_i \qquad \sum_i \delta x_{ji} = \delta x_j \qquad (15)$$

$$\sum_{ij} w_{ij} \delta x_{ij} = 0 \qquad (16)$$

$$\sum_i \delta x_i = 0. \qquad (17)$$

§ 15.—**Solution of the problem.**—I solve this problem by the method of Lagrangian multipliers; I multiply equation 17 by $-\log S$, equations 15 by $(\frac{1}{2} + \log \mu_j)$ and $(\frac{1}{2} + \log \lambda_j)$, respectively, and equation 16 by α. I add and obtain:

$$\frac{p}{N} \delta \log W + \alpha \frac{p}{N} \delta U + F(\lambda_j, \mu_j, S, \delta x_i, \delta x_{ij}) = 0 \qquad (18)$$

I put the coefficients of the δx_i and δ_{ij} equal to zero and obtain the following equations:

$$\log x_i - \log S = \log \lambda_i + \log \mu_i \qquad i = 1, 2 \ldots q \qquad (19)$$

$$\log x_{ij} - \alpha w_{ij} = \log \lambda_i + \log \mu_j \qquad i = 1, 2 \ldots q \qquad (20)$$
$$j = 1, 2 \ldots q$$

I have $q^2 + 3q + 2$ equations between the $q^2 + 3q + 2$ unknowns: $x_i, x_{ij}, \lambda_i, \mu_j, S, \alpha$.

Before proceeding further, I shall give the physical significance of the coefficient α. From the equations given above, it can be seen that if W represents the probability of the most probable complexion for the energy U, the variation δW of that probability, when the energy changes by δU, everything else being equal, is given by:

$$\delta \log W = -\alpha \delta U.$$

The entropy is by definition (Boltzmann):

$$E = K \log W$$

and from Carnot's principle*:

$$\delta E = \frac{\delta U}{T}.$$

From these three relations I find:

$$\alpha = -\frac{1}{KT}. \qquad (21)$$

The coefficient α defines the temperature of the system. Putting $a_{ij} = \exp(-w_{ij}/KT)$, equations 19 and 20 become:

$$x_i = S\lambda_i \mu_i \qquad (22)$$

$$x_{ij} = a_{ij} \lambda_i \mu_j \qquad (23)$$

and putting these values of x_i, x_{ij} in equations 11 and 13 one has:

$$\sum_i S\lambda_i \mu_i = 1 \qquad (24)$$

$$\sum_i a_{ij} \lambda_i = S\lambda_j \qquad j = 1, 2, \ldots q \qquad (25)$$

$$\sum_i a_{ji} \mu_i = S\mu_j \qquad j = 1, 2, \ldots q \qquad (26)$$

25 and 26 form two sets of linear, homogeneous equations in λ_i and μ_i. In order to have a non-zero set of solutions, the common determinant of the two systems must be zero:

$$\Delta(S) = \begin{vmatrix} a_{11} - S & a_{12} & \cdots & a_{1q} \\ a_{21} & a_{22} - S & \cdots & a_{2q} \\ \vdots & & & \vdots \\ a_{q1} & \cdots & & a_{qq} - S \end{vmatrix} = 0 \qquad (27)$$

This is an equation of degree q in S. If we denote by A_{ij} the minor of Δ relating to the ith line and j^{th} column, the following relations hold between the λ_i and μ_i:

$$\frac{\lambda_1}{A_{11}} = \frac{\lambda_2}{A_{12}} = \cdots = \frac{\lambda_q}{A_{1q}} = K \qquad (28)$$

$$\frac{\mu_1}{A_{11}} = \frac{\mu_2}{A_{21}} = \cdots = \frac{\mu_q}{A_{q1}} = K' \qquad (29)$$

---

*The system is put in contact with a heat bath and there are assumed to be only changes of heat.

Putting into 25 and 26 the values of λ_i, μ_i from the preceding relations, one has:

$$KK' \sum_i A_{1i} A_{i1} = \frac{1}{S} \qquad (30)$$

but from the known properties of minors of determinants:

$$\frac{A_{11}}{A_{i1}} = \frac{A_{1i}}{A_{ii}}.$$

Equation 30 can be written in the more symmetric form:

$$KK' A_{ii} \sum_i A_{ii} = \frac{1}{S}. \qquad (31)$$

Using 28, 29 and 31, the values of x_i, x_{ij} given by equations 22 and 23 can be written:

$$x_i = \frac{A_{ii}}{\sum A_{ii}} \qquad (32)$$

$$x_{ij} = \frac{a_{ij}}{S} \frac{A_{ij}}{\sum A_{ii}}. \qquad (33)$$

These equations solve the problem. By noting that one can write:

$$A_{ij} = \frac{\partial \Delta}{\partial a_{ij}} \qquad -\frac{\partial \Delta}{\partial S} = \sum_i A_{ii}$$

then the determinant Δ is homogeneous in a_{ij} and in S

$$\sum_{ij} a_{ij} \frac{\partial \Delta}{\partial a_{ij}} + S \frac{\partial \Delta}{\partial S} = q\Delta = 0.$$

Equations 32 and 33 can be transformed into

$$x_i = -\frac{\frac{\partial \Delta}{\partial a_{ii}}}{\frac{\partial \Delta}{\partial S}} \qquad x_{ij} = -\frac{a_{ij}}{S} \frac{\frac{\partial \Delta}{\partial a_{ij}}}{\frac{\partial \Delta}{\partial S}}. \qquad (35)$$

§ 16.—**The case of two different types of carriers.**—The preceding results can be easily extended to the case of a mixture of several carriers. Each group of p carriers will contain, for example, α carriers of type P, and β of type Q. Equation 13 is replaced by the following:

$$\sum_i \alpha_i x_i = P \qquad \sum_i \beta_i x_i = Q \qquad (36)$$

with:

$$P + Q = 1 \qquad \alpha_i + \beta_i = p \qquad (37)$$

P and Q are the proportions of atoms of the two types present in the mixture.

Equation 17 is replaced by the following:

$$\sum_i \alpha_i \delta x_i = 0 \qquad \sum_i \beta_i \delta x_i = 0.$$

Multiplying these by $-\log S$ and $-\log \sigma$ and adding them to the same equations that previously were multiplied by the same coefficients, and cancelling the coefficients of the δx_i and δx_{ij} one has:

$$x_i = S^{\alpha_i} \sigma^{\beta_i} \lambda_i \mu_i$$

$$x_{ij} = a_{ij} \lambda_i \mu_j.$$

Putting these values into the conditional equations, we obtain two sets of linear equations in $\lambda_i \mu_i$, the common determinant of which Δ must be zero:

$$\Delta = \begin{vmatrix} a_{11} - S^{\alpha_1}\sigma^{\beta_1} & a_{12} & & a_{1q} \\ a_{21} & a_{22} - S^{\alpha_2}\sigma^{\beta_2} & \cdots & a_{2q} \\ \vdots & & & \vdots \\ a_{q1} & & & a_{qq} - S^{\alpha_q}\sigma^{\beta_q} \end{vmatrix} = 0 \quad (38)$$

For a given set of values of S, σ, which make the determinant vanish, the values of the x_i, x_{ij} are:

$$x_i = K S^{\alpha_i} \sigma^{\beta_i} A_{ii} \qquad x_{ij} = K a_{ij} A_{ij}.$$

Putting these values into 36 gives:

$$KS \frac{\partial \Delta}{\partial S} + P = 0 \qquad (39)$$

$$K\sigma \frac{\partial \Delta}{\partial \sigma} + Q = 0. \qquad (40)$$

Equations 38, 39 and 40 allow calculation of S, σ, K.

§ 17.—**Internal energy, Entropy, Magnetization.**—In the case of one carrier, from 12:

$$\frac{p}{N} U = \sum_{ij} x_{ij} w_{ij} = -\frac{1}{S \frac{\partial \Delta}{\partial S}} \sum_{ij} a_{ij} w_{ij} \frac{\partial \Delta}{\partial a_{ij}}.$$

but:

$$\frac{\partial a_{ij}}{\partial \frac{1}{KT}} = -w_{ij} a_{ij},$$

and since $\Delta = 0$:

$$\sum_{ij} \frac{\partial \Delta}{\partial a_{ij}} \cdot \frac{\partial a_{ij}}{\partial \frac{1}{KT}} + \frac{\partial \Delta}{\partial S} \frac{\partial S}{\partial \frac{1}{KT}} = 0$$

then

$$U = -\frac{N}{p} \frac{\partial}{\partial \frac{1}{KT}} \log S. \qquad (41)$$

The entropy is given by the formula

$$E = K \log W.$$

For the magnetization one has:

$$\mathfrak{J} = \frac{N}{p} \sum_{ij} x_{ij} \mu_{ij}$$

μ_{ij} being the magnetic moment of a bond.*

In the same way as before, this relation can be transformed into the following:

$$\mathfrak{J} = \frac{N}{p} \frac{\partial}{\partial \frac{H}{KT}} \log S, \qquad (42)$$

since the potential energy for the bond ij is of the form:

$$w_{ij} = w'_{ij} - \mu_{ij} H$$

on writing out the magnetic energy explicitly.

In the case of a mixture of two carriers, one is led to the analogous relations:

$$U = -\frac{N}{p} \frac{\partial}{\partial \frac{1}{KT}} \log [S^P \sigma^Q] \qquad (43)$$

$$\mathfrak{J} = \frac{N}{p} \frac{\partial}{\partial \frac{H}{KT}} \log [S^P \sigma^Q]. \qquad (44)$$

§ 18.—**Study of the case of two neighbours.**†—I shall apply the preceding equations to the particularly simple case of two neighbours. The case of eight or

---

* The bond, of course, does not have a magnetic moment, but to facilitate the calculation it is convenient to locate the moment of the group on its two bonds, as has already been done for the energy (see § 14).

† During proof correction I have become aware of a paper by Ising (Zts. f. Phys. *31*, 1925, 253) which treats this case by another method and finds equation 47.

twelve neighbours is closer to reality, but the corresponding equation for S, given by the preceding theory, will be of the 8th or 12th degree. Moreover, it is only obtained by expanding a determinant of order 8 or 12. The complexity of the calculation prohibits any attempt at a numerical solution.

On the other hand, as the number of neighbours increases, the results approach those obtained by neglecting fluctuations. The extreme case of two neighbours is interesting, since it accentuates the modifications of the classical results brought about by fluctuations.

The four different bonds that have to be considered are shown schematically below:

$$\rightarrow\rightarrow \qquad \rightarrow\!\!\leftarrow \qquad \leftarrow\!\!\rightarrow \qquad \leftarrow\!\!\leftarrow$$

We assume that the externally applied magnetic field H is directed along the row of carriers.

To within a constant, the respective potential energies per bond are:

$$-w-\mu H \qquad 0 \qquad 0 \qquad -w+\mu H.$$

The determinant involving S can be written:

$$\Delta = \begin{vmatrix} e^{\frac{w+\mu H}{KT}} - S & 1 \\ 1 & e^{\frac{w-\mu H}{KT}} - S \end{vmatrix} = 0$$

that is

$$S^2 - 2Se^{\frac{w}{KT}} \operatorname{ch}\frac{\mu H}{KT} + e^{\frac{2w}{KT}} - 1 = 0. \quad (45)$$

This second degree equation for S has two positive roots, one lying between 0 and $\exp\{(w-\mu H)/KT\}$ and the other greater than $\exp\{(w+\mu H)/KT\}$. From the physical significance of the minors of the determinant Δ, they must all have the same sign; since the adjoint determinant of Δ is:

$$\begin{vmatrix} e^{\frac{w-\mu H}{KT}} - S & -1 \\ -1 & e^{\frac{w+\mu H}{KT}} - S \end{vmatrix}$$

Hence the only acceptable root is the largest:

$$S = e^{\frac{w}{KT}} \operatorname{ch}\frac{\mu H}{KT} + \sqrt{1 + e^{\frac{2w}{KT}} \operatorname{sh}^2\frac{\mu H}{KT}}. \quad (46)$$

From equation 42, I have:

$$\mathcal{J} = N\frac{\partial}{\partial\frac{H}{KT}}\log S = \frac{N\mu}{\sqrt{1 + \dfrac{1}{e^{\frac{2w}{KT}} \operatorname{sh}^2\frac{\mu H}{KT}}}} \quad (47)$$

This equation gives the magnetization of a row of carriers, in a magnetic field, when only interactions between adjacent carriers are considered.

§ 19.—**The case of four neighbours.**—As an example of the method, I shall also treat the case of four neighbours. In this case, $p=2, q=4$, there are 16 bonds ij to consider and the equation for S can be written:

$$\Delta(S) = \begin{vmatrix} 1 - Se^{\frac{2\mu H}{KT}} & e^{-\frac{3}{2}\frac{w}{KT}} & e^{-\frac{5}{2}\frac{w}{KT}} & e^{-\frac{3w}{KT}} \\ e^{-\frac{5}{2}\frac{w}{KT}} & e^{-\frac{2w}{KT}} - S e^{-\frac{3w}{KT}} & e^{-\frac{3}{2}\frac{w}{KT}} \\ e^{-\frac{3}{2}\frac{w}{KT}} & e^{-\frac{3w}{KT}} & e^{-\frac{2w}{KT}} - S & e^{-\frac{5}{2}\frac{w}{KT}} \\ e^{-\frac{3w}{KT}} & e^{-\frac{5}{2}\frac{w}{KT}} & e^{-\frac{3}{2}\frac{w}{KT}} & 1 - Se^{-\frac{2\mu H}{KT}} \end{vmatrix} = 0$$

This equation can be expanded in the form:

$$AB - S^2 \operatorname{sh}^2 \frac{\mu H}{KT} = 0,$$

with:

$$A = 1 + e^{-\frac{3w}{KT}} + \frac{\left(e^{-\frac{3}{2}\frac{w}{KT}} + e^{-\frac{5}{2}\frac{w}{KT}}\right)^2}{S - e^{-\frac{2w}{KT}} - e^{-\frac{3w}{KT}}} - S\operatorname{ch}\frac{2\mu H}{KT}$$

$$B = 1 - e^{-\frac{3w}{KT}} - \frac{\left(e^{-\frac{3}{2}\frac{w}{KT}} - e^{-\frac{5}{2}\frac{w}{KT}}\right)^2}{S - e^{-\frac{3w}{KT}} + e^{-\frac{3w}{KT}}} - S\operatorname{ch}\frac{2\mu H}{KT}$$

In the absence of an external field (H=0), the only acceptable root is the largest root of A=0:

$$S_0 = \frac{1}{2}\left[1 + e^{-\frac{2w}{KT}} + 2e^{-\frac{3w}{KT}} + \right. \quad (47a)$$

$$\left. + \sqrt{1 - 2e^{-\frac{2w}{KT}} + 4e^{-\frac{3w}{KT}} + 9e^{-\frac{4w}{KT}} + 4e^{-\frac{5w}{KT}}}\right]$$

When the external field is weak, one can write:

$$S = S_0(1 + \epsilon).$$

ϵ is small and is determined by

$$B_0 \left[\epsilon \frac{\left(e^{-\frac{3w}{2KT}} + e^{-\frac{5w}{2KT}}\right)^2}{\left(S_0 - e^{-\frac{2w}{KT}} - e^{-\frac{3w}{KT}}\right)^2} + \frac{2\mu^2 H^2}{K^2 T^2} + \epsilon \right] = S_0 \frac{4\mu^2 H^2}{K^2 T^2}$$

B_0 is the value of B for $H = 0$. The equation which gives the magnetization is then

$$\mathcal{J} = 2 \frac{N\mu^2 H}{KT} \frac{B_0 + 2S_0}{B_0} \frac{1}{1 + \left(\frac{e^{-\frac{2w}{2KT}} + e^{-\frac{2w}{2KT}}}{S_0 - e^{-\frac{2w}{KT}} - e^{-\frac{3w}{KT}}}\right)^2}$$

At high temperatures, one can write:

$$S_0 = 4 - \frac{8w}{KT} + 10\frac{w^2}{K^2 T^2} + \ldots$$

$$B_0 = -4 + 11\frac{w}{KT} - \frac{59}{4}\frac{w^2}{K^2 T^2} + \ldots$$

and consequently:

$$\mathcal{J} = \frac{N\mu^2 H}{KT\left(1 - \frac{2w}{KT} + \frac{w^2}{K^2 T^2}\right)},$$

putting $(2w/KT) = \theta/T$, I have

$$\mathcal{J} = \frac{N\mu^2 H}{KT\left(1 - \frac{\theta}{T} + \frac{1}{4}\frac{\theta^2}{T^2}\right)}. \tag{48}$$

Similarly, *at low temperatures*, I have:

$$S_0 = 1 + 2e^{-\frac{3w}{KT}} + \ldots$$

and:

$$\mathcal{J} = \frac{N\mu^2 H}{KT} e^{+\frac{3}{2}\frac{\theta}{T}}. \tag{49}$$

These latter calculations show the extent to which the complexity of the problem increases, when the interaction of more than two neighbours is taken into account.

CHAPTER III

STUDY OF THE INITIAL SUSCEPTIBILITY

§ 20.—Purpose of this chapter.—In the preceding chapter, by using certain simplifying assumptions, I have made a rigorous study of a lattice with two neighbours and of a lattice with four neighbours, and I have obtained expressions for the magnetization, the internal energy, etc. I shall now go on to discuss the results from a physical point of view.

I shall introduce an approximate treatment of the problem of the initial susceptibility, when there are any number, $2p$, of neighbours. The results obtained above in the simple cases $p = 1$, $p = 2$ will help to determine the limits within which the approximations of my new calculation are valid.

§ 21.—Initial susceptibility in the case of two neighbours.—When $\mu H/KT$ is small, equation 47 becomes:

$$\mathcal{J} = \frac{N\mu^2 H}{KT} e^{\frac{w}{KT}} \tag{50}$$

I let $w = K\theta$, introduce the susceptibility χ and have:

$$\frac{1}{\chi}\frac{N\mu^2}{K\theta} = \frac{T}{\theta} e^{-\frac{\theta}{T}}$$

I have shown in figure 2, curve A, the variation of $(1/\chi)(N\mu^2/K\theta)$ as a function of (T/θ).

The curve has as asymptote the line θF with equation:

$$\frac{1}{\chi}\frac{N\mu^2}{K\theta} = \frac{T}{\theta} - 1.$$

The distance from the curve to the asymptote varies as $(\theta/2T)$, the approach is not very rapid. At absolute zero, the curve is tangential to the temperature axis: $1/\chi$ vanishes only at the absolute zero. There is no spontaneous magnetization. It is only that the system becomes easier to magnetize as the temperature is lowered.

On the other hand, if I had neglected fluctuations, I would have obtained, in place of curve A, the line θF. In that case, $1/\chi$ vanishes at θ; there is a temperature region 0θ where there is a spontaneous magnetization.

When there are no interactions, $w=0$, and equation 47 reduces to:

$$\mathcal{J} = N\mu \, \text{th} \, \frac{\mu H}{KT} \tag{51}$$

This is the same law of magnetization as that given by Lenz.

§ 22.—Approximate calculation of the initial susceptibility, valid for $T > \theta$.—I shall now develop an approximate method, valid for any number of neighbours. This is, basically, a first approximation in the calculation of fluctuations. The approximation will consist of assuming that the carriers neighbouring a central carrier have independent orientation probabilities.

In the material, each carrier, of moment μ, is surrounded by $2p$ neighbours. The relative interaction energy of two parallel carriers is $-w/2$ and that of two anti-parallel carriers is $w/2$.

In an external magnetic field H, the material has a magnetization $\mathcal{T}$, and, using Lenz's hypothesis, there are $N(1+\varepsilon)/2$ carriers oriented in the direction of the field and $N(1-\varepsilon)/2$ in the opposite direction, so that,

$$\mathcal{J} = N\mu\varepsilon \tag{52}$$

Amongst the $2p$ neighbours of the carrier considered, there are $2p!/(2p-q)!\,q!$ different ways of taking $2p-q$ carriers parallel to the field and q anti-parallel. The probability $\bar{w}$ of any one of these distributions is, if one assumes the independence of the probabilities of orientation of the neighbouring carriers*:

$$\bar{\omega} = \left(\frac{1}{2}\right)^{2p} (1+\varepsilon)^{2p-q} (1-\varepsilon)^q.$$

For this configuration of neighbouring carriers, the magnetization of the central carrier is

$$\bar{\mu} = \mu \, \text{th} \, \frac{\mu H + w(p-q)}{KT}.$$

When it is parallel to the magnetic field, its energy is effectively:

$$-\mu H - \frac{w}{2}(2p-q) + \frac{w}{2}q = -\mu H - w(p-q)$$

and this expression with the sign changed when it points in the opposite direction.

---
*This assumption becomes more valid as the temperature increases.

I obtain the total magnetization, by taking the sum of the partial magnetizations, related to the different configurations of the neighbours, each with their statistical weights:

$$\mathcal{J} = N\mu \sum_{q=0}^{2p} \frac{2p!}{(2p-q)!} \left(\frac{1}{2}\right)^{2p} (1+\varepsilon)^{2p-q} (1-\varepsilon)^q$$

$$\text{th} \, \frac{\mu H + w(p-q)}{KT}$$

This relation, together with (52), can be solved to give ε and hence the complete law of magnetization.

I intend to study only the initial susceptibility: ε is then as small as one wishes, and one can reduce $(1+\varepsilon)^{2p-q}(1-\varepsilon)^q$ to the first two terms in the series expansion. Similarly, the magnetic field H being small, one can expand the hyperbolic tangent in a Taylor series:

$$\text{th} \, \frac{\mu H + w(p-q)}{KT} = \frac{w(p-q)}{KT} + \frac{\mu H}{KT}\left(1 - \text{th}^2 \frac{w(p-q)}{KT}\right) + \ldots$$

I thus obtain:

$$\mathcal{J} = \frac{N\mu^2 H}{KT} \cdot \frac{1 - \left(\frac{1}{2}\right)^{2p} \sum_{q=0}^{2p} \frac{2p!}{(2p-q)!\,q!} \text{th}^2 \frac{2w}{KT}(p-q)}{1 - \left(\frac{1}{2}\right)^{2p} 2\sum_{q=0}^{2p} \frac{2p!}{(2p-q)!q!}(p-q)\text{th}\frac{w}{KT}(p-q)}$$

In order to simplify this, I expand the hyperbolic tangent in a series and neglect terms in $w^5(p-q)^5/K^5T^5$ and higher terms; the series will only be valid if T is large, as I shall explain in more detail later.

I use the known relations:

$$\left(\frac{1}{2}\right)^{2p} \sum_{q=0}^{2p} \frac{2p!}{(2p-q)!q!}(p-q)^2 = \frac{p}{2}$$

$$\left(\frac{1}{2}\right)^{2p} \sum_{q=0}^{2p} \frac{2p!}{(2p-q)!q!}(p-q)^4 = \frac{p(3p-1)}{4}$$

I put $\theta = pw/K$, and, after completing the calculation, the inverse susceptibility, $1/\chi$, is given by the following expression:

$$\frac{1}{\chi} = \frac{K\theta}{N\mu^2}\left[\frac{T}{\theta} - 1 + \frac{1}{2p}\frac{\theta}{T} - \frac{1}{6p^2}\frac{\theta^2}{T^2} + \frac{2-3p}{12p^3}\frac{\theta^3}{T^3} + \ldots\right]$$

When $p=1$, it can be seen that this expansion coincides up to the term in θ^3/T^3 with the following expansion obtained by the exact method of Chapter II:

$$\frac{1}{\chi} = \frac{K\theta}{N\mu^2}\left(\frac{T}{\theta} - 1 + \frac{1}{2}\frac{\theta}{T} - \frac{1}{6}\frac{\theta^2}{T^2} + \frac{1}{24}\frac{\theta^3}{T^3} + \ldots\right)$$

Equally, for $p=2$ it agrees with the first terms of the exact expansion.

It is interesting to note that the preceding argument does not involve the position of the neighbours, but only their number.

§ 23.—**Approximate calculation of the initial susceptibility valid for $T<\theta$.**—The preceding series are only valid if $\theta/T<1$. Below the Curie point, it is necessary to develop a different argument.

When the thermal agitation is weak, the carriers tend to become parallel to each other. Going along the line of carriers, one meets n carriers aligned in one direction, then n' carriers aligned in the opposite direction, etc. ... n and n' are large: put another way, the number of carriers is large compared with the number of reversals,* so that two consecutive reversals are separated by a large number of carriers. From the point of view of the energy, it is as if two neighbouring parallel carriers interact with an energy zero and two neighbouring antiparallel carriers with an energy $p(p+1)w/2$, the potential energy of a reversal.

To find the law of magnetization under these conditions, it is sufficient to replace w by $p(p+1)w/2$ in equation 50. At low temperature I thus have:

$$\frac{1}{\chi} = \frac{KT}{N\mu^2} e^{-\frac{(p+1)}{2}\frac{\theta}{T}} \quad \text{valable pour } T<\theta.$$

§ 24.—**Discussion.**—To summarize, the results obtained in the study of the initial susceptibility are, as a function of the number of neighbours, $2p$, as follows:

$$T > \theta \qquad \frac{1}{\chi} = \frac{K\theta}{N\mu^2}\left[\frac{T}{\theta} - 1 + \frac{1}{2p}\frac{\theta}{T} + \ldots\right] \quad (53)$$

$$T < \theta \qquad \frac{1}{\chi} = \frac{KT}{N\mu^2} e^{-\frac{p+1}{2}\frac{\theta}{T}} \quad (54)$$

These results are in agreement, both at high and low temperatures, with the complete expressions obtained for $p=1$ and $p=2$. Moreover, for $p=\infty$ the formulae above reduce to:

$$T > \theta \qquad \frac{1}{\chi} = \frac{K\theta}{N\mu^2}\left[\frac{T}{\theta} - 1\right] \quad (55)$$

$$T < \theta \qquad \frac{1}{\chi} = 0 \quad (56)$$

These are the expressions obtained by the classical theory, without taking account of fluctuations. Since it is obvious, a priori, that with an infinite number of neighbours these are no fluctuations, the results are consistent.

In figures 2 and 3 I have drawn, from equations 53 and 54, the curves corresponding to different values of p. Figure 3 has a scale five times larger than figure 2.

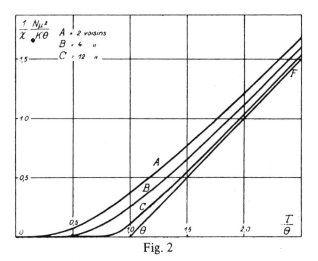

Fig. 2

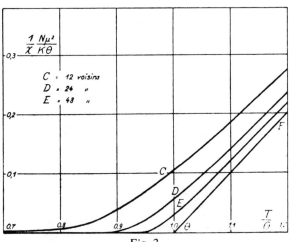

Fig. 3

---

*By a reversal I mean a bond between two antiparallel carriers.

As p increases the curves approach more and more closely to the broken line OθF. They all have the same asymptote:

$$\frac{1}{\chi}\frac{N\mu^2}{K\theta} = \left(\frac{T}{\theta} - 1\right)$$

The fundamental fact, however, is the following: whatever the value of p, $1/\chi$ vanishes at the absolute zero; there is no spontaneous magnetization.

Hence, the short range interatomic interactions, taken by themselves, do not give spontaneous magnetization. The mechanism for that must be sought elsewhere; the short range interactions only have the effect of making the process of magnetization very much more easy: the term $\exp\{-(p+1)\theta/2T\}$ tends to zero very rapidly as p increases.

§ 25.—The case of a negative bond energy.

In the preceding chapters I assumed that w was positive. This is the case for ferromagnetic bodies. The same arguments and the same calculations apply for a negative value of w and lead to the equation (for two neighbours)

$$\frac{1}{\chi}\frac{N\mu^2}{K\theta} = \frac{T}{\theta} e^{\frac{\theta}{T}} \qquad (57)$$

I have shown in figure 4 the variation of $(N\mu^2/K\theta\chi)$ as a function of (θ/T). At high temperatures, the curve is asymptotic to the line BF. That line represents the variation of $1/\chi$ if fluctuations are neglected. At low temperatures the behaviour is completely different: $1/\chi$ is a minimum at $T = \theta$. The susceptibility goes to zero at the absolute zero.

There are no examples of such behaviour.

The assumption of Lenz is much too drastic here. I shall show in Chapter VIII that, if the carriers of the magnetic moment can take up any orientation with respect to the magnetic field, one is led to a constant paramagnetism at low temperatures.

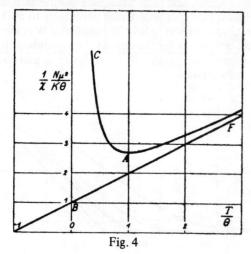

Fig. 4

CHAPTER IV

FERROMAGNETISM

§ 26.—Effect of distant carriers.

In Chapter III I have studied the temperature variation of the initial susceptibility in the simple schematic case of equal interactions between a carrier and its $2p$ nearest neighbours.

In reality more distant carriers also have an effect, and a second approximation consists of dividing into two groups the carriers which interact with a central carrier: a first group will be formed by the $2p$ neighbours which I have already taken into account and a second group; neighbours in the second shell and beyond. As this second shell contains, generally, four times as many carriers as the first, one can see, from the form of the curves already given, that it is legitimate to neglect fluctuations, all the more so since the total interaction with them is much weaker than for the first.

Hence, I shall replace the effect of shells beyond the first by a molecular field proportional to the magnetization,

$$h_m = n\bar{\mu}.$$

Strictly, this assumption is only valid when the ferromagnetic material is in an external magnetic field large compared with the coercive field.

Ferromagnetic materials exhibit very different effects.

In weak magnetic fields, of the order of a few gauss, hysteresis is shown.

Now, at a given temperature, if the field is increased up to several thousand gauss, saturation is approached according to the following law of approach [6]:

$$\sigma = \sigma_T \left(1 - \frac{a}{H}\right),$$

a is independent of the temperature: one is dealing with a crystalline phenomenon: inside an elementary crystal there are structural fields [7], equivalent to magnetic fields of some hundreds of gauss, which make magnetization easier along one or other of the crystal axes. In order to completely align the magnetization, these structural fields must be overcome. I have deliberately kept away from these phenomena. I simply note that an interpretation of them can be given naturally in terms of the ideas developed above. In fact, according to my basic assumptions, the two pairs of carriers shown below:

have the same potential energy. I made this assumption just so that the properties in different directions in the lattice should not be different. This is certainly not so and the introduction of a difference between the two potential energies introduces a structural field.

This is not the place to develop this point further, I shall also assume that measurements are made along the easy direction and I propose simply to account for the spontaneous magnetization.*

§ 27.—Demagnetization energy.

Anticipating the results of Chapter V, the energy per carrier required to demagnetize the material, or, rather, to go from absolute zero to a temperature much higher than θ, is:

$$W_f = \frac{pw}{2} = \frac{K\Theta}{2} \quad (58)$$

this is for a molecular field with fluctuations. In the case of a molecular field without fluctuations it is [8]:

$$W_n = \frac{n\mu^2}{2}. \quad (59)$$

Then, from the superposition of the two types of molecular field, the total demagnetizing energy per carrier is:

$$W = W_f + W_n. \quad (60)$$

Consider a material in which there are any number of neighbours. Assume, for the moment, that h_m does not exist, then $1/\chi$ as a function of temperature is shown in curve C of figure 5.

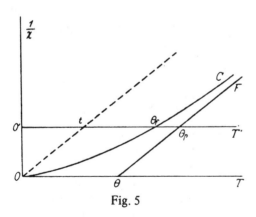

Fig. 5

If I now introduce h_m, one has:

$$\mathcal{J} = \frac{H}{\frac{1}{\chi} - n}$$

and the new value of the susceptibility χ' is given as a function of the old by the relation:

$$\frac{1}{\chi'} = \frac{1}{\chi} - n.$$

It is sufficient to keep the curve C already drawn, and to take a new temperature axis, O'T', such that O O' $= n$. It meets the asymptote θF at θ_p, which is the paramagnetic Curie point of the material and, since the slope of the asymptote is (K/μ^2), I find:

$$O'\theta_p = O\Theta + n\frac{\mu^2}{K}$$

which, from (58) and (59), can be written:

$$\frac{K\theta_p}{2} = \frac{n\mu^2}{2} + \frac{K\Theta}{2} = W_f + W_n = W. \quad (61)$$

The paramagnetic Curie point depends only on the total demagnetizing energy. One could also say that the demagnetizing energy per carrier of a material with a paramagnetic Curie point θ_p is equal to the energy of one degree of freedom at that temperature.

The axis O'T' meets the curve C at θ_f. It is easy to see, following the argument of Weiss [2], that below θ_f the material has a spontaneous magnetization, although it does not have this above θ_f. The region O'θ_f is the region of ferromagnetism, the region $\theta_p T'$ that of paramagnetism, θ_f is the ferromagnetic Curie point.

When the number of neighbours $2p$ is low, an extremely weak molecular field (n very small) is sufficient to produce a Curie point θ_f very close to θ_p. Under these conditions, θ_p is practically coincident with Θ: taking $p=2$ one finds for iron that a thousandth of a gauss is sufficient to give $\theta_f = 0.94\ \Theta$.

To illustrate the completely different significance of the two Curie points, one can imagine two materials, with the same total demagnetizing energy and thus the same paramagnetic Curie point θ_p, one having a ferromagnetic Curie point close to absolute zero, and the other a ferromagnetic Curie point near to θ_p. For the first it is sufficient to take W_n small and W_f close to W, and conversely for the second. I have shown these two extreme cases schematically in figure 6a and 6b. In both cases I have taken $p=1$.

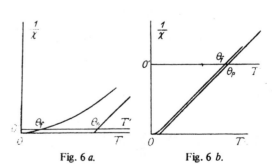

Fig. 6 a.   Fig. 6 b.

---

*Spontaneous magnetization, not remanent magnetization.

Thus the difference between the two Curie points is a function of both the number of neighbours in the first shell, and the ratio between the interaction which arises from the first shell and that from the following shells.

§ 28.—Approach to saturation in the general case.

I shall now study, by an approximate method, the magnetization in a strong field, below the Curie point. I shall treat this problem for any value of p by a particular method and, as a justification of the method, I shall compare the results with those which I have already obtained by a rigorous method for $p=1$.

I shall suppose that conditions are such that the great majority of the carriers will be pointing in the direction of the magnetic field. In this state there will be, distributed along the length of the line of carriers pointing in the field direction, carriers in groups of 1, 2, ... q, pointing in the opposite direction, but, if the magnetization reaches 90% of saturation, the carriers, and, consequently, the reversed groups will be separated by an average of at least ten carriers pointing in the field direction: hence one can assume that on both sides of a group of reversed carriers there are at least p carriers pointing in the field direction.

Under these conditions, the ptoential energy of two parallel carriers being taken as 0, the potential energy of an isolated reversed carrier is $2pw$. The potential energy of a group of q reversed carriers, isolated in a line of carriers pointing in the field direction, is equal to:

$$q(2q - q + 1)w, \text{ si}: q \leqslant p, \text{ et à}: p(p+1)w, \text{ si}: q > p.$$

Taking the probability of a particular carrier being parallel to the general direction of magnetization to be 1, the probability of that carrier being antiparallel to the field in the middle of parallel carriers is $\exp(-2pw/KT)\exp(-2\mu H/KT)$, since the difference in potential energy between the two orientations is $2pw + 2\mu H$. Similarly, the probability of that same carrier being in an isolated group of antiparallel carriers is $q\exp(-w_q/KT)$, where w_q is the potential energy of the group. I have multiplied by q to take account of the q positions which the initially chosen carrier can occupy in the group. In addition I have:

$$w_q = q(2p - q + 1)w + 2q\mu H \quad \text{pour } q \leqslant p$$
$$w_q = p(p+1)w + 2q\mu H \quad \text{pour } q > p.$$

I put:

$$W = \sum_{q=1}^{\infty} q e^{-\frac{w_q}{KT}}.$$

The magnetization is then:

$$\mathcal{J} = N\mu \frac{1-W}{1+W} \qquad (62)$$

This equation is only valid if W is small. In the case where $p=1$, one has:

$$W = e^{-\frac{2w}{KT}} \sum_{q=1}^{\infty} q e^{-\frac{2q\mu H}{KT}} = \frac{e^{-\frac{2w}{KT}}}{4 \operatorname{sh}^2 \frac{\mu H}{KT}}$$

which gives:

$$\mathcal{J} = N\mu \left[1 - \frac{e^{-\frac{2w}{KT}}}{2 \operatorname{sh}^2 \frac{\mu H}{KT}} \right]$$

This equation agrees exactly with that which can be deduced from equation 47, in the case where $\exp(2w/KT)\operatorname{sh}^2(\mu H/KT)$ is large. Hence the method used is correct.

The general expression for W can also be written:

$$W = \sum_{q=1}^{p-1} q e^{-\frac{q(2p-q+1)w}{KT}} e^{-\frac{2q\mu H}{KT}} + e^{-\frac{p(p+1)w}{KT}} \frac{1}{4 \operatorname{sh}^2 \frac{\mu H}{KT}} \quad (63)$$

Let us assume first that the number of neighbours $2p$ is sufficiently large, with $(\mu H/KT)$ non-zero. Writing out the first terms under the Σ sign and replacing pw by $K\theta$, one has:

$$W = e^{-\frac{2\theta}{T}} e^{-\frac{2\mu H}{KT}} + 2e^{-\frac{4\theta}{T}} e^{\frac{2\theta}{pT}} e^{-\frac{4\mu H}{KT}} + \dots$$

The second term is much smaller than the first; the law of approach is approximately of the form:

$$\mathcal{J} = N\mu \left[1 - 2e^{-\frac{K\theta + \mu H}{KT}} - \varepsilon \right] \quad (63a)$$

with ε positive and very small.

This limiting law of approach is exactly that which would be obtained by neglecting fluctuations in the molecular field.

However, it is important to note the presence of the term $\exp\{-\theta(p+1)/T\}\{4 \operatorname{sh}^2(\mu H/kT)\}^{-1}$ in the expression for W. It shows that the material becomes demagnetized when the external field is zero. In fact, whatever the value of the factor $\exp\{-\theta(p+1)/T\}$ one can find a sufficiently small value of H so that $\exp\{-\theta(p+1)/T\}\{4\,\mathrm{sh}^2(\mu H/kT)\}^{-1}$ becomes as large as one wishes. In that case, however, W is no longer small and equation 62 is not valid.

These results, together with those I have already obtained in Chapter III in the study of the initial susceptibility, can be summarized in the following expression for the law of magnetization:

$$\mathcal{J} = \frac{N\mu}{\sqrt{1 + \dfrac{e^{-\frac{\theta}{T}(p+1)}}{\mathrm{sh}^2 \dfrac{\mu H}{KT}}}} (1 - 2V) \qquad (64)$$

with

$$V = \sum_{q=1}^{p-1} q\, e^{-\frac{q(2p-q+1)}{KT}}\, e^{-\frac{2q\mu H}{KT}}$$

This expression is valid when the magnetization is small or in the region of saturation, whatever the value of p.

Whatever the magnetization, it is rigorous for $p=1$, hence I think that it provides a useful approximation for any magnetization and for any value of p.

The character of this law of magnetization is very interesting: $\exp\{-\theta(p+1)/T\}$ is very small if p is not very high (for example, equal to $1/22{,}000$ for $p=4$ and $T=\theta/2$) so that a very weak external magnetic field is sufficient to give the expression,

$$\frac{1}{\sqrt{1 + \dfrac{e^{-\frac{\theta}{T}(p+1)}}{\mathrm{sh}^2 \dfrac{\mu H}{KT}}}},$$

a value close to unity. Thus, the relative saturation $N\mu(1-2V)$ is reached very quickly at the temperature considered. On the other hand, in order to reach the absolute saturation $N\mu$, very intense fields, of the order of $\theta K/\mu$, i.e. some million gauss, are required.

Unfortunately, it is impossible to verify equation 64 experimentally. All the phenomena in relatively weak fields, i.e. the only fields that can be produced experimentally, are profoundly modified by crystalline effects due to structural fields of the order of 1,000 gauss.

Nevertheless, note that the preceding theory allows one to avoid the idea of elementary domains in the interpretation of magnetization curves.

§ 29.—Spontaneous magnetization near the curie point.

The general form of the magnetization law, valid for small values of μ is: (a and b are functions of T)

$$\frac{\bar{\mu}}{\mu} = a\frac{\mu H}{KT} + b\left(\frac{\mu H}{KT}\right)^3 + \ldots \qquad (65)$$

In the case of spontaneous magnetization the material is subject to the single molecular field:

$$h_m = n\vec{\mu} \qquad (66)$$

Comparing 65 and 66 one sees that the spontaneous magnetization vanishes for $T=\theta$ such that:

$$a_0 \frac{n\mu^2}{K\theta} = 1$$

where a_0 is the value of a at $T=\theta$. One can easily deduce the spontaneous magnetization μ_s for any value of T in the region of θ:

$$\frac{\mu_s^2}{\mu^2} = \frac{a_0^3}{b}\frac{T^3}{\theta^3} - \frac{aa_0^2}{b}\frac{T^2}{\theta^2} \qquad (67)$$

In the study of magnetism, it is usual to consider the square of the magnetization as a function of temperature. It is of interest to know the slope of that curve at the Curie point, i.e. the quantity:

$$\mathrm{tg}\, V = \left[\frac{d\left(\dfrac{\mu_s^2}{\mu^2}\right)}{d\left(\dfrac{T}{\theta}\right)}\right]_\theta$$

a simple calculation gives:

$$\mathrm{tg}\, V = \frac{a_0^3}{b_0}\left(1 - \frac{1}{a_0}\frac{da}{dT}\right) = \frac{a_0^3}{b_0}\frac{d}{dt}\log\frac{1}{\chi}. \qquad (68)$$

Using equation 64, one can put as a first approximation:

$$\frac{\bar{\mu}}{\mu} = a\frac{\mu H}{KT} - \left(\frac{a^3}{2} - \frac{a}{3}\right)\frac{\mu^3 H^3}{K^3 T^3} + \ldots \qquad (69)$$

with

$$a = e^{\frac{\theta}{T}\frac{p+1}{2}}.$$

I can neglect $a/3$ in comparison with $a^3/3$ and I thus have:

$$\mathrm{tg}\, V = -2\left(1 + \frac{p+1}{2}\right). \qquad (70)$$

For $p=0$ one finds the classical value -3 [9].

If p is non-zero, the absolute value of the slope is larger: the curve is straighter.

I recall that the value found in Langevin's theory is:

$$\text{tg V} = -\frac{5}{3}.$$

Assuming that there is spatial quantization with a quantum number j, one obtains the following formula given by Stoner [9]:

$$\text{tg V} = -\frac{10(j+1)^2}{3[j^2+(j+1)^2]}.$$

As j varies from $\frac{1}{2}$ to infinity, the corresponding values of tg V vary from -3 to $-5/3$. If one applied the theory of fluctuations to these more complicated cases, one would similarly obtain values for tg V larger in absolute magnitude. However, experiment gives values of |tg V| larger than 3. From the values of $-d\sigma^2/dT$ obtained by Weiss and Forrer [10] one can deduce the value -5.3 for tg V.

Hence, whichever mechanism is adopted, it is necessary to introduce fluctuations.

§ 30.—**Conclusions.**—In summary, the curve of spontaneous magnetization, obtained taking account of fluctuations is always below the curve obtained by neglecting them. This is a result of the presence of ε in equation 63a. But it is above the spontaneous magnetization curve of a material, with a molecular field without fluctuations, which would have its Curie point at θ_f. I have shown these properties schematically in figure 7, the dotted curves represent the spontaneous magnetization for substances with a molecular field without fluctuations, the full curve that of a substance having a molecular field with fluctuations.

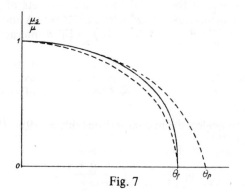

Fig. 7

CHAPTER V

INTERNAL ENERGY AND SPECIFIC HEAT

§ 31.—**The case where there is neither external magnetic field nor molecular field h_m.**—Under the assumption of two neighbours ($p=1$) the following expression is obtained for the internal energy (§ 18):

$$U = \frac{Nw}{1+e^{\frac{w}{KT}}}, \qquad (71)$$

to within a constant.

To go from absolute zero to a very high temperature I must give to the material the energy $w/2$ or $K\Theta/2$ per carrier.

Even above the paramagnetic Curie point, it is still necessary to give energy to the material, i.e. there is an anomaly in the specific heat, above the ferromagnetic Curie point.

In figure 8 I have shown $U = f(T)$ (curve A). With the assumption of four neighbours, I have:

$$U = -\frac{\partial S_0}{\partial \frac{1}{KT}},$$

where S_0 is given by equation 47a.

I have used that expression to calculate curve B.

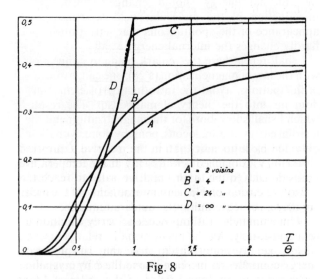

Fig. 8

Finally, in D, I have drawn the curve obtained by neglecting fluctuations in the molecular field. In that case, I have the following relation between the square of the spontaneous magnetization and the internal energy [11]:

$$U = -n\frac{\mu_s^2}{2}.$$

When the number of neighbours increases the curves of internal energy approach curve D.

By using the results of the method of paragraph 22, it is easy to obtain a power series for U valid above the Curie point; I have:

$$U = \frac{wp}{2} - \frac{1}{2}\sum_{q=0}^{2p}\frac{2p!}{(2p-q)!\,q!}w(p-q)\,\text{th}\,\frac{w(p-q)}{KT}$$

Using the same notation, and expanding:

$$U = \frac{pw}{2}\left(1 - \frac{1}{p}\frac{\Theta}{T} + \frac{1}{6}\frac{3p-1}{p^3}\frac{\Theta^3}{T^3} + \cdots\right) \quad (72)$$

In this equation U is taken to be zero at absolute zero.

§ 32.—**General case.**—In the preceding I have not taken into account the superimposed uniform molecular field h_m; but showed in the course of section 28, that it was very small since it represents the effect of carriers far from the central carrier. This implies that below the Curie point when one goes from the state of zero magnetization to apparent saturation $N\mu(1-2V)$ the internal energy should change a little. The introduction of the molecular field h_m causes the appearance of the spontaneous magnetization but it hardly changes the internal energy at all.

I shall repeat here the remark made in connection with the law of magnetization (§ 28); one can, by means of fluctuations, avoid the introduction of elementary domains into the theory of magnetism. The results which I shall show, both for the magnetization and the internal energy, are analogous, point by point, with those obtained by supposing that, in the material, there are elementary domains in which the magnetization is uniform and equal to the relative saturation $N\mu(1-2V)$. The idea of elementary domains is a useful language which interprets and summarizes the experimental facts, without necessarily corresponding to a physical reality.

It is equally important to note that the internal energy continues to increase above the paramagnetic Curie point, up to $T = \infty$.

The specific heat of a ferromagnetic body, above the Curie point, is again greater than if the material were not ferromagnetic. When the number of neighbours is infinite, one observes a discontinuity in the specific heat; when it is finite, there is a spreading out of that discontinuity. The spreading becomes larger as the number of neighbours decreases. It is directly related to the difference between the Curie points.

From equation 72, it can be seen that the total demagnetizing energy per carrier is equal to $pw/2$, i.e. $K\Theta/2$, Θ being the paramagnetic Curie point of the material.

This is a quantity accessible to experiment, it is sufficient to measure the area between the curve giving the atomic specific heat of a ferromagnetic and that for the atomic specific heat of an analogous non-ferromagnetic material. This is a means of determining the number of carriers, analogous to that of the classical theory.

CHAPTER VII*

ALLOYS

§ 36.—**Basic assumptions.**—I shall now, following the guiding principles already developed, consider a metallic solid solution.

Consider such a solution, consisting of a total of N carriers, PN of which belong to the species A and have a moment μ_A, the other QN belong to the species B and have a moment μ_B. By definition, one has

$$P + Q = 1.$$

I assume that there is no superlattice, i.e. that the two kinds of carriers are distributed at random on the lattice points. Furthermore, I shall assume that this distribution is not modified by the introduction of an external magnetic field.

As in Chapter II, I shall only take into account interactions between contiguous carriers. I have defined (§ 9) the potential energy of interaction between two carriers of the same type, for example A. I shall now denote this by w_{AA}; similarly, for the carriers of type B I define an energy w_{BB}.

However, in the crystal there are also sets of two neighbouring carriers of which one is of type A and the other of type B, for which there is a third interaction energy w_{AB}.

The quantities μ_A, μ_B, w_{AA}, w_{BB} are given by study of the pure materials; thus it is necessary to account for the properties of the solid solution by means of a single parameter w_{AB}.

I shall consider the consequences of these assumptions from the double point of view of ferromagnetism and paramagnetism.

---

*Chapter VI (§ 33–§ 35) has been omitted.

§ 37.—Ferromagnetism.

In a ferromagnetic body the saturation at absolute zero provides the result which can be most easily interpreted.

Consider a ferromagnetic material A($w_{AA} > 0$), with a moment μ_A, to which I add small quantities of another material B, with moment μ_B.

In the pure material A, at absolute zero, all the carriers are aligned parallel, so that the observed magnetization per atom is μ_A.

§ 38.—Case of low concentrations.

I now replace some of the A carriers by B carriers so as to make alloys of progressively increasing concentration in B.

The first B carriers added are surrounded by only A carriers; their orientation with respect to the latter is thus determined by the sign of w_{AB}.

If w_{AB} is positive, the new B carriers arrange themselves parallel to the general magnetization. The atomic magnetization, as a function of the atomic concentration, is given by the equation:

$$\mu_0 = P\mu_A + Q\mu_B \qquad w_{AB} > 0 \qquad (74)$$

This is a straight line; extrapolated it gives the moment μ_B of the pure material B.

If w_{AB} is negative, the B carriers arrange themselves antiparallel to the general magnetization. The magnetization as a function of the concentration, is given by the equation:

$$\mu_0 = P\mu_A - Q\mu_B \qquad w_{AB} < 0 \qquad (75)$$

This is a straight line; extrapolated it gives the moment $-\mu_B$ for the pure B material, i.e. an apparent negative moment.

Hence, when the concentration of B is low the results are independent of the sign of w_{BB}.

In any case, study of low concentrations allows determination of the moment μ_B.

§ 39.—Case of high concentrations.

The effects here are more complicated; in order to make the ideas more concrete, I assume that there are twelve neighbours, as for example, in the face-centred cubic lattice. The probability of any carrier having α neighbours of type A, and β neighbours of type B is:

$$\bar{\omega}_{\alpha\beta} = \frac{12!}{\alpha!\,\beta!} P^\alpha Q^\beta \qquad (\alpha + \beta = 12).$$

When w_{AB} and w_{BB} are both positive, the most stable situation corresponds to all the carriers being parallel. One then has, for all concentrations, a linear variation of the atomic saturation.

But one can also meet the following important case: w_{AB} is positive, w_{BB} is negative.

I assume that all the carriers neighbouring a given carrier B, i.e. α A carriers and β B carriers, are arranged parallel to the general magnetization (1).

It is obvious that the orientation of the central B carrier is determined by the sign of the quantity $H = \alpha w_{AB} + \beta w_{BB}$.

If H is positive, the central carrier lies parallel to its neighbours, while if H is negative it lies antiparallel. Denoting by α_0 the value of α for which H is zero, the probability of a B carrier being reversed is:

$$\Pi = \sum_{\alpha=0}^{\alpha_0} \bar{\omega}_{\alpha\beta}$$

and the saturation at absolute zero is given by the equation:

$$\mu_0 = P\mu_A + Q(1 - 2\Pi)\mu_B \qquad (76)$$

This equation is correct, in so far as one can neglect Π^2 in comparison with Π.

In table I, I give, for $\alpha + \beta = 12$ and different values of P, the value of $\bar{\omega}_{\alpha\beta}$. That table allows the calculation, point by point, of the curve of atomic saturation as a

TABLE I
Values of $\bar{\omega}_{\alpha\beta}$

P =	0,50	0,45	0,40	0,35	0,30	0,25	0,20	0,15	0,10	0,05
β = 0	0,000	0,000	0,000	0,000	0,000	0,000	0,000	0,000	0,000	0,000
1	003	001	000	000	000	000	000	000	000	000
2	016	007	002	000	000	000	000	000	000	000
3	054	028	012	005	001	000	000	000	000	000
4	120	076	042	020	008	002	000	000	000	000
5	193	149	101	059	029	011	003	000	000	000
6	225	212	177	128	079	040	016	004	000	000
7	193	222	227	204	158	103	053	019	004	000
8	120	170	213	237	231	194	133	068	021	002
9	054	092	142	195	240	258	236	172	085	017
10	016	034	064	109	168	232	283	292	230	099
11	003	008	018	037	071	127	206	301	377	341
12	000	001	002	006	014	032	069	142	282	540

function of concentration. In particular, one sees that, in the case where μ_B is much larger than μ_A, one can have, as the proportion of B increases, an increase in the saturation, followed by a maximum and by a decrease.

§ 40.

The curves calculated in this way are profoundly modified if the atoms of the two materials are not distributed at random in the crystal, but have a regular arrangement, in a word—if there is a superstructure; the curve may then be replaced by a line with bends in it. This is shown immediately by the calculation.

---

*This assumption is only valid provided that the number of B carriers reversed is small compared with the total number of B carriers.

§ 41.—**Study of the paramagnetism.**—The theory which I have outlined, in section 16, allows the solution of the problem, but it is very complicated. In order to simplify, I shall assume that the number of neighbours $2p$ is very large, i.e. in short, neglect the troublesome region near the Curie point.

Consider a crystal containing PN carriers A and QN carriers B; with the given conditions, the magnetization is of the form:

$$\bar{\mu} = \bar{\mu}_A + \bar{\mu}_B \quad (76 \text{ bis})$$

$\bar{\mu}_A$ and $\bar{\mu}_B$ are the contributions to the magnetization from the carriers A and B, respectively. I consider, for example, an A carrier; it is surrounded by PN carriers A and QN carriers B. The effect of the B carriers on the A carrier can be replaced by a fictitious magnetic field:

$$h_A^B = v_A^B \bar{\mu}_B \quad (77)$$

the coefficient v_A^B being chosen suitably. At saturation, their effect is equivalent to that of a fictitious field equal to Ppw_{AB}/μ_A. For a general magnetization, this fictitious field must be multiplied by the ratio $\bar{\mu}_B/P\mu_B$ of that magnetization to the saturation value; one thus obtains:

$$h_A^B = \frac{Ppw_{AB}}{\mu_A} \frac{\bar{\mu}_B}{P\mu_B}. \quad (78)$$

Comparing (77) and (78), I write:

$$v_A^B = \frac{pw_{AB}}{\mu_A \mu_B} = b.$$

Similarly one finds:

$$v_A^A = \frac{pm_{AA}}{\mu_A^2} = a \quad v_B^A = \frac{pw_{AB}}{\mu_A \mu_B} = b \quad v_B^B = \frac{pw_{BB}}{\mu_B^2} = c$$

(76)

I have introduced the symbols a, b, c to simplify the writing.

In the absence of interactions, the magnetization for the carriers of type A is of the form:

$$\bar{\mu}_A = \frac{PC_A}{T} H. \quad (80)$$

In order to take account of interactions, the molecular field $h_A^B + h_A^A$, which represents the effects of carriers B and carriers A, must be added to the external magnetic field H.

Thus, I can write

$$\bar{\mu}_A = \frac{PC_A}{T} [H + a\bar{\mu}_A + b\bar{\mu}_B]. \quad (81)$$

Similarly,

$$\bar{\mu}_B = \frac{QC_B}{T} [H + b\bar{\mu}_A + c\bar{\mu}_B]. \quad (82)$$

Eliminating μ_A and μ_B from the three equations 76a, 81 and 82, I have:

$$\bar{\mu} = \frac{T(PC_A + QC_B) - PQC_AC_B(a + c - 2b)}{T^2 - T(PaC_A + QcC_B) + PQC_AC_B(ac - b^2)} H. \quad (83)$$

The inverse susceptibility $1/\chi$ as a function of temperature is no longer represented by a straight line but by an hyperbola.

§ 42.—**Ferromagnetic Curie Point.**—The ferromagnetic Curie point, i.e. the point where $1/\chi$ becomes zero, is given by the largest root of the equation:

$$T^2 - T(PaC_A + QcC_B) + PQC_AC_B(ac - b^2) = 0$$

that is

$$\theta_f = \frac{1}{2}$$

$$[PaC_A + QcC_B + \sqrt{(PaC_A - QcC_B)^2 + 4PQC_AC_Bb^2}]. \quad (84)$$

The variation of θ_f is parabolic. When $a<0, c<0$ the representative curve is such as AB (figure 11); the tangents at A and B are sufficient to give an indication

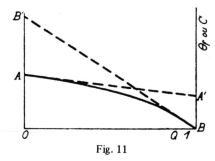

Fig. 11

of the variation. A simple calculation shows that the equation of the tangent AA' is:

$$\theta_f = PaC_A + Q\frac{b^2}{a} C_B \quad (85)$$

and that of the tangent BB':

$$\theta_f = P\left(a - \frac{b^2}{c}\right) C_A. \tag{86}$$

§ 43.—Apparent Curie constant.—In the neighbourhood of the ferromagnetic Curie point the form of the curve $1/\chi = f(T)$ is determined by the tangent at the Curie point, i.e. by an apparent Curie constant C at that point.

$$C = \frac{1}{\partial \frac{1}{\chi} / \partial T} \quad \text{pour } T = \theta_f$$

and one immediately finds:

$$C = \frac{1}{2}(PC_A + QC_B)$$
$$+ \frac{(PC_A + QC_B)(PaC_A + QcC_B) - 2PQ(a + c - 2b)C_A C_B}{2\sqrt{(PaC_A - QcC_B)^2 + 4PQb^2 C_A C_B}}. \tag{87}$$

With a curve such as AB (figure 11), the tangents AA', BB' are given, respectively, by the equations:

$$C = PC_A + QC_B\left(2\frac{b}{a} - \frac{b^2}{a^2}\right)$$
$$C = PC_A\left(1 - 2\frac{b}{c} + \frac{b^2}{c^2}\right). \tag{88}$$

§ 44.—Very high temperatures.—At high temperatures the curve $1/\chi = f(T)$ is asymptotic to the line:

$$\frac{1}{\chi} = \frac{C'}{T - \Theta'} \quad \text{avec :}$$
$$C' = PC_A + QC_B$$

and

$$\Theta' = \frac{P^2 a C_A^2 + 2PQb C_A C_B + Q^2 c C_B^2}{PC_A + QC_B}. \tag{89}$$

The Curie constants obey the law of mixtures. The variation of $\Theta' C'$ is parabolic.

§ 45.—Conclusions.—From the preceding calculations, note that in an alloy the variation of $1/\chi$ with temperature is no longer linear.

Moreover, the variation of the ferromagnetic Curie point has a parabolic form, in agreement with experiment.

In the third part, in connection with solid solutions of cobalt and of platinum, I shall use the preceding results.

CHAPTER VIII

CONSTANT PARAMAGNETISM

§ 46.—General.—In the course of the preceding chapters, I have assumed in order to simplify the calculations, that the carriers of magnetic moment could take up only two orientations with respect to the external field: parallel or antiparallel.

I have thus arrived at the magnetization law of Lenz:

$$\bar{\mu} = \mu \,\text{th}\, \frac{\mu H}{KT}.$$

This method of proceeding was legitimate, since I was not looking for the law of magnetization itself, but solely for the influence of fluctuations in the molecular field on the law of magnetization. Now, I have shown that the influence of fluctuations became weaker as the temperature increased. All this study led only to the corrections which had to be applied to the classical law of magnetization, which itself formed a first approximation. It was sufficient to study the modifications which that concept made to any law of magnetization to have an idea of the effects in other cases.

However, towards absolute zero the effect of fluctuations becomes very large: they no longer form small corrections: the general form of the phenomena changes completely.

For example, in a material with a negative molecular field, the susceptibility tends to a finite non-zero value when fluctuations are neglected; on the other hand, it tends to 0 when they are taken into account (see § 25).

It is necessary to abandon the hypothesis of Lenz and admit that a carrier can adopt all possible orientations relative to a magnetic field.

§ 47.—Law of approach.—There are two principal cases for which it will be essential to study the modifications which this new point of view brings to the preceding results: (a) in ferromagnetics, the approach to saturation as a function of temperature: (b) in materials with negative molecular fields, the study of the initial susceptibility.

§ 48.—Materials with negative molecular fields.—I

shall study the second case. I assume that the relative potential energy of two neighbouring carriers is equal to $(w/2)\cos\alpha$, w being positive and α the angle between the magnetic moments of the two carriers. This is the simplest and most natural assumption that can be made, for the given symmetry conditions. Moreover, it will be seen that, in broad terms, the results are independent of the form chosen.

Consider, for example, a material crystallizing in the body-centred cubic form: this can be treated as the superposition of two simple cubic lattices. At absolute zero, the stable equilibrium orientation of the carriers is as follows: the magnetic moments of the carriers situated at the lattice points of one of the lattices will be aligned in a certain direction, while the moments of the carriers on the other lattice will be oriented in the opposite direction. The material is magnetically neutral. If I apply a magnetic field, this distribution is modified; there is a magnetization in the direction of the field. If the magnetic field is perpendicular to the initial direction of the carriers, I have:

$$\bar{\mu} = \frac{\mu^2 H}{2pw} \qquad (90)$$

$2p$ being the number of neighbours.

In a first approximation, the effect of temperature is negligible, as we shall see in the following paragraph. I therefore have a constant paramagnetism.

It is necessary to take into account the fact that the initial direction of the carriers is arbitrary and to multiply the second term in (90) by 2/3, and, noting that $pw = 3K\Theta$ (1), one finally has:

$$\bar{\mu} = \frac{1}{9} \frac{\mu^2 H}{K\Theta} \qquad (91)$$

§ 49.—Effect of temperature.—The rigorous

treatment of the general case leads to calculations which are intractable. Following our usual method, we shall devise a simple case for which the calculations can be carried through to the end.

Consider a material in which the carriers are associated in pairs and where the interaction energies between pairs is neglected. In a pair there are two magnetic moments OA_1, OA_2, equal in absolute value, with an angle α between them. I assume that the internal energy of a pair is simply a function of that angle α.

The carrier OA_1 has a probability proportional to $d\omega_1$ of lying within a small solid angle $d\omega_1$. I denote the total potential energy of a pair by w, and the component of its magnetic moment along the direction of the external magnetic field by M. By applying Boltzmann's theorem, the average magnetization, per carrier, is then:

$$\bar{\mu} = \frac{\int M e^{-\frac{w}{KT}} d\omega_1 d\omega_2}{2 \int e^{-\frac{w}{KT}} d\omega_1 d\omega_2}$$

the integrals are taken over a complete sphere.

In order to perform the integrations, it is convenient to take the direction of one of the magnetic moments as fixed and to let the direction of the magnetic field vary. Let Ox (figure 12) be the direction of that magnetic moment OA_1. The external magnetic field OH has a probability $(1/4\pi)\sin\omega\, d\omega$ of making an angle with Ox, of between ω and $\omega + d\omega$. Let Oy be the projection of OH on the plane perpendicular to Ox. The position of the second moment OA_2 of the pair is completely defined by the angle α which it makes with Ox, and the angle V between the plane yox and the plane zOx which contains OA_2. The probability of those angles having values in the ranges α to $\alpha + d\alpha$ and V to $V + dV$ is $(1/4\pi)\sin\alpha\, d\alpha\, dV$.

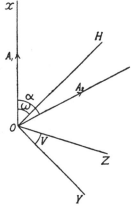

Fig. 12.

The potential energy of OA_1 is $-\mu H \cos\omega$, that of OA_2 is $-\mu H[\cos\omega\cos\alpha + \sin\alpha\sin\omega\cos V]$, so that the average magnetization of a carrier is given by the equation:

$$\bar{\mu} = \frac{1}{2} \frac{\partial}{\partial \frac{H}{KT}} \log \int_0^\pi d\alpha \int_0^\pi d\omega \int_0^{2\pi} \sin\alpha \sin\omega$$

$$\exp\left\{ \frac{-w'}{KT} + \frac{\mu H}{KT}(\cos\omega + \cos\omega\cos\alpha + \sin\alpha\sin\omega\cos V) \right\} dV.$$

---

(1) At very high temperatures the material obeys the Langevin–Weiss law with a Curie point $-\Theta$

w' represents the internal energy of a pair, a function of α alone. If I assume that $\mu H/KT$ is sufficiently small so that the series expansion of the exponential can be limited to two terms, it is easy to integrate with respect to V and ω and one finds:

$$\bar{\mu} = \frac{1}{3} \frac{\mu^2 H}{KT} \frac{\int_0^\pi (1+\cos\alpha) e^{-\frac{w'}{KT}} \sin\alpha\, d\alpha}{\int_0^\pi e^{-\frac{w'}{KT}} \sin\alpha\, d\alpha} \quad (92)$$

In order to proceed further, it is necessary to have an explicit expression for w': I put $w' = w\cos\alpha$ (1), with w a constant.

Equation (92) gives immediately:

$$\bar{\mu} = \frac{1}{3} \frac{\mu^2 H}{w} \left[1 - \frac{2w}{KT} \frac{1}{e^{\frac{2w}{KT}} - 1}\right] \quad (93)$$

At high temperatures, this reduces to:

$$\bar{\mu} = \frac{1}{3} \frac{\mu^2 H}{KT + \frac{w}{3}}. \quad (94)$$

and with $w=0$, to the classical formula of Langevin. Our fictitious material obeys a Weiss law with a Curie point $-\theta = -w/3K$.

Putting θ into equation (93) I have:

$$\bar{\mu} = \frac{1}{9} \frac{\mu^2 H}{K\theta} \left[1 - 6 \frac{\theta}{T} \frac{1}{e^{+6\frac{\theta}{T}} - 1}\right] \quad (95)$$

When T is small, I have a temperature independent paramagnetism. The value of that paramagnetism, expressed as a function of the Curie point of the material at high temperatures, is the same as the value just calculated at absolute zero; the result justifies the method.

For $T = \theta$, the difference between the susceptibility at that temperature and the susceptibility at absolute zero is again only 1.5% (figure 13).

In order to obtain the preceding results, I have adopted a particular form for the function w. It is possible, for example, that it varies much more rapidly

---

(1) In fact, previously I have taken $w=(w/2)\cos\alpha$, but as I now group the carriers in pairs I omit half of the bonds; it is necessary to double the value for those which remain.

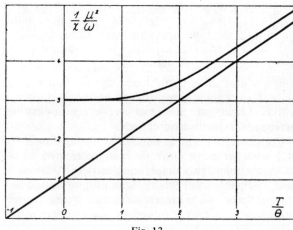

Fig. 13.

with the angle α. In that case, calculation shows that one again obtains at low temperatures a constant paramagnetism, but that for the same Curie point it is weaker at high temperatures.

§ 50.—Effect of distant carriers.

It now remains for me to investigate the influence of carriers which are not nearest neighbours. I can divide these into successive spherical shells, as I have already done in the case of ferromagnetism.

Already in the second shell the number of neighbours is sufficiently great that I could legitimately replace their effect by a uniform molecular field proportional to the magnetization: $h_m = -n\bar{\mu}$; n is a positive coefficient.

Consider C, the curve (figure 14) which represents the variation of $1/\chi$ as a function of T when only the first shell of neighbouring carriers is taken into account. In order to take into account the effect of the following shells, it is sufficient to displace the temperature axis OT downwards, to O'T', by an amount $OO' = n$. One then sees that, for the same value

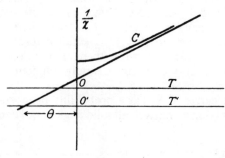

Fig. 14

of θ, the constant paramagnetism becomes smaller and that the temperature range over which it remains constant, to the same approximation, becomes wider. In the limit when all the molecular fields are considered as being without fluctuations one gets back to the straight line of the Langevin–Weiss theory.

§ **50a.—Effect of the number of neighbouring carriers.**—Under the assumption of a single neighbour, I have noted that the susceptibility changes only by 1.5% when the temperature increases from 0 to θ. When the number of neighbours is large, the results approach those of the classical theory. One knows that in the classical theory when the temperature increases from 0 to θ the susceptibility decreases by a half. The effect of the number of neighbours is great.

REFERENCES

[1] LANGEVIN, *Ann. Chimie. Phys.*, 4, 1905, p. 70
[2] WEISS, *J. Phys.*, 6, 1907, p. 661.
[3] HEISENBERG, *Z für Phys.*, 49, 1928, p. 619.
[4] FOWLER, *Statistical Mechanics*, Cambridge Univ. Press, 1929.
[5] LENZ, *Phys. Zeits.*, 21, 1920, p. 613.
[6] P. WEISS, *J. Phys.*, 9, 1910, p. 373.
[7] P. WEISS, *J. Phys.*, 4, 1905, p. 49 and 829.
[8] WEISS and BECK, *J. Phys.*, 7, 1908, p. 249.
[9] STONER, *Phil. Mag.*, 10, 1930, p. 27.
[10] WEISS and FORRER, *Ann. de Phys.*, 5, 1926, p. 211
[11] WEISS, *J. Phys.*, 8, 1930, p. 1.

A17 (1934)
THE EQUATION OF STATE AND THE ELEMENTARY CARRIERS OF MAGNETISM OF NICKEL

Summary: In the introduction a review is given of the principal interpretations of the saturation magnetization and the Curie constant of nickel, and it is shown that, in order to reach a definitive conclusion, the magnetic equation of state of nickel must be interpreted in its entirety. A model is then proposed for a ferromagnetic material, using the law of Lenz as a starting point, which allows a complete calculation of the effect of fluctuations of the molecular field. The properties of the model are compared with those of nickel, in particular the initial susceptibility (in the paramagnetic state), the spontaneous magnetization, the specific heat etc. One then introduces some corrections to the proposed model suggested by quantum mechanics which allow a quantitative interpretation of the law of approach to saturation and of the parasitic magnetization term of Weiss and Forrer. In the last part, the arguments in favour of the correctness of the idea of fluctuations of the molecular field are brought together. The overall magnetic equation of state of nickel suggests the following interpretation, which best fits the experimental facts: the magnetic properties of nickel are due to a variable number of moment carriers, each having a resultant spin $S = \frac{1}{2}$. That number is a function of the magnetization. At saturation there is 0.607 carrier per nickel atom and at low magnetizations there is 0.867.

INTRODUCTION

(1) Interpretation of the absolute saturation and of the Curie constant: Nickel is the only ferromagnetic material for which the magnetic equation of state is known accurately over a wide range, thanks to the experiments of Weiss and Forrer [1][2]. One also has very accurate data for the specific heat anomalies which are closely related to the magnetic properties [3]. Nevertheless, this information has not led to a general interpretation of ferromagnetism, since the explanation of the specific properties of nickel presents difficult problems, for which several solutions have been proposed, but none which have proved definitive.

Following the work of Weiss on the theory of ferromagnetism [4], a ferromagnetic body is considered as a paramagnetic body subject to a magnetic field equal to the sum of an external field and a molecular field proportional to the magnetization. Two fundamental constants for nickel, are, at low temperatures, the absolute saturation $\sigma_0 = 57.5$, at high temperatures the Curie constant $C = 0.0055$: these two constants are referred to one gram. Taking the atom as the carrier of the magnetic moment, one can deduce that a gram-atom of nickel has 3 Weiss magnetons[1] at low temperatures and 8 magnetons in the paramagnetic state; Langevin's law of paramagnetism is used in calculating the latter value. These two moments are, to within one part in a thousand, simple integer multiples of the magneton. It may be recalled here that experiments on paramagnetic salts in certain cases confirm the existence of the magneton as an elementary unit. The two multiples 3 and 8 appear all the more remarkable as their difference is very close to 4.94, the Bohr magneton. The same method of interpretation leads to an analogous increase in the moment from 11 to 15.79 μW for iron and from 9 to 15 μW for cobalt.

Unfortunately one is led in this way to a dead end: the hypothesis is not susceptible to any experimental verification and the magnetic equation of state, still so rich in detail, particularly in the region of the Curie point, remains without a complete interpretation. Moreover, theoretical and experimental study of salts of the rare earths shows that Langevin's formula in general leads only to an apparent moment: there will not then be any reason to compare the moments 3 and 8.

For a long time it has been suggested [6] that the spin of the electron should be taken as the carrier of magnetic moment. The absolute saturation gives 0.607 carrier per atom.

As I am here looking simply for a formal interpretation of ferromagnetism, I shall not attempt to justify the existence of such a number of carriers, which has no simple relation with the number of atoms;

---
[1] 1 Weiss magneton = 1 μW = 1125.6 C.G.S.

similarly for the molecular field, I shall not consider the question of its origin.

The law of paramagnetic magnetization is given by the formula of Lenz [7]:

$$\sigma = \sigma'_0 \, \text{th} \, \frac{\mu H}{kT}. \tag{1}$$

If the number of carriers of the moment were the same at high and at low temperatures, σ'_0 would be equal to the absolute saturation σ_0: on the contrary:

$$\sigma'_0 \doteq \frac{kC}{\mu} = 82.2$$

that is, 0.867 carrier per atom. From the ferromagnetic state to the paramagnetic state the number of carriers increases by 43%.

The spontaneous magnetization calculated by this hypothesis agrees much better with experiment than that calculated from Langevin's formula, but irregularities occur in the region of the Curie point.

In the preceding hypotheses, interpretation of the observations requires an increase in the number of carriers at the Curie point: one could claim that this increase was only an apparent one due to the use of an incorrect law of magnetization. It is thus that Wolf [9] suggests that 30% of the nickel atoms are in the ^3F state and the other 70% in the ^1D state. This assumption gives a correct absolute saturation, but the apparent paramagnetic moment calculated from those data is 7.6 in place of the observed moment of 8 magnetons. Nevertheless, the possibility of an hypothesis of this kind must be borne in mind.

(2) The need for an interpretation of the overall magnetic equation of state: The absolute saturation and the Curie constant thus only give information which is quite insufficient to be conclusive. The study of the overall magnetic equation of state is essential. It is in the neighbourhood of the Curie point that one will obtain the most important results, since one can then investigate an isotherm over a large range of magnetization, which is of the same interest as the study of a paramagnetic, without a molecular field, in the region of the absolute zero. Unfortunately, it is in this region that one observes phenomena which do not fit indisputably into the simple picture of ferromagnetism introduced at the beginning of this article; namely molecular field paramagnetism. First, let us mention the existence of a paramagnetic Curie point different from the ferromagnetic Curie point and, correspondingly, the more and more accentuated curvature of the function $1/\chi = f(T)$ as the Curie point is approached[2]. Thus, in the paramagnetic state, the apparent moment[3] decreases from the Curie point, in disagreement both with the first hypotheses of an increase in the moment and with the hypothesis of a constant moment. Finally, the spontaneous magnetization curve is much straighter in the region of the Curie point than required by the theory.

Apart from the intrinsic interest in the interpretation of these particular phenomena, their prior interpretation is essential for the extraction from the magnetic isotherms of information of a general kind on the nature of the elementary carrier of moment in nickel.

I
STUDY OF THE INFLUENCE OF FLUCTUATIONS ON A SEMI-CLASSICAL MODEL

(3) Principles of the theory of fluctuations of the molecular field: In an earlier work [10], I have shown that fluctuations of the molecular field produce a difference between the two Curie points, but I have not attempted any numerical applications. The main difficulties of the theory are mathematical: in order to obtain numerical results it is necessary to resort to schematic models of magnetic materials, but the results thus obtained have a general relevance nevertheless.

The principle of the explanation which I suggest is the following: I divide the interactions between carriers of the magnetic moment into two groups. First, the long range interactions for which the fluctuations in time and space are negligible and which can be legitimately replaced by a fictitious, molecular magnetic field, proportional to the magnetization, and completely analogous to the Weiss molecular field. The second group of interactions consists of short range interactions which fluctuate in a significant way: it is necessary to include them in the calculation from the outset. At least in the model used, this latter type of interaction does not lead to ferromagnetism, the model is still paramagnetic, but the classical law of paramagnetism: $\chi = C/T$ is replaced by a more complicated law, shown schematically by curve Γ in Fig. 1. This curve is tangential to the temperature axis at the origin and asymptotic to the line $A\Theta$. Introducing now the long range interactions in the form of a molecular field

$$H_m = n\sigma \tag{2}$$

---
[2] χ = susceptibility per gram.

[3] I call the moment calculated from the formula: $\mu = \sqrt{3RMC}$ the apparent moment.

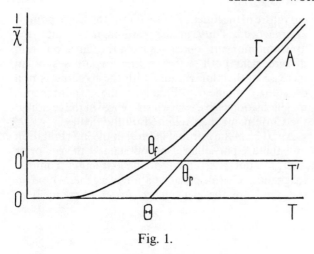

Fig. 1.

the new diagram is obtained [10, Fig. 5] by shifting the OT axis by: $OO' = n$. The new axis $O'T'$ cuts the curve at θ_F and the asymptote at θ_p. These two points are the ferromagnetic and paramagnetic Curie points, respectively. If the molecular field of this same model were entirely without fluctuations, θ_f would become coincident with θ_p and $1/\chi$ would be represented by the straight line $A\theta_p$. Hence it is θ_p which is the true Curie point, a function of the total molecular field and independent of fluctuations. Just as the actual curve which represents $1/\chi$ as a function of T only deviates from the ideal line $A\theta_p$ in the neighbourhood of the Curie point, so the actual curve of spontaneous magnetization should deviate progressively from the ideal magnetization curve for a material without fluctuations θ_p and finish up at θ_p.

This is effectively what is shown by experiment. Assuming the law of magnetization (1), with a molecular field such that the Curie point would be 378°, the paramagnetic Curie point of nickel, I have calculated the spontaneous magnetization as a function of temperature. I have shown that curve as a dotted line in Fig. 2, and the experimental data of Weiss and Forrer as a full line. The curves come together unexpectedly well.

This result justifies the idea of attributing the anomalies in the region of the Curie point to fluctuations in the molecular field.

(4) Model of a ferromagnetic material and calculation: In order to calculate the effect of fluctuations, I treat the material as a collection of groups, each containing N carriers of magnetic moment[4]: each of the carriers can orientate itself either parallel or antiparallel to the magnetic field. Each

[4] In § 16 and the following sections I shall justify the idea of limiting N to a finite value and I shall indicate the modifications which the new mechanics brings to the semi-classical scheme suggested here.

group has $N!/(p!q!)$ distinct configurations in which p carriers are oriented parallel to the magnetic field and q antiparallel. In the absence of interactions, this model is identical with that of Lenz: it applies to carriers with a resultant spin $S = \frac{1}{2}$. I have chosen it because it leads to simple calculations: moreover, the experimental curves of spontaneous magnetization agree more closely with those obtained from the law of Lenz than those obtained from the law of Langevin.

Short range interactions will be taken into account by introducing a potential energy between carriers. Let w be the mutual potential energy of any two antiparallel carriers within a group, and 0 the energy of any two parallel carriers. With respect to any one carrier, all the other carriers in its group play the same role. Between two distinct groups, we will assume that there are no interactions other than those which can be represented on average by a molecular field of the Weiss type.

If μ is the magnetic moment of a carrier, the potential energy of a group (p, q)[5] is:

$$W = pqw - (p - q)\mu H \qquad (3)$$

H is the sum of the external magnetic field and the molecular field.

From the formula of Boltzmann, the probability of occurrence of a configuration (p, q) amongst all the possible configurations can be written:

[5] p carriers parallel to the magnetic field, q in the opposite direction.

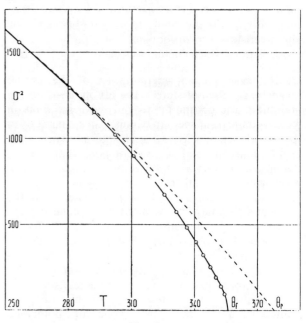

Fig. 2.

$$\bar{\omega}(p,q) = \frac{\frac{N!}{p!q!} e^{-\frac{W}{kT}}}{\Sigma \frac{N!}{p!q!} e^{-\frac{W}{kT}}} \quad (4)$$

the sign Σ extends over all positive integers, subject to the condition: $p+q=N$.

The magnetic moment of a configuration (p, q) is $(p-q)\mu$, and the average magnetization of a group is then $\Sigma(p-q)\mu\bar{\omega}$. For one gram, and denoting by σ_0 the magnetization of the whole material when all the carriers are aligned parallel to the field, the magnetization σ can be written:

$$\sigma = \sigma_0 \Sigma \frac{p-q}{N} \bar{\omega}. \quad (5)$$

The number of carriers in a group plays the role of a fluctuation coefficient. When N is small the relative importance of fluctuations is large. Conversely, when N tends to infinity the properties of the model approach more and more those of a material where the molecular field is purely of the Weiss type.

Even if this model gives a good representation of the experimental facts, it cannot be concluded that in a real magnetic material the carriers are arranged in groups of N units, or that the interactions between a carrier and its N nearest neighbours are all equivalent. This aspect of the theory is only a device used in the calculation and it is necessary to guard against attributing to the coefficient N its literal meaning. An infinity of other types of interactions would also give magnetic equations of state showing the same general characteristics. I propose simply to study here the modifications which fluctuations make to the equation of state: the actual interaction laws are outside the scope of the present subject.

(5) Initial susceptibility [11]: In the following I shall compare equation (5) with experiment in the case of nickel. I shall study particularly the initial susceptibility, the spontaneous magnetization and the specific heat. Study of the initial susceptibility is particularly valuable for the determination of the constants. Equation (5), expanded in increasing powers of H up to the term in H^3, can be written:

$$\sigma = \sigma_0(aH - \lambda a^3 H^3) \quad (6)$$

a and λ have the following significance:

$$a = \frac{A_2 N \mu}{A_0 kT} \quad \lambda = \frac{A_0^2}{A_2^2}\left(\frac{A_4}{6A_2} - \frac{A_2}{2A_0}\right) \quad (7)$$

putting

$$A_i = \Sigma \frac{N!}{p!q!} \frac{(p-q)^i}{N^i} e^{-\frac{pqw}{kT}} \quad (8)$$

The initial susceptibility χ_0 is equal to $a\sigma_0$.

We now look for the asymptote to the curve $1/\chi_0 = f(T)$. At high temperatures, the exponential which appears in the expression for the A_i reduces to

$$\left(1 - \frac{pqw}{kT}\right)$$

and a simple calculation gives

$$\frac{1}{\chi} = \frac{1}{\sigma_0 \mu}\left[kT - \frac{w(N-1)}{2}\right].$$

In the coordinate system $1/\chi$, T, this is the equation of a straight line which cuts the T axis at the point:

$$\Theta = \frac{w(N-1)}{2k};$$

this is the required asymptote.

This result agrees with that which I have obtained in previous work [10, equation 55]: $w(N-1)/2$ represents, in effect, the average bond energy per carrier.

To take account of the long-range molecular field: $H_m = n\sigma$, it is sufficient to subtract n from the values of $1/\chi_0$, and finally:

$$\frac{1}{\chi} = \frac{k}{\sigma_0 \mu} \frac{A_0 T}{A_2 N} - n \quad (9)$$

this is a function of the temperature and the four constants:

$$\frac{\sigma_0 \mu}{k}, \quad \Theta, \quad n \text{ and } N.$$

(6) Determination of the constants: From 750 K, experiment shows that $1/\chi$ is a linear function of temperature, i.e. the curve is practically coincident with its asymptote. Then, denoting the experimental Curie constant by C, one has

$$\frac{\sigma_0 \mu}{k} = C. \quad (10)$$

In order to eliminate the constant n and reduce the amount of trial adjustments, it is advantageous to

TABLE I.

T	$\partial \dfrac{1}{\chi} / \partial T$	$\dfrac{T}{\Theta}$	$C \partial \dfrac{1}{\chi} / \partial T$
637,5	105,1	0,990	0,565
640,3	109,4	0,995	0,588
643,9	115,7	1,000	0,622
647,6	121,8	1,006	0,655
651,5	127,2	1,012	0,684
655,0	133,1	1,018	0,716
662,3	141,3	1,029	0,760
673,2	152,6	1,046	0,820
685,0	161,9	1,064	0,870
697,5	165,7	1,084	0,891
709,2	170,9	1,102	0,918
726,7	177,6	1,129	0,955
784,4	181,9	1,163	0,978
774,2	186,9	1,203	1,005
820,2	186,0	1,274	1,000
854,0	185,3	1,327	0,996

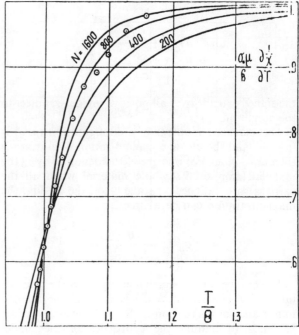

Fig. 3.

compare the derivative curves, $\partial(1/\chi)/\partial T$ as a function of T.

A value of N is chosen and the whole of the corresponding derivative curve is derived. The value of Θ is chosen so as to give agreement with the experimental curves. If the agreement between the two curves is not satifactory the derivative curve is calculated for another value of N. In Table I, I give the values of $\partial(1/\chi)/\partial T$ as a function of T, derived from the experiments of Weiss and Forrer (columns 1 and 2), which bring together in a unified way the paramagnetic and ferromagnetic data. In Fig. 3, I have shown the theoretical curves

$$\frac{\sigma_0 \mu}{k} \frac{\partial}{\partial T} \frac{1}{\chi} = f\left(\frac{T}{\Theta}\right),$$

calculated for the four values: N = 200; 400; 800 and 1600.[6]

Assuming the provisional values $\Theta = 643.6°$ and C = 0.00536, I have calculated the reduced experimental values of $C\partial(1/\chi)/\partial T$ as a function of T/Θ

[6] I have started by calculating the values of $1/\chi$: then I have differentiated graphically with respect to T. The calculation of $1/\chi$ required that of A_0 and A_2, which are sums extending over a large number of terms, at least 200. To within a good approximation one can replace them by definite integrals calculated by Simpson's method. It is sufficient to divide the interval of integration into about ten parts to have a relative accuracy of 0.1%. The calculations do not present any difficulty apart from their length.

(columns 3 and 4 of Table I), which are directly comparable with the theoretical values from relation (10).

They are represented by the circles in Fig. 3. These points are very close to the curve N = 800. A further calculation shows that with N = 750 the deviations are of the order of magnitude of the experimental errors. In Table II, I give the values of $(\sigma_0\mu/k) \partial(1/\chi)/\partial T$ as a function of T/Θ for N = 750. The preceding comparisons provide a correct value for N and an approximate value of Θ. The final constant n is obtained by a direct comparison between the calculated and experimental values of the susceptibility. If necessary the assumed values of Θ and $\sigma_0\mu/k$ are modified further.

The definitive values adopted are as follows:

$$N = 750 \quad \Theta = 642°8 \quad n = 2400 \quad \frac{k\Theta}{\sigma_0\mu} = 120\,000. \quad (11)$$

The significance of $k\Theta/\sigma_0\mu$ is obvious: it is the coefficient of the Weiss molecular field, fluctuations being neglected, which is equivalent to the short-range interactions. One thus sees that, from the point of view of the energy, the long-range interactions represent only one fiftieth of the short-range interactions.

$$n = 0{,}02 \frac{k\Theta}{\sigma_0\mu} \qquad (12)$$

TABLE II. ($N = 750$)

$\dfrac{T}{\Theta}$	$\dfrac{\sigma_0\mu}{k}\dfrac{\partial\frac{1}{\chi}}{\partial T}$	$\dfrac{T}{\Theta}$	$\dfrac{\sigma_0\mu}{k}\dfrac{\partial\frac{1}{\chi}}{\partial T}$
1,6037	1,000	1,0064	0,647
1,4855	0,996	0,9988	0,603
1,3828	0,993	0,9939	0,560
1,2934	0,990	0,9890	0,527
1,2148	0,978	0,9841	0,491
1,1453	0,954	0,9793	0,455
1,0976	0,917	0,9745	0,416
1,0683	0,874	0,9698	0,379
1,0458	0,820	0,9651	0,341
1,0297	0,765	0,9605	0,307
1,0166	0,701	0,9559	0,272

TABLE III.

T	$\dfrac{1}{\chi}$ calc.	$\dfrac{1}{\chi}$ obs.	diff. p. 100
919°5	49 900	50 208	− 0,6
858°2	38 530	38 763	− 0,6
804°6	28 622	28 883	− 0,9
757°23	19 990	19 963	+ 0,1
715°17	12 290	12 423	− 1,0
695°84	9 184	9 115	+ 0,7
677°53	6 197	6 172	+ 0,4
667°00	4 586	4 572	+ 0,3
656°79	3 129	3 134	− 0,2
650°15	2 260	2 273	− 0,6
643°65	1 475	1 491	− 1,1
640°45	1 114	1 132	− 1,5
637°28	783	789	− 0,7
634°14	474	468	+ 1,2

(7) Theory and experiment: In Table III I give, for different temperatures (column 1), values of $1/\chi$ calculated using the coefficients in (11) (column 2). In column 3 I have entered the experimental values of $1/\chi$, corresponding to the same temperatures, interpolated from the experiments of Weiss and Forrer.

I have chosen this method of comparison because the authors have made a very large number of measurements: the interpolations are thus simple and accurate. The values shown have been reduced to zero field and corrected for the demagnetizing field. Finally, in column 4, I give the percentage difference between the calculated and observed values. The two curves agree to within almost 1% over the whole range. The differences, sometimes negative sometimes positive, vary sufficiently randomly. It is possible that a further fine adjustment of the constants — in particular N — would improve the agreement, but the result would be out of proportion with the length of the calculation involved: on the one hand, the model used is certainly too far removed from reality for an agreement to within 0.1% to be truly significant; on the other hand, the experiments of Weiss and Forrer at high temperatures are a little less precise than in the region of the Curie point, the method is applied with difficulty to paramagnetic measurements. In fact, the apparent moment thus calculated is 7.9 μW, while the paramagnetic measurements normally give 8 μW. Nevertheless, in the following, in order to discuss a consistent set of experiments, I shall always take $C = 0.00535$, the value of Weiss and Forrer.

The agreement between theory and experiment is particularly satisfying when one notices that, in the temperature range considered, the susceptibility varies in the ratio 1 to 265. The theoretical curve is shown in Fig. 4, the circles represent the experimental points. On the scale used the agreement is complete.

Using the constants determined above, the paramagnetic Curie point θ_p is at $T/\Theta = 1.02000$, that is 382.45°, while experiment gives 378°. The curve, although parallel to its asymptote, is still a certain distance from it.

By interpolation, the ferromagnetic Curie point θ_f of the model is calculated to be at $T/\Theta = 0.97830$, that is 355.85°, and the corresponding value of $(\sigma_0\mu/k)\partial(1/\chi)/\partial T$ is 0.446. The difference between the two Curie points is: $\theta_p - \theta_f = 26.6°$.

It appears already that the introduction of fluctuations provides the key to anomalies observed in the region of the Curie point: moreover, the characteristic constants of the molecular field, Θ, N and n have been determined accurately. We shall now go on to show that a more extensive comparison between theory and experiment again leads to consistent results.

(8) Calculation of the second term in the law of magnetization: The law of magnetization of any paramagnetic material can be written:

$$\sigma = \alpha'H + \beta'H^3 + \ldots \quad (13)$$

H denotes the external field, α', β' are functions of temperature. In general, the term in H^3 is not experimentally accessible. If the material follows the Curie law, one might determine it by experiments near the absolute zero. If it has a Curie point, it is in that

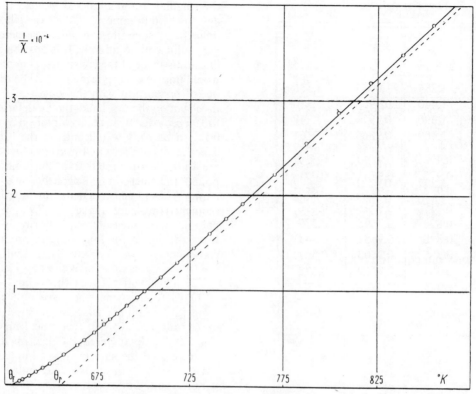

Fig. 4.

region that β' can be calculated from experimental results.

In the model which I have adopted, in order to calculate the coefficients α', β' of the expansion (13) it is necessary to take the law of paramagnetic magnetization which takes into account the short-range interactions and which, for brevity, I write:

$$\sigma = \alpha H' + \beta H'^3 + \ldots \qquad (14)$$

H' represents the total internal field: $H' = H + H_m$. One then has

$$\sigma = \alpha(H + n\sigma) + \beta(H + n\sigma)^3 + \ldots$$

Replacing σ in this equation by its value taken from equation (13), and comparing the coefficients of H and H^3 on the two sides of the equation, I have:

$$\alpha' = \frac{\alpha}{1 - n\alpha} \quad \beta' = \beta \frac{\alpha'^4}{\alpha^4}$$

Already at this stage, one notices the great advantage of working close to the Curie point: the ratio β'/α' of the two coefficients in the expansion (13) can be written:

$$\frac{\beta'}{\alpha'} = \frac{\beta}{\alpha} \frac{\alpha'^3}{\alpha^3}$$

i.e. it is equal to the ratio β/α of the coefficients for a purely paramagnetic material, multiplied by the ratio $(\alpha'/\alpha)^3$ which becomes very large at the Curie point: neglecting the fluctuations one has:

$$\frac{\alpha'}{\alpha} = \frac{T}{T - \Theta}.$$

In practice, this indicates that the susceptibility varies with the magnetic field according to a magnetization all the weaker as one is closer to the Curie point.

(9) Calculation of $\partial(1/\chi)/\partial H^2$: To determine the coefficient β' from experimental results, it is useful to represent $1/\chi$ graphically as a function of the square of the magnetization. One obtains a straight line with an intercept $1/\alpha'$ on the ordinate axis. The slope of that line is:

$$\frac{\partial \dfrac{1}{\chi}}{\partial H^2} = -\frac{\beta'}{\alpha'^2} = -\beta \frac{\alpha'^2}{\alpha^4}$$

TABLE IV.

$\dfrac{T}{\Theta}$	λ	$\dfrac{T}{\Theta}$	λ
1,33510	1,284	1,00133	3,336
1,25166	1,440	0,99633	3,310
1,17803	1,890	0,99142	3,248
1,11259	2,409	0,98653	3,161
1,08252	2,768	0,98169	3,063
1,05403	3,126	0,97691	2,952
1,03765	3,297	0,97216	2,829
1,02177	3,410	0,96281	2,564
1,01144	3,415	0,95365	2,294

but by definition from equation (6):

$$\alpha = \sigma_0 a \quad \beta = -\sigma_0 \lambda a^3,$$

whence:

$$\beta = -\sigma_0 \frac{\lambda \alpha^3}{\sigma_0^3} = -\frac{\lambda \alpha^3}{\sigma_0^2} \quad \text{et} \quad \frac{\partial \frac{1}{\chi}}{\partial H^2} = \frac{\lambda}{\sigma_0^2} \frac{\alpha'^2}{\alpha}. \quad (15)$$

The coefficient λ in equation (15) is given, in terms of the quantities A_0, A_2 and A_4, by the expression in (7). It is a function of N and T/Θ. I have calculated the values of λ for $N = 750$ and different values of T/Θ.

They are collected in Table IV. On the other hand, α', the susceptibility in an infinitely weak field, is known by extrapolation of the curve $1/\chi = f(H^2)$. α is calculated from the relation: $1/\alpha = (1/\alpha') + n$. Hence equation (15) finally allows the calculation of σ_0 as a function of $\partial(1/\chi)/\partial H^2$.

(10) Experimental values of $\partial(1/\chi)/\partial H^2$: In the experiments of Weiss and Forrer, 4 isotherms are suitable to give a sufficiently accurate value of $\partial(1/\chi)/\partial H^2$: they correspond to the following Centigrade temperatures: 376° 47; 380° 15; 383° 83 and 387° 50. The isotherms corresponding to lower temperatures are not suitable, because, in the magnetic fields used, $1/\chi$ is no longer a linear function of H^2: it is necessary, in the expression for the magnetization, to take account of terms in H^3 and beyond. On the other hand, at temperatures above 387° 50, the determination of $\partial(1/\chi)/\partial H^2$ cannot be made sufficiently accurately. Already, at 387° 50, the relative accuracy is only 10%. In the first three columns of Table V I give the temperatures of the isotherms together with the corresponding values of $1/\alpha'$ and $\partial(1/\chi)/\partial H^2$. Column 4 gives the relative accuracy in the determination of $\partial(1/\chi)/\partial H^2$ as a percentage.

TABLE V.

T	$\dfrac{1}{\alpha'}$	$\dfrac{\partial \frac{1}{\chi}}{\partial H^2}$	p. 100	σ_0	$H_{\text{sup.}}$	$\sigma_{\text{sup.}}$
1	2	3	4	5	6	7
369°07				57,0	21 315	11,2
372°77				58,0	21 315	9,68
376°47	2 162	0,820	2	63,5	17 775	7,21
380°15	2 661	0,584	2	64,6	21 315	7,21
383°83	3 135	0,428	4	67,0	21 315	6,34
387°50	3 654	0,318	10	69,5	17 775	4,68
358°90	253,8	24,6	2	71,9	758	2,8

I have added to these data a result taken from a series of measurements, which I have made and which will be published shortly, on the magnetization of nickel in the neighbourhood of the Curie point, in magnetic fields of 100 to 1000 gauss. This measurement has given the following results:

H gauss.	179.9	293.8	443.0	577.8	757.6
$1/\chi$...	255.0	255.7	258.7	261.8	268.0

The temperature remained constant to within (1/200)th of a degree during the measurements. Plotting $1/\chi$ as a function of H^2 (Fig. 5) one obtains an excellent straight line.

(11) Study of the variations of σ_0: Thus I finally have 5 values of $\partial(1/\chi)/\partial H^2$ which allow the calculation of the 5 values of σ_0 shown in column 5 of Table V. Note

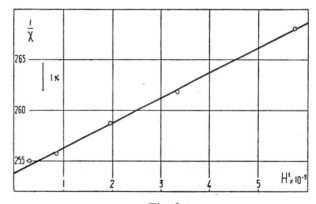

Fig. 5.

that the relative error in σ_0 is half the relative error in $\partial(1/\chi)\partial H^2$.

Finally, anticipating the results of the following sections, I have determined σ_0, by another method, at temperatures of 372.77°C and 369.07°C namely by comparing the complete theoretical and experimental isotherms.

If the proposed model of a ferromagnetic material were entirely satisfactory σ_0 would be constant and equal to the absolute saturation 57.5. This is not so: σ_0 varies within wide limits. The preceding results also show that σ_0 does not bear any simple relation to the temperature. One could then consider the effect of the molecular field. In order to be more general, we look for a relation between σ_0 and the magnetization. In the final analysis, σ_0 is determined by the slope of a line drawn on a graph which has the square of the magnetization as abscissa: hence it is the point obtained in the highest field which plays the principal part in the drawing of the line. I shall therefore study σ_0 as a function of σ_{sup}, the magnetization attained in the highest magnetic field (columns 6 and 7 of Table V). Plotting these points on a graph (Fig. 6), one finds an obvious relation between σ_0 and σ_{sup}: σ_0 decreases as the magnetization increases. This effect is in agreement with the hypothesis of an increase in the moment or the number of carriers at the Curie point.

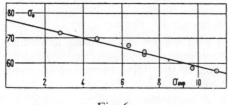

Fig. 6.

Since one is led to conclude that σ_0 is a function of the magnetization, $1/\chi$ is no longer a linear function of H^2, and the preceding arguments have to be considerably modified. I shall not go over them again, in order not to make this paper too long. I have simply calculated, for the isotherms studied, the values of $1/\chi$ as a function of H^2, by assuming that σ_0 is correctly represented as a function of the magnetization by the straight line in Fig. 6. The curves thus calculated are shown by the full lines in Fig. 7. These are almost straight lines except in the neighbourhoof of $H^2 = 0$.

This small correction improves the agreement with experiment. Careful examination of the experimental points, apart from any theoretical interpretation, shows that the convexity of the curve $1/\chi = f(H^2)$ is directed towards the H^2 axis, in such a way that one can write, for weak magnetizations at a given temperature T:

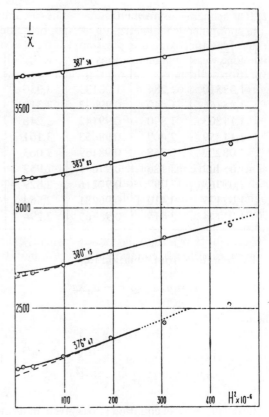

Fig. 7.

$$\frac{1}{\chi} = a + bH^2 + cH^4 + \cdots$$

a, b, c being all three *positive*. This result will be used in the conclusion.

For $\sigma = 0$, one has $\sigma_0 = 77 \pm 2$ (from Fig. 7). But the quantity σ_0 which occurs in the Curie constant $C = \sigma_0\mu/k = 0.00535$ is equally related to weak magnetizations, hence it is the value 77 which should be given to it, from which $\mu = 5782$. This value is not very far from 5561, the Bohr magneton. Thus, experiment shows that the resultant spin of a carrier is $\frac{1}{2}$.

The preceding facts suggest the following picture: in 1 g of nickel there is a certain number of groups of N carriers. The number N of carriers in a group will be taken as always fixed and equal to 750.

On the other hand, the number of groups will be taken to vary with the magnetization. A first approximation to the values of the spontaneous magnetization and of the specific heat will thus be obtained by taking as the number of groups that which corresponds to the saturation magnetization.

(12) The spontaneous magnetization: The calculation of the spontaneous magnetization σ_{sp} can be made by means of the approximate expression (13), but it is valid only for weak magnetizations — one tenth of saturation at the most — also it is preferable to use the rigorous expression (5):

$$\sigma = \sigma_0 \Sigma \frac{p-q}{N} \bar{\omega}.$$

The method consists of determining point by point a small portion of the isotherm in weak external magnetic fields and interpolating in such a way as to calculate the spontaneous magnetization σ_{sp} in zero external field.

The quantity W is the only unknown which occurs in the expression for $\bar{\omega}$, equation (4), since N and T are given. But, from (3):

$$W = pqw - (p-q)\mu(H + n\sigma)$$

The coordinates of a point on an isotherm are obtained by assuming, *a priori*, a value m for the auxiliary variable $\mu(H + n\sigma)$. One can then calculate W, $\bar{\omega}$ and then σ/σ_0 as a function of m. One then has:

$$\mu H = m - \mu n\sigma = m - 0{,}02 k\Theta \frac{\sigma}{\sigma_0},$$

from relation (12) of §7. By giving different values to m, one calculates other points and one can draw an isotherm in terms of the coordinates σ/σ_0 and μH: the intersection with the line $\mu H = 0$ gives the ratio σ_{sp}/σ. The values in the following table have been obtained in this way.

T	$\frac{\sigma_{sp}}{\sigma_0}$
354°96	0,1037
351°90	0,2078
348°89	0,2665
345°89	0,3090
340°01	0,3738
334°22	0,4189
322°97	0,4820
312°14	0,5286

In Fig. 8, I have drawn as a full line the theoretical curve giving the square of the spontaneous magnetization as a function of temperature. I have assumed the value $\sigma_0 = 57.5$ for the spontaneous

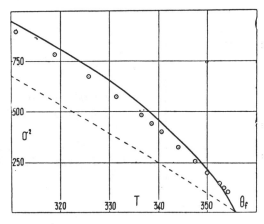

Fig. 8.

magnetization at absolute zero. The circles are the experimental points of Weiss and Forrer. The broken line represents the curve of spontaneous magnetization obtained with a hyperbolic tangent magnetization law, but without any fluctuations in the molecular field. The introduction of fluctuations greatly improves the agreement between theory and experiment.

(13) The isotherms in the neighbourhood of the Curie point: The results of § 11, in relation to weak magnetizations, show that the quantity σ_0, which we have assumed constant, in reality varies from $\sigma_0 = 77$ for very weak magnetization to $\sigma_0 = 57.5$ for saturation.

For intermediate magnetizations, it appears, *a priori*, impossible to use the formulae established above since the assumed law which relates the number of carriers to the magnetization is unknown. Equation (5) has a precise meaning only for a fixed number of carriers, each with a moment of one Bohr magneton. I have shown in § 11 that it was valid for small magnetizations ($\sigma < 1$), with $\sigma_0 = 77$, and the characteristic constants of the molecular field having the following values: $\Theta = 642°8$, $n = 2400$, $N = 750$.

I shall now attempt to represent the experimental isotherms empirically by equation (5), keeping the same constants of the molecular field, but this time treating σ_0 and μ as adjustable constants which are functions of the temperature. Trial shows that this is possible to a very high accuracy.

A point of an isotherm is determined by the method indicated in the preceding section. Table VI gives the values of σ/σ_0 as a function $N\mu(H + n\sigma)/kT$ for 8 different temperatures. By trial, one very quickly finds the values of σ_0 and μ which allow a good representation of the chosen isotherm. Table VII gives the values of σ_0, $N\mu/kT$ and μ (columns 2, 3 and 4) which correspond to the values of the temperature

TABLE VI.

$\dfrac{N\mu(H+n\sigma)}{kT}$	372°77	369°07	365°36	361°84	358°03	354°96	351°09	346°9
1	0,03877	0,04296	0,04781	0,05326				
2	0,07640	0,08439	0,09367	0,10377				
3	0,11192	0,12303	0,13578	0,14946	0,16601	0,18148	0,1971	
4	0,14464	0,15810	0,17326	0,18925	0,20827	0,2250	0,2427	0,2739
5	0,17427	0,18928	0,20593	0,2232	0,2432	0,2606	0,2785	0,3093
					0,2718	0,2890	0,3067	0,3359
						0,3120	0,3287	0,3561

TABLE VII.

T	σ_0	$\dfrac{kT}{N\mu}$	μ
372°77	58	9 320	7 690
369°77	57	9 118	7 810
363°36	57,5	9 190	7 700
361°84	58	9 230	7 630
358°03	58,5	9 390	7 450
354°96	58,5	9 462	7 360
351°9	58,5	9 551	7 250
346°9	59,2	10 050	6 840

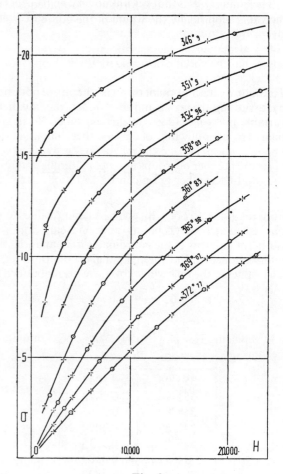

Fig. 9.

shown in column 1. In Fig. 9 I have shown by circles the values calculated in this way. The crosses represent the experimental points of Weiss and Forrer[7].

Inspection of Table VII shows that σ_0 remains very close to 58 for all the isotherms, while μ varies systematically. Since μ is here only an adjustable coefficient, the physical significance of which is not obvious, I shall not dwell on it. The constant σ_0 is also only an adjustable coefficient, but it is, at the same time, the magnetization of the material in an infinite external field.

One now sees by inspection of Fig. 9 that the interpolation formula is a good one, since it is sufficient to determine two constants, i.e. one point and a tangent, in order for all the points of an isotherm to be represented accurately. Moreover, since σ_0 is practically independent of temperature, with a value equal to the absolute saturation of nickel, one can say that the formula is applicable for magnetizations much stronger than those which have been attained by Weiss and Forrer in the range of temperatures considered (345°C to 372°C). In fact, these coincidences are not due to chance. To summarize this discussion, we can say that a single adjustable constant μ, which is a function of temperature, is sufficient for the empirical representation of the experimental isotherms. The

---

[7] The calculation has been made for a substance having the same demagnetizing field as the specimen studied by Weiss and Forrer.

second constant σ_0 is independent of temperature: it has the significance, equally well in principle as in fact, of an absolute saturation.

It is quite interesting and instructive that the correct absolute saturation of nickel should thus be obtained from isotherms which correspond to temperatures above the Curie point, i.e. from data which all correspond (in particular the constants of the molecular field) to the properties of nickel in the paramagnetic state. This illustrates well the continuity and unity of the para and ferromagnetic states. Thus equation (5), used as an interpolation formula, with μ varying with temperature, is correct for strong magnetizations: but § 11 shows that it is not valid, at least not with the same value of σ_0, for weak magnetizations of the order of a few C.G.S. units. At a given temperature a single formula is not adequate to represent empirically the whole of the curve $\sigma = f(H)$.

(14) Calculation of the specific heat: It remains to show that the specific heat anomalies in the proposed model agree with experiment. The energy of a given group (p, q), with N carriers, is the sum of the internal energy pqw of the group and the magnetic energy of the carriers in the magnetic field: $-(p-q)\mu H$. If one restricts oneself to study of the spontaneous magnetization, $H = n\sigma$: the magnetic energy is $-\tfrac{1}{2}n\sigma^2$.

As the number of groups in 1 g of materials is $\sigma_0/N\mu$, the total energy per gram can be written:

$$\frac{\sigma_0}{N\mu} \sum pqw\omega - \frac{1}{2} n\sigma_{sp}^2$$

Taking account of the relations:

$$n = 0{,}02 \frac{k\Theta}{\sigma_0 \mu} \qquad w = \frac{2k\Theta}{N-1}$$

and replacing $N-1$ by N, the energy per gram-atom becomes:

$$U = 2\nu R\Theta \left[\sum \frac{pq\omega}{N^2} - \frac{1}{200}\left(\frac{\sigma_{sp}}{\sigma_0}\right)^2 \right];$$

R is the gas constant and ν the number of carriers per atom. The quantity in brackets, say V, can be calculated as we have calculated analogous quantities above. The atomic specific heat, in units of Dulong and Petit R/2J, is:

$$c = -4\nu \frac{dV}{d\frac{T}{\Theta}}$$

TABLE VIII.

T	$\dfrac{c}{\nu}$	T	$\dfrac{c}{\nu}$
558°3	0,020	361°1	1,108
507°9	0,032	355°9	1,400
463°1	0,048		5,064
432°5	0,092	353°4	4,420
413°6	0,172	348°5	4,092
399°2	0,240	347°4	3,836
388°9	0,324	340°0	3,468
380°5	0,516	328°6	2,952
373°8	0,652	317°6	2,700
367°4	0,856		

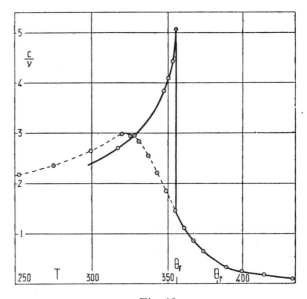

Fig. 10.

The energy U depends on the molecular field constants n, Θ, N, the values of which are known, the temperature and the ratio ν of the number of carriers to the number of atoms.

I have calculated U/ν as a function of T/Θ, and by differentiation I have deduced C/ν. The results from the subject of Table VIII. Figure 10 shows, in the full line, the temperature variation of the specific heat.

If the coefficient n of the long-range molecular field were zero, there would still be a specific heat anomaly represented by the broken line. Naturally, above the Curie point the two curves coincide since the long-range interactions vanish for zero magnetization.

The most striking feature of the specific heat curve is the persistence, after the discontinuity, of a specific heat anomaly which then disappears gradually. One also notices that the discontinuity, here equal to 3.66v, is larger than the discontinuity for a material with a molecular field without fluctuations which is only 3v. This large discontinuity is to be compared with the temperature variation of the square of the spontaneous magnetization, in the neighbourhood of the Curie point, which is more rapid with the fluctuations than in the classical case.

(15) Determination of v: One knows that Weiss has shown [12] the existence in the specific heat of nickel of a term of uncertain origin which he has called 'terme inconnue'. Below the Curie point this term is a function of the temperature, and appears constant above. One also knows that the specific heat of an ideal nickel, freed of its ferromagnetism, would not be constant as suggested by the classical theory: at low temperatures there is quantum degeneracy: at high temperatures the oscillators are no longer harmonic, which leads to the increasing term of Born and Brody. For these different reasons it is better to compare theory and experiment over a temperature range sufficiently small that the perturbing terms only vary a little: that is why I shall at first compare the discontinuities. The theory gives 3.66v and experiment 2.105 (this number is obtained after reduction to constant volume and after correction for quantum degeneracy). The comparison gives $v = 0.575$.

This method seems crude: the discontinuity obtained by Mme Lapp [3] is the result of an extrapolation, because the nickel used in her experiments did not show a sharp discontinuity: the fall in the specific heat was spread over an interval of 5° to 6°, this was probably due to a mixture of several varieties of nickel with sligltly different Curie points. However, the necessary extrapolation has been made by Mme Lapp under the assumption that there is no anomaly above the Curie point. Given the difficulties which would be involved in restoring the correct discontinuity, I have judged it preferable to compare the specific heats at two points, situated on the two sides of the Curie point, at a distance sufficiently large that the errors due to the spread of the Curie point should be very attenuated, but not so large that the perturbing terms vary appreciably.

I have chosen the temperatures 350° and 362°; the results are summarized in the following table:

	350°	362°	Diff.
$c_{calc.}$	4,04 v	1,05 v	2,99 v
$c_{obs.}$	8,797	7,132	1,665

Equating the two values given in the last column, one obtains:

$$v = 0.556$$

This value is a little smaller than that which has been calculated from the discontinuity. Mme Lapp has estimated the maximum relative error in the total specific heat as $\pm 2\%$. The absolute error near the Curie point will then be ± 0.16. I shall assume that the absolute error in the difference between the two specific heats is also ± 0.16, assuming a partial compensation in the errors. One then deduces:

$$v = 0.556 \pm 0.056$$

It is probable that the number v, which occurs in the specific heat, is related to the carriers effectively aligned at saturation. The assumption $S = \frac{1}{2}$ leads at low temperatures to 0.607 carrier per atom (§ 1). This number agrees with the value of v calculated above.

II
QUANTUM MECHANICAL CORRECTIONS

(16) Ferromagnetism treated by quantum mechanics: The calculations which I have made in the first part of this account are of a semi-classical nature: they have the disadvantage of neglecting the origin of the molecular field, which can only be correctly interpreted by means of quantum mechanics. Of course, fluctuations are a general phenomenon which arise equally well in quantum statistics as in classical statistics, but it is necessary to specify them more precisely.

I shall use the quantum model of Heisenberg, with the method of calculation given by Pauli in his report to the 6th Solvay Congress[8]. This solution, suggested by the methods of Dirac and Slater, appears in a form suitable for the treatment of the problem of fluctuations in a similar manner to that which I have used above. I shall not reproduce the calculations and I shall limit myself to quoting what is strictly essential for my purpose.

Consider a group of N carriers, each in an S state, with a single electron; the unperturbed eigenfunction of each carrier is known. The mutual interactions are introduced as a perturbation, and the perturbed eigen functions of the group are expanded in terms of the unperturbed eigenfunctions, neglecting, in particular, heteropolar states.

---

[8] Le Magnétisme. Rapports et discussions du 6ᵉ Congrès Solvay (Gauthier-Villars), Paris, 1932

Let m be the quantum number which represents[9], for a possible state of the system, the component of the resultant angular momentum parallel to the magnetic field. The Pauli exclusion principle limits to $r = C^{(N/2)+m}$,[10] the number of distinct solutions having a quantum number m. In order to write the partition function Z in a rigorous form, from which one could deduce the total magnetization of the system by means of the relation:

$$\sigma = kT \frac{\partial}{\partial H} \log Z,$$

it is necessary to know the energy corresponding to each solution, i.e. to solve an equation of the rth degree. Faced with the impossibility of attaining that result rigorously, one simply calculates the sum of the characteristic energies of the r solutions, which can be done easily, and one makes use of the following observations.

The resultant S of the angular momentum, in a state of energy E, is, in the absence of an external magnetic field, oriented in any direction in space: however, if a magnetic field is introduced, the component of S parallel to the magnetic field can only take the values S, $S-1, S-2, \ldots, -S$. Naturally, the energy of each of these quantized states remains equal to E. Conversely, the states which correspond to a component m of the angular momentum parallel to the magnetic field arise from various states, of which the resultant angular momenta and energies are different.

One can now calculate the average energy of the states for which the resultant angular momentum is S:

$$\bar{E}_s = H_i - \frac{N(N-4) + 4s(s+1)}{2N(N-1)} \sum_{i<k} H_{ik};$$

H_i represents the unperturbed energy and H_{ik} the exchange integral between two atoms of row i and k.

In the later calculations, the different states of resultant S, having different energies, are replaced by a single state of energy $\bar{E}_s$, with a statistical weight:

$$f(s) = C_N^{\frac{N}{2}+s} \frac{2s+1}{\frac{N}{2}+s+1}.$$

These formulae are valid whatever the values of H_{ik} the exchange integrals as functions of the i and k. But their use in writing a partition function Z, provides only an approximate solution, the more exact as the energies of the different states of resultant S deviate less from the average energy E_s. This deviation would be a minimum if the exchange energies were distributed uniformly over all the possible pairs of two atoms taken amongst the N of the group. That is, if all the atoms played the same role with respect to each other.

Thus the treatment of this quantum mechanical problem requires in practice, because of the inadequacy of any mathematical solutions, the implicit assumption that the H_{ik} are independent of the i and k. Expressed otherwise, one considers as equivalent all problems in which the average value of the H_{ik} is the same. These conditions are identical with those which I have adopted in the semi-classical model used in this paper.

One thus studies the laws of magnetization of a group of N atoms between which the interactions are identical. However, although in the semi-classical model I have taken a finite value for N, *in order to produce the fluctuations in a form which is mathematically tractable*, in the quantum model N tends to infinity, the effect of fluctuations is thus eliminated and one is back at the theory of Weiss.

In order to take account of fluctuations, omitting the implicit introduction of exchange integrals which are all equal whatever the distance between the atoms, it would be necessary, for example, to improve the calculation, taking account of the mean square of the difference $E_s - \bar{E}_s$. However, the analogous calculations carried out on the semi-classical model show that the approximation thus obtained is still very bad at the Curie point.

(17) Comparison of the semi-classical model and the quantum model: Since the semi-classical model, with N finite, leads to results which agree very well with experiment, the use of an analogous trick in wave mechanics should be fruitful.

Be that as it may, the partition functions, Z for the wave model[11] and Z' for the semi-classical model, can be written, to the same order of approximation, putting $\alpha = \mu H/kT$:

$$Z = \sum_{p=\frac{N}{2}}^{N} C_N^p \left(1 - \frac{q}{p+1}\right) e^{\frac{pqA}{kT}}$$

$$[e^{(p-q)\alpha} + e^{(p-q-2)\alpha} + \ldots + e^{(q-p)\alpha}]$$

$$Z' = \sum_{p=\frac{N}{2}}^{N} C_N^p e^{\frac{pqA'}{kT}} [e^{(p-q)\alpha} + e^{(q-p)\alpha}]$$

---

[9] In units of $h/2\pi$.

[10] I put, as usual,

$$C_N^{(N/2)+m} = \frac{N!}{[(N/2)+m]![(N/2)-m]!}$$

[11] This expression is deduced directly from equations (52) and (54) of the report of Pauli already cited, with minor changes of notation and the approximations valid when N is sufficiently large to neglect N in comparison with N^2.

with $p+q=N$.

A is calculated, in principle, by means of the exchange integrals:

$$A = -\frac{2}{N^2} \sum_{i<k} H_{ik}$$

and A' by means of the interaction energies defined at the beginning of this article[12]:

$$A' = -w = -\frac{2}{N^2} \sum_{i<k} w_{ik} .$$

These partition functions Z and Z' are exact if the H_{ik} and the ω_{ik} are all equal. Otherwise they are only approximate.

At high temperatures, for weak magnetizations Z and Z' again give asymptotically the same hyperbolic tangent law of magnetization. It is at low temperatures, near saturation, that they differ most.

In this last case, $q/(p+1)$ is small and the coefficients of the exponentials are practically equal in Z and Z'. The only fundamental difference between the two models is the following: in the semi-classical model, I have assumed that the component of the magnetic moment parallel to the magnetic field takes only the values $(p-q)\mu$ and $-(p-q)\mu$, while quantum mechanics tells us that in reality it can take the series of values $(p-q)\mu$, $(p-q-2)\mu$, ..., $-(p-q)\mu$.

As is shown by the comparison of calculation and experiment, the semi-classical model is adequate in the region of the Curie point, but it is completely incorrect with respect to the experimental laws of approach to saturation.

(18) Corrections to be made to the semi-classical calculations: Rather than starting again with lengthy and tedious calculations using the quantum model, I have preferred to modify the classical model, which can be made rigorous in the region of saturation.

Close to the saturation σ_0, the influence of fluctuations has become negligible, and the magnetization of the group of N atoms can be written from the hyperbolic tangent law:

$$S = N\mu \left(1 - 2e^{-\frac{2\theta_p}{T}}\right) \quad (16)$$

This magnetization arises from some states with groups parallel to the magnetic field and some antiparallel states, but the probability of the latter is completely insignificant, so that S also represents the average resultant of the group, to which one can apply the results of the quantum calculation. The component of S in the direction of the magnetic field, is quantized with a large number of equispaced possible values. The correct saturation is thus obtained by applying the Langevin law to carriers of moment S: and the magnetization arising from a group of N carriers is:

$$\sigma = S\left(\coth a - \frac{1}{a}\right), \quad (17)$$

putting:

$$a = \frac{S(H_m + H)}{kT}$$

H_m is equal to the long-range molecular field: H is the external magnetic field. Comparison of equations (16) and (17) gives, neglecting the terms of second order, equating the hyperbolic tangent to 1, and replacing S by N in the correction term[13]:

$$\sigma = \sigma_0 \left[1 - \frac{kT}{N\mu(H + H_m)} - 2e^{-\frac{2\theta_p}{T}}\right]. \quad (18)$$

This equation gives the law of approach to saturation as a function of the temperature and the magnetic field. All the constants which occur in this equation have been determined in sections 5 to 7, by studying paramagnetic nickel. I shall therefore use:

$$\theta_p = 655°, \quad N = 750, \quad H_m = 2400\,\sigma_0 .$$

An earlier discussion has led to the adoption as the moment of a carrier the Bohr magneton, 5564 C.G.S. Finally, the study of the isotherms in the neighbourhood of the Curie point gives a value close to 58 for σ_0. The precision with which that value is known is, of course, not sufficient for the close study of the laws of approach to saturation. I have therefore chosen the value of σ_0 which gives the best agreement under these conditions: namely 58.94.

(19) Law of approach as a function of temperature [13]: I take the external field to be zero. I have calculated at different temperatures, the values of $\sigma_0 kT/(N\mu H_m)$, $2\sigma_0 \exp(-2\theta_p/T)$ and σ (columns 1, 2, 3 and 4 in Table IX). The last column gives the experimental values of Weiss and Forrer [2, p. 322]. The calculated values agree with the experimental values to within 7 parts in 10 000. In Fig. 11 I have

---

[12] The potential energy of two atoms i and k is taken as ω_{ik} when they are antiparallel and as zero when they are parallel (cf. 10, p.8).

[13] Study of the numerical values given later shows that all these approximations are justified.

TABLE IX.

T	$\frac{\sigma_0 kT}{N\mu H_m}$	$2\sigma_0 e^{-\frac{2\theta_p}{T}}$	σ calc.	σ obs.
100°	0,83	0,00	58,11	58,13
150°	1,24	0,02	57,68	57,71
200°	1,65	0,17	57,12	57,08
250°	2,07	0,63	56,24	56,20
288°	2,38	1,25	55,31	55,34

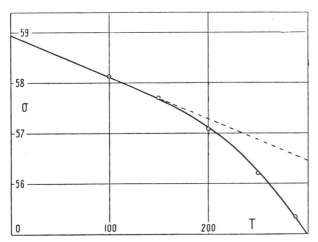

Fig. 11.

shown the calculated curve by a full line. The circles are the experimental points. The broken straight line is the tangent at the origin. The law of approach as T is not valid above 100 K. This law of approach as T is related to the model used, in respect of the finite number of atoms. For an infinite number, one knows that the rigorous quantum law of approach is as $T^{3/2}$. Here, because of the method of approximation used, the term in $T^{3/2}$ is replaced by two terms, one in T and the other exponential.

The unhoped for agreement of the theory with experiment shows that the model used is adequate for temperatures above 100 K. Below, there are no further experimental data to make the comparison.

(20) Law of approach as a function of magnetic field [13]: At saturation the long-range molecular field is some 139 000 gauss. When the external field is comparatively small, one can write:

$$\sigma = \sigma_0 \left[1 - \frac{kT}{N\mu H_m} \left(1 - \frac{H}{H_m} \right) \right]$$

where

$$\sigma = \sigma_T(1 + bH)$$

denoting by σ_T the relative saturation at temperature T in zero field. One obviously has:

$$b = \frac{kT}{N\mu H_m^2} = 1{,}02\ T \times 10^{-9}$$

This term corresponds exactly to the parasitic magnetization of Weiss and Forrer. The numerical value of b, calculated at 288 K, is 0.29×10^{-6}, while the same authors find 0.22×10^{-6}. If one notices that in a field of 10 000 gauss, bH only represents 0.3% of the total magnetization, the agreement seems satisfactory, given the difficulty of the experiments[14].

CONCLUSIONS

(21) Results of the study of the magnetic equation of state of nickel: The results obtained from this general study of nickel can be divided into two categories: one set, related to fluctuations in the molecular field, are of a purely formal kind and are of interest only for the part of the system of magnetic isotherms near the Curie point: the other set, of a more general kind, provide us with information on the nature of the elementary carrier in nickel.

(22) Fluctuations of the molecular field: The arguments in favour of this hypothesis are the following:
1. The relation with temperature of the susceptibility of nickel in the paramagnetic state can be interpreted quantitatively. Moreover, no other interpretation has been given.
2. The calculated spontaneous magnetization no longer shows large discrepancies with the observed values.
3. The specific heat anomalies are correctly represented: the conclusions that one can draw from their study agree with those which arise from the study of the usual magnetic properties.
4. The theory provides an empirical formula, which contains only two arbitrary constants, and which

---

[14] One knows that the complete law of approach to saturation at temperature T is of the form

$$\sigma = \sigma_1[1 - (a/H) + bH]$$

b is nearly constant for all the specimens but a varies over a wide range. The term a is probably connected with crystalline and intercrystalline effects. However, Forrer [14] has given it a different, and attractive, interpretation.

provides an accurate representation of the isotherms in the neighbourhood of the Curie point.

5. Certain quantum mechanical corrections which must be applied to the calculations, near saturation, allow a quantitative treatment to be given of the law of approach to saturation and give the parasitic magnetization term correctly. The data used in these latter calculations are taken entirely from the study of paramagnetic nickel.

These arguments taken together show, in an indisputable way, the important role of fluctuations in the molecular field.

It should be noted that the hypothesis of fluctuations in the molecular field truly constitutes a new hypothesis only if one keeps strictly to the formal point of view of the molecular field of Weiss. If, on the other hand, one follows Heisenberg and starts from the origin of ferromagnetism in the exchange interaction, it is only in a first approximation that one obtains the molecular field of Weiss, and it is the modifications obtained by taking the approximations further that, using a language that seems to me useful, I have attributed to fluctuations in the molecular field. Hence I have not introduced any new hypotheses, but only some methods of calculation designed to increase the accuracy of results provided by the same fundamental hypotheses.

(23) The nature of the elementary carrier in nickel: The gyromagnetic effect shows that the orbital moment plays no role in ferromagnetism: the orbits are quenched and the moment is due only to spin. The calculations developed above, apply only to the case where the resultant angular momentum of a carrier is equal to $\frac{1}{2}$: for example, when there is only a single electron per carrier. It is almost impossible to perform the corresponding calculations when S has the value of 1 or a higher value. When one keeps to the assumption $S = \frac{1}{2}$ to interpret both the saturation σ_0 and the Curie constant C, it is necessary to vary the number of carriers. Does this variation correspond to reality or is it only an artifice due to the use of an incorrect value for S?

In favour of a real variation in the number of carriers and of the value $S = \frac{1}{2}$ one can assemble the following arguments:

1. In this way one calculates correctly the discontinuity in the specific heat and the initial curvature of the isotherms in the neighbourhood of the Curie point.
2. Experiment shows that near the Curie point, at a given temperature, $1/\chi$ can be expanded as a function of H^2 in the following form (see § 11):

$$\frac{1}{\chi} = a + b H^2 + c H^4 + \ldots$$

a, b, c being all three positive. It does not seem possible to obtain a series of this kind by means of a law of magnetization related to a fixed number of carriers: one obtains the coefficients a and b positive but a coefficient c negative. For example, the calculation is easy to carry out for a paramagnetic material without a molecular field. Conversely, if the number of carriers decreases when the magnetization increases, a positive coefficient c can be justified.

In summary, only the hypothesis of a number, varying with the magnetization, of carriers with resultant spin $S = \frac{1}{2}$, seems compatible with the experimental facts.

REFERENCES

[1] Weiss and Forrer, *Ann. Physique*, **5**, 1926, p. 153.
[2] Weiss and Forrer, *Ann. Physique*, **12**, 1929, p. 279.
[3] Weiss, Piccard and Carrard, *Arch. Sc. phys. et. nat.*, **42**, 1917, p. 378; **43**, 1917, pp. 22, 113, 199; Lapp (Mme) *Ann. de Physique*, **12**, 1929, p. 442.
[4] Weiss, *J. Phys.*, **6**, 1907, p. 666.
[5] Langevin, *Ann. de Chim. et de Phys.* **4**, 1905, p. 70.
[6] Stoner, *Phil. Mag.*, **10**, 1930, p. 27.
[7] Lenz, *Physik, Z.*, **21**, 1920, p. 613.
[8] Tyler, *Phil. Mag.*, **9**, 1930, p. 1036; **11**, 1931, p. 596.
[9] Wolf, *Z. Physik*, **70**, 1931, p. 519.
[10] Néel, *Ann. Physique*, **17**, 1932, p. 5.
[11] Néel, *C.R.*, **197**, 1933, p. 1195.
[12] Weiss, *C.R.*, **187**, 1928, p. 12.
[13] Néel, *C.R.*, **197**, 1933, p. 1310.
[14] Forrer, *J. Phys.*, **10**, 1929, p. 247.

A40 (1939)

NOTE ON THE MAGNETIC PROPERTIES OF A GAS OBEYING BOSE–EINSTEIN STATISTICS

Consider a gas of N particles, with mass m and spin J, contained in a volume V and obeying Bose–Einstein statistics (J integer). Below a temperature Θ defined by

$$24{,}8\,\Theta = \left(\frac{N}{V}\right)^{2/3} \frac{h^2}{m\mathrm{K}},$$

a 'condensation' occurs; this was noted by Einstein and the mechanism has recently been analysed by London (*Phys. Rev.*, **54**, 1938, p. 947) with application to helium II. The 'condensation', i.e. the separation of the gas into two phases, takes place only in momentum space; within the ordinary space of volume V, the gas is homogeneous. The energy of the particles which are in the 'condensed' state is negligible. The number N′ of particles in that state is:

$$N' = N(\Theta^{3/2} - T^{3/2})/\Theta^{3/2}.$$

At the absolute zero all the particles are in the condensed state.

If the particles have a magnetic moment, the gas has some remarkable magnetic properties.

Above Θ, it is paramagnetic with a susceptibility χ, referred to N particles, which becomes infinite at the point Θ. Putting

$$\chi_0 = N\mu^2 g^2 j(j+1)/3k\Theta;$$

one finds that χ_0/χ as a function of T/Θ is given by the following table:

χ_0/χ	T/Θ
0	1
0,537	1,366
0,817	1,580
1,066	1,781
1,305	1,979
1,542	2,182
1,782	2,392
2,029	2,610
2,283	2,840
2,549	3,082
2,826	3,228
3,422	3,895
4,086	4,523
4,827	5,236

When T/Θ is greater than 5 one can use

$$\chi_0/\chi = [1 - 0{,}9236\,(\Theta/T)^{3/2}]\,T/\Theta.$$

Below Θ, because of the condensed fraction, the gas becomes *ferromagnetic*, i.e. an infinitesimally small magnetic field is required to produce a relative saturation equal to $N'gj\mu$, which tends, for zero absolute temperature, towards the absolute saturation $Ngj\mu$ with a law of approach in $T^{3/2}$.

Thus the gas in question possesses a paramagnetism with a Curie point and a ferromagnetism, *without a molecular field*.

For the known particles Θ is extremely small, but these considerations may present a certain interest if one imagines that, in the transition metals, the 3d or 4s electrons combine to give complexes with, for example: J = 1; Θ would then be an accessible temperature.

A22 (1935)
THE NUMBER OF ELECTRONS WHICH CONTRIBUTE TO THE PARAMAGNETISM OF NICKEL

Summary (1977): Contrary to certain earlier suggestions, it is shown that the number of electrons in nickel which are magnetically active does not change in passing from the ferromagnetic to the paramagnetic state.

In a paper which will appear shortly Manders has studied the temperature variation of the magnetic susceptibility of a series of solid solutions based on nickel. He found generally that the susceptibility χ was well represented by the formula $\chi = a + C/(T-\theta)$, which gives evidence for a temperature-independent paramagnetism a superimposed on a molecular field paramagnetism obeying the Weiss law and defined by a Curie constant C and a Curie point θ.

The experiment shows that C varies linearly as a function of the concentration of the alloy, provided that the concentration of the metal alloyed with nickel is not too large (from 0 to 10 at %, for example). Let ΔC be the slope of the line showing the variation of the Curie constant with atomic concentration. Then $-\Delta C$ represents the reduction in the Curie constant corresponding to a variation in the atomic concentration equal to 1. The following table gives the values of $-\Delta C$ for a number of metals. One can see at once that the reductions become larger as the number of valence electrons in the metal increases, and that they are proportional to the number n of electrons in the incomplete shells of the metals considered. Also, dividing $-\Delta C$ by n gives the numbers c shown in the third column of the table.

Métal.	$-\Delta C$.	n.	c.
Cu	0,429	1	0,429
Al	1,41	3	0,470
Ti	1,75	4	0,438
Sn	2,29	4	0,572
V	2,73	5	0,546
Sb	2,65	5	0,530
Mo	3,04	6	0,507
W	3,10	6	0,516

Bearing in mind the difficulties involved in the determination of $-\Delta C$, one could say, as a first approximation, that c is a constant and is equal to the mean value 0.512.

Some of the values show significant deviations. A more detailed discussion can account for these. Thus the value of c for copper has not been calculated from the results of Manders but from the older results of Alder (Thesis, Zurich, 1916). Alder assumed that $a = 0$, which does not seem legitimate; his Curie constants were then too large and the reduction which one deduced too small. The excessively large value of c found for tin is to be attributed to the very irregular results obtained in the study of the corresponding alloys. In general, the alloys for which the results are more consistent and inspire more confidence also give the values of c which approach the average more closely.

Whichever metal is alloyed with nickel, each electron in the incomplete shells produces the same reduction in the Curie constant. This result is in agreement with remarks made independently by Stoner [1] and by Dorfman [2] on the decrease in the saturation magnetization of the same alloys [experiments of Sadron [3]]. One is led to the following result. Everything occurs as if: (a) in the alloy the atoms of the foreign metal replace the nickel atoms devoid of any moment; (b) the electrons of the foreign metal are aligned antiparallel to the magnetic electrons of nickel, for example, when entering the cells of phase space already occupied by an electron of nickel.

Thus *one electron from the foreign metal cancels the magnetic contribution of one electron of nickel.* On adopting this conclusion deduced from the study of ferromagnetism, it would follow that if nickel had one magnetic electron per atom its Curie constant would be equal to 0.512 since that is precisely the reduction produced by one electron per atom. Since the Curie constant of nickel is 0.323 the number of electrons which contribute to the paramagnetic properties is equal to $0.323/0.512 = 0.63$ electron per atom.

This number is identical, within the experimental

error, with the number 0.60 of electrons which produce ferromagnetism. Hence there is complete continuity between ferro- and paramagnetism, *they are the same electrons which are involved in both cases*. Hence in the interpretation of the magnetic properties of nickel it is necessary to reject hypotheses which make an appeal to an increase in the number of magnetic carriers when the Curie point is passed.

REFERENCES

[1] Stoner, *Phil. Mag.*, **15**, 1932, p. 1018.
[2] Dorfman, *Phys. Zts. d. Sowjet Union*, **3**, 1933, p. 399.
[3] Sadron, *Ann. de Physique*, **17**, 1932, p. 371.

A35 (1938)
PARAMAGNETISM OF ELECTRONS IN A RECTANGULAR BAND

Summary (1977): A study of the magnetic properties of electrons in a rectangular band in order to make a comparison with the classic results of Stoner for a parabolic band and thus to evaluate the influence of the shape of the band.

According to current ideas on the electron theory of metals, the electrons are distributed in *bands*. In the iron group the magnetic properties are attributed to the incompletely filled 3d band. In order to calculate the magnetic properties it is necessary to know the shape of the band, i.e. the number of allowed electronic states as a function of energy. For the metals in which we are interested, situated at the end of the group, it is only the shape of the top of the band which is important, since the band is almost completely filled. In their calculations some authors, for example Stoner [1], represent the band by a parabola by analogy with the case of free electrons: the calculations are tedious. However, examination of those cases where the shape of the band has in fact been determined [2] shows that there is no reason to represent a band by a parabola rather than a rectangle or a triangle. Since it is most important at present from the point of view of magnetism to compare the new concepts with the old, it is more sensible to adopt as a working hypothesis a *rectangular* band shape which leads to very simple calculations.

Let us consider a gram-atom. The empty part of the band is completely determined by its width $k\Phi$ (k = Boltzmann's constant), which has the dimensions of an energy, and hence Φ is a temperature, and by the number νN (N = Avogadro's number) of electrons which are necessary to fill it, allowing two electrons per cell in phase spaces. If μ is the Bohr magneton the saturation magnetization is $\mathscr{J}_0 = \nu N \mu$. In a magnetic field H, at an absolute temperature T application of Fermi statistics gives the magnetization

$$\mathscr{J} = \mathscr{J}_0 \frac{T}{\Phi}\left[y - \text{arg sh}\left(e^{-\frac{\Phi}{T}} \text{sh } y \right) \right]$$

where $y = \mu H / kT$.

The initial susceptibility χ_0 can be written

$$\chi_0 = \frac{\mathscr{J}_0 \mu}{k\Phi}\left(1 - e^{-\frac{\Phi}{T}} \right).$$

The inverse, $1/\chi_0$, of the initial susceptibility as a function of T is represented by a curve, tangential at low temperatures to the horizontal line $1/\chi_0 = k\Phi/\mathscr{J}_0\mu$ (constant paramagnetism) and asymptotic at high temperatures to the line $1/\chi_0 = (T+\Phi)k/\mathscr{J}_0\mu$. This law of magnetization for a metal will then replace the Curie law. It can be seen that the existence of a negative Curie point for a substance obeying the Weiss law does not necessarily imply the existence of a molecular field. This comment could explain the finite Curie points obtained in very dilute solid solutions of Mn or Cr [3].

As has been shown by L. Brillouin [4], the introduction of a coupling between electrons can be interpreted by the appearance of a molecular field H_m, proportional to the magnetization, $H_m = n' \mathscr{J}$, and the new initial susceptibility χ, is related to the original χ_0 by $1/\chi = 1/\chi_0 - n'$. For $n' > k\Phi/\mathscr{J}_0$ there will be a ferromagnetism characterised by the existence of a spontaneous magnetization J_s below some Curie point Φ. Putting

$$n' = n\frac{\mu \mathscr{J}_0}{k\Phi}, \quad z = \frac{\mathscr{J}_s}{\mathscr{J}_0}, \quad \tau = \frac{T}{\Phi},$$

one finds that the reduced spontaneous magnetization z is related to the reduced temperature τ by the equation

$$e^{-\frac{1}{\tau}} \text{sh}\frac{nz}{\tau} = \text{sh}(n-1)\frac{z}{\tau}.$$

The Curie point is given by the formula

$$\frac{\Phi}{\Theta} = L \frac{n}{n-1}.$$

These simple formulae can facilitate the discussion of the magnetic properties of the iron group of metals, but in doing that it is also necessary to take account of the complications introduced by fluctuations and by the temperature variation of the molecular field. By comparison with the results of Stoner (loc. cit.), relative to a parabolic band, they allow a discussion of the influence of the shape of the band on the magnetic properties.

REFERENCES

[1] Stoner, *Proc. Roy. Soc. London* **165A**, 1938, p. 372.
[2] For copper: Slater, *Phys. Rev.*, **49**, 1936, p. 537; for calcium: Manning and Krutter, *Phys. Rev.*, **51**, 1937, p. 761.
[3] Neel, *J. Phys. Rad.*, **3**, 1932, p. 160.
[4] Brillouin, *C.R. Acad. Sci.*, **194**, 1932, p. 255.

C5 (1939)
INTERIONIC EQUILIBRIUM

Summary (1977): The first large international conference devoted to magnetism was held in Strasbourg from 21 to 25 May 1939. The three volumes which recorded the proceedings, edited in Paris in 1940 by the 'Institut international de Coopération intellectuelle', are practically unobtainable today; most of the copies were destroyed at the beginning of the war.

Under the title 'Molecular field, saturation magnetization and Curie constants of transition elements and their alloys', L. Néel presented a report of 100 pages (Vol. II, Ferromagnetisme, pp. 65–164), part of which was entirely original. In particular, in order to reconcile the hypothesis of localised magnetic moments and that of itinerant moments he made the suggestion that the metals of the first transition series should be considered as mixtures of localized 3d ions, in equilibrium in a bath of 4s electrons treated by band theory.

The seven pages which describe this study are given here.

(1) Some suggestions: If a treatment such as that of Mott–Slater is adequate to represent properly the properties of the 4s electrons, it is not so for the 3d electrons. Thus the band shapes calculated by Slater [1] and later by Jones and Mott [2] do not agree with the experimental curves of Farineau.

Basically, the 4s wave functions are associated with the lattice as a whole and not with a particular individual atom; on the other hand the 3d wave functions overlap each other very little; for the ferromagnetics they *must* overlap relatively little in order for the exchange integral to have the proper sign and they thus keep their individuality very strongly. Their magnetic properties resemble more closely those of ions than those of a collective assembly of electrons.

Here and in the rest of this report, the term ion will be understood to mean the core of the atom including the 3d levels, but without any 4s electrons.

Thus a transition metal will consist of *a mixture of ions in equilibrium* immersed in an atmosphere of 4s electrons. In summary, one adopts ideas analogous to those of Stoner, Wolf, Dorfman, Vogt, but makes them more precise in order to find the equilibrium conditions: there is a dynamic equilibrium in which, *on average*, all the lattice points are identical and consequently all the atomic polyhedra are neutral.

(2) Equilibrium conditions at low temperatures: A rigorous theory is not given here, but rather the development of a language which will be useful for the classification and prediction of the experimental facts.

Suppose that in the metal in question, nickel, for example, there are two types of ions A and B, in the sense of the preceding section, then we shall show, by means of a simple hypothesis, that the conditions for equilibrium between two solid phases (A and B) and a gas phase (the atmosphere of 4s electrons), with two independent and consequently invariant constituents: the electronic pressure or, what is equivalent, the electron concentration *must have* a well defined value for a fixed temperature.

Proof —

Let x and y be the atomic concentrations of A and B ($x+y=1$), e and e' the charge of the ions A and B (in elementary units of positive charge), S the number of 4s electrons per atom. Since the total system is neutral:

$$S = xe + ye' \qquad (1)$$

The energy of the system is the sum of the following terms:

(1) A term $xW_A + yW_B$ which represents the energy of the ions: W_A and W_B are constants which depend on the nature of the ions A and B;

(2) the energy of the atmosphere of electrons, which, if we consider a given volume, is a function of the number of electrons S, say $f(S)$;

(3) the electrostatic interaction energy between the ions and the electronic atmosphere contained within their associated polyhedra; I shall assume that this is proportional to the number of ions and their charge, i.e.

$$(xe + ye')g(S) \qquad (2)$$

(4) the electrostatic interaction energy between the ions is compensated by that between the ions and the 4s electrons situated outside their associated polyhedra, since I have assumed that the polyhedra are, on average, neutral.

The total energy can thus be written:

$$W = xW_A + yW_B + f(S) + (xe + ye')g(S) \qquad (3)$$

and on writing the equilibrium condition $dW/dS = 0$, one obtains the following equation, independent of x and y:

$$(W_A - W_B)/(e - e') + d/dS[Sg(S) + f(S)] = 0 \qquad (4)$$

Equilibrium between the ions A and B is achieved only when the number of 4s electrons has the particular value S which is the solution of the preceding equation. Figuratively, one could also speak of an electronic equilibrium pressure p; this has the advantage of reminding us that the volume plays a part in the equilibrium.

Thus, in any metal or alloy, where by hypothesis ions A and B exist in equilibrium, the relative proportions of A and B establish themselves in such a way that the electronic pressure is just equal to the equilibrium pressure p.

The equilibrium is moreover almost independent of temperature, since the atomic volume varies little with T and the electronic gas, being almost completely degenerate, has an energy and pressure which do not depend appreciably on the temperature; this explains the language adopted above.

(3) **Application to nickel and to copper:** Two very different cases arise:
(1) that of a metal such as copper where one of the possible ions, Cu^+ ($3d^{10}$) in this case, is so much more stable than the other, *that it exists alone*, in the presence of an atmosphere of 4s electrons, with one electron per atom, i.e. $S = 1$;
(2) that of a metal such as nickel where two varieties of nickel ions, A and B, exist in equilibrium. It is evident *a priori* that for such an equilibrium to be possible it is necessary that the variety which is more stable should be that which has the fewer electrons in the 3d level. This is precisely the case in the transition metals, from nickel to manganese for which the most stable spectroscopic states are the states $d^{n-2}s^2$. In the solid, these states would give too many 4s electrons, with two large an energy, and equilibrium is restored by the formations of the ions $d^{n-1}s$ or d^ns^0. With $S = 1$, the energy of the 4s atmosphere is 5 to 7 electron-volts, while the difference in energy between states such as the two cited above is of the order of 1 electron-volt. The energies are of the required order of magnitude.

Since the saturation magnetic moment of nickel is less than one Bohr magneton, it is necessary that one of the varieties has a zero moment: that is, the neutral ion $3d^{10}$. If x is the atomic concentration of the other variety $3d^{10-r}$, of which the magnetic moment is $r\mu_B$, then:

$$rx = 0{,}606 = S. \qquad (5)$$

By the nature of the equilibrium, *only* two varieties A and B can exist in equilibrium in the presence of each other, I shall choose one of these: B($3d^{10}$). The other, characterised by the index r has not been fixed.

(4) **Application to the experiments of Sadron:** Consider dissolving into nickel at atomic concentration z of a metal which has q valence electrons. These valence electrons will be incorporated into the 4s atmosphere of nickel, since they are themselves s electrons: the proportions of ions A and B of nickel will be modified in such a way as to re-establish the equilibrium pressure: S must remain unchanged.

In pure nickel one has

$$rx = S$$

In the alloy, if x' is the new concentration of the magnetic nickel ions A, then

$$rx' + qz = S \qquad (6)$$

and combining these two relations one obtains:

$$r(x - x') = -qz \qquad (7)$$

which shows precisely that *the reduction of the saturation moment is equal to one Bohr magneton for each valence electron introduced.*

It is obvious that the effect continues until all the magnetic carriers A of nickel have disappeared: it is necessary for that to have added about 0.6 electron per atom. Beyond that point the electronic concentration S ceases to be constant, it increases beyond 0.6, there is then in the mixture only the non-magnetic variety B of nickel.

Analogous phenomena occur in palladium, which at the same concentration as nickel, may be considered as an equilibrium mixture which is changed by the addition of s electrons. Svensson [3] and others have observed, in fact that the susceptibility of palladium decreases linearly with the addition of hydrogen until the value zero is obtained by the addition of 0.65 atom of hydrogen per atom of palladium. Since one knows in other ways that the paramagnetic moment of palladium is about the same as that of nickel, S for palladium must have approximately the same value as for nickel: 0.6, in excellent agreement with the result quoted above.

(5) Transition elements dissolved in nickel: This case is appreciably more complicated than the preceding one, since generally these metals when in the ionised state have an incomplete d shell, i.e. a characteristic magnetic moment.

As above, provided that the dissolved metal is dilute, the electronic pressure will remain constant and equal to that of pure nickel. From the point of view of the dissolved metal, it is placed in an electronic atmosphere with a well-defined pressure; it will assume the configuration which is most stable in those conditions: its state will be essentially different from its average state in its pure form where it is concerned with an equilibrium.

The dissolved transition metal may be characterised by the number n which defines its configuration $d^{10-n}s^0$, i.e. the number of positive holes in its neutral ion: Pd, Co, Fe, Mn are characterised by $n = 0, 1, 2, 3$, respectively. The most stable state in nickel will be denoted by d^{10-q}, with q positive holes. The z ions of this nature dissolved in nickel will introduce $z(q-n)$ 4s electrons. If the number of A ions in nickel in the new equilibrium state is denoted by x', one will have:

$$rx' + z(q - n) = S \qquad (8)$$

Several cases now arise: the moment of the ion d^{10-q} is effectively equal to q Bohr magnetons when q is smaller than or equal to S, i.e. when the ion is in the second half of the periodic table, while it is equal to $10 - q$ Bohr magnetons when q is greater than 5, when the ion is in the first half of the periodic table. From a macroscopic point of view, the contribution of the dissolved ion to the total magnetization will be positive or negative according to the sign of the coefficient of the molecular field of interaction between the dissolved metal and nickel. Let γ be that coefficient. We will examine briefly the four possible cases:

(1) $\qquad q \leqslant 5; \quad \gamma > 0$

the moment of the solution, expressed in Bohr magnetons, is $rx' + zq$ and the rate of change can be written:

$$\frac{1}{z}[rx - (rx' + zq)] = n \qquad (9)$$

The rate of change is independent of q: hence magnetic experiments cannot give any information on the nature of the dissolved ion.

Palladium $n = 0$ — The rate of change is zero: the magnetic moment must remain constant. This is in good agreement with experiment. Following the early comment, it is possible, for example, that the dissolved palladium has the d^{10} configuration, without a magnetic moment, so that a Ni-Pd alloy with 50 atomic percent of palladium will contain as magnetic carriers only the A ions of nickel, in exactly the same proportion as in pure nickel.

Cobalt, iron — $n = 1$ and 2; the experimental rates of change are 1.2 and 2.2.

(2) $\qquad q > 5; \quad \gamma > 0$

in this case the magnetic moment of the solution is $rx' + z(10 - q)$ Bohr magnetons: one thus deduces a rate of change

$$n + 10 - 2q \qquad (10)$$

manganese could be in this category.

(3) $\qquad q \leqslant 5; \quad \gamma < 0$

The magnetic moment of the solution is $rx' - zq$ Bohr magnetons, from which there is a rate of change:

$$n - 2q \qquad (11)$$

platinum could be in this category; in fact the experimental rate of change is -0.6, very far from any expected value: perhaps the platinum-nickel molecular field is too small to produce a definite orientation in one sense or the other; moreover, the experimental results are very irregular and depend strongly on the heat treatment.

(4) $\qquad q > 5; \quad \gamma < 0$

the magnetic moment of the solution, expressed in Bohr magnetons, is $rx' - (10 - q)z$, which gives a rate of change

$$n = 10 \qquad (12)$$

again independent of q.

It is very important to note in this last case that $n - 10$ is the total number of 3d and 4s electrons in the dissolved metal, so that the result obtained above is the same as one would obtain by supposing that all the outer electrons of the dissolved metal were transferred into the 4s state, which is equivalent to treating these metals in the same way as copper or zinc, by the method of section 4.

Titanium–vanadium — They certainly fall into this category. One would then have rates of change of 4 and 5, while experiment gives 3.8 and 5.2. This agreement, which is satisfactory bearing in mind the difficulty of the measurements, does not give any certain information about what actually occurs in the solid solution.

Chromium–tungsten ($n=4$) — The intermolecular field with nickel is negative, but, depending on the state of ionization of the dissolved ion, one will be in either the third or fourth category.

In the case of tungsten experiment gives a rate of change of -5.8, compatible with the rate of change -6 predicted in the fourth category: the stable tungsten ion in nickel will thus be $5d^5$ or $5d^4$.

In the case of chromium experiment gives the rate of change -4.4 and, taking the fourth category, one has:

$$4-2q = -4.4$$

from which $q=4.2$; this tends to show that chromium is not ionized in nickel and is in the $3d^6$ state.

The experimental results for molybdenum are too uncertain for the discussion to present the least interest.

REFERENCES

[1] Slater, *Phys. Rev.*, **49**, 1936, pp. 437, 932.
[2] Jones and Mott, *Proc. Roy. Soc., London*, **162**, 1937, p. 49.
[3] Svensson, *Ann. der Physik*, **18**, 1933, p. 299.

Appendix to C5 (1939)

SATURATION MAGNETIC MOMENT OF THE ELEMENTS OF THE FIRST TRANSITION SERIES AND THEIR ALLOYS

Note (1977): This is concerned with a figure taken from Néel's Report presented at the 1939 Strasbourg conference (see summary of the previous paper). By assuming that the number D of the positive holes in the $3d$ band is given by the saturation magnetic moment of an element or alloy one can deduce the number S of electrons in the $4s$ band. If one plots S as a function of D for Ni, Co, Fe and their alloys one obtains the graph shown here.

It can be seen that when D reaches a certain critical value D_0 of about 2.3 there is often a sharp break in the curves, as if the $3d$ band could not accept a greater number of holes than D_0.

This representation seems to be the best way of showing this effect.

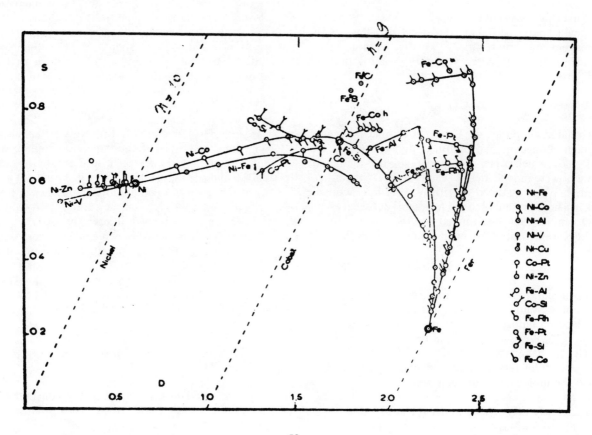

C29 (1957)
THE WEISS MOLECULAR FIELD AND THE LOCAL MOLECULAR FIELD

Note (1977): The first 16 pages of a report presented by L. Neel to the national conference commemorating the work of P. Weiss (Strasbourg 8–10 July 1957) are reproduced here. The last 7 pages, relating to the interpretation of the magnetic properties of pyrrhotite by means of the theory of phases, are given in Ch. VI (C29).

Summary: In a first part the author gives an historical account of the work of P. Weiss on the theory of the molecular field and spontaneous magnetization; he emphasizes the successes in the interpretation of the magnetic and energetic properties of ferromagnetic materials and the subsequent improvements: the energy molecular field and the molecular field to correct the equation of state.
  In a second part, after examining the difficulties which the theory has come against, the author shows that they have been resolved by introducing the concept of a local molecular field, which has provided the key to the interpretation of the properties of antiferro- and ferrimagnetic materials.

FIRST PART
The WEISS MOLECULAR FIELD

(1) First enunciation of the theory: It was in 1905 that Langevin published [1] the electron theory of diamagnetism and the kinetic theory of paramagnetism. The latter, in particular, accounted for the Curie law

$$J = CH/T$$

and related the constant C to the atomic magnetic moment. Langevin had assumed that the atoms or molecules which were the carriers of the magnetic moment were independent of each other. On the other hand P. Weiss assumed that there were interactions, and, in a note presented to the Academie de Sciences on 10 December 1906, he assumed that the effects of these could be represented by means of an *internal field*

$$H_m = NJ$$

proportional to the magnetization J, in the same direction as it and augmenting the applied field [2]. A little later, in a more detailed paper which has become a classic [3], Weiss gave H_m the name *molecular field* which it has retained since. The constant N, independent of temperature and magnetization, was called the molecular field coefficient.
  Weiss then showed that a material could become sponaneously magnetized under the action of its own molecular field, in the absence of any external field. This *spontaneous magnetization*, the direction of which is distributed at random in a demagnetized ferromagnetic material, aligns itself along the direction of an applied field, so that the *saturation magnetization* is nothing other than the spontaneous magnetization made observable to us. It was also shown that the spontaneous magnetization, corresponding at the absolute zero to alignment of all the elementary magnets, decreases as the temperature increases and tends to zero at the Curie point.
  In spite of its remarkable fertility, this audacious and completely new idea of spontaneous magnetization took a long time to be accepted, although, at the outset Weiss had made the comment "that it should not cause any surprise since, in an analogous manner, a liquid can exist, with its high density, under zero external pressure, i.e. under its own internal pressure".
  Ferromagnetism was thus explained in a very simple way, in particular the temperature variation of the saturation magnetization and its disappearance at the Curie point θ. The theory also showed [4] the appearance of paramagnetism above the Curie point with a magnetization given by the expression

$$J = \frac{C'H}{T - \theta}$$

where the constant C' is equal to the constant C in Langevin's theory, and thus co-ordinated the properties of ferromagnetics over the whole temperature range.

At about the same period, P. Weiss and his collaborators showed that this law, called molecular field paramagnetism or the Curie–Weiss law, provided an interpretation of the properties of paramagnetic anhydrous salts and their concentrated solutions: it was found in this case that the coefficient N, which is essentially positive for ferromagnetics, could also take negative values.

In December 1907, Weiss showed [5] that the molecular field hypothesis led to the prediction of an anomaly in the specific heat of ferromagnetics equal to $\frac{1}{2}N(dJ^2/dT)$ and in particular a discontinuity at the Curie point. Experiment showed quantitative agreement with the theory.

Notably later, Weiss and Piccard discovered the magnetocaloric effect [6] and interpreted it completely in the framework of the molecular field theory within which it could be very naturally described.

The success of the molecular field theory was thus impressive. It was further strengthened by the experiments of Perrier and Balachowsky [7] who provided a further argument in favour of the reality of spontaneous magnetization by showing that, in the absence of any demagnetizing field, the remanent magnetization of a ferromagnetic substance varied reversibly with the temperature and was proportional to the spontaneous magnetization.

(2) Elementary domains: A little later, the experiments of Barkhausen [8] led to the concept of *elementary domains*, regions of the material, independent of the crystalline grains, in which the magnetization is uniform and equal to the spontaneous magnetization. This new concept showed its full value when, after the technique of producing single crystals had been perfected, it became possible to study in detail their magnetic properties and to show that these domains arose from the simultaneous occurrence of a spontaneous magnetization and of directions of easy magnetization.

We shall not dwell on the extraordinary fertility of this concept of elementary domains which forms the basis of the current interpretation of all the properties of ferromagnetic materials, and particularly the magnomechanical and magnetoelectric effects.

Thus, the molecular field theory coordinates a wide variety of phenomena and, as with any good theory, stimulated a large amount of new experimental work, the results of which were extremely useful.

(3) Atomic moments: For its full development the molecular field theory required clear ideas about the atomic moment. It is obviously unfortunate that at the time of its development the only known ferromagnetic materials were the metals: iron, nickel and cobalt, for which the notion of an atomic moment makes little sense, while even today the interpretation of their saturation moment is not well understood.

It is thus that the quite natural idea of identifying the carrier of magnetic moment with the atom was bound to lead to the attribution to the nickel atom of a magnetic moment increasing in the ratio 3 to 8, in going from the ferromagnetic to the paramagnetic state above the Curie point.

The success of the theory as well as the number and variety of materials whose properties could be interpreted by the Langevin–Weiss formalism induced the idea of a universal theory, and according to the statement of P. Weiss, a body which did not follow the theory had to be considered as a body which did not keep the same numbers of magnetons [see below] in the temperature interval studied [10]. It is thus that the curve representing the temperature variation of the inverse susceptibility of magnetite, which we now know to be parabolic, was analysed into 5 successive straight lines, corresponding to discontinuous variations in the number of magnetons [11].

This was the origin of the *experimental magneton* or the *Weiss magneton* as the elementary unit of magnetic moment [12]; this concept was in part inspired by a very careful and most ingenious investigation by Ritz [13], which was aimed at interpreting the laws of spectral series as well as the principle of combination: it involved an atomic model, which is now completely forgotten, consisting of identical elementary cylindrical permanent magnets assembled together in a variable number. The introduction of the Bohr magneton, the development of quantum theories and the increase in accuracy due to progress in techniques eventually led to the disappearance of the experimental magneton, which had at least had the great merit of stimulating numerous and varied experiments which have been most useful in the progress of science.

(4) Crystalline properties: The position of molecular field theory in relation to magnetocrystalline effects deserves our attention all the more because, as Weiss himself noted, it was the properties of a crystal of pyrrhotite which led to the concept of a molecular field.

The pyrrhotite crystal has a magnetic plane, outside of which it is very difficult to create a significant component of the magnetization. In the magnetic plane itself, one can distinguish a direction OX of easy magnetization and a perpendicular direction OY of difficult magnetization. After overcoming great difficulties due to the inevitable presence of imperfections and after studying in detail the variation of magnetization as a function of the magnitude and orientation of the magnetic field, P. Weiss concluded [14] that "everything happens as if the material were demagnetized by a demagnetizing effect due to its structure: a component of the field proportional to the component J'_y of the magnetization in the direction OY of difficult magnetization". From this it can be deduced that there is a kind of demagnetizing field $H_D = -N_s J'_y$, which was later given the name *structural field*, the law of magnetization was the same for all azimuths in the magnetic plane XOY.

In the currently accepted theory, simple symmetry considerations show that, in the plane of easy magnetization of an orthorhombic crystal (which is approximately the case for pyrrhotite [15]), the magnetocrystalline energy E_c takes the form

$$E_c = A\alpha^2 + B\beta^2 \quad (A < B), \tag{1}$$

where α and β denote the direction cosines with respect to the axes XOY of the vector $\mathbf{J}_s$ representing the *spontaneous magnetization*. Apart from a constant which is of no interest here, this expression is equivalent to

$$E_c = +\frac{1}{2} N_s J_y^2 \tag{2}$$

where J_y represents the component along OY of the spontaneous magnetization and N is given by

$$N_s = 2(B - A) : J_s^2. \tag{3}$$

However, the conclusions of Weiss summarized above lead to the implication that there is a coupling energy with the lattice of the form

$$E_c = +\frac{1}{2} N_s J'^2_y \tag{4}$$

where J'_y this time represents the component along OY of the *macroscopic magnetization*.

The expressions (2) and (4) are equivalent only in the case where there is only one type of elementary domain, i.e. a single *phase*, in the sense of the general theory of the subdivision of crystals into elementary domains [16]. In orthorhombic crystals, two phases, in which the spontaneous magnetization is oriented differently, are generally present, so that the expressions (2) and (4) are not equivalent.

Since the interpretation of P. Weiss is in excellent agreement with the experimental facts and there is no reason to doubt the sound basis of the form which is currently given by the theory of magnetocrystalline effects, only one conclusion can be drawn: J'_y and J_y are equal. Thus, when a crystal of pyrrhotite contains two phases, the components along OY of the spontaneous magnetization of those two phases are equal. This is exactly what is predicted [16] by the general theory of equilibrium between phases, so that the experiments of Weiss in fact provide a very beautiful verification of that theory, of which we shall give a summary in the Appendix at the end of this article (cf. Ch. VI, C29).

Taking everything into account, if the interpretation of crystalline phenomena by means of the structural field agrees with the facts in the case of pyrrhotite, this is to some extent accidental. One knows from elsewhere that in cubic materials, the magnetocrystalline energy

$$E_c = K(\alpha^2 \beta^2 + \beta^2 \gamma^2 + \gamma^2 \alpha^2)$$

is of the fourth order in the direction cosines of the spontaneous magnetization and cannot be translated into the language of the structural field. We thus see why P. Weiss, starting from the fact that the law of approach to saturation of iron in a magnetic field showed a relationship between the energy of the system and the orientation of the spontaneous magnetization, deduced that a crystal of iron must be composed of elementary crystallites, with properties analogous to those of a crystal of pyrrhotite and characterised by three different coefficients of the structural field: N_1, N_2 and N_3 along the three axes [3]; we know today that this is not so.

The concept of a structural field and the attempt at a general synthesis of exchange forces and coupling with the crystal lattice solely in terms of the language of molecular and structural fields is thus of no great interest. It is necessary to stress here the basic reason for the success of the molecular field theory: it is the fact that the corresponding energy

$$E_w = -\frac{1}{2} N J_s^2$$

is invariant with respect to the rotation group; this invariance is one of the essential characteristics of the Heisenberg exchange interaction which is at the foundation of ferromagnetism. It is not the same for other interactions, such as those which arise from magnetic dipole coupling.

(5) Ewing's model: The success of the theory of Weiss has put into the shade, at least in France, theories of ferromagnetism proposed earlier and notably the theory, or more properly the model of Ewing [17] which had the undeniable merit of providing the first reasonable representation of the phenomena of hysteresis. Unfortunately, the ideas of Weiss and of Ewing were taken to be incompatible and the progress of the theories of hysteresis and of technical magnetization were consequently impeded. In fact, far from being mutually exclusive these two concepts are complementary. On the one hand, the concept of isotropic coupling developed by Weiss and Heisenberg

[18], Bloch [19], etc., provides an account of the spontaneous magnetization; while, on the other hand, the concept of anisotropic coupling, such as magnetic dipole coupling, developed by Ewing and Mahajani [20], Akulov [21], etc., allows an interpretation of the magnetocrystalline phenomena: in particular, it is the very existence of the spontaneous magnetization which justifies the use of the small magnetized needles in the model of Ewing.

(6) The origin of the molecular field: From the beginning P. Weiss recognized that magnetic dipolar interactions were far too weak to explain the enormous magnitude (several million gauss) of the molecular field of iron and considered it simply as a fictitious magnetic field.

In 1914, based on the variation with concentration of the Curie point of solid solutions of nickel and of cobalt, together with the experiments of Maurain [22] on thin electrolytic deposits, P. Weiss showed [23] that the molecular field interaction must decrease as the inverse sixth power of the distance, thus excluding a magnetic or electrostatic origin. This conclusion depends on the fact that as in a solution in a non-magnetic metal the Curie point tends to zero at the same time as the concentration of the magnetic metal, the molecular field constant N must also tend to zero. In fact, one has $N = \theta/C$; as θ and C tend to zero together there is an indeterminacy. As to the experiments of Maurain, their interpretation is not clear even today; the thin films in question perhaps had gaps in them, and that would entirely change the nature of the problem.

Later, after the theory of Debye and experiments determining the order of magnitude of electric dipole moments, P. Weiss took up again [24] the idea of an electrostatic origin of the molecular field by assuming that a magnetic moment and an electric moment were superimposed on the same atom.

In a more subtle form, and taking into account the effect of the Pauli exclusion principle, the theory of Heisenberg shows that it is indeed electric forces which lead to ferromagnetism.

(7) Energy molecular field and molecular field to correct the equation of state: P. Weiss always sought to give a more perfect and complete form of the molecular field theory. After having shown [23] in 1914 that writing in a more general form

$$h_m = N_1 J + N_2 J^3 + \cdots$$

did not lead to anything new but rather was in disagreement with experiment, and, inspired on the other hand by his researches on the equation of state of fluids, P. Weiss [25] was led to distinguish an *energy molecular field* defined in terms of the internal energy U by

$$H_m = -\frac{\partial U}{\partial J}$$

and a *molecular field to correct the equation of state* defined by the relation

$$J = f\left(\frac{H + h_m}{T}\right),$$

giving the expression for the magnetization of a paramagnetic subject to a molecular field in terms of the law of magnetization

$$J = f\left(\frac{H}{T}\right)$$

for a corresponding simple paramagnetic. He obtained the following relation between the two

$$H_m = h_m - T\frac{dh_m}{dT}.$$

This form of exposition, which is very attractive, allows a very elegant treatment of the energy properties of ferromagnetic bodies.

Weiss, who attached great importance to the linear character of the inverse susceptibility of paramagnetic materials, showed thermodynamically that this characteristic implied the proportionality of the energy molecular field to the magnetization and its independence of the temperature:

$$H_m = N_0 J \quad \text{with} \quad N_0 = \theta/C'.$$

He concluded from this that it was the same as the coefficient of molecular field to correct the equation of state. In fact a more complete analysis shows [26] that h_m is more generally a linear function of the temperature

$$h_m = H_m(1 + \lambda T).$$

The constant C' in the Curie–Weiss law is then related to the Curie law constant C by

$$C' = C(1 + \lambda\theta).$$

In the same paper Weiss showed that in the case of nickel the molecular field coefficient deduced from the specific heat anomalies in the ferromagnetic region had the same value as the molecular field coefficient deduced from the ratio θ/C' in the paramagnetic region. It corresponds in both cases to the coefficient N_0 of the energy molecular field and it is overall a beautiful confirmation of the molecular field theory. It has been wrongly criticized by Potter [27] following considerations deduced from a study of magnetocaloric phenomena in the region of the Curie point, in the neighbourhood of which the molecular field theory is not applicable as we shall see later, whereas Weiss's verification omitted precisely that region.

It does not seem that one can reconcile these results with the temperature independence of the coefficient N_s of the structural field of pyrrhotite. We have, in fact, shown that N_s cannot be considered as the difference of two coefficients N_x and N_y of the molecular field relative to two perpendicular axes, but rather as a specific constant representing a particular anisotropic interaction. One can easily verify that purely dipolar magnetic interactions give a coefficient N_s independent of T.

(8) Some unexplained facts: It is obviously not necessary to deal with those which resulted from an imperfect knowledge, at the time of the development of molecular field theory, of atomic moments and of the statistics which were applicable to them.

Amongst the other most troublesome facts were: the existence of materials with a *constant paramagnetism*, such as manganese and several other metals characterised by a susceptibility independent of temperature and field; the hyperbolic form of the temperature variation of the inverse susceptibility of ferrites, magnetite in particular, above their Curie point; the fact that for the common ferromagnetics: iron, nickel, cobalt and their alloys the susceptibility could not be represented exactly, in the region of the Curie point, by a Curie–Weiss law, so that it is convenient to distinguish between a ferromagnetic Curie point θ_F, at which the ferromagnetic properties disappear, and a paramagnetic Curie point θ_p obtained by extrapolation of the Curie–Weiss law; the related fact that the specific heat anomalies of ferromagnetics do not disappear sharply at the Curie point as predicted by the theory but continue to decrease up to about a hundred degrees above the Curie point, as has been shown by more precise measurements than those carried out originally [28].

SECOND PART
THE LOCAL MOLECULAR FIELD

(9) Fluctuations of the molecular field: The theory of Heisenberg showed that the ferromagnetic interactions must have a short range of action and are only effective between first and at the most second nearest neighbours. From then it was no longer possible to treat the molecular field as constant in time and space and we have proposed to take account of its fluctuations [29]: we have thus explained the difference between the para and ferromagnetic Curie points, and the anomalies in the specific heat above the ferromagnetic Curie point.

In fact the mathematical treatment used on that occasion did not consist specifically of a calculation of fluctuations but was more a statistical treatment of the kind which has been pursued with such vigour since then. The difficulties of an exact treatment are considerable and they have not been overcome even at the present time.

It was clear that to obtain directly applicable results it was best to take over the ideas of Weiss concerning the representation of interactions by fictitious fields, but to modify them appropriately to take account of their local characters.

(10) The local molecular field: In the examples treated by Weiss, the molecular field theory was applied to materials composed of atoms which on average were magnetically identical with each other. Let us now consider a material formed of atoms of type A and of type B and denote by J_A and J_B their respective contributions to the overall magnetization; the hypothesis of Weiss amounts to writing the energy in the form

$$E_c = -\frac{1}{2} n(J_A + J_B)^2.$$

In fact, since this energy is the sum of contributions from pairs of neighbouring atoms such as A–A, A–B and B–B, it should rather be written as

$$E_c = -\frac{1}{2}(n_{AA} J_A^2 + 2n_{AB} J_A J_B + n_{BB} J_B^2).$$

In brief, one should no longer consider a common and uniform molecular field but *local molecular fields*: a local molecular field $H_A = n_{AA}J_A + n_{AB}J_B$ acts on the atoms A and a local molecular field $H_B = n_{AB}J_A + n_{BB}J_B$ acts on the atoms B. The simplicity of the original theory is lost; in particular the inverse of the paramagnetic susceptibility χ is no longer represented by a straight line but by the hyperbola

$$\frac{1}{\chi} = \frac{T}{C} - n - \frac{\sigma}{T - \theta};$$

the three coefficients n, σ, and θ are complicated functions of n_{AA}, n_{AB} and n_{BB}.

The theory was first applied to alloys of platinum and of cobalt [30].

(11) Ferrites: The most striking application of these ideas is in relation to ferrites [31] in which the coefficient n_{AB} is *negative* and with an absolute value clearly larger than those of n_{AA} and of n_{BB}. This leads to the formation at low temperatures of two sub-lattices A and B, corresponding to different crystallographic sites with spontaneous magnetizations J_A and J_B aligned in opposite directions to each other: the macroscopic magnetization is equal to the difference between them. Thus a magnetically ordered state is established, although in some cases all the interactions may be negative: we have called this ferrimagnetism. We recognize for such materials that their saturation moment at absolute zero is much smaller than the moments of the ions present and that the curves of $(1/\chi, T)$ are hyperbolae.

The temperature variation of the saturation magnetization of ferrimagnetics often show very different features from that of ordinary ferromagnetics. It sometimes happens, for example, that at a certain temperature T_c, called the compensation temperature, the spontaneous magnetization of the two sub-lattices may be exactly equal and opposite: the apparent spontaneous magnetization is zero, although there is no question of a Curie point. It is in fact a temperature at which the spontaneous magnetization changes sign.

Ferrites of the garnet type, with a formula $5Fe_2O_3 \cdot 3M_2O_3$, provide an example of an even more complicated example of ferrimagnetism, with three sub-lattices and three local molecular fields: H_A, H_B and H_C, bringing into play six molecular field coefficients: the $(1/\chi, T)$ curve is then of the third degree. Application of the local molecular field method to such complicated problems nevertheless provide excellent results which agree quantitatively with the experimental data [32].

(12) Antiferromagnetism: The same general ideas about the localization of the molecular field have led us to suppose [33] that in a material containing only one kind of magnetic ion and characterized by a negative molecular field, two different but equivalent sub-lattices, magnetized in opposite senses could occur at low temperatures. In the absence of an applied field, the spontaneous magnetizations J_A and J_B are equal and opposite so that the macroscopic spontaneous magnetization is zero. The material behaves as a paramagnetic: when a magnetic field is applied the antiparallel assembly of J_A and J_B becomes deformed and one obtains a susceptibility independent of the magnetic field and of temperature: one is then dealing with *antiferromagnetism*.

Application of the method of the local molecular field shows [34] that these materials must have an antiferromagnetic transition point which marks the disappearance of magnetic order. Below that point and in the absence of magnetocrystalline coupling, the susceptibility remains independent of temperature; above, the susceptibility obeys the Curie–Weiss law with a negative Curie point. A short time after the publication of this theory, the first antiferromagnetic substances were discovered [35].

Antiferromagnetics have specific heat anomalies in the neighbourhood of their transition point analogous to those in ferromagnetics. This is not surprising, since each sub-lattice has a spontaneous magnetization. Effects which depend on an odd power of the spontaneous magnetization, such as the saturation magnetization, are hidden, while effects which depend on an even power, such as energy effects, are similar to those in ferromagnetics.

The intervention of magnetocrystalline coupling complicates the phenomena by creating preferred *directions of antiferromagnetism* with minimum energy: the direction of antiferromagnetism is defined as the exterior bisector of the spontaneous magnetizations of the two sublattices. When the applied field is perpendicular to the preferred direction of ferromagnetism the susceptibility varies in the way we have described above. On the other hand, when the field is parallel to the direction of antiferromagnetism, the theory shows immediately that the susceptibility, zero at the absolute zero, increases with temperature; it also shows that there is a certain critical coupling field H_c for which the direction of antiferromagnetism changes its orientation suddenly and becomes perpendicular to the direction of the applied field, as soon as the value H_c is passed [33]. Almost twenty years after their announcement [36], these theoretical predictions received experimental confirmation in the course of a study of the antiferromagnetism of $CuCl_2 \cdot 2H_2O$.

Finally, it may be noted that the method of neutron diffraction has provided confirmation of the reality of the decomposition into two sub-lattices and determination of the structure.

(13) The Weiss molecular field and the local molecular field: The rapid review given in the preceding paragraphs clearly illustrates the fertility of the concept of a local molecular field. It leads to the question of whether it is an independent concept or if, on the contrary, it is only a simple generalization of the Weiss molecular field to apply to two or more sub-lattices. We think that in fact there are two different concepts. Weiss always insisted, for example, on the uniformity of the molecular field within the material and he again wrote in 1932 that "the minimum domain required for the production of a molecular field contains a large number of elementary magnets which interact through their average orientation and in which, in the molecular field mechanism, the fluctuations were ineffective" [37]. On the other hand, the concept of a local molecular field calls on the idea of fluctuations.

This difference arises even in the application to the most simple case of a material containing only one type of magnetic ion and a single magnetic lattice. In the picture of paramagnetism in a Weiss molecular field, the decrease of spontaneous magnetization with temperature increasing from absolute zero arises from the thermal oscillations, of increasing amplitude, of the atomic magnets in the molecular field in a fixed direction: that of the macroscopic magnetization. In particular, there is no correlation between the motion of the magnetic moments associated with two neighbouring atoms.

In the local field picture, much greater importance is attached to the parallelism of the atomic moments of neighbouring atoms. In effect, when the magnetization is no longer uniform an increase occurs in the Heisenberg exchange energy, given for cubic materials by the well-known formula of Landau–Lifshitz:

$$E_c = \frac{1}{12} Na^2 [(\text{grad } J_x)^2 + (\text{grad } J_y)^2 + (\text{grad } J_z)^2],$$

(5)

where N is the molecular field coefficient, a the distance between nearest-neighbour atoms and J_x, J_y and J_z the projections of the spontaneous magnetization on three rectangular axes. The presence of the factor a shows that a local field theory is indeed involved. Consider now a *wave of spontaneous magnetization* corresponding to deviations, periodic, in x, y, z, of the direction of the spontaneous magnetization with respect to an average direction. Equation (5) shows that the energy of a wave is inversely proportional to the square of the wavelength. Each wave, corresponding to one degree of freedom, has an energy of $kT/2$. One is thus led to replace the statistics of independent atoms, characteristic of the Langevin–Weiss theory, by statistics taking the waves as units.

Using classical mechanics, for a three-dimensional lattice one thus finds a law of approach to absolute saturation as a function of temperature which is expressed by the relatiohn

$$J_{sT} = J_{s0}(1 - aT - \cdots)$$

the form of which is identical with that in the Langevin–Weiss theory, but in which the coefficient a is different: it now depends strongly on the nature of the crystal lattice. For 1 and 2 dimensional lattices, the local field theory shows that spontaneous magnetization cannot occur; properly speaking, such bodies will not be ferromagnetic.

It is sufficient to quantize semi-classically the spontaneous magnetization waves of which we are speaking in order to obtain a law of approach to saturation in $T^{3/2}$, in agreement with Bloch's [19] quantum theory of spin waves. The concept of a local field thus leads to a more complete interpretation of the observations than is obtained with the Weiss field and allows a simpler connection with statistical theories.

(14) The value of molecular field theories: The theories of the molecular field, the two successive forms of which have been reviewed, provide, in an intuitive way and by means of very simple mathematics, an excellent representation of the magnetic properties of ferromagnetic, ferrimagnetic and antiferromagnetic bodies: more recently, they have been also adopted to the case of metamagnetic bodies [38] and to certain curious materials, which, like dysprosium [39], show a transition from antiferro- to ferro-magnetism. They provide a guide which is almost always reliable and an invaluable tool for research.

In fact, to the present time no rigorous statistical theory of ferromagnetism has been carried through completely even in the simple case of a simple cubic lattice with identical atoms, with spin $\frac{1}{2}$, placed at the lattice points. Consequently, and this is most disagreeable, we do not know the accuracy of the molecular field approximation or what confidence can be placed in the interaction coefficients which it enables us to calculate.

Be that as it may, we should pay tribute to Pierre Weiss for having introduced into science an idea, the usefulness and fertility of which are far from being exhausted, and for having founded, along with Pierre Curie and Paul Langevin, a French school of magnetism of which the vitality and prestige have never ceased to be maintained.

REFERENCES

[1] P. Langevin, *Ann. de Chim. et de Phys.*, **5**, 1905, p. 70.
[2] P. Weiss, *CR. Acad. Sci.*, **143**, 1906, p. 1137.
[3] P. Weiss, *J. Phys.*, **6**, 1907, p. 666.
[4] P. Weiss, *CR. Acad. Sci.*, **144**, 1907, p. 25.
[5] P. Weiss, *C.R. Acad. Sci.*, **145**, 1907, p. 1417.
[6] P. Weiss and A. Piccard, *C.R. Acad. Sci.*, **166**, 1918, p. 352.
[7] A. Perrier and Balachowsky, *Arch. Sc. Phys. Nat. Geneve*, **2**, 1920, p. 5.
[8] H. Barkhausen, *Physik, Z.*, **20**, 1919, p. 401.
[9] G. Urbain, P. Weiss and F. Trombe, *C.R. Acad. Sci.*, **200**, 1935, p. 2132.
[10] P. Weiss, *J. de Phys.*, **2**, 1911, p. 965.

[11] P. Weiss, *C.R. Acad. Sci.*, **152**, 1911, p. 79.
[12] P. Weiss, *C.R. Acad. Sci.*, **152**, 1911, p. 367.
[13] W. Ritz, *C.R. Acad. Sci.*, **145**, 1907, p. 178.
[14] P. Weiss, *J. de Phys.*, **4**, 1905, p. 469.
[15] F. Bertaut, *C.R. Acad. Sci.*, **234**, 1952, p. 295.
[16] L. Neel, **J. de Phys.**, **5**, 1944, pp. 241, 265.
[17] J. A. Ewing, *Magnetic induction in iron.*
[18] W. Heisenberg, *Z. Physik*, **39**, 1929, p. 619.
[19] F. Bloch, *Z. Physik*, **61**, 1930, p. 206.
[20] G. S. Mahajani, *Proc. Cambridge Phil. Soc.*, **23**, 1925, p. 136; *Trans. Roy. Soc.*, **228A**, 1929, p. 63.
[21] N. S. Akulov, *Z. Physik*, **59**, 1930, p. 264; **69**, 1931, pp. 78, 822.
[22] C. Maurain, *Eclair. Elect.*, **26**, 1901, p. 212; *J. de Phys.*, **1**, 1902, pp. 90, 151.
[23] P. Weiss, *Ann. de Phys.*, **1**, 1914, p. 134; *Arch. Sc. Phys. Nat., Geneve*, **37**, 1914, pp. 105, 201.
[24] P. Weiss and G. Foex, *Le Magnetisme* (Armand Colin, Paris, 1926).
[25] P. Weiss, *J. de Phys.*, **7**, 1917, p. 129; **1**, 1930, p. 3; *Ann. de Phys.*, **17**, 1932, p. 97.
[26] L. Neel, *Ann. de Phys.*, **8**, 1937, p. 237.
[27] H. H. Potter, *Proc. Roy. Soc.*, **146**, 1934, p. 362.
[28] L. Neel, *C.R. Acad. Sci.*, **207**, 1938, p. 1384.
[29] L. Neel, *Ann. de Phys.*, **17**, 1932, p. 5.
[30] L. Neel, *C.R. Acad. Sci.*, **198**, 1934, p. 1311.
[31] L. Neel, *Ann. de Phys.*, **4**, 1948, p. 137.
[32] R. Pauthenet, *Ann. de Phys.*, **3**, 1958, p. 424.
[33] L. Neel, *Ann. de Phys.*, **5**, 1936, p. 232.
[34] L. Neel, *C.R. Acad. Sci.*, **203**, 1936, p. 304.
[35] C. Squire, H. Bizette and B. Tsai, *C.R. Acad. Sci.*, **207**, 1938, p. 449.
[36] J. van den Handel, H. Gijsman, and J. Poulis, *Physica*, **18**, 1952, p. 862.
[37] P. Weiss, *Ann. de Phys.*, **17**, 1932, p. 97.
[38] L. Neel, *Antiferromagnetisme et metamagnetisme* (Rapport presente devant le 10^e Conseil Solvay, 1954, p. 251; Stoops, editeur, Bruxelles).
[39] L. Neel, *C.R. Acad. Sci*, **242**, 1956, p. 1549; **242**, 1956, p. 1824.

Chapter II
FERRIMAGNETISM
A61, A70, A72, A77, A94, C42

A61 (1948)
MAGNETIC PROPERTIES OF FERRITES: FERRIMAGNETISM AND ANTIFERROMAGNETISM

Remark (1977): The distribution of the orientations of atomic magnetic moments, which is one of the fundamental hypotheses at the basis of this work, was confirmed three years later by Shull, by means of neutron diffraction.

Summary: After a short introduction (sections 1–5), in the first part of this paper (sections 6–15) the author develops and discusses, as a function of different parameters, the theory of the magnetic properties of a substance made up of carriers of magnetic moment, identical to each other, distributed between two categories of sites A and B which are crystallographically different. For this the molecular field approximation has been generalized by characterizing the magnetic interactions by three coefficients of the molecular field relative to carriers situated on sites of the same category, or of different categories. In certain cases one obtains substances which, at low temperature, present the same essential characteristics as ordinary ferromagnetic substances although the three coefficients of the molecular field are *negative* and therefore all the exchange integrals are negative. The author proposes calling such substances *ferrimagnetic*. These are characterized by:
(a) a saturation magnetization, at absolute zero, well below the absolute saturation which corresponds to all the carriers being parallel,
(b) a variation with temperature of the spontaneous magnetization which can differ considerably from those of ordinary ferromagnets and
(c) a hyperbolic variation of the inverse of the susceptibility, above the Curie point, which is quite different from Weiss' Law. When the carriers are equally distributed between sites A and B, one finds in a more general form the theory of antiferromagnetism previously proposed by the author (sections 16–19).

After discussion of a case which is intermediate between ferro- and ferri-magnetism (sections 20–22) it is shown in a second part that, within the accuracy of the experiments, it is possible to account for the paramagnetic and ferromagnetic properties of the ferrites by using the theory developed above (sections 23–26). All the characteristic parameters of copper ferrite are calculated (sections 27 and 18). This shows magnetically that the ferric ions of the ferrites are distributed on two different categories of sites, in agreement with the ideas of Barth and of Posnjak. An interpretation of the abnormal properties of the ferrites of zinc and of cadmium is given (section 29).

In the same way one can account for the value of the saturation magnetization of magnetite as well as the variation with temperature of its susceptibility above the Curie point (sections 30 and 31). When the values of the unknown parameters have been completely determined from the paramagnetic properties it is possible to calculate the variation with temperature of the spontaneous magnetization, in agreement with the experimental results of Weiss. In the same way, it is possible to explain the value of the saturation magnetization of the cubic variety of ferric oxide.

The same general system of explanation is applicable to the compound of manganese and antimony Mn_2Sb studied by Guillaud; its paramagnetic properties especially are considered and a variation with temperature of the spontaneous magnetization has been deduced which agrees with the experimental data (sections 32 and 33).

In the third part, the application of the theory to the experimental data of the ferrites and the antiferromagnetic substances, respectively, allows the calculation, for each bond, of the characteristic coefficient of the molecular field, proportional to the corresponding exchange integral (sections 34–37). The values obtained are not consistent and do not agree with those deduced from the study of pure metals. The necessity is thus reached for important interactions between atoms very distant from each other and separated, for example, by oxygen ions (section 38–39). A more consistent interpretation of the experimental facts is obtained by admitting the possibility of indirect interactions, with anions as intermediaries, according to the superexchange mechanism studied by Kramers. In all the substances studied, these superexchange interactions would always be negative whereas the ordinary exchange actions would be positive, in agreement with the data provided by the study of pure metals (sections 40–43). The presence of excited states of anions, without which there could not be a superexchange, is confirmed by the existence of the constant paramagnetism of ions with a magnetic band either empty or full, to which G. Foex has already drawn attention.

INTRODUCTION

(1) Magnetic properties of ferrites: Ferrites have the general formula Fe_2O_3MO where M is a bivalent metal such as Cu, Ag, Mg, Pb, Ni, ..., etc. Magnetite itself can be considered as iron ferrite; Fe_2O_3FeO. These bodies are ferromagnets at ordinary temperatures, with the exception of zinc ferrite which is paramagnetic and of cadmium ferrite which is either paramagnetic or ferromagnetic, depending on the conditions of preparation: their saturation magnetic moment is weak, of the order of 1 to 2 Bohr magnetons per iron atom. Above their Curie point θ, their magnetic susceptibility χ does not obey Weiss' Law

$$\chi = \frac{C}{T-\theta} \quad (1)$$

even in an approximate manner. The curve $(1/\chi, T)$ on the contrary shows a strong concavity towards the temperature axis. These remarkable magnetic properties present great difficulties of interpretation. In fact, with the exception of magnetite, iron occurs in ferrites only in the form of trivalent ions whose magnetic band has the configuration $3d^5$, with a moment of 5 Bohr magnetons quite different from the moment provided by the magnetization at saturation. Besides, it would be expected that these substances would follow Weiss' Law, in the paramagnetic region, with a Curie constant equal to that of the ferric salts, which is far from being the case.

(2) The interpretation by Mlle. Serres: In her fundamental work on ferrites [1], Mlle. Serres supposed that, in addition to a paramagnetism obeying Weiss' Law, there would be superimposed a paramagnetism independent of temperature and equal to that of the ferric oxide $Fe_2O_3\alpha$; the observed curves are reproduced in this way. Theoretically the objection must be made that since the temperature independent paramagnetism is a property linked to the distances and the relative positions of the iron atoms, rather than an atomic property, there is no reason why $Fe_2O_3\alpha$ should conserve in the ferrites, which are generally cubic, the same constant paramagnetism as in the sesquioxide which is rhombohedric. This question therefore remains open.

(3) Crystal structure of the ferrites: Some recent crystallographic results appear able to provide the key to the magnetic properties of the ferrites.

Magnetite, as well as many other ferrites, such as those of magnesium, zinc, etc, have the same crystal structure as the spinels (type H_{11} of Ewald). The oxygen ions, being much larger than the others, form very approximately a compact cubic structure and the metal ions take up positions in the interstices. The metallic ions have a choice of two sorts of sites: the tetrahedric sites, or A sites, surrounded by 4 oxygen atoms; and the octahedric sites, or B sites, surrounded by 6 oxygen atoms. Per molecule there is one A site and two B sites.

We shall call the group of A sites, sub-structure A and the group of B sites, sub-structure B.

Normal structure of the spinels. In the *normal structure*, the B sites are occupied by the irons of the trivalent metal and the A sites by the ions of the divalent metal. The spinels Al_2O_3MO, with M being Mn, Fe, Co, Ni or Zn, belong to this kind [2].

Inverse structure. However it has been noticed [3] that for bodies such as Al_2O_3MgO this structure contradicted a generally verified crystallographic rule, according to which the cations have a coordination number, which is larger as the cations are bigger. Here on the contrary, the ion Al^{III} which is much smaller than the ion Mg^{II} is surrounded by six oxygen ions while Mg^{I} is surrounded by 4 atoms. In fact, Barth and Posnjak [4] have shown that certain spinels like Fe_2O_3MgO or Fe_2O_3TiO possess the *structure called "inverse"* in which the A sites are occupied by trivalent ions while the rest of the divalent and trivalent ions are shared equally and at random on the B sites [5]. Magnetite belongs to this type [6] as both X-rays and electrical conductivity show.

Mixed structure. These two structures must correspond to free energies which are quite close. We shall suppose, in a more general manner, that ferrites possess a *mixed structure* in which a certain fraction λ of trivalent ions occupy A sites while the other fraction μ occupy the B sites[1]. Here, the former ions will be called A ions and the latter B ions.

By definition

$$\lambda + \mu = 1 \quad (2)$$

Following Machatschki [7] Ga_2O_3MgO would belong to this mixed type.

The normal structure corresponds to $\lambda = 0$ and the inverse structure to $\lambda = 0.5$.

We intend to examine the magnetic properties of such structures as a function of λ and of the nature of the interactions between the magnetic ions. For the sake of simplicity, we shall first study in detail the case where only one of the two ions is magnetic, for example the trivalent ion, and we shall consider later more complicated cases.

---
[1] Following the notation of Verwey and Heilmann, the mixed ferrites, corresponding to the ferrite Fe_2O_3MO, would, for example, be written

$$Fe_{2\lambda}M_{1-2\lambda}(Fe_{2-2\lambda}M_{2\lambda})O_4$$

(4) State of the problem: It is therefore necessary in a general manner to study the magnetic properties of a substance which consists of magnetic ions of the same kind, divided into two groups A and B on sites which are crystallographically different. These ions are characterised magnetically by the angular momentum quantum number j and the Landé factor g, so that their basic magnetic moment is equal to $gj\mu_B$, where μ_B is the Bohr magneton[2].

We shall call M_j, or simply M, the absolute saturation of one ion-gramme, corresponding to all the elementary magnets being parallel. With Avogadro's number, N,[3] one obtains:

$$M = Ngj_B \quad (3)$$

For example, for the trivalent ion Fe^{III}, $j = 5/2$ and $g = 2$, so that $M = 27920$.

The case where the interactions are negligible. The magnetic ions behave in these conditions like free ions and the magnetization of an ion-gramme is written as a function of the magnetic field H as follows

$$\mathcal{J} = M_j B_j[z], \quad (4)$$

where $B_j[z]$ is the Brillouin function

$$B_j[z] = \frac{2j+1}{2j} \operatorname{ctgh} \frac{2j+1}{2j} z - \frac{1}{2j} \operatorname{cth} \frac{1}{2j} z, \quad (5)$$

and z is the Langevin variable

$$z = \frac{M_j H}{RT}, \quad (6)$$

where T is the absolute temperature and R the gas constant[4]. When z is small compared with 1 then the formula can be developed as

$$B_j[z] = \frac{j+1}{3j} z - \frac{[(j+1)^2 + j^2](j+1)}{90 j^3} z^3 + \ldots, \quad (7)$$

If the second term is neglected then the Curie Law is obtained.

$$\mathcal{J} = \frac{C_j H}{T} \quad \text{avec} \quad C_j = \frac{j+1}{3j} \frac{M^2}{R} \quad (8)$$

For the ferric ions, one thus obtains $C_j = 4.377$. This value is confirmed by the experimental study of ferric salts such as $FeCl6H_2O$ or $Fe(NO_3)_3 9H_2O$ in which the Fe^{III} ions are relatively far from each other.

The case where the interactions are appreciable. In this case, as the A and B sites have different neighbours, the magnetization $\vec{\mathcal{J}}_a$ of an ion-gramme of A ions differs from the magnetization $\vec{\mathcal{J}}_b$ of an ion-gramme of B ions. For the whole of the substance, the magnetization of an ion-gramme is therefore given by[5]

$$\vec{\mathcal{J}} = \lambda \vec{\mathcal{J}}_a + \mu \vec{\mathcal{J}}_b, \quad (9)$$

where λ and μ are the proportion of the magnetic ions distributed on each of the two kinds of sites.

The interactions to be considered arise from the Heisenberg exchange actions and decrease very rapidly with distance, so that only immediate neighbours need to be taken into account. They are defined by three exchange integrals A_{aa}, A_{bb} and A_{ab}. The first relates to two neighbouring ions both on A sites, the second to two neighbouring ions both on B sites and the third to two neighbouring ions, one on an A site and the other on a B site.

(5) The molecular field approximation: In a simple substance, characterized by a single exchange integral it has not been possible up till now to calculate rigorously the effect of interactions so that the Weiss approximation of the molecular field remains the most simple method of calculation and nevertheless gives a better agreement with the experimental results than more sophisticated methods of calculation. In the much more complicated case which we consider here, the only method applicable is thus to carry over the hypothesis of the molecular field.

As we have done previously for alloys of two constituents [8] we shall proceed by assuming that the action of the neighbours of an A ion is equivalent to the sum $\vec{h}_a$ of the molecular fields $\vec{h}_{aa}$ and $\vec{h}_{ab}$. The first, $\vec{h}_{aa}$ comes from the action of the A neighbours and is proportional to the average magnetic moment of these A ions and to their number λ. The second $\vec{h}_{ab}$ comes from the action of the B ions and is proportional to the average magnetic moment of the B ions and to their number μ. In a general way, we can therefore write $\vec{h}_a$ in the form:

$$\vec{h}_a = n(\alpha\lambda\vec{\mathcal{J}}_a + \epsilon\mu\vec{\mathcal{J}}_b), \quad (10)$$

where *n is always positive* and where ε is either equal to $+1$ or -1. We shall establish later[6] the relationships which link the characteristic coefficients α and n of the molecular field, acting on the A ions, to the exchange

---

[2] $\mu_B = 0.9273 \times 10^{-20}$
[3] $N = 6.023 \times 10^{23}$
[4] $R = 8.314 \times 10^7$ erg K^{-1}

[5] Here, and from now on, a vector will be presented by $\vec{X}$ and its magnitude by X.
[6] Cf. Section 34.

integrals A_{aa} and A_{ab} which define the interactions of an A ion with its neighbours. In the same way the molecular field $\vec{h}_b$, equal to the action on a B ion of the atoms which surround it, is given by the formula:

$$\vec{h}_b = n(\beta\mu\vec{\mathcal{J}}_b + \epsilon\lambda\vec{\mathcal{J}}_a) . \tag{11}$$

To calculate the magnetization of the A ions we shall assume that they are subject to a total magnetic field equal to the geometric sum of the applied field $\vec{H}$ and the molecular field $\vec{h}_a$ given by (10); similarly for the B ions. In the paramagnetic region, the following equations are obtained, where C is the atomic Curie constant of the ions in question

$$\left. \begin{array}{l} \vec{\mathcal{J}}_a = \dfrac{C}{T}(\vec{H} + \vec{h}_a) \\[6pt] \vec{\mathcal{J}}_b = \dfrac{C}{T}(\vec{H} + \vec{h}_b) \end{array} \right\} \tag{12}$$

Together with equations (8), (10) and (11) these allow the calculation of the macroscopic magnetization.

Remark It is very important to note that the hypothesis of the molecular field on which the validity of the preceding equations rest has a wider implication than in simple substances such as pure metals, since we assume in addition here that all the ions have the same effective surroundings. That is, a number of neighbours of each type equal to the average number. This approximation certainly becomes false when λ or μ becomes very small, when the number of neighbours belonging to the corresponding type also becomes very small and it is known that [9] the approximation of the molecular field becomes less valid as the number of neighbours becomes smaller. It is also implicitly assumed that the ions are distributed randomly on the sites of each kind without any correlation with the position of neighbouring ions. That is to say there is no superstructure.

It will therefore be necessary to consider the magnetic properties of the system with respect to different values of the coefficient λ which determines the distribution of the ions between the two sub-lattices, and the values of n, ϵ, α and β which characterize the interactions. In the first part, IA and IB, we shall assume that the interactions between different sub-lattices are negative, when the coefficient ϵ will be equal to -1. In the part IC we shall consider the case where these interactions are positive, when the coefficient ϵ will be equal to $+1$.

FIRST PART

A. WHEN INTERACTIONS BETWEEN TWO SUB-LATTICES ARE NEGATIVE, ($\epsilon = -1$; $\lambda \neq \mu$)

(6) The law of paramagnetism: When $\vec{\mathcal{J}}_a$, $\vec{\mathcal{J}}_b$, $\vec{h}_a$ and $\vec{h}_b$ are eliminated from the 5 equations (9), (10), (11), and (12), one obtains

$$\vec{\mathcal{J}} = \frac{T - nC\lambda\mu(2 + \alpha + \beta)}{T^2 - nC(\lambda\alpha + \mu\beta)T + n^2C^2\lambda\mu(\alpha\beta - 1)} \vec{H} . \tag{13}$$

The atomic susceptibility χ is equal to $\vec{\mathcal{J}}/\vec{H}$. It is more convenient to use its inverse $1/\chi$. That is

$$\frac{1}{\chi} = \frac{T}{C} + \frac{1}{\chi_0} - \frac{\sigma}{T - \theta} , \tag{14}$$

where

$$\frac{1}{\chi_0} = n(2\lambda\mu - \lambda^2\alpha - \mu^2\beta) \tag{15}$$

$$\sigma = n^2 C\lambda\mu[\lambda(1 + \alpha) - \mu(1 + \beta)]^2 \tag{16}$$

$$\theta = nC\lambda\mu(2 + \alpha + \beta) . \tag{17}$$

The form given in equation (14) differs from Weiss' classical law of paramagnetism with a molecular field, by the term in $\dfrac{\sigma}{T-\theta}$: it represents a hyperbola instead of a straight line. As will be shown later this form (14) fits well the paramagnetism of the ferrites, but before discussing this it is convenient to discuss the form of the curve $\left(\dfrac{1}{\chi}, T\right)$ as a function of the values of the 4 independent parameters n, λ, α and β. Since, apart from C which is known and is equal to the atomic Curie constant of the ions considered, the expression (14) only includes three parameters χ_0, σ and θ, there exist an infinite number of sets of values of n, λ, α and β which give identical curves for the thermal variation of the susceptibility. Also, since λ, α and β are coefficients with no dimension, while nC and n have the dimensions of temperature and the inverse of susceptibility respectively, then nC and n simply fix the scales of the axes (abscissa and ordinate) without changing the form of the curve $\left(\dfrac{1}{\chi}, T\right)$. It is therefore necessary to study the form of this curve as a function of α and β, for each value of λ.

(7) Asymptotic Curie point, θ_a: The equation (14) represents a hyperbola, asymptotic to the straight line

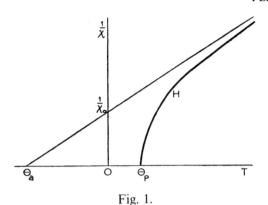

Fig. 1.

$$\frac{1}{\chi} = \frac{T}{C} + \frac{1}{\chi_0} . \quad (18)$$

When extrapolated the straight line cuts the temperature axis at a point

$$\theta_a = -\frac{C}{\chi_0} \quad (19)$$

which we shall call the asymptotic Curie point.

(8) Paramagnetic Curie point or ordering Curie point θ_p: The hyperbola (14) which is schematically represented as H on Fig. 1, cuts the temperature axis at a point θ_p given by

$$\theta_p = \frac{nC}{2}[\lambda\alpha + \mu\beta + \sqrt{(\lambda\alpha - \mu\beta)^2 + 4\lambda\mu}], \quad (20)$$

This we shall call the *paramagnetic Curie point* or the *ordering Curie point*. If θ_p is negative the substance remains paramagnetic to absolute zero, but if θ_p is positive the susceptibility becomes infinite at this point. As we shall make clear later, a spontaneous magnetization appears which remains finite when the external field goes to zero; there is a sort of ferromagnetism. However, in the case studied here when the interactions between sub-lattices are negative ($\varepsilon = -1$), the properties of the substance are sufficiently different from those of a ferromagnet to justify the attribution of a special name: we shall say there is ferrimagnetism.

In order for there to be ferrimagnetism, θ_p must be greater than zero. Now, from equation (20) θ_p is zero for

$$\alpha\beta = 1 \quad (21)$$

As a result, if we consider α and β as rectangular

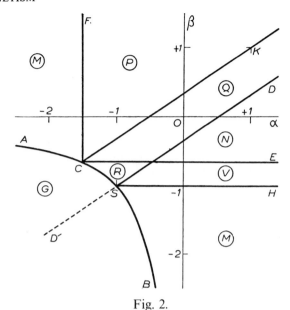

Fig. 2.
Diagram giving the different types of magnetism as a function of α and β for negative interactions between sub-networks ($\varepsilon = -1$). The figure is drawn for $\lambda/\mu = 2/3$.

coordinates of a point M, the negative section ACSB (Fig. 2) of the hyperbola given by equation (21) divides the graph (α, β) into two regions. On the side which includes the origin, θ_p is positive and ferrimagnetism appears at a temperature sufficiently low. On the other side, θ_p is negative and the substance always remains paramagnetic.

(9) Contingent validity of Weiss' law: When σ is zero, the equation (14) takes the form (18) and the substance obeys Weiss' law up to the Curie point.

$$\lambda(\alpha + 1) = \mu(\beta + 1). \quad (22)$$

For a given value of λ, this equation represents the straight line D'SD (Fig. 2) which passes through the summit S of the hyperbola and which has a slope equal to λ/μ. The curve $(1/\chi, T)$ is then represented by Figs. 3a or 3b, depending on whether the point M is on D'S or on SD.

(10) The appearance of spontaneous magnetization: By analogy with the behaviour of simple ferromagnets we deduce that above the ordering Curie point θ_p, ferrimagnets possess a certain spontaneous magnetization $\vec{\mathcal{F}}_s$ which is equal to the resultant of the partial spontaneous magnetizations, $\lambda \vec{\mathcal{F}}_{as}$ and $\mu \vec{\mathcal{F}}_{bs}$, of the ions on the A sites and of the ions on the B sites. These spontaneous magnetizations are

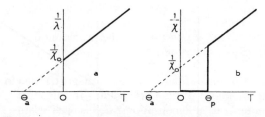

Fig. 3.
Variations with temperature of the inverse of the susceptibility corresponding to the points (α, β) situated (a) on the segment D'S of the straight line, (b) on the segment DS of the straight line.

created and maintained by the sole action of the molecular fields $\vec{h}_a$ and $z\vec{h}_b$,[7] so that $\vec{\mathcal{J}}_{as}$ and $\vec{\mathcal{J}}_{bs}$ are the solutions of the two simultaneous equations

$$\vec{\mathcal{J}}_{as} = MB_j \left[\frac{M\vec{h}_a}{RT} \right] \quad ; \quad \vec{\mathcal{J}}_{bs} = MB_j \left[\frac{M\vec{h}_b}{RT} \right] \quad (23)$$

As the interactions between the two sub-lattices are negative, it necessarily follows that the two partial spontaneous magnetizations are oriented *antiparallel*, the position of stable equilibrium.

The observable values $\mathcal{J}_s$ of the saturation magnetization correspond to the arithmetic differences between the absolute values of the two partial spontaneous magnetizations. Then the relatively small magnitude of the experimental values is easily explained. In this situation the partial spontaneous magnetizations are solutions of the equations

$$\left. \begin{array}{l} \mathcal{J}_{as} = MB_j \left[\dfrac{Mn(\alpha\lambda\mathcal{J}_{as} + \mu\mathcal{J}_{bs})}{RT} \right] \\[2mm] \mathcal{J}_{bs} = MB_j \left[\dfrac{Mn(\beta\mu\mathcal{J}_{bs} + \lambda\mathcal{J}_{as})}{RT} \right] \end{array} \right\} \quad (24)$$

and the resultant spontaneous magnetization is given in absolute value by the equation

$$\mathcal{J}_s = |\lambda \mathcal{J}_{as} - \mu \mathcal{J}_{bs}|. \quad (25)$$

The preceding equations permit the calculation of the thermal variation of the spontaneous magnetization by using, for example, the method described in appendix 1. The curves obtained present an extraordinary variety of forms depending on the values of α, β and λ as is shown by the schematic Figs. 7 and 8L, and by the curves of Figs. 15, 18 and 20. In order to orient the comparisons with experimental

[7] $\vec{h}_a$ and $\vec{h}_b$ are derived by equations (10) and (11) by replacing $\vec{\mathcal{J}}_a$ and $\vec{\mathcal{J}}_b$ by $\vec{\mathcal{J}}_{as}$ and $\vec{\mathcal{J}}_{bs}$.

results it is necessary to classify the forms obtained. Three criteria allow this classification:

(a) the value of the saturation magnetization at absolute zero.

(b) the comparison of the values of the spontaneous magnetization in the neighbourhood of the Curie point and at absolute zero.

(c) the shape of the thermal variation of the spontaneous magnetization at very low temperature.

We shall examine successively these different points in paragraphs 11 to 14 and will give a summary of the results in paragraph 15.

(11) Spontaneous magnetization at absolute zero: This is determined directly by writing that the energy of the molecular field W is a minimum at absolute zero. For an ion-gramme this is written

$$W = -\frac{1}{2} \lambda \vec{\mathcal{J}}_{as} \vec{h}_a - \frac{1}{2} \mu \vec{\mathcal{J}}_{bs} \vec{h}_b, \quad (26)$$

and since $\vec{\mathcal{J}}_{as}$ has an opposite orientation to $\vec{\mathcal{J}}_{bs}$, this expression can be transformed following (10) and (11) into

$$W = -\frac{n}{2} (\alpha\lambda^2 \mathcal{J}_{as}^2 + 2\lambda\mu \mathcal{J}_{as} \mathcal{J}_{bs} + \beta\mu^2 \mathcal{J}_{bs}^2), \quad (27)$$

where now $\mathcal{J}_{as}$ and $\mathcal{J}_{bs}$ are always positive, and at the maximum are equal to the absolute saturation M. The values of $\mathcal{J}_{as}$ and of $\mathcal{J}_{bs}$ which make the preceding equation a minimum belong to one of the four following groups of solutions I, II, III and IV.

Solution I

$\mathcal{J}_{as} = \mathcal{J}_{bs} = 0$. The energy W_I is then $W_I = 0$. (28)

Solution II

$\mathcal{J}_{as}$ and $\mathcal{J}_{bs}$ are both equal to the maximum value M. The energy is written:

$$W_{II} = -\frac{nM^2}{2} (\alpha\lambda^2 + 2\lambda\mu + \beta\mu^2). \quad (29)$$

Solution III

$\mathcal{J}_{as}$ takes the maximum value M, while $\mathcal{J}_{bs}$ is determined by the condition $\partial W_{III}/\partial \mathcal{J}_{bs} = 0$ which gives

$$\mathcal{J}_{bs} = -\frac{\lambda}{\mu\beta} M. \quad (30)$$

The corresponding energy is then written

$$W_{III} = -\frac{nM^2}{2} \lambda^2 \left(\alpha - \frac{1}{\beta}\right). \tag{31}$$

Solution IV

$\mathcal{J}_{bs}$ takes the maximum value M; $\mathcal{J}_{as}$ is determined by the condition $\partial W_{IV}/\partial \mathcal{J}_{as} = 0$, which gives

$$\mathcal{J}_{as} = -\frac{\mu}{\lambda\alpha} M \tag{32}$$

with an energy

$$W_{IV} = -\frac{nM^2}{2} \mu^2 \left(\beta - \frac{1}{\alpha}\right). \tag{33}$$

Notice that in the cases III and IV, the calculated values of $\mathcal{J}_{as}$ and of $\mathcal{J}_{bs}$ must fall between O and M. On the other hand, one can easily see that there can be no solution for the minimum energy corresponding to values of $\mathcal{J}_{as}$ and of $\mathcal{J}_{bs}$ falling at the same time between O and M.

An elementary discussion allows, for a given value of λ, and as a function of α and β, the determination of the particular preceding solution corresponding to the minimum energy. The results follow.

Suppose first that $\lambda < \mu$. The graph (α, β), (see Fig. 2), is then divided into 4 regions: by the section ASB of the hyperbola of equation $\alpha, \beta = 1$, by the half of the straight line CF of the equation $\alpha = -\mu/\lambda$, and by the half of the straight line CE of the equation $\beta = -\lambda/\mu$. Also, the straight line SH, of the equation $\beta = -1$, divides into two parts the region BCE. Table I summarizes the types of solution corresponding to each of the regions of the graph, as well as values of $\mathcal{J}_{as}$, of $\mathcal{J}_{bs}$ and of the total spontaneous magnetization $\mathcal{J}_s$ in each case.

When λ is greater than μ, the preceding discussion applies with the interchange of α and β.

(12) Variation of the spontaneous magnetization in the neighbourhood of the Curie point: In this region, the spontaneous magnetization, $\mathcal{J}_s$, as well as the two partial spontaneous magnetizations, $\mathcal{J}_{as}$ and $\mathcal{J}_{bs}$, are small. One can then substitute for the function $B_j(z)$, a function of z, a series expansion in the system of equations (24). Then, neglecting all but the first two terms, the system can be solved by successive approximations.

The first term in the series expansion of $\mathcal{J}_s$ is then written (cf. appendix II)

$$\mathcal{J}_s = \lambda \mathcal{J}_a - \mu \mathcal{J}_b =$$

$$= FM \sqrt{\frac{\theta_p - T}{\theta_p}} \left(\lambda\sqrt{k} - \frac{\mu}{\sqrt{k}}\right) \sqrt{\frac{\lambda k + \frac{\mu}{k}}{\lambda k^2 + \frac{\mu}{k^2}}}, \tag{34}$$

where k is the positive root of the equation:

$$\lambda k^2 + (\beta\mu - \lambda\alpha) k - \mu = 0, \tag{35}$$

and where F is a function of j which is equal to 1.486 for $j = 5/2$.

In the neighbourhood of the Curie point, the spontaneous magnetization varies as $\sqrt{\theta_p - T}$ as in ordinary ferromagnets. Also, more interestingly, this magnetization has the same direction as $\mathcal{J}_{as}$ when $\Delta = \lambda\sqrt{k} - (\mu/\sqrt{k})$ is positive. However, it has the same direction as $\mathcal{J}_{bs}$ when Δ is negative. Now Δ changes sign for $k = \mu/\lambda$ and in substituting this value of k in the equation (35) one finds that Δ changes sign when

$$\lambda(\alpha + 1) - \mu(\beta + 1) = 0. \tag{36}$$

For λ and μ constant, this equation represents the straight line SD already drawn, in the graph (α, β) (cf. section 9). Thus Δ is positive in the upper half of this graph, which contains the origin O, while Δ is negative in the lower half.

From the comparison of these results with the results which relate to the spontaneous magnetization at absolute zero, it follows that in the region ASD (Fig. 2) the spontaneous magnetization always remains in the same direction as $\mathcal{J}_{bs}$, at absolute zero as well as in the neighbourhood of the Curie point. In the same way, in the region BSH, the spontaneous magnetization remains always in the same direction as $\mathcal{J}_{as}$.

On the contrary, in the triangular region HSD, the spontaneous magnetization has the same direction as

TABLE I.

Area of graph	Minimum energy soln.	$\frac{\mathcal{J}_{as}}{M}$	$\frac{\mathcal{J}_{bs}}{M}$	$\frac{\mathcal{J}_s}{M}$	Type of magnetization law
GASB	I	\multicolumn{3}{c|}{always paramagnetic}			G
ACF	IV	$-\frac{\mu}{\lambda\alpha}$	1	$\mu\left(1 + \frac{1}{\alpha}\right)$	M
FCE	II	1	1	$\mu - \lambda$	P, Q, N
ECSH	III	1	$-\frac{\lambda}{\mu\beta}$	$-\lambda\left(1 + \frac{1}{\beta}\right)$	R, V
HSB	III	1	$-\frac{\lambda}{\mu\beta}$	$\lambda\left(1 + \frac{1}{\beta}\right)$	M

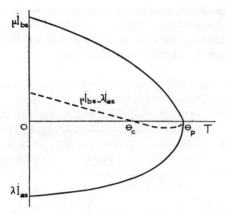

Fig. 4.
Variation with temperature of the partial spontaneous magnetizations and of the resultant spontaneous magnetization, in the case where a compensation temperature θc exists.

$\vec{\mathcal{I}}_{bs}$ at absolute zero while in the neighbourhood of the Curie point it has the same direction as $\vec{\mathcal{I}}_{as}$. Fig. 4 represents schematically the thermal variations of $\lambda \mathcal{I}_{as}$ and of $\mu \mathcal{I}_{bs}$ in this case, as well as the values of $\mathcal{I}_s$, and allows one to understand what happens. There exists notably a certain temperature θ_c, which we shall call the *compensation temperature*, where the partial spontaneous magnetizations of the sublattices are equal and opposite. The resultant observable spontaneous magnetization is therefore zero. Finally in this region HSD, the curve representing the thermal variation of the spontaneous magnetization will have a shape analogous to that represented in the figures 7N or 7V while in the regions ASD and BSH this will not be the case since the curves of the thermal variation will have the familiar shape of the curves which relate to ordinary ferromagnetism.

(13) Variation of the spontaneous magnetization in the neighbourhood of absolute zero: To complete this brief study, we shall again examine the shape of the thermal variation of the spontaneous magnetization at low temperatures. For this, it is convenient to make a distinction between the region FCE (Fig. 2), in which the two sub-lattices are saturated at absolute zero, and the two regions ACF and BCE where only one of the sub-lattices is saturated.

(A) Only one of the sub-lattices is saturated. Take for example the region ACF, where only sub-lattice B is saturated. At absolute zero the curve of magnetization corresponding to this sub-lattice is OMT (Fig. 5), and P_0 is the point which represents the magnetization under the action of the molecular field h_0. At a temperature T close to absolute zero the

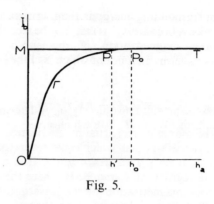

Fig. 5.

magnetization curve is deformed and is shown as Γ with an initial slope of the order of $1/T$. The molecular field takes a certain value h' close to h_0, so that the point representing the magnetization becomes P: the variations of the spontaneous magnetization of this sub-lattice, B, are thus at least of second order in T, or in $h_0 - h'$, because the curve Γ is so close to its asymptote, and are negligible compared with the variations in the spontaneous magnetization of sub-lattice A, as we shall see.

In fact, at absolute zero the curve of magnetization corresponding to this is the curve OMT (Fig. 6). On the

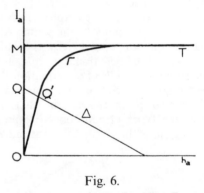

Fig. 6.

other hand, the molecular field acting on this sub-lattice is given, following formula (10) by

$$h_a = n(\mu \mathcal{I}_b + \alpha \lambda \mathcal{I}_a), \qquad (37)$$

where α is basically negative, since we are in the region ACF and where $\mathcal{I}_b$ is practically constant and equal to M, as was shown above.

The equation (37) is represented on Fig. 6 by a straight line Δ which cuts the curve OMT in a point Q. The ordinate of this point, $OQ = -\mu M/\lambda\alpha$ represents the magnetization of the sub-lattice A at absolute zero. At a temperature T close to O, the magnetization curve is shown as Γ. $\mathcal{I}_b$ remains almost constant, so that the new point representing the spontaneous magnetization of sub-lattice A becomes Q'. It can thus be seen that the spontaneous magnetization of the sub-lattice A becomes smaller by an amount which is of the order of T:

$$\mathcal{I}_a = \mathcal{I}_{a0}(1 - cT + \cdots). \quad (38)$$

Thus the spontaneous partial magnetization of that one of the two sub-lattices which is not saturated starts to diminish directly with temperature. It is deduced that in the regions ACF and BSH where the partial spontaneous magnetization of the unsaturated sub-lattices is smaller in absolute value than the partial spontaneous magnetization of the saturated sub-lattices, the total spontaneous magnetization $\mathcal{I}_s$ begins by increasing directly with temperature so that, qualitatively, the curves of the thermal variation of the resultant spontaneous magnetization have the shape shown in Fig. 7M.

In the region ECSH, on the contrary, the partial spontaneous magnetization of the unsaturated sub-lattice has a higher absolute value than the partial spontaneous magnetization of the saturated sub-lattice so that the total spontaneous magnetization starts to decrease with temperature. The curves have the shapes shown in Figs. 7R and 7V.

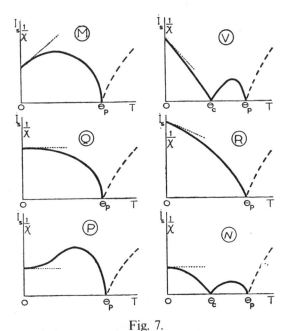

Fig. 7.
The principal possible schematic types of the laws of magnetization. (Thermal variations of the spontaneous magnetization and of the inverse of the susceptibility.)

(B) Both of the sub-lattices are saturated. The two partial spontaneous magnetizations are so close to saturation that one can use the following approximate expression for the Brillouin function

$$B_j[z] = 1 - \frac{1}{j} e^{-\frac{z}{j}}. \quad (39)$$

On the other hand, the molecular fields have values which are infinitely close to their values at absolute zero; that is, $Mn(\lambda\alpha + \mu)$ for the A ions and $Mn(\mu\beta + \lambda)$ for the B ions. Therefore

$$\mathcal{I}_s = M\left[\mu - \lambda + \frac{\lambda}{j}\exp\left\{-\frac{Mn(\lambda\alpha + \mu)}{RT}\right\} - \frac{\mu}{j}\exp\left\{-\frac{Mn(\mu\beta + \lambda)}{RT}\right\}\right]. \quad (40)$$

The approach to saturation, as a function of T, depends therefore on the two exponentials in this expression, which have opposite signs. For very small values of T, the term with the greatest absolute value is that which has the smallest absolute value of the argument of the exponential. These two arguments are equal when:

$$\lambda\alpha + \mu = \mu\beta + \lambda. \quad (41)$$

On the graph (α, β) this equation is represented by the straight line CK (Fig. 2) passing through the point C, through the point k ($\alpha = 1, \beta = 1$) and parallel to the straight line SD. It can easily be deduced that in the region FCK the spontaneous magnetization *increases with temperature*, in the neighbourhood of absolute zero (cf. Fig. 7P), while, in the region KCE the spontaneous magnetization *decreases with temperature* (cf. Fig. 7Q or 7N). However, in both cases the tangent at absolute zero must be horizontal.

(14) Comparison of the forms of the thermal variation of the spontaneous magnetization of normal ferromagnets and of ferrimagnets in the region FCE: The straight line CK defined in the preceding paragraph, has a remarkable property: when the point M(α, β) is situated on this straight line, the thermal variation of the spontaneous magnetization is identical to that of a normal, simple ferromagnet. In fact, if r is the common value of $\alpha\lambda + \mu$ and of $\beta\mu + \lambda$ we see that the two equations (24) have a solution given by $\mathcal{I}_{as} = \mathcal{I}_{bs} = \mathcal{I}'$, where $\mathcal{I}'$ is given by the implicit relationship

$$\mathcal{I}' = MB_j\left[\frac{Mnr\mathcal{I}'}{RT}\right]. \quad (42)$$

The resultant spontaneous magnetization

$$\mathfrak{J}_s = \mu \mathfrak{J}_{bs} - \lambda \mathfrak{J}_{as} = (\mu - \lambda) \mathfrak{J}'$$

is proportional to $\mathfrak{J}'$ which, as equation (42) shows, can be considered as the spontaneous magnetization of a normal ferromagnet with the coefficient of the molecular field equal to nr.

In the region FCK of the graph (α, β) the reduced curves[8] which represent the thermal variation of the spontaneous magnetization are situated *above* the curve of a normal ferromagnet while in the region KCE they are situated *below*.

(15) The different possible types of magnetization laws: To summarize the results of the preceding paragraphs, we are thus led to distinguish between 6 principal types of thermal variation of spontaneous magnetization. These are distinguished according to the shape of this variation at low temperatures and according to the eventual existence of a compensation temperature, lower than the Curie point, at which the spontaneous magnetization is zero since the partial spontaneous magnetizations of the two sub-lattices exactly cancel each other out. These different types of variation are schematically represented in Fig. 7 and are labelled with the same letters as in Table 1, and which indicate in Fig. 2 the regions of the graph (α, β) where these types are observable.

These types of variation correspond to the region of the graph (α, β) situated on the same side of the hyperbolic arc ASB as the origin O. On the other side there is always paramagnetism: type G, with a simple hyperbolic variation of $1/\chi$ throughout the whole temperature range (Fig. 8G).

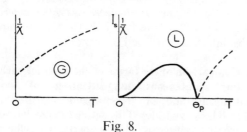

Fig. 8.

G shows the thermal variation of the inverse of the susceptibility corresponding to points (α, β) which are situated on the negative side of the branch of the hyperbola ASB. L shows a particular type of the magnetization law which corresponds to $\lambda = \mu = 0.5$ and to $\varepsilon = -1$ (cf. section 16).

---

[8] That is to say, the curves which represent the reduced magnetization $\mathfrak{J}_s / \mathfrak{J}_{s0}$ as a function of the reduced temperature T/θ_p.

Of course, for other pairs of values of λ and μ the straight lines which limit the different regions of the graph (α, β) take different positions. (The straight line SH is an exception and is fixed.) Thus the relative extent of these different regions varies.

It would be convenient to finish this study by examining the transitions between the different types of laws. We shall not do so as the interest for the moment would be comparatively small.

B. ANTI FERROMAGNETISM ($\varepsilon = -1$. $\lambda = \mu$)

(16) The two coefficients λ and μ, are equal: Up to this point, λ has been supposed to be smaller than μ. When λ is greater than μ this becomes again like the preceding case if α and β are interchanged. Therefore, the case must be studied where $\lambda = \mu = 0.5$ which corresponds to the equal sharing of the magnetic ions between the two sub-lattices. In this case the two straight lines CK and SD become identical to the line bisecting the diagram while the line CE becomes SH. One thus obtains the diagram of Fig. 9.

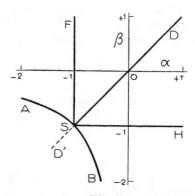

Fig. 9.

Diagram to determine, as a function of α and of β, the different types of the magnetization laws relative to negative interactions between the sublattices and to equal values of λ and μ (cf. section 16).

In the two regions ASF or BSH the curves of spontaneous magnetization are of type M with a spontaneous magnetization at absolute zero equal to $M/2(1 + 1/\alpha)$ or equal to $M/2(1 + 1/\beta)$. In the quadrant FSH, on the contrary, the spontaneous magnetization at absolute zero is always zero by cancelling of the partial spontaneous magnetizations of the two sub-lattices. A spontaneous magnetization then appears with rising temperature, provided that α is different from β, and type L is obtained, represented by Fig. 8L.

Finally, if α is equal to β, that is if the point M is situated on the straight line SD, the spontaneous magnetization always remains zero, whatever the temperature. This is because the partial spontaneous magnetizations of the sub-lattices always remain equal and opposite although both continually decrease with temperature. One recognizes here the essential properties of *antiferromagnetism* which have been the subject of theoretical work by Néel [10], [9], by Bitter [11] and by Van Vleck [12]. There is just one difference which, however, has no effect on the magnetic properties. Here the two sub-lattices are in principle crystallographically different while, in the antiferromagnets studied by the above authors, the two sub-lattices are fictitious and are obtained by sharing sites basically identical between two different categories, in such a way as to make the greatest possible number of negative interactions figure under the form of interactions between the two sub-lattices.

Note: It is to be noted that in antiferromagnets at least two configurations exist which are macroscopically indistinguishable. For example one can attribute spin + to sub-lattice A and spin − to sub-lattice B, or vice versa. Instead of supposing that the substance possesses one or other of these configurations it is probably more exact to imagine that they exist simultaneously with the same weight so that in the absence of an external field the average magnetic moment of any atom should always be zero.

(17) The case of antiferromagnetism ($\lambda = \mu = 0.5$; $\alpha = \beta$): We shall rapidly take up again the study of this particularly important case. Equations (14) and (17) give the value of the paramagnetic susceptibility

$$\frac{1}{\chi} = \frac{T}{C} + \frac{n}{2}(1 - \alpha) \ . \qquad (43)$$

A straight line of Weiss is obtained which is extrapolated for $1/\chi = 0$ up to a point θ_a

$$\theta_a = -\frac{nC}{2}(1 - \alpha) \ . \qquad (44)$$

But the law (43) is only valid up to the Curie point θ_p defined by the relation (20) and which is here written[9]:

$$\theta_p = \frac{nC}{2}(1 + \alpha) \ . \qquad (45)$$

At this point, $1/\chi$ is equal to n.

Below this temperature, the two sub-lattices take partial spontaneous magnetizations $\mathfrak{J}'_{as}$ and $-\mathfrak{J}'_{as}$

[9] In the literature of antiferromagnetism, this Curie point is also known as the *transition point* and written θ_λ

which are equal and opposite. The two equations (22) then combine in one.

$$\mathfrak{J}_{as} = MB_j \left[\frac{Mn(\alpha + 1)\mathfrak{J}_{as}}{2RT} \right] \qquad (46)$$

which allows the determination of $\mathfrak{J}_{as}$. The *resultant spontaneous magnetization is zero* so that below the point θ_p the substance is not ferromagnetic, but remains paramagnetic as we shall see. However, because of the existence of partial spontaneous magnetizations it must present those anomalies of the ferromagnets which depend on an even power of the spontaneous magnetization such as the anomalies of specific heat and of thermal expansion. In particular the specific heat will undergo a discontinuity at the point θ_p.

(18) Calculation of the susceptibility below the Curie point: The two partial spontaneous magnetizations are parallel, and antiparallel to a certain common direction Δ. Let us first suppose that the magneto-crystalline energy of the coupling between the partial spontaneous magnetizations and the crystal lattice is zero. The direction Δ then turns freely with respect to the crystal axes. It can then be shown, [10], that when a field H is applied to the system the direction Δ is oriented at right angles to H, while the two antiparallel vectors turn slightly with respect to each other, producing a component proportional to the field. The susceptibility is independent of the temperature and is equal to $1/n$

$$\frac{1}{\chi} = n \ , \qquad (47)$$

that is to say, it is exactly equal to the susceptibility given by formula (43) at the point θ_p. The whole of the curve $(1/\chi, T)$ is shown by A in Fig. 10.

However, in the *a priori* much more probable case where a magneto-crystalline energy exists (cf. section 19) of the same order of magnitude as in the ferromagnets, then the direction Δ remains fixed with respect to the crystal axes, at least while H remains fairly small. Then there are parts of the substance in which the direction Δ is parallel to the field H. In this case, as Bitter has shown [11], the partial spontaneous magnetization which has the same direction as the field increases by the amount $\delta \mathfrak{J}_{as}$ while the other spontaneous magnetization which has the opposite orientation decreases by $\delta \mathfrak{J}_{as}$. $\delta \mathfrak{J}_{as}$ is easily found to be the solution of the equation

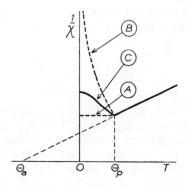

Fig. 10.
Thermal variation of the inverse of the susceptibility for antiferromagnets. For A the direction of antiparallel orientation is perpendicular to the applied field; in B it is parallel and in C it is distributed at random.

$$\mathcal{J}_{as} + \delta\mathcal{J}_{as} = MB_j\left[\frac{2MH + Mn(\alpha+1)\mathcal{J}_{as} + Mn(\alpha-1)\delta\mathcal{J}_{as}}{2RT}\right], \tag{48}$$

where $\mathcal{J}_{as}$ was defined by equation (46).

The part of the substance under consideration thus takes a certain magnetization $\delta\mathcal{J}_{as}$ in the direction of the field, which is proportional to H. The corresponding susceptibility, which is zero at absolute zero, increases and tends towards n as T tends towards θ_p, as represented schematically by B in Fig. 10. In order to obtain the susceptibility of the whole of the substance, with the distribution at random of the orientations of Δ, it is necessary [12] to take the average of the two curves drawn, with a weighting of 2 in the case where Δ is perpendicular to H and a weighting of 1 when Δ is parallel to H. The curve C of Fig. 10 is obtained. Notice especially that the susceptibility at absolute zero χ_0 is equal to 2/3 of the susceptibility χ_p at θ_p.

A certain number of antiferromagnets are known in which the susceptibility varies as shown by C in Fig. 10, such as MnO, FeO, FeFe, MnF$_2$ studied by Bizette and Tsai [13]. We shall say more on this subject later (cf. sections 36 and 37). The values corresponding to the relation χ_0/χ_p are given in Table XI. They vary from 0.69 to 0.82 and are therefore systematically larger than the theoretical value of 0.67. This disagreement is perhaps due to fluctuations in the molecular field which have been ignored in the theory given here.

(19) Elementary domains of antiferromagnets: The direction of antiparallel orientation, Δ, plays the same role for antiferromagnets as the direction of spontaneous magnetization does for ferromagnets. In particular, as the energies of coupling with the crystal lattice for a ferromagnet, called magnetocrystalline energies, only depend on the *direction* of the spontaneous magnetization and not on its sense, it results that in antiferromagnets the direction of antiparallel orientation, which is only the direction of two spontaneous magnetizations of opposite sense, is coupled with the crystal network exactly in the same way as a simple direction of spontaneous magnetization. The formulae which give the coupling energy as a function of the components α, β, γ of the direction of antiparallel orientation are then the same as the familiar formulae of ferromagnetism relative to the magnetocrystalline energy and the coefficients k which figure in them must be of the same order of size as for ferromagnets.

There exist, in particular, a certain number of directions of minimum energy which we shall call *preferred directions*. It will be seen that an antiferromagnet must be subdivided below its transition point into elementary domains analogous to Weiss domains, inside which the direction of antiparallel orientation follows the same preferred direction. The different elementary domains are separated from each other by transition regions where the direction of antiparallel orientation gradually turns. They are thus analogous to Bloch walls; their theory is the same, and, as a consequence, their thickness and their superficial energy are of the same order of size, that is, 0.1 micron and a few ergs per square centimetre respectively, as the thickness and energy of the Bloch walls of ferromagnets.

However, the observable phenomena are presented in a different way for it is necessary to consider antiferromagnets like ferromagnets with a spontaneous magnetization which is variable and proportional to the applied magnetic field H. Therefore it is by studying the variations of the susceptibility $\mathcal{J}/H$, as a function of the field, of an antiferromagnet that analogies must be found with the curves which represent the variations with the field of the magnetization of ferromagnets. For example, the variations of the residual susceptibility of antiferromagnets must correspond to the variations of the residual magnetization of ferromagnets.

Antiferromagnets must therefore show variations of susceptibility and phenomena of hysteresis, which are on one hand linked to reversible rotations of the directions of antiferromagnetism and on the other hand linked to reversible and irreversible displacements of the separation walls between the elementary domains. An essential difference arises because the energies provoking these phenomena are, for one cubic centimetre, of the order of $\chi_0 H^2$ for antiferromagnets, where χ_0 is the susceptibility at absolute zero, while they are of the order of magnitude of $\mathcal{J}_s H$ for ferromagnets. It is therefore deduced that the displacements of the walls which are seen in an almost pure ferromagnet at fields of 1 to 10 gauss will appear here at fields of 3,000 to 10,000 gauss[10]. Also the rotations, which are produced at 100 and 1000 gauss in ferromagnets must only be produced here between 30,000 and 100,000 gauss. These figures are, of course, only given as a guide to orders of magnitude.

It is, no doubt, with an interpretation of this kind that one must explain the variation of susceptibility with the magnetic field observed by Bizette and Tsai for antiferromagnets, such as MnO and MnF_2, below their transition point. However, the experimental data are still too fragmentary at present to develop a systematic theory concerning them.

C. THE INTERACTIONS BETWEEN DIFFERENT SUB-LATTICES ARE POSITIVE ($A_{ab} > 0$; $\varepsilon = +1$)

(20) Paramagnetic properties: Study of formulae (10) and (11) show that one can pass from the case where ε is equal to -1 to the case where it is equal to $+1$ by replacing n, α and β in the formulae by $-n$, $-\alpha$ and $-\beta$. In particular, the calculations of section 6 are valid and the expression for the paramagnetic magnetization as a function of the field is written in the following form which replaces equation (13)

$$\vec{\mathcal{J}} = \frac{T + nC\lambda\mu(2 - \alpha - \beta)}{T^2 - nC(\lambda\alpha + \mu\beta)T + n^2C^2\lambda\mu(\alpha\beta - 1)} \vec{H}. \quad (49)$$

Naturally, the general expression for $1/\chi$ as a function of T keeps the form (14) but the coefficients of this formula are now given by the following formulae

$$\frac{1}{\chi_0} = -n(2\lambda\mu + \lambda^2\alpha + \mu^2\beta), \quad (50)$$

$$\sigma = n^2 C\lambda\mu[\lambda(1-\alpha) - \mu(1-\beta)]^2, \quad (51)$$

$$\theta = -nC\lambda\mu(2 - \alpha - \beta). \quad (52)$$

(21) Ferromagnetic properties: When the sub-lattices have spontaneous magnetizations, the fact that the sign of A_{ab} changes, simply changes the relative orientation of the two partial spontaneous magnetizations. In the present case, the resultant spontaneous magnetization

[10] As a basis for comparison, for 1 cm³, we have taken $\mathcal{J}_s = 1,700$ and $\chi_0 = 300 \times 10^{-6}$.

$\mathcal{J}_s$ then becomes the *sum* of the partial spontaneous magnetizations so that equation (25) which defines $\mathcal{J}_s$ is replaced by

$$\mathcal{J}_s = \lambda \mathcal{J}_{as} + \mu \mathcal{J}_{bs} \quad (53)$$

but the equations (24) remain valid, so that the values of $\mathcal{J}_{as}$ and of $\mathcal{J}_{bs}$ do not depend on the sign of A_{ab}, if all other things are equal.

In particular, the ordering Curie point θ_p, considered as the point of annulation of the partial spontaneous magnetizations, is always given by the formula (20), *whatever the sign of A_{ab}*.

The discussion of the general properties of the system as a function of α and β, for a given value of λ, immediately results in a transposition of the results of sections 11 to 15. The graph (α, β) is divided into 4 regions (Fig. 11) by the negative branch ACB of the hyperbola $\alpha\beta = 1$ and the two half straight lines CE, of the equation $\beta = -\lambda/\mu$ and CF, of the equation $\alpha = -\mu/\lambda$. On the negative side of ACB, the substance is

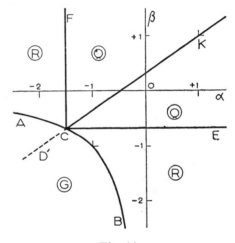

Fig. 11.
Diagram giving the different possible kinds of magnetization laws, as a function of α and β, for positive interactions between the sublattices ($\varepsilon = +1$). The figure is drawn for $\lambda/\mu = 2/3$.

always paramagnetic; on the positive side it becomes ferromagnetic below a certain temperature θ_p, given by equation (20). For each region of the graph, the values of the partial spontaneous magnetizations and of the resultant spontaneous magnetization, as well as the kinds of magnetization laws corresponding to the diagrams of Fig. 7, are given in Table II. It will be noticed that in the region FCE, the resultant spontaneous magnetization is equal to the absolute saturation M.

TABLE II.

Area of graph	Minimum energy soln.	$\dfrac{\mathscr{T}_{as}}{M}$	$\dfrac{\mathscr{T}_{bs}}{M}$	$\dfrac{\mathscr{T}_{s}}{M}$	Type of magnetization law
GACB	I	\multicolumn{3}{c}{always paramagnetic}			G
ACF	IV	$-\dfrac{\mu}{\lambda\alpha}$	1	$\mu\left(1-\dfrac{1}{\alpha}\right)$	R
FCE	II	1	1	1	Q
ECB	III	1	$-\dfrac{\lambda}{\mu\beta}$	$\lambda\left(1-\dfrac{1}{\beta}\right)$	R

(22) Possible validity of Weiss' law of paramagnetism: In order for the law of paramagnetism (14) to take the form of Weiss' law, the coefficient σ must be zero, that is:

$$\lambda(1-\alpha) = \mu(1-\beta). \quad (54)$$

For a given value of λ, this equation is represented by the straight line D'CK (Fig. 11) which passes through the point C ($\alpha = -\mu/\lambda$; $\beta = \lambda/\mu$) and the fixed point K($\alpha = \beta = 1$). Also, when the point M(α, β) is on this straight line, $1/\chi$ is a linear function of the temperature. In the ferromagnetic region, that is, if the point M is on the straight line CK, the substance also possesses a notable property: $\mathscr{J}_{as}$ and $\mathscr{J}_{bs}$ are equal, so that the resultant spontaneous magnetization is a solution of an equation of the type (42). With reduced coordinates $\mathscr{J}_{s/M}$ and T/θ the curve representing the thermal variation of the spontaneous magnetization is then identical to that of a normal ferromagnet. When M is outside this straight line, the curve of the spontaneous magnetization lies below the curve corresponding to a normal ferromagnet.

SECOND PART
INTERPRETATION OF THE EXPERIMENTAL DATA
A. THE FERRITES

(23) The role of fluctuations in the molecular field; the two Curie points: It is now necessary to compare the theory developed above with the experimental results relating to the magnetic properties of the *ferrites*. To begin with, we shall be concerned with the paramagnetic properties for which we have the most complete and accurate experimental data, thanks to the work of Mlle. Serres [1]. The theory has shown that the magnetic susceptibility, for one ion-gram of ferric iron, is expressed as a function of the absolute temperature T, by the formula:

$$\frac{1}{\chi} = \frac{T}{C} + \frac{1}{\chi_0} - \frac{\sigma}{T-\theta}, \quad (14)$$

where χ_0, σ, θ are constants and C is the atomic Curie constant of the ferric ion, which is approximately 4.4. A brief examination shows that such a formula can only represent the experimental results on the condition that the results obtained in the immediate neighbourhood of the ferromagnetic Curie point of the substance considered, are not included.

A priori, such an observation is not surprising. Indeed, it is known that in classic ferromagnetic substances, the theory of the molecular field predicts that the linear variation of the inverse of the susceptibility as a function of temperature extends from the Curie point up to the highest temperatures. However experiments show that this linear variation only really begins at at least a hundred degrees above the Curie point. Closer to the Curie point, the observed susceptibility is weaker than the expected susceptibility. Thus the Weiss formula:

$$\chi = \frac{C}{T-\theta_p} \quad (1)$$

is only valid at a certain distance above the Curie point, while the constant θ_p which occurs in this formula, called the paramagnetic Curie point is appreciably different from the experimental Curie point θ_f, or ferromagnetic Curie point, defined as the point where the spontaneous magnetization disappears. The difference $\theta_p - \theta_f$ is of the order of 20° for nickel and of 50° for iron. In the language of the molecular field, this difference is attributed to the *fluctuations* of this field. This difference thus makes clear the difference between the molecular field approximation and a rigorous theory. Now, the theory developed above is in fact only a generalisation of the molecular field theory to substances with a crystal structure which is more complex than that of pure metals. It is then necessary to anticipate the existence of anomalies which are analogous to those of ordinary ferromagnets. In particular, the analogue of the paramagnetic Curie point will be the temperature θ_p at which the susceptibility χ, extrapolated by means of formula (14), becomes infinite. It is then necessary to expect that this temperature θ_p will be considerably higher than the temperature θ_f where the spontaneous magnetization disappears. The difference $\theta_p - \theta_f$ must be here of the same order of magnitude as for pure ferromagnetic metals.

(24) Calculation of the coefficients χ_0, σ and θ from the experimental data: The most simple procedure consists in writing equation (14), satisfied for three pairs χ_i, T_i of experimental values, and by solving the three simultaneous equations thus obtained, which presents no difficulty. The three pairs must, of course, be chosen to be as far distant from each other as possible, while taking care to choose the first pair at a hundred degrees above the Curie point in order to avoid the region which is disturbed by the fluctuations. This procedure has the inconvenience of leaving aside most of the experimental points and does not allow the evaluation of the accuracy obtained nor the determination of the limit where the fluctuations begin to be felt.

It is preferable, after the experimental curve has been carefully drawn in the high temperature region to choose the point (χ_a, T_a) which is at the highest possible temperature and which is also known accurately. On combining the two equations (14) which correspond to the point (χ_a, T_a) and to the current point (χ, T) the following is obtained

$$\frac{1}{\dfrac{\dfrac{1}{\chi_a}-\dfrac{1}{\chi}}{T_a-T}-\dfrac{1}{C}} = \frac{1}{\sigma}(T-\theta)(T_a-\theta). \quad (55)$$

If a graph is drawn where y is the inverse of

$$\frac{\dfrac{1}{\chi_a}-\dfrac{1}{\chi}}{T_a-T}-\frac{1}{C}$$

and x is T_a-T, a straight line is obtained whose intercepts with the axes yield $(T_a-\theta)$ and $\dfrac{(T_a-\theta)^2}{\sigma}$. Thus σ and θ can be determined. When these values of σ and θ are substituted in (14) one then obtains as many determinations of $1/\chi_0$ as there are pairs χ, T of the experimental values. The average is taken and at the same time one can assess the spread of values compared with the calculated curve.

As an example, Fig. 12 shows the straight line in question for a variety of cadmium ferrite studied by Mlle. Serres [1]. Close to $T-T_a=300°$, the appearance of differences due to fluctuations can be noticed.

(25) Results of the study of experimental data for the ferrites: Table III gives the values of χ_0, σ and θ corresponding to a certain number of ferrites studied by Mlle. Serres. The range for which formula (14) is valid has also been given. In the sixth and seventh columns the values are given for the paramagnetic

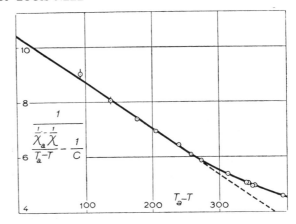

Fig. 12.
Graph to determine the characteristic constants of equation (14) for a ferrite of cadmium. ($T_a=732°$).

Curie point θ_p and the ferromagnetic Curie point θ_f when the latter is known. The differences $\theta_p-\theta_f$ are from $40°$ to $60°$ in the most accurate cases, which is of the expected order of magnitude.

In order to estimate how accurately a formula of type (14) represents the experimental results, we have, in Tables IV, V and VI, compared the experimental values with the values calculated by formula (14), using the values for the constants χ_0, σ and θ given in Table III. In the last columns of these tables the relative differences between these values has been indicated as a percentage. For this comparison substances have been chosen for which there are the most numerous measurements and the most extended temperature range. The Tables show well the temperature below which deviations from formula (14) appear, which are produced by fluctuations of the molecular field. Above this temperature, the formula represents the experimental results with an accuracy of a few thousandths, comparable with the accuracy of the measurements. Thus, in the paramagnetic domain, the theory proposed allows the complete interpretation of the form of the experimental curves.

(26) Calculation of the values of the characteristic coefficients α, β, and n: In order to progress further it is necessary to go from the values of χ_0, σ and θ to the values of the quantities α, β, n and λ which will finally physically define the substance considered by the proposed theory. We already know that there is an infinite number of solutions since the system of equations (15), (16) and (17) or (50, 51) and (52) only give three relationships between these four quantities. We shall then suppose that ε is negative and shall calculate α, β and n as a function of the experimental data χ_0, σ, θ and with λ as parameter. In order to

FERRIMAGNETISM

TABLE III.
Constants for the law of paramagnetism, for some ferrites studied by Mlle. Serres.

Substance	$\dfrac{1}{\chi_0}$	σ	θ	Range in which (14) is valid	θ_p	θ_f	$\theta_p - \theta_f$
Fe_2O_3MgO I	296,7	14 700	601,8	420-720°C	635°K	588°K	47°
Fe_2O_3MgO II	259,3	2 486	705,2	480-720	711	612 ?	99 ?
Fe_2O_3PbO a	298,2	10 610	750,0	550-720	772	708	64
Fe_2O_3PbO b	305,8	19 630	708,6	550-750	750	?	
Fe_2O_3CuO	292,7	9 970	744,1	530-710	766	728	38
Fe_2O_3NiO	239,9	10 600	881,0	680-740	905	863	42
Fe_2O_3CdO If	139,8	18 600	555,3	450-750	622	523 ?	99 ?
Fe_2O_3CdO Ip	158,7	37 100	384,0	450-750			
$2Fe_2O_3, CuO, ZnO$	291,5	79 200	366,7	400-710			
$4Fe_2O_3, CuO, 3ZnO$	238,4	223 000	− 419	370-700			
Fe_2O_3ZnO	197,0	300 000	− 1 210	30-700			

TABLE IV.
Fe_2O_3MgO, I.

T °K	$\dfrac{1}{\chi_m}$ calc.	$\dfrac{1}{\chi_m}$ obs.	diff. %
722,1	485,5	481,0	0,9
651,6	461,5	460,8	0,1
607,1	443,9	444,0	0,0
598,4	440,2	439,9	0,1
583,0	433,4	432,9	0,0
573,2	428,9	428,8	0,0
558,8	421,8	421,9	0,0
544,4	414,3	414,1	0,0
521,6	401,1	400,6	0,1
497,5	384,7	385,2	− 0,1
489,0	378,1	378,7	− 0,2
474,4	365,6	366,4	− 0,2
463,7	355,1	355,9	− 0,2
447,9	337,1	336,6	0,1
422,8	298,5	301,6	− 1,0
400,0	243,2	259,5	− 6,7

TABLE V.
Fe_2O_3PbO.

T °K	$\dfrac{1}{\chi_m}$ calc.	$\dfrac{1}{\chi_m}$ obs.	diff. %
747,5	474,8	474,6	0,0
718,7	461,9	462,3	− 0,1
704,1	454,6	455,6	− 0,2
686,1	445,4	445,8	− 0,1
665,6	433,8	434,8	− 0,2
646,4	421,7	422,1	− 0,1
644,0	420,0	420,3	− 0,1
633,1	412,3	412,0	0,1
620,8	402,9	401,9	0,2
612,7	396,7	395,9	0,1
592,7	377,5	377,5	0,0
578,4	361,8	360,6	0,3
561,7	339,8	339,3	0,1
546,7	315,4	317,8	− 0,8
524,7	266,8	280,0	− 5,0

resolve the system of equations (15), (16) and (17) it is better, for simplification, to introduce two quantities ρ and τ, with no dimensions, defined by the relationships

$$\rho = \left(1 + \frac{C}{\chi_0 \theta}\right)\lambda\mu \; ; \quad \tau = \frac{\eta}{\theta}\sqrt{\lambda\mu C\sigma}, \quad (56)$$

where η is equal to ± 1.

After eliminating n, one then obtains two first degree equations in α and β which give, when resolved

$$\left.\begin{aligned}\alpha &= \frac{-\rho + \mu(1 - 2\tau)}{\rho + \tau(\mu - \lambda)} \\ \beta &= \frac{-\rho + \lambda(1 + 2\tau)}{\rho + \tau(\mu - \lambda)}\end{aligned}\right\} \quad (57)$$

TABLE VI.
Fe$_2$O$_3$CdO.

T °K	$\dfrac{1}{\chi_m}$ calc.	$\dfrac{1}{\chi_m}$ obs.	diff. %
732,0	327,3	327,3	0,0
638,8	295,1	295,8	− 0,2
594,9	279,0	279,1	0,0
555,4	263,1	263,1	0,0
528,4	251,7	251,7	0,0
494,0	236,4	236,2	0,1
476,1	228,1	227,1	0,4
459,7	220,0	219,1	0,4
420,3	199,2	198,2	0,1
394,1	183,9	183,9	0,0
390,5	181,7	182,1	− 0,2
382,9	177,0	176,7	0,2
379,9	175,1	176,9	− 1,0
339,5	147,1	152,5	− 3,7
289,4	104,2	117,6	− 12,8

α and β being thus determined, n can be deduced by means of equation (17).

The curves Γ. The possible values of α and of β correspond, on a graph (α, β) to a curve consisting of two branches Γ and Γ', one for $\eta = +1$ and the other for $\eta = -1$. As an example, Fig. 13 shows the curve drawn for the ferrite Fe$_2$O$_3$CuO; we limit ourselves to showing the parts of the curves corresponding to values of λ less than, or equal to, 2 because, from the physical nature of the problem, the sub-lattice A can contain at most only half of the ferric ions. Also, the curve has been marked with the values of λ corresponding to each point.

Thus, the study of the paramagnetic properties simply allows the definition in the form of a curve Γ, Γ' the place of possible positions of the representative point M(α, β, λ). In order to progress further, it is necessary to include information given by the study of the substance below the ferromagnetic Curie point and particularly of the saturation magnetization at absolute zero, equal to the resultant spontaneous magnetization at the same temperature. In fact, as we saw above, this spontaneous magnetization depends on λ and on the type of the magnetization law.

(27) The choice of the values of α and of β from the values of the saturation magnetization; application to Fe$_2$O$_3$CuO: We have discussed above the different possible types of magnetization laws following the values of α, β and λ. When the results of this discussion are applied here, it is found, for the branch of the curve Γ, that when λ falls between 0.5 and 0.265, between the points A and B (Fig. 13) the magnetization curve is of the type Q. When λ is smaller than 0.265, beyond the point B, it is of type R. Considering the curve Γ', the magnetization curve is of type N when λ is greater than 0.38, between the points A' and B'. It is of type V when λ is smaller than 0.38, beyond the point B'.

Recognizing this, the data of Table I allow the calculation, as a function of λ, of the spontaneous magnetization at absolute zero, corresponding to the different points (α, β) of the curves Γ and Γ'. Therefore one finds the curves abc and ade of Fig. 14, corresponding respectively to the curves Γ' and Γ. The

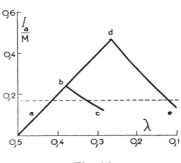

Fig. 14.

arcs ab, bc, ad, de correspond respectively to the types N, V, Q and R of the magnetization laws. In this figure, the magnetizations are expressed by taking as unity the absolute saturation M[11], which corresponds to the parallel alignment of all the spins of the ferric ions present.

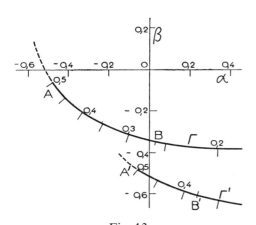

Fig. 13.
Characteristic curves Γ, Γ' of the ferrite of copper, marked according to the values of λ.

[11] Not to be confused with the spontaneous magnetization at absolute zero.

On the other hand, what do the experiments give? M. Fallot has kindly provided me with unpublished measurements, taken by him in 1935 at Strasbourg, of the saturation magnetization of a ferrite of copper slowly cooled from a high temperature, consequently analogous to the ferrite studied by Mlle. Serres. By a double extrapolation at absolute zero and in an infinite field, he has found 0.85 Bohr magneton, that is, 0.17 M.

The possible solutions correspond to the intersections of the curves of Fig. 14 with the straight line of ordinate 0.17. There are four solutions of which the characteristic constants are summarized in Table VII.

TABLE VII.

Type of magnetisation law	λ	μ	α	β
Q	0,415	0,585	− 0,36	− 0,19
N	0,415	0,585	+ 0,14	− 0,58
R	0,12	0,88	+ 1,18	− 0,41
V	0,325	0,675	+ 0,46	− 0,66

By using the method of calculation indicated in Appendix I, the thermal variation of the spontaneous magnetization is calculated for these four solutions. The curves obtained are reproduced in adjusted coordinates in Fig. 15.

The four curves are very different from each other. Unfortunately, we do not possess the complete corresponding experimental curve, but only two measurements, one at 77 K with $\mathcal{J}/M = 0.165$ and the other at 291 K with $\mathcal{J}/M = 0.142$. Since the Curie point of the ferrite is at 728 K, the corresponding reduced temperatures are respectively equal to $0.106 = \frac{77}{728}$ and $0.4 = \frac{291}{728}$; hence the two points drawn on Fig. 15 and surrounded by a small circle. These two points are close to the curve Q, but have no relation at all to the other curves. Therefore, solution Q must be adopted and solutions R, V and N rejected.

The determination of the constant n still remains. For that, formula (17) is used with $\theta = 744$. Then $n = 480$ is obtained.

Thus a satisfactory representation is obtained of all the magnetic properties of the slowly cooled copper ferrite by attributing them to the unequal division of the ferric ions between the two possible categories of sites. That is, on average 0.415 ions on sublattice A, the tetrahedric positions, and 0.585 ions on sublattice B, the octahedral positions.

The preceding discussion relates to the case where ε is negative, that is to say, the case where the

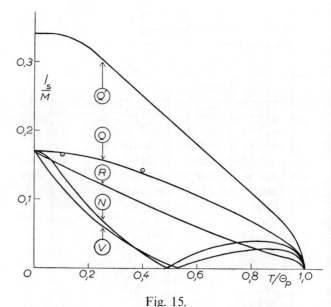

Fig. 15.
Q, R, V, and N show the different possible thermal variations of the spontaneous magnetization of copper ferrite, determined from the paramagnetic data and the value of the saturation magnetization at absolute zero. The circles show the experimental points. Q' shows the calculated curve for the same quenched ferrite.

interactions between different sub-lattices are negative. An analogous discussion shows that it is not possible to obtain an acceptable solution for the case where ε is positive. Therefore the sets of information provided by the paramagnetic properties are incompatible.

(28) The role of heat treatments: *A priori*, this distribution ($\lambda = 0.415$, $\mu = 0.585$) is fairly surprising. Indeed, if w is the increase in the energy of the system connected with the transfer of a ferric ion from an A site to a B site, and supposing w to be constant and independent of λ, there are two possible alternatives. At absolute zero, if w were negative all the ferric ions would have to occupy the sites of the sub-lattice B; one would thus have $\lambda = 0$, $\mu = 1$. If w were positive, there would have to be the maximum possible number of ions on sub-lattice A; that is $\lambda = 0.5$, $\mu = 0.5$. In fact an intermediate case is observed.

The situation shows close analogies with the situation of the superstructures. It is known that many superstructures exist, $FeNi_3$ for example, which are very difficult to obtain in a perfect state, even using prolonged annealing, because the differences of energy which correspond to the different possible distributions of atoms are too small. One could suggest that it is the same here; w would be positive and small. In this case, it must then be expected that different annealings lead to different values of λ. That is, to different values of the spontaneous magnetization at absolute zero which is equal, as we have seen to $M(1-2\lambda)$. This is precisely the case: experiment shows that the saturation magnetization of ferrites in general, and of copper ferrite in particular greatly depends on the heat treatment. For example, suppose that a ferrite is taken to a very high temperature: if the two possible positions of the ions are almost equivalent, energetically, the ions will be distributed at random on all the possible sites. Since the B sites are twice as numerous as the A sites, there will be twice as many ferric ions on sub-lattice B as on sub-lattice A. There, $\lambda = 0.333$ and $\mu = 0.667$. By a violent quenching, it must be possible to conserve such a distribution in false equilibrium at ordinary temperatures which will be characterized by a spontaneous magnetization, at absolute zero, equal to 0.333 M. Experiment shows, in complete agreement with these predictions, that the magnetization of quenched ferrites is well above that of annealed ferrites.

M. Fallot has obtained spontaneous magnetizations at absolute zero which have reached 1.72 Bohr magnetons per ferric ion, that is 0.344 M, which is very close to the value predicted above.

The considerations above are also confirmed by the shape of the thermal variation of the spontaneous magnetization at low temperature. Experiment shows that the relative diminution of the spontaneous magnetization from the temperature of liquid nitrogen (77 K) to room temperature (291 K) is definitely bigger (22.5% of the saturation magnetization at absolute zero) in the quenched copper ferrite than in the ferrite which was slowly cooled (13% of saturation). Now if the values of α and of β are taken to be independent of λ, as a first approximation, the thermal variation of the spontaneous magnetization can be calculated for $\lambda = 0.333$ and it is then found that when the adjusted temperature passes from 0.106 to 0.400 the relative diminution of the spontaneous magnetization is 26% (cf. Fig. 15 Q') while it is only 17% for $\lambda = 0.415$ under the same conditions. Thus a simple interpretation of the experimental results is obtained. This argument supposes that the reduced temperatures are equal in both cases; that is, that the Curie points are equal. In fact, from the experimental data it does not seem that the relative difference between the Curie points reaches 10%. On the other hand, if n is supposed to be constant, the calculated variation in the Curie point also does not reach 10%, so that qualitatively the preceding arguments keep their value.

(29) The other ferrites: *The ferrites of magnesium and of lead.* In the same way one can interpret the paramagnetic properties of magnesium ferrite and of

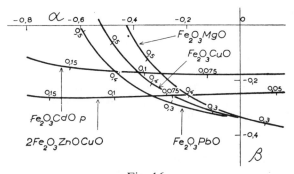

Fig. 16.
Characteristic curves, Γ, Γ' of different ferrites.

lead ferrite which are very close to those of copper ferrite and which have been studied by Mlle Serres. Fig. 16 allows their characteristic curves, Γ, to be compared. Unfortunately we do not possess any precise data on their saturation magnetization so that it is not possible to specify the point on the characteristic curve which corresponds to their real state but as it seems that this saturation magnetization is of the order of magnitude of that of copper ferrite, the distribution parameter λ must be, as in the former case, between 0.33 and 0.50.

The ferrites of cadmium and of zinc. Experiments show that a certain number of ferrites exist whose properties are clearly different from those of the ferrites studied above. In particular, their Curie point is much lower. For example, these are cadmium ferrite, zinc ferrite and mixed ferrites containing zinc. Also, for cadmium ferrite, the properties are very different depending on the origin of the sample and the heat treatment. On Fig. 16 we have shown, for example, the characteristic curve relating to a sample of cadmium ferrite studied by Mlle Serres. It can be seen that it is very different from the curves studied above. This arises perhaps from the considerable differences between the values of the constants α, β and n, corresponding to the various categories of ferrites. However, this interpretation does not seem very likely as zinc ferrite, for example, has the same crystal structure as copper ferrite and the interatomic distances are almost the same in both. It seems more natural to attribute the differences in paramagnetic properties to the different distributions of the ferric ions. If we suppose that the values of α, β and n corresponding to copper ferrite and to cadmium ferrite are similar, then the point representing the state of cadmium ferrite on the graph (α, β) must be placed close to the point representing the state of copper ferrite; for example, the point corresponding to $\lambda = 0.1$. One then obtains $\alpha = -0.362$, $\beta = -0.175$ and $n = 524$. 90% of the ferric ions would then be on B sites, that is, on places occupied by trivalent ions in normal spinels. An analogous conclusion is reached in the case of the mixed ferrite $2Fe_2O_3 \cdot CuO \cdot ZnO$ whose characteristic curve has been shown on Fig. 16.

In spite of this comforting agreement, it seems difficult to go further here with the interpretation of the magnetic properties of the ferrites of zinc and of cadmium ferrite. On one hand, in fact, we lack certain data on the properties of these bodies at low temperatures. On the other hand, and especially, when λ is small the approximations which are the basis of the theory are no longer valid, as was specified in section 5, because of the magnitude of the molecular field fluctuations.

Also it must not be expected that the theory set out above can be applied to substances such as $Fe_2O_3 \cdot ZnO$ or $3Fe_2O_3 \cdot 2ZnO \cdot Cu \cdot O$ in which it is expected, *a priori*, that most of the ferric ions will be found placed on B sites.

At most, one can obtain from formula (14) the approximate form of the phenomena at *high temperature*, since one knows, *a priori*, that it is in this region that the influence of the molecular field fluctuations is most weak. In this spirit, an attentive study shows that the magnetic properties of the two ferrites mentioned above ($Fe_2O_3 \cdot ZnO$, $3Fe_2O_3 \cdot 2ZnO \cdot CuO$ seems to agree suitably with values of λ of order 0 to 0.5 and values of α, β and n close to those found for copper ferrite.

(29)(a) Conclusion of the study of ferrites: Our results thus agree qualitatively with those of Verwey and Heilmann [5] which were chiefly based on studies using X-rays. In fact, following these authors, all the ferrites would possess inverse structures (cf. section 3) with the exception of the ferrites of zinc and of cadmium whose structures would be normal and the ferrites of zinc and of copper which would possess a mixed structure. However, a close examination shows some differences.

The case of zinc ferrite. In this case we find, for example, that a small, but not zero, number of ferric ions exist on A sites while, according to the authors quoted above, there are none. It is possible that this difference arises from the fact that the particular ferrite of zinc whose magnetic properties we have interpreted contains a slight excess of Fe_2O_3 compared with the ideal composition. This produces considerable modifications in the magnetic properties as Snoëk [14] has shown.

The case of inverse ferrites. The differences here are more profound, for our interpretation seems to show that at high temperature the ferric ions of the inverse spinels are distributed at random on the A sites and B sites, while according to Verwey and Heilmann, all the B sites would be occupied by ferric ions. Also, we attribute the diminution of the magnetization at saturation, produced by the anneal at 750°C of quenched ferrites, to a displacement of the ferric ions from the B sites to the A sites, while according to the authors quoted, it would consist of the segregation of a new phase consisting of non-magnetic $Fe_2O_3\alpha$.

This last interpretation seems indisputable when it concerns ferrites containing an excess of Fe_2O_3 compared with the ideal composition Fe_2O_3MO but seems less certain when it concerns the correct composition and, in all cases, it does not exclude the possibility of a simultaneous displacement of ferric ions. In fact quite recently Verwey, Haymann and Romeijn [15] have interpreted the variations of the electrical conductivity of the spinel Fe_2AlO_4, produced by an anneal at 850°C, by the passing of ferric ions from B sites to A sites.

For the complexity of these phenomena, the magnetic data which we possess are too incomplete, too imprecise, and relate to substances which are not sufficiently well defined for us to be able to give a definite solution to these problems at present.

B. MAGNETITE AND CUBIC IRON SESQUIOXIDE

(30) Magnetization at saturation: One of the essential magnetic characteristics of copper ferrite is the preponderance of the interactions between the two sub-lattices: α and β are small compared with unity. It seems that this is a general property of the ferrites. These interactions are negative and finally bring about, towards absolute zero (at least if λ, or μ, is not very small), the saturation magnetization of the two sub-lattices in opposite directions.

Application to magnetite. If, with Verwey and de Boer [6] one assumes that in a molecule $Fe_2O_3 \cdot FeO$ the ferrous ion is placed on sub-lattice B while one of the ferric ions is placed on sub-lattice A and the other on sub-lattice B, it follows that at saturation, sub-lattice B contains a ferric ion with 5 Bohr magnetons and a ferrous ion[12] with 4 Bohr magnetons both in the same direction, while sub-lattice A contains one ferric ion oriented in the opposite direction. That is, there is a total resultant moment of 4 Bohr magnetons per Fe_3O_4 molecule. Experiment [16] gives 22,740 U.C.G.S. which is 4.07 Bohr magnetons, thus agreeing well with theoretical predictions. The small excess of 0.07 μ_B perhaps arises from the orbital moment.

Application to cubic iron sesquioxide. Cubic iron sesquioxide, or $Fe_2O_3\gamma$, gives the same rays as magnetite under X-rays. In general it is assumed that

---

[12] According to D. Ray-Chaudhuri (*Nature* 1932, **130**, 891), the value of g for magnetite is equal to 1.97. Practically, the spin moment acts alone.

the crystal structure is the same as that of magnetite, allowing for the fact that one atom of iron in every nine is displaced, while the oxygen lattices are identical in both cases. As we have seen above, the ferric ions tend to occupy A sites. The eight ferric ions which are distributed among the 3 A sites and 6 B sites occupy the 3 A sites and 5 of the B sites while the last B site remains empty. The resultant moment will then be 10 Bohr magnetons for 8 ferric ions. That is, 2.5 Bohr magnetons per Fe_2O_3 molecule. Experiment [16] gives 13,330 C.G.S. units which is 2.39 Bohr magnetons. P. Weiss and R. Forrer estimate their measurements are accurate to the nearest decimal point. The agreement between calculation and experiment is therefore satisfactory.

(31) Paramagnetic properties: Cubic iron sesquioxide is unstable and decomposes in the neighbourhood of its Curie point so that it has not been possible to study it in the paramagnetic region, but magnetite is stable and has been studied by Kopp [18]. The curve $(1/\chi, T)$ does not obey Weiss' law but rather is similar to the susceptibility curves of the ferrites.

The theory of the paramagnetic properties of magnetite is also much more complicated than that for the ferrites since each molecule contains three magnetic ions: two ferric ions of atomic Curie constant $C_{5/2} = 4.377$ and one ferrous ion of atomic Curie constant $C_{4/2} = 3.001$ instead of only two ferric ions as for the ferrites. Therefore, it is necessary, at least, to introduce 5 exchange integrals instead of 3.[13]

For simplicity's sake, we shall be content to use the same exchange integrals as in the case of the ferrites. That is to say, the same molecular field coefficients, $n\alpha$, $n\beta$ and $-n$. We thus suppose that the molecular field coefficients which define the interactions between two ions only depend on the nature of the two sub-lattices on which these ions are placed, and do not depend on the nature of the ions.

If $\lambda_1, \lambda_2, \ldots, \lambda_i$, and $\mu_1, \mu_2, \ldots, \mu_i$, designate the proportions of ions of Curie constant $C_1, C_2, \ldots, C_i$, distributed respectively on sub-lattice A and on sub-lattice B, then one has

$$\sum_i (\lambda_i + \mu_i) = 1 . \qquad (58)$$

It is easy to show that, concerning the paramagnetic properties[14], the susceptibility is calculated as if the sub-lattices A and B contained proportions λ' and μ' of the same kind of ions of Curie constant C'. λ', μ' and C'

[13] If there were ferrous ions and ferric ions at the same time on the A sites and the B sites, 9 exchange integrals would have to be introduced.

[14] Of course, this is no longer the case when the calculation of the spontaneous magnetization is concerned.

are given by the formulae

$$C' = \sum_i (\lambda_i + \mu_i) C_i ; \quad \lambda' C' = \sum_i \lambda_i C_i ;$$

$$\mu' C' = \sum_i \mu_i C_i . \qquad (59)$$

With the distribution of ions indicated in paragraph 30, one obtains, for magnetite

$$C' = \frac{2}{3} C_{5/2} + \frac{1}{3} C_{4/2} = 3{,}92 , \qquad (60)$$

as well as

$$\lambda' = 0{,}372 ; \quad \mu' = 0{,}628 . \qquad (61)$$

There are some excellent series of measurements of the paramagnetic susceptibility, taken by Kopp [18], which can be represented with accuracy by formulae of type (14) with $C = 3.92$, as shown by Fig. 17, which relates to his third series of measurements.

On Fig. 17 the influence of the molecular field fluctuations is clearly shown. The paramagnetic Curie point is at about 596° and the ferromagnetic Curie point at 575°. The difference is 21° which is very close to that of nickel (about 20°). On taking the average of the different series of measurements, it is found that the susceptibility χ_m of magnetite, relative to the atom-gram of iron, is expressed in the form

$$\frac{1}{\chi_m} = \frac{T}{3{,}92} + 295 - \frac{57\,000}{T - 760} . \qquad (62)$$

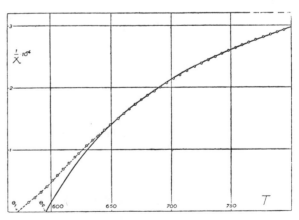

Fig. 17.
Magnetite. Variation with temperature of the inverse of the susceptibility (adjusted theoretical curve and experimental points of W. Kopp).

TABLE VIII.

	1st solution	2nd solution
α ...	− 0,51	+ 0,81
β ...	+ 0,01	− 0,86
n ...	553	424

When the numerical coefficients of this formula are substituted in equations (56) and (57), two possible groups of solutions are found for α, β and n (see Table VIII).

In order to determine the correct set it is necessary to consider the thermal variation of the spontaneous magnetization and compare it with the experimental curve of P. Weiss [19].

In order to avoid too complicated calculations, we have supposed that all ions possess the same spin $j=5/2$ and have put $\lambda=0.357$ and $\mu=0.643$ in order to find the same connection between the spontaneous magnetization at absolute zero and the absolute saturation M. We have used the method of Appendix 1, as the error thus involved can not be too great. It can then be seen that the second solution is not suitable at all while the first gives a curve which is very close to the experimental curve as shown in Fig. 18.

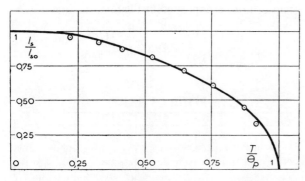

Fig. 18.
Magnetite. Variation with temperature of the spontaneous magnetization. (Curve calculated from the paramagnetic data and experimental points of P. Weiss.)

C. MANGANESE ANTIMONIDE Mn_2Sb

(32) Magnetic properties of Mn_2Sb: This compound Mn_2Sb possesses, per manganese atom, a saturation moment of 0.936 μ_B [20] which is well below the usual moment of manganese. It is also known that, in this substance, the manganese atoms occupy two kinds of sites which are different crystallographically. As the distances between neighbouring atoms belonging to different kinds of sites correspond to negative interactions, C. Guillaud has supposed that there exists a mixture of $3d^5$ ions of 5 Bohr magnetons and of $3d^7$ ions of 3 Bohr magnetons, of almost equal quantities and oriented parallel but opposite to each other. If the quantities of these two sorts of ions were equal, the moment at saturation would be 1 Bohr magneton per atom of manganese. The measurements of A. Serres [2] on the same substance above the Curie point show an accentuated curvature of the curve $(1/\chi, T)$, and thus confirms qualitatively the hypothesis of a mixture. However, in order to interpret the anormal shape of the thermal variation of the spontaneous magnetization, Guillaud also supposes that a progressive liberation of electrons is produced as the temperature rises. In other words, that the proportion of $3d^5$ ions increases with temperature. We shall show that this additional hypothesis is unnecessary and that it is possible to quantitatively interpret the paramagnetic data and the thermal variation of the spontaneous magnetization at the same time.

(33) Interpretation of the paramagnetic properties: We shall thus suppose that this substance is made up of λ ions of 5 Bohr magnetons situated on a sub-lattice A and of μ ions of 3 Bohr magnetons on a sub-lattice B. If the interactions between the two sub-lattices are negative and if the representative point M is placed in the region FCE of the graph (α, β), then there are two possible ways of finding the experimental value of the saturation magnetization by putting either $\lambda=0.258$ and $\mu=0.742$ or $\lambda=0.492$ and $\mu=0.508$. The calculation shows that the first solution does not allow values of n, α and β to be obtained which are compatible with the paramagnetic properties of the temperature variation of the spontaneous magnetization. Therefore, we shall reject this solution and adopt the second.

The paramagnetic properties are the same as if the substance was made up of identical ions, of Curie constant C′ distributed in the proportions of λ′ and μ′ on the two sub-lattices (cf. section 31). λ′, μ′ and C′ are defined by:

$$C' = \lambda C_{5/2} + \mu C_{3/2} = 3,11 \; ; \lambda' = \lambda \frac{C_{5/2}}{C'} = 0,693 \; ;$$

$$\mu' = \mu \frac{C_{3/2}}{C'} = 0,307$$

(63)

On the other hand, if the second method of paragraph 24 is applied to the experimental data of A. Serres, the following formula is obtained to represent the atomic susceptibility above the Curie point.

$$\frac{1}{\chi_m} = \frac{T}{3{,}11} + 295{,}2 - \frac{14\,500}{T - 542{,}4} \qquad (64)$$

Fig. 19 shows that this formula represents well the experimental results, with the exception of the usual confused region near the Curie point. The paramagnetic Curie point is about 297°C while the ferromagnetic Curie point is at 277°C. The difference of 20° is of the usual order of magnitude.

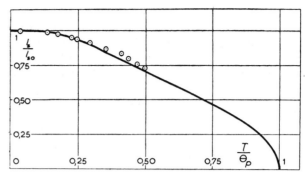

Fig. 20.
Variation with temperature of the spontaneous magnetization. (Theoretical curve calculated from the paramagnetic data and experimental points of Guillaud.)

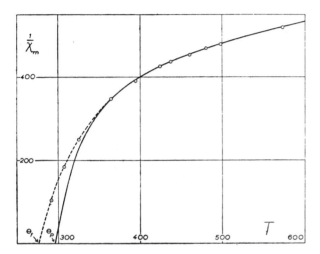

Fig. 19.
Manganese Antimonide Mn_2Sb. Variation with temperature of the inverse of the susceptibility. (Adjusted theoretical curve and experimental points of A. Serres.)

When the numerical coefficients of equation (64) are substituted in equations (56), (57) and (17), two sets of solutions are obtained for α, β and n. The first leads to a thermal variation of the spontaneous magnetization which is incompatible with the experimental results. On the contrary, the second:

$$\alpha = -0{,}241\,;\ \ \beta = -0{,}202\,;\ \ n = 526 \qquad (65)$$

allows the calculation, by means of the method described in Appendix II, of the curve of the thermal variation of the spontaneous magnetization represented in Fig. 20, in reduced coordinates, taking as unity the saturation magnetization and the temperature θ_p of the paramagnetic Curie point. On the same figure have been drawn the experimental points of Guillaud. The agreement with experiment is satisfactory, especially if one thinks that small variations in the values of α and β can greatly modify the values of the spontaneous magnetization. The particular shape of this thermal variation, compared with that of normal ferromagnets, receives a satisfactory interpretation, as can be seen.

Conclusion. Thus one interprets satisfactorily all of the magnetic properties of Mn_2Sb by supposing it to be made up of 0.492 ions at 5 μ_B and of 0.508 ions at 3 μ_B, with negative interactions between these two sorts of ions. There is reason to bring these numbers closer, since the two sub-lattices of manganese each contain the same number of sites so that one would rather expect to find $\lambda = \mu = 0.5$. I do not think it is possible to discuss this difference at greater extent. It could simply arise from the orbital moment of the ion $3d^7$ which, being added to the spin moment of that ion, would diminish the resultant moment at saturation by giving 0.936 μ_B instead of 1 μ_B.

FOURTH PART
THE INTERACTIONS

(34) Relationships between the molecular field coefficients and the exchange integrals: In this fourth part, we shall be concerned with the calculation, by means of the theory explained above applied to the experimental data, of the numerical values of the exchange integrals and to relate them to the nature of the interacting ions and their mutual distance.

First of all, we shall establish the relationships which link the exchange integrals with the molecular field coefficients. Let the two neighbouring ions be A and B and $\vec{S}_a$ and $\vec{S}_b$ be the matrices which represent their *resultant spin*, with $h/2\pi$ as unity[15]. Following Van Vleck [22], their mutual magnetic energy is written

$$W_{ab} = -2A_{ab}\vec{S}_a\vec{S}_b \qquad (66)$$

[15] The corresponding magnetic moments are then $g\mu_B\vec{S}_a$ and $g\mu_B\vec{S}_b$

where A_{ab} is the classic exchange integral relating to two carriers of spin $\frac{1}{2}$. As Stoner has already shown, the molecular field approximation consists of replacing, in expression (66), the instantaneous value of $\vec{S}_b$ by its average value in time, which is equal by definition (cf. section 4) to $\vec{\mathfrak{J}}_b/Ng\mu_B$.

In these conditions, one bears in mind that the magnetic moment of ion A is equal to $g\mu_B\vec{S}_a$, then the mutual energy W_{ab} has the same value as if this ion were placed in an imaginary magnetic field $u_{ab}\vec{\mathfrak{J}}_b$ where the coefficient u_{ab} is defined by

$$u_{ab} = \frac{2A_{ab}}{Ng^2\mu_B^2} \qquad (67)$$

u_{ab} then possesses the significance of a *reduced molecular field coefficient*. In the same way the coefficients u_{aa} and u_{ab} would be defined as a function of A_{aa} and A_{bb}.[16]

Value of the total molecular field. In the ferrites let p_{aa} and p_{ab} be the number of ions A and B which are neighbours of an ion A and let p_{ba} and p_{bb} be the corresponding quantities relating to neighbours of an ion B. From the point of view of energy, everything is as if the ions A and B were subjected to magnetic fields $\vec{h}_a$ and $\vec{h}_b$ respectively, defined by the following relationships.

$$\left.\begin{array}{l} \vec{h}_a = p_{aa}u_{aa}\vec{\mathfrak{J}}_a + p_{ab}u_{ab}\vec{\mathfrak{J}}_b \\ \vec{h}_b = p_{ba}u_{ab}\vec{\mathfrak{J}}_a + p_{bb}u_{bb}\vec{\mathfrak{J}}_b \end{array}\right\} \qquad (68)$$

(35) Application to ferrites and to magnetite: As each molecule Fe_2O_3MO contains $2Fe^{III}$ ions for 2 B sites and 1 A site, on average there exist $2\lambda Fe^{III}$ ions on each A site and μFe^{III} ions on each B site.

As an A site is surrounded by 4 A sites and 12 B sites [23], then

$$p_{aa} = 8\lambda, \quad p_{ab} = 12\mu ; \qquad (69)$$

and similarly one B site is surrounded by 6 B sites and 6 A sites

$$p_{ba} = 12\lambda, \quad p_{bb} = 6\mu . \qquad (70)$$

When formulae (10), (11), (68), (69) and (70) are compared, one obtains

$$u_{aa} = \frac{n\alpha}{8}, \quad u_{ab} = -\frac{n}{12}, \quad u_{bb} = \frac{n\beta}{6} . \qquad (71)$$

[16] The coefficients u_{aa}, u_{ab}, u_{bb} have the same significance and are directly comparable to the coefficient u or to the quantity $W/N\mu^2$, which are defined in the preceding publications of the author: *Ann. de Phys*, 1937, **8**, 238, or *Ann. de Phys*. 1936, **5**, 232.

The same formulae are applicable to magnetite, on the condition that the values obtained are multiplied by 2/3 to take account of the fact that there are 3 magnetic ions per molecule instead of 2.

The results of Table IX are thus obtained. These values are to be compared to the corresponding values relating to pure metals [10] which are given in Table X.

TABLE IX.

	u_{aa}	u_{ab}	u_{bb}
Copper ferrite	−22	−40	−15
Magnetite	−24	−31	−0.6
Interatomic distances[17] in Å	3.64	3.49	2.97

[17] These distances are practically the same for Fe_3O_4 and Fe_2O_3CuO.

TABLE X.

Metal	Cr	Mn	Fe_α, I	Co	Ni	Fe_α, II
u	−179	−79	0	102	167	138
d in Å	2,49	2,58	2,48	2,49	2,49	2,86

(36) Application to antiferromagnets. Calculation of n and α: We shall now determine the values of the reduced molecular field coefficients which relate to antiferromagnets. The following formulae deduced from formulae (44) and (45) first of all allow the calculation of n and α as a function of the experimental data, the asymptotic Curie point θ_a and the transition point θ_p.

$$n = \frac{\theta_p - \theta_a}{C} ; \quad \alpha = \frac{\theta_p + \theta_a}{\theta_p - \theta_a} . \qquad (72)$$

Table XI drawn up from the data of Bizette [13] gives the values of these different quantities for MnO, MnS, MnF_2, FeO, FeF_2.

To these data we have added those relating to rhombohedric iron sesquioxide $Fe_2O_3\alpha$. In fact, from the work of A. Serres [1], K. Endô [24] and R. Chevallier [25], this substance behaves like an antiferromagnet with a transition point at 950 K. Above this temperature, its susceptibility is compatible with the atomic Curie constant of 4.4, characteristic of ferric salts, and an asymptotic Curie point of −2000 K. Below 950 K, $Fe_2O_3\alpha$ also possesses a very weak ferromagnetism, some hysteresis and a saturation

TABLE XI.

Substance	χ_0/χ_p	θ_a °K	θ_p °K	C	n	α
MnO	0,69	− 610	122	4,4	166	− 0,67
MnS	0,82	− 528	165	4,3	161	− 0,52
FeO	0,75	− 570	198	6,24	123	− 0,48
MnF$_2$	0,76	− 113	72	4,08	45	− 0,22
FeF$_2$	0,72	− 117	79	3,88	51	− 0,19
Fe$_2$O$_3$	?	− 2 000	950	4,4	670	− 0,36

magnetization which is of the order of one hundredth that of ordinary ferromagnets and varies according to the samples. As Snoek [26] has shown, this ferromagnetism disappears gradually by prolonged heating in oxygen at 1200°C. It is therefore a secondary phenomenon which is probably due to a slight excess of iron atoms compared with the ideal composition Fe$_2$O$_3$.

(37) Determination of the reduced molecular field coefficients: Since antiferromagnets are concerned, we have

$$\epsilon = -1 \; ; \quad \lambda = \mu = 0{,}5 \; ; \quad p_{aa} = p_{bb} \; .$$

If, for simplification, we only take account of the actions, supposed to be identical to each other, of the close neighbours, then all the reduced molecular field coefficients are equal

$$u_{aa} = u_{ab} = u_{bb} = u \qquad (73)$$

and when the equations (10), (11), (68) and (73) are compared, the following relationships are obtained

$$-n = 2p_{ab}\,u_{ab} = 2p_{ab}\,u \; ; \quad n\alpha = 2p_{aa}\,u_{aa} = 2p_{aa}\,u \; . \qquad (74)$$

From these two formulae, one deduces

$$\alpha = \frac{p_{aa}}{p_{ab}} \; . \qquad (75)$$

In order to go further, it is now necessary to specify the method of splitting up into sub-lattices (cf. section 16). The method to be chosen is that corresponding to the greatest possible value of p_{ab}, that is, the methods I, III and V indicated in Table XIII. The same table shows the corresponding values of p_{aa} and p_{ab}, from which is deduced: $\alpha = -0{,}5$ for MnO, MnS, FeO and $\alpha = 0$ for MnF$_2$, FeF$_2$, Fe$_2$O$_3$. The reason for this difference comes from the fact that in the first case the cations are coordinated by 12 cations while in the second case they are only coordinated by 2 or 4 cations.

These values of α only show small agreement with the experimental values given in Table XI.

When the first of formulae (74) is used, the values of n are calculated and given in Table XII and accompanied by the corresponding values of the distances d between neighbouring cations.

The values in Table XII are of an order of magnitude comparable with that of the values obtained for ferrites.

TABLE XII.

Substance	MnO	MnS	FeO	MnF$_2$	FeF$_2$	Fe$_2$O$_3\alpha$
d in Å	3,12	3,68	3,03	3,30	3,36	2,88 / 2,94
u	− 10,4	− 10,1	− 7,7	− 11,2	− 12,7	− 84

(38) Appraisal of the preceding results: The preceding results call for a number of observations.

(A) The exchange integrals, which are proportional to u (cf. equation (67)) are thus all negative although one would expect to find them positive from the results of previous studies on the molecular field of metals [10]. For example, in Fe$_2$O$_3\alpha$, one finds $u = -84$ for distances d between iron atoms of 2.88Å and 2.94Å, while in α iron, one finds $u = +138$ for $d = 2.86$Å. In the ferrites, large values of u are found, going from -20 to -40 which relate to interatomic distances of 3.49Å and 3.64Å beside smaller values, going from 0 to -15 for a distance of 2.97Å. It is the same for the compounds of manganese. In MnO, $u = -10.4$ for $d = 3.12$Å while the preceding results, notably those relating to the ferromagnetism of MnBi, MnAs, MnSb [20], would rather lead to the prediction that u is positive for distances greater than 2.80Å. All these results are inconsistent and would remain so if the distance between magnetic bands took the place in the discussion of the distance between the atoms, which would be more rigorous.

(B) In FeF$_2$ and MnF$_2$ on one hand and Fe$_2$O$_3\alpha$ on the other hand, the iron atoms only possess 2 and 4 close neighbours respectively. In spite of this weak coordination, α is not zero but equal to -0.19, -0.22, or -0.36.

(C) In several cases the anions which are much bigger than the cations, surround the latter so closely that interactions between cations seem impossible. This is especially the case for the ferrites where an oxygen ion separates the ferric ions A and B and where, however, u_{ab} is still equal to -40, a considerable value.

(39) Hypothesis of a deformation of the magnetic band: The objections (A) can be countered by the hypothesis of a deformation of the magnetic 3d band, from spherical symmetry, produced by the positive charges of the anions. This deformation would bring the magnetic bands closer to each other, following the direction of the links Fe–Fe or Mn–Mn, and would be equivalent to the atoms drawing closer together. However, the deformations which would be necessary appear too large to be very likely. Also the objections (B) and (C) would still exist. Without completely rejecting this mechanism, which in any case must play a certain part, it is necessary to look for another explanation.

(40) The hypothesis of superexchange: Faced with these contradictions, it seems necessary to adopt a more revolutionary point of view by allowing that there exist some interactions between the magnetic cations, with anions of oxygen, sulphur or fluorine as intermediaries. A long time ago H. A. Kramers [27] showed that such interactions, called super-exchange, exist theoretically and H. Bizette [13] has put forward arguments of a chemical order which justify this hypothesis.

The paramagnetism of ions in an empty or complete band. It seems that in such interactions the anion which serves as a "relay", such as F^- or Cl^- or O^{--} must take, at least partially, an unsaturated configuration like $(n-1)p^5 ns$, or lose an electron to give $(n-1)p^5$. Such an effect, which we shall call a 'deformation' for the sake of brevity, can be shown very directly in a certain number of other phenomena; it concerns the constant *paramagnetism* of compounds such as TiO_2, V_2O_5, CrO_3, etc. to whose properties G. Foëx has attracted attention several times [28] and in which the ion possess an *empty or incomplete magnetic band*. This constant paramagnetism is higher as the degree of ionisation of the cation is higher, which shows that the 'deformation' of the anion is larger as the positive charges which surround it are more concentrated. Also, this deformation must be weaker if the anion is more electronegative. WF_6 is effectively diamagnetic while WCl_6 possesses a constant paramagnetism. In the same way, the superimposed constant paramagnetism which nickel possesses above its Curie point [29], or when it is diluted in copper, and which one is led to attribute to the $3d^{10}$ configuration, must arise from the same mechanism. This paramagnetism which is 130×10^{-6} per gram atom, is much higher than those of the substances quoted by Foëx which are of the order of 20×10^{-6}, conforming to the lesser stability of the $3d^{10}$ configuration of nickel compared with the $2p^6$ configuration of the ion O^{--}.

(41) Direct and indirect bonds: The preceding arguments show that it is not unreasonable to allow that interactions exist between the cations M by the intermediate action of the anions O which link them.

We shall designate them by the notation M–O–M and we shall call them *indirect bonds* to distinguish them from ordinary bonds M–M, which we shall call *direct bonds*. Just as the exchange interactions were represented by a coefficient u of the molecular field, the super exchange actions will be represented by a coefficient V of the molecular field.

Unfortunately, the theory does not specify the order of magnitude of these interactions and the factors on which this depends; among which factors must, no doubt, be counted the electric charge, the geometric disposition and distance of neighbouring cations, as well as the angle of the two bonds M–O–M. There possibly also exist some super exchange actions of a higher order, of the kind M–O–O–M, M–O–O–O–M, etc. If the great number of variables is considered together with the scarcity of the usable data, the hope of settling this question experimentally can no longer exist.

Nevertheless, in order to appreciate the progress brought about by these new concepts, we shall propose a very simplistic, and certainly inexact, hypothesis: we shall suppose that all the simple indirect interactions, of the type M–O–M which can be established between a cation and its neighbours, by the intermediary of the anions which link them, to be all equal to each other, and all the other indirect interactions will be ignored.

We see at once that objection (B) disappears, for the introduction of indirect bonds links a cation with a greater number of neighbours than if the direct bonds existed alone, so that it is not surprising that the experimental values of α should always be considerably different from zero. Objection (C) also disappears. As for objection (A), in order to remove it, we shall allow that the direct bonds of the ferrites and of the antiferromagnets correspond to positive values of the coefficient u of the molecular field. Since the asymptotic Curie points of the substances in question are always negative, it then follows necessarily that the coefficient v, which relates to the indirect bonds, is *negative*.

(42) Determination of u and v: Let p_{aa} and q_{aa} designate respectively the number of direct and indirect bonds between a cation and its neighbours belonging to the same sub-lattice, and let p_{ab} and q_{ab} designate the number of direct and indirect bonds between a cation and its neighbours belonging to another sub-lattice. The formulae (74) can then be written in the form:

$$-n = 2p_{ab}u + 2q_{ab}v\ ; \quad n\alpha = 2p_{aa}u + 2q_{aa}v, \quad (76)$$

and allow the calculation of u and v from the experimental data, once the manner of splitting into sub-lattices is known. As u is positive and v negative,

TABLE XIII.
The number of the direct or indirect bonds relating to the splitting up
of the antiferromagnets in sub-lattices

Type of lattices (Ewald)	B1		C4		D51	
Corresponding antiferromagnets	MnO FeO MnS		FeF_2 Mn_2		$Fe_2O_3\alpha$	
Method of splitting up into sub-lattices	I	II	III	IV	V	VI
Axes Δ	quatern.	tern.	quatern.	quatern.	tern.	tern.
p_{aa}	4	6	0	2	0	3
p_{ab}	8	6	2	0	4	1
q_{aa}	14	12	4	4	6	6
q_{ab}	16	18	8	8	12	12

TABLE XIV.

Substance	MnO	MnS	FeO	FeF_2	MnF_2	$Fe_2O_3\alpha$
u ...	0	5,9	5,7	4,0	3,2	18,8
v ...	− 4,6	− 6,4	− 5,4	− 3,2	− 2,8	− 30
d (A) .	3,12	3,68	3,03	3,36	3,30	2,91(1)
d' (A) .	2,21	2,61	2,14	2,12	2,11	2,03(1)

(1) Average value

the most stable manner will correspond to the greatest possible values of p_{aa} and of q_{ab}.

Table XIII indicates the principal ways of splitting up, corresponding to the antiferromagnets studied as well as the corresponding values of p and of q. In order to describe the ways of splitting up, the cations are arranged in successive numbered levels, perpendicular to a certain axis Δ indicated in the table.

In the methods I, II, IV and V, the sub-lattice A is made up of levels of uneven order and the sub-lattice B is made up of levels of even order. In the methods III and VI, sub-lattice A is made up of levels $4n$, $4n+1$ and sub-lattice B of levels $4n+2$, $4n+3$.[18]

The methods to be adopted will therefore be methods II, IV and V for they are more stable respectively than the methods I, III and VI.

By using the formulae (76) the values of u and v are obtained which are given in Table XIV and which are completed by showing the distances d between neighbouring cations and the distances d' between neighbouring anions and cations. These values are of a very acceptable order of magnitude. For $Fe_2O_3\alpha$, u possesses a relatively large value which is probably connected with the greater proximity of the iron atoms. As to the large value of $-v$, perhaps this must be connected to the fact that the oxygen ions are surrounded by trivalent ions while in the other antiferromagnets they are surrounded by divalent ions. Their deformation must therefore be greater.

The case of copper ferrite. In the ferrites, the oxygen is similarly surrounded by trivalent ions whose indirect bonds must be strong. In fact, the study of the crystalline structure of the ferrites shows that the bond between the two iron atoms A and B, equal to -40 in Fe_2O_3CuO, is made up of a direct bond Fe–Fe and an indirect bond Fe–O–Fe. As the value u of the direct bond is probably very small because of the large

[18] In method III, all the levels are equidistant. In method VI, the levels $4n+1$ and $4n+2$ are more separated than the levels $4n+2$ and $4n+3$.

distance Fe–Fe (3.49Å), the value v_{ab} of the indirect bond must be close to -40, a high value compared to that of $Fe_2O_3\alpha$. As for the bonds between two iron atoms B, equal to -15 (Fe_2O_3CuO) or to 0.6 (Fe_3O_4), they are equal to the difference between a positive direct bond u_{bb} and two negative indirect bonds v_{bb}. As the surroundings of the ions O^{--} are not at all symmetrical, it is not likely that v_{bb} will be equal to v_{ab}. Finally, it seems that it is necessary to attribute the interactions AA to indirect bonds of the type Fe–O–O–Fe which would be of the order of -4.

(43) Conclusions: To sum up, the introduction of super-exchange and of indirect bonds bring some light to the interpretation of the magnetic properties of ferrites and of antiferromagnets, with the adoption of values of u and of v of an order of magnitude which is reasonable and compatible with the known facts. However, progress cannot be made in a discussion of these phenomena unless a more detailed theory of this kind of interaction is available.

All the indirect super-exchange bonds which have been met so far were negative bonds, favouring the antiparallel orientation of the magnetic moments of the two cations. However it is possible to doubt the generality of this result: in fact it is known that a Heusler alloy $MnAlCu_2$ exists in which the ferromagnetism is linked to the existence of a super-structure where each atom of manganese is surrounded directly by 8 neighbours of copper, situated at a distance of 2.57Å, while the first neighbours of manganese are situated at a distance of 4.20Å, a distance such that the direct bonds Mn–Mn are certainly negligible. The ferromagnetism therefore arises from *positive* indirect bonds Mn–Cu–Mn, in which the copper ions serving as relays are in the $3d^{10}$ configuration.

APPENDIX I
CALCULATION OF THE SPONTANEOUS MAGNETIZATION

We shall calculate numerically, as a function of T, the solutions of the system (cf. section 10)

$$\left. \begin{array}{l} \mathscr{I}_a = M_j B_j \left[\dfrac{M_j n(\alpha\lambda\mathscr{I}_a + \mu\mathscr{I}_b)}{RT} \right] \\ \\ \mathscr{I}_b = M_k B_k \left[\dfrac{M_k n(\beta\mu\mathscr{I}_b + \lambda\mathscr{I}_a)}{RT} \right] \end{array} \right\} \quad (77)$$

relating, for a more general case, to atoms A and B which are different, and of internal quantum number j and k.

We shall put

$$u = \frac{M_j n}{RT}(\alpha\lambda\mathscr{I}_a + \mu\mathscr{I}_b); \quad v = \frac{M_k n}{RT}(\beta\mu\mathscr{I}_b + \lambda\mathscr{I}_a).$$

We propose to find a point for which

$$\frac{\mathscr{I}_a}{\mathscr{I}_b} = \frac{M_j B_j[u]}{M_k B_k[v]} = \varphi, \quad (79)$$

where φ is a number given *a priori*.

When $\mathscr{I}_a$, $\mathscr{I}_b$, $B_j[u]$ and $B_k[v]$ are eliminated from the preceding equations, we obtain

$$\frac{u}{v} = \frac{M_j}{M_k}\frac{\alpha\lambda\varphi + \mu}{\beta\mu + \lambda\varphi} = \Psi. \quad (80)$$

and so ψ is known as a function of φ.

On taking logarithms, these two last equations can be written

$$\log B_j[u] - \log B_k[v] = \log \frac{M_k \varphi}{M_j}$$

$$\log u - \log v = \log \Psi. \quad (81)$$

If we have previously drawn on a graph the two curves C and C' of the equations

$$(C) \begin{cases} x = \log B_j[u] \\ y = \log u \end{cases} \quad (C') \begin{cases} x = \log B_k[v] \\ y = \log v \end{cases} \quad (82)$$

we will have to find two points M and M', placed respectively on curves C and C' such that the difference between the ordinates may be equal to $\log(M_k\varphi/M_j)$ and the difference between the abscissae equal to $\log\psi$. This can be done very easily using a transparency. When v and $B_k[v]$ have been thus determined, T can then be calculated from the following formulae, deduced from (78) and (79)

$$\frac{RT}{M_k^2 n} = (\beta\mu + \lambda\varphi)\frac{B_k[v]}{v}, \quad (83)$$

as well as the resultant spontaneous magnetization $\mathscr{I}_s$

$$\mathscr{I}_s = \lambda\mathscr{I}_a - \mu\mathscr{I}_b = (\lambda\varphi - \mu)M_k B_k[v]. \quad (84)$$

By choosing successively different values of φ, the curve required can thus be constructed point by point.

Curie point At the Curie point, u and v are infinitely small, so, from equations (7) and (79)

$$\varphi = \frac{M_j(j+1)ku}{M_k(k+1)jv}, \qquad (85)$$

and when this value is taken into (80), it is seen that φ is a root of the second degree equation

$$\frac{\lambda C_k}{C_j}\varphi^2 + \left(\frac{\beta\mu C_k}{C_j} - \lambda\alpha\right)\varphi - \mu = 0. \qquad (86)$$

APPENDIX II
CALCULATION OF THE SPONTANEOUS MAGNETIZATION IN THE NEIGHBOURHOOD OF THE CURIE POINT

In the fundamental equations (24), $\mathcal{I}_a$ and $\mathcal{I}_b$ being very small, this can be developed as a series B_j in the form given in (7). Let us then put

$$U = \frac{R}{Mn}(\alpha\lambda\mathcal{I}_a + \mu\mathcal{I}_b); \quad V = \frac{R}{Mn}(\beta\mu\mathcal{I}_b + \lambda\mathcal{I}_a). \qquad (87)$$

Being limited to the terms in U^3 and V^3, equations (24) are then transformed to

$$\left.\begin{array}{l} \dfrac{V - \beta U}{1 - \alpha\beta} - \dfrac{\lambda C U}{T} = -\dfrac{\lambda D U^3}{T^3} \\[2mm] \dfrac{U - \alpha V}{1 - \alpha\beta} - \dfrac{\mu C V}{T} = -\dfrac{\mu D V^3}{T^3} \end{array}\right\} \qquad (88)$$

putting, for the sake of brevity

$$C = \frac{M^2 n}{R}\frac{j+1}{3j}; \quad D = \frac{M^2 n}{R}\frac{[(j+1)^2 + j^2](j+1)}{90j^3}. \qquad (89)$$

When T tends towards the Curie point θ, the ratio U/V tends towards a certain limit k which is a solution of the two equations (88) written without the second term

$$\frac{1 - \beta k}{1 - \alpha\beta} - \frac{\lambda C k}{\theta} = 0 \; ; \quad \frac{k - \alpha}{1 - \alpha\beta} - \frac{\mu C}{\theta} = 0 \; ; \qquad (90)$$

k is then the positive root of the following equation obtained by eliminating θ between the two equations (90)

$$\lambda k^2 + (\beta\mu - \lambda\alpha)k - \mu = 0. \qquad (91)$$

At a temperature T, lower than the Curie point and very close to it, we can put

$$T = \theta - t \; ; \quad U : V = k + \epsilon , \qquad (92)$$

where t and ε are infinitely small.

When these expressions are substituted in equations (88), we obtain, on neglecting the infinitely small quantities of an order greater than t, two linear and homogeneous equations in t, ε and V^2, from which we obtain, by eliminating ε and taking account of (91)

$$\lambda k + \frac{\mu}{k} = \frac{DV^2}{C\theta t}k\left(k^2\lambda + \frac{\mu}{k^2}\right). \qquad (93)$$

But on the other hand, from equations (87), (90) and (91), when the infinitely small quantities of order $t^{3/2}$ are ignored, the spontaneous magnetization is written

$$\mathcal{I}_s = \lambda\mathcal{I}_a - \mu\mathcal{I}_b = \frac{RVC\sqrt{k}}{Mn\theta}\left(\lambda\sqrt{k} - \frac{\mu}{\sqrt{k}}\right), \qquad (94)$$

and when V is replaced by its value obtained from (93)

$$\mathcal{I}_s = M\frac{j+1}{3j}\left(\lambda\sqrt{k} - \frac{\mu}{\sqrt{k}}\right)\sqrt{\frac{Ct\left(\lambda k + \dfrac{\mu}{k}\right)}{D\theta\left(\lambda k^2 + \dfrac{\mu}{k^2}\right)}}. \qquad (95)$$

For $j = 5/2$, we find that

$$\frac{j+1}{3j}\sqrt{\frac{C}{D}} = 1{,}486, \qquad (96)$$

from which comes the expression (34) given above.

REFERENCES

[1] A. Serres, *Ann. de Phys.*, **17**, 1932, 53.
[2] R. C. Evans, An introduction to crystal chemistry, Cambridge, 1946.
[3] T. F. W. Barth and E. Posnjak, *J. Wash. Ac. Sc.*, **21**, 1931, 255.
[4] T. F. W. Barth and E. Posnjak, *Z. Krist.*, **82**, 325.
[5] Verwey and Heilmann, *J. Chem. Phys.*, **15**, 1947, 174.
[6] E. J. W. Verwey and J. H. de Boer, *Rec. Trav. Chim. Pays-Bas.*, **55**, 1936, 531.
[7] F. Machatschki, *Z. Krist.*, **82**, 1932, 348.
[8] L. Neel, *Ann. de Phys.*, **17**, 1932, 61; *C.R. Acad. Sci.*, **198**, 1934, 1311.
[9] L. Néel, *Ann. de Phys.*, **17**, 1932, 5.
[10] L. Néel, *J. Phys.*, **3**, 1932, 160; *Ann. de Phys.*, **5**, 1936, 232; *C.R. Acad. Sci.*, **203**, 1936, 304.
[11] F. Bitter, *Phys. Rev.*, **54**, 1938, 79.
[12] J. H. Van Vleck, *J. Chem. Phys.*, **9**, 1941, 85.

[13] H. Bizette, *Ann. de Phys.*, **1**, 1946, 223 (cf. bibliographic).
[14] J. L. Snoek, New developments in ferromagnetic materials, p. 98 (Elsevier, Amsterdam, 1947).
[15] E. J. Verwey, P. W. Haymann and F. C. Romeijn, *J. Chem. Phys.*, **15**, 1947, 181.
[16] P. Weiss and R. Forrer, *Ann. de Phys.*, **12**, 1929, 279.
[17] F. Kordes, *Z.f. Krist.*, **91**, 1935, 193; E. J. H. Verwey, *Ibid*, 65; G. Hagg, *Ibid*, 114.
[18] W. Kopp, Thesis, Zurich, 1919.
[19] P. Weiss, *J. de Phys.*, **6**, 1907, 661.
[20] C. Guillaud, Thesis, Strasbourg, 1943.
[21] A. Serres, *J. de Phys.*, **8**, 1947, 146.
[22] J. H. Van Vleck, Electric and magnetic susceptibilities, Oxford, 1932.
[23] Ewald and Hermann, *Strukturbericht*, **1**, 350.
[24] K. Endo, *Sc. Rep. Tohotu Univ.*, **25**, 1937, 879.
[25] R. Chevallier, *Ann. de Phys.*, **18**, 1943, 258.
[26] J. L. Snoek, *Physica*, **3**, 1936, 463.
[27] H. A. Kramers, *Physica*, **1**, 1934, 182; *Réunion d'Etudes sur le Magnétisme*, Strasbourg, **3**, 1939, 45.
[28] G. Foex, *J. de Phys.*, **9**, 1938, 37; *Réunion d'Etudes sur le Magnétisme*, Strasbourg, **3**, 1939, 187.
[29] L. Neel, *Réunion d'Etudes sur le Magnétisme*, Strasbourg, **2**, 1939, 65; M. Fallot, *J. de Phys.*, **5**, 1944, 153.

A70 (1950)
SATURATION MAGNETIZATION OF CERTAIN FERRITES

Summary (1977): Experimental verification of the theory of ferrimagnetism based on the determination of the saturation magnetization of ferrites Fe_2O_3MO, with M = Mn, Fe, Co, Ni, Zn; and on the influence of heat treatment on the saturation magnetization of Fe_2O_3CuO.

X-rays as well as study of magnetic properties show [1] that in the spinels of the type Fe_2O_3MO, the divalent metal ion M occupies the tetrahedral (A) sites, surrounded by 4 oxygen ions (normal spinels) or the octahedral (B) sites, surrounded by 6 oxygen ions (inverse spinels). We will indicate these differences in affinity more precisely by the energy W required for an ion M to move from a B-site to an A-site. This movement is of course, accompanied by an inverse movement of an Fe^{+++} ion from an A- to a B-site. We call $\theta = W/k$ (k = Boltzmann's const.) the characteristic temperature of the M-ion. We assume that θ is, to a first approximation, independent of the distribution of the M-ions over the A- and B-sites and that it is on the whole determined by the stiffness of the O^{--} ion position.

Let us assume that all arrangements of n M-ions on the disposable n A-sites and $2n$ B-sites in n molecules are *a priori* equally probable. An elementary consideration of statistical mechanics gives *the equilibrium distribution at a temperature T* of the M-ions on the A and B sites. If the proportions of the ions on the 2 sites are x and $1-x$ respectively, we have:

$$\frac{x(1+x)}{(1-x)^2} = e^{-\frac{\theta}{T}}. \qquad (1)$$

A ferrite which is normal at absolute zero corresponds to a negative θ, ($x=1$) and one which is inverse ($x=0$) to a positive θ. It seems *a priori* possible to conserve at absolute zero in *metastable equilibrium* any state with values of x between 0 and 1, by rapid quenching. Thus, in the series M = Mn, Fe, Co, Ni, Cu, Zn, in order of atomic numbers, the first five ferrites are inverse, the last one is normal. It is hence logical to think that θ, at first positive, decreases regularly as the atomic number grows, and becomes negative with zinc. In particular one will expect that the positive value of θ for copper, is small.

Moreover, according to the ideas which I have developed elsewhere [1], the ferromagnetism of the ferrites, or *ferrimagnetism*, is essentially due to the *negative exchange interaction* between the spins of the A-ions and the spins of the B-ions, which arrange themselves in opposing directions to each other. Let us, therefore, take a magnetic moment of m Bohr magnetons for the M-ion and 5 for the Fe^{+++} ion; the saturation moment μ at absolute zero of a molecule Fe_2O_3MO, becomes

$$\mu = m + 2x(5-m). \qquad (2)$$

For an inverse ferrite one finds simply $\mu = m$, and, neglecting the orbital moments, assumed quenched, one predicts the moments given in the second line of Table I.

TABLE I.

M.	Mn.	Fe.	Co.	Ni.	Cu.	Zn.
μ_{calc}	5	4	3	2	1	0
μ_{obs}	5,0	4,06	3,35	2,25	1,37	0

The third line of the table gives the values found experimentally by E. W. Gorter for Mn, P. Weiss and

R. Forrer [2] for Fe, E. W. Gorter and R. Pauthenet [3] for Co, E. W. Gorter, C. Guillaud and M. Roux [4] for Ni and finally R. Pauthenet [3] for Cu. The general agreement is good. The excess of the observed over the theoretical values, as concerns cobalt and nickel are probably due to the effect of the orbital moment, in agreement with the ferromagnetic resonance experiments of H. G. Beljers and D. Polder on nickel-ferrite.

For copper, the values of μ depend on the temperature from which the sample was quenched, after having been kept a long time at that temperature (Table II). This was measured by R. Pauthenet on samples prepared and heat treated by L. Weil and L. Bochirol.

TABLE II.

T°K.	573.	673.	773.	968.	1073.	1123.	1173.	1268.
μ	1,37	1,69	1,99	2,01	2,08	2,18	2,26	2,36
θ	1700°	1500	1300	1600	1700	1600	1600	1550

For each value of T one can calculate the corresponding value of θ, using equations (1) and (2). If the explanation given is correct, these values should be identical. One can see that this is approximately the case. The variations of the moment of copper ferrite should thus arise from differences in distribution of Cu-ions on the A- and B-sites, rendered particularly important by the small value of θ which, as we have indicated above, one must normally infer.

These facts give solid experimental confirmation of the theory of *ferrimagnetism*.

REFERENCES

[1] For the bibliography see A69.
[2] *Ann. de Physique*, **12**, 1929, p. 279.
[3] Unpublished. The samples were prepared by L. Weil and L. Bochirol.
[4] C. Guillaud, *Comptes rendus*, **229**, 1949, p. 1133.

A72 (1950)

SATURATION MAGNETIZATION OF THE MIXED NICKEL–ZINC FERRITES

Summary (1977): Experimental verification of the theory of ferrimagnetism based on the variation with zinc-concentration of the saturation magnetization of mixed nickel–zinc ferrites.

In order to interpret the molecular saturation moment M_s of the mixed nickel–zinc ferrites Fe_2O_3, x NiO, $(1-x)$ZnO, we assume, in accord with the work by Vervey et al [1] that the nickel is situated on the octahedral B-sites and the zinc on the tetrahedral A-sites, and that the magnetic moments of the ions on the A-sites point in the opposite direction to those on the B-sites [2]. The result is that replacement of a Ni^{++} ion, of moment to $2\mu_B$ (μ_B = Bohr magneton) neglecting the orbital moment, by a Zn^{++} ion of zero moment, accompanied by a move of a Fe^{+++} ion from A to B with reversal of its moment of $5\mu_B$, gives a total change in moment of $(2 \times 5) - 2 = 8\mu_B$. The tangent of the curve $M_s = f(x)$ should therefore extrapolate to $10\mu_B$ at $x=0$, in accordance with experiment as shown in the figure in which are plotted as G and P the experimental values of C. Guilland [3] and E. W. Gorter [4]. The latter author has found analogous verifications for several other series of mixed ferrites.

At the zinc-rich side there is no longer a sufficient number of Fe^{+++} ions on the A-sites to align completely in opposition to each other the A and B sublattices in the presence of the negative B–B interactions. Thus one passes from Q-type to R-type magnetization, described in paragraph 11 of a previous publication [2], with $M_s = -M_a(1+(1/\beta))$ where M_a is the saturation moment of the ions on A, $5 \times \mu_B$ in this case and β the reduced molecular field coefficient relating to the B–B interaction. The study of the paramagnetic properties of the same mixed ferrites has already given $\beta = 0.15$ [5], giving the theoretical straight line BC. The trend of the broken line ABC represents the general form of the experimental variation of M_s.

At medium concentrations the deviations arise from the inadequacy of the molecular field approximation. The moments of the ions on B sites are, in effect, pointing in the direction of M_s through the influence of the six neighbouring A-ions. If, by chance, all these six ions are non-magnetic zinc, the B-ion is only exposed to the action of its six B ion neighbours, which tend to turn its moment against M_s. It can also be shown that, taking account of the values of n, α, β [5], the orienting effect of a single magnetic ion amongst the six A-

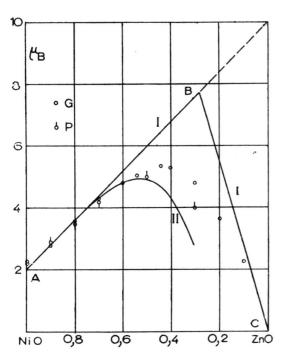

neighbours is nearly compensated by the opposing effect produced by the six B-neighbours, in the sense that the moment of the B-ion in question behaves like that of a free ion not saturated by the magnetizing field. The saturation becomes normal when at least two of the six nearest neighbour A-ions are magnetic. Assuming that the Zn ions are distributed at random over the A-sites, one can calculate the probabilities of the various configurations and the corrections which apply to the values of M_s. In this way one obtains the curve II which is very close to the observed points.

To sum up, the theory of *ferrimagnetism* allows a satisfactory interpretation of the saturation magnetization of the mixed nickel–zinc ferrites.

REFERENCES

[1] E. J. W. Verwey and E. H. Heilmann, *J. Chem. Physics*, **15**, 1947, p. 174.
[2] L. Néel, *Ann. de Physique*, **3**, 1948, p. 137. (A61)
[3] *Comptes rendus*, **229**, 1949, p. 1133.
[4] *Comptes rendus*, **230**, 1950, p. 190.
[5] L. Néel and P. Brochet, *Comptes rendus*, **230**, 1950, p. 280.

A77 (1951)
EFFECT OF THERMAL EXPANSION ON THE VALUE OF THE CURIE CONSTANT OF FERRITES

Summary: The author shows that the existence of great anomalies of thermal expansion entails necessarily an important temperature variation of the molecular field coefficient. The result is that the asymptotic Curie constant should be greater than the theoretical classical value. For magnetite the observed increase is very close to the value calculated using dilatometric data.

Recent studies by A. Serres [1] and P. Maroni [2] show that the inverse molecular magnetic susceptibility $1/\chi_n$ of ferrites above their Curie point varies with absolute temperature according to:

$$\frac{1}{\chi_M} = \frac{T}{C} + \frac{1}{\chi_0} - \frac{\sigma}{T - \theta} \quad (1)$$

which follows from the theory of ferrimagnetism [3], provided that T is not too close to the ferromagnetic Curie point: the table gives corresponding values of C for some ferrites as well as for $\alpha\text{-}Fe_2O_3$ an antiferromagnetic substance.

The theory [3] [5] indicates that C should be equal to the sum C', also given in the table, of the usual values of the atomic Curie constants of the corresponding ions measured on salts with Curie points near absolute zero. But one sees on the contrary that C is noticeably greater than C'.

It is possible, as we have suggested before [5], that this disagreement arises from the thermal variation of the molecular field, the importance of which we have emphasized long ago [6]. For simplicity, suppose that amongst the three coefficients which characterize the molecular field of ferrites [7], only the coefficient n is a function of temperature, α and β being constant, and put

$$n = n_0(1 + \gamma T). \quad (2)$$

Under these conditions the expression of the temperature variation of the susceptibility retains the form [equation (1)] but the significance of the coefficients changes: in particular the constant C is connected with the theoretical Curie constant C' by the relation:

$$\frac{1}{C} = \frac{1}{C'} + \frac{\gamma}{\chi_0}. \quad (3)$$

Substance	C	C'	$\frac{1}{\chi'_0}$	$-\gamma \cdot 10^4$	References
$Fe_2O_3 MgO$	14,0	8,8	153	2,76	[1]
$Fe_2O_3 NiO$	19,5	10,0	186	2,77	[2]
$Fe_2O_3 CoO$	13,9	11,8	109	1,17	[2]
$Fe_2O_3 FeO$	14,2	11,9	98	1,37	[2]
Fe_2O_3	13,6	8,8	316	1,27	[4]

This equation has allowed the calculation of the values of γ given in the table; they vary from -1.2 to -2.8×10^{-4}. This is a reasonable order of magnitude, because, if we attribute the decrease of the molecular field to the progressive increase in the distance d between two neighbouring atoms because of thermal expansion, one finds that, with a linear expansion coefficient λ of 16×10^{-6} (near 1000 K) n varies as d^{-p} where $p = 7.3$ to 17.3 according to the material. This decrease is comparable with that of interatomic forces of repulsion where p lies between 8 and 14 [8].

This temperature variation of n is accompanied by an anomaly of length at absolute zero. In fact when a magnetically saturated substance undergoes a relative change in length r, the exchange energy per gram-molecule is given by

$$E = E_0 \left(1 + \frac{\gamma r}{\lambda}\right) \quad (4)$$

Passing from a state of ordered spins to one of disorder, such a substance undergoes a relative change in length given by

$$r_0 = \frac{E_0 k_0 \gamma}{9 V \lambda},$$

where V is the atomic volume and k_0 the cubic compressibility coefficient. For magnetite $V = 45$ cm^3, $k_0 = -0.54 \times 10^{-12}$, $\lambda = 16 \times 10^{-6}$ and the results of Maroni [2] allow to calculate $E_0 = 2.55 \times 10^{11}$ ergs. Thus one finds $v_0 = 2.9 \times 10^{-3}$, while the thermal expansion measurements of Chevenard [9] give a value of 3×10^3. This excellent agreement is no doubt just accidental. For magnesium and nickel ferrites the observed anomaly of length is smaller than the calculated value. This shows that the problem has been oversimplified. Indeed, on the one hand, oscillation terms have been neglected [6], on the other hand, the same coefficient γ of thermal variation has been given to the three characteristic molecular fields of ferrites.

Let us mention that Pauthenet [10] has calculated the thermal variation of the spontaneous magnetization of ferrites of Fe, Co and Ni and found values agreeing with experiment. He has used the values of n_0, α and β, corrected for expansion, obtained by Maroni [2] from experiments in the paramagnetic region.

To sum up, the temperature variation of the molecular field due to expansion plays an important role in the interpretation of the magnetic properties of ferrites and very probably also of antiferromagnets, notably as far as comparison of the asymptotic Curie point and transition point, of which the relative values are changed, is concerned.

REFERENCES

[1] G. Foex and A. Serres, *C.R. Acad. Sci.*, **230**, 1950, 729.
[2] P. Maroni, *Colloques de Ferromagnétisme et d'Antiferromagnétisme*, Grenoble, 1950 (*J. Phys.*, **12**, 1951, 256).
[3] L. Néel, *Ann. Phys.* **3**, 1948, 137. (A61)
[4] K. Endo, *Sc. Rep. Tôhoku Imp. Univ.*, **25**, 1937, 879.
[5] L. Néel, *Ann. Inst. Fourier*, **1**, 1949, 163.
[6] L. Néel, *Ann. Phys.*, **8**, 1937, 237. (A30)
[7] For the symbols see Refs [3] and [5].
[8] R. H. Fowler and E. A. Guggenheim, *Statistical Thermodynamics*, Cambridge, 1939.
[9] P. Chevenard, *C.R. Acad. Sc.* **172**, 1921, 320.
[10] R. Pauthenet, *C.R. Acad. Sc.*, **230**, 1950, 1842.

A94 (1954)

ON THE INTERPRETATION OF THE MAGNETIC PROPERTIES OF THE RARE EARTH FERRITES

The magnetic properties of the rare earth ferrites are interpreted with the help of the molecular field method, by subdividing the principal lattice into a ferrimagnetic sublattice of the iron ions which is weakly coupled to a paramagnetic sublattice of the rare earth ions. At very low temperatures the paramagnetic sublattice in its turn becomes probably ferrimagnetic.

The rare earth ions recently discovered [1] and studied [2] by H. Forestier and G. Guiot-Guillain have extremely curious and complicated magnetic properties. A more profound analysis of these has just been taken up for the case of gadolinium ferrite by R. Pauthenet and P. Blum [3].

We propose to show that one can understand these properties by dividing the lattice of the magnetic ions (Fe^{+++} and M^{+++} where M is the rare earth ion) into two primary sublattices, the one, A, belonging to the Fe-ions the other, B, to the M-ions. First we will examine the properties of each of these sublattices separately, supposing it to be isolated from the other and then we will introduce a coupling in the form of a molecular field between A and B. Magnetization will always refer to a gram molecule $Fe_2O_3M_2O_3$. The notations of the transformation points are those of Pauthenet and Blum [3].

As far as the primary A-sublattice is concerned we assume it is subdivided below the ferromagnetic Curie point θ_f into two secondary sublattices A' and A", with antiparallel spontaneous magnetization a little different from each other, so that we get a weak ferrimagnetism near perfect antiferromagnetism. In these conditions the magnetization J of A in a field H becomes:

$$J = \sigma_A + aH.$$

The susceptibility a is nearly that corresponding to a perfect antiferromagnetic arrangement of the Fe ions with supposedly the same interactions: more precisely we are dealing with the susceptibility measured parallel to the direction of the antiferromagnetism, since that is automatically aligned in the direction of the field H, under the action of H on σ_A. The result is that a, starting from zero at absolute zero [4] begins to increase till the Curie point θ_f is reached, and then decreases obeying a Curie–Weiss law.

According to the theory of ferrimagnetism [5], the temperature variation of σ_A can behave as type Q, characteristic of normal ferromagnetics, or as type N, characterized by a compensation point θ_c at which σ_A goes through zero on changing sign, or as type P for which σ_A begins to grow, passing through a maximum before vanishing at θ_f. Besides it must be said that types N and P only apply to ferrimagnets near antiferromagnetism, as shown by theory as well as the experiments by E. W. Gorter [6].

Let us now consider the primary sublattice B. We regard it as paramagnetic with a susceptibility b given by

$$b = \frac{C}{T - nC}, \quad (1)$$

where C is the molecular Curie constant of the two M ions (15.76 for M=Gd) and n a molecular field coefficient, probably small, representing the M–M interactions in B.

Finally we introduce the interactions between A and B by a molecular field coefficient m. The magnetizations J_A and J_B of A and B become

$$J_A = \sigma_A + a(H + mJ_B),$$
$$J_B = b(H + mJ_A).$$

The resulting magnetization $J = J_A + J_B$ takes the form

$$J = \sigma_s + \chi_s H, \qquad (2)$$

where the spontaneous magnetization σ_s and the susceptibility χ_s are given by:

$$\sigma_s = \sigma_A \frac{1 + bm}{1 - abm^2}, \qquad (3a)$$

$$\chi_s = b + \frac{a(1 + bm)^2}{1 - abm^2}. \qquad (3b)$$

Equation (3a) shows that σ_s becomes zero at the same points as σ_A, i.e. at θ_f and eventually at θ_c, but also, if m is negative at the temperature θ'_c corresponding to $1 + bm = 0$. This temperature θ'_c is also a compensation point: it occurs when the Fe ions induce in the M ions an equal and opposite magnetization to their own.

Equations (2) and (3) can form the basis of interpretation of the magnetic properties of the rare earth ferrites. To do this we accept that the temperature variation of the spontaneous magnetization σ_A of the A sublattice belongs to type N and the two corresponding characteristic temperatures θ_c and θ_f must be identified with the two temperatures θ_1 and θ_2 of Forestier and Guiot-Guillain [2] [7]. These two authors have shown that θ_1 and θ_2 vary very regularly with the ionic diameter and do not visibly depend on the magnetic properties of the M-ions. These facts can easily be explained in the framework of our hypothesis, since the causes of variations of θ_1 and θ_2 are the changes in lattice parameter and not the nature of M. In addition, as θ_1 is a compensation temperature, no thermoremanent magnetization is connected with it, in agreement with experimental results.

The θ_3 point in $Fe_2O_3Gd_2O_3$ at 306 K [3] must be identified with θ'_c. If n is negligible one deduces the value $m = -20$. This explanation implies that a remanent magnetization given initially to the ferrite must change in direction if, in zero field, the temperature goes through θ_3. This phenomenon has not yet been observed in gadolinium ferrite, probably because of inhomogeneities. However Guillain and Forestier have told me that they have seen such an inversion in dysprosium ferrite with similar properties. Equation (3a) shows that σ_s varies almost as $1/T$ below θ_3, in agreement with experimental results.

According to (3b) χ_s is obtained by adding a coercive term to the susceptibility of the ions M, supposed to be not coupled to the Fe-ions. When a is small one obtains, at low temperatures, the approximate expression

$$\chi_s = \frac{C}{T - C(n + am^2)} + \cdots$$

This Curie—Weiss law represents the experimental results well. The Curie point θ_5 is at -8 K for M = Gd, from which one deduces $n = -0.5$, neglecting am^2 with respect to n, which is justified since a tends to 0 at absolute zero as we have seen above.

As n is negative one must expect that at very low temperatures the primary B sublattice also breaks up into two secondary sublattices B' and B", magnetized in opposition. As the sublattices A' and A" are slightly assymmetric, the same must be the case for B' and B" so that the Gd-ions are slightly ferrimagnetic. In these conditions the two directions of antiferromagnetism of the A and B lattice must align themselves parallel to the applied field so that the susceptibility goes to 0 towards absolute zero, in accordance with the experimental results. In this interpretation, the temperature θ_4 is the Curie point of the ferrimagnetic Gd-sublattice ions. It lies between 4 and 14 K. One can also suppose that B' and B" form a perfect antiferromagnetic lattice, but this assumption does not allow as good an explanation as the former one for the fact that the susceptibility at 4 K is at least 5 times smaller than at 14 K according to Pauthenet and Blum.

To conclude, it seems that the proposed model gives a satisfactory explanation of the set of magnetic properties of gadolinium ferrite. It remains to understand why the Fe ions, and perhaps at lower temperatures the Gd-ions adopt a slightly ferrimagnetic arrangement rather than a perfect antiferromagnetic one. Are we dealing with a lattice imperfection or a specific property of the lattice? The question is open and also applies to the magnetic properties of αFe_2O_3 [8]. One cannot give a precise answer before knowing much better than now the crystal structure of these rare earth ferrites: one knows only that it belongs to the structure of the perovskites [9].

REFERENCES

[1] *Comptes rendus*, **230**, 1950, p. 1844.
[2] *Comptes rendus*, **235**, 1952, p. 48.
[3] *Comptes rendus*, **239**, 1954, p. 33.
[4] F. Bitter, *Phys. Rev.*, **54**, 1938, p. 79.
[5] L. Néel, *Ann. Phys.* **3**, 1948, p. 137.
[6] *Thesis*, Leyden, 1954, to be published in *Philips Research Reports*.
[7] G. Guiot-Guillain, *Comptes rendus*, **237**, 1953, pl 1534.
[8] L. Néel and R. Pauthenet, *Comptes rendus*, **234**, 1953, p. 2172; L. Néel, *Ann. Phys.*, **4**, 1949, p. 249.
[9] G. Guiot-Guillain, *Comptes rendus*, **232**, 1951, p. 1832.

C42 (1964)
THE RARE EARTH GARNETS

Summary (1977): This brief history of the discovery of the rare-earth garnets formed Néel's introduction to a review article on these compounds by R. Pauthenet and B. Dreyfus published in the 1964 volume of "Progress in Low Temperature Physics".

The rare earth garnets, of general formula $Fe_5M_3O_{12}$ where M is a trivalent rare earth ion, form an important class of magnetic compounds whose properties can be remarkably simply and quite accurately interpreted by a theory of three ferrimagnetic sub-lattices. Theoretically, their study is of interest because of the large number of atomic substitutions that can be realized in the lattice and one hopes that this may lead to a solution of the problem of magnetic interactions. Their practical interest, especially at high frequencies, is considerable, as they are excellent electrical insulators, preparable as monocrystals and characterized by very narrow resonance lines.

The story of the rare earth garnets started in 1950, at Strasbourg. Forestier and Guiot-Guillain showed [1] that, on heating an equi-molecular mixed-precipitate of Fe_2O_3 and M_2O_3 where M is a rare earth such as Nd, Er, Y, Sm, Pr, La, they obtained strongly ferromagnetic products which they identified as $FeMO_3$ and whose Curie points ranged from 520 to 740°K. In 1951, Guiot-Guillain [2] showed that the ferrites of praseodymium and lanthanum had a perovskite-type structure.

Continuing their studies in 1952 and 1953, Forestier and Guiot-Guillain [3] brought to light a curious fact: with M = Yb, Tm Y, Gd and Sm, the aforementioned products had two Curie-points about a hundred degrees apart. These points varied regularly from one metal to the next as a function of the atomic radius of M. On passing through the upper Curie-point one obtained a strong thermoremanent magnetization.

These results attracted the attention of the "Laboratoire de Magnétisme" at Grenoble: Pauthenet and Blum [4] prepared a gadolinium ferrite and showed in 1954 that, besides the two Curie-points $\theta_1 = 570°K$ and $\theta_2 = 678°K$ already mentioned, there was a third temperature $\theta_3 = 306°K$ at which the spontaneous magnetization was zero. It certainly seemed that this was a compensation temperature, in the sense of Néel's theory of ferrimagnetism; that is, a temperature where the magnetization changes sign.

At the same time, Néel [5] proposed a likely explanation of these facts on the supposition that the Fe^{3+} ions in the lattice formed a ferrimagnetic ensemble A having a spontaneous magnetization opposite in direction to that of the sublattice B of M^{3+} ions. The molecular coupling field was sufficiently weak such that for temperatures above that of liquid air the magnetization of the B ions was proportional to the field. In particular, Néel suggested that the point θ_3 corresponded to an exact compensation of the magnetizations of the ferrimagnetic A ions and B ions. A few days later [6] the existence of such compensation points was confirmed for the ferrites of dysprosium and erbium, at 246° and 70°K respectively; at these temperatures the remanent magnetization changed sign.

In spite of this success the supposed ferrimagnetic distribution of the Fe^{3+} ions was inconsistent with the recently confirmed perovskite structure of the compound $FeMO_3$. At the request of L. Néel, Remeika at the Bell Telephone Laboratory prepared monocrystals of $FeMO_3$ which, after study by S. Geller [7], were sent to Grenoble.

Starting from the idea that the observed phenomena were due to a new type of compound, Bertaut and Forrat showed in January 1956 [8] that one was dealing with a cubic compound $Fe_5M_3O_{12}$, space-group O_h^{10} Ia 3d, with eight molecules per unit cell — an entirely unexpected discovery. The ferrimagnetic distribution of the Fe^{3+} ions was thus 24 ions on the tetrahedral sites 24d, oppositely magnetized to the remaining 16 Fe^{3+} ions on the octahedral sites 16a.

In the following months, many experiments quantitaatively verified the accuracy of this model. In particular, Pauthenet, after his work on pure $FeGdO_3$ and $Fe_3Gd_5O_{12}$, showed [9] that gadolinium ferrite (the object of his first investigations [4]) was in fact a mixture of 0.26 g of garnet, 0.64 g of perovskite and

0.1 g of Gd_2O_3; the points θ_1 and θ_2 were respectively the Curie-points of the garnet and perovskite. He soon completed his results with a study of the complete series of rare earth garnets [10], while, with Aléonard and Barbier he studied [11] the properties of the isolates sublattice of Fe^{3+} ions in yttrium ferrite. All these results were communicated to the Congress of Physics at Moscow [12] (25–31 May, 1956).

The same research teams also did the first work on substitutions [13], the accurate determination of the parameters [14] and, with Herpin and Mériel [15], structure determination by neutron diffraction.

Following this, Bell Telephone Labs. conducted a beautiful series of experiments on substitutions in the garnets, originating in the work of S. Geller [16].

The two following articles are devoted to the garnets, the first by R. Pauthenet on their magnetic properties, and the second by B. Dreyfus on their optical and thermal properties, especially at low temperatures. In view of the numerous articles on the garnets since their discovery, it has been necessary to make a limited choice of the subject treated.

REFERENCES

[1] H. Forestier and G. Guiot-Guillain, *C.R. Paris*, 1950, **230**, 1844.
[2] G. Guiot-Guillain, *C.R. Paris*, 1951, **231**, 1832.
[3] H. Forestier and G. Guiot-Guillain, *C.R. Paris*, 1952, **235**, 48; G. Guiot-Guillain, *C.R. Paris*, 1953, **237**, 1654.
[4] R. Pauthenet and P. Blum, *C.R. Paris*, 1954, **239**, 33.
[5] L. Néel, *C.R. Paris*, 1954, **239**, 8.
[6] G. Guiot-Guillain, R. Pauthenet and H. Forestier, *C.R. Paris*, 1954, **239**, 155.
[7] S. Geller, *Phys. Rev.*, 1955, **99**, 1641.
[8] F. Bertaut and F. Forrat, *C.R. Paris*, 1956, **242**, 382.
[9] R. Pauthenet, *C.R. Paris*, 1956, **242**, 1859.
[10] R. Pauthenet, *C.R. Paris*, 1956, **243**, 1499.
[11] R. Aleonard, J. C. Barbier and R. Pauthenet, *C.R. Paris*, 1956, **242**, 2531.
[12] L. Néel, F. Bertaut, R. Pauthenet and F. Forrat, Comptes Rendus de l'Académie des Sciences d'U.R.S.S., 1957, **21**, 6.
[13] F. Bertaut and F. Forrat, *C.R. Paris*, 1956, **243**, 1219.
[14] F. Bertaut and F. Forrat, *C.R. Paris*, 1956, **244**, 96.
[15] F. Bertaut, F. Forrat, A. Herpin and P. Meriel, *C.R. Paris*, 1956, **243**, 989.
[16] S. Geller and M. A. Gilleo, *Acta Cryst.*, 1957, **10**, 243 and 787.

Chapter III

ANTIFERROMAGNETISM
A29, A112, A113, A114, A115

A29 (1936)

THEORY OF CONSTANT PARAMAGNETISM. APPLICATION TO MANGANESE

Editorial note (1977): This is a return, in a more rigorous way, to a problem already raised in A7 (Chapter I). In substances with negative magnetic interactions it is shown that the antiferromagnetic order which exists at low temperatures disappears at a certain temperature Θ_N, analogous to the Curie point. It is nowadays known as the Néel temperature. In particular, it is marked by a discontinuity in the specific heat. The term "antiferromagnetism" itself was only suggested later, in 1939, by F. Bitter.

The example of antiferromagnetism which has been chosen here, manganese, is not a particularly happy one, but it must be remembered that well-characterized antiferromagnets such as NiO were only discovered two years later by Bizette in 1938.

A more complete account of antiferromagnetism is given in Chapter II in A61, from paragraph 16 onwards.

On several occasions [1] I have shown that a substance with an *atomic moment* and a *negative molecular field* possesses, at low temperatures, a susceptibility which is independent of temperature. But since these proofs have been given for special cases where too much importance is attributed to fluctuations in the molecular field, I think it worthwhile to take up the question again in a more general and rigorous way.

At absolute zero, the atomic moments orient themselves in a position of minimum potential energy, all parallel to a certain direction, one half in one sense and the other half in the opposite sense. From now on, let us mentally isolate one of these halves and treat it as a substance A which is magnetized to saturation at absolute zero. The other half will constitute a substance B. In an external field $\vec{H}$ at temperature T, the magnetization of the two halves will be represented by the vectors $\vec{\sigma}_A$ and $\vec{\sigma}_B$. These magnetizations are related to the fields actually acting $\vec{H}_A$ and $\vec{H}_B$ by the two identical laws of paramagnetism:

$$\vec{\sigma}_A = f\left\{\frac{\vec{H}_A}{T}\right\}, \quad \vec{\sigma}_B = f\left\{\frac{\vec{H}_B}{T}\right\}. \quad (1)$$

The effective field $\vec{H}_A$ is the geometric sum of the external field $\vec{H}$ and of a molecular field $n\vec{\sigma}_A$, due to the influence on an A atom of the other A atoms, and of a molecular field $n'\vec{\sigma}_B$ due to the influence of the B atoms. We have

$$\vec{H}_A = \vec{H} + n\vec{\sigma}_A + n'\vec{\sigma}_B \quad \text{et} \quad \vec{H}_B = \vec{H} + n\vec{\sigma}_B + n'\vec{\sigma}_A \quad (2)$$

The coefficients n and n' are negative. Equations (1) and (2) allow a complete calculation of the magnetic properties if the coupling of the vectors $\vec{\sigma}_A$ and $\vec{\sigma}_B$ with the lattice may be neglected.

When H = 0, and if $n - n' > 0$, the two vectors $\vec{\sigma}_A$ and $\vec{\sigma}_B$ are oppositely directed and their common magnitude $\vec{\sigma}_T$, the solution of the equation

$$\sigma_T = f\left\{\frac{(n-n')\sigma_T}{T}\right\},$$

is a true *spontaneous magnetization*. The temperature of disappearance of this spontaneous magnetization is given by the formula $\Theta = (n-n')C$ where C is the Curie constant of the paramagnetism of formulae (1) ($\sigma_A = CH_A/T$ when H_A/T is small). With this spontaneous magnetization is associated a total magnetic energy $-(n-n')\sigma_T^2$, hence on excess specific heat $-(n-n')d\sigma_T^2/dT$ with a discontinuity at the Θ-point.

For $H \neq 0$, the susceptibility is determined as follows. If $T < \Theta$ and $M < -2n'\sigma_T$, the susceptibility χ is *independent of field and temperature*, equal to $-1/n'$. If $T > \Theta$, χ follows the Weiss law

$$\chi = \frac{2C}{T - (n+n')C}.$$

At the Θ-point, the two lines join and form a cusp; there is no discontinuity in χ.

For all metals with constant paramagnetism, except for manganese, the Θ-point is too high to be experimentally accessible. Let us show how the preceding considerations apply to manganese. The susceptibility of pure manganese has recently been determined by Y. Shimizu [2]. It has a constant value 7.53×10^{-6} up to 643 K and thereafter decreases up to the $\alpha \rightarrow \beta$ transformation at 1070 K. On the inverse susceptibility plot this decrease can be represented by a rather poorly defined Weiss line, with an atomic Curie constant $C_M = 0.68$. The true specific heat of manganese has been measured very recently by Ashworth [3]. The curve obtained has the same appearance as the curve of the specific heat of nickel. In particular, the author finds, at 628 K, a discontinuity of 0.125 cal/degree. This quite unexpected phenomenon is immediately explained by the above theory. In particular, it is noteworthy that the temperature of the discontinuity in specific heat should coincide, within experimental error, with the temperature where the susceptibility begins to vary with temperature. It is also noteworthy that above this temperature there should be an excess specific heat while the susceptibility remains completely constant. These are precisely the essential features of the theory.

It remains to show that the predicted orders of magnitude are correct. From the atomic susceptibility of αMn at low temperature one calculates $n' = -2400$. In general, depending on the nature of the lattice, n is included between 0 and $n'/2$. The complexity of the manganese lattice does not allow an exact calculation to be made, but the formula $C_M = 2C = 2\Theta/(n-n')$ allows us to calculate $C_M = 0.52$ for $n = 0$ and $C_M = 1.04$ for $2n = n'$ whereas the experiment gives 0.68. The discontinuity in specific heat calculated on the assumption that $2n = n'$ and supposing that σ_T varies as the spontaneous magnetization of nickel in reduced coordinates, is equal to 0.13. The hypothesis that $n = 0$ yields essentially the same value. Experiment gives 0.125. The agreement between theory and experiment is better than could be hoped for.

These agreements provide a solid experimental basis for the theory proposed for constant paramagnetism which, in its crude form, already satisfactorily related a number of phenomena concerned with the curve of interaction energy as a function of distance [4].

REFERENCES

[1] Néel, *Ann. de Physique*, 1932, **17**, 63; 1936, **5**, 232; *J. de Physique*, 1932, **3**, 160.
[2] *Sc. Rep. Tôhoku Imp. Univ.*, 1930, **19**, 411.
[3] *Proc. Phys. Soc.*, 1936, **48**, 456.
[4] Néel, *Ann. de Physique*, 1936, **5**, 232.

A112 (1961)
SUPERPARAMAGNETISM OF VERY SMALL ANTIFERROMAGNETIC PARTICLES

Summary: It is shown that in the ordered region, antiferromagnetic fine particles must generally show a sort of superparamagnetism due to the permanent magnetic moment resulting from imperfect compensation of the two sublattices. The superparamagnetic susceptibility may reach or exceed the order of magnitude of the normal antiferromagnetic susceptibility.

From a theoretical viewpoint, the subdivision of a bulk antiferromagnetic material into very fine particles, say of order 100 Å or less, should lead to the appearance of a certain number of new properties below the transition temperature Θ_N.

First, there is the fact that the magnetocrystalline energy, or more precisely the variations of energy connected with changes in the antiferromagnetic direction Δ, become of the same order of magnitude as the energy of thermal agitation kT when they are related to the mass of a given particle. This direction Δ can therefore no longer be considered as invariant and it is necessary from now on to take into account its time dependence.

A second important fact comes about because there is no good reason why the two antiferromagnetic sublattices (assuming for simplicity, that there are only two) should contain exactly the same number of magnetic atoms. The particle should, in practice, exhibit a slight ferrimagnetism. The corresponding permanent magnetic moment M, equal to the difference in the spontaneous magnetizations of the two sublattices, becomes relatively greater the smaller the particle. It is parallel to the antiferromagnetic direction. In large particles, for which Δ is fixed, M preserves its direction practically invariant so that in an ensemble of these particles the individual moments M mutually cancel and therefore escape macroscopic observation. But when the particle is small enough for Δ to become mobile, it develops a genuine paramagnetism in the sense studied by Langevin. The moment of the carrier of the magnetism is M, acquired in the way indicated above. This phenomenon has much in common with *superparamagnetism* of very fine ferromagnetic properties [1]. We give it the same name. It is a question then of both a surface effect and a volume effect at the same time, because the magnitude of M depends on the structure of the particles' surface and the coherence between the elementary moments which make up M is assured by the antiferromagnetic structure of the whole particle.

Finally, the antiferromagnetic particles should show pure surface effects such as *surface magnetocrystalline anisotropy* [2]. The consequence of this anisotropy is to introduce into the expression for the part of the energy of the particle depending on the orientation of Δ some new terms associated with the particles' shape. Another surface effect arises from the difference in the properties of the surface and bulk atoms arising from the differences in their surroundings.

The antiferromagnetism of the particle—A very small particle is thus antiferromagnetic and superparamagnetic at the same time. From an antiferromagnetic point of view it is characterized by a parallel susceptibility $S_\parallel$ and a perpendicular susceptibility $S_\perp$ corresponding respectively to orientations of the magnetic field H parallel and perpendicular to Δ. The quantities $S_\parallel$ and $S_\perp$ are specific to a given particle, and they are proportional to its volume. They depend on temperature. To study them in isolation it is necessary to have a bulk single crystal. We set $S = S_\perp - S_\parallel$ and introduce a quantity b defined by the relation $b = SH^2/2kT$ where kT is the thermal agitation energy and $SH^2/2$ the change in magnetic energy in the field H when Δ turns from the direction parallel to H to a perpendicular direction. When b is small, Δ remains oriented at random on the average, and the average antiferromagnetic susceptibility of a given grain is equal to $S_\parallel + (2/3)S$. When b is large, Δ aligns itself perpendicular to the applied field and the average susceptibility of the particle tends towards $S_\parallel + S$, in other words towards $S_\perp$.

It is worth remarking here that for an ensemble of

identical large particles with fixed directions Δ, the average initial susceptibility is also $S_\parallel + (2/3)S$, but this is an effect of the average over different particles.

The superparamagnetism of the particle — The important parameter is Langevin's variable a defined by the relation $a = MH/kT$. If a is small, the particle has an average magnetic moment m_s parallel to H given by the classical formula $m_s = M^2H/3kT$. We will study rigorously the superposition of superparamagnetic and antiferromagnetic magnetization in a subsequent paper. In particular we will justify the intuitive idea that these two magnetizations are superposable when a and b are much less than 1. The total average magnetization of a given particle is therefore given by

$$m = m_s + S_\parallel H + \frac{2}{3} SH \qquad (1)$$

The permanent moment M — Let us suppose, for simplicity, that we are concerned with a normal antiferromagnetic particle composed of just two sublattices A and B, with a total number n of magnetic atoms. The determination of M is closely linked to that of the difference p between the number of magnetic atoms contained in the two sublattices.

The evaluation of p as a function of n is a virtually intractable problem whose solution requires a precise knowledge of the architecture of the particle, supposing it to be monocrystalline.

Let us first suppose that the particle is built up by some sort of random draw of the n atoms for the two sublattices. The average value of p^2 is then equal to n, and p is of order $n^{1/2}$.

Suppose now, on the contrary, that the particle is built up very regularly by adding successive lattice planes. From the magnetic point of view these planes can be sorted into different categories. They are *neutral* when they contain equal numbers of atoms of both sublattices. They are *active* when they contain only atoms belonging to one or other of the two sublattices. They can be A active or B active. When the planes which delimit a particle, assumed to be entirely formed, are all neutral or else can be grouped in parallel pairs of different activity, p is zero. But when the two parallel planes have the same activity, p is equal to the number of magnetic atoms contained in the planes. It is therefore of order $n^{2/3}$.

One can imagine plenty of other situations, for example a crystal bounded by neutral planes which are incompletely filled, with a random distribution of surface atoms among the two sublattices, p is then of order $n^{1/3}$.

As we are simply interested here in taking an example and working out an order of magnitude, let us adopt the intermediate hypothesis, $p = n^{1/2}$. Insofar as the atomic moment is concerned, we suppose provisionally that all the magnetic atoms in the particle possess the same magnetic moment μ, corresponding to the spontaneous magnetization of each of the two antiferromagnetic sublattices. This moment μ decreases regularly from its saturation value of $gj\mu_B$ (g is the Landé factor, j is the total quantum number and μ_B is the Bohr magneton) down to zero as the temperature increases from absolute zero to the Néel temperature Θ_N. The average value of M^2 is equal to the average value of $p^2\mu^2$, that is to say $n\mu^2$ with the hypothesis we decided to adopt.

The superparamagnetic magnetization — It can be written

$$m_s = \frac{M^2 H}{3kT} = \frac{n\mu^2 H}{3kT}. \qquad (2)$$

It is worthwhile comparing formula with the one which gives m_P, the magnetization of the particle in the paramagnetic region above Θ_N. Applying molecular field theory and denoting by Θ_p the paramagnetic Curie temperature, generally negative, one finds

$$m_p = \frac{n\mu_B^2 g^2 j(j+1) H}{3k(T - \Theta_p)} \qquad (3)$$

With the same theory, the value of $S_\perp H$ or that of SH at absolute zero is obtained on replacing T by Θ_N in the expression for m_P. It follows that, close to absolute zero in sufficiently small magnetic fields,

$$\frac{m_s}{SH} = \frac{j}{j+1} \frac{\Theta_N - \Theta_p}{T} \qquad (4)$$

Hence, it may be seen that the superparamagnetic magnetization may become much larger than the antiferromagnetic magnetization.

Formula (2) shows that m_s is proportional to the number of atoms in the particle. The superparamagnetic susceptibility of an ensemble of particles expressed per unit mass therefore does not seem to depend on the particle size. In reality it is also necessary to take account of the saturation of the superparamagnetism which begins to appear when the variable a in the Langevin function is greater than unity. Now a varies as $n^{1/2}$. In other words, the larger the particles, the smaller the values of magnetic field for which saturation appears. At the same time, for the large particles this saturation magnetization M becomes small compared with the antiferromagnetic magnetization SH induced by magnetic fields of a few thousand gauss.

Furthermore, the fact that the superparamagnetic

susceptibility per unit mass is independent of particle size is simply a consequence of the starting hypothesis, that $p = n^{1/2}$. Now, we have shown that other hypotheses are just as plausible. If p varies as $n^{2/3}$, one finds that the specific susceptibility varies as $n^{1/3}$, in other words as the linear dimensions of the particles. If p varies as $n^{1/3}$, the specific susceptibility is inversely proportional to the particle diameter.

Temperature dependence of the superparamagnetic magnetization — This dependence is substantial since the effect of the $1/T$ factor is added to that of the μ^2 factor. It can be studied with some precision if μ varies with T like the spontaneous sublattice magnetization. Unfortunately the above supposition, that all the atoms in a particle have the same moment, is open to serious objections. The molecular fields acting on internal and external atoms are surely quite different.

In general, it seems logical to suppose that the external atoms are subject to a weaker molecular field than the internal atoms. Although it is weaker, this field is sufficient to saturate magnetically the external atoms when the temperature is close to absolute zero, but when the temperature approaches Θ_N this is no longer the case. The decrease in average magnetic moment with increasing temperature should be greater for the external atoms than for the internal atoms.

It even seems possible that M changes sign because the decrease in magnetic moment of atoms belonging to an external lattice plane of an antiferromagnetic particle must produce effects analogous to the progressive removal of external planes. If these are active, the planes removed belong alternatively to one or other of the two sublattices so that the moment M oscillates between two limits of opposite sign.

Role of the anisotropy — We have so far supposed that we are dealing with an isotropic antiferromagnet. In other words we have neglected the variation in energy of the particle associated with rotations of the direction Δ. To tell the truth, we do not know of any isotropic antiferromagnetic substance, but we do know of some which approximate to this condition, those where the direction Δ can turn freely in a plane. This is the case, for example, for NiO, where Δ turns freely in the plane perpendicular to the rhombohedral axis. The theory of the magnetic properties of these substances greatly resembles that which we have just developed, and only differs insofar as the saturation is concerned for values of a greater than unity.

Another situation is likely to arise where Δ can take up a certain number of discreet positions of minimum energy, called *spontaneous antiferromagnetic directions*. They are separated by potential barriers the height of which we denote by w for a particular particle.

In bulk material or in large particles it is sometimes possible to pass from one spontaneous antiferromagnetic direction to another without change of energy by a mechanism of domain wall movement, but this is impossible for small particles. It is necessary there to surmount the potential barrier. Moreover, to achieve this, the influence of a field is far less effective than in a ferromagnet because the energy available, which is of order $1/2\, SH^2$ in an antiferromagnet, is much less than the energy $J_s H$ in a ferromagnet. (J_s is the spontaneous magnetization.) Finally, for small particles the only possible mechanism for changing the antiferromagnetic direction and thus reorienting the associated moment M is that of thermal fluctuations.

One can now develop a theory for antiferromagnetic small particles analogous to that for ferromagnetic small particles. The result is the following: the direction Δ passes continually from one orientation to the antiparallel orientation, or from one spontaneous antiferromagnetic direction to another carrying the moment M with it. The mean dwell time in a particular orientation is

$$\tau = \frac{1}{\nu} \exp \frac{w}{kT} , \qquad (5)$$

where ν is a frequency of order w/h, where h is Planck's constant.

Since, on the one hand w is proportional to the volume of the particle and on the other hand tends to zero as T tends to Θ_N, a certain critical temperature T_B can be defined for each size of particle so that above T_B, τ is less than the duration of experiments which it is possible to carry out in the laboratory with standard equipment. Thermal equilibrium is then established. On the contrary, below T_B, τ is longer than the duration of possible experiments, thermal equilibrium cannot be established, and the permanent moment M preserves the orientation which it had when it was cooled for the first time below T_B. This is why this temperature is called the blocking temperature.

So long as the temperature remains higher than T_B, the particle exhibits superparamagnetism essentially as we have described above. Below T_B the moment remains in some sense *frozen*, and superparamagnetism ceases.

Nevertheless, notwithstanding this disappearance, a remanent magnetization can appear when the particles have been cooled from a temperature above T_B in such a way that a field H is acting while the temperature passes T_B. Except for the thermal variation of M, the remanent moment at low temperature is equal to the superparamagnetic magnetization produced by the field H at the temperature T_B. We are concerned here with a mechanism in every way analogous to that for the acquisition of thermoremanent magnetization by rocks which we analyzed previously. It is evidently possible to modify this thermoremanent magnetization during an isothermal process, but to do this it is necessary to use magnetic fields much greater than those which were necessary to create the magnetization by field cooling.

Let us note that an antiferromagnet, perfectly isotropic in the bulk, whether in a plane or in all directions in space, is susceptible to become anisotropic and acquire a blocking temperature T_B when it is subdivided into very small particles. It is sufficient for the particle not to have spherical symmetry so that there will be surface magnetocrystalline anisotropy analogous to the anisotropy suspected to exist in ferromagnetic materials [2].

REFERENCES
[1] L. Néel, *Ann. Géophys.*, 1949, **5**, 99.
[2] L. Néel, *Comptes rendus*, 1953, **237**, 1468; *J. Phys. Rad.*, 1954, **15**, 225.

A113 (1961)

SUPERPOSITION OF ANTIFERROMAGNETISM AND SUPERPARAMAGNETISM IN A VERY SMALL PARTICLE

Summary: This is a calculation of the average magnetic moment of a very small antiferromagnetic particle placed in a magnetic field and having a permanent magnetic moment, in the case where variations in coupling energy of the antiferromagnetic direction with the crystal lattice are negligible compared with the thermal agitation energy.

We have shown in a previous communication [1] that, in addition to the usual antiferromagnetic properties, a very small particle of antiferromagnetic material below the transition temperature Θ_N is characterized by a certain permanent magnetic moment M parallel to the antiferromagnetic direction Δ. The moment M depends on temperature T and on the surface structure of the particle. Furthermore the particles considered are sufficiently small for Δ to change direction under the influence of thermal agitation.

We propose to calculate the average magnetization assumed by the particle in question when it is placed in a magnetic field H. The antiferromagnetic properties are defined by the two susceptibilities $S_\parallel$ and $S_\perp$ corresponding to magnetic fields parallel and perpendicular to Δ respectively. These susceptibilities are related to the whole particle under consideration. For simplicity, we treat the problem in the case where variations in coupling energy of the direction Δ with the crystal lattice are negligible compared with kT.

The energy W of the particle depends on the angle α of the direction with the field H, and is written

$$W = -\frac{1}{2} SH^2 \sin^2 \alpha - \frac{1}{2} S_\parallel H^2 - MH \cos \alpha, \quad (1)$$

where we have put $S = S_\perp - S_\parallel$. S is a quantity which usually decreases smoothly as the temperature increases from absolute zero up to the Néel temperature Θ_N, where $S = 0$.

The two first terms on the right hand side of equation (1) correspond to the induced antiferromagnetic magnetization and the last term corresponds to the permanent moment M. As for the component m of the total magnetic moment in the direction of the field, it may be written

$$m = SH \sin^2\alpha + S_\parallel H + M \cos \alpha. \quad (2)$$

Hence, in the expressions for w and m, there appear terms in $S_\parallel$, independent of the orientation of Δ. Taking this point into consideration and supposing that M is large compared with the Bohr magneton, the average magnetization m of the particle undergoing thermal agitation is obtained by applying classical Boltzmann statistics. It may be written

$$\overline{m} = S_\parallel H + \frac{\int_0^\pi (M \cos \alpha + SH \sin^2 \alpha) \exp(a \cos \alpha + b \sin^2 \alpha) \sin \alpha \, d\alpha}{\int_0^\pi \exp(a \cos \alpha + b \sin^2 \alpha) \sin \alpha \, dx} \quad (3)$$

an expression in which we have introduced the following notation

$$a = \frac{MH}{kT}; \quad b = \frac{1}{2} \frac{SH^2}{kT} \quad (4)$$

In order to calculate the integrals which appear on the right hand side of (3) we put $x = \cos \alpha$ and

$$S_n = \int_{-1}^{+1} x^n \exp(ax - bx^2) dx. \quad (5)$$

We then easily obtain the relation

$$aS_0 - 2bS_1 = 2 e^{-b} \operatorname{sh} a,$$

from which S_1 is deduced. It is also equal to dS_0/da. By deriving S_1 with respect to a under the integral sign one deduces

111

$$S_2 = \left(\frac{1}{2b} + \frac{a^2}{4b^2}\right) S_0 - \left(\frac{\operatorname{ch} a}{b} + \frac{a \operatorname{sh} a}{2b^2}\right) e^{-b}.$$

Finally, one obtains

$$\bar{m} = S_\| H + M \frac{S_1}{S_0} + SH \left(1 - \frac{S_2}{S_0}\right). \tag{6}$$

As for S_0, it may be simply expressed in terms of the error function $\Phi(x)$:

$$\Phi(x) = \frac{2}{\sqrt{\pi}} \int_0^x e^{-x^2} dx,$$

which may be found in tables. Depending on the case, one finds

$$S_0 = \frac{1}{2}\sqrt{\frac{\pi}{b}} e^{\frac{a^2}{4b}} \left[\Phi\left(\sqrt{b} + \frac{a}{2\sqrt{b}}\right) \right.$$
$$\left. + \Phi\left(\sqrt{b} - \frac{a}{2\sqrt{b}}\right)\right], \quad a < 2b; \tag{7}$$

or

$$S_0 = \frac{1}{2}\sqrt{\frac{\pi}{b}} e^{\frac{a^2}{4b}} \left[\Phi\left(\frac{a}{2\sqrt{b}} + \sqrt{b}\right) \right.$$
$$\left. - \Phi\left(\frac{a}{2\sqrt{b}} - \sqrt{b}\right)\right], \quad a > 2b. \tag{8}$$

The problem posed at the outset is therefore solved, but a brief discussion of equation (6) is of some interest.

The case where $S_\|$ and S are zero. Equation (6) reduces to the classical Langevin formula

$$\bar{m} = M \left(\operatorname{coth} a - \frac{1}{a}\right).$$

The case where the permanent moment M is zero. Equation (6) reduces to

$$\bar{m} = S_\| H + SH \left(1 - \frac{1}{2b} + \frac{1}{\sqrt{\pi b}} \frac{e^{-b}}{\Phi(\sqrt{b})}\right), \tag{9}$$

The asymptotic expressions for values of b very much greater or very much less than unity are

$$\bar{m} = S_\| H + \frac{2}{3} SH\left(1 + \frac{2b}{15} - \frac{4b^2}{315} + \cdots\right); \quad b \ll 1; \tag{10}$$

and

$$\bar{m} = S_\| H + SH\left(1 - \frac{1}{2b} + \frac{e^{-b}}{\sqrt{\pi b}} + \cdots\right); \quad b \gg 1. \tag{11}$$

Formula (10), valid for weak fields, corresponds to a random distribution of orientations of Δ whereas for strong fields (formula (11)) Δ is oriented in a direction perpendicular to H.

It is important to note that formula (10), calculated on the supposition that the direction Δ is not coupled to the crystalline lattice, is also valid for any strength of coupling on condition that m is taken as an average value over all possible orientations of the particle in question.

The case where a and b are both smaller than unity. Limiting the expression to third order terms, one finds

$$\bar{m} = S_\| H + \frac{2}{3} SH\left(1 + \frac{2b}{15} - \frac{4b^2}{315} + \cdots\right) +$$
$$+ \frac{M^2 H}{3kT}\left(1 - \frac{a^2}{15} + \cdots - \frac{8b}{15} + \frac{8b^2}{105} + \cdots\right). \tag{12}$$

Examination of the formula shows that everything behaves as if there were simply superposition of the effects of bulk antiferromagnetism and Langevin paramagnetism with a second order correction equal to

$$-\frac{8}{45} Mab \left(1 - \frac{b}{7} + \cdots\right).$$

The same remark can be made as in the preceding section. The validity of formula (12), established by neglecting the coupling of Δ with the crystal lattice, extends to the case of coupling of arbitrary strength on condition that $\bar{m}$ is interpreted as an average value over all possible orientations of the particle in question.

Case when b is greater than unity. It is useful to distinguish two cases depending on the relative values of a and b.

If a is greater than $2b$, one obtains

$$\bar{m} = S_\| H + M - \frac{kT}{H} + \cdots; \tag{13}$$

which is simply an approximation to the paramagnetic Langevin function. The anisotropic part of the bulk antiferromagnetism is suppressed.

If *a* is smaller than 2*b*, one obtains on the contrary

$$\bar{m} = SH - \frac{kT}{H} + \cdots \qquad (14)$$

The influence of the permanent moment has disappeared, at least to first order.

Final remark — The evolution of *m* as a function of temperature will generally be very complicated since the quantities M, S, *a* and *b* depend enormously on temperature, following laws which are quite different from each other.

REFERENCE

[1] L. Néel, *Comptes rendus*, 1961, **252**, 4075.

A114 (1961)
SUPERANTIFERROMAGNETISM IN SMALL PARTICLES

Summary: It is shown that very small antiferromagnetic particles in the ordered region, for a range of field which becomes larger the smaller the particle, show an increase in susceptibility which is of the same order of magnitude as the normal susceptibility, the increase susceptibility is related to a progressive rotation of the antiferromagnetic direction from one end of the particle to the other.

Let us consider an antiferromagnetic substance with two sublattices A and B such that the magnetic atoms are arranged on successive lattice planes, equally spaced a distance e apart, parallel to a plane II. The A sublattice is composed of even-numbered planes and the B sublattice is composed of odd-numbered planes. Furthermore, let us suppose that the antiferromagnetic direction Δ can turn freely in the plane II without change of energy. This is approximately what happens in NiO or in αFe_2O_3 above $-15°C$.

Now let us consider a small particle composed of n of these planes, supposed perfect and identical apart from a lattice translation. When the temperature is sufficiently low, in the absence of an external magnetic field, each lattice plane possesses a magnetic moment parallel to Δ equal to $+M$ or $-M$ according to whether it belongs to A or B.

If the number of planes n is odd, the two extreme lattice planes which delimit the particle possesses moments with the same sign. The particle is therefore endowed with a permanent magnetic moment of magnitude M, and it shows a sort of superparamagnetism whose effects are superposed on those of the antiferromagnetism. We have studied this phenomenon in two preceding communications [1].

Henceforth we consider the case where n is even. The magnetic moments of the two extreme lattice planes are then of opposite sign and the total magnetic moment of the particle is zero. Let us take as reference direction in the plane II the initial direction of Δ, Oz, and apply a field H, parallel to π and perpendicular to Oz. It is known that the field causes the spontaneous magnetization of the two sublattices to deviate from their initial directions. The angles they make with Δ become equal to ω and $\pi - \omega$. The antiferromagnetic direction Δ, defined as the external bisector of the directions of spontaneous magnetization of the two sublattices, is not affected and remains parallel to Oz throughout the particle. But besides the classical effect, it is also necessary to consider a surface effect. The field H exerts a couple on the magnetic moment of the external lattice planes which tends to align the direction Δ locally along the direction of H. This couple is easily evaluated on observing that it should change in sign but not in magnitude when one extra lattice plane is added to the particle because the magnetic moment of the plane is of opposite sign to the moment of the plane on which it has been placed. Since the variation of the couple in the course of this operation is equal to $\vec{M} \wedge \vec{M}$, it follows that the couples acting on the ends are equal to $\pm 1/2\, \vec{M} \wedge \vec{H}$.

Hence, the antiferromagnetic direction should turn progressively within the particle. Let us identify each lattice plane by its distance x from the central plane, and denote the thickness of the particle by $2L$ ($ne = 2L$). Let us call φ the angle between Δ and H. This angle is equal to $\pi/2$ in the middle of the particle and it varies towards edges tending to φ_0 for $x = L$ and $\pi - \varphi_0$ for $x = -L$. Furthermore, the larger H is, the smaller is φ_0. In order to determine the magnetic state of the particle it is still necessary to know the angles ω and $\pi - \omega$ which the moments of the lattice planes of A and B atoms make with Δ. ω is also a function of x. By symmetry, $\varphi(x) + \varphi(-x) = \pi$ and $\omega(x) = \omega(-x)$.

We will now go on to calculate ω and φ as a functution of H, supposing that the number of planes n is large enough to be able to consider ω and φ as continuous functions of x. We also suppose that ω is always small.

Setting up the equations — The magnetic moment of any lattice plane M is in equilibrium under the influence of the couples exerted by the external field and by the neighbouring lattice planes. Let us restrict ourselves to the influences of the first and second neighbour planes, and represent them by molecular fields. We therefore allow a plane with moment M to exert a molecular field of $-n_1 \vec{M}$ on the two adjacent planes and a molecular field of $-n_2 \vec{M}$ on the two next nearest neighbour planes.

Let us now consider the five successive lattice planes I, II, III, IV, and V, centred around plane III which we suppose, for the sake of argument, to belong to sublattice A. The angles ψ which the planes make with H are given by the following expansions in terms of increasing powers of e,

$$\begin{cases} \psi_I = \varphi - 2e\varphi' + 2e^2\varphi'' - \omega + 2\omega'e \; ; \\ \psi_{II} = \pi + \varphi - e\varphi' + \dfrac{1}{2} e^2\varphi'' + \omega - \omega'e \; ; \\ \psi_{III} = \varphi - \omega \; ; \\ \psi_{IV} = \pi + \varphi + e\varphi' + \dfrac{1}{2} e^2\varphi'' + \omega + \omega'e \; ; \\ \psi_V = \varphi + 2e\varphi' + 2e^2\varphi'' - \omega - 2\omega'e \; ; \end{cases} \quad (1)$$

where φ', φ'', ω' are the successive derivatives of φ and ω with respect to x. We truncate the expansions for φ with the term in e^2 and those for ω with the term in e because $e\varphi'$ is of order ω, as we can show *a posteriori*.

Let us now write the condition that the magnetic moment of plane III is in equilibrium.

$$\begin{aligned} MH\sin(\varphi + \omega) = \; & n_1 M^2 [\sin(\psi_{II} - \psi_{III}) + \\ & + \sin(\psi_{IV} - \psi_{III})] \\ & + n_2 M^2 [\sin(\psi_I - \psi_{III}) + \\ & + \sin(\psi_V - \psi_{III})] \; . \end{aligned} \quad (2)$$

Replacing the ψ_s by their values given by (1) and expanding the sine as a series, one obtains

$$n_1 M^2 (e^2 \varphi'' + 4\omega) - 4 n_2 M^2 e^2 \varphi'' = \\ = MH(\sin\varphi - \omega \cos\varphi) \; . \quad (3)$$

When the central plane III belongs to the B sublattice, a new equilibrium equation can be deduced from the previous one by replacing ω by $-\omega$ and φ by $\varphi + \pi$ in (2), giving

$$n_1 M^2 (e^2 \varphi'' - 4\omega) - 4 n_2 M^2 e^2 \varphi'' = \\ = MH(-\sin\varphi - \omega \cos\varphi) \; . \quad (4)$$

Adding and subtracting equations (3) and (4) we obtain the two new equations.

$$H \sin\varphi = + 4 n_1 M \omega \; , \quad (5)$$

$$- H\omega \cos\varphi = (n_1 - 4 n_2) M e^2 \varphi'' \; , \quad (6)$$

the first is nothing more than the classical relation giving the deviations $\pm \omega$ with respect to Δ of the spontaneous magnetizations of the two sublattices as a function of the field $H \sin\varphi$ normal to Δ. Substituting now the value of ω given by (5) into (6), one obtains a second order differential equation for φ

$$\varphi'' + \frac{H^2}{4 n_1 (n_1 - 4 n_2) e^2} \cos\varphi \sin\varphi = 0 \; . \quad (7)$$

Boundary conditions — Let us imagine that the crystal lattice is cut between planes II and III. The first part of the lattice exerts a couple on the second equal to the sum of the couple $-n_1 M^2 \varphi'$ produced by plane II on plane III, the couple $2 n_2 M^2 e\varphi'$ produced by plane I on plane III, and the couple equal to the preceding one are produced by plane II on plane VI. The total couple is finally $-(n_1 - 4 n_2) M^2 e \varphi'$, to first order. If we had cut the lattice between planes III and IV, thus interchanging the sublattices of the two planes on either side of the cut, we would have obtained exactly the same result.

For the two external lattice planes, it is necessary that the total couple calculated in this manner be balanced by the couple $1/2 \vec{M} \wedge \vec{H}$ produced by the external field on the ends. Denoting as φ_0 the value of φ for $x = \pm L$, this gives

$$(n_1 - 4 n_2) M^2 e \varphi'_0 = \frac{1}{2} MH \sin\varphi_0 \; . \quad (8)$$

the problem is therefore solved.

Calculation of φ — To simplify the notation, let us set

$$\text{tg}^2 \theta = \frac{n_1}{n_1 - 4 n_2} \quad \text{and} \quad h = \frac{LH}{2 n_1 e M} \; , \quad (9)$$

where h is a reduced field. Equation (7) is written

$$\varphi'' + \frac{h^2 \, \text{tg}^2 \theta}{L^2} \sin\varphi \cos\varphi = 0 \; , \quad (10)$$

and can be integrated once after multiplying through by $2\varphi'$.

$$\varphi'^2 = \frac{h^2 \, \text{tg}^2 \theta}{k^2 L^2} (1 - k^2 \sin^2 \varphi) \; , \quad (11)$$

where k is a constant of integration. A second integration gives

$$x = \frac{Lk}{h \, \text{tg}\, \theta} \int_{\varphi_0}^{\frac{\pi}{2}} \frac{d\varphi}{\sqrt{1 - k^2 \sin^2 \varphi}} = \\ = \frac{Lk}{h \, \text{tg}\, \theta} \left[F\left(k, \frac{\pi}{2}\right) - F(k, \varphi) \right] \; , \quad (12)$$

where $F(k, \varphi)$ is an elliptical function of the first kind of argument k. To determine k we write the boundary conditions using the new notation.

$$k \sin \varphi_0 = \cos \theta. \tag{13}$$

Also, for $x = L$, equation (12) gives

$$\frac{h \, \text{tg} \, \theta}{k} = F\left(k, \frac{\pi}{2}\right) - F(k, \varphi_0). \tag{14}$$

Given a value for φ_0, one can thus calculate k from (13), and then determine the corresponding value of h from (14).

When H increases from 0 to $+\infty$, k increases from an initial value of

$$\cos \theta = (n_1 - 4n_2)^{\frac{1}{2}} (2n_1 - 4n_2)^{-\frac{1}{2}}$$

up to 1. At the same time $\sin \varphi_0$ decreases from 1 to $\cos \theta$. The variation of φ with x is of the form $\varphi = \sin ax$ where a is an appropriate constant.

Induced magnetic moment — The magnetic moment induced in the particle by the field H is the sum of the antiferromagnetic moment m_Q induced in the bulk of the particle and the moment m_P of the two ends. That is

$$m_P = 2 \frac{M}{2} \cos \varphi_0, \tag{15}$$

$$m_Q = \int_{-L}^{+L} \frac{H \sin^2 \varphi \, dx}{4n_1 e} = \frac{Mk}{\text{tg} \, \theta} \int_{\varphi_0}^{\frac{\pi}{2}} \frac{\sin^2 \varphi \, d\varphi}{\sqrt{1 - k^2 \sin^2 \varphi}}, \tag{16}$$

or

$$m_Q = \frac{M}{k \, \text{tg} \, \theta} \left[F\left(k, \frac{\pi}{2}\right) - F(k, \varphi_0) - E\left(k, \frac{\pi}{2}\right) + E(k, \varphi_0) \right], \tag{17}$$

where E is an eliptical function of the second kind.

In the absence of the end effects, the particle would have a normal antiferromagnetic susceptibility given by $S_0 = L(2n_1 e)$.

Now the real susceptibility is the sum of the two susceptibilities $S_P = m_P/H$ and $S_Q = m_Q/H$ which for small values of H are written as

$$S_P/S_0 = \text{tg}^2 \theta - \frac{1}{3} h^2 \, \text{tg}^4 \theta (1 + 2 \, \text{tg}^2 \theta) + \cdots, \tag{18}$$

$$S_Q/S_0 = 1 - \frac{1}{3} h^2 \, \text{tg}^4 \theta + \cdots \tag{19}$$

When h becomes very large, S_P/S_0 decreases to zero and S_Q/S_0 increases to unity. The following table brings together some intermediate values calculated for $\tan \theta = 1$, that is for $n_2 = 0$.

h.	k^2.	S_P/S_0.	S_Q/S_0.
0,455	0,587	0,845	0,951
0,700	0,671	0,721	0,915
0,934	0,750	0,618	0,894
1,178	0,821	0,531	0,880
1,453	0,883	0,453	0,874
1,783	0,933	0,382	0,876
2,220	0,970	0,314	0,885
2,935	0,992	0,240	0,905

We therefore observe that S_Q, initially equal to S_0, decreases to $0.87 S_0$ before tending again to S_0. As for the surface susceptibility, it is also initially equal to S_0 (for $n_2 = 0$). The total initial susceptibility is then twice as great for particles as for the bulk material. It may even be very much larger when n_2 is non-zero. It tends to infinity as $4n_2$ tends to n_1.

When $4n_2$ is greater than n_1, the substance takes on the helical antiferromagnetic structure described by Yoshimori [2] and by Herpin and Villain [3].

It is somewhat odd that the total initial susceptibility $S_P + S_Q$ per unit mass does not depend on the grain size even though S_Q arises from a surface effect. On the other hand, the range of magnetic field for which an increase in susceptibility beyond its normal value S_0 may be detected in inversely proportional to $2L$, the thickness of the particle. In effect it is of order of the reduced field, that is $2n_1 eM/L$. This is a consequence of the fact that the total additional magnetization associated with the existence of the surface susceptibility S_Q is equal to $M \cos \varphi_0$, and is independent of the thickness $2L$ of the particle.

To this increase in susceptibility which appears in certain very small antiferromagnetic particles we propose to give the name *superantiferromagnetism*. Together with the superparamagnetism described in the previous communications [1], the new process should allow an interpretation of the remarkable magnetic properties of fine particles of NiO, αFe_2O_3, Cr_2O_3 etc which have recently been found by various authors [4].

Similar phenomena should occur in bulk antiferromagnets containing dislocations. A ring of dislocations should behave in the same way as the surface lattice plane of a particle.

REFERENCES

[1] L. Néel, *Comptes rendus*, 1961, **252**, 4075; 1961, **253**, 9.
[2] A. Yoshimori, *J. Phys. Soc.* (*Japan*), 1959, **14**, 807.
[3] Herpin, Meriel, and Villain, *Comptes rendus*, 1959, **249**, 1334; Herpin and Meriel, *Ibid.*, 1960, **250**, 1450.
[4] J. T. Richardson and W. O. Milligan, *Phys. Rev.*, 1956, **102**, 1289; J. C. Creer, J. Cohen and K. Gopal Srivastava (Private communication).

A115 (1961)

ON THE CALCULATION OF THE ADDITIONAL SUPER-ANTIFERROMAGNETIC SUSCEPTIBILITY OF SMALL PARTICLES AND ITS TEMPERATURE DEPENDENCE

Summary: The initial extra superantiferromagnetic susceptibility of small particles is calculated by a rigorous method. Having established the idea of an end moment, the type of differential equation established previously governing the variation of φ, the angle between the antiferromagnetic direction within the particle and the magnetic field, is justified. Then the relation between the end moment and the moment of the internal lattice planes is studied, and the temperature dependence of the extra susceptibility is deduced.

In a previous communication [1], which we will henceforth refer to as paper A, we have studied the problem of superantiferromagnetism of very small particles of an antiferromagnetic substance. It is a question of antiferromagnets with two sublattices formed from a family of successive parallel lattice planes called *active lattice planes*. The odd-numbered planes form one of the two sublattices and the even-numbered planes from the other. The particles are supposed to be made up of an even number n of these planes. The interplanar spacing is e, and they possess a permanent and invariant moment M which is identical for all the planes. The orientations of the moments depend on the applied field H and on the magnetic interactions between planes defined by the energies $n_1 M^2 \cos\psi$ or $n_2 M^2 \cos\psi$ associated respectively with a pair of nearest neighbour planes or any pair of second neighbour planes. ψ designates the angle between the moments of the pair of planes considered. The energies associated with the distant pairs of planes are neglected, as are magnetocrystalline energies.

Assuming in particular that the angles defining the orientations of the moments are continuous functions of the distance x of the active lattice planes from the centre of the particle, we have shown that for $n_2 = 0$, the initial susceptibility of these particles was twice as great as that of the corresponding bulk antiferromagnet. The range of field in which this increase in susceptibility was detectable reaches $4n_1 M/n$.

Given the interest of the result and the objections which are likely to be raised against the method of calculation used, it is worthwhile taking up the question using a rigorous method. We will now write down the condition that moment of each one of the planes is in equilibrium under the influence of the field H and the forces exerted by the neighbouring planes. We simply suppose that H is small enough for all the planes to be approximately perpendicular to the field direction so that the couple exerted by the field on the pth plane is equal to $(-1)^p MH$. This corresponds to values of H which are small compared with $4n_1 M/n$.

Now let $\vec{M}_p$ be the moment of the pth plane and $\vec{D}_p$ a unit vector parallel to $\vec{M}_p$ when p is odd and antiparallel to p when p is even. We designate the angle between $\vec{D}_{p+1}$ and $\vec{D}_p$ as ε_p. The $p-1$ values of ε_p are, according to our hypothesis, small compared with unity. The equilibrium condition for M_p is written

$$\epsilon_p - \epsilon_{p-1} - \rho(\epsilon_{p+1} - \epsilon_{p-2}) = (-1)^{p-1} \mu, \quad (1)$$

After simplifying the notation by writing

$$\rho = \frac{n_2}{n_1 - n_2} \quad \text{et} \quad \mu = \frac{H}{M(n_1 - n_2)}. \quad (2)$$

Equation (1) is valid for all values of p on condition that $\varepsilon_{-1} = \varepsilon_0 = \varepsilon_n = \varepsilon_{n+1} = 0$. The system of n equations thereby obtained is symmetric in the sense that one finds the same equations on substituting $n-p$ for p.

Taking this symmetry into account, the general solution of (1) is

$$\epsilon_p = (-1)^{p-1} A + B(\lambda^{p-1} + \lambda^{n-p-1}) + D. \quad (3)$$

After substituting in (1), one finds that A is given by

$$2A(1 + \rho) = \mu, \quad (4)$$

and that λ is the smaller of the two roots of the equation

$$\lambda^2 + \lambda\left(1 - \frac{1}{\rho}\right) + 1 = 0. \quad (5)$$

For the roots to be real. it is necessary that ρ be less than 1/3, that is to say that $4n_2$ be less than n_1. We recall that if $4n_2$ is greater than n_1 we are no longer dealing with an antiferromagnetic material but a helimagnetic material of the type described by Yoshimori and Villain, Meriel and Herpin [2].

The two constants B and D are obtained on writing that $\varepsilon_1, \varepsilon_2, \varepsilon_3$ satisfy two equations (1) corresponding to $p=1$ and $p=2$ which are given by the following:

$$\begin{cases} D(1-\rho) + B(1 - \rho\lambda - \rho\lambda^{n-3} + \lambda^{n-2}) = \dfrac{\mu}{2}, \\ \\ D\rho + B(1 - \lambda + \rho\lambda^2 + \rho\lambda^{n-4} - \lambda^{n-3} + \lambda^{n-2}) = \\ \qquad\qquad\qquad\qquad\qquad\qquad\qquad = \dfrac{\mu\rho}{2(1+\rho)} \,. \end{cases} \quad (6)$$

The value of ε_p being therefore determined completely, the moment M induced in the particle by the field H is obtained by adding the moments together in pairs, 1 and 2, 2 and 3, etc. The resultant moment of each of these pairs is parallel to H. Hence

$$\bar{M} = (\epsilon_1 + \epsilon_3 + \epsilon_5 + \cdots + \epsilon_{n-1}) M \quad (7)$$

and finally

$$\bar{M} = \frac{1}{2} nM(A + D) + 2 BM \frac{1-\lambda^n}{1-\lambda^2}. \quad (8)$$

The induced moment is therefore the sum of three terms, the last of which, involving B, is independent of n and represents an end effect localized at the surface of the particle which is negligible when ρ is small or when n is a few tens or greater. One can now simply write $M = n(S_0 + S_a)H$ after putting $S_0 = MA/2H$ and $S_a = MD/2H$. Replacing A by the value given in (4), one finds that S_0, equal to $1/(4n_1)$, is just the bulk susceptibility for a single plane. The second term $M_a = nS_aH$ then represents an additional induced moment related to the division into particles. S_a is the extra susceptibility.

When ρ is small, we obtain

$$S_0 = \frac{1}{4n_1}$$

and

$$S_a = \frac{1}{4n_1}\left[1 + 2\rho + 2\rho^2\left(3 - \frac{4}{n}\right) + \cdots\right].$$

For $\rho = 0$, the result is identical to that of paper A, which in the present notation gives $S_a = 1/4n_1[(1+\rho)/(1-3\rho)]$. When ρ is different from zero, the result is slightly different.

The end moment — Notice that, always neglecting the end effects (the terms in B), we obtain

$$(n-1)\,D = \epsilon_1 + \epsilon_2 + \epsilon_3 + \ldots + \epsilon_{n-1},$$

where the right hand side is simply equal to the angle through which the antiferromagnetic direction Δ turns in the thickness of the particle. Now let φ be the angle between Δ and H. In the centre of the particle, $\varphi = \pi/2$ by symmetry, and at the two ends $\varphi = \varphi_0$ and $\varphi = \pi - \varphi_0$. To within $1/n$, $\pi/2 - \varphi_0 = nD/2$.

The extra moment M_a can then be written $M_a = M \sin(\pi/2 - \varphi_0)$ and it can be considered as the projection on the direction of the applied field of two moments $+1/2M$ and $-1/2M$ associated with the two ends of the particle and parallel to the local antiferromagnetic direction. We give these two moments the name *end moments*. A direct argument allows us to account for the existence and magnitude of such a moment, which must obviously change in sign without changing in magnitude when an additional active lattice plane is added to the particle. The end moment is therefore equal to half the total change in moment of the particle produced by adding an extra lattice plane. As this change is equal to M, the end moment is equal to $1/2M$.

The case of high fields — The preceding method no longer applies when $\pi/2 - \varphi_0$ is not much less than 1. We now use the idea of end moment. The twist in the antiferromagnetic direction, measured by the value of the derivative $\varphi' = d\varphi/dx$ when φ is a continuous function of x, corresponds to a couple Γ_1, which is balanced at the two ends of the particle by the couples $\pm 1/2\,MH$ due to the effect of the field on the end moments. Since in half the thickness of the particle $L = 1/2(n-1)e$ the direction Δ turns in the negative sense through an angle $\pi/2 - \varphi_0$, one can write the sequence of equalities establishing a relation between φ' and Γ_1

$$-\varphi' = -\frac{d\varphi}{dx} = \frac{2}{en}\left(\frac{\pi}{2} - \varphi_0\right) = \frac{2\,M_a}{en\,M} = \frac{2\,S_a H}{e\,M} =$$

$$= \frac{4\,S_a}{e\,M^2}\,\Gamma_1\,. \quad (9)$$

This relation of proportionaltiy, established when $\pi/2 - \varphi_0$ is small, remains valid for values of order $\pi/2$ because of the smallness of the ε_p, $n/2$ times smaller than $\pi/2 - \varphi_0$, which is really the essential condition.

An element dx containing dx/e planes and subject to a field H experiences a couple $\Gamma_2 dx$, when φ is different from $\pi/2$, due to the susceptibility being equal to $S_0 dx/e$ in the direction perpendicular to Δ and equal to zero in the parallel direction.

$$\Gamma_2 = \frac{S_0 H^2 \sin \varphi \cos \varphi}{e} \qquad (10)$$

The element dx is in equilibrium when the sum of the couple $\Gamma_2 dx$ and the two couples $-\Gamma_1$ and $\Gamma_1 + (d\Gamma_1/dx)$ acting on the extremities is zero, this gives

$$\frac{d^2 \varphi}{dx^2} + \frac{4H^2 S_0 S_a}{e^2 M^2} \sin \varphi \cos \varphi = 0 ; \qquad (11)$$

at the two ends of the particle. The solution must satisfy the boundary conditions deduced from (9) which take the form

$$-\varphi'_0 \doteq \frac{2S_a H}{eM} \sin \varphi_0 \qquad (12)$$

The two equations (11) and (12) are identical to equations (7) and (8) of paper A, when $n_2 = 0$. They differ slightly when n_2 is non-zero. Nevertheless, equation (10) and the following ones in the same paper A remain valid if one replaces equations (9) which define θ and h by the following ones

$$\operatorname{tg}^2 \theta = \frac{S_a}{S_0} ; \quad h = \frac{n S_0 H}{M}, \qquad (13)$$

with values of S_a and S_0 given above. All the conclusions of paper A naturally remain valid.

The twist of the antiferromagnetic direction measured by φ' only involves the internal properties of the particle, and particularly the moment M_0 of the active internal lattice planes. The same is true of the couple Γ_1. Let us now denote by $1/2 M_e$ the end moment. We have so far supposed that M_e was equal to M_0, but if this is not the case and if, for example, M_e is q times smaller than M_0 a field q times greater is needed to balance the same twist since the induced moment, proportional to M_e, is q times smaller. In the general case, the extra susceptibility is proportional to M_e^2, that is

$$S'_a \doteq \frac{S_a M_e^2}{M_0^2}. \qquad (14)$$

Temperature dependence of the end moment — The assumption made above of the equality of all values of M for whatever value of p is only valid close to absolute zero. It is not valid in the neighbourhood of the Né'el temperature Θ_N since in that region the moment depends greatly on the molecular field and the surface planes, having only one nearest-neighbour plane, are subject to a weaker molecular field than the inner planes.

The value S_a of the extra susceptibility calculated above is therefore the one that applies at absolute zero. To determine the extra susceptibility S_a at a temperature T, it is therefore necessary to calculate M_e as a function of T. This is a complex problem because M_e depends on the values of n_1 and n_2 and on the interactions among the atoms of the same active lattice plane, determined by a coefficient n_0. We content ourselves by examining the case where only the coefficient n_1, is non-zero. The analytical determination of the moments of successive planes M_p starting from the surface plane $p = 1$ is not easy in the general case. It is preferable to have recourse to a numerical method. For this it is sufficient to have a good table giving the values of the moment of an isolated active lattice plane as a function of the magnetic field acting on it. A value of M_1 is then chosen *a priori*, from it the field h_1 having created M_1 is deduced which is due to the moment M_2 of the second plane which can thus be calculated. From it h_2 is deduced, which is due to the already known influence of M_1 and that of M_3 which can thus be calculated, and so on. If M_1 is equal to the required value, the values of M_P will tend to M_0, the spontaneous magnetic moment corresponding to the bulk state, which may be easily calculated otherwise. In practice, the convergence is very rapid so that it is only necessary to work out four or five terms to find out whether the chosen value of M_1 is too small or too large.

Once the values of M_P are known, the values of the end moment are deduced from them, applying the definition given above, that is the limit as p tends to infinity of the expression

$$\frac{1}{2} M_e = M_1 - M_2 + M_3 - \ldots - M_{2p} + \frac{1}{2} M_{2p+1}.$$

We thus have calculated the values of M_0/M_s, M_1/M_s, M_e/M_s, S'_a/S_a given in the following table as a function of T/Θ_N for

$$j = 1/2, n_0 = n_2 = 0,$$

We designate as M the absolute saturation moment of an active lattice plane.

T/Θ_N	0	0,20	0,25	0,31	0,35	0,39	0,47	0,52	0,60	0,67	0,76
M_0/M_s	1	1,00	1,00	1,00	0,99	0,99	0,97	0,95	0,91	0,86	0,76
M_1/M_s	1	0,99	0,96	0,93	0,89	0,85	0,77	0,70	0,61	0,52	0,39
M_e/M_s	1	0,97	0,93	0,85	0,79	0,72	0,58	0,51	0,40	0,33	0,23
S'_a/S_a	1	0,95	0,86	0,73	0,63	0,53	0,36	0,29	0,20	0,15	0,09

It is possible to complete the table when the temperature is close to Θ_N because one can then linearize the problem and treat it analytically, expanding as a series the magnetic moment as a function of field. Having put $f^2(\Theta_N - T)/\Theta_N$, it is found that the moment of the pth plane is equal to $(-1)^{p-1} M_0 \tanh pf$ and that the end moment is given approximately by $1/2 M_e = 1/4 M_0 f$. In this region of temperature, one then obtains $S'_a/S_a = f^2/4$.

From this ensemble of results obtained from the molecular field theory it is most important to remember that the temperature dependence of the extra susceptibility is considerable. It is already reduced by half for $T = 0.4\Theta_N$ and by 4/5 for $T = 0.6\Theta_N$.

REFERENCES

[1] L. Néel, *Comptes rendus*, 1961, **253**, 203.
[2] Herpin, Meriel and Villain, *Comptes rendus*, 1959, **249**, 1334; Herpin and Meriel, *ibid.*, 1960, **250**, 1450; A. Yoshimori, *J. Phys. Soc.* (Japan), 1959, **14**, 807.

Chapter IV
MAGNETIC INTERACTIONS
A18, A23, A33, A30

A18 (1934)
ON THE INTERPRETATION OF MAGNETIC PROPERTIES OF ALLOYS

Summary (1977): Use of a formula, already established in A7 (Chapter I), to determine, in a solid solution AB, the interaction energies of different atom pairs A–A, A–B, and B–B.

A formal interpretation of the magnetic properties of solid solutions of ferromagnetic metals can be based on the following hypotheses:

(1) The constituent atoms preserve their identity and their magnetic moment in solution: in principle, the contributions J_A and J_B of the constituents A and B to the total magnetization can be distinguished;

(2) the distribution of atoms is statistical: all the atoms in solution have the same local environment;

(3) the effect of the atoms adjacent to an atom A is formally equivalent to an effective magnetic field h_A, which adds to the external field, and which is the sum of two terms, one proportional to J_A and the other to J_B: $h_A = aJ_A + bJ_B$. Similarly, the effect of atoms adjacent to an atom B is equivalent to an effective field: $h_B = b'J_A + cJ_b$; h_A and h_B can be different, a, b and c are independent of temperature and concentration. A more complete study shows that $b' = b$.

Under these conditions, C_A and C_B being the atomic Curie constants, P and Q the atomic concentrations of the constituents A and B, the atomic susceptibility of the alloy is [1]

$$\frac{1}{\chi_A} = \frac{T^2 - T(PaC_A + QcC_B) + PQC_AC_B(ac - b^2)}{T(PC_A + QC_B) - PQC_AC_B(a + c - 2b)},$$

(1)

a and C_A, C and C_B can be determined from the Curie points and the Curie constants of the pure constituents. A single interaction constant b represents mutual actions between the two constituents. For an alloy, the straight line of the Curie–Weiss law will be replaced by a curve. At first sight, this seems in disagreement with the experimental results of ferromagnetic alloys which obey the Weiss law. However, study of the formula (1) shows that in the temperature interval of measurement, the differences between a straight line and the curve (1) are smaller than experimental errors. The straight lines observed are merely the tangents at the Curie point and *the corresponding Curie constants do not have a simple meaning.*

The first hypothesis, conservation of the magnetic moment, is not verified for a pure metal when the structure changes. I would therefore only apply formula (1) in a face-centred cubic system. It is also very probable that change in the structure parameter would change the interaction energies through a, b and c. We can consider them as constants only in a first approximation.

On the other hand, to replace the interactions by a molecular field means that the *effects of fluctuations* are neglected. For a ferromagnet, these effects are only important in a region within about a hundred degrees of the Curie point and are negligible. The theory should therefore apply particularly well to Ni–Co, for which the constituents have positive molecular fields. In effect, formula (1) fits, within experimental uncertainties, the results of the experiments of Bloch on Ni–Co alloys (thesis, Zurich 1912). However, for an atom with a negative molecular field, like Fe_γ, the effects of fluctuations become predominant, as they transform a variable paramagnetism into temperature independent paramagnetism: formula (1) only gives the general form of the variations of the magnetic properties of Ni–Fe_γ and of Co–Fe_γ.

For a pure ferromagnet, the interaction constant a is linked to the Curie constant and to the Curie point by the formula $\theta = aC$. For a "constant" paramagnet (Mn or Cr) the constant is given as a function of the susceptibility χ by the formula $a = -\lambda/\chi$; λ being a numerical constant between 0.5 and 1. Taking $\lambda = 0.75$, however, does give an order of magnitude. Finally, the study of Fe_γ–Ni, Fe_γ–Co, and Ni–Co alloys allows three other interaction constants between different metals to be evaluated.

It is interesting, as Slater and Stoner [2] did in a slightly different form, to compare the interaction constants with the ratio R of the distance between neighbouring atoms and the effective radius of the unfilled 3d orbital. The values of R are taken from Stoner. The results are for the γ phase (face centred cubic). The table shows that the interaction constant, a, varies very regularly as a function of R.

These results constitute an extension and a confirmation of the ideas of Slater and of Stoner. In the γ phase, the change from a negative molecular field to a positive molecular field takes place at a value of R near 1.65.

REFERENCES

[1] L. Néel, *Ann. de Physique*, **17**, 1932, p. 61.
[2] Slater, *Phys. Rev.*, **36**, 1930, p. 57; Stoner, *Proc. Leeds Phil. Soc.*, **2**, Part IX, 1933, p. 391.

	Cr-Cr	Mn-Mn	Fe_γ-Fe_γ	Fe_γ-Co	Fe_γ-Ni	Co-Co	Ni-Co	Ni-Ni
R	1,30	1,47	1,63	1,72	1,79	1,82	1,89	1,97
$a \times 10^{-5}$	−1,75	−0,76	−0,06	0,42	0,93	0,73	1,17	1,19

A23 (1936)

MAGNETIC PROPERTIES OF THE METALLIC PHASE AND INTERACTION ENERGY BETWEEN MAGNETIC ATOMS

Remarks (1977): The article given here, taken from a more extensive paper, concerns the variation of the interactions as a function of the interatomic distances. The interaction energy is first calculated by various procedures. It is then shown that a general relation exists between the magnitude of the interactions and the shortest distance between the magnetic orbitals (in general 3d). At first negative (antiferromagnetism) when this shortest distance is small, the interactions become positive (ferromagnetism) for about 1.05A and go through a maximum near 1.30A.

At the same time, in dealing with the calculation of the interactions starting from the susceptibility of temperature independent paramagnetic substances, it is shown that when a field H is applied parallel to the direction D of antiferromagnetism, the magnetic susceptibility shows, for a certain critical value H_0 of the field H, a discontinuity that corresponds to a sudden rotation of the direction of antiferromagnetism. This phenomenon, called sometimes 'flip-flop'[1] was only observed 16 years later, in Copper Chloride $CuCl_2$, by C. J. Gorter and N. J. Poulis.

(1) Different methods for calculating the interaction energies: I have used the following procedures:

(a) Use of the Curie constant and the Curie point for materials that obey the Weiss law of paramagnetism: Ni, Co, Fe, Pt, Pd.

(b) By an extension of the previous method to alloys, to obtain the mixed interaction energies- Ni–Co, Ni–Fe and Fe–Co.

(c) By the susceptibility of temperature independent paramagnetic metals such as: Cr, Mn, V, Ti, etc.

(d) By the study of the superimposed constant paramagnetism of the alloys of nickel to calculate mixed interaction energies: Ni–V, Ni–Ti.

(2) General basis of the calculation: As there is no quantitative theory of magnetism for a system of magnetic carriers, whatever the sign of the interaction, I will use a classical model that I have already used [1] and which gives coherent results. I will take the atoms as carriers, and I will consider each atom to have q unpaired electrons coupled rigidly in parallel, so that the resultant magnetic moment is equal to $q\mu_B$.[2] In the case of nickel, one is led to assume that all the atoms are not identical. I will consistently apply Langevin statistics [2].

I will take $-w\cos\alpha$ as the interaction energy between two unpaired electrons on two neighbouring atoms on which the magnetic moments are at a relative angle α. All unpaired electrons of a given atom are linked to each electron of a different atom in an identical way. Hence, between two neighbouring atoms, each with q electrons, the total interaction energy is $-q^2 w\cos\alpha$. w can be positive or negative; it is zero for two atoms that are not near neighbours.

I consider the preceding hypotheses to be capable, in all cases, of completing the calculation so as to explore the subject as thoroughly as possible. Clearly these calculations will have to be resumed by a more rigorous method which will permit a closer study of the phenomena.

(3) Formulae valid at high temperature: In this case, whatever the sign of w, the magerial obeys the Weiss law. For a structure comprising a total of N atoms of moment $q\mu_B$, each having $2p$ neighbours, the magnetization of the system can be written[3]

$$\mathcal{I} = \frac{1}{3} \frac{Nq^2\mu_B^2 H}{kT - \frac{2pq^2 w}{3}} \qquad (1)$$

The Curie constant and the Curie point are written

$$C = \frac{1}{3}\frac{Nq^2\mu_B^2}{k} \qquad \Theta = \frac{2pq^2 w}{3k} \qquad (2)$$

The formula (1) is the same as formula 94 of my thesis [1] except for the factor $2p$ introduced here to take

---
[1] Also known as spin–flip.
[2] $\mu_B = 1$ Bohr magneton.

account of the 2p neighbours. I have established this in a particular case but it is very easy to generalize.

From the formulae (2) one can derive

$$\frac{w}{N\mu_B^2} = \frac{1}{2p} \frac{\Theta}{C} \qquad (3)$$

a relation which allows w to be calculated from experimental data for θ and C.

To apply formula (3), it is necessary that all atoms are identical. In fact instead of distributing the Nq magnetic electrons equally, we can distribute them equally between $\bar{\omega}N$ atoms $(0<\bar{\omega}<1)$; so that $(1-\bar{\omega}N)$ atoms have no magnetic moment. The moment of the magnetic atoms is then $q\mu_B/\bar{\omega}$ and the interaction energy between these neighbouring atoms will be $q^2\omega/\bar{\omega}^2 \cos\alpha$. The number of neighbours will be reduced to $2p\bar{\omega}$. Formulae (2) will be replaced by the following

$$C = \frac{1}{3}\frac{N\bar{\omega}q^2\mu_B^2}{k\bar{\omega}^2} \quad \text{and} \quad \Theta = \frac{2p\bar{\omega}wq^2}{3\bar{\omega}^2 k} \qquad (4)$$

However formula (3) is still valid as C and θ are both divided by $\bar{\omega}$. The formula can thus be applied with confidence in particular to the case of nickel where there are strong reasons to believe that all atoms are not identical.

(4) Application of formula (3) to metals obeying the Weiss law: The parameters and results of the calculation are shown in Table I. The metals considered all have a face centred cubic crystal structure; the number of near neighbours is therefore consistently twelve. The Curie constants used in Table I refer to a gram-atom, N is therefore equal to Avogadro's number: 6.06×10^{23}.

I have quoted two values for Fe_γ; their agreement is poor. The value corresponding to ω therefore only indicates an order of magnitude.

(5) Extension of the theory to alloys: Consider a solid solution of two metals A and B, for which the moments, per atom, are μ_A and μ_B; with interaction energies ω_{AA} and ω_{BB}, corresponding respectively to the interaction between two A atoms and two B atoms. It is necessary to add an interaction energy ω_{AB} corresponding to the interaction between an atom A and an atom B. In the case of a statistical distribution of atoms, any atom is surrounded by $2pP$ A atoms and $2pQ$ B atoms, where P

---

[3] I arrive at the Langevin law as I consider two neighbouring carriers may occupy any relative position, a hypothesis necessary in order to find, with ω negative, a temperature independent paramagnetism near absolute zero.

TABLE I.

Element	C	Θ	Author	$\frac{\omega}{N\mu_B^2} \times 10^{-2}$
Ni	0,323	650°		1,67
Co	1,14	1400	Preuss [2]	1,02
Fe_γ	3,3	− 730	,,	− 0,18
Fe_γ	9,3	−5500	Peschard [3]	− 0,49
Pt	0,318	−1096	Kopp [4]	− 2,88
Pd	0,321	− 228	,,	− 0,59

and Q designate the proportions of atoms A and B $(P+Q=1)$.

If the atoms of the alloy were not coupled, the total magnetization would be the sum of partial magnetizations $\mathcal{J}_A$ and $\mathcal{J}_B$ of the two consistuents; with

$$\mathcal{J}_A = \frac{C_A H}{T} \quad \text{and} \quad \mathcal{J}_B = \frac{C_B H}{T}. \qquad (5)$$

To take account of mutual interactions, ignoring the fluctuations, a molecular field must be added to the external field. The molecular field that acts on the A atoms is of the form: $a\,\mathcal{J}_A + b\,\mathcal{J}_B$, and that acting on the B atoms is of the form $b\,\mathcal{J}_A + c\,\mathcal{J}_B$. One finds

$$a = \frac{2pw_{AA}}{N\mu_A^2} \quad b=b'=\frac{2pw_{AB}}{N\mu_A\mu_B} \quad c = \frac{2pw_{BB}}{N\mu_B^2} \qquad (6)$$

with

$$\mathcal{J}_A = \frac{C_A}{T}(H + a\mathcal{J}_A + b\mathcal{J}_B)$$

and

$$\mathcal{J}_A = \frac{C_B}{T}(H + b\mathcal{J}_A + c\mathcal{J}_B)$$

from which one can easily derive $\mathcal{J} = \mathcal{J}_A + \mathcal{J}_B =$

$$\frac{T(PC_A + QC_B) - PQC_AC_B(a+c-2b)}{T^2 - T(PaC_A + QcC_B) + PQC_AC_B(ac-b^2)} H. \qquad (7)$$

I have already shown (16) that this formula represents well the set of experimental results for Ni–Co, the only case where it can be validly applied in the complete range of concentrations. It also allows the general trends of the magnetic properties of Pt–Co to be interpreted [1, p.100].

In the representation $(1/\chi, T)$, formula (7) describes a curve; experimentally, the region studied extends to about a hundred degrees above the Curie point, and what is interpreted as a straight line is only the tangent to the curve 7 at the Curie point[4] for which the equation can be written in the form

$$\frac{1}{\chi} = \frac{T - \theta'}{C'}. \quad (8)$$

(6) Calculation of w_{AB} from experimental data: If the concentration of metal B is small, then

$$\theta' = P a C_A + Q \frac{b^2}{a} C_B$$

and

$$C' = P C_A + Q C_B \left(2\frac{b}{a} - \frac{b^2}{a^2}\right) \quad (9)$$

which show, as a function of concentration, linear variations of the apparent Curie point and Curie constant. Extending the straight lines obtained as far as $Q = 1$; with θ'' and C'' as the values of θ' and C' at $Q = 1$; then from (9) we have

$$\frac{C''}{\theta''} = \frac{2}{b} - \frac{1}{a} \quad \text{or} \quad b = \frac{2}{\frac{C''}{\theta''} + \frac{C_A}{\theta_A}} \quad (10)$$

noting that for the pure metal A, with Curie constant C_A and Curie point θ_A, we have: $a = \theta_A / C_A$. C'' and θ'' can be determined experimentally by extrapolating the initial tangents to the curve of the variation of the Curie constant and Curie point as a function of concentration. I have used this method to calculate the interaction energies of the cross interactions ω_{AB}: Ni–Co, Ni–Fe, Co–Fe, using the experimental data of Preuss [2], of Peschard [3] and of Bloch [5].

In the previous calculation, w_{AB} represents the total interaction energy between two moments μ_A and μ_B. To have comparable values with w of §4, it is necessary to describe w_{AB} in terms of the interaction energy w_{AB} of two electrons, carried, one by atom A, and the other by atom B. Let us set $\mu_A = q\mu$, $\mu_B = q'\mu$, where μ is the Bohr magneton. This gives immediately: $w_{AB} = qq' w_{AB}$. From this, using formula (6), and as the factor qq' cancels:

$$b = \frac{2p w'_{AB}}{N\mu^2}. \quad (11)$$

[4] With the reservation, naturally, that anomalies produced by fluctuations in the molecular field will be superimposed on the phenomena described above, but which fortunately only extend over a limited range of temperature (they disappear at 100° above the Curie point).

TABLE II.

Mixture	θ''	C''	$\frac{w'_{AB}}{N\mu^2} \times 10^{-2}$
Ni-Co (a)	1 800°	1,06	1,53
Ni-Co (b)	1 100	0,235	1,62
Co-Fe$_\gamma$	850	1,67	0,60
Ni-Fe$_\gamma$	2 300	2,23	1,13

Table II gives values of C'', θ'', ω'_{AB} corresponding to different mixtures. The crystal structure being face-centred cubic, there is always $2p = 12$.

In Ni–Co, the tangents to the experimental curves can be taken, either starting with nickel (a), or starting with cobalt (b); the two values for ω'_{AB} are consistent. For solid solutions of iron, I limited myself to taking tangents starting with either cobalt or nickel, since if iron is taken as the starting point, the alloy may be either in the α or γ phase, depending on the temperature, and in the latter case, particularly, the observed Curie constant is rather uncertain.

(7) The case of low temperature: Formula (1) is no longer valid[5] when T is smaller than the absolute value of θ. When w is positive, the material is ferromagnetic, but the temperature can always be raised sufficiently to observe the paramagnetic region; on the contrary when ω is negative, ω is generally large in absolute terms, and whatever the temperature, one is always in the region where formula (1) is not valid, from which comes the need for a special study of this region, which for materials with a negative molecular field is the equivalent of the ferromagnetic region for materials with a positive molecular field.

At absolute zero, each atom is aligned antiparallel to its neighbours in order to obtain a minimum potential energy for the system, as shown in Fig. 1. The moments are all aligned in the same direction D, but they point in different directions instead of all in the same direction as in a ferromagnet. A magnetic field h, perpendicular to the direction D, will disturb this structure and magnetize it. All atoms will turn through the same small angle α (Fig. 2). With $2p$ neighbours, N the number of atoms, the total energy; sum of the deformation and magnetic energies, is equal to:

$$W = +p N w (1 - 2\alpha^2) - \mu N h \alpha \quad (12)$$

[5] In fact, formula (1) is never rigorously valid. With the proposed model, when w is negative, calculation shows that x varies with temperature in a complicated way. At high temperature, the curve is asymptotic to the straight line represented by formula (1), while at absolute zero one obtains a horizontal tangent (x constant). Generally, one can say that the temperature, $T = -\theta$, separates the curve into two parts, based on the distance that separates it from one or the other of these straight lines.

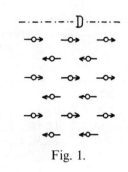

Fig. 1.

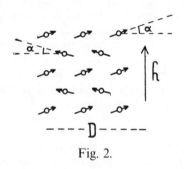

Fig. 2.

This expression is a minimum for

$$\alpha = \frac{\mu h}{-4pw}$$

From which the total magnetization

$$\mathcal{J} = \frac{N\mu^2 h}{-4pw}. \tag{13}$$

This is the expression for temperature independent paramagnetism.

(8) The role of the crystal field: Following the previous considerations, if the direction D is not initially perpendicular to the external magnetic field h, the carriers will turn to give a distribution for which the potential energy is a minimum. Because of this rotation of carriers, it is unlikely that the direction D will be able to change without any variation in energy. In fact, as in ferromagnets, we can attribute the moment to the electron spin. These electrons are in fixed orbitals, for which the orientations are determined relative to the crystal structure. A weak spin–orbit coupling is sufficient to orient the moments in a preferred direction relative to the crystal axes. This is how, following Powell [6], the properties of single crystals of nickel and cobalt can be interpreted.

To introduce these effects in as simple a way as possible, I will consider the material as a collection of domains, each having a preferred direction D; I will write the coupling energy between the magnetic moment and the crystal structure in the form $w'' \cos 2\theta$, designating by θ the angle between the moment and the preferred direction. w'' will be much smaller than ω, as is the case for ferromagnetic materials.

Let β be the angle of the external magnetic field with the preferred direction D. In equilibrium, half of the moments (Fig. 3) will make an angle $\psi + \alpha$ with H and the other half the angle $-(\psi - \alpha)$.

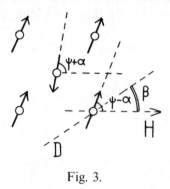

Fig. 3.

As the magnetic energy is always very small compared with the interaction energy, the angle α will be very small, and the coupling energy per atom between the moment and the crystal will be $w'' \cos 2(\psi - \beta)$, and as the component of the magnetic field perpendicular to the mean moment is $H \sin \psi$, the deformation angle is $\alpha = \dfrac{\mu H \sin \psi}{-4p\omega}$ and the magnetic energy $-\alpha\mu H \sin \psi$. The total energy is

$$W = -w'' \cos 2(\psi - \beta) - \frac{\mu^2 H^2 \sin^2 \psi}{-4pw} \tag{14}$$

Let:

$$\lambda = \frac{\mu^2 H^2}{8pww''} \qquad \text{tg } 2\varphi = \frac{\sin 2\beta}{\cos 2\beta - \lambda} \tag{15}$$

Then $0 < 2\varphi < \pi/2$ when the second term in (15) is positive and $\pi/2 < 2\varphi < \pi$ when the second term is negative. Under these conditions, the energy is a minimum for $\psi = \varphi$ and the relative magnetisation for N atoms can be written as

$$\mathcal{J} = \frac{N\mu^2 H}{-4pw} \sin^2 \varphi. \tag{16}$$

(9) Discussion of formula (16): This formula shows that the susceptibility is a function of the magnetic field

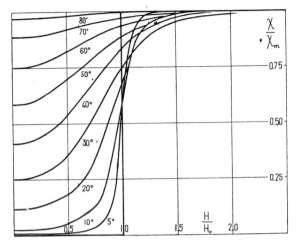

Fig. 4.
Susceptibility as a function of field for different orientations of the preferred direction.

H and the angle β. Let H_0 be a critical field defined by the equation

$$H_0^2 = \frac{-8pww''}{\mu^2} \qquad (17)$$

and let

$$\chi_m = \frac{N\mu^2}{-4pw}. \qquad (18)$$

Figure 4 shows the variations of χ/χ_m as a function of H/H_0, for different values of β spaced at 10° intervals.

When the preferred direction is perpendicular to the external magnetic field the susceptibility is independent of the field and equal to χ_m. However, when the direction of alignment, D, is parallel to the external field ($\beta = 0$), the susceptibility is zero for fields smaller than the critical field H_0, and equal to χ_m for larger fields, with a discontinuity at $H = H_0$. For intermediate directions the variation is more complicated.

It is interesting to observe that the susceptibility remains practically independent of the field, both in the case where the external field is larger than twice the critical field, and in the case where it is smaller than one fifth of the critical field, whatever the preferred orientation. For polycrystalline material two extreme cases can be considered:

(a) *A very small external field* In the different elementary domains the preferred directions are distributed at random.

For each direction, the susceptibility is given by formula (16) in which $\varphi = \beta$ as $\lambda = 0$. For the average susceptibility, the average value of $\sin^2 \beta$, (2/3) must be taken, then the susceptibility of the system can be written:

$$\chi_m = \frac{N\mu^2}{-6pw}. \qquad (19)$$

(b) *A very large external field* In this case the initial orientation of D is not relevant, and the average susceptibility is equal to the maximum susceptibility.

The preceding arguments are valid at absolute zero. For temperatures small compared with θ, the preceding conclusions are unchanged. However, at high temperatures, in the region where Weiss' law is valid, the effect of the crystal field is negligible, exactly as for ferromagnets.

(10) Calculation of the interaction energy of metals with constant paramagnetism: It is to the mechanism described in sections (7) *et seq* that I attribute the temperature independent paramagnetism of most metals. Essentially these show an atomic moment and a negative molecular field. Depending on the moment and on the interaction energy, the susceptibility at low temperatures is given by formulae (18) or (19) depending on the relative values of the external magnetic field and the critical field H_0. It is probable that in most cases the critical field is greater than the fields used in susceptibility measurements; hence the following formula should be used

$$\chi = \frac{N\mu^2}{-6pw}.$$

If the atomic moment is q Bohr magnetons, the interaction energy is equal to $q^2 w$ and

$$\chi = \frac{Nq^2\mu^2}{-6pq^2w} = \frac{N\mu^2}{-6pw}. \qquad (20)$$

The susceptibility at low temperatures is independent of the number of parallel coupled electrons on each atom. It gives merely the value ω of the interaction energy between two electrons on neighbouring atoms.

It is important to determine the number, $2p$, of effective neighbours which appears in formula (20).

(11) The number of effective neighbours: Two simple cases can be treated.

(a) *Body centred cubic lattice.* This lattice can be represented as two interlaced simple cubic lattices: each atom of one of the lattices has 8 neighbours which all belong to the other lattice. It is reasonable to

TABLE III.

Configuration	Metal	$-\chi_A \times 10^6$	Metal	$-\chi_A \times 10^6$	Metal	$-\chi_A \times 10^6$
d^8s^2	Ni	43	Pd	72	Pt	93
d^7s^2	Co	46	Rh	77	Ir	99
d^6s^2	Fe	50	Ru	83	Os	106
d^5s^2	Mn	53				
d^4s^2	Cr	57				
d^3s^2	V	61				
d^2s^2	Ti	65				

suppose that at absolute zero all the atoms of one of the lattices are oriented parallel to the preferred direction, and all the atoms of the other lattice are oriented in the opposite direction. This is also the situation of minimum potential energy. In these conditions the number of neighbours is 8 and

$$\chi = \frac{N\mu^2}{-24w}. \quad (21)$$

(b) *Face centred cubic lattices.* The problem is more difficult. One arrangement that appears to give the minimum potential energy is the following: we can divide the atoms into successive planes, perpendicular to one of the diagonal axes; we can then orient the moments contained in one of these planes parallel to the preferred direction, and the moments contained in the adjacent plane antiparallel and so on. In this way each atom will have four neighbours oriented in the same direction as itself in the same plane, and in the neighbouring planes, 8 neighbours oriented in the opposite direction. In the small deformation produced by the action of the magnetic field, the potential energy between neighbours oriented in the same direction does not vary as the moments remain parallel; it is as if these neighbours did not exist and it was only necessary to consider the 8 antiparallel neighbours, as in the body centred cubic structure. Formula (21) will also serve to calculate the energy.

Since metals that do not crystallise in one of the two preceding systems, but crystallise in a lattice that nevertheless differs little, I will use formula (21) in all cases.

(12) Diamagnetic corrections: The constant paramagnetism of metals is generally rather weak compared in absolute magnitude to the atomic diamagnetism of the ions, from which arises the necessity for a diamagnetic correction intrinsic to the atoms.

The semi-empirical formulae of Slater [7] allow the atomic diamagnetism to be calculated for a given electronic configuration. Table III gives some values for transition elements.

These data refer to free neutral atoms. In the metallic state the *s*-electron contribution in particular is substantially changed. The only paramagnetic metal for which the diamagnetism is known experimentally with precision is palladium, the experiments of Kopp giving $\chi_A = -58 \times 10^{-6}$. The value given for platinum is much less certain. The calculated value for palladium being $= -72 \times 10^{-6}$, it is necessary to multiply it by 0.81 to obtain the experimental value. I have estimated the order of magnitude of the atomic diamagnetism by multiplying the values in Table III by the same factor as for palladium; values calculated in this way are given in Table IV.

TABLE IV.

Metal	Mn	Cr	V	Ti	Rh	Ru	Ir	Os
$-\chi_A \times 10^6$	43	46	49	52	62	67	80	85

(13) Numerical values of w for some metals with constant paramagnetism: The second column of Table V gives the atomic susceptibility of some metals. Values corrected for diamagnetism using figures from Table IV are given in the fourth column. Finally, in the last column I have calculated $w/N\mu^2$ using formula (21).

For iridium and osmium the diamagnetic correction is so large that the corresponding values of ω are very uncertain. For magnesium I have chosen the value for the susceptibility of Mn_γ as the corresponding crystal structure is tetragonal, with a simple and well known arrangement of neighbouring atoms, which is not the case for the α phase for which four types of atoms with different environments must be distinguished.

(14) Superposed paramagnetism of nickel alloys: I have attributed this superposed paramagnetism to the

TABLE V.

Metal	$\chi_A \times 10^6$ obs.	Author	$\chi_A \times 10^6$ corr.	$-\dfrac{w}{N\mu^2} \times 10^{-2}$
Cr	187	I.C.T. [8]	233	1,79
Mn	484	Shimizu [9]	527	0,79
V	71	I.C.T.	120	3,47
Ti	60	"	112	3,72
Rh	111,1	Guthrie	173,1	2,40
Ru	43,4	et	110,4	3,77
Ir	25,7	Bourland	105,7	3,93
Os	9,9	[10]	94,9	4,40

TABLE VI.

Interaction	m	$-\dfrac{w}{N\mu^2} \times 10^{-2}$
Ni-Al	177	3,76
Ni-Mo	404	1,67
Ni-W	401	1,66
Ni-V	390	1,61
Ni-Ti	172	3,87
Ni-Cr	410	1,62

formation of antiparallel links between the nickel moment and that of the foreign atoms. The corresponding interaction energy can be calculated if a is known; in fact, for a rather dilute solution with all foreign metal atoms, X having only nickel atoms as neighbours, the superposed constant paramagnetism of the solid solution is due to distinct groupings, separated from each other, and consisting of the following: an atom X at the centre, with q electrons oriented parallel to a given direction, surrounded by q electrons of the nickel atoms oriented antiparallel. The interaction energy of two electrons is then $-w\cos\alpha$; at low temperature the susceptibility of this formation is

$$\chi = \frac{2q^2\mu^2}{-3q^2 w}$$

(This is formula (20), with $2p = 1$ and $N = 2$.)

Let τ be the atomic concentration of the metal X, N Avogadro's number, then the atomic susceptibility is written

$$\chi_A = \frac{2\tau N\mu^2}{-3w}.$$

This formula is only applicable to dilute concentrations since the number of links per atom increases with the concentration because of the interactions between different groupings, which eventually lock all the atoms into a complete structure. The general formula is written

$$\chi_A = \frac{2\tau N\mu^2}{3w(1 + 11\tau)}. \quad (22)$$

I have calculated the average value m of $\dfrac{\chi_A(1+11\tau)}{\tau}$ for the different series of alloys studied, and the results are given in Table VI together with the corresponding values of $w/N\mu^2$.

(15) The interaction energy and size of the magnetic orbitals: I have assembled in Table VII the values of ω calculated in the preceding paragraphs.

It can be noted that for elements in the first transition series, the molecular field increases regularly from titanium to nickel; as it does also in the series Ru, Rh, Pt and Os, Ir, Pt.

Slater [7] has compared the ferromagnetic properties of iron, nickel and cobalt with the smallness of the diameter of their magnetic shells — the 3d band — compared with the distance between two neighbouring atoms in the crystal. Let δ be the diameter of this band, d the distance between two neighbours: Stoner [7] has pointed out that the interaction energies vary regularly with the ratio d/δ. I have mentioned this comment [11] concerning the interactions Cr–Cr, Mn–Mn, Ni–Co, Ni–Fe, Co–Fe. However, I have since noticed that the interaction energies could just as well be considered as a function of the single variable $d - \delta$, the smallest distance between the magnetic shells.

The formulae of Slater allow calculation of the diameters of the different electronic shells for a given atomic configuration. I have calculated the diameter of the d-shell, the magnetic shell, using the configurations given in Table III. These are the values of δ given in the fourth column of Table VII, in units of angstroms. In the third column I give the distances, d, in angstroms, between two neighbouring atoms in the crystal structure. The differences, $d - \delta$ are given in the last column. For mixed interactions I have taken the average of the diameters of the magnetic shells of the two atoms as the value δ.

Fig. 5 shows the values of the interaction energy as a function of $d - \delta$. The systematic trend of the results is as good as can be expected in view of the uncertainty attached to the calculation of the size of the magnetic shells.

This systematic trend should not, however, be taken as too significant as, in the difference $d - \delta$, while δ varies in the range 1 to 2, d only varies in the range from

TABLE VII.

Interaction	$\frac{w}{N\mu^2} \times 10^{-2}$	d	δ	$d - \delta$
Ti-Ti	− 3,72	2,92	2,46	0,46
V-V	− 3,47	2,63	2,09	0,54
Cr-Cr	− 1,79	2,49	1,82	0,67
Mn-Mn	− 0,79	2,58	1,61	0,97
Fe-Fe	− 0,33	2,57	1,44	1,13
Co-Co	+ 1,02	2,49	1,30	1,19
Ni-Ni	+ 1,67	2,49	1,19	1,30
Ru-Ru	− 3,77	2,69	2,19	0,50
Rh-Rh	− 2,40	2,69	1,98	0,71
Pd-Pd	− 0,59	2,74	1,82	0,92
Os-Os	− 4,40	2,71	2,54	0,17
Ir-Ir	− 3,93	2,76	2,32	0,44
Pt-Pt	− 2,88	2,76	2,12	0,64
Ni-Co	+ 1,58	2,49	1,25	1,24
Co-Fe	+ 0,60	2,49	1,37	1,12
Ni-Fe	+ 1,13	2,49	1,30	1,19
Ni-Ti	− 3,87	2,49	1,82	0,67
Ni-V	− 1,61	2,49	1,64	0,85

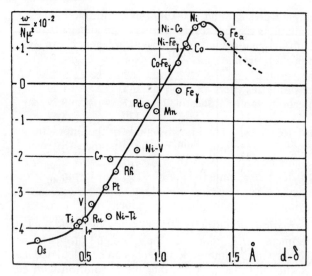

Fig. 5.
The molecular field as a function of the distance between magnetic shells.

1 to 1.1. In fact, if ω is presented as a function of d/δ the curve obtained has exactly the same shape as in the previous case. The systematic trend observed shows above all that the molecular field depends on the diameter of the magnetic shells.

However, the following arguments show that ω varies equally with distance:

(a) Normally in manganese, the distance between neighbouring atoms is 2.58Å; the molecular field is negative. If a dilute solution is made of manganese in copper, where $d = 2.55$Å, two accidentally neighbouring manganese atoms will have a negative molecular field; on the other hand, a solution of manganese in silver ($d = 2.88$Å) will give $d - \delta = 1.27$Å, corresponding to a positive interaction. In fact, only for dilute Mn–Cu solid solutions is the molecular field negative, while it is positive for solutions of Mn–Ag.

(b) Sadron has found [13] that in Mn–Ni, the manganese is oriented parallel to the nickel. In fact, the distance between the magnetic shell of a manganese atom and the magnetic shells of the neighbouring nickel atoms is $2.49 − 1.40 = 1.09$ Å, corresponding to a positive interaction; the opposite to solutions of manganese in cobalt: $d − \delta = 1.03$ Å, corresponding to a negative interaction; when Mn is oriented antiparallel to Co [13].

(c) In the α phase of iron (iron at room temperature), each iron atom has 8 neighbours at a distance of 2.46Å, and 6 neighbours at a distance of 2.86Å. The corresponding values of $d - \delta$ are 1.04Å and 1.42Å. The first value, from Fig. 12, corresponds to a near zero interaction. The positive molecular field of iron must come from the 6 more distant neighbours. The corresponding value of $w/N\mu^2$ is 1.38×10^2, which is not in disagreement with the curve in Fig. 5. This seems to show that, for distances between magnetic shells greater than 1.40Å, the interaction energy decreases with distance, a very natural result. It is very remarkable that M. Forrer [14], by very different considerations, has already shown that the molecular field of iron is produced by the neighbours at a distance of 2.86Å.

(d) For iron dissolved in gold ($d = 2.88$Å), two accidentally neighbouring iron atoms have $d - \delta = 1.44$Å, hence a positive interaction. Solid solutions of Fe–Au have in fact a positive Curie point [15].

(e) From Fig. 5 it can be deduced that thermal expansion increases the molecular field for all metals except iron α. When the molecular field is negative, its absolute value decreases with temperature, and the susceptibility should increase. It is perhaps in this way that the slight increase with temperature of the susceptibility of the metals such as V, Ti, Os, Ir, Ru, Rh should be explained.

Unfortunately, the exact magnitude of this effect cannot be calculated; the temperature increases not only the lattice constants, but also the amplitude of atomic oscillations about their equilibrium position, from which arises a change in the average interaction energy since it is not a linear function of distance. At temperatures around 1000 K, the amplitude of these oscillations reaches several tenths of an angström; the resulting change in the interaction energy can be important. In materials with positive molecular fields, the variations of the molecular field affect the Curie constant; this effect must be taken into account in the interpretation of values of the Curie constants.

(16) Remarks: The known values of the interaction energy are reliable when the distance between magnetic orbitals is in the range from 0.1Å to 1.5Å. Outside these limits the curve is difficult to pin down. When $d - \delta$ increases, ω should tend towards zero after having taken slightly negative values.

For values of $d - \delta$ smaller than 0.1Å, molybdenum can be used ($d - \delta = -0.04$Å), but the diameter of its magnetic orbital is so large (2.55Å) that the interactions between atoms that are not nearest neighbours are no longer negligible. The results obtained can only be illusory.

We should remember above all from these results that the domain of ferromagnetism is rather tightly restricted to distances between magnetic shells in the range from 1.1 to 1.5Å.

BIBLIOGRAPHY

[1] L. Neel, *Annl. de Phys.*, **17**, 1932, p. 5.
[2] A. Preuss, Thesis, Zurich, 1912.
[3] M. Peschard, Thesis, Strasbourg, 1925.
[4] W. Kopp, Thesis, Zurich, 1919.
[5] O. Bloch, Thesis, Aurich, 1912.
[6] F. C. Powell, *Proc. Roy. Soc.*, **130**, 1930, p. 167; *Proc. Camb. Phil. Soc.*, **27**, 1931, p. 561.
[7] J. C. Slater, *Phys. Rev.*, **36**, 1930, p. 57.
[8] International Critical Tables, New York.
[9] Y. Shimizu, *Sc. Rep. Tohoku Imp. Univ.*, **19**, 1930, p. 411.
[10] A. N. Guthrie and L. T. Bourland, *Phys. Rev.*, **37**, 1931, p. 303.
[11] L. Neel, *C.R. Acad. Sci.*, **198**, 1934, p. 1311.
[12] L. Neel, *J. Phys.*, **3**, 1932, p. 160.
[13] C. Sadron, *Ann. de Phys.*, **17**, 1932, p. 371.
[14] R. Forrer, *J. Phys.*, **4**, 1933, p. 109.
[15] J. W. Shih, *Phys. Rev.*, **38**, 1931, p. 2051.
[16] L. Neel, *Comm. Soc. Franc. Phys.*, Bull. no. 355, 1934.

A33 (1938)

INTERPRETATION OF THE PARAMAGNETIC CURIE POINT OF ELEMENTS IN THE RARE EARTH GROUP

Note (1977): It is shown here that the magnetic coupling energy for atoms in this group is proportional to the product of the number of spins of the two interacting atoms. It is implicitly assumed that the coupling is direct, but DeGennes has recently shown (*J. Phys*, **23**, 1962, p. 510) that the coupling is in fact indirect, through the intermediary of the 5s conduction band electrons.

The remarkable magnetic properties of the rare earth metals have above all been spotlighted by the discovery of the ferromagnetism of gadolinium by Urbain, Weiss and Trombe [1]. Very recently Klemm and Bommer [2] have studied the whole group. In particular they have shown that, from gadolinium to thulium, and in the metallic state, these elements have the same magnetic moment as the equivalent tri-valent ion: their fundamental state is hence perfectly determined. The same authors have measured the paramagnetic Curie points given in Table 1. Their results, relative to gadolinium are in agreement with those of Trombe.

I propose to interpret the values of these Curie points. In order to do this, I will assume that the only coupling force between different atoms corresponds to the interaction between the vectors $\vec{S}$ as $\vec{L}$ and $\vec{S}$ are rigidly coupled, with the resultant $\vec{J}$, it follows that the $\vec{J}$ of two different atoms are also coupled, but through the intermediary of the $\vec{S}$ vectors only.

Let L, S, and J be the quantum numbers of an atom; for the elements under consideration $J = L + S$. The atomic moment at saturation is $\sigma = g^J \mu_B = (L + 2S)\mu_B$, where g and μ_B represent the Landé factor and the Bohr magneton respectively. Let ω be the coupling energy between two spins on two neighbouring atoms. At saturation, the coupling energy between two atoms (the magnetic energy), is obviously proportional to the square of the number of spins carried by each atom; which is $4S^2 w$ assuming that all interactions between spins are equivalent. The elements under consideration all crystallize in a close-packed hexagonal structure; each atom having therefore twelve near neighbours and a total magnetic energy per atom of $24S^2 w$, being careful not to count any interactions twice. This energy is also equal to $n\sigma^2/2$ where n is a coefficient of the molecular field. Equating these two expressions, I obtain $n = 48 \omega S^2 / g^2 J^2 \mu_B^2$. However, the Curie point, θ, is related to the Curie constant, $C = g^2 \mu_B^2 J(J+1)/3k$, by the formula $\theta = nC$. Finally the Curie point is:

$$\Theta = \frac{16w}{k} \frac{S^2(J+1)}{J}.$$

The energy w is unknown, however I have shown [3] that in the group of transition elements, it depends only on the shortest distance, $d - \delta$, between the magnetic shells, and I will assume that this result is applicable here. Fortunately, for the elements under consideration, $d - \delta$ is constant to within a few hundredths of an angstrom. Hence ω equally is constant. The distances d between neighbouring atoms, given in Table II, are taken from Klemm and Bommer. They are average distances, since the neighbours fall into two groups of 6 for which the distances are very similar: for example in gadolinium each atom has 6 neighbours at 3.622Å and 6 at 3.554Å. The diameters δ of the 4f magnetic shells have been calculated using the screening constants of Slater [4].

The Curie point of these elements is therefore simply proportional to $S^2(J+1)/J$. In Table I, I give the values of L, S and J currently accepted for the trivalent ions (Hund), and which allow their magnetic moments to be found with complete success. I have then calculated $S^2(J+1)/J$. The Curie point of gadolinium being taken as 302 K. I have calculated the Curie points given in the

TABLE I.

Metal	2S	L	2J	$\frac{4S^2(J+1)}{J}$	Θ observed	Θ calculated
Gd	7	0	7	63	302°K	302°K
Tb	6	3	12	42	205	202
Dy	5	5	15	28,3	150	136
Ho	4	6	16	18		86
Er	3	6	15	10,4	40	50
Tm	2	5	12	4,7	10	22

TABLE II.

Metal	Gd.	Tb.	Dy.	Ho.	Er.	Tm.
d	3,588	3,546	3,546	–	3,496	3,484
δ	0,86	0,83	0,80	0,77	0,74	0,72
$d-\delta$	2,73	2,72	2,74	–	2,76	2,76

last column of Table I. The agreement with the experimental values is satisfactory.

Briefly, this shows that the molecular field of rare earth metals comes only from interactions between spins. These interactions are undoubtedly a sort of exchange interaction as for the iron group of metals. There are no interactions between the orbital moments.

BIBLIOGRAPHY

[1] *Comptes rendus*, **200**, 1935, 2132; **201**, 1935, 642.
[2] *Z.f. anorg. allgm. Chemie*, **231**, 1935, 138.
[3] *Ann. de Physique*, **5**, 1936, 232.
[4] *Phys. Rev.* **36**, 1930, 57.

A30 (1937)
STUDY OF THE MOMENT AND MOLECULAR FIELD OF FERROMAGNETS

Summary: 1st and 2nd parts. The author develops a general theory of the volume anomalies of the ferromagnets: the anomaly is the sum of two terms; one, already known, is independent of temperature, the other is proportional to the temperature: it arises from the anomalous value of the expansion coefficients of ferromagnets. A theoretical formula is obtained which coincides with the formula previously found by Chevenard to accurately fit the experimental results.

The magnetic forces between atoms have been introduced in the form of an interaction energy between neighbouring atoms; a function of the distance between these atoms. Other work, based on the comparison of the magnetic properties of metals have allowed the author to define this dependence, and the present theory confirms these results: the experimental data on the volume anomalies provide the first derivative of the interaction energy versus distance, as well as a relationship between the second and third derivatives.

3rd part. Two completely independent sources allow a determination of the variation of the interaction energy with distance between atoms. The important problem of the variation of the molecular field with temperature becomes possible to treat, and has been solved. Two phenomena affect the interaction energy: (a) an expansion due to a progressive increase, by raising the temperature, of the distance between interacting atoms; (b) an oscillation effect; the thermal agitation separates and brings together the atoms, from which comes a change in the average energy. Finally, it is found that the molecular field is a linear function of the temperature, which can be approximately calculated from data previously obtained. Above the Curie point, the inverse of the susceptibility of a paramagnet with a molecular field is always a linear function of temperature, but the coefficient which plays the role of the Curie constant is not the same as the Curie constant of the corresponding pure paramagnet. The proposed theory allows this coefficient to be calculated. In this way it is found that the true Curie constants of nickel and cobalt are those corresponding respectively to 1 and 2 uncompensated spins per atom.

FIRST PART
THEORY OF THE EXPANSION ANOMALIES OF FERROMAGNETS

(1) Definition of length anomalies: The study of ferromagnetic materials leads to a distinction between the spontaneous magnetization[1] — which is the magnetization within an elementary domain — and the bulk magnetization, the geometrical resultant of the spontaneous magnetizations of the different elementary domains [1].

The largest part of the magnetic energy depends on the spontaneous magnetization: for the same spontaneous magnetization the true specific heat only varies slightly with the bulk magnetization, while the spontaneous magnetization shows up important anomalies.

Similarly the relative variations in length, as a function of the bulk magnetization and direction relative to the crystal axes are small, of the order of 10^{-6}. I will leave aside these phenomena which constitute the so called magnetostriction. On the contrary, the length anomalies due to the spontaneous magnetization are much larger, of the order of 10^{-3}.

Definition of the length anomaly. Above the Curie point, the spontaneous magnetization is zero, which allows the normal expansion of the material to be determined as a function of temperature, and the length l_A which the material would have at a temperature t for zero spontaneous magnetization to be calculated by extrapolation below the Curie point. Let l_B be the true length at the same temperature for a zero bulk magnetization. I will call the difference, the length anomaly

$$l = l_B - l_A$$

In a similar way, the volume and dilation anomalies can be defined.

(2) The origin of length anomalies: It is the interactions between the carriers of magnetic moment that produce the length anomalies. In a ferromagnet such as nickel, the magnetization is only due to the

---
[1] The spontaneous magnetization should not be confused with the remanent magnetization.

spin, and the molecular field comes from the interactions between spins. The corresponding energy is proportional to the square of the spontaneous magnetization, σ. To these interactions should be added the coupling forces between spin and orbital angular momentum as well as the forces due to the magnetic field of the spin: dipole and quadrupole terms, etc. These last forces play a major role in the phenomena of the bulk magnetization, hysteresis, magnetocrystalline properties and corresponding magnetostriction, but have only a minor influence on the spontaneous magnetization.

Hence, the length anomalies of a ferromagnetic material, placed in zero external field, with no net magnetization, and a polycrystalline isotropic structure, can be attributed almost exclusively to the interactions between spins, as already noticed by Powell [2].

The length anomalies depend on the variations of magnetic energy as a function of volume, or more evidently, the variations, as a function of the interatomic distance, of the coupling forces between spins. Many authors have already drawn attention to the theoretical significance of these anomalies. Bauer [3] has calculated the length anomaly at absolute zero on a purely thermodynamic basis, assuming that the molecular field coefficient was inversely proportional to the volume. With this hypothesis, Verschaffelt [4] calculates the expansion anomaly at the Curie point, without however making any comparison with experiment. Bauer [5] himself recognised that the nickel anomaly could be better interpreted with a molecular field inversely proportional to the square root of the volume. Finally, Fowler and Kapitza [6] and Powell [2] have made analogous calculations from the standpoint of the quantum theory of ferromagnetism. Generally the agreement between theory and experiment is only fragmentary. In fact, the experiments of Chevenard[2] on the expansion anomalies of ferromagnets and their alloys, which constitute a systematic study of the question, provide the following basic result: the relative length anomaly $\delta l/l$ can be expressed as a function of the temperature T and the spontaneous magnetization σ by the formula:

$$\frac{\delta l}{l} = \sigma^2(\lambda + \mu T + \nu T^2). \qquad (1)$$

The preceding theories give only the temperature independent term in λ.[3] It is therefore necessary to

[2] For a description of the work of Chevenard on this question, and a complete bibliography, the report of Weiss to the Solvay meeting in 1930 should be consulted: L'anomalie de volume des ferromagnetiques. Paris, Fauthier-Villars (1932).

[3] Chevenard has already shown by a fairly general argument the proportionality of the length variations with the square of the spontaneous magnetization.

complete and then elaborate them. This seems possible since the comparative study of the magnetic properties of a sufficiently large number of materials has allowed the curve representing the interaction energy as a function of the interatomic distance to be defined.

I will show that to the volume anomaly at absolute zero, it is necessary to add a thermal anomaly which is essentially separate and which comes from the difference between the expansion coefficients of a magnetic material and the same material without its spontaneous magnetization.

Firstly, I will calculate the elastic energy and the magnetic energy of a material as a function of an isotropic deformation and derive from this the expansion coefficient.

(3) Energy as a function of an elastic isotropic deformation: The potential energy is the sum of an elastic energy W_E and a magnetic energy W_M, these quantities being for a gramme-atom.

Elastic energy. Let V be the atomic volume of the material at absolute zero, and d the distance between the two near neighbour atoms. For an isotropic deformation the volume becomes $V + \delta V$ and the distance $d + x$: the atomic deformation energy can be put into the form

$$W_E = Ax^2 + Bx^3 + Cx^4 + \cdots \qquad (2)$$

I will restrict the expansion to the first terms in the series, assuming that x is small compared with d.

The term A can be calculated from the compressibility x_0 at absolute zero. It is known in fact that for a very small deformation, the stored energy in compression can be written: $1/2\chi_0(\delta V/V)^2 V$ and as in (2) the term in x^3 and following are negligible; in these conditions one has

$$\frac{1}{2\chi_0}\left(\frac{\delta V}{V}\right)^2 V = Ax^2 ; \qquad (3)$$

By combining (3) with the obvious relationship

$$\frac{x}{d} = \frac{1}{3}\frac{\delta V}{V} \qquad (4)$$

one can calculate A. Extrapolation of experimental values of compressibility to absolute zero is uncertain, since the thermal variation is large, and also the values in Table I cannot be considered to be accurate to better than 10%. In formula (2), x is expressed in angstroms and W in ergs.

In principle, B can be calculated from the pressure coefficient of the compressibility, but the experimental values of this coefficient are much too uncertain for any

TABLE I.

Metal	Ni	Co	Fe
$\chi_0 \times 10^{12}$	0,50	0,49	0,55
$A \times 10^{-13}$	0,96	0,99	0,94
$-B \times 10^{-13}$	2,4	2,4	1,94

use to be made of this data. It is preferable to calculate B from the expansion coefficient. Paragraph 5 is devoted to this question, and I will briefly discuss the accuracy obtained in paragraph 12.

The magnetic energy. It is known that, provided one neglects the effects of fluctuations, the interactions between carriers of moment can be replaced by a molecular field h_m, proportional to the spontaneous magnetisation σ

$$h_m = n\sigma ; \qquad (5)$$

n is the molecular field coefficient; the magnetic energy can then be written

$$W_M = -\frac{1}{2} n\sigma^2 . \qquad (6)$$

To calculate n as a function of the interaction energy, the magnetic energy corresponding to absolute saturation, i.e. all carriers oriented parallel, can be evaluated in two different ways.

Let two atoms, each carrying q parallel coupled spins, be separated by a distance l_i at absolute zero; when their magnetic axes make an angle α, the magnetic energy can be written as

$$-q^2 w_i \cos \alpha$$

where w_i is a function of l_i.[4]

Let $2p_i$ be the number of neighbours of each atom at a distance l_i; when the spins of all the atoms are oriented parallel, the magnetisation can be written as:

$$\sigma = Nq\mu \qquad (\mu = \text{Bohr magneton}), \qquad (7)$$

where N is Avogadro's number; the magnetic energy can then be written, being careful not to count each interaction more than once

$$W_M = -Nq^2 \sum_i p_i w_i ; \qquad (8)$$

[4] I have already used several times this method of decomposition of the magnetic energy. It leads to simple results. See [7], [9].

The summation is over all layers of neighbours surrounding a given atom. Comparing formulae (6), (7) and (8), one obtains the relation

$$n_0 = \frac{2}{N\mu^2} \sum_i p_i w_i \qquad (9)$$

in which I have written n_0 as a reminder that this is the molecular field coefficient at absolute zero.

To simplify, I put[5]

$$u_i = w_i \frac{1}{N\mu^2} . \qquad (10)$$

It is now a question of calculating n as a function of x. For an isotropic deformation in which d becomes $d + x$, the distance l_i becomes $l_i + xl_i/d$ and the corresponding interaction energy can be written as w'_i

$$w'_i = w_i + \frac{xl_i}{d}\frac{\partial w_i}{\partial l_i} + \frac{1}{2}\frac{x^2 l_i^2}{d^2}\frac{\partial^2 w_i}{\partial l_i^2} + \frac{1}{6}\frac{x^3 l_i^3}{d^3}\frac{\partial^3 w_i}{\partial l_i^3} , \qquad (11)$$

From (8) and (10) the molecular field coefficient can be written in the form

$$n = n_0 + n_1 x + n_2 x^2 + n_3 x^3 \qquad (12)$$

having put for brevity

$$n_0 = \sum_i 2p_i u \qquad n_1 = \sum_i \frac{2p_i l_i}{d}\frac{\partial u_i}{\partial l_i} \\ n_2 = \sum_i \frac{p_i l_i^2}{d^2}\frac{\partial^2 u_i}{\partial l_i^2} \qquad n_3 = \sum_i \frac{p_i l_i^3}{3d^3}\frac{\partial^3 u_i}{\partial l_i^3} \qquad (13)$$

As a function of x, the magnetic energy takes the form

$$W_M = -\frac{1}{2}\sigma^2(n_0 + n_1 x + n_2 x^2 + n_3 x^3) . \qquad (14)$$

The expression for the magnetic energy only has a clear meaning if all the atoms are stationary: for example, when the deformation x is produced by compression at absolute zero. On the contrary, when the temperature makes the atoms oscillate, the ordinary thermo-elastic energy and the part of the magnetic energy which varies with temperature cannot be separated[6].

[5] It is the quantities u_i for which I have studied the variations as a function of the distance between magnetic shells [10] (in the last column of the table, 10^{-2} should be read as 10^{-25}; also [7].
[6] This point is dealt with in detail in the 3rd part, para. 20.

In principle, to calculate the coefficients $n_0, n_i, \ldots$, it is necessary to extend the summations Σ_i of the expressions (13) to all the successive layers of atoms surrounding the central atom. In practice, it is sufficient to restrict this to the first two layers for a body-centred cubic lattice, and to the first layer alone for a face-centred cubic lattice.

Total energy. This is, apart from a constant, the sum of the elastic and magnetic energies

$$W = W_E + W_M = -\frac{1}{2} n_0 \sigma^2 - \frac{1}{2} n_1 \sigma^2 x \\ + \left(A - \frac{1}{2} n_2 \sigma^2\right) x^2 + \left(B - \frac{1}{2} n_3 \sigma^2\right) x^3 \quad (15)$$

Examination of this expression shows that the magnetization has a double effect:
(a) to change the equilibrium position of the atoms at absolute zero;
(b) to modify the expansion coefficient.

(4) The length anomaly at absolute zero: The new equilibrium position, δx, is defined by $(\partial W/\partial x) = 0$. Let

$$\delta x = \frac{n_1 \sigma^2}{2(2A - n_2 \sigma^2)}$$

to first order, neglecting $(B - \frac{1}{2} n_3 \sigma^2) \delta x$ relative to $A - \frac{1}{2} n_2 \sigma^2$. This can be further simplified when $n_2 \sigma^2$ and $n_3 \sigma^2$ are small compared with A and B respectively. The relative length anomaly, $\delta l/l$, linked to δx by the relationship $\delta l/l = \delta x/x$, can be written, at absolute zero

$$\left(\frac{\partial l}{l}\right)_0 = \frac{n_1 \sigma^2}{4Ad}. \quad (16)$$

(5) Basic theory of the expansion of a solid: Before going further, it is necessary to express the expansion of a solid as a function of the coefficients A and B and the elastic energy.

Expansion coefficient of a diatomic molecule. Consider a molecule composed of two atoms. We will take into account only the vibrations along the line of centres. At absolute zero the equilibrium distance is d, corresponding to a minimum in the potential energy. When the distance between the centres of the two atoms becomes $d + x$, the potential energy of the molecule can be written as

$$w = ax^2 + bx^3 + \cdots \quad (17)$$

At a temperature T, the average kinetic energy and the average potential energy are each equal to $kT/2$.[7] As equation (17) includes a term in x^3, the vibrations are anharmonic, and the average distance between two atoms becomes $d + \delta x$. Neglecting effects of quantum degeneracy, one finds

$$-\delta x = \frac{3}{4} \frac{b}{a^2} kT \quad (k = \text{Boltzmann constant}) \quad (18)$$

and the expansion coefficient, α, of the molecule, exactly analogous to the expansion coefficient, is written

$$-\alpha = \frac{3}{4} \frac{b}{a^2} \frac{k}{d}. \quad (19)$$

A linear chain of atoms. In the same way, a linear chain of atoms can be considered assuming that the interaction energy between two consecutive atoms is given by formula (17) and that the interaction energy between two nonconsecutive atoms is negligible. The expansion coefficient, α, of such a chain is

$$-\alpha = \frac{1}{2} \frac{b}{a^2} \frac{k}{d}. \quad (20)$$

This formula is taken from Damköhler [11]. It can be derived by a simple change of notation in formula 48 of his article.

The general case. For the general case, let me take atoms distributed in a body-centred cubic lattice, such that only the interaction energies between an atom and its $2p$ nearest neighbours at a distance d need to be considered. When this distance becomes $d + x$, the energy is given by formula (17). The expansion coefficient should take the form

$$-\alpha = D \frac{b}{a^2} \frac{k}{d} \quad (21)$$

D being a numerical coefficient which only depends on the geometric nature of the crystal lattice. This formula can also be justified on purely dimensional arguments, remembering that T can only enter in the form kT which has the dimensions of energy. It remains to relate a and b to A and B. Now each link in the chain, for which the energy is represented by formula (17), corresponds to one degree of freedom, while there are 3N degrees of freedom in a gram-atom. I can therefore write

$$A = 3Na \quad \text{and} \quad B = 3Nb \quad (22)$$

[7] The deviations from equipartition, as a result of anharmonic vibrations, are negligible.

The only difficulty now is the knowledge of D. By analogy to the case of chains of atoms, I have taken D=0.5. It is possible, in fact, to imagine a fictitious lattice composed of 3 categories of chains, each at right angles to the others. Anyway, to formulate a theory of expansion anomalies, it is sufficient to know that D is a pure number. Its numerical value is only important if it is intended to calculate numerical values for n_2 and n_3.

In the preceding paragraphs, the calculation has been restricted to the first term in the expansion series defined by

$$\frac{\delta l}{l} = \alpha T + \beta T^2 + \cdots \quad (23)$$

At the expense of greater complexity, β could have been linked to the coefficients A, B and C of formula (2).

(6) A complete expression for the length anomaly of a solid: Take a non-magnetic material for which the elastic energy is given by formula (2). At sufficiently high temperatures, the vanishing of the expansion coefficient at low temperatures, which is uniquely associated with the vanishing of the specific heat, can be ignored. In these conditions, the relative expansion from $0°$ at T K is given by formula (23). Taking D=0.5, and using (21) and (22) one can write

$$-\alpha = \frac{3}{2} \frac{B}{A^2} \frac{R}{d} \quad (24)$$

R, as usual, represents the ideal gas constant. Let us now consider a magnetic material; expression (2) for the energy must be replaced by the more complete form (15), and the new expansion coefficient, α', is given by[8]

$$\alpha' = \frac{3}{2} \frac{R}{d} \frac{B - \frac{1}{2} n_3 \sigma^2}{\left(A - \frac{1}{2} n_2 \sigma^2\right)^2}. \quad (25)$$

When $n_2\sigma^2$ and $n_3\sigma^2$ are small relative to A and B respectively, and taking into account (24), one can write

$$\alpha' = \alpha \left(1 + \frac{n_2 \sigma^2}{A} - \frac{n_3 \sigma^2}{2B}\right). \quad (26)$$

The result is that at a temperature T, the magnetic material experiences an expansion in length, of which

[8] In fact, one should first make a change of variables to remove the term in x in equation (15), and to use new coefficients A' and B' in the formula so obtained. However, the error from simplification, as above, is negligible in the case of iron and nickel.

the excess relative to that experienced by a corresponding non-magnetic material is given by the formula

$$\delta l_T = l\sigma^2 \left(\frac{n_2}{A} - \frac{n_3}{2B}\right) T. \quad (27)$$

The total length anomaly, which is the excess length of the magnetic material, is obtained by adding to the anomaly at absolute zero the thermal anomaly given by the preceding formula; from which

$$\frac{\delta l}{l} = \left(\frac{\delta l}{l}\right)_0 + \left(\frac{\delta l}{l}\right)_T =$$
$$= \sigma^2 \left[\frac{n_1}{4Ad} + \alpha T\left(\frac{n_2}{A} - \frac{n_3}{2B}\right)\right]. \quad (28)$$

The formula of Chevenard [12], given in para. 2 (relation (1)), differs from formula (28) simply in the presence of an additional term in T^2 which has been neglected here in order not to complicate the explanations.

Comparing expression (28) with formulae (13) which define the quantities n_0, n_1, n_2, n_3, it can be noticed that the anomaly at absolute zero allows the first derivative of the interaction energy as a function of distance to be derived, while the thermal part of the anomaly depends on the higher derivatives. There is therefore a great interest in the study of expansion anomalies, both for the interpretation of the phenomenon itself and for the information it can provide about the molecular field.

Formula (28) shows that there is no discontinuity in length at the Curie point; it is known, anyway, that the loss of ferromagnetism cannot be associated with a change of state. On the contrary, there is a discontinuity in the expansion coefficient; the anomaly in the expansion coefficient is in fact written

$$\Delta\alpha = \frac{d}{dT}\frac{\delta l}{l} = \frac{d\sigma^2}{dT}\left[\frac{n_1}{4Ad} + \alpha T\left(\frac{n_2}{A} - \frac{n_3}{2B}\right)\right] + \sigma^2 \alpha\left(\frac{n_2}{A} - \frac{n_3}{2B}\right). \quad (29)$$

At the Curie point, the discontinuity, $\Delta\alpha_\theta$, in the expansion coefficient is therefore

$$\Delta\alpha_\theta = \left(\frac{d\sigma^2}{dT}\right)_\theta \left[\frac{n_1}{4Ad} + \alpha\Theta\left(\frac{n_2}{A} - \frac{n_3}{2B}\right)\right]. \quad (30)$$

TABLE II.

Data for Nickel.

1 t°C.	2 σ^2 $\times 10^{-6}$	3 $-\dfrac{d\sigma^2}{dt}$ $\times 10^{-4}$	4 α exp. $\times 10^6$	5 α corr. $\times 10^6$	6 $\Delta\alpha$ exp. $\times 10^6$	7 δl $\times 10^4$	8 $\dfrac{\delta l}{\sigma^2}$ exp. $\times 10^{11}$	9 $\dfrac{\delta l}{\sigma^2}$ calc. $\times 10^{11}$	10 $\Delta\alpha$ calc. $\times 10^6$	11 α calc. $\times 10^6$
0°	10,45	0,90	12,48	13,18	−0,69	−0,74	−0,71	−0,70	−0,58	12,59
50	9,85	1,20	13,18	13,72	−0,46	−1,04	−1,06	−0,99	−0,46	13,18
100	9,15	1,60	13.76	14,17	−0,32	−1,26	−1,38	−1,29	−0,31	13,77
150	8,30	2,05	14,4	14,76	−0,04	−1,33	−1,60	−1,56	−0,14	14,32
200	7,10	2,65	15,0	15,29	+0,18	−1,31	−1,85	−1,84	+0,11	14,93
250	5,60	3,40	15,69	15,96	+0,54	−1,15	−2,06	−2,10	+0,43	15,58
300	3,60	4,55	16,53	16,76	+1,03	−0,77	−2,20	−2,34	+0,88	16,38
325	2,32	5,70	16,90(1)	17,12	+1,20			−2,48	+1,31	17,01
350	0,65	8,50	17,8 (1)	18,02	+1,98			−2,60	+2,17	17,99
Θ	0	9,30	18,05(1)	18,25	+2,15			−2,63	+2,45	18,33
400			16,29	16,49						
500			16,83	16,98						
600			17,44	17,58						

(1) Unlike the others, these 3 values are taken from Williams.

(7) Experimental data on the expansion anomaly of nickel: *The experiments of Chevenard.*

From Chevenard, I will check that the length anomalies of ferromagnets follow exactly a formula of the type:

$$\frac{\delta l}{l} = \sigma^2 (\lambda + \mu T + \nu T^2), \qquad (30a)$$

which will constitute a confirmation of the proposed theory. I think it is useful to review, in the case of iron and nickel, the verification of Chevenard as the corresponding publications are rather inaccessible and little known.

Chevenard studied, as a function of temperature, the expansion of iron, nickel and chromium and a very large number of their alloys[9]. From the excellent curves given in his article, I have calculated the values of the true expansion coefficient given in column 4 of Tables II and III.

Other data on nickel. The expansion of nickel has been the subject of studies by Colby [13], prior to those of Chevenard, but made on samples of uncertain

[9] Experimental research on the alloys of iron, nickel and chromium. Travaux et mémoires du Bureau International des poids et mesures, 17, Paris, Gauthier-Villars, 1927; see also: P. Weiss, report of the 1930 Solvay meeting.

purity, and which I therefore shall not reproduce. Amongst other work, we can cite those of Hidnert [14] on very pure samples — 99.94% nickel — and especially the experiment of Williams [15] on a single crystal of nickel with 99.90% purity. He made precise measurements in the neighbourhood of the Curie point and found on the expansion curve a discontinuity at 358°, in excellent agreement with the Curie point of pure nickel — 357.9° — determined by magnetic techniques in a small magnetic field [16].

Fig. 1 shows data from the different experimenters. It shows that below the Curie point, the measurements of Williams are in excellent agreement both with those of Chevenard and those of Hidnert.

At the Curie point, Williams finds a more striking anomaly than that found by Chevenard. This is partly explained by the fact that the sample used by Chevenard was less pure than the sample used by Williams, as the latter found that the height of the anomaly is reduced by 40% for 1% of impurities. Perhaps also the discontinuity in the expansion coefficient at the Curie point is different in a polycrystalline material and only shows its theoretical value in a single crystal. Anyway the larger value is more reliable.

Above the Curie point, in a temperature interval of about fifty degrees, it is seen that the expansion

TABLE III.

Data for Iron.

1	2	3	4	5	6	7	8	9	10	11
t°C	σ^2 $\times 10^{-8}$	$-\dfrac{d\sigma^2}{dt}$ $\times 10^{-5}$	α exp. $\times 10^6$	α corr. $\times 10^6$	$\Delta\alpha$ exp. $\times 10^6$	δl $\times 10^3$	$\dfrac{\delta l}{\sigma^2}$ exp. $\times 10^{11}$	$\dfrac{\delta l}{\sigma^2}$ calc. $\times 10^{11}$	$\Delta\alpha$ calc. $\times 10^6$	α calc. $\times 10^6$
0°	1,46	0,40	11,44	12,25	+ 2,64	− 2,08	− 1,42	− 1,415	+ 2,67	11,47
100	1,42	0,53	12,85	13,29	+ 2,82	− 1,81	− 1,27	− 1,264	+ 2,95	12,98
200	1,37	0,60	14,16	14,41	+ 3,16	− 1,50	− 1,09	− 1,09	+ 3,16	14,16
300	1,29	0,85	15,18	15,44	+ 3,41	− 1,17	− 0,905	− 0,905	+ 3,30	15,07
400	1,18	1,20	16,07	16,29	+ 3,46	− 0,83	− 0,705	− 0,702	+ 3,38	15,99
500	1,05	1,80	16,66	16,84	+ 3,25	− 0,405	− 0,470	− 0,48	+ 3,29	16,70
600	0,82	2,7	17,03	17,19	+ 2,82	− 0,188	− 0,230	− 0,24	+ 2,69	16,90
650	0,67	3,3	16,94	17,08	+ 2,31	− 0,065	− 0,097	− 0,11	+ 2,09	16,72
700	0,45	4,5	16,63	16,76	+ 1,61	+ 0,030	+ 0,067	+ 0,02	+ 1,11	16,13
750	0,18	6,6	14,46	14,57	− 0,98	+ 0,056	+ 0,310	+ 0,16	− 0,58	14,86
Θ	0	8,0	12,56	12,65	− 3,02	+ 0,020	„	+ 0,20	− 1,6	13,98
800			15,64	15,75						
850			16,22	16,33						
900			16,62	16,72						

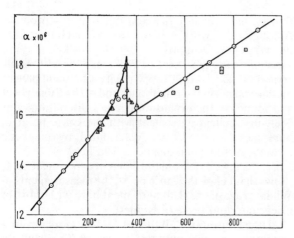

Fig. 1.
The expansion coefficient of nickel as a function of temperature.

coefficient does not immediately return to its normal behaviour, namely a steady increase with temperature as is usual in a non-magnetic material. This residual anomaly is connected with the phenomenon of the two Curie points — the paramagnetic Curie point slightly different from the ferromagnetic Curie point — and to the analogous phenomenon for the specific heat [17], observed by Ahrens [18]. The curves of thermoelectric power as a function of temperature also show a similar form [19]. At high temperature, the experiments of Hidnert agree well with those of Chevenard if the greater difficulty of the measurements is taken into account.

In general, this discussion shows that other work has confirmed the results of Chevenard obtained with his differential expansion meter. As only the results of Chevenard cover a sufficient range of temperature, and as he has studied many elements and alloys, I have preferred to use systematically the results of this author in the following analysis.

(8) Comparison of theory and experiment in the case of iron and nickel: Experimental determination of the length anomaly — Fig. 1 — shows that below the Curie point, the true expansion coefficient is a linear function of temperature, which facilitates extrapolation to low temperatures and gives in this way the expansion coefficient of the non-magnetic substance. These are the expansion formulae which are most appropriate for nickel and iron when the spontaneous magnetization is zero

$$\text{Nickel}: \alpha = 12{,}2 \cdot 10^{-6} + 6{,}2 \cdot 10^{-9}\,T$$
$$\text{Iron}: \alpha = 7{,}55 \cdot 10^{-6} + 7{,}8 \cdot 10^{-9}\,T \tag{31}$$

By subtracting the values given by the preceding formulae, from the actually observed values, the expansion anomaly can be obtained.

At room temperature, the expansion coefficient is already affected by quantum degeneracy. To make the corresponding correction, I assume that the relative change in the expansion coefficient is the same as that for the specific heat[10]. Let α be the true expansion coefficient, and α' the expansion if there was no degeneracy, then one has

$$\alpha = \alpha' \Phi \left(\frac{T}{\Theta} \right)$$

$\Phi(x)$ being the Debye function [20] and Θ the Debye temperature. I assume [21]: $\Theta = 375°$ for nickel and $\Theta = 420°$ for iron. The corrected values are given in column 5 of Tables II and III. In columns 2 and 3 I have given the values of σ^2 of the atomic spontaneous magnetization and its derivative $d\sigma^2/dT$ from Weiss and Forrer [22] for nickel, and Hegg [22] for iron.

In column 6 are given the expansion coefficient anomalies, $\Delta\alpha$, from which the corresponding relative length anomalies are calculated (column 7). Dividing these latter values by σ^2 gives an expression $\delta l/l\sigma^2$ (column 8) which should be a quadratic function of temperature. I found, in fact, that the expressions

$$\text{Nickel}: \frac{\delta l}{l\sigma^2} = 1{,}1 \cdot 10^{-11} - 7{,}1 \cdot 10^{-14}\,T + \\ + 1{,}88 \cdot 10^{-17}\,T^2 \quad (32)$$

$$\text{Iron}: \frac{\delta l}{l\sigma^2} = -1{,}74 \cdot 10^{-11} + 0{,}95 \cdot 10^{-14}\,T + \\ + 0{,}88 \cdot 10^{-17}\,T^2 \quad (33)$$

reproduce well the experimental values, as shown by the comparison of columns 8 and 9. Finally, in columns 10 and 11, the values of $\Delta\alpha$ and α calculated from (32) and (33) are found. I have added to the values of α so calculated the negative quantum degeneracy correction, to allow a direct comparison of the figures in column 11 with the gross experimental data.

I have drawn, as a continuous line, on Figs. 1 and 2, the calculated curves giving the expansion coefficient as a function of temperature. The agreement with experiment is remarkable for both nickel and iron, despite the complexity of the experimental curve especially in the case of iron.

[10] It is easily shown that this property holds for a diatomic oscillator, which has already been considered (para. 5). Grüneisen (Ann der Phys. 39 (1912) 287) has shown that this relationship is nearly exact for a solid.

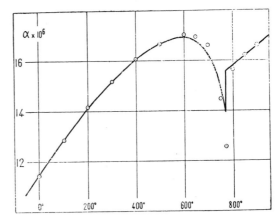

Fig. 2.
True expansion coefficient of iron as a function of temperature.

Briefly, from a formal standpoint, the proposed theory gives a true representation of the experimental results.

Formulae (32) and (33) allow the length anomalies at absolute zero to be calculated

$$\left. \begin{array}{l} \text{Nickel}: \dfrac{\delta l}{l} = +1{,}25 \times 10^{-4} \\[6pt] \text{Iron}: \dfrac{\delta l}{l} = -2{,}66 \times 10^{-3} \end{array} \right\} \quad (34)$$

This latter value agrees reasonably well with the value calculated by Chevenard, while there is no agreement for nickel as the quantum degeneracy correction which was not made by Chevenard is relatively large in this case.

SECOND PART
THE INTERACTION ENERGY CURVE DEFINED BY THE EXPANSION ANOMALIES

(9) The significance of the interaction energy curve as a function of the distance between magnetic shells: The comparison of the magnetic properties of different materials, ferromagnets, paramagnets with a molecular field, temperature independent paramagnets, has allowed, to show in a first approximation, that the interaction energy depends only on the distance, $d - 2r$, between the magnetic shells of the atoms. As a function of $d - 2r$, the energy is represented by the curve Γ in Fig. 3.

This curve has been obtained on the whole from different types of atoms placed at distances relative to each other fixed by the dimensions of the

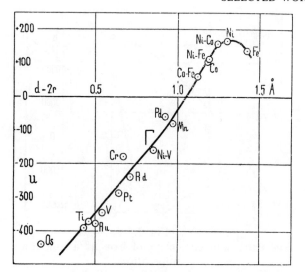

Fig. 3.
The interaction energy as a function of the distance between magnetic shells.

corresponding crystal lattice spacing. In $d-2r$, d represents the distance between the centres of two atoms, and r the radius of the magnetic shells. Hence, to obtain the curve Γ, I have varied r, taking different atoms while d varied very little, as the interatomic distances in the iron group of metals are all very similar.

Conversely, one can take an atom of a well-defined type; an atom of nickel for example. At absolute zero the distance between the magnetic shells of two neighbouring atoms in the crystal is 1.30Å. The corresponding point on the curve Γ is that marked Ni. If this distance is artificially changed, the variations in interaction energy will be represented by the curve Γ. In other words, I can trace a portion of the curve Γ by leaving r constant and varying d.

However, it is reasonable to suppose that for atoms of a given type, the interaction energy can vary with distance following a well defined law, even if it is not possible to reduce to a unique formulation the laws corresponding to atoms of different types. The theory of expansion anomalies developed in the first part of this article is selfconsistent: it allows, in principle, for a given material (for example nickel), the derivative of the interaction energy to be determined relative to the distance. The validity of this theory, directly checked by experiments, is hence completely independent of the greater or lesser exactitude of the ideas which will be developed now.

I propose to show that the values of the derivative[11]

of the experimental curve Γ agree with the values $\partial u/\partial l$, $\partial^2 u/\partial l^2$ ... etc, deduced from the expansion anomalies; a necessary consequence of the notion of the existence of a unique curve Γ for all the magnetic elements.

Hence, the question now is what is the real significance of this curve: if there is no agreement, then the points representing the interaction energy as a function of distance fall accidentally on a smooth curve. On the contrary, if the two methods of obtaining the derivatives give consistent results, this provides a strong confirmation for the existence of this curve, and the careful study of expansion anomalies will take on a greater interest since it will be possible to draw general conclusions about the molecular field.

I will first define the general relationships which connect the successive derivatives of the energy to the length anomaly at absolute zero and its thermal variation. I will then pass to a more detailed study of iron and nickel.

(10) Information provided by the length anomaly at absolute zero: Formulae (13) and (16) give

$$\sum_i \frac{2 p_i l_i}{d} \frac{\partial u_i}{\partial l_i} = \frac{4 A d}{\sigma^2} \left(\frac{\partial l}{l}\right)_0. \qquad (35)$$

When only the effect of the nearest group of neighbouring atoms is important, this formula can be simplified and written

$$\frac{\partial u}{\partial l} = \frac{2 A d}{\sigma^2 p} \left(\frac{\partial l}{l}\right)_0 \qquad (36)$$

The sign of the length anomaly at absolute zero is also the sign of the first derivative of the interaction energy as a function of distance.

The curve Γ (Fig. 3) has a maximum near $d-2r=1.31$Å. Hence, when the distance between magnetic shells is less than 1.31Å, the length anomaly is positive; and conversely, it is negative for distances greater than 1.31Å.

I have assembled, in Table IV, for several substances, the signs of the volume anomalies[12] and the distances between magnetic shells, $d-2r$. The preceding deductions are completely confirmed.

The case of ferro-nickels is particularly interesting since it is possible to follow, with increasing proportions of iron, the relative increase in the length anomaly due to the shift of the point representing the interaction energy towards the left on the curve Γ, from the point for Ni to the point for Fe–Ni.

---

[11] In the theory of expansion anomalies, the derivatives are taken relative to d; they are identical to the derivatives relative to $d-2r$ since r is constant.

[12] The corresponding expansion measurement studies have been done by Chevenard (loc. cit.) except for AsMn which was studied by Smits, Gerding and Vermast. Z.f. physik. Chemie; Bodenstein, Fetschrift (1931), 357.

TABLE IV.

Composition	Distance between magnetic atoms	Distance between magnetic shells	Sign of the length anomaly
Fe^3C	2,60 Å	1,16 Å	+
Ni-Fe	2,49	1,19	+
Mn-As	2,89	1,28	+
Ni	2,49	1,30	+
Fe	2,86	1,42	−
Fe^2O^3	2,88	1,44	−
FeS	2,89	1,45	−
Fe^3O^4	2,97	1,53	−

The average radius of the magnetic shell is an indispensable parameter. These radii have been calculated using the semi-empirical formula of Slater [24]. In absolute terms, the precision obtained in this way is no doubt poor; however one can hope that the errors included are of the same order and in the same sense when the average radii of the magnetic shells of atoms of similar structure are calculated, as is the case for the metals in the iron group, such that the slope of the curve Γ is not greatly affected. Hence, it is not surprising that the sutdy of these metals gives a regular curve for the interactions. These phenomena can be expected to be more complicated for compounds with polar characteristics. In fact, in a metal the magnetic shell has a close to spherical symmetry, while in a compound like magnetite, Fe_3O_4, because of the electrostatic interactions, the diameter of the magnetic shell of the Fe^{+++} ion must be greater in the direction of neighbouring Fe^{+++} ions, than it is in the direction of the O^{--} ions, however, it is precisely the diameter in the direction of the Fe^{+++} ions that is important, not the average radius of the shell.

This is why I have considered it unnecessary to discuss quantitatively the expansion properties of such compounds; I have merely given in Table IV some qualitative information.

(11) Information provided by the study of the temperature dependence of the length anomaly: Consider formulae (28) which give the full expression for the length anomaly. When the temperature rises from absolute zero, one observes either an increase or a decrease in the numerical value of the anomaly. As $d\sigma^2/dT$ is zero at absolute zero (if the approach to saturation follows a $T^{3/2}$ law), the coefficient of the term in T of the anomaly is

$$\sigma^2 \alpha \left(\frac{n_2}{A} - \frac{n_3}{2B} \right).$$

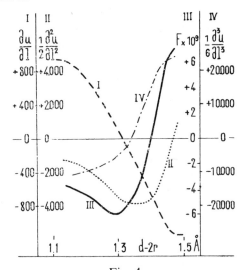

Fig. 4.
Derivatives of the interaction energy.

The sign of the temperature variation of the anomaly is then the sign of $n_2/A - n_3/2B$.

I have included in Fig. 4, as a function of $d-2r$, the approximate values of n_1, n_2 and n_3 as I have been able to calculate them in the following sections. The ordinate scales are different for each curve. In addition, with the hypothesis: $p=6$ and $-B=2A=10^{+13}$, I have drawn the curve $F = n_2/A - n_3/2B$ as a continuous line. It can be seen that F is initially negative for small values of $d-2r$, passes through a minimum near $d-2r=1.30$Å, then rises rapidly, ending at $d-2r=1.40$Å.

The results are already very complicated, and the curve F is not known with great accuracy. Nevertheless, the few substances for which we know the sign of the temperature variation of the anomaly give results in agreement with the indications given below.

The results would be even more complicated if one wished to describe, even schematically as a function of $d-2r$, the expansion anomaly at the Curie point. It can

TABLE V.

Compound	$d - 2r$	Sign of the thermal variation of the anomaly
Fe^3C	1,16 Å	−
Fe-Ni	1,19	−
Ni	1,30	−
Fe	1,42	+
Fe^3O^4	1,53	+

be noted in particular that the sign of the anomaly at absolute zero is completely independent of the sign of the discontinuity in the expansion coefficient at the Curie point.

(12) The accuracy of the method: The few qualitative results just obtained underline the interest in a detailed study of these phenomena. Unfortunately, such a study is not easy.

To deduce the length anomalies, the interaction energy and its derivatives as a function of distance, the quantities A and B must be known; the coefficients of the equation describing the energy as a function of the deformation (formula (2)). A is directly connected to the compressibility by relations (3) and (4); this is poorly known, and in the case of a polycrystal the significance of what is measured is not clear: at the high pressures used by Bridgman [25] — 10,000 atmospheres and above — compression is without doubt accompanied by reorganisation of the polycrystalline structure. In fact, Grüneisen [26] who worked by an indirect method — standing waves — with low pressures, gives compressibilities which differ by more than 10% from those of Bridgman. A is probably not known to better than 10%.

For B the situation is even worse. From (24), and with D' as a numerical coefficient, one can write

$$B = D'A^2 \alpha d$$

Already the presence of the factor A^2 introduces a relative error, double that on A. α is the expansion coefficient at absolute zero if there is no quantum degeneracy: the difficulties in accurate determination are obvious. D' is a numerical coefficient that depends on the geometry of the lattice. The value I have taken is not very reliable. Finally, the presence of d, the distance between two neighbouring atoms, shows that the formula is only valid in the case when only the energy between nearest neighbours is relevant. If, as for iron, it is necessary to include the next group, new difficulties arise from this fact.

In brief, great care is necessary in the interpretation of the results.

(13) Special study of nickel: Comparison of equations (28) and (32) allows the following to be written

$$\frac{n_1}{4Ad} = 1{,}1 \times 10^{-11} \qquad (38)$$

$$\alpha\left(\frac{n_2}{A} - \frac{n_3}{2B}\right) = -7{,}1 \times^{-14} . \qquad (39)$$

I will adopt the following numerical values

$$A = 0{,}96 \times 10^{+13} \quad ; \quad B = -2{,}4 \times 10^{+13} \; ;$$

$$d = 2{,}49 \text{ Å} \quad ; \quad \alpha = 12{,}2 \cdot 10^{-6} \quad ; \quad p = 6.$$

The first derivative of u. Using the value of n, given by (13), equation (38) gives:

$$\frac{\partial u}{\partial l} = 88 . \qquad (40)$$

The corresponding tangent agrees well with what is known of other parts of the interaction curve.

Higher derivatives of u. A relationship between the two unknowns, $\partial^2 u/\partial l^2$ and $\partial^3 u/\partial l^3$ is provided by (39).

An infinite number of solutions are possible, but there is no proof *a priori* that any of them will agree with the interaction curve Γ as we know it. I will try three solutions, (a), (b) and (c), corresponding to values of n_3 given below:

(a) $\qquad n_3 = 0$

(b) $\qquad n_3 = 60{,}000$

(c) $\qquad n_3 = 120{,}000$

The three solutions are characterized by the numerical values given in Table VI.

TABLE VI.

Solution	u	$\dfrac{\partial u}{\partial l}$	$\dfrac{1}{2}\dfrac{\partial^2 u}{\partial l^2}$	$\dfrac{1}{6}\dfrac{\partial^3 u}{\partial l^3}$
a	167	88	− 4 650	0
b	167	88	− 3 700	− 5 000
c	167	88	− 2 650	− 10 000

The three curves shown in Fig. 5 correspond to these three solutions. The right hand branches of these curves are almost identical and are compatible with the only known point on this side of the interaction curve; that for Fe. This point is however rather poorly known as it has been determined assuming that the energy U, is zero at small distances — 2.49Å — which is perhaps not rigorously valid[13].

---

[13] Once the interaction energy curve is drawn, one sees that it gives the value $u=0$ for $d-2r=1.03$A, which is precisely the distance between the magnetic shells of two iron atoms placed 2.49A apart.

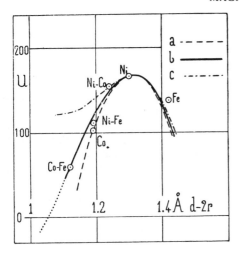

Fig. 5.
Curves of interaction energy deduced from expansion anomalies of nickel.

Examination of the left hand branch of the curves shows that the only solution that is appropriate is b, or a solution close to it. For example, for this solution, the corresponding curve crosses the axis of the abscissa near $d - 2r = 1.05$Å, which is in good agreement with what is known of the curve Γ from Fig. 3.

It would be illusory to wish to be more exact, as the abscissa values of the points on the curve are certainly not known to better than about 0.01Å: the uncertainty in the determination of the radius of the magnetic shell is of this order. In addition, I have represented the curve Γ, over a rather large range, by a series expansion, limited at the term in x^3, from which arise supplementary errors.

In any case, the results provided by the study of length anomalies seem not only compatible with the curve Γ determined by basically different procedures, but also allows its shape near the abscissa to be pinned down for nickel.

These conclusions will be reinforced by the study of iron.

(14) Special study of iron: Comparison of equations (28) and (33) allows us, this time, to write

$$\frac{n_1}{4Ad} = -1{,}74 \times 10^{-11} \quad (41)$$

$$\alpha \left(\frac{n_2}{A} - \frac{n_3}{2B} \right) = 0{,}95 \times 10^{-14}. \quad (42)$$

I will adopt the following numerical values

$$A = 0{,}94 \cdot 10^{+13} \quad ; \quad B = -1{,}94 \cdot 10^{+13} \quad ;$$
$$\alpha = 7{,}55 \cdot 10^{-6}.$$

For B particularly, the uncertainty is very large; it is only known to about 40%.

Two magnetic interatomic distances must be taken into account for iron; that corresponding to neighbours along the ternary axes, and that corresponding to neighbours along the cubic axes. I will identify them respectively by indices 1 and 2. One has

Distance 1 : $l_1 = 2{,}49$ Å $u_1 = 0$ $p_1 = 4$
Distance 2 : $l_2 = 2{,}86$ Å $u_2 = 138$ $p_2 = 3$

For the distance 1 the representative point of the interaction energy is clearly on the straight line representing the negative interaction energies of a large number of temperature independent paramagnetic metals or metals with a negative Curie point. The slope of this line gives

$$\frac{\partial u_1}{\partial l} = 600, \quad (43)$$

and the higher derivatives are obviously zero

$$\frac{\partial^2 u_1}{\partial l^2} = 0 \qquad \frac{\partial^3 u_1}{\partial l^3} = 0. \quad (44)$$

From the relationships (13) one has

$$n_1 = \frac{p_1 l_1}{d} \frac{\partial u_1}{\partial l} + \frac{p_2 l_2}{d} \frac{\partial u_2}{\partial l},$$

and from (41) and (43), replacing p_1, l_1, etc. by their numerical values, I have

$$\frac{\partial u_2}{\partial l} = -930. \quad (45)$$

Making the same comment now as for nickel: there is only the one relationship (42) to determine $\partial^2 u_2 / \partial l^2$ and $\partial^3 u_2 / \partial l^3$. I will therefore try three solutions: a, b and c corresponding to the following values of $\frac{1}{6} \partial^3 u_2 / \partial l^3$: 10,000, 20,000 and 30,000. Under these conditions, one notes that $n_3 \sigma^2$ is no longer negligible compared with B. The simplified formula (26) representing the expansion coefficient of the material is no longer useable; it is necessary to return to the

TABLE VII.

Solution	u_2	$\dfrac{\partial u_2}{\partial l}$	$\dfrac{1}{2}\dfrac{\partial^2 u_2}{\partial l^2}$	$\dfrac{1}{6}\dfrac{\partial^2 u_2}{\partial l^3}$
a	138	−930	−1 100	10 000
b	138	−930	−3 100	20 000
c	138	−930	−5 000	30 000

complete formula (25): numerical calculations take a little longer.

The three solutions correspond to the numerical values of Table VII.

The three corresponding curves are shown in Fig. 6. The right hand branches are very similar whatever solution is adopted. The interaction energy must go to zero, or at least become very small when $d-2r$ falls between 1.55Å and 1.60Å.

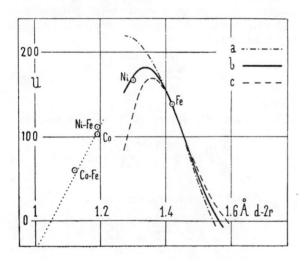

Fig. 6.
Curves of interaction energy deduced from the expansion anomalies of iron.

The curve corresponding to solution b passes near the point representing nickel. However, if this curve were prolonged, it would be noted that it no longer agreed with experimental points further to the left. This is not surprising, as a series expansion limited to the third term is not valid far from $x=0$, especially as the curve here has a complicated shape.

In this way, while it is not possible to obtain great accuracy, the set of coherent results obtained so far shows that the schematic representation made of the magnetic phenomena is rather realistic, and that the order of magnitude of the different quantities which enter into the theory are rather well known.

(15) The Curie point as a function of pressure: It is possible to obtain in a very direct way modifications in the interaction distance by compressing isotropically — hydrostatic pressure — a ferromagnetic material. The Curie point will move, which allows in principle the calculation of the first derivative of the interaction energy. The preceding theory allows a complete calculation of the shift of the Curie point.

Let P be the pressure, k_0 the compressibility; then the change δl in the distance between two atoms separated by a distance l when the pressure rises from 0 to P is:

$$\delta l = \frac{1}{3} l \kappa_0 P \qquad (36)$$

and, all other things being unchanged, the relative shift of the Curie point $\Delta\theta/\theta$ is equal to the relative change in the interaction energy, hence

$$\frac{\Delta\theta}{\theta} = \frac{\dfrac{\delta l}{l}\sum_i p_i l_i \dfrac{\partial u_i}{\partial l}}{\sum_i p_i u_i}. \qquad (47)$$

When only one magnetic distance is involved formula (47) can be simplified and written

$$\frac{\Delta\theta}{\theta} = -\frac{\partial u}{\partial l}\delta l. \qquad (48)$$

The necessary data for the calculation are given in Table VIII. The compressibilities, k_0, included are the compressibilities at absolute zero obtained by extrapolation.

In the last column of the table are the shifts of the Curie point calculated for a pressure increase of 10^9 bars, about 1000 atmospheres.

These shifts seem difficult to measure. Those for iron and nickel are too small. For cobalt the Curie point is at too high a temperature. Nevertheless, for iron–nickel γ, the values of $1/u$, $\partial u/\partial l$ should exceed 8 — the value for cobalt — when the proportion of iron reaches 50%. As the Curie points of these alloys are easily accessible, it could be possible to observe the expected shift of more than two degrees for 1000 atmospheres.

TABLE VIII.

Metal	$\kappa_0 \times 10^{12}$	p	u	$\dfrac{\partial u}{\partial l}$	l	θ	$\dfrac{\Delta\theta}{\theta P} \times 10^{12}$	$\dfrac{\Delta\theta}{P = 10^9}$
Ni	0,50	6	167	88	2,49	630°	0,21	0°13
Co	0,49	6	103	800	2,49	1 400°	3,3	4°6
Fe	0,55	(4 (3	0 138	600 −935	2,49) 2,86)	1 040°	0,82	−0°79

PART THREE
THE INFLUENCE OF THE TEMPERATURE VARIATION OF THE MOLECULAR FIELD ON PARAMAGNETIC PROPERTIES

(16) The temperature variation of the interaction energy: The study of the length anomalies of ferromagnets confirms the importance of the variation of the molecular field with the distance between atoms. Hence, the effect of temperature on the coefficient of the molecular field should not be negligible. Let us study first the effect on a single isolated interaction between two atoms at a distance l_i at absolute zero. The interaction energy for a distance $l_i + y_i$ becomes:

$$u = u_i + y_i \frac{\partial u_i}{\partial l} + \frac{1}{2} y_i^2 \frac{\partial^2 u_i}{\partial l^2} + \cdots \quad (49)$$

y_i is a function of time and temperature since the atoms oscillate and their average separation increases as the temperature is raised. It is known, moreover, that for two neighbouring atoms the amplitude of oscillation is large compared with the average additional separation corresponding to the expansion. It can be shown that it is permissible to calculate separately the effect on the molecular field of the expansion and of the oscillation.

The effect of expansion. Letting α be the average expansion coefficient between O and T, gives:

$$y_i = \alpha l_i T \quad (50)$$

and the new value of the energy, neglecting terms in T^2, can be written:

$$u = u_i + \alpha l_i \frac{\partial u_i}{\partial l} T \quad (51)$$

The effect of oscillation. To simplify, let us study what happens at constant volume: the two atoms under consideration oscillate about their equilibrium positions, and as we assume that there is no expansion, the average value of y_i is zero; however the average value of y_i^2 from now on designated $\bar{y}_i^2$ is not zero. The average interaction energy is then:

$$u = u_i + \frac{1}{2} \bar{y}^2 \frac{\partial^2 u_i}{\partial l^2}. \quad (52)$$

I will show in the following paragraph that $\bar{y}_i^2$ is proportional to T, so that one can write, in combining the two preceding equations

$$u'_i = u_i + \left[\alpha l_i \frac{\partial u_i}{\partial l} + \frac{1}{2T} \bar{y}_i^2 \frac{\partial^2 u_i}{\partial l^2} \right] T, \quad (53)$$

the term within brackets being independent of temperature.

Multiplying equation (53) by $2p_i$, and adding all the equations corresponding to different indices, i, one has

$$\sum_i 2p_i u'_i = \sum_i 2p_i u_i + \\ + \left[2\alpha \sum_i l_i p_i \frac{\partial u_i}{\partial l} + \frac{1}{T} \sum_i p_i \bar{y}_i^2 \frac{\partial^2 u_i}{\partial l^2} \right] T. \quad (54)$$

Now, the first term in the preceding equation is in fact n, the coefficient of the molecular field at temperature T; putting n_0 explicitly as the coefficient of the molecular field at absolute zero, one obtains

$$n = n_0 (1 + \lambda T), \quad (55)$$

and putting

$$\lambda n_0 = 2\alpha \sum_i l_i p_i \frac{\partial u_i}{\partial l} + \frac{1}{T} \sum_i p_i \bar{y}_i^2 \frac{\partial^2 u_i}{\partial l^2}. \quad (56)$$

In a first approximation, the molecular field is shown in this way to follow a linear function of temperature.

Before studying the consequences of formula (55), I will say a few words about the determination of $\bar{y}_i^2$.

(17) Relative oscillations of two atoms: The problem is to determine the mean square amplitude of oscillation, $\bar{y}_i^2$, along the line joining two atoms at a distance l_i.

The case of nearest neighbour atoms. If the atoms under consideration are nearest neighbours, the problem can be resolved using the following hypothesis: assume that for the elastic energy the interaction between nearest neighbours is dominant. At absolute zero, two neighbouring atoms are in equilibrium at a distance d, and when the distance between two atoms becomes $d+y$, the potential energy increases by

$$w = a'y^2 + b'y^3 + \cdots \qquad (57)$$

The coefficients a' and b' corresponding to a real interaction should not be confused with the coefficients a and b of paragraph 5 which correspond to interactions in a schematic linear model.

Starting from absolute zero, let us increase the temperature, but at constant volume. In a first approximation the atoms undergo harmonic oscillations about their equilibrium position, and the average potential energy of the interaction can be reduced to

$$\bar{w} = a'\bar{y}^2 ,$$

and for the pN bonds in a gram–atom, the total average potential energy is

$$W = pNa'\bar{y}^2 . \qquad (58)$$

However, this total energy is also equal to $3RT/2$ so that

$$\bar{y}^2 = \frac{3RT}{2pNa'} ; \qquad (59)$$

a' can be immediately calculated from the compressibility at absolute zero: in fact, the total potential energy stored in a static isotropic deformation, in which d becomes $d+y$ is

$$W' = pNa'y^2 , \qquad (60)$$

However, it is also equal to

$$W' = \frac{1}{2\kappa_0}\left(\frac{\delta V}{V}\right)^2 V = \frac{9}{2\kappa_0}\left(\frac{y}{d}\right)^2 V \qquad (61)$$

V being the atomic volume, as $\delta V/V = 3y/d$ and the comparison of formulae (60) and (61) gives

$$a' = \frac{9V}{2pN\kappa_0 d^2}$$

and, incorporating this formula in equation (59) one finally obtains

$$\bar{y}^2 = \frac{R\kappa_0 d^2}{3V} T \qquad (62)$$

The initial hypothesis plays a role only in the derivation of equation (58), which can be written only under the following condition: the elastic energy due to other than the first set of neighbours is negligible. For a close packed metal like nickel this condition is certainly satisfied to high accuracy, and the value of $\bar{y}^2$ given by (62) is correct.

Using this technique, neither the absolute amplitude of the oscillations, nor their relative phases enter into consideration. In fact the elastic and magnetic energies have been directly compared, since the same interactions are involved in both cases.

For a metal like iron, with a less compact structure, formula (62) only gives an approximate estimate if p is given a value corresponding to the nearest set of neighbours.

The case of distant atoms. In contrast to the previous case, it seems interesting to calculate now the mean square relative amplitude of oscillation of two distant atoms. The phases of their movement will then be independent, and if $\bar{z}^2$ is the mean square amplitude of oscillation of one atom in a given direction — relative to fixed axes — one will have

$$\bar{y}^2 = 2\bar{z}^2$$

Assuming, in a first approximation that the atoms oscillate with the same average frequency v, and the same average amplitude z_0, the movement of one atom will be represented by the equation

$$z = z_0 \sin 2\pi v t$$

The average kinetic energy is $m2\pi v^2 z_0^2$; equal to $kT/2$, hence:

$$\bar{z}^2 = \frac{z_0^2}{2} = \frac{kT}{4\pi^2 v^2 m} = \frac{RT}{4\pi^2 v^2 M} \qquad (63)$$

M being the atomic mass.

The average frequency can be determined from the Einstein temperature which best represents, by

Einstein's law[14], the specific heat at low temperature for the material under consideration. Let θ_E be the characteristic Einstein temperature determined in this way, then one has

$$k\theta_E = h\nu$$

In Table IX I give the values of θ_E and the values of the mean square relative amplitude of neighbouring atoms and distant atoms. It can be seen that the relative values for distant atoms are a little larger than those corresponding to neighbouring atoms. In any case, the order of magnitude of the amplitudes is well established.

TABLE IX.

	Ni	Co	Fe
θ_E	281°	289°	315°
$\dfrac{\overline{y^2}}{T} \times 10^5$; neighbouring atoms	1,32	1,28	1,35
$\dfrac{\overline{y^2}}{T} \times 10^5$; distant atoms	2,08	1,96	1,78
e en Å	0,163	0,160	0,165

To be more precise, I have calculated at 1000 K the maximum separation, e, relative to the equilibrium distance, of two neighbouring atoms. The total amplitude of motion is greater than 0.3Å; corresponding to considerable variations in the interaction energy.

(18) Temperature dependence of the susceptibility: The susceptibility of an ideal paramagnet, without a molecular field, obeys Curie's law:

$$\chi = \frac{C_0}{T} \quad (64)$$

where C_0 is the Curie constant. Introducing a molecular field proportional to the magnetization:

$$h_m = n\sigma, \quad (65)$$

the susceptibility is given by the Weiss law

$$\chi = \frac{C_0}{T - nC_0}. \quad (66)$$

However, from equation (55), the coefficient, n, of the molecular field is a linear function of temperature and making this dependence explicit, I obtain

$$\chi = \frac{C_0}{T - n_0 C_0 (1 + \lambda T)} = \frac{\dfrac{C_0}{1 - \lambda n_0 C_0}}{T - \dfrac{n_0 C_0}{1 - \lambda n_0 C_0}} = \frac{C}{T - \Theta}. \quad (67)$$

The experimental susceptibility of a material still obeys the Weiss law in the sense that the inverse susceptibility is still a linear function of temperature. Putting $\Theta_0 = n_0 C_0$; Θ_0 is the Curie point the material would show, if it kept the same molecular field coefficient as at absolute zero when the temperature varied. Relations (67) show the experimental Curie constant and the experimental Curie point, Θ, are given as a function of C_0 and Θ_0 by the relationships

$$\begin{cases} C_0 = (1 - \lambda\Theta_0) C \\ \Theta_0 = (1 - \lambda\Theta_0) \Theta \end{cases} \text{ or } \begin{cases} C = (1 + \lambda\Theta) C_0 \\ \Theta = (1 + \lambda\Theta) \Theta_0 \end{cases} \quad (68)$$

(19) The energetic molecular field and the corrective molecular field of the equation of state: Formula (67) shows that with an interaction energy which is a linear function of temperature, the susceptibility formally follows the Weiss law; the interaction energy is a linear function of temperature. To demonstrate this proposition I will follow an argument developed by P. Weiss [27].

P. Weiss distinguishes

(1) *An energetic molecular field* defined by

$$H_m = -\frac{\partial U}{\partial \sigma}$$

with U representing the magnetic energy of the material.

Thermodynamics shows that when a material obeys the Weiss law[15]:

---

[14] It is known that Einstein's theory of specific heats assumes monochromatic vibrations.

[15] It is known that for many ferromagnetic materials this law is carefully verified. See for example para. 7 of the article by P. Weiss: The molecular field hypothesis. Ann. de Physique, **17**, 1932, 97.

$$\chi = \frac{C}{T - \theta}, \quad (70)$$

The energetic molecular field is proportional to the magnetization and is independent of temperature:

$$H_m = n_0 \sigma \quad (71)$$

and the coefficient of proportionality, n_0, is given by:

$$n_0 = \frac{\theta}{C}. \quad (72)$$

(2) *A corrective molecular field of the equation of state* h_m, so that, if we denote by

$$\sigma = f\left(\frac{H}{T}\right), \quad (73)$$

the law of magnetization of an ideal paramagnetic, the law of magnetization with a molecular field is represented by the equation:

$$\sigma = f\left(\frac{H + h_m}{T}\right). \quad (74)$$

P. Weiss finds between H_m and h_m the following relationship:

$$H_m = h_m - T\frac{dh_m}{dT}. \quad (75)$$

If the energetic molecular field, H_m, is independent of temperature, for instance if the material obeys the Weiss law, integration of the differential equation (75) gives[16]:

$$h_m = H_m(1 + \lambda T), \quad (76)$$

λ being a constant; the corrective molecular field of the equation of state is a linear function of temperature.

---

[16] P. Weiss only considers the solution $H_m = h_m$. However, the complete solution shows that the reduction of the experimental equation to the ideal gas form is possible in several ways. In conformity with (67), and writing:

$$\chi = \frac{C_0}{T(1 - \lambda n_0 C_0) - n_0 C_0}$$

it can be seen that if a reduction is possible, then another can also be obtained by changing the temperature scale, which is obvious. However, this does not diminish the consequences of the introduction of the variability of n.

P. Weiss has shown that in the case of nickel, the coefficient, n_0, of the energetic molecular field is the same at high and low temperatures, which demonstrates the internal consistency of the molecular field theory. Formula (72) shows that at high temperatures, the coefficient n_0 of the energetic molecular field is obtained, and formula (69) shows that the energy phenomena, following from the specific heat, also gives H_m, hence n_0. However, from these results, no information can be drawn about the temperature coefficient of the corrective molecular field of the equation of state. This is completely inaccessible.

Potter thought of disproving the constancy of the energetic molecular field coefficient by the study of the magneto-caloric effect near the Curie point. He detected a strong reduction in n_0 at the Curie point. In fact this measurement assumed that it was legitimate to use the concept of a molecular field near the Curie point, while I have shown that in this region the fluctuations in the molecular field are such that the average value loses all meaning, and the results of Potter far from weakening, as the author thinks, the molecular field theory and its fluctuations actually underline their essential character. The study of the magnetic properties of a material near the Curie point can never give in a simple way either the moment or the molecular field, since it is a fundamentally perturbed region.

(20) The apparent paradox of a temperature independent energetic molecular field: The magnetic energy U, mentioned in the previous paragraph, has the form

$$U = -\frac{1}{2} n_0 \sigma^2, \quad (77)$$

if the material obeys the Weiss law; it is independent of temperature.

Elsewhere, in paragraph 3, equation (14), I had a magnetic energy, W_M, composed of a first term equal to U, and additional terms dependent on temperature. This difference arises solely from the definitions.

The meaning of U — at constant temperature the variations in U represent the variations in the total internal energy due to changes in magnetization.

The meaning of W_M — at constant temperature and at constant volume, when the magnetisation, σ, is changed, the magnetic energy W_M is changed, but what I call the elastic energy W_E — see paragraph 3 — is also changed. In fact the forces which tend to keep the atoms in their equilibrium position are changed, and as the average kinetic energy per degree of freedom remains constant at $kT/2$, the mean square amplitude changes and hence also the elastic energy.

In other words, at constant volume, there is in the elastic energy W_E as well as in the magnetic energy W_M, a term which depends on temperature. This term can also be called the oscillation term as it arises from oscillations of atoms about their equilibrium positions. Whatever the magnetization the sum of these two oscillation terms is always equal to $3RT/2$, following the law of equipartition of energy. This sum is equal to the elastic oscillation energy of a non-magnetic material.

Everything behaves as if the total magnetic energy was independent of temperature. Thermodynamics gives the result very directly but analysis of the mechanism shows that the process is complex. Variations in the elastic oscillation energy exactly compensate the variations in the magnetic oscillation energy. Study of the specific heat cannot therefore give any information about W_M, and the total energy of demagnetization which is easily accessible experimentally is simply equal to

$$\tfrac{1}{2} n_0 \sigma^2$$

(21) The experimental Curie constant and the true Curie constant:
In this way the corrective molecular field of the equation of state is not identical to the energetic molecular field. As a result the experimental Curie constant given by the Curie–Weiss law (70) is not identical to the Curie constant that a material would have if the interaction energy was removed without anything else being changed: it is only an apparent Curie constant.

The relationships (68) show that the true Curie constant, C_0, can be calculated from the formula

$$C_0 = \frac{C}{1 + \lambda \theta}. \quad (78)$$

This correction, as I will show in the following paragraphs, is far from negligible.

The relationships (68) allow moreover to be written

$$n_0 = \frac{\theta_0}{C_0} = \frac{\theta}{C}. \quad (79)$$

n_0 can be determined independently by the ratio of the apparent Curie point to the Curie point, or by the ratio θ_0/C_0 which is in fact the definition of n_0. This remark is important as it justified the method of determination of n_0 that I used in a previous article [7] in which I assumed that the molecular field was independent of temperature.

The Curie constant, C_0, of ideal paramagnets can be written

$$C_0 = \frac{1}{3} \cdot \frac{\mu_A^2}{R} \text{ with: } \mu_A = g\sqrt{j(j+1)}\,\mu_B. \quad (80)$$

In the case of iron, nickel and cobalt, the gyromagnetic effect shows that essentially $g=2$, hence only the spin is involved. If the atom has 1, 2 or 3 coupled spins, then $j = 1/2, 2/2$ or $3/2$ and the corresponding atomic Curie constants, calculated with $\mu_B = 5.554$ are

$j =$	$\dfrac{1}{2}$	$\dfrac{2}{2}$	$\dfrac{3}{2}$
C_0	0,372	0,993	1,862

I will determine the true Curie constants of nickel and iron and compare them with these figures.

(22) Calculation of the true Curie constant of nickel:
The correction formula is formula (78); θ is known, so only λ must be determined. It is hence necessary to determine the first derivative and the second derivative of the interaction energy as a function of distance.

The first derivative is given unambiguously by the length anomaly at absolute zero, and from (40) one has

$$\frac{\partial u}{\partial l} = 88 \; ;$$

the angstrom always being taken as the unit of length. To determine the second derivative the results of the study of the thermal variation in the length anomaly can be used. In this way I have obtained a value near -3700 for $1/2\,\partial^2 U/\partial l^2$. This result presents the inconvenience of being based on a prior determination of B which is very poorly known, much worse than A. This difficulty can be avoided in the following manner: a parabola can be fitted to the already known points on the interaction energy curve, taken sufficiently near that of nickel for the terms in x^3 to be neglected. The tangent determined above will be used. I have found that the following parabola

$$u = 167 + 88(x - 1.30) - 3210(x - 1.30)^2,$$

gives a good fit, as is shown in Fig. 7.

The second derivative determined in this way agrees to within 15% with the value calculated from the

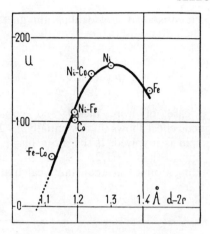

Fig. 7. The parabolic representation of the interaction energy curve near the point representing nickel.

thermal expansion anomaly. I will adopt the following values

$$\frac{\partial u}{\partial l} = 88 \qquad -\frac{\partial^2 u}{\partial l^2} = -3\,210\,; \qquad (80c)$$

and find after calculation

$$\alpha l \frac{\partial u}{\partial l} \theta = 2{,}03 \qquad \frac{1}{2} \bar{y}^2 \frac{\partial^2 u}{\partial l^2} \theta = -26{,}6\,,$$

from which

$$\lambda \theta = -0{,}146\,.$$

The experimental Curie constant of nickel for a gram–atom is 0.323 [31], from which

$$C_0 = \frac{0\,323}{1 - 0{,}146} = 0{,}379\,. \qquad (81)$$

It can be noted that if the moment of a nickel atom were 1 Bohr magneton, the Curie constant would be 0.372. The difference between these two values is no larger than the possible errors. In particular, the correction can be affected by an error of 10 to 15%. I do not think that this quasi-coincidence is an accident. The study of the magnetic equation of state of nickel, and in particular the study of the curvature of the magnetic isotherms near the Curie point has already allowed me to deduce that in the paramagnetic state, the moment of the magnetic carriers was 1 Bohr magneton to within 4%. It is also known that the spontaneous magnetization curve of nickel agrees well with the assumption of carriers with 1 magneton.

Elsewhere, I believed, in studying the magnetic properties of nickel [7], that the number of electrons active in the paramagnetic state was 0.6 per atom, but when that article was written I did not suspect the importance of the corrections to be made to the Curie constant, and hence some of the conclusions must be revised.

Hence the distribution of 1 Bohr magneton per nickel atom does not present problems. The electronic configuration of the incomplete shells of nickel would be: $3d^9$, $4s$. It is the $3d$ shell with 9 electrons which plays the role of paramagnetic carrier with one Bohr magneton. The 4s electron, as in copper for which the configuration is $3d^{10}$, $4s$, does not play any role[17].

(23) Calculation of the true Curie constant of cobalt: There is no information about the expansion anomalies of cobalt. In addition the derivatives of the interaction energy curve can only be determined from a series expansion which best reproduces the interaction energy curve near the point for cobalt. This point being near the point of inflexion of the curve, it is necessary to introduce terms in the third power. By iterations, the following expansion is obtained:

$$u = 106 + 800\,(x - 1{,}19) - \\ - 1\,400\,(x - 1{,}19)^2 - 800\,(x - 1{,}19)^3\,.$$

This expression is satisfactory (Fig. 8); in particular u goes to zero at $x = 1.06$Å, near the value $x = 1.05$Å which was adopted earlier. I obtain in this way:

$$\frac{\partial u}{\partial l} = 800 \qquad \frac{1}{2}\frac{\partial^2 u}{\partial l^2} = -1\,400\,.$$

The average expansion coefficient, between 0 K and the Curie point (1400 K) is not known. I have adopted 14.5×10^{-6}, the average value of the expansion coefficients of nickel and iron between 0 K and 1400 K: nickel and iron lie either side of cobalt in the periodic table of elements, which justifies this procedure. With these numerical values, one can calculate

$$\alpha l \frac{\partial u}{\partial l} \Theta = 40{,}5 \qquad \frac{1}{2}\bar{y}^2 \frac{\partial^2 u}{\partial l^2} \Theta = -25{,}2\,,$$

$$\lambda \Theta = 0{,}148\,.$$

---

[17] More precisely, the electrons of this shell would participate in the Pauli type constant paramagnetism.

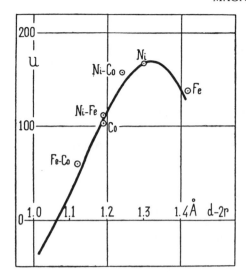

Fig. 8.
Third degree polynomial curve representing the interaction energy near the abscissa value for cobalt.

From the experiments of Bloch [31] the atomic Curie constant of cobalt is 1.14; the true value is

$$C_0 = \frac{1{,}14}{1 + 0{,}148} = 0{,}99 \qquad (82)$$

It is interesting to note that the correction is the opposite sign to that for nickel.

The Curie constant corresponding to 2 unpaired spins per atom is 0.993. The agreement between these two values is significant: the following electronic configuration for the incomplete shells must be attributed to cobalt: $3d^8.4s$. Taking into account the atomic number of cobalt, this is in fact the configuration that could logically be expected in the light of our knowledge of those for nickel and copper.

These consistent results obtained for nickel and cobalt are mutually reinforcing and give strong backing to the theory.

Taking these results as a basis, iron which has one less electron than cobalt should have a configuration $3d^7.4s$, with 3 Bohr magnetons per atom.

(24) The case of iron: The experimental atomic Curie constant is equal to 1.23 [32]. However, the correction indicated by the preceding theory cannot be calculated with confidence. In fact, for distances between magnetic shells greater than 1.42Å — the magnetic distance in iron — there are no experimental points to fix the slopes of the interaction curve and to calculate the terms in the corresponding polynomial expansion.

I will, in addition, tackle the problem from the opposite direction; I will show that the values of the derivatives of the interaction energy deduced from the correction to be applied to the experimental Curie constant in order to give the theoretical value, 1.862, which corresponds to 3 Bohr magnetons, taken together with the results of the study of the expansion anomalies, allows the part of the interaction curve as a function of the distance to be calculated, and to be remarkably consistent with what is independently known of this curve.

In fact, one has

$$1 + \lambda\theta = \frac{1.23}{1.862} = 0.660; \qquad \text{where } \lambda\theta = -0.34$$

and $\lambda = 3.25 \times 10^{-4}$. Taking as the average expansion coefficient between 0 and θ, $\alpha = 11.6 \times 10^{-6}$, and using the value of the first derivative (see para. 14), one can deduce

$$\sum_i p_i u_i = 414 \qquad \sum_i p_i l \frac{\partial u_i}{\partial l_i} = 2\,020;$$

from which

$$\sum_i \frac{\overline{y_i^2}}{2} p_i \frac{\partial^2 u_i}{\partial l^2} = 0.114 \quad \text{and} \quad \frac{1}{2} \frac{\partial^2 u^2}{\partial l^2} = -2820.$$

The value, calculated in this way is very near the value -3100 derived from study of the temperature dependence of the expansion anomaly.

Hence, it seems quite justified to attribute to iron a moment of 3 Bohr magnetons.

(25) General comments: This way of interpreting the Curie constants of ferromagnetic metals allows us to understand why cobalt and iron have such similar experimental Curie constants — 1.14 and 1.23 — while their ferromagnetic moments are substantially different: it is the corrections for the temperature variation of the molecular field which are of opposite sign and large.

A fourth ferromagnetic metal has recently been discovered, gadolinium [33]. Its experimental Curie constant is very near the theoretical value. This could be expected, since, taking a structure with twelve neighbours, gadolinium is found to have $u = 3/2$, a value more than 50 times smaller than that of nickel. In addition, the distance between magnetic shells — the 4f shell — is not well known, but it is certainly very large: between 2.5 and 3.0Å. The point representing gadolinium is therefore on a part of the interaction

TABLE X.

Metal	d	$d-2r$	u	$\dfrac{\partial u}{\partial l}$	$\dfrac{1}{2}\dfrac{\partial^2 u}{\partial l^2}$	$\dfrac{1}{6}\dfrac{\partial^3 u}{\partial l^3}$
Ni	2,49	1,30	167	87	−3 200	−5 000
Co	2,49	1,19	103	800	−1 400	−8 000
Fe$_1$	2,49	1,05	0	600	0	0
Fe$_2$	2,86	1,42	138	−935	− 000	+20 000

curve very near the axis of $u=0$, the tangent is small, the curvature weak, λ is therefore very small and the correction to the Curie constant is negligible.

The diameters of the magnetic shells used in this work have been calculated [7] assuming that the magnetic shells of nickel, cobalt and iron contain 8, 7 and 6 electrons respectively. The preceding paragraphs have shown that these numbers should be replaced by 9, 8 and 7. There is no fundamental change to the interaction energy curve; it must simply be shifted, as a whole, to the left by about 0.07Å, which does not present any important problem.

I have summarised in Table X the most probable values for the interaction energy and its derivatives, deduced from the preceding discussions.

BIBLIOGRAPHY

[1] Weiss and Foex, Le Magnétisme, Paris, Armand Colin.
[2] F. C. Powell, Proc. Phys. Soc., **42**, part 5, 1930, 390.
[3] Bauer, J. Phys., **10**, 1929, 345.
[4] Verschaffelt, Conseil Solvay, 1930, Paris, Gauthier-Villars (1932), p. 370.
[5] Bauer, ibid., p. 368.
[6] Fowler and Kapitza, Proc. Roy. Soc., **124**, 1929, 1.
[7] L. Neel, Ann. de Phys., **5**, 1936, 232.
[8] P. Weiss, Ann. de Phys., **17**, 1932, 97.
[9] L. Neel, J. Phys., **5**, 1934, 104; Ann. de Phys., **5**, 1936, 232.
[10] L. Neel, Bull. Soc. Franc. Phys., no. 374, 1935, p. 93.
[11] Damkohler, Ann. de Phys., **24**, 1935, 1.
[12] Chevenard, C.R. Acad. Sci., **172**, 1921, 1655.
[13] Colby, Phys. Rev., **30**, 1910, 506.
[14] Hidnert, Bur. of Stand., **5**, 1930, 1305.
[15] Williams, Phys. Rev., **46**, 1934, 1011.
[16] L. Neel, J. Phys., **6**, 1935, 27.
[17] L. Neel, J. Phys., **5**, 1934, 103.
[18] Ahrens, Ann. der Phys., **21**, 1934, 169.
[19] Dorfman, Janus and Kikoin, Z.f. Phys., **54**, 1929, 277.
[20] Debye, Ann. der Phys., **39**, 1912, 789.
[21] Tables de Landolt, 2e Suppl.
[22] Weiss and Forrer, Ann. de Phys., **5**, 1926, 153.
[23] Hegg, Diss., Zurich, 1912.
[24] Slater, Phys. Rev., **36**, 1930, 57.
[25] Bridgman, Proc. Nat. Acad. Amer., **8**, 1922, 361; Proc. Amer. Acad., **58**, 1923, 163.
[26] Gruneisen, Ann. der Phys., **33**, 1910, 1329.
[27] P. Weiss, J. Phys., **1**, 1930, 3.
[28] P. Weiss, Congrès Solvay, 1930, p. 291, Fauthier-Villars, Paris (1932); J. de Phys., **1**, 1930, 163.
[29] Potter, Proc. Roy. Soc., **146**, 1934, 362.
[30] L. Neel, **17**, 1932, 5.
[31] Bloch, Thesis, Zurich, 1912; Alder, Thesis, Zurich, 1916; Safranek, Rev. Métall. (février 1924); Peschard, Thesis, Strasbourg, 1925.
[32] L. Neel, C.R. Acad. Sci., **193**, 1931, 1325.
[33] Urbain, Weiss and Trombe, C.R. Acad. Sci., **200**, 1935, 2132; Trombe, C.R. Acad. Sci., **201**, 1935, 1591.

Chapter V
APPROACH TO SATURATION
A62, A63, A93

A62 (1948)

THE LAW OF APPROACH IN a/H AND A NEW THEORY OF MAGNETIC HARDNESS[1]

Abstract: P. Weiss found experimentally that in fields greater than 1 kOe, the magnetization of a ferromagnetic material approaches a saturation value $\mathcal{J}_s$ given by

$$\mathcal{J} = \mathcal{J}_s \left(1 - \frac{a}{H}\right).$$

The coefficient a has been termed the magnetic hardness. After reviewing various pertinent experimental data the author shows that no satisfactory interpretation has yet been given for the existence of the term in a/H. It is proposed that the term be attributed to the presence of voids (cavities) or non-magnetic inclusions within the ferromagnetic medium. The results of some experiments of Lorin are discussed in which the hardness of sintered porous iron specimens increases as the relative volume of the cavities increases.

The author then develops a new calculation which allows of a rigorous determination of the law of approach as a function of the volume of the voids. This law is complex: in very strong fields, greater than 100 kOe, the approach follows $1/H^2$ but in the field regime accessible to experiment (between 2 kOe and 10 kOe), the approach approximates to $1/H$, confirming the experimental results. The results of Lorin can be explained quantitatively in this way.

The same theory provides a qualitative interpretation of the large magnetic hardness observed in multiphase alloys (associated with the segregation of a non-magnetic phase) and inhomogeneous solid solutions. The temperature variations observed by Sadron are similarly explained. The rôle played by cavities and non-magnetic inclusions in the approach to saturation compares with their rôle in recent theories of coercivity and shows their importance in the theory of ferromagnetism.

I. INTRODUCTION

(1) Approach to saturation; magnetic hardness: *Definitions* — It is known that the magnetization $\mathcal{J}$ of a ferromagnetic body when subjected to an ever increasing magnetic field H, tends to a limit, $\mathcal{J}_s$, termed the *relative saturation*, at the temperature considered. Weiss showed [1] that in sufficiently strong fields, the approach to saturation may be expressed by the following law:

$$\mathcal{J} = \mathcal{J}_s \left(1 - \frac{a}{H}\right) ; \qquad (1)$$

a is conventionally termed the coefficient of hardness or more simply the magnetic *hardness*. The $1/H$ form of the law of approach has been confirmed by the work of Weiss and Forrer [2], Peschard [3], Sadron [4], Fallot [5, 6] at Strasbourg and also by Steinhaus, Kussman and Schoen [7] at the Physikalische-Technischen Reichsanstalt using an altogether different technique. Czerlinski [8] has contested this conclusion but his own results should not be considered decisive because they were obtained in too weak a field.

Magnetic hardness of metals — The hardness of pure materials possessing a cubic structure is generally smaller than 10 and varies significantly from one specimen to another: from 2.5 to 8 for iron [2, 7], and from 0.3 to 20 for nickel [9, 2]. It appears [7] that the hardness is not related to coercive field but increases with impurity content: generally speaking the purer and better annealed the specimen, the lower the hardness.

The hardness of non-cubic materials is considerably higher: values reported [2] for iron–boron, cementite and cobalt are 255, 460 and >1000 respectively. Magnetite is particularly noteworthy in this respect having a transition temperature at 120 K. Above the transition it is cubic with $a = 17$ whereas below 120 K it loses symmetry (although retaining the same atomic moment) and has a value of $a = 160$.

The hardness of alloys — Two distinct cases occur for alloys namely solid solutions and multiphase alloys. The hardness of solid solutions is small, comparable

---

[1] The experimental and theoretical work described in this paper was done and completed during the Occupation. The publication was delayed on account of material difficulties. In view of printing delays already apparent in 1945, a short note on the same subject was published in 1945 (*C.R. Acad. Sci.*, 1945, **220**, 738).

with that of pure metals. On the other hand, as shown by Fallot [6] in a study of iron alloyed with platinum type metals, multiphase alloys are characterized by large magnetic hardness generally of the order of several hundreds.

Temperature variation of hardness — This has been mostly studied between room temperature and liquid air temperature. According to Weiss and Forrer [2] and confirmed by work of Polley [9] on nickel, the hardness of iron and nickel are constant over this temperature range.

According to Sadron [4] and Fallot [5], the hardness of solid solutions with an iron base is also independent of temperature except for solutions rich in aluminium, chromium and gold where it increases with temperature.

Finally, and again according to Sadron [4], the hardness of solid solutions possessing a nickel base rises with temperature, especially as the alloy composition approaches the solubility limit. The hardness is often several times higher at room temperature than at liquid air temperature. Marian [10] has studied nickel alloys made as homogeneous as possible by suitable heat treatment. His results show that hardness diminishes with increasing temperature, regardless of composition: compared with liquid air temperature, the value at room temperature is frequently reduced by a factor of two. Undoubtedly, it is inhomogeneities which are responsible for this increase in a with temperature.

(2) Theoretical interpretations: According to present thinking, the changes in magnetization of a polycrystalline material produced by medium and strong fields originate from the *rotation* of the spontaneous magnetization away from the magnetocrystalline easy axes towards an alignment with the applied field. To these magnetocrystalline forces, which are related to the symmetry of the crystal lattice, must be added other internal directional forces arising from lattice imperfections. Several years ago, Weiss commented [2] that when these internal forces constrain the atomic moment to lie along directions different from that of the applied field, the approach to saturation must be of the second order in $1/H$.

$$\mathcal{J} = \mathcal{J}_s \left(1 - \frac{b}{H^2}\right). \quad (2)$$

In fact Akulov [11], Gans [12], and Becker [13] predicted theoretically that the law of approach involving saturation against magnetocrystalline forces is of the form (2) rather than (1).

Nevertheless, in a recent paper, Brown [14] has suggested that certain deformations, arising from dislocations of the lattice, would give rise to an approach law varying as $1/H$. These dislocations result from the decrease (negative dislocation) or increase (positive dislocation) by one unit of the number of atoms which constitute successive rows of the crystal. According to a theory of Becker [13] these defects give rise to easy directions of magnetization; rotation of the magnetization from such an easy axis requires energy expenditure proportional to both the elastic stress and the magnetostriction constant λ. Brown's result is not compatible with the comment of Weiss. Moreover we may adduce the following simple but relevant argument. The internal energy of magnetization is measured by the area included between the magnetization curve and its asymptote — this is infinite for an approach of the form[2] $1/H$. An infinite energy would therefore be required to overcome the disturbing but finite effects of the dislocations and so align the magnetization with the field direction. Brown's result would thus appear invalid[3].

If models based on the existence of internal perturbing forces are rejected another mechanism must be found e.g. thermal agitation, to which Weiss's comment obviously does not apply. In this case the body is divided into polyatomic domains each characterized by a resultant moment M of constant magnitude but variable orientation. The domains can be considered as discrete moments to which Langevin's paramagnetic theory may be applied. This yields an effective law of approach of the form given by (1) viz,

$$\mathcal{J} = \mathcal{J}_s \left(1 - \frac{kT}{MH}\right), \quad (3)$$

where T is the absolute temperature and k is Boltzmann's constant. But there is no other evidence for the existence of such a subdivision which has been totally fabricated for the purpose of the model. Besides, contrary to the prevailing experimental data, this theory predicts a large increase of a with temperature.

Finally, Forrer [2] proposed that the $1/H$ approach law owes its origin to an intra-atomic distortion of the moment caused by the applied field. The large variation of the hardness as a function of impurity content and the nature of the heat treatment excludes all models of this type.

---

[2] The area approximates to $\int_0^\infty \frac{a\mathcal{J}_s}{H} dH$.

[3] Brown's result holds when, in order to avoid the problem of a finite displacement produced at infinity by an isolated dislocation, the dislocations are arranged in *pairs*, comprising a positive and a negative dislocation. This introduces a singularity at the origin which provokes the appearance of an $1/H$ term if the integration includes the origin of the pair. The integration should be stopped at a distance from the origin which is of the order of magnitude of the distance between the components of the pair.

Faced with the impossibility of finding a reasonable explanation it is legitimate to question the validity of equation (1). It has been established only with certainty up to 10 kOe; beyond, corrections for magnetic images make the measurements misleading. On the other hand, directional forces arising from a field of 10 kOe are certainly weak compared with the perturbing forces found, for example, in the neighbourhood of dislocations. At these places therefore the spontaneous magnetization cannot be considered perfectly aligned with the applied field and the conditions under which the $1/H^2$ law is valid are not fulfilled. *The experimental law of approach in l/H must be thought of as an approximation, transforming gradually at high fields (20 kOe–50 kOe) into a l/H^2 law.* Unfortunately accurate measurements in this field regime can be made only with extreme difficulty.

(3) Rôle of cavities: It is hardly necessary to mention how far this new aspect of the matter complicates the theoretical task: it is no longer a question of finding the actual approach law in very high fields, still relatively easy, but of unravelling the phenomena in an intermediate region where the applied field is neither large nor small compared with the effects of inhomogeneities. Confronted with these difficulties it would seem preferable to approach the problem via the experimental route but we have no way of preparing materials with a *known* average density of randomly distributed defects.

Quite apart from deformations, perturbing influences may be linked to a more commonplace origin and whose importance has so far remained undetected. This is the presence of simple *cavities* or non-magnetic inclusions located within the body and unaccompanied by distortion. The internal magnetic fields created by the fictitious poles which appear on the surfaces of the cavities play the same rôle as perturbing fields. However, it is relatively easy to prepare specimens containing a given proportion of randomly distributed cavities and to study their magnetic properties. The corresponding theory is also relatively easy to develop, all of which serves to make the experiments of interest.

II.
APPROACH TO SATURATION OF POROUS IRON
(Néel and Lorin)

(4) Materials studied and experimental set-up: We have studied specimens of porous iron of different densities, obtained by sintering. They were prepared[4]

[4] The specimens were prepared at the Laboratoire Central des Industries mecaniques. We would like to acknowledge its director, M. l.Ingenieur general Nicolau and M. Girschig for their help.

from iron powder obtained either by hydrogen reduction or by decomposition of iron carbonyl. The specimens, suitably compressed into cylinders of diameter 27 mm and height 10 mm were then annealed in hydrogen for one hour at 880°C. Their densities ranged between 4.92 and 7.19. Ellipsoids of revolution of length 10 mm and diameter between 2 and 3 mm were cut from the cylinders either with the major axis parallel (position V) or perpendicular (position H) to the compression direction.

Measurement of coercive force — This was done by an attractive method using a translation balance constructed by Weil [15]. The coercive force varied between 4 and 7 Oe and showed no marked dependence on density or orientation.

Measurement of magnetization — The magnetization in medium and high fields was determined by a technique due to Weiss and Forrer [2, 16]: it consists of removing the specimen from a system of coils placed within the field of an electromagnet and then noting the deflection on a ballistic galvanometer. We have used the apparatus installed at Grenoble under the direction of Forrer. The accuracy, 0.2–0.3%, is inferior to that obtainable with the apparatus at Strasbourg [2] because the only electromagnet available (kindly placed at our disposal by M. Fortret) is small with 9 cm diameter pole pieces. Thanks to hollow pole pieces the specimen was extracted along the field axis; this procedure has an advantage over a removal perpendicular to the field direction because the corrections required for magnetic images are far less significant: in fact they are negligible in the field regime (up to 10 kOe) used here.

Care was always exercised to operate the electromagnet on the descending branch of the hysteresis loop. The magnetic field was determined as a function of current using the ballistic method. The apparatus was calibrated by means of a bulk specimen of pure iron.

In order to determine the demagnetizing factor N, we have used the fact that, in weak fields, the magnetization curve is practically straight with a slope of $1/N$.

(5) Results: Several magnetization curves are reproduced in Fig. 1. Iron becomes more difficult to magnetize as it becomes more porous. The curves depend on the orientation of the ellipsoid with respect to the direction of compression (curves 9H, 9V); it would therefore appear that the form and distribution of the cavities are not isotropic, a fact which is hardly surprising. Magnetization curves have also been plotted as a function of $1/H$ in order to reveal the nature of the approach to saturation (Fig. 2). *Within experimental error the behaviour is described by an approach law of the form $1/H$ valid between a lower limit*

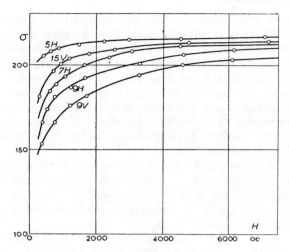

Fig. 1.
Magnetization curves of specimens of sintered iron of different densities (cf. Table I).

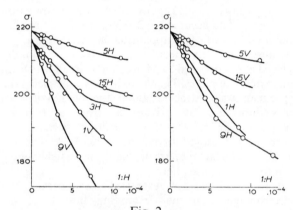

Fig. 2.
Magnetization curves in strong fields, shown as a function of 1/H, for various specimens of sintered iron (cf. Table I)

close to 2 kOe and 10 kOe, the upper limit of our experiment. Table I gives the corresponding coefficients a for the various samples and the two chosen orientations as well as the corresponding densities d.

For each density we have also calculated a weighted average, a_m, for a. Weightings of 2 and 1 were given for a specimen cut perpendicular and parallel respectively to the compression axis since, amongst the three orthogonal positions, there are two H positions for one V position.

Table I also gives the saturation magnetization σ_s per gram, obtained by extrapolation and assuming that the 1/H approach law holds. Values between 216 and 219 are found, very close to the saturation value of pure iron (217.8).

TABLE I.

Specimen	d	σ_s	a	a_m
5 H	7,19	218,5	46	47
5 V	"	218,5	49	
15 H	6,75	218	115	110
15 V	"	218	101	
3 H	6,34	216	128	132
3 V	"	216	141	
7 V	6,12	217	145	146
7 H	"	218	148	
1 H	5,74	218	168	170
1 V	"	217	172	
9 H	4,92	217,5	242	280
9 V	"	217,5	357	

III.
THEORY OF THE LAW OF APPROACH FOR A MATERIAL WITH A RANDOM DISTRIBUTION OF CAVITIES

(6) Analytical representation of an inhomogeneous distribution of spontaneous magnetization: Before discussing the theory of the approach of law for a material containing cavities, we treat first the case of a body, assumed for simplicity to be a sphere of volume V, whose spontaneous magnetization $\mathfrak{J}_s$ varies randomly from one point to another. We assume that the body is basically isotropic and that the mean wavelength of inhomogeneities is small enough to make the average magnetization value $\mathfrak{J}_m$ within a cube of side L (L is large compared with the wavelength but small compared with the sphere), arbitrarily centred about a point O, independent of the position of O. In a more general way, the fraction of the body contained within such a cube represents a faithful image of the whole and, to all intents and purposes, possesses the same properties.

Let an initial arbitrary cube be chosen with edges parallel to Oxyz. Consider now the fictitious material obtained by completely filling the sphere with cubes derived from the initial one by successive translations of L parallel to its edges. This fictitious material has properties practically identical to the original material and its spontaneous magnetization is representable by the Fourier series:

$$\mathfrak{J}_s = \mathfrak{J}_m + \sum i_{pqr} \exp\left\{\frac{2i\pi}{L}(px + qy + rz)\right\}, \quad (4)$$

where the summation sign includes, now and hereafter, all positive and negative integral values of p, q, r and where i_{000} is zero. The coefficients i_{pqr} are complex quantities defined through the following integral which extends over all the fictitious material contained within the sphere:

$$i_{pqr} = \frac{1}{V} \int_V (\mathcal{J}_s - \mathcal{J}_m) \times \exp\left\{-\frac{2i\pi}{L}(px + qy + rz)\right\} dx\, dy\, dz \tag{5}$$

Since $\mathcal{J}_s$ is real, if the complex conjugate is denoted by an asterisk *, we have

$$i_{-p,-q,-r} = i^*_{pqr} \tag{6}$$

Obviously, if another origin for the initial cube is chosen an expansion is obtained with different coefficients i_{pqr} but which corresponds to the same average properties. It is more important, therefore, to know the probable values of i_{pqr} for some arbitrary position of the initial cube than the actual values for some specific position.

It can be shown that the probability law which governs i_{pqr} depends upon one variable only

$$\rho = \sqrt{p^2 + q^2 + r^2}.$$

The co-ordinate axes are now transformed in order to make the new OX axis perpendicular to the plane $px + qy + rz = 0$. Equation (5) then becomes

$$i_{pqr} = \frac{1}{V} \int_V [\mathcal{J}_s(X, Y, Z) - \mathcal{J}_m] \times \exp\left\{-\frac{2i\pi}{L}\rho X\right\} dx\, dy\, dz. \tag{7}$$

Since, by hypothesis, the substance is isotropic, the result of this integration does not depend on the orientation of OX with respect to the sphere — it depends only on ρ. In particular the average value of the product $i_{pqr} i^*_{pqr}$, which is positive and real, depends only on ρ: we can therefore put simply,

$$\overline{i_{pqr} i^*_{pqr}} = i^2_\rho. \tag{8}$$

Note also that according to equation (5) the average value of i_{pqr} is zero:

$$\overline{i_{pqr}} = 0. \tag{9}$$

Thus, to provide a complete analytic description of a random distribution of spontaneous magnetization requires in principle a knowledge of the probability laws which give i_{pqr} as a function of ρ. By good fortune there exists a certain number of properties, including the approach to saturation, which depend only on S, the mean square difference between the spontaneous magnetization and its mean value, or, and which comes to the same thing, on the mean value $\overline{\mathcal{J}^2_s}$ of the square of the spontaneous magnetization. In fact we have,

$$\overline{\mathcal{J}^2_s} = \mathcal{J}^2_m + S. \tag{10}$$

To calculate $\overline{\mathcal{J}^2_s}$, equation (4) is first squared then averaged with respect to x, y, z, and then averaged with respect to all possible positions of the initial cube. Having regard to equations (6) and (8) we have finally

$$\overline{\mathcal{J}^2_s} = \mathcal{J}^2_m + \sum i^2_\rho. \tag{11}$$

As the number of terms included in the sum is very large, it can be replaced without appreciable error by an integral; each element of unit volume of pqr space corresponds to one term only of the sum. If in addition a change of variable is made,

$$r = \rho \cos\varphi, \quad p = \rho \sin\varphi \cos\theta, \quad q = \rho \sin\varphi \sin\theta,$$

$$dp\, dq\, dr = \rho^2 \sin\varphi\, d\varphi\, d\theta\, d\rho,$$

we obtain:

$$S = \sum i^2_\rho = \int_{-\infty}^{+\infty} \int_{-\infty}^{+\infty} \int_{-\infty}^{+\infty} i^2_\rho\, dp\, dq\, dr$$

$$= 4\pi \int_0^\infty \rho^2 i^2_\rho\, d\rho. \tag{12}$$

(7) Calculation of the law of approach for the preceding material: Under the action of a strong field H parallel to Oz, the magnetization at every point approaches Oz and its projections on the xOz and yOz planes make only small angles α and β with Oz. If we assume further that the functions of the spontaneous magnetization are always small with respect to the average magnetization $\mathcal{J}_m$, then to a first approximation, we obtain finally for the projections $\mathcal{J}_x$, $\mathcal{J}_y$, $\mathcal{J}_z$, of the spontaneous magnetization along the three axes:

$$\mathcal{J}_x = \alpha \mathcal{J}_m, \quad \mathcal{J}_y = \beta \mathcal{J}_m,$$

$$\mathcal{J}_z = \mathcal{J}_m + \sum i_{pqr} \exp\left(\frac{2i\pi}{L}(px + qy + rz)\right) \tag{13}$$

It must now be noted that the spontaneous magnetization is in equilibrium at every point under the action of the total magnetic field, magnetocrystalline forces and Heisenberg exchange forces. If it is assumed that no correlation exists between the orientation of the crystallites and local variations of the spontaneous magnetization, both considered as perturbations, then these two phenomena can be treated independently: we therefore reserve for later study the influence of magnetocrystalline forces. Exchange effects are negligible as long as the average wavelength of the perturbation remains large (as assumed here) compared with the boundary thickness (0.1 μm) separating the domains.

Finally we note that the spontaneous magnetization is in equilibrium with the total field i.e. the geometric sum of the external field H and the internal demagnetizing field arising from the non-uniformity of the magnetization: the latter is derivable from a certain potential V. Hence

$$\alpha H = -\frac{\partial V}{\partial x}, \quad \beta H = -\frac{\partial V}{\partial y}. \tag{14}$$

The potential V satisfies Poisson's equation,

$$\Delta V + 4\pi m = 0$$

with

$$m = -\text{div } \mathcal{J} = -\frac{\partial \mathcal{J}_x}{\partial x} - \frac{\partial \mathcal{J}_y}{\partial y} - \frac{\partial \mathcal{J}_z}{\partial z},$$

which gives

$$\frac{\partial^2 V}{\partial z^2} + \left(1 + \frac{4\pi \mathcal{J}_m}{H}\right)\left(\frac{\partial^2 V}{\partial x^2} + \frac{\partial^2 V}{\partial y^2}\right)$$
$$= -\frac{8i\pi^2}{L} \sum r i_{pqr} \exp\left\{\frac{2i\pi}{L}(px + qy + rz)\right\} \tag{15}$$

This equation is satisfied by a potential V of the form,

$$V = \sum v_{pqr} \exp\left\{\frac{2i\pi}{L}(px + qy + rz)\right\}, \tag{16}$$

where the coefficients v_{pqr} are given by the relation

$$v_{pqr} = \frac{2iLr}{r^2 + a(p^2 + q^2)} i_{pqr} \tag{17}$$

in which, to simplify the notation we have written

$$a = 1 + \frac{4\pi \mathcal{J}_m}{H}. \tag{19}$$

Having thus determined V, it can be used to find α and β (from equation (14)) and thence the value of $\mathcal{J}$, the magnetization along the field direction

$$\mathcal{J} = \bar{\mathcal{J}}_s \left(1 - \frac{\alpha^2 + \beta^2}{2}\right). \tag{20}$$

The average value of this component, taken with respect to x, y, z, and all possible positions of the initial cube, can be expressed, all calculations having been done, as

$$\bar{\mathcal{J}} = \mathcal{J}_m \left\{1 - \frac{8\pi^2}{H^2} \sum \frac{(p^2 + q^2)r^2}{[r^2 + a(p^2 + q^2)]^2} i_\rho^2\right\}. \tag{21}$$

As was done in the analogous case at the end of the preceding section the sum can be replaced by an integral. After making the same change of variable, we have

$$\sum \frac{(p^2 + q^2)r^2}{[r^2 + a(p^2 + q^2)]^2} i_\rho^2$$
$$= \int_{-\infty}^{+\infty} \int_{-\infty}^{+\infty} \int_{-\infty}^{+\infty} \frac{(p^2 + q^2)r^2}{[r^2 + a(p^2 + q^2)]^2} \tag{22}$$
$$\times i_\rho^2 \, dp \, dq \, dr = SM$$

where S is the mean square, given by (12), of the difference between the spontaneous magnetization and its average value. M is a quantity given by,

$$M = \int_0^{\frac{\pi}{2}} \frac{\sin^2 \varphi \cos^2 \varphi \sin \varphi \, d\varphi}{(\cos^2 \varphi + a \sin^2 \varphi)^2} \tag{23}$$

A simple integration gives,

$$M = \frac{2a+1}{(2a(a-1)^2} \frac{\text{arth } \varepsilon}{\varepsilon} - \frac{3}{2(a-1)^2}$$
$$\text{with } \varepsilon = \sqrt{\frac{a-1}{a}} \tag{24}$$

Finally, the approach law (21) can be completely determined and in the last analysis depends only on the value of S. Using the value of S given by (10), the law of approach takes the form:

$$\bar{\mathcal{J}} = \mathcal{J}_m \left[1 - \frac{1}{2} \frac{\bar{\mathcal{J}}_s^2 - \mathcal{J}_m^2}{\mathcal{J}_m^2} (a-1)^2 M\right], \tag{25}$$

which is suitable for applications.

(8) Application to a material containing voids: We now apply the equation above to a specimen of iron containing a large number of randomly, but isotropically, distributed cavities much as a sponge. Let v be the volume per cm³ occupied by cavities and $\mathcal{J}_0$ the spontaneous magnetization of bulk iron. By definition,

$$\mathcal{J}_m = (1-v)\mathcal{J}_0 \quad \text{and} \quad \overline{\mathcal{J}_s^2} = (1-v)\mathcal{J}_0^2 \qquad (26)$$

and after substituting these values into equation (25) we obtain the law of approach as

$$\mathcal{J} = \mathcal{J}_m \left[1 - \frac{v}{2(1-v)}(a-1)^2 M \right]. \qquad (27)$$

In order to compare this theoretical law with the experimental results discussed above from numerous specimens of different values of v, it is convenient to represent

$$y = \frac{\mathcal{J}_m - \mathcal{J}}{\mathcal{J}_m} \frac{2(1-v)}{v}$$

as a function of

$$x = \frac{4\pi \mathcal{J}_m}{H} = \frac{4\pi \mathcal{J}_0 (1-v)}{H}.$$

According to equation (27) y is a function of x, independent of v, given in parametric representation by

$$y = (a-1)^2 M, \quad x = (a-1) \qquad (28)$$

which is shown as the solid line in Fig. 3.

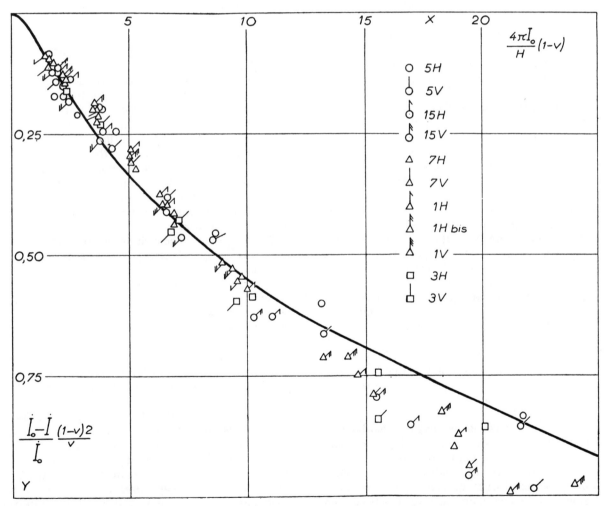

Fig. 3.

The approach to saturation. A comparison of the results from "void" theory with experimental points from the various specimens studied (reduced coordinates). The theoretical curve is shown as the solid line.

This curve displays a point of inflexion such that in the region between $x=0.4$ and $x=4$ (which for dense alloys corresponds to fields between 50 kOe and 5 kOe) the approach law is clearly of the form $1/H$. Only in fields greater than 100 kOe, does the approach law become $1/H^2$. In the same figure are shown the experimental points for the specimens, of widely differing densities, already included in Table I. The agreement between theory and experiment is excellent, especially in strong fields. In weak fields, i.e. <1 kOe, the observed magnetization is systematically less than the theoretical value: the most probable reason for this is the effect of magnetocrystalline forces which have been ignored in the theory.

IV.
THE RÔLE OF CAVITIES IN THE LAW OF APPROACH FOR METALS AND ALLOYS

(9) The approximate nature of the 1/H law: We may summarize as follows. Theory and experiment both agree in showing that a random array of holes or non-magnetic inclusions provokes the appearance of an approach law of the form $1/H$ over a range of fields extending from thousands of oersteds to 50 kOe. It is precisely in this region that a $1/H$ law has been observed in common materials.

Since the perturbing effects of randomly distributed internal demagnetizing fields can give rise to, over a suitable field range, an approximate law of the form $1/H$ it seems not unreasonable to suppose that irregular perturbations of any sort, perhaps resulting from a random stress distribution, should also give, over a limited field range, an approximate $1/H$ law.

Speaking generally, therefore, it seems valid to relate any approach in $1/H$ to the effects of a random distribution of internal perturbing forces. Two examples of such forces have just been given but there are others. Thus in non-cubic materials the magnetocrystalline forces, which are considerable, tend to align the magnetization along a direction which varies from one crystal grain to the next. These forces play the same rôle as random internal perturbations and are responsible for the considerable hardness of *hcp* cobalt and cementite, etc.

(10) The rôle of cavities in the magnetic hardness of common materials: Metals and alloys certainly contain real cavities: it is known that forging changes the density although necessarily not by much. However, in the absence of cavities, non-magnetic inclusions or inclusions more weakly magnetized than the principal phase play an exactly similar or analogous rôle. Steels, for example, contain a considerable proportion of non-magnetic constituents or material less magnetic than ferrite: austenite, cementite, etc. It is sufficient for the volume occupied by these inclusions to attain a value of about 10% in order to give a hardness of 100, which is the magnitude observed in steel. The great hardness of multiphase alloys is satisfactorily explained in a similar way.

It should not be forgotten that certain materials, for example quenched steels, contain internal stresses which also contribute to the hardness.

(11) Temperature variation of the hardness: In our present state of knowledge it is difficult to distinguish the influence of stresses from that of inclusions. In this respect the temperature variation of a does give some interesting pointers. *In the inclusion theory* the demagnetizing fields which oppose saturation are proportional to the spontaneous magnetization; the temperature variation of a must reflect this behaviour. As a result therefore, the temperature variation of any hardness due to inclusions should be insignificant in the case of iron since the temperature variation (below room temperature) of the spontaneous magnetization is negligible.

On the contrary, *according to the stress theory*, the hardness varies approximately as the magnetostriction constants since the temperature variation of the elastic constants is comparatively small. According to Takai [17] the magnetostriction constants decrease rapidly as the temperature increases, of the order of 4% near room temperature, and therefore a rapid diminution in a would be expected as the temperature rises. Since the magnetic hardness of iron is independent of temperature its origin must be attributed to inclusions rather than stresses.

Nickel alloys — The inclusion theory can be used in a similar way to clarify the results of Sadron summarized at the end of section 1. In all likelihood we are concerned here with solid solutions which are far from homogeneous and possess concentration gradients. As a result their properties are analogous to a strongly magnetized matrix containing a sizeable fraction of second phase which is less magnetic and has a lower Curie point. The magnetization of both phases as a function of temperature is illustrated schematically in Fig. 4. As far as the approach law is concerned, the influence of the lesser magnetized inclusions becomes more marked, all other things being equal, as the difference between the magnetization of the two phases increases. Fig. 4 shows that this difference is virtually independent of temperature at points far removed from the Curie point but grows more rapidly as the Curie point of the lesser magnetized phase is approached. Consequently, a increases at an increasing rate with temperature. This explains the observations of Sadron who found that the temperature increases in a were more marked as the Curie point was closer to room

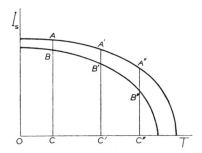

Fig. 4.
The relative difference AB/AC, A'B'/A'C', etc of the spontaneous magnetizations of two bodies with neighbouring Curie points increases as the region of the Curie points is approached.

temperature. Such effects do not occur in the homogeneous alloys prepared by Marian. The model would also explain why no temperature variation is seen in iron-based solid solutions: the Curie point is too high.

Summarizing the results above we can say that the study of the temperature variation of *a* shows that the hardness arises from voids and inclusions rather than internal stresses.

(12) Magnetic hardness and coercive field: The important rôle played by voids and inclusions in the approach to saturation compares with their rôle in the question of coercivity. For a long time it was thought that the coercive force originated exclusively from a random distribution of internal stresses [18]. Recently, however, Kersten [19] and later Néel [20] have shown that cavities and inclusions can give rise to considerable coercivity. In this matter also it is difficult to separate the effects of inclusions and stress. There are, nevertheless, differences in the rôle of voids in the two situations. As far as the coercivity is concerned holes some 0.1 μm in size are the most influential while large cavities are unimportant. On the other hand, holes both large and small are equally significant for the hardness.

It seems, therefore, that a comparative study of the coervice force and magnetic hardness could furnish interesting results about the structure of metals and alloys. This field of research has scarcely yet been explored.

REFERENCES

[1] P. Weiss, *J. de Physique*, 1910, **9**, 373.
[2] P. Weiss and R. Forrer, *Ann. de Physique*, 1929, **12**, 279.
[3] Peschard, *Rev. de Métallurgie*, 1925, **22**, 490, 581 and 663.
[4] C. Sadron, *Ann. de Physique*, 1932, **17**, 371.
[5] Fallot, *Ann. de Physique*, 1936, **6**, 305; 1937, **7**, 420.
[6] Fallot, *Ann. de Physique*, 1938, **10**, 291.
[7] Steinhaus, Kussmann and Schoen, *Phys. Z.*, 1937, **38**, 777.
[8] Czerlinski, *Ann. der Physik*, 1932, **13**, 80.
[9] Polley, *Ann. der Physik*, 1939, **36**, 625.
[10] Marian, *Ann. de Physique*, 1937, **7**, 459.
[11] Akulov, *Z. Physik*, 1931, **69**, 822.
[12] Gans, *Ann. der Physik*, 1932, **15**, 28.
[13] R. Becker and W. Doring, Ferromagnetismus (Berlin Springer, 1939).
[14] W. F. Brown Jr, *Phys. Rev.*, 1941, **60**, 139.
[15] *Thèse Strasbourg*, 1941 (Grenoble Allier, éd.).
[16] P. Weiss and R. Forrer, *Ann. de Physique*, 1926, **5**, 153.
[17] H. Takaki, *Z. Physik*, 1937, **105**, 92.
[18] Cf. Becker, Ferromagnetismus (*Loc. cit.*), p. 203 and 213.
[19] Kersten, *Physik Z.*, 1943, **44**, 63.
[20] Néel, *Cahiers de Physique*, 1944, **25**, 21–44; *Annales Université de Grenoble*, 1946, **22**, 299–343.

A63 (1948)

RELATION BETWEEN THE ANISOTROPY CONSTANT AND THE LAW OF APPROACH TO SATURATION IN FERROMAGNETIC MATERIALS[1]

Abstract: In an increasing field the magnetization, $\mathcal{J}$, of a cubic polycrystalline ferromagnetic approaches its saturation value, $\mathcal{J}_s$, according to the law

$$\mathcal{J} = \mathcal{J}_s \left(1 - \frac{b}{H^2}\right) \qquad (A)$$

and a current theory states that $b = 8K^2/105\,\mathcal{J}_s^2$ where K is the anisotropy constant. The author shows this theory is erroneous because it neglects the considerable magnetic interaction between crystallites. The experimental vindications that have been claimed for it are specious either because of inadequate interpretation or because specimens containing undue residual internal stresses were used.

A new calculation is then developed which allows for magnetic interactions and results in a rigorous approach law which does not have the form (A). However, in the region of most practical interest, i.e. between 200 and 1000 Oe, it approximates to the form (A) but the value of b is about half as big as the figure quoted above. Some new experiments performed on well annealed, high purity nickel have confirmed the results. The formulae obtained allow a determination of K from the experimental data gathered from these polycrystalline specimens.

I. INTRODUCTION

(1) Determination of the anisotropy constant K from a study of the approach to saturation: *Single crystal case* — We know that the free energy F of a ferromagnetic crystal is the sum of various terms including one, F_m, the magnetocrystalline energy, which depends on the orientation of the saturation magnetization $\mathcal{J}_s$ which respect to the crystal axes. In a cubic crystal F_m can be satisfactorily expressed by the following relation

$$F_m = K(\alpha^2\beta^2 + \beta^2\gamma^2 + \gamma^2\alpha^2). \qquad (1)$$

where α, β, γ, are the direction cosines of the magnetization $\mathcal{J}_s$ with respect to the cubic axes.

In an external field, $\vec{H}$, the free energy includes an additional second term $F_h = -\vec{H}\vec{\mathcal{J}}_s$.

For a perfect crystal the equilibrium orientation of $\vec{\mathcal{J}}_s$ is obtained when the free energy $F = F_m + F_h$ is a minimum. As the field intensity increases, still maintaining its direction with respect to the crystal, the spontaneous magnetization tends to parallelism with the field and its component $\mathcal{J}$ along this direction approaches the saturation value, $\mathcal{J}_s$. It can easily be shown that the approach to saturation follows a law of the form

$$\mathcal{J} = \mathcal{J}_s\left(1 - \frac{b}{H^2}\right), \qquad (2)$$

where the coefficient b depends on the orientation of H with respect to the crystal axes.

Polycrystalline case — The relation (2), applicable to a single crystal, has never been tested experimentally but over the years several authors have applied it to a polycrystal. The first, Akulov [1] calculated the value of b pertinent to a polycrystal by averaging over the values obtained on allowing a single crystal all conceivable orientations: the result is

$$b = \frac{8}{105}\frac{K^2}{\mathcal{J}_s^2} \qquad (3)$$

Gans [2] and later Becker [3] followed the same reasoning and arrived at the same result.

Experimental verification — In the meantime Czerlinski [4] showed experimentally that

---

[1] The experimental and theoretical work described in this paper was done and completed during the Occupation. The publication was delayed on account of material difficulties. In view of printing delays already apparent in 1945 a short note on the same subject was published in 1945 (*C.R. Acad. Sci.*, 1945, **220**, 814).

polycrystalline samples of iron and nickel obey an approach law of the form (2) and thence, from the values of b so obtained together with equation (3), deduced values of K which, between $-150°C$ and $150°C$, agreed with literature values obtained by the methods usually applied to single crystals. For his part Polley [5] repeated the study of polycrystalline nickel and has shown that, in fact, the approach law (2) must be replaced by a more complex one of the form

$$J = J_s \left(1 - \frac{a}{H} - \frac{b}{H^2}\right) + \chi H ; \quad (4)$$

Polley assumed that the term in a/H arising from some unknown mechanism did not alter the term in b/H^2 and so was able to calculate values of K at different temperatures: once again these agreed with the literature.

It seems therefore that we are in possession of a proven method for the determination of anisotropy constants based on measurements taken from polycrystals. This is of obvious experimental interest. Unfortunately, serious objections can be raised against this corpus of data, as much from the theoretical point of view as from the experimental.

(2) Theoretical objections: It seems questionable whether the coefficient b for a polycrystal can be calculated by averaging values of b for supposedly isolated single grains because this neglects the internal demagnetizing fields associated with free magnetic poles. In turn these poles arise from the non-uniformity of the magnetization of the polycrystal.

As far as an isolated grain is concerned, the presence of the demagnetizing field offers no particular difficulty: the crystallite has an average spherical shape with a demagnetizing factor independent of direction and equal to $4\pi/3$. Moreover, the theory tacitly assumes that the external field is sufficiently intense to make each grain a single domain particle. As a result the crystal is always magnetized to saturation and the only changes in magnetization that occur arise solely[2] from the rotation of a vector with constant magnitude J_s. During the rotation the magnetostatic energy remains constant at $\frac{1}{2}(4/3)\pi J^2$ erg cm^{-3} and hence there is no need to take it into account.

This is no longer the case for the assembly of juxtaposed grains which constitute the polycrystal. Let us suppose in fact that the assembled crystallites possess the same state of magnetization that they would have if isolated. Since the direction of spontaneous magnetization varies from one grain to another, fictitious magnetic poles appear on the surfaces separating them so creating magnetic fields which increase the free energy. However, unlike the case of the isolated grain where this energy remains unchanged when the magnetization aligns with the field, here, the extra energy vanishes at saturation i.e. when all the particles have the same parallel magnetization. (It is assumed that the polycrystal is a long rod lying in the field direction.) Consequently, forces exist which favour saturation.

The effect of these interactions may be appreciated by the following argument: if an elementary grain is extracted from the polycrystal, the field which remains in the corresponding cavity is, *on average*, equal to $H + (4/3)\pi J$, where J is the average magnetization of the polycrystal. It is this total field which effectively acts on the crystallite. Since J is virtually equal to J_s near saturation, the term b/H^2 in the relation (2) must be replaced by a term in $b/(H + (4/3)\pi J_s)^2$. For the case of nickel, where $J_s \sim 500$ G, it can be established that in a field of 1 kOe the new term is nine times smaller than the original. Undoubtedly the treatment sketched here is too rough to give an exact result but it suffices to indicate that crystallite interactions are significant and cannot be neglected.

Finally, the law (2) with the value of b given by (3) is very much in error.

(3) Experimental objections: We are concerned to explain under what conditions the authors cited above were able to obtain correct values of K using incorrect formulae.

Czerlinski's experiments — As far as these experiments are concerned, a close examination of the author's paper, which will not be reproduced here, shows that his results are far better represented by a formula of the type

$$J = J_s \left(1 - \frac{a}{H} - \frac{c}{H^3}\right), \quad (5)$$

with no b/H^2 term than by the formula which he actually indicates (this lacks a term in a/H but contains one in b/H^2) and which presents systematic deviations from experiment. Moreover, this new formula gives a value of a equal to 5.2 which is of the same order of magnitude as values obtained by other authors for pure iron [8]. Thus Czerlinski's experiments, far from supporting the Akulov–Gans–Becker theory, show rather, that the value of the b/H^2 term predicted by it is too large. However it should be noted that the presence of the extra term in a/H, due probably to defects in the crystallites, makes the interpretation uncertain because it is possible that these imperfections will be manifest through an extra term in b'/H^2.

---

[2] Concerning this question see the note by Néel: Laws of magnetization and subdivision into elementary domains of an iron single crystal. *J. de Physique*, 1944, **5**, 241–251.

Polley's experiments — These experiments are more favourable from this point of view because the a/H term is very small. Nevertheless the sample of this author must still be far from perfect as the following technique shows.

(4) Calculation and significance of the work of magnetization: At the present time it is not possible to calculate the magnetization curve of a polycrystalline material, but a quantity does exist whose experimental determination is easy and which, independent of all theory, can be calculated as a function of the anisotropy constant: this is the work required to magnetize a ferromagnetic substance.

From the experimental aspect, the work required is equal to the area included by the magnetization curve (magnetization as a function of field), the asymptote at saturation and the ordinate axis. This method assumes that there is no hysteresis. However, as shown by Kersten [7] in an analogous case, the reversible work of magnetization can be evaluated when the hysteresis is small by measuring the area contained between the descending branch of the limiting cycle and the asymptote at saturation (Fig. 1). This is because the branch SR of the loop is sensibly reversible.

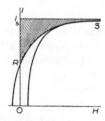

Fig. 1.
The descending branch SR of the major loop is more or less reversible: the hatched area corresponds therefore to the work of magnetization.

From the theoretical point of view, the work required for magnetization is equal to the difference between the internal energy at saturation and the internal energy of the initial state when the magnetization is zero.

In the initial state, the spontaneous magnetization of the various elementary domains is aligned with the so-called easy directions corresponding to a minimum in the magnetocrystalline energy. When K is positive (the case of iron) the minimum in the magnetocrystalline energy, F_m, is zero; when K is negative (the case of nickel) the minimum equals $K/3$. On the other hand at saturation, the spontaneous magnetization is parallel to the applied field so that the average value of the energy of a polycrystalline material equals the mean value of F_m given by expression (1) when the unit vector α, β, γ, takes on all possible orientations: an elementary calculation shows that this average equals $K/5$.

Finally, we have the result that the work of magnetization for a polycrystal is equal to $K/5$ erg cm^{-3} when K is positive and $K/5 - K/3 = -2K/15$ erg cm^{-3} when K is negative.

Application to Polley's experiments — According to data from the literature concerned with single crystals of nickel (cf. Becker *loc. cit.* p. 123), $K \sim -5 \times 10^4$ erg cm^{-3}, from which, at room temperature, the work required to magnetize a polycrystalline nickel sample should be $\sim 6.6 \times 10^3$ erg cm^{-3}.

From the graphical and numerical data contained in Polley's paper we obtain a value of 13×10^3 erg cm^{-3} for the work to magnetize a nickel sample at 14°C. This represents a doubling of the theoretical value. Despite the annealing treatment this specimen was in all probability unduly affected by internal stresses which altered the magnetization curve. In fact it is known that nickel is particularly sensitive to internal stresses on account of its large magnetostriction.

(5) New measurements on nickel: In these circumstances it was of interest to undertake again the measurement of b for stress-free nickel.

Method — To this end we investigated a wire of pure nickel at room temperature. The wire, of 0.4 mm diameter and 130 cm long was held inside a solenoid of length 95 cm. The flux coil suitably compensated to indicate directly variations in $4\pi \mathcal{J}_s$ on a ballistic galvanometer was 20 cm in length. All arrangements were made to trace the loops, point by point, starting from an arbitrarily chosen origin. The wire had been annealed for 3 hours at 1000°C in an atmosphere of carefully purified hydrogen. Subsequent anneals at temperatures as high as 1200°C did not perceptibly modify the shape of the magnetization curve.

Results — They are presented in Figs. 2 and 3. The coercive force of this nickel was very low, equal to 0.66 Oe. It was hoped therefore that the specimen was essentially stress-free. We found that the area contained between the descending branch of the major loop and its asymptote was equal to 6.4×10^3 erg cm^{-3} at a temperature of 27°C: this value is very close to the theoretical one obtained above.

In order to establish the law of approach to saturation for nickel treated in this way, the magnetization is plotted in Fig. 3 as a function of $1/H^2$. Between 150 and 750 Oe a clear straight line is obtained which extrapolates to a saturation magnetization of 497.5 G. The terms a/H and χH in the series (4) are not apparent in this field interval. The relation (2) with a coefficient $b = 290$ is a satisfactory representation of the experiments at 27°C. If Polley's results are interpolated for this same temperature, a having suffered very similar heat treatment; suffice it to say that the Polley sample was more massive than ours, 5 mm in diameter, and the impurities would have had more difficulty in diffusing to the surface before being eliminated. In addition, Polley does not seem to have taken due precaution about purifying the hydrogen.

Be that as it may, the main experimental conclusion is that for nickel, as free as possible from parasitic stresses, the term b/H^2 in the approach law is — between 200 and 700 Oe — a factor of two smaller than the theoretical term calculated on the basis of negligible crystallite interaction.

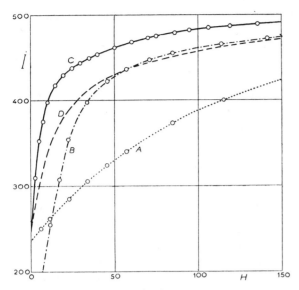

Fig. 2.
Magnetization curves in intermediate fields for different nickel specimens: A, wire straight from the reel ($H_c = 8.7$ Oe); B, the same wire annealed in air for 15 min at 900°C ($H_c = 5.4$ Oe); C, the same wire annealed in hydrogen for 3 h at 1000°C; D, the wire studied by Polley.

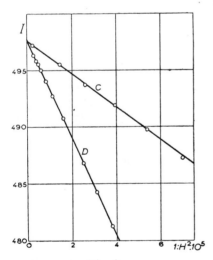

Fig. 3.
Approach to saturation in strong fields as a function of the inverse square of the field: C, nickel wire annealed in hydrogen for 3 h at 1000°C; D, wire studied by Polley (taken from his results).

value of $b = 550$ is obtained, a factor of two greater than the previous figure.

It is difficult to account for the difference in properties of the two pure nickel specimens both

II.
THEORY OF THE LAW OF APPROACH FOR A POLYCRYSTALLINE MATERIAL

(6) Analytical representation of magnetocrystalline forces in a polycrystal: We now demonstrate a calculation applicable to a polycrystal when placed in a magnetic field of sufficient intensity to ensure that the magnetization at every point is close to the saturation value. The polycrystal comprises elementary crystallites with orientations distributed at random. The magnetization is everywhere in equilibrium under the action of the external field H and the magnetocrystalline forces. As long as the study is limited to the approach to saturation the latter are formally equivalent to a magnetic field F of appropriate intensity perpendicular to the field H: naturally the magnitude and direction of F change from one grain to another.

Let a cube of side L be cut out from the polycrystal such that L is very large compared with the dimensions of the crystallites. If cubes identical to the first one are stacked together in places determined by translations along three mutually perpendicular axes then an extended periodic polycrystal is obtained which has practically identical properties with the initial polycrystal. This accomplished, the periodic polycrystal is now referred to Cartesian co-ordinates OXYZ. The component of the field F along an arbitrary direction Δ can be expressed as a Fourier series.

$$F_\delta = \sum M_{pqr} \exp\left\{\frac{2i\pi}{L}(px + qy + rz)\right\} \quad (6)$$

where the summation sign extends here, and in what follows, over all integral values of pqr between $-\infty$ to $+\infty$. The coefficients M_{pqr} are complex quantities which, F_δ being real, satisfy the relation

$$M_{-p,-q,-r} = M^*_{pqr} \quad (7)$$

where an asterisk (*) denotes the complex conjugate.

If the initial cube is cut from another site in the polycrystal then, naturally, a completely different set of coefficients M_{pqr} is obtained. As a result it is more important to know their probable value than the actual values corresponding to a particular expansion. As we have shown already in an analogous case [6], the fact that the polycrystal is, on average, isotropic, leads to the result that the probability law which fixes the value of M_{pqr} depends upon one parameter only, viz, $\rho = (p^2 + q^2 + r^2)^{1/2}$. This is because the expansion must preserve the same form for any arbitrary rotation of the axes. In particular, the mean value $\overline{M_{pqr} M^*_{pqr}}$ of the product $M_{pqr} M^*_{pqr}$ is a positive function of ρ. Assume now that the axes are rearranged so that OZ is parallel to the external field H; the components F_x and F_y along OX and OY can be represented by two series of the type (5), identical on average by reasons of symmetry. The corresponding coefficients, M_{pqr} and M'_{pqr} obey the following relations in which quantities beneath bars have been averaged with respect to all possible positions of the initial cube

$$\overline{M_{pqr}} = \overline{M'_{pqr}} = 0, \quad \overline{M_{pqr} M^*_{pqr}} = \overline{M'_{pqr} M'^*_{pqr}} = F_\rho \quad (8)$$

where F_ρ is a positive quantity which depends only on ρ. In addition, since the coefficients M and M' are independent, we have

$$\overline{M_{pqr} M'^*_{pqr}} = 0, \quad \overline{M^*_{pqr} M'_{pqr}} = 0. \quad (9)$$

(7) The approach law without interaction between crystallites: Let α and β be the angles made by the projections of the magnetization $\mathcal{J}_s$ at some point onto the XOZ and YOZ planes with the OZ direction. Since the angles α and β are small, the equilibrium condition for the magnetization is simply

$$\alpha = \frac{F_x}{H}, \quad \beta = \frac{F_y}{H}; \quad (10)$$

and to second order, the macroscopic magnetization $\mathcal{J}$ which is equal to the mean value of the magnetization on OZ, is

$$\mathcal{J} = \mathcal{J}_s \left(1 - \frac{\overline{\alpha^2} + \overline{\beta^2}}{2}\right) = \mathcal{J}_s \left(1 - \frac{\overline{F_x^2} + \overline{F_y^2}}{2H^2}\right) \quad (11)$$

To find $\overline{F_x^2}$ the relation (6) must be squared and then successively averaged with respect to xyz and all possible values of the coefficients M: taking into account relations (7) and (8) the final result is

$$\overline{F_x^2} = \overline{F_y^2} = \sum \overline{M_{pqr} M^*_{pqr}} = \sum F_\rho, \quad (12)$$

from which the law of approach may be obtained as

$$\mathcal{J} = \mathcal{J}_s \left(1 - \frac{\Sigma F_\rho}{H^2}\right), \quad (13)$$

This law must be identical with that obtained by the conventional methods. So, on comparing (2), (3) and (13) we may deduce that

$$\sum F_\rho = \frac{8}{105} \frac{K^2}{\mathcal{J}_s^2} \quad (14)$$

(8) Introduction of interactions: Allowance will now be made for the crystallite interactions ignored by Akulov and his successors, Gans and Becker. These interactions are made manifest by the existence of an internal demagnetizing field, derivable from a potential V, which has as origin the non-uniformity of the magnetization in the polycrystal. The equilibrium condition (10) must be replaced by the following:

$$\alpha = \frac{1}{H}\left(F_x - \frac{\partial V}{\partial x}\right); \quad \beta = \frac{1}{H}\left(F_y - \frac{\partial V}{\partial y}\right). \quad (15)$$

In addition the potential V satisfies Poisson's equation: $\nabla^2 V + 4\pi m = 0$ with $m = -\text{div } \mathcal{J}$. However, since the local components of magnetization along the three axes, are, to a second order, given by $\alpha \mathcal{J}_s, \beta \mathcal{J}_s$ and $\gamma \mathcal{J}_s$ respectively, we may write

$$-m = \frac{\partial \mathcal{J}_x}{\partial x} + \frac{\partial \mathcal{J}_y}{\partial y} + \frac{\partial \mathcal{J}_z}{\partial z}$$
$$= \frac{\mathcal{J}_s}{H}\left(\frac{\partial F}{\partial x} + \frac{\partial F}{\partial y} - \frac{\partial^2 V}{\partial x^2} - \frac{\partial^2 V}{\partial y^2}\right) \quad (16)$$

and Poisson's equation takes the form

$$\left(1 + \frac{4\pi \mathcal{J}_s}{H}\right)\left(\frac{\partial^2 V}{\partial x^2} + \frac{\partial^2 V}{\partial y^2}\right) + \frac{\partial^2 V}{\partial z^2}$$
$$= \frac{4\pi \mathcal{J}_s}{H}\left(\frac{\partial F}{\partial x} + \frac{\partial F}{\partial y}\right). \quad (17)$$

From expansions of type (6) which define F_x and F_y, we get

$$\frac{\partial F}{\partial x} + \frac{\partial F}{\partial y} = \frac{2i\pi}{L} \sum (pM_{pqr} + qM'_{pqr})$$
$$\times \exp\left\{\frac{2i\pi}{L}(px + qy + rz)\right\} \quad (18)$$

Equation (17) is thus satisfied by a series of the form

$$V = \sum v_{pqr} \exp\left\{\frac{2i\pi}{L}(px + qy + rz)\right\}. \quad (19)$$

in which the coefficients v are given by

$$v_{pqr} = \frac{bL(pM_{pqr} + qM'_{pqr})}{2i\pi[b(p^2+q^2)+p^2+q^2+r^2]} \quad (20)$$

with

$$b = \frac{4\pi J_s}{M}.$$

After substituting this expression for V into relation (15) an average value of α^2 with respect to xyz can be deduced as

$$\overline{\alpha^2} = \frac{1}{H^2}\sum\left[\frac{bp(pM+qM')}{b(p^2+q^2)+\rho^2} - M\right]$$
$$\times\left[\frac{bp(pM^* + qM'^*)}{b(p^2+q^2)+\rho^2} - M^*\right] \quad (21)$$

with a symmetrical expression for β^2.

The average of $\overline{\alpha^2} + \overline{\beta^2}$ over all possible values of M and M' is now sought. Taking note of relations (8) and (9) the mean value, $\overline{\overline{\alpha^2}} + \overline{\overline{\beta^2}}$ turns out to be

$$\overline{\overline{\alpha^2}} + \overline{\overline{\beta^2}} = \frac{1}{H^2}\sum F_\rho \left\{1 + \left[\frac{\rho^2}{\rho^2 + b(p^2+q^2)}\right]^2\right\} \quad (22)$$

Because the number of terms inside the summation sign is very large and the probability of finding one in the volume $dpdqdr$ is everywhere simply equal to $dpdqdr$, the sum may be replaced by an integral. By introducing now a change of variable: $p = \rho\sin\varphi\cos\theta$, $q = \rho\sin\varphi\sin\theta$, $r = \rho\cos\varphi$, $dpdqdr = \rho^2\sin\varphi d\varphi d\rho d\theta$, it can be shown that the variables are separable. Integration over θ yields the following

$$\overline{\overline{\alpha^2}} + \overline{\overline{\beta^2}} = \frac{1}{H^2}\int_{-\infty}^{+\infty}\int_{-\infty}^{+\infty}\int_{-\infty}^{+\infty}$$
$$F_\rho\left\{1 + \left[\frac{\rho^2}{\rho^2 + b(p^2+q^2)}\right]^2\right\}dp\,dq\,dr =$$
$$= \frac{4\pi G}{H^2}\int \rho^2 F_\rho d\rho, \quad (23)$$

where G is the definite integral

$$G = \int_0^{\frac{\pi}{2}}\left[1 + \frac{1}{(1+b\sin^2\varphi)^2}\right]\sin\varphi\,d\varphi \quad (24)$$

which is easily calculated as

$$G = 1 + \frac{1}{2(b+1)}$$
$$+ \frac{1}{2(b+1)^2}\sqrt{\frac{b+1}{b}}\,\text{argth}\sqrt{\frac{b}{b+1}}. \quad (25)$$

The same transformation from sum to integral, when applied to the sum ΣF_ρ gives

$$\Sigma F_\rho = 4\pi\int_0^\infty \rho^2 F_\rho\,d\rho. \quad (26)$$

Finally, after substituting expression (23) into equation (11) and using relation (26) and (14) the law of approach is obtained as

$$J = J_s\left(1 - \frac{8K^2}{105 J_s^2 H^2}\frac{G}{2}\right). \quad (27)$$

where G is given by equation (25) and $b = 4\pi J_s/H$.

(9) Discussion of the law of approach (27): Let us first assume that the external field is very large; in this case b is very small and, if second order terms in b are neglected, G reduces to $1/(1+b/3)^2$. The approach law (27) becomes

$$J = J_s\left[1 - \frac{8K^2}{105 J_s^2\left(H + \frac{4}{3}\pi J_s\right)^2}\right] \quad (28)$$

Equation (28) confirms that the effect of crystallite interaction is to augment the external field by a field equal to $(4/3)\pi J_s$: this is the classic Lorentz field whose origin has been discussed briefly above. However, from a practical aspect, it is impossible to attain the conditions under which equation (28) is valid. Consider the case of iron for which $4\pi J_s = 21,500$: here a field of at least 100 kOe is required to make b sufficiently small. In the event we know that the fields involved in the experimental study of the $1/H^2$ approach law did not exceed 2 kOe. In these conditions b must be considered as large. Evaluation of expression (25) shows that G/2 (which equals 0.5 when $1/b = 0$) increases almost linearly with $1/b$ up to $1/b = 0.3$. The figures in the table below are limited to the region amenable to experiment. In this regime G varies but little with H

TABLE

$\dfrac{1}{b}$	0,022	0,050	0,116	0,276
$\dfrac{G}{2}$	0,506	0,513	0,531	0,576

which means that the actual law of approach becomes to a first approximation a rigorous $1/H^2$ law, as found experimentally. However, the coefficient of the term in $1/H^2$ is clearly half as big as that obtained if interactions are ignored.

In the experiments described above (§ 5) on the polycrystalline nickel wire the study of the approach law in $1/H^2$ was made in fields ranging from 150 Oe to 750 Oe: hence $1/b$ varies from 0.024 to 0.1 which, according to the table, corresponds to a mean value of $G/2 \sim 0.515$. Having found the coefficient of the $1/H^2$ term to be 290, equation (27) now yields a value of $K = 4.3 \times 10^4$ erg cm^{-3} at $t = 27°C$. The measurements of Brukhatov and Kirensky [8] on a single crystal at the same temperature give $K = 4.16 \times 10^4$ erg cm^{-3}. There is excellent agreement between the two values.

(10) Conclusions: As far as nickel is concerned, the introduction of interactions between crystallites leads to a formulation which is in agreement with experiment.

Conversely, it seems that equation (27) may be used to deduce the anisotropy constant from the law of approach; as a first approximation, G may be taken as unity, i.e. corresponding to the Akulov–Becker law but with a coefficient half as big.

K can also be obtained from the energy required for magnetization, as outlined in § 4, but this method appears less certain and less convenient.

Speaking generally it would be good to check, for example with X-rays, that the material is genuinely isotropic with a random crystallographic distribution of grains, something that is not always true of drawn wires. In addition the material should be as free as possible of internal stresses and contain neither voids nor non-magnetic inclusions: as we have shown previously [6] the existence of cavities greatly affects the approach law. If these precautions are taken then measurements on polycrystalline specimens could yield useful information about the magnetic anisotropy of cubic materials.

REFERENCES

[1] Akulov, *Z.f. Physik*, 1931, **69**, 822.
[2] Gans, *Ann. der Physik*, 1932, **15**, 28.
[3] Becker and Doring, Ferromagnetismus, p. 168 (Springer, Berlin, 1939).
[4] Czerlinski, *Ann. der Physik*, 1932, **13**, 80.
[5] Polley, *Ann. der Physik*, 1939, **36**, 625.
[6] L. Néel, *J. de Phys.*, 1948, **9**, 184.
[7] Kersten, *Z.f. Physik*, 1932, **76**, 505.
[8] Brukhatov and Kirensky, *Phys. Z. Sowjetunion*, 1937, **12**, 602.

A93 (1954)
THE APPROACH TO SATURATION OF THE MAGNETOSTRICTION

The law of approach to saturation of the longitudinal magnetostriction, λ, of a cubic polycrystalline ferromagnetic has been studied by Schlechtweg [1] who showed that the first term in a series expansion is of the form $1/H$. Recently, Lee [2] has calculated the subsequent term in $1/H^2$, a term which vanishes except in the case of isotropic magnetostriction. Starting with the existence of a magnetocrystalline energy F_m given by

$$F_m = K(\alpha_1^2 \alpha_2^2 + \alpha_2^2 \alpha_3^2 + \alpha_3^2 \alpha_1^2)$$

and following the path used by Becker [3] in his work on the approach to saturation of the magnetization, Lee shows that the longitudinal magnetostriction of a single crystal magnetized along a direction with direction cosines $\alpha_1, \alpha_2, \alpha_3$ (referred to the cube edges as co-ordinates) approaches saturation as

$$\lambda = \frac{3}{2}(\lambda_{100} - \lambda_{111})\left[S_4 - \frac{1}{3} + 2\rho(S_6 - S_4^2) + \rho^2(7S_8 - 17S_4 S_6 + 10S_4^3)\right]$$
$$+ \lambda_{111}\left[1 - \frac{3}{2}\rho^2(S_6 - S_4^2)\right]. \quad (1)$$

In this formula we have put

$$\rho = \frac{K}{HJ}, \quad S_n = \alpha_1^n + \alpha_2^n + \alpha_3^n,$$

while λ_{100} and λ_{111} represent the saturation magnetostriction along the cube edges and cube diagonals respectively; J is the spontaneous magnetization and H, the magnetic field. On averaging over all spatial directions Lee finds the following approach law for an assembly of randomly oriented crystallites:

$$\lambda = \frac{2\lambda_{100} + 3\lambda_{111}}{5}$$
$$+ (\lambda_{100} - \lambda_{111})\left[\frac{8}{5.7}\rho - \frac{32}{7.11.13}\rho^2\right]$$
$$- \lambda_{111}\frac{8}{5.7}\rho^2 + \cdots \quad (2)$$

Expression (2) applies only to independent particles. However, we know [4, 5] that in the case of magnetization saturation, allowance can be made for particle interaction by multiplying the $1/H^2$ term in the approach law for "independent particles" by a factor $G/2$ which is less than unity. To take account of interactions Lee proposes, therefore, to replace H in expression (2) by an internal field H' related to the external field through

$$\left(\frac{H}{H'}\right)^2 = \frac{G}{2}.$$

It seems to me preferable to effect a correction for interactions by multiplying all the terms in ρ and ρ^2 of equation (2) by $G/2$. This procedure can be justified by a scrutiny of the method used previously to resolve the problem of interactions [5]. Close to saturation an assembly of N atomic moments possesses $2N$ degrees of oscillational freedom. By a suitable linear combination of these degrees of freedom an equivalent system can be obtained, also containing $2N$ terms, but separable into two groups each with N terms having the following properties. In the first group, the divergence of the vector representing fluctuations in magnetization is zero: the oscillations are therefore not accompanied by disperse fields and so can occur freely. In the second group on the other hand, the divergence is non-zero with the result that oscillations of the atomic moments engender a magnetic field together with a significant increase in potential energy,

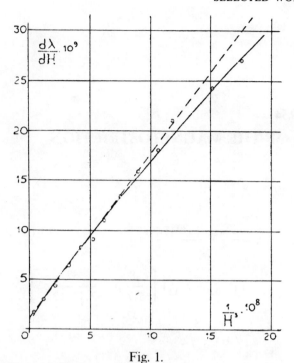

Fig. 1.
The approach to saturation of the longitudinal magnetostriction of nickel. The circles represent the experimental points of Lee: the dashed line is that traced by Lee to represent his results. The solid curve corresponds to the theoretical formula of the text with a correction factor of $\frac{1}{2}$.

although, in fact, the amplitude of such oscillations can only be small. Thus over a large range of fields half the degrees of freedom are effectively blocked: the presence of interactions has thereby halved the deviations from saturation. Calculation actually shows [5] that in the field region of interest here $G/2$ is always close to $1/2$. Obviously this argument applies equally as well to the magnetostriction as to the magnetization. In practice the correction can be effected to a first approximation by multiplying all ρ and ρ^2 terms of equation (2) by a half.

The theoretical results obtained in this way for iron and nickel have been compared with experiment. For iron we have used

$$K = 5.10^5 \text{ ergs/cm}^3, \quad \lambda_{100} = 20.10^{-6},$$
$$\lambda_{111} = -20.10^{-6}, \quad J = 1700 \text{ e.m.u.}$$

and for nickel,

$$K = -5,1.10^4 \text{ ergs/cm}^3, \quad \lambda_{100} = -54.10^{-6},$$
$$\lambda_{111} = -27.10^{-6}, \quad J = 470.$$

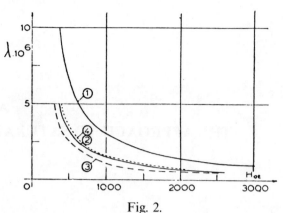

Fig. 2.
The approach to saturation of the longitudinal magnetostriction of iron. 1. Theoretical curve without correction for interactions. 2. Theoretical curve with a correction factor of $\frac{1}{2}$. 3. Experimental results of Weil and Reichel. 4. Experimental results of Kornetzki.

As far as experimental results are concerned we have made use of the excellent measurements of Lee [6] on nickel as well as adopting his graphical technique of plotting $d\lambda/dH$ as a function of $1/H^3$. For iron we have used the results of Kornetzki [7] suitably corrected for the demagnetizing field ($N = 0.135$) and the volume term proportional to the field as well as the results of Weil and Reichel [8] obtained at Grenoble using a strain gauge technique. Figs. 1 and 2 show satisfactory agreement between theory and experiment. In particular for the case of iron we have shown the theoretical curve (curve 1, Fig. 2) uncorrected for interactions: it is very much less satisfactory than the corrected curve. This would be still more striking for the case of nickel.

It may also be concluded from these results that variations in the behaviour of the magnetostriction at high fields can easily be interpreted by current theories and seem to present no anomalies.

REFERENCES

[1] O. Rudiger and H. Schlechtweg, *Ann. Physik*, 1941, **39**, 1; 1942, **41**, 1; 1942, **5**, 87.
[2] E. W. Lee, Private communication. *Proc. Phys. Soc.*, in the press.
[3] R. Becker and W. Doring, Ferromagnetismus, Springer, Berlin, 1939.
[4] T. Holstein and H. Primakoff, *Phys. Rev.*, 1941, **59**, 388.
[5] L. Néel, *C.R. Acad. Sci.*, 1945, **220**, 814; *J. Physique Rad.*, 1948, **9**, 193.
[6] E. W. Lee, *Proc. Phys. Soc.*, B, 1952, **65**, 162.
[7] M. Kornetzki, *Z. Physik*, 1933, **87**, 560.
[8] L. Weil and K. Reichel, *Comm. Soc. franc. Physique*, Grenoble, 11 février 1954; *J. Physique Rad.*, 1954, **15**, 72s.

Chapter VI

PHASES AND MODES IN FERROMAGNETIC CRYSTALS
A52, C29, A58, A127, A85

52 (1944)

THE LAWS GOVERNING THE MAGNETIZATION PROCESS AND THE SUBDIVISION INTO ELEMENTARY DOMAINS OF MONOCRYSTALLINE IRON

Abstract: *First part*: This first section deals with the laws governing the magnetization process in a *pure undeformed iron monocrystal*, the basic supposition being that there are no internal magnetic charges. The elementary domains, using the nomenclature of Weiss, are first grouped into *phases* each of which corresponds to a well determined direction of the spontaneous magnetization. It is shown that, in general, the magnetization curve of an iron monocrystal is composed of four successive parts each corresponding to a different magnetization *mode*. These modes, I, II, III and IV correspond respectively to the presence of 6, 3, 2 and 1 phases in equilibrium.

In mode I the material is magnetically isotropic, the internal field is zero and the overall magnetization depends only on the external geometry of the crystal. In mode II the internal field is always oriented along a three-fold axis and the spontaneous magnetization directions of the three phases all make equal angles to this axis. Simple conditions are also satisfied by the internal field and the spontaneous magnetization directions of the two phases of mode III.

The most important examples experimentally are then dealt with in detail and give the following original results: the laws governing the magnetization process according to mode II are established for randomly oriented thin bars and a comparison is made with the experimental results of Kaya and Sizoo; the process according to mode III is also determined for bars whose long axis is oriented approximately perpendicular to a fourfold axis of the crystal. The magnetization processes according to modes I, III and IV are also established for the case of discs sectioned from one of the cube basal planes and a comparison is made with the experimental results of Webster and Honda.

The last two sections concern original work on the determination of the form and dimensions of elementary domains.

Second part: It is shown that the elementary domains have the form of plane sheets stacked together with their plane perpendicular to a threefold axis in mode II and to a twofold axis in mode III. This allows the interpretation of the experimental results of Sixtus and Kaya on the orientation of powder lines (Bitter patterns).

Third part: The theory of the secondary surface structure in the elementary domains is established (notion of closure domains). The theory is used to calculate the thickness of the sheet-like domains composing the main structure. The variations in the spacing of the periodic powder lines as a function of applied field is examined in a simple case leading to a quantitative interpretation of the experimental results of Sixtus and Kaya.

INTRODUCTION

(1) Elementary domains and the magnetization laws: After establishing the theory of ferromagnetism which still remains valid today in its main concepts, P. Weiss was obliged to reconcile the apparent ease of demagnetization with the strength of the molecular field: he proposed therefore that an apparently demagnetized ferromagnetic material is in reality subdivided into *elementary domains* each magnetized to saturation but along different directions. This hypothesis, universally accepted nowadays, provoked a good deal of doubt at first but Barkhausen's experiments, which revealed the magnetization discontinuities produced during a hysteresis cycle, were subsequently considered as a direct proof of the existence of domains. In fact these discontinuities, at least in the region of low magnetization and weak fields, concern only irreversible displacements of a relatively small fraction of walls separating adjacent domains. Many other less direct but more convincing observations lead to the present day situation in which the existence of elementary domains is an absolute certainty.

A certain number of points are now clear: it is known that, in weak fields, the spontaneous magnetization can only be oriented along a finite number of directions, called *privileged or easy magnetization directions*, for iron these directions are the three fourfold crystal axes. The expression for the energy, the so-called magnetocrystalline energy, necessary to bring the spontaneous magnetization out of a privileged into an arbitrary direction is also known: thus to bring the magnetization from a fourfold into a threefold direction requires an energy expenditure of around 1.4×10^5 ergs/cm^3.

Since the work of Bloch [1] it is also known that the wall separating two elementary domains is characterized by a certain surface energy and has a finite thickness which results from the equilibrium between the Heisenberg exchange forces which tend to make the wall as wide as possible and the magnetocrystalline energy which tends on the contrary to reduce the wall width. In the case of iron[1] the wall thickness varies from 500 to 2000 Å and the corresponding surface energy is in the range 0.7 to 1.4 ergs/cm^2.

It is also known that magnetization changes in a material occur by two different mechanisms: (a) a *rotation*, usually reversible, of the spontaneous magnetization within each domain whose boundaries remain fixed; (b) a *displacement*, reversible or irreversible, of the *walls separating* adjacent domains whose magnetization directions remain fixed[2].

In the general case of an arbitrary overall magnetization little is known about the directions taken up by the spontaneous magnetization within the domains, about the different possible types of domain, their shape, size and relative spatially distributions.

The macroscopic magnetization laws alone have allowed only two cases to be successfully treated: that of a monocrystal under a field directed along a symmetry axis (four, three or two-fold axis) and that of a crystal in a high field. In reality in these two cases only the validity of the mathematical expression for the magnetocrystalline energy as a function of the direction cosines of the magnetization is verified since all phenomena involving subdivision into elementary domains are completely ignored. In the first case this omission has no serious repercussions since, by symmetry, the different categories of domains possess the same proprieties so that as far as the calculation is concerned it is as if there were only one type of domain. The same situation applies to the second case since due to the strength of the applied field, there is indeed only one type of domain.

This brief summary shows that there is no existing theory for the magnetization process in a ferromagnetic material which takes any real explicitly detailed account of the subdivision into elementary domains.

In this article we try to lay the foundations for such a theory from purely energetic considerations by noting that as far as the subdivision into elementary domains is concerned the demagnetizing field plays a fundamental role the importance of which has been recognized by only a few authors, de Waard [2], Landau and Lifshitz [3], Gorter [4] among others. Since this subdivision has no direct relation with hysteresis effects it appears advisable to investigate the process for an ideal, pure and undeformed iron crystal. It so happens that the real crystals which have been experimentally studied by various investigators are sufficiently near to this ideal state to allow meaningful comparison with the theory. From this point of view iron is a very good material of much greater interest than nickel in which accidental internal stresses provoke serious perturbations because of the large magnetostriction.

The first part of this article will be devoted to a purely formal study of the decomposition into domains. We will determine the *number of distinct categories of elementary domains, their relative proportions* as well as the corresponding *directions* of the *spontaneous magnetization*. In the second part we investigate the *form* and *orientation* of the domains. Finally in the third part we determine their *absolute dimensions* in relation to certain complicated surface effects in which special types of domain called closure domains come into play.

FIRST PART
DIFFERENT MAGNETIZATION MODES FOR A MONOCRYSTAL

(2) Magnetic energy and internal charges: Consider any ferromagnetic material divided into elementary domains whose size is such that one can take volume elements $d\tau$ within the material which are large with respect to the domains but sufficiently small compared to the crystal dimensions. The mean internal field $\vec{H}_i$ and mean magnetization $\vec{J}$ at any point are then defined as the mean field and mean magnetization of volume element $d\tau$ centred on this point. The mean field $\vec{H}_i$ is the sum of the applied field $\vec{H}$ and of a mean demagnetizing field $\vec{H}_d$ which is created by a fictitious charge distribution defined by the mean volume charge density $\rho = -\text{div}\,\vec{J}$ and the mean surface charge density $\vec{J}_N$.

At a point taken within the element $d\tau$ the local magnetization is equal to $\vec{J} + \vec{i}$ and the local field is given by $\vec{H} + \vec{H}_d + \vec{h}$: the field $\vec{h}$ is created by the charge distribution on the surfaces separating the elementary domains. The mean values of $\vec{h}$ and $\vec{i}$ within the element $d\tau$ are zero by definition

---

[1] For details refer to an article by the author concerning a detailed study of walls in iron (Cahiers de Physique — to be published 1945). This article will be referred to several times in the present article under the symbol A.

[2] See the experiments of Elmore (*Phys. Rev.* 1938, **53**, p. 757) who has directly observed the discontinuous wall displacements under the microscope. The detailed theory of wall movements in weak fields is established by Néel (Cahiers de Physique 1942, no. 12, p. 1 and 1943, no. 13, p. 18).

$$\int_{d\tau} \vec{h}\, dv = 0, \quad \int_{d\tau} \vec{i}\, dv = 0. \tag{1a}$$

The magnetic energy W_M of the system is given by the integral

$$W_M = -\int \vec{H}(\vec{J} + \vec{i})\, dv + \int \frac{(\vec{H}_d + \vec{h})^2}{8\pi}\, dv \tag{1b}$$

taken over all space.

By decomposing the integration space into volume elements $d\tau$ for which one can write

$$\left. \begin{array}{l} \int_{d\tau} \vec{H}\vec{i}\, dv = \vec{H} \int_{d\tau} \vec{i}\, dv = 0, \\[6pt] \int_{d\tau} \vec{H}_d \vec{h}\, dv = \vec{H}_d \int_{d\tau} \vec{h}\, dv = 0, \end{array} \right\} \tag{1c}$$

one finally obtains the expression

$$W_M = -\int \vec{H}\vec{J}\, dv + \int \frac{H_d^2}{8\pi}\, dv + \int \frac{h^2}{8\pi}\, dv \tag{1d}$$

The presence of surface charges on the domain walls introduces therefore a supplementary term $h^2/8\pi$ which is always positive and which can reach very large values: h can take on the value $4\pi J_s$ which, for iron, gives an energy term of 1.8×10^7 ergs/cm³. The importance of such a value can be appreciated by remarking that the total energy of certain subdivisions into domains which are investigated later hardly exceed 100 ergs/cm³.

Any subdivision into elementary domains not carrying internal free charges must therefore have a particularly small potential energy since the supplementary energy term must then be zero because h is zero throughout the material.

Furthermore, if the spontaneous magnetization of domains making up an internal charge free subdivision is in equilibrium under the combined influence of the applied field, the demagnetizing field and the magnetocrystalline forces, and if the different domains are in equilibrium with respect to each other, then the subdivision in question will be stable with respect to all the parameters which characterize it: namely the relative proportions of the domains, the directions of the spontaneous magnetization and the orientation of the walls. We will suppose that such a subdivision corresponds to the most stable state of the crystal in the field considered.

(3) Objectives of this investigation and fundamental hypotheses: We propose to establish the magnetization laws for an ideal and perfect iron monocrystal, limited by a *second order surface*[3] *and placed in a uniform magnetic field* $\vec{H}$. The nomenclature ideal and perfect crystal is taken to mean a crystal whose lattice is identical to the theoretical one without any vacancies, inclusions or any deformation.

(A) Following the conclusions of the previous section we suppose that the structure of the elementary domains, into which the crystal is subdivided, is such that the internal free charges are everywhere zero. We will show in detail in the second part of the present article that in the case of a monocrystal it is always possible to find such a structure (this is not possible in the case of polycrystals). For the moment it suffices to recognize the existence of this type of structure.

(B) Since there are no internal charges the internal field H_i is uniform and equal everywhere to the vectorial sum of the applied field H and a uniform demagnetizing field H_d which is the same as that which would exist in the crystal if the magnetization were uniform and equal to the overall mean magnetization.

(C) In addition we accept the possibility of decomposing the crystal into a *finite number of phases* each possessing the same uniform magnetization. A phase i thus includes all domains whose magnetization is parallel to a certain direction. The modulus of this spontaneous magnetization is of course equal to the saturation magnetization at the ambient temperature, that is $\vec{J}_s = 1710$ cgs at room temperature. If x_i represents the fraction of the total volume occupied by the phase i we obtain, by definition:

$$\sum x_i = 1, \quad \sum x_i \vec{J}_i = \vec{J}. \tag{1e}$$

(D) Finally, we suppose that a reversible mechanism exists allowing the growth of any phase at the expense of another, in fact this will occur by the movement of domain walls, and we will examine the equilibria existing between the different phases. In this first section we will take account only of the magnetic and of the magnetocrystalline energies neglecting the surface energy of the wall separating the domains.

(4) The equilibrium positions of the spontaneous magnetization in a given field: The first problem is to determine the possible phases. The following remark simplifies the solution: in the equilibrium state the

---

[3] It is known that a uniformly magnetized body limited by a second order surface has a uniform demagnetizing field: in a uniform field it therefore takes up a uniform magnetization. The same situation holds when the demagnetizing field is negligible, that is when an isotropic material sectioned into a very long cylinder of arbitrary cross-section is magnetized along the direction of one of the generating lines.

spontaneous magnetization $\vec{J}_i$ of each phase is in equilibrium under the combined action of the internal field $\vec{H}_i$ and the magnetocrystalline forces which couple the magnetization to the crystal lattice. It is necessary then to begin by examining this equilibrium.

In cubic ferromagnetic crystals the potential energy which corresponds to the magnetocrystalline forces can, to a sufficient approximation be expressed in the following form

$$W_C = K(\alpha^2 \beta^2 + \beta^2 \gamma^2 + \gamma^2 \alpha^2), \qquad (2)$$

where (α, β, γ) are the direction cosines of the spontaneous magnetization with respect to the fourfold crystal axes taken as coordinate axes. In the case of iron K is positive and equal to 4.3×10^5 ergs/cm^3. When the field H_i is zero there are six distinct equilibrium positions:

$$\alpha = \pm 1, \quad \beta = \gamma = 0; \quad \beta = \pm 1, \quad \alpha = \gamma = 0;$$
$$\gamma = \pm 1, \quad \alpha = \beta = 0.$$

These six positions all correspond to the same total energy, equal to zero: they are therefore rigourously equivalent. *Consequently when the internal field is zero six different phases can coexist in equilibrium within the crystal.*

Apply now a field $\vec{H}_i$ with direction cosines (p, q, r) with respect to the fourfold axes of the crystal OX, OY, OZ $(p \geq q \geq r)$. If H_i is sufficiently weak the different spontaneous magnetization directions are only slightly deflected from their initial equilibrium positions and the corresponding potential energies take on the following values, to within the second order:

$$\mp p H_i \mathcal{J}_s, \quad \mp q H_i \mathcal{J}_s, \quad \mp r H_i \mathcal{J}_s.$$

The six positions so obtained are in general no longer equivalent: the most stable position is that having the lowest energy, in the present case this would be the position originating from the initial position OX.

To summarize, when p is greater than q one of the phases is energetically more favourable than the others and this corresponds to the phase whose initial position OX was the closest to the positive direction of the field. We will suppose that this somewhat self evident proposition, remains true for any value of H_i.

From the preceeding proposition we immediately deduce that *if the internal field is non-zero and if its direction is such that p is bigger than q and r then it is impossible to establish an equilibrium between the different phases: the crystal contains a single phase.*

On the other hand when the direction cosines are such that $p = q > r$ there are two equivalent equilibrium positions which are symmetrical with respect to the plane $x + y = 0$, and finally when $p = q = r$ there are three equivalent equilibrium positions, symmetrical with respect in the three-fold axes. *For equilibrium between three phases the internal field must be parallel to a threefold axis: for equilibrium between two phases the internal field direction should be such that two of its direction cosines are equal, the third having a smaller modulus than the other two.*

These simple remarks allow four possible magnetization modes to be distinguished depending on the number of coexisting phases and the nature of the internal field. Table I summarizes the characteristics of these different phases. We will show

TABLE I.
The different magnetization modes for an ideal iron monocrystal.

Magnetization mode	Internal field		Number of phases
	Intensity	Direction	
I	$H_i = 0$		6
II	$H_i > 0$	$p = q = r$	3
III	$H_i > 0$	$p = q > r$	2
IV	$H_i > 0$	$p > q > r$	1

later that in general the magnetization curve of a single crystal is subdivided into four parts with a change in slope from gfe to the next and that each segment corresponds to a different magnetization mode.

(5) Geometrical representation of the mean magnetization: Consider the case in which six phases coexist (mode I) and let x and $\bar{x}$ represent the fractions of the total volume occupied by the phases magnetized respectively along the direction OX and the reverse direction OX'. Similarly let $y, \bar{y}$ and $z, \bar{z}$ be the fractions occupied by the phases magnetized along OY, OY' and OZ and OZ'. By definition

$$x + \bar{x} + y + \bar{y} + z + \bar{z} = 1, \qquad (3)$$

whilst the mean magnetization $\vec{J}$ is written as a function of the unit vectors $\vec{i}, \vec{j}, \vec{k}$ along the fourfold axes

$$\vec{J} = [(x - \bar{x})\vec{i} + (y - \bar{y})\vec{j} + (z - \bar{z})\vec{k}] \mathcal{J}_s. \qquad (4)$$

From a point O take three rectangular axes OX, OY, OZ parallel to the fourfold crystal axes and consider a vector OM equal to the mean magnetization. It is clear that the point M will always lie within the regular octahedron AA'BB'CC' limited by the eight planes

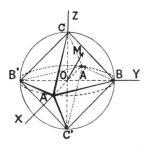

Fig. 1.

$\pm x' \pm z' \pm y' = \mathcal{J}_s$ (Fig. 1). This octahedron is included within the sphere with centre O and of radius $\mathcal{J}_s$. Each point M within this octahedron represents a possible magnetization state for the mode I.

Similarly, elementary geometry shows that all the points representing magnetization states belonging to the modes II and III are to be found within the volume enclosed between the octahedron and the circumscribed sphere. The point representing the magnetization of a demagnetized body is at O. When this specimen is magnetized to saturation the point M representing the magnetization will move from O towards the surface of the sphere of radius $\mathcal{J}_s$. It will cross the octahedron: the initial magnetization process will therefore take place according to the mode I.

(6) The initial magnetization phase (mode I): Since in this magnetization mode the internal field is zero the magnetization taken up by the crystal in an applied field H will be found by setting the demagnetizing field H_d equal and opposite to the applied field H. *For the sake of simplicity we will suppose henceforth that the applied field is always parallel to one of the three principal axes of the ellipsoid which delimits the crystal.* Despite this restriction our treatment will cover all experiments so far made with single crystals whilst a treatment of the more general case would not reveal anything essentially new. In these conditions, since the field is directed along a principal axis, the demagnetizing field is also directed along this axis and is equal to $-N\mathcal{J}$ where N represents the demagnetizing factor relative to the corresponding principal axis. The mean magnetization is therefore both parallel and proportional to the applied field

$$\mathcal{J} = \frac{1}{N}\vec{H}. \qquad (5)$$

It is independent of the orientation of the crystal axes with respect to those of the ellipsoid ... Experiment shows that cubic crystals are indeed magnetically isotropic in weak applied fields [5].

The magnetization and the field maintain the proportionality defined by equation (5) as long as the extremity M of the vector OM is within the octahedron. The passage across the surface of the octahedron is accompanied by a change in the magnetization mode due to the fact that the internal field becomes non-zero. We will call the magnetization value $\mathcal{J}_r$ corresponding to this transition the *ideal remanent magnetization*. Its value is obtained by taking the intersection of the face $x' + y' + z' = \mathcal{J}_s$ of the octahedron with a straight line parallel to the applied field and passing through O. Let l, m, n be the direction cosines of the field and $x\mathcal{J}_s$, $y\mathcal{J}_s$, $z\mathcal{J}_s$ be the coordinates of the intersection, then

$$\frac{x\mathcal{J}_s}{l} = \frac{y\mathcal{J}_s}{m} = \frac{z\mathcal{J}_s}{n}$$

$$= \frac{\mathcal{J}_s}{l+m+n} = \sqrt{\frac{(x^2+y^2+z^2)\mathcal{J}_s^2}{l^2+m^2+n^2}} = \mathcal{J}_r. \qquad (6)$$

The ideal remanent magnetization is thus related to the direction cosines of the applied field according to the relation

$$\mathcal{J}_r = \frac{\mathcal{J}_s}{l+m+n}$$

Kaya [6], from purely empirical considerations, was the first to propose this formula to describe the results of his experiments on monocrystalline iron bars, but the underlaying theoretical signification was only revealed at a later date by Gorter [7].

The magnetization given by formula (5) is well defined; nevertheless the relative proportions of the six coexisting phases are not determined since the six quantities $x, \bar{x}, y, \bar{y}, z, \bar{z}$ are only given by four relations: equation (3) and the three equations deduced from the vectorial equation (4). The situation changes when the remanent magnetization is reached: at this point the proportions $\bar{x}, \bar{y}, \bar{z}$ of the phases magnetized along OX', OY', OZ' decrease to zero and only the three phases magnetized along OX, OY, OZ remain with relative proportions given according to (4) and (6) by

$$\left. \begin{array}{c} x = \dfrac{l}{l+m+n}, \quad y = \dfrac{m}{l+m+n}, \\[6pt] z = \dfrac{n}{l+m+n}. \end{array} \right\} \qquad (7)$$

If none of the direction cosines is zero then three phases coexist in the material so that by continuity a subsequent increase in the applied field must change the overall magnetization according to mode II.

(7) Examination of the magnetization process according to mode II in the case of bars: We have just shown that in the general case the first region of the magnetization curve, in which the magnetization proceeds according to mode I with six phases, is followed by a region within which the magnetization proceeds according to mode II with three phases. We now examine this second region in the particularly important experimental case of monocrystalline bars. These bars are always sufficiently elongated that the transverse demagnetizing field coefficient is very close to 2π. We will suppose, in addition, in order to simplify the presentation, that the longitudinal demagnetizing field coefficient is zero: in practice this is never exact but it is easy to reduce the situation to this case by a simple correction which consists of subtracting the longitudinal demagnetizing field, which in any case is always weak, from the values of the applied field.

The direction cosines of the long axis of the bar with respect to the fourfold crystal axes, chosen with appropriate positive directions, will be called l, m, n ($l > m > n > 0$). The external field will likewise be parallel to this long axis. The remanent magnetization $\mathcal{J}_r = \mathcal{J}_s/(l+m+n)$ is attained for an infinitesimally small field beyond which the magnetization proceeds according to the mode II: the internal field $\vec{H}_i$ must therefore be directed along the threefold crystal axis which is closest to the applied field, that for which the three direction cosines are equal to $1/\sqrt{3}$. Since the

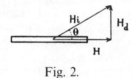

Fig. 2.

longitudinal demagnetizing field is zero the internal field $\vec{H}_i$ is reduced to the geometrical sum of the external field $\vec{H}$, parallel to the axis of the bar and of a transverse demanetizing field $\vec{H}_d$ perpendicular to the axis of the bar (Fig. 2). Taking θ as the angle between $\vec{H}$ and $\vec{H}_i$

$$\cos\theta = (l + m + n) : \sqrt{3},$$

shence

$$H_i = H : \cos\theta = \sqrt{3}H : (l + m + n). \qquad (8)$$

The spontaneous magnetizations OA, OB, OC of the three phases, initially directed along OX, OY, OZ, rotate under the effect of this internal field H_i and

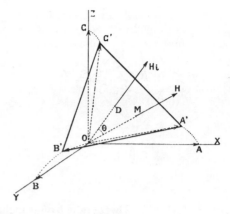

Fig. 3.

approach symmetrically the threefold axis following OA', OB', OC' (Fig. 3). The plane A'B'C' normal to the threefold axis intersects this axis at D and the length of OD is a function of H_i

$$OD = F(H_i) \qquad (9)$$

If we denote by ξ, η, ζ, the proportions of the phases magnetized along OA', OB', OC' ($\xi + \eta + \zeta = 1$) and by $\vec{i}'$, $\vec{j}'$, $\vec{k}'$ the corresponding unit vectors the mean magnetization of the system is then given by

$$\vec{OM} = (\xi\vec{i}' + \eta\vec{j}' + \zeta\vec{k}')\mathcal{J}_s. \qquad (10)$$

The point M is therefore situated in the plane, and within the triangle, A'B'C'. But we also know that *this magnetization OM must be almost exactly parallel to the axis of the bar*, since because of the large value of the transverse demagnetization field coefficients an extremely small deviation of the magnetization away from the long axis will produce the required transverse demagnetizing field $\vec{H}_d$. As a result the magnetization $\mathcal{J} = OM$ can be obtained to within a sufficient approximation by taking the intersection of the plane A'B'C' with a line through O parallel to the axis of the bar. This gives

$$QM = OD/\cos\theta$$

whence finally

$$\mathcal{J}\frac{l+m+n}{\sqrt{3}} = F\left(\frac{H\sqrt{3}}{l+m+n}\right) \quad \text{(mode II)} \qquad (11)$$

Thus the curve obtained by plotting $\mathcal{J}(l+m+n)$ as a function of $H/(l+m+n)$ is identical for all bars whatever their orientation.

This curve is easy to calculate since OD is the

projection of the vector OA' onto a threefold axis: it is one of the equilibrium positions of the spontaneous magnetization under the combined effect of the magnetocrystalline forces and a field H_i parallel to the threefold axis. The curve is therefore the classical magnetization curve along a threefold axis as calculated by Akulov [8]. Putting

$$\eta = OD : \mathcal{J}_s,$$

we find

$$3H_i\mathcal{J}_s : K = (\eta + \sqrt{2}\sqrt{1-\eta^2}) \times (2\eta\sqrt{2} + \sqrt{1-\eta^2})(\eta\sqrt{2} - \sqrt{1-\eta^2}), \quad (12)$$

whence by inversion we find the function (9).

The numerous measurements of Kaya [9] and Sizoo [10] on diversely oriented bars are plotted in Fig. 4,

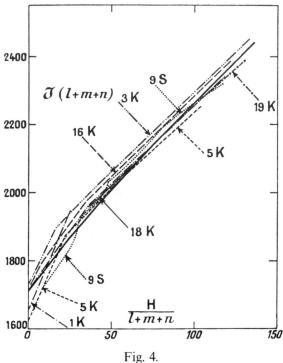

Fig. 4.

Mode II magnetization curves for variously oriented monocrystalline iron bars. The full line corresponds to the theoretical curve, the broken lines are the experimental curves. The suffixes K and S refer respectively to the experiments of Kaya and of Sizoo.

according to the representation indicated above, together with the theoretical curve deduced from the relationship (12). The deviations from the theoretical curve which are rarely as much as 2% appear to be mainly due to small errors, of the order of a degree, in the orientation of the axes. The general agreement of the calculation with experiment is therefore very satisfactory and leaves no doubt as to the correctness of the mechanism which we have described.

The process governed by equation (11) remains valid as long as three phases coexist. But according to the relationship (10) one of these phases will be eliminated when the vector OM has one of its components along the three directions OA', OB', OC' equal to zero: this is exactly what happens when, under the influence of the increasing field H_i, the triangle A'B'C' shrinks so that the direction OM rotates into the plane A'OB'. Henceforth only two phases continue to coexist and the magnetization process must necessarily continue according to the mode III. An elementary calculation in spherical trignometry shows that the magnetization $\mathcal{J}_c$ corresponding to this transition is given by the formulae

$$\mathcal{J}_c = \mathcal{J}_s \cos\lambda \quad \text{with} \quad \tan\lambda = \frac{2(l+m) - 4n}{\sqrt{2}(l+m+n)}. \quad (13)$$

This magnetization, $\mathcal{J}_c$, beyond which the law (11) is no longer valid varies according to the orientation of the bar. The relations (13) are used to find the appropriate cut-off points for the curves of Fig. 4.

(8) Magnetization of a bar in mode III: When the magnetization value is greater than $\mathcal{J}_c$ only two phases remain, these are magnetized along the directions OA' and OB' which correspond to the two fourfold axes closest to the long axis of the bar. The calculation of the magnetization becomes rather difficult except in the case where $n = 0$ that is to say when the long axis of the bar is situated in one of the basal planes of the cube: in this case the mode II does not exist and there is a direct transition from mode I to mode III. In this situation one of the symmetry planes of the bar is also a plane of magnetic symmetry. Since the applied field is in this plane so are the mean magnetization and the internal field: the internal field is therefore parallel to that twofold axis which internally bisects the two axes OX and OY. The problem reduces to two dimensions and by following exactly the same arguments as for the three dimensional problem examined in the previous paragraph it can be shown that in mode III, if n is zero, then the magnetization is related to the field by the formula

$$\mathcal{J}\frac{l+m}{\sqrt{2}} = G\left(\frac{H\sqrt{2}}{l+m}\right) \quad \text{(mode III)}. \quad (14)$$

In this case the relationship $\mathcal{J} = G(H)$ represents the classical magnetization curve along the twofold axis. It is derived from (11) by inverting the function

$$H_i \mathcal{J}_s = 2K\eta(2\eta^2 - 1) \quad \text{avec} \quad \eta = \mathcal{J} : \mathcal{J}_s. \quad (15)$$

when H is small, the first terms in the series development of (14) are written

$$\frac{\mathcal{J}}{\mathcal{J}_s} = \frac{1}{l+m} + \frac{\mathcal{J}_s H}{2K(l+m)^2} + \cdots \quad \text{(mode III)} \quad (16)$$

The initial slope is, therefore,

$$\mathcal{J}_s \Big/ 2K(l+m)^2$$

As we will see later the relationship (14) is well confirmed by experiment but since it is extremely difficult to obtain bars which are exactly oriented along one of the basal planes of the cube, n is never exactly zero and it is useful to examine what happens in this case.

We let n take a very small value whilst maintaining the direction cosines, l and m fixed at their values for $n=0$ to within the second order in n. By this operation the remanent magnetization (Fig. 5) is reduced from the value:

$$OA = \mathcal{J}_s/(l+m)$$

to the value

$$OB = \mathcal{J}_s/(l+m+n).$$

AD shows the magnetization curve for a bar oriented exactly along one of the cube basal planes, the curve BCD' indicates the effect of a slight misorientation.

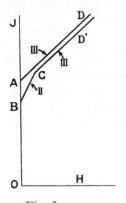

Fig. 5.

Since the magnetization process follows the mode II from the point B the formula (11) must be used, the series development close to the remanence position is in this case

$$\frac{\mathcal{J}}{\mathcal{J}_s} = \frac{1}{l+m+n} + \frac{\mathcal{J}_s H}{K(l+m+n)^2} + \cdots \quad \text{(mode II)}. \quad (17)$$

The initial slope is therefore equal to $\mathcal{J}_s/K(l+m+n)^2$; its value is nearly twice the value corresponding to $n=0$. The slope of BC (Fig. 5) is therefore twice that of AD. In addition since n is small the part of the curve corresponding to the mode II is also very small: there must therefore be a elbow at C beyond which the magnetization will increase following mode III. In the figure we have shown the segment CD' to be very close to AD: the calculation shows that in fact the difference is of the second order in n. The equation of the straight line CD' is found to be

$$\frac{\mathcal{J}}{\mathcal{J}_s}\left[1 + \frac{n^2}{(l+m)^2}\right] = \frac{1}{l+m} + \frac{\mathcal{J} H}{2K(l+m)^2} + \cdots \quad (n \ll m < 1), \quad (18)$$

whilst the equation for AD is given by equation (16). This is an important point since it shows that the law (14) which is only rigorously true for $n=0$ represents to a relative precision of the order $n^2/(l+m)^2$ those parts of the magnetization curve which corresponds to mode III for bars which are not exactly oriented along one of the cube faces. The error introduced is less than 0.5% when n is less than 0.1.

We have exploited this property to plot the theoretical values of $\mathcal{J}(l+m)$ as a function of $H/(l+m)$, these are compared to the experimental values of Sizoo and of Kaya (loc. cit.) for a series of bars of various orientations but such that n was less than 0.1, Fig. 6. We have however also used a bar for which $n=0.13$ with the correction given by equation (8). We have, of course, only plotted the sections of the curve corresponding to mode III. As in the examples of the previous section the general agreement between theory and experiment is extremely satisfactory.

When the external field attains a sufficient value one of the two phases reduces to zero and all the domains are now magnetized along a single direction, the magnetization process now continues according to the mode IV. In order to complete our analysis we will examine this process in detail for the experimentally

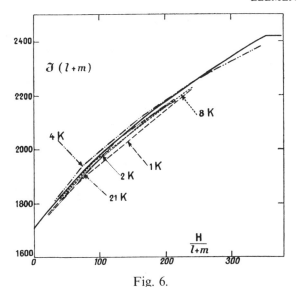

Fig. 6.

Mode III magnetization curves of bars approximately oriented along one of the basal planes of the cube. The theoretical curve is shown as the full line. The dashed curves indicate the bars experimentally studied by Kaya, the numbers correspond to those of his article.

important case of discs with a symmetry of revolution (flattened ellipsoids) whose equatorial plane is perpendicular to a fourfold crystal axis.

(9) Magnetization of discs following modes III and IV: We will suppose that the external field H is uniform and parallel to the equatorial plane of the disc and that it makes an angle φ with the bisector OC of the two fourfold axes OX and OY in this plane (Fig. 7).

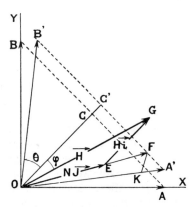

Fig. 7.

Diagram of the magnetizations and fields for a flattened ellipsoid of revolution whose small axis coincides with a fourfold crystal axis.

The general results of section 5 show that in the initial phase (mode I) the magnetization is both parallel and proportional to the field until it reaches the value

$$\mathcal{J}_r = \frac{\mathcal{J}_s}{l+m} = \frac{\mathcal{J}_s}{\sqrt{2}\cos\varphi}. \quad (19)$$

This occurs when the external field takes the value $H_r = N \mathcal{J}_r / \sqrt{2}\cos\varphi$ where N represents the demagnetizing factor of the disc in a direction perpendicular to the short axis. When $\vec{H}$ is larger than this value the internal field $\vec{H}_i$ is no longer zero and, since the equilibrium is now between two phases (mode III), its direction is in the plane perpendicular to the external bisector of the two axes OX and OY. By symmetry it must therefore be directed along OC. Since H and H_i are in the plane XOY the relationship

$$\vec{H}_i = \vec{H} - N\vec{\mathcal{J}}$$

shows that the magnetization $\vec{\mathcal{J}}$ is also in this plane, but is not parallel to the applied field $\vec{H}$: it has a component $\mathcal{J}_N$ perpendicular to the field which we must calculate.

The following segments are represented in Fig. 7

$$\overline{OE} = N\vec{\mathcal{J}}; \quad \overline{EG} = \vec{H}_i; \quad \overline{OG} = H.$$

Because of the effect of the internal field $\vec{H}_i$ the spontaneous magnetization of each of the two phases rotates slightly from OA to OA' and from OB to OB' symmetrically with respect to OC. The vector OB' is in equilibrium under the action of the magnetocrystalline forces and the field H_i. Representing by θ the angle made by OB' with OC we have on the one hand

$$OC' = OB' \cos\theta$$

whilst the total energy of the system is written as

$$W = \frac{K}{4}\cos^2 2\theta - \mathcal{J}_s H_i \cos\theta.$$

This has a minimum for $\partial W/\partial\theta = 0$ which gives finally equation (20)

$$H_i = \frac{2K}{\mathcal{J}_s}\cos\theta (2\cos^2\theta - 1), \quad (20)$$

this relationship defines θ as a function of H_i.

Furthermore the field H_i causes the disc to subdivide into elementary domains magnetized along OA' or OB'. According to the relative proportions of these two categories the extremity M of the vector OM, representing the mean magnetization, follows the straight line A'B' as described previously in the case of formula (10). But, by hypothesis, the mean magnetization is directed along OE. It is therefore equal to OF where F represents the intersection of OE with A'B'. Projecting the contour OG = OE + EG onto OC we obtain the relationship

$$H \cos \varphi = N \overline{OC'} + H_i \qquad (21)$$
$$= N \mathcal{I}_s \cos \theta + \frac{2K}{\mathcal{I}_s} \cos \theta \, (2 \cos^2 \theta - 1),$$

which relates the external field H to the angle θ. By projecting the same contour onto a perpendicular to OG we obtain the perpendicular magnetization component

$$\mathcal{I}_N = -\frac{1}{N} H_i \sin \varphi \qquad (22)$$
$$= -\frac{2K}{N \mathcal{I}_s} \sin \varphi \cos \theta \, (2 \cos^2 \theta - 1),$$

These two formulae relate $\mathcal{I}_N$ to H through the parameter θ; the problem is thus resolved. When the internal field is weak OA' is close to OA: one can set $\cos \theta = (\sqrt{2}/2) + \varepsilon$ where ε is very small; neglecting ε^2 and eliminating ε we obtain from (21) and (22)

$$\mathcal{I}_N = -\left(H - \frac{N \mathcal{I}_s}{\sqrt{2} \cos \varphi}\right) \frac{\cos \varphi \sin \varphi}{N \left(1 + \frac{N \mathcal{I}_s^2}{4K}\right)} \qquad (23)$$

We have seen above that the perpendicular magnetization remains zero until the field H reaches the value $H_r = N \mathcal{I}_s / \sqrt{2} \cos \varphi$. The formula (23) shows that beyond this point the perpendicular magnetization is initially proportional to $H - H_r$.

It would now be easy to calculate the parallel magnetization component and the relative proportions of the two phases as well but this would be of no real interest. We simply remark that the magnetization variations observed here are due both to a real rotation of the spontaneous magnetization within the domains and to a change in the proportions of the domains magnetized along OA' and OB'. As H increases the point F approaches, and finally touches A'. All the domains are then magnetized along OA' and from this point onwards the magnetization changes come uniquely from a reversible rotation of the vector OA' under the influence of the field H; the magnetization process therefore proceeds according to the mode IV.

The calculation of the perpendicular component in this last magnetization mode is a classical problem [12] which we now briefly recapitulate. The total energy is the sum of the magnetocrystalline energy term

$$-K \cos r\theta/8$$

and the applied field energy

$$-H \mathcal{I}_s \cos (\varphi - \theta).$$

It is not necessary to take account of the demagnetizing energy since we are only concerned with the rotation of a vector of constant modulus following directions whose demagnetizing field coefficient is constant, the corresponding energy is therefore constant. The total energy is minimum when θ satisfies the relation

$$K \sin 4\theta = 2 H \mathcal{I}_s \sin (\varphi - \theta).$$

The perpendicular component is then

$$\mathcal{I}_N = \mathcal{I}_s \sin (\varphi - \theta).$$

To sum up, the curve describing the normal magnetization component as a function of the applied field consists of three regions in the case of a disc: the first region in which the normal magnetization component is zero corresponds to mode I; the second region, corresponding to mode III, in which the normal magnetization component increases rapidly from zero as the field increases, according to the equations (21) and (22); and finally the region, corresponding to the mode IV, in which the normal magnetization decreases with the field following the equation established in the previous section. We have compared these theoretical results with the experiments of Webster [13] and of Honda and Kaya [14]. The curves in Fig. 8 were calculated using $K = 4.3 \times 10^5$ and $\mathcal{I}_s = 1\,710$ and taking the theoretical demagnetizing factors $N = 0.734$ and $N = 0.190$ indicated respectively by these authors. The segments of the curve corresponding to mode I are represented by OA and OB and those corresponding to mode III by AC and BD: they depend on the value of the demagnetizing field whilst the common segment CDE, corresponding to the mode IV is independent of the demagnetizing field. The experimental points fall reasonably close to the theoretical curves and the discrepancies may equally well be due to impurities and deformations of the crystal as to the inevitable imperfections in the machining of the ellipsoid.

(10) The subdivisions of the magnetization curve: All the preceding results prove that in the general case

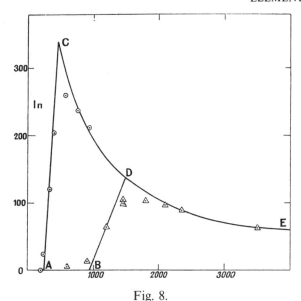

Fig. 8.
Magnetization component perpendicular to the field for discs sectioned following one of the cubic basal planes. The theoretical curves are the full lines; the circles correspond to the experiments of Honda and Kaya, the triangles to those of Webster. The magnetic field is the abscissa.

(arbitrary orientation of the crystal axes with respect to the ellipsoid axes) the magnetization curve of an ellipsoidal iron monocrystal is subdivided into four magnetization regions corresponding to the four modes predicted in Table I of section 4; they follow each other in the indicated order as the applied field increases. The magnetization is not discontinuous at the critical field values corresponding to the passage from one mode to the next: the curve shows simply a sharp elbow indicating a discontinuity in the first derivative $d\mathcal{J}/dH$. The relative extent of the different regions varies considerably with the orientation of the crystal axes. For certain privileged orientations one or more modes may even disappear completely.

Each of these modes is characterized by the existence of a well determined number of phases in equilibrium (6, 3, 2 or 1), whose respective volumes vary with the field; these volumes can be calculated by the indicated methods. The direction of the spontaneous magnetization of each phase also varies with the field except in mode I where it coincides with a fourfold axis.

All these results, stated and described for the first time in the present article, are obtained without any preconceptions concerning the form and relative dispositions of the elementary domains: this is the problem which will now be examined starting from the experimental observations. We will examine in particular the domains corresponding to the modes II and III above the first elbow in the magnetization curve. In fact as we have demonstrated in section 5, in mode I above the first elbow the relative proportions of the six types of elementary domains are not determined in an ideal crystal even though the resultant magnetization is: minor variations of stress, inclusions, etc., will finally fix the relative proportions which may be extremely variable from crystal to crystal: the surface polish for example plays an important role and considerably modifies the superficial domain structure. For this reason we limit ourselves for the present[4] to an analysis of the domain structures corresponding to modes II and III which are completely determined by the theory enunciated above and which are much less sensitive to accidental perturbations. As for the mode IV the problem is inexistent since by definition only a single type of elementary domain exists in this mode.

SECOND PART
THE FORM AND ORIENTATION OF ELEMENTARY DOMAINS

(11) Powder patterns: The most useful experimental information about the form and orientation of elementary domains is given by the powder patterns first observed by Hamos and Thiessen [15] and first studied in detail by Bitter [16]; these patterns are regular lines, observable under the microscope, which form on the polished surface of a ferromagnetic material by deposition from a colloidal suspension of ferrous oxide (Fe_2O_3) in an appropriate liquid (water, alcohol, mineral oil).

The detailed appearance of these patterns depends on the nature of the material and on the region examined in the magnetization curve (Figs. 9 and 10). In the case of iron, in the region of concern here, above the first bend in the magnetization curve, there exist two beautiful series of independent observations by Sixtus [17] and by Kaya [18] obtained by magnetizing the crystal parallel to the surface under observation. To be more explicit these are the type II patterns observed by Sixtus and the third type of pattern in the observations of Kaya: for the sake of brevity we will call such patterns *remanence patterns*.

(12) The characteristics of remanence patterns: These patterns consist of parallel lines which are often very long and fine, approximately regularly spaced and *aligned along crystal lattice axes*: a rotation of the magnetic field in the plane of the observation surface sometimes causes the disappearance of the initial set of

---

[4] An article will subsequently be devoted to the study of mode I.

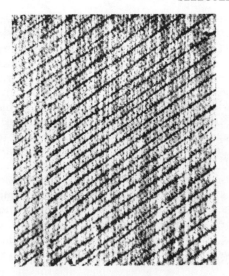

Fig. 9.
Powder pattern lines (Δ lines) observed by Sixtus (*Phys. Rev.*, 1937, **5**, p. 870) on the surface of a silicon–iron ribbon (magnification 15 ×).

Fig. 10.
Powder pattern lines (θ lines) observed by Kaya (*Z. f. Physik*, 1934, **10**, p. 551) on the (110) face of an iron crystal magnetized along a threefold axis (magnification 35 ×).

lines which is replaced discontinuously by a second system of differently oriented lines. The lines of each system always conserve a fixed orientation with respect to the crystal. The degree of surface polish does not modify their appearance.

The observations reveal two types of remanence patterns, in the first type, which we will call type Θ, the lines are, according to Kaya, parallel to the line of intersection of the surface of observation and the plane normal to the threefold axis closest to the magnetic field: as for the second type, which we will call type Δ, according to Kaya and Sixtus the lines are parallel to the intersection of the plane normal to the binary axis closest to the field direction with the surface of observation. These distinctive orientations have not been interpreted up to the present.

Table II shows the types of pattern observed by Kaya depending on the magnetization and on the

TABLE II.

	Observation plane	Field direction	Observed lines	Magnetisation range
α	(001)	[110]	Δ	
β	(001)	10° de [100]	Δ	
γ	(1$\bar{1}$0)	[111]	Θ	
δ	(1$\bar{1}$0)	[110]	Θ	Very weak
δ′	(1$\bar{1}$0)	[110]	Δ	$J < 1200$
ε	(1$\bar{1}$0)	23° de [110]	Θ	$J > 1200$
ε′	(1$\bar{1}$0)	23° de [110]	Δ	$J < 1400$
η	(1$\bar{1}$1)	15° de [110]	Θ	$J > 1400$
η′	(1$\bar{1}$1)	15° de [110]	Δ	$J < 1300$
				$J > 1300$

orientation of the applied field and the observation plane with respect to the crystal axes.

The Θ and Δ lines show characteristic differences in appearance; the Θ lines are relatively short, of irregular thickness, and rather unequally spaced (spacing roughly 10 to 20 μm, Fig. 20), the Δ lines often show beautiful regularly spaced lines crossing the entire field of observation (Fig. 9). These have a larger spacing and are more regular than the Θ lines; their mean separation diminishes as the applied field increases; thus on a surface close to the (100) plane the separation of the lines decreases from 350 to 90 μm when the field increases from a few gauss to 200 gauss (Sixtus). Kaya, in his experiments, found a decrease from 400 to 50 μm when the field increased.

(13) General interpretation of the patterns: We will suppose first of all that the powder pattern lines represent the traces on the observation plane of the surfaces separating the domains and that, particularly in the case of the remanence patterns, the patterns are related to the internal domain structure and not merely to some superficial structure[5]. All the observations of remanence patterns show parallel arrays of lines whatever the orientation of the surface of observation with respect to the crystal axes and the magnetization direction: if our initial supposition is correct this must indicate that the *elementary domains are in the form of*

---

[5] We will return to this question later (§ 23).

plane layers piled the one on top of the other like the pages of a book. We will give an explanation of this structure later.

Furthermore the observations of Kaya and Sixtus show that the remanence patterns only appear after the first bend in the magnetization curve and that they disappear near to saturation: these conditions show that these patterns are related to the magnetization modes II and III. To be more explicit since in the same crystal in an increasing applied field the Θ and Δ lines are observed successively it appears normal to associate the first set, the Θ lines, with a magnetization process following mode II with three phases in equilibrium and the second set, the Δ lines, with a magnetization according to the mode III with two phases in equilibrium. According to this interpretation a Θ *type structure is made up of a pile of plane layers alternatively magnetized along three different directions, whilst in a Δ type structure the stack of layers are magnetized alternatively along two different directions.*

(14) Orientation of the walls in the Θ structure: In order that there shall be no free internal charges, in accord with the conclusions of paragraph 2, it is necessary and sufficient that *the components perpendicular to the wall of the spontaneous magnetizations in the two adjacent phases separated by the wall be equal.* We apply this result to the case of a magnetization process following mode II with three phases in equilibrium: since the separating walls are all parallel the normal components of the spontaneous magnetization are pairwise equal, therefore they must all be equal. However we have already shown (§ 7) that the spontaneous magnetization directions of the three phases are equally inclined with respect to the internal field which itself is parallel to the threefold axis closest to the applied field. *The walls, themselves parallel to the plane of the layers, are therefore perpendicular to* this same threefold axis. We can thus readily interpret Kaya's observations concerning the orientation of the Θ lines which were described at the beginning of section 11. This first result is thus a very encouraging success.

Moreover, we can now understand why the domains take the form of plane layers: it is indeed the only manner by which the crystal can be decomposed without provoking the appearance of internal charges within the crystal.

To sum up, in the case of a magnetization process according to mode II, we know the form and the orientation of the three types of domain: they are all layers with their plane perpendicular to the threefold axis closest to the applied field. Following the methods developed in the first part of this article we know in principle how to calculate the corresponding spontaneous magnetization directions as well as the relative volumes occupied by the three domain types; on the other hand we know nothing about the absolute thickness of the constituent layers: this question is reserved for the third section.

(15) The orientation of the walls in the Δ-type structures: In this case where the layers are alternatively magnetized along only two different directions the condition for zero internal free charge simply shows that there is an infinity of possible orientations given by the set of planes passing through the external bisector of the two directions of spontaneous magnetization. But, in an article concerning an analysis of walls in iron (cf. A), we have already shown that the most stable among all the possible orientations is perpendicular to the internal bisector of the two possible spontaneous magnetization directions; for all other orientations magnetostrictive effects, which deform the initially cubic crystal lattice, provoke a mutual deformation of two juxtaposed domains and thereby lead to an increase in the potential energy. This increase is relatively large since it reaches 200 ergs/cm^3 for walls which are parallel to the two spontaneous magnetization directions and this value doubles the energy associated with the system of walls.

So, when the two privileged directions are two fourfold axes the walls between the layers must be perpendicular to the binary axis closest to the field direction since this axis is the internal bisector of the two spontaneous magnetization directions. This gives a simple and direct interpretation of the observations of Δ-lines made by Kaya and by Sixtus.

(16) The interpretation of Table II: It now becomes easy to in interpret the set of observations made by Kaya which is summarized in Table II.

When the field is directed out of the basal planes of the cube the magnetization process after the first elbow in the curve follows the mode II, this means that Θ lines should be observed and this is seen to be what happens in the γ, ε and η cases. When the overall magnetization is sufficiently high the number of phases is reduced to two and the magnetization proceeds following mode III as was shown in part I of the present article, Δ lines should be observed in this region and this is confirmed by the experimental results at ε' and η'. In the γ case mode III should be forbidden by symmetry since the field is applied along a threefold axis: this again is confirmed by experiment.

When the applied field direction is in one of the basal planes of the cube we know (§ 8 and § 9) that the mode II disappears and that the magnetization beyond the first bend in the curve directly follows mode III: therefore only Δ lines should be observed: this is exactly what happens in the cases α and β. The very weak Δ lines observed in the case δ probably come from a slight misorientation of the applied field.

The set of experimental observations due to Kaya are seen to entirely confirm our general theory and verify in particular the theoretical predictions for modes II and III and the transition from one to the other. Other observations also support the theory: we cite the example of a disc, cut from the basal plane of the cube, in a field in this plane along a direction making an angle of 10° with a fourfold axis (β, Table II). An analogous case was investigated theoretically in section 8: whilst the magnetization proceeds according to mode III the binary axis closest to the applied field always remains as the internal bisector of the spontaneous magnetizations of the two phases. The walls between the leaflets, perpendicular to this bisector maintain therefore a fixed orientation until the walls disappear. This explains why Kaya observed powder pattern lines of invariant orientation a result which appears very mysterious at first view, especially when it is realized that in the extensive observation range the true rotation of the spontaneous magnetization within the elementary domains is considerable.

THIRD PART
SECONDARY SURFACE STRUCTURE AND LAYER THICKNESS

(17) The need for a secondary surface layer: The plane parallel layer structures, examined in the second part of this article, create a demagnetizing field which is identical to that which would occur throughout the material if its magnetization were uniform and equal to the mean magnetization of the layers, *except in a surface layer* whose thickness is of the same order of magnitude as that of the layers. In fact magnetic charges appear on the external surface of the crystal with a density equal to the normal component of the spontaneous magnetization which varies from one leaflet to the next so that the magnetic fields at two neighbouring points can differ by several thousand gauss. It is unlikely that such charges, so widely variable from point to point, can exist in iron since they would considerably increase the potential energy of the system[6] and the resulting magnetic fields would be sufficiently strong as to locally reverse the spontaneous magnetization of the original domains.

The surface layer should therefore be the site of a reorganization of the domain structure: in contrast to the primitive plane layer structure we will suppose that the formation of a secondary surface structure is characterized by the complete or nearly complete disappearance of the periodically variable magnetic charges.

(18) The secondary structures without any free internal charges are a good approximation to the real structure: We will now attempt to determine the form of the charge-free surface structures by using the technique which worked so well in the case of the primary structure.

These secondary surface structures will be characterized by systems of magnetic charges throughout the material, and particularly in the surface layer, *identical to those which would be associated with a uniform magnetization equal to the mean overall magnetization of the system of plane layers*. We will show further on that such structures always exist and we will try to find those whose energy is minimum.

We first note that the structures obtained in this way are not really equilibrium structures, but that small modifications suffice to make them so. As we will see later the secondary domains which form near the external surface of the crystal have the form of thin prisms whose spontaneous magnetization directions are such as to give the appropriate magnetic densities on the walls of the prism. There is therefore no reason why this spontaneous magnetization be in equilibrium under the influence of the internal field and the magnetocrystalline forces: the spontaneous magnetization will therefore rotate but since the transverse demagnetizing factor is considerable, 2π on the average, this rotation can only be very small since it is limited by the appearance of the demagnetizing field: in this way, in the most unfavourable case in which the spontaneous magnetization is parallel to the generators of the prism, the application of a transverse field of 100 gauss, which is quite a large field, will only turn the spontaneous magnetization of the prism through an angle of $H/2\pi J_s$, less than 1/100 radians, whilst the corresponding reduction of the potential energy, equal to $H^2/2N$, is less than 1 000 ergs/cm³. This energy is negligible compared with the total energy of formation of the prisms which reaches 17 000 ergs/cm³ for the same field in an example examined further on (cf. § 24).

To summarize, structures without internal free charges usually constitute a very good second approximation to equilibrium structures: it will be important therefore to study them in detail. It should not be forgotten however in the examination of some problems, in particular that of the mechanism of the deposition of powders, that the spontaneous magnetization so determined for the surface layers

---

[6] As an example, calculation shows that parallel bands of width $d/2$ carrying alternatively densities $+J_s$ and $-J_s$ possess a magnetic energy of $0.426\, d\, J_s^2$ ergs/cm². For $d = 100$ μm one obtains 12,400 ergs/cm², whilst the energy of the secondary structures studied later are generally less than 200 ergs/cm².

undergoes in reality small deviations whilst weak magnetic charges appear on the walls of the prisms.

(19) The search for secondary structures. Closure domains: Csondier a basic periodic structure, with two phases (mode III), made up of plane leaflets piled one on top of another: the odd leaflets have the thickness e, and the spontaneous magnetization $\vec{J}_1$, whilst the even ones have thickness e_2 and magnetization $\vec{J}_2$. This basic structure is characterized by a periodicity d and a mean magnetization $\vec{J}$ defined by the relations

$$e_1 + e_2 = d \; ; \quad e_1\vec{J}_1 + e_2\vec{J}_2 = d\vec{J}. \tag{33}$$

Moreover, to avoid free internal charges, the unit vector n normal to the plane of the leaflets must obey the conditions

$$\vec{n}\vec{J}_1 = \vec{n}\vec{J}_2 = \vec{n}\vec{J}_3. \tag{34}$$

We now isolate from this structure two consecutive leaflets which we delimit by an arbitrary plane S defined by a unit vector s normal to the plane and directed towards the exterior. Consider a plane perpendicular to the plane of the leaflets and to the limiting plane S (Fig. 11). In Fig. 11b the surface element AB carries a charge density of $\vec{s}\cdot\vec{J}$, whilst the element BC has an associated charge density $\vec{s}\cdot\vec{J}_2$.

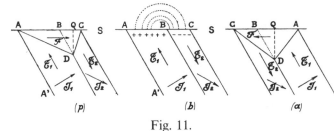

Fig. 11.
Secondary surface structure in the case of two isolated leaflets. The two principal closure positions (a) and (p).

The problem is to balance the charges: we must therefore find a prismatic domain ACD whose spontaneous magnetization direction is such that the walls AD and DC carry no charge and that the charge density on AC is equal to the density $\vec{s}\cdot\vec{J}$ which would be produced by the mean magnetization[7].

---

[7] In their investigation of the properties of cobalt, Landau and Lifschitz (*Physik Z. der Sowjetunion*, 1935, **8**, p. 153) were the first to use prismatic domains to produce closed magnetic circuits but they did so without establishing any general doctrine concerning this type of domain. Their conclusions as to the periodicity of cobalt powder patterns do not agree with the results of Elmore (*Phys. Rev.* 1938, **53**, p. 757): the secondary structure of cobalt is probably much more complicated.

With this in mind we decompose the spontaneous magnetization $\vec{J}_1$ of the first layer into three vectors, the first $\vec{J}'$ is parallel to the mean magnetization $\vec{J}$, the second $\vec{\mathscr{T}}_1$ is parallel to the line AA' along which the wall plane cuts the plane of the figure, and the third $\vec{\mathscr{P}}_1$ is parallel to the intersection of the walls with the plane S, that is perpendicular to the plane of the figure. This gives

$$\vec{J}_1 = \vec{J}' + \vec{\mathscr{T}}_1 + \vec{\mathscr{P}}_1, \tag{35}$$

but on multiplying both sides of this equation by n and taking account of the relation (34) it is found that $\vec{J}'$ is identical to $\vec{J}$. In the same way $\vec{J}_2$ is decomposed into three vectors $\mathscr{J}', \vec{\mathscr{T}}_2, \vec{\mathscr{P}}_2$

$$\vec{J}_2 = \vec{J} + \vec{\mathscr{T}}_2 + \vec{\mathscr{P}}_2. \tag{36}$$

By substituting in the second equation (33) the values of $\vec{J}_1$ and $\vec{J}_2$ deduced from (35) and (36) one finds

$$e_1 \vec{\mathscr{T}}_1 + e_2 \vec{\mathscr{T}}_2 = 0 \tag{37}$$

this equation expresses the fact that the flux of the vector $\mathscr{T}$ leaving the first layer is equal to the flux which enters the second. This flux closure is achieved *outside* the crystal and produces the magnetic charges which we wish to annihilate. This becomes possible by closing the flux *within* the crystal by means of the prismatic domain ACD, whose corners are perpendicular to the plane of the figure: we will therefore call this domain a *closure domain*. Its spontaneous magnetization must have a component $\vec{\mathscr{T}}$ parallel both to the limiting surface S and to the plane of the figure, that is to say parallel to AC, such as to ensure flux closure; this gives

$$e_1 \vec{\mathscr{T}}_1 = h\vec{\mathscr{T}} = - e_2 \vec{\mathscr{T}}_2, \tag{38}$$

where $h = DQ$ is the height of the prism. This height h is thus determined by $\vec{\mathscr{T}}$.

In summary, this component $\vec{\mathscr{T}}$ is used to cancel out the charges created on the wall AD by the component $\vec{\mathscr{T}}_1$ and on the wall DC by the component $\vec{\mathscr{T}}_2$. To eliminate the charges created on these same walls by the common component $\vec{\mathscr{J}}$ of $\vec{J}_1$ and $\vec{J}_2$ a component equal to $\vec{\mathscr{J}}$ must also be attributed to the spontaneous magnetization of the closure domain. As for the components $\vec{\mathscr{P}}_1$ and $\vec{\mathscr{P}}_2$ they do not produce charges on AD and DC; it would seem therefore that the spontaneous magnetization of the closure domain can be given a component $\mathscr{G}$ perpendicular to the plane of the figure and of arbitrary intensity, but since the resultant is equal in modulus to $\mathscr{J}_1$ the following condition must be satisfied

$$(\vec{\mathscr{F}} + \vec{\mathscr{I}} + \vec{\mathscr{G}})^2 = \vec{\mathscr{I}}_s^2, \qquad (39)$$

and this determines $\vec{\mathscr{G}}$ as a function of $\vec{\mathscr{F}}$.

It is easily seen that in the structure described the only free charges come from the component of $\vec{\mathscr{I}}$ perpendicular to the limiting surface S and these charges are the same as those which would be produced if the material had a uniform magnetization equal to the mean magnetization since $\vec{\mathscr{I}}$ is in fact equal to this mean magnetization.

(20) The two principal flux closure positions for two leaflets: The closure vector $\vec{\mathscr{F}}$ is defined in *direction*: it is parallel to AC, Its *sense* depends on the relative positions of the two leaflets: it changes sign when they are inversed. This means that two closure positions must be distinguished: we will call position A that for which the order of the layers is such that the vectors $\vec{\mathscr{I}}$ and $\vec{\mathscr{F}}$ are directed along opposite senses in the wall plane (Fig. 11a) and position P that for which the order of the layers is such that $\vec{\mathscr{I}}$ and $\vec{\mathscr{F}}$ are directed in the same sense in the wall plane (Fig. 11p).

In each of these two positions the *magnitude* of $\vec{\mathscr{F}}$ is not as yet determined: we will finally retain that value which corresponds to the minimum energy.

A given value of $\vec{\mathscr{F}}$ corresponds to a well defined height h of the closure prism (equation (38)) and a well defined orientation of its spontaneous magnetization: this allows the calculation of the energy value w (per cm^3) necessary to turn the spontaneous magnetization out of its equilibrium position in one or other of the layers into the position which it takes up in the closure prism, because in these two positions the field acting is the same and equal to the resultant of the applied field and the demagnetizing field due to the uniform magnetization $\vec{\mathscr{I}}$. The energy required to form the closure prism for a pair of layers of given thickness is therefore proportional to hw; we need to find therefore the minimum value of this product.

We note that all other things being equal the formation energy of the closure prism for a pair of domains of total thickness d is proportional to the surface of the prism base, that is to d^2. The energy of formation of the closure prisms in positions A and B can therefore be denoted by $d^2 W_a$ and $d^2 W_p$ respectively, the quantities W_a and W_p being independent of the absolute dimensions of the leaflets[8].

(21) Flux closure for an unlimited structure with one plane surface: We wish to know which solution should be adopted in the case of the structure resulting from the periodic and unlimited repetition of the elementary pattern made up of the two layers treated above: one might be tempted to retain that closure position, either A or P, corresponding to the smallest of the two energies W_a and W_p. However in reality the lowest energy is obtained by combining the two positions as schematized in Fig. 12, in which the flux of a portion xd of the elementary pattern is closed in position A whilst the flux of the remaining fraction $(1-x)d$ is closed in position P.

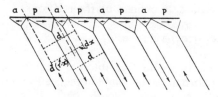

Fig. 12.
Secondary surface structure in the general case of an unlimited number of pairs of layers.

The energy of the two closure prisms with respect to just one elementary motif is equal to $x^2 d^2 W_a + (1-x)^2 d^2 W_p$. This expression, considered as a function of x, is minimum for

$$x = \frac{W_p}{W_a + W_p}. \qquad (40)$$

It takes the value

$$d^2 \frac{W_a W_p}{W_a + W_p}. \qquad (41)$$

The closure structure corresponding to an unlimited succession of elementary patterns is thus completely determined.

(22) Flux closure in a real material and the determination of the absolute dimensions of the domains: (1) *The case of a cylindrical limiting surface* — We now apply the preceding results to a real material in which the layers are not limited by a plane but by a closed surface: a second order cylinder is first taken as the limiting surface. We will simply suppose that the radii of curvature of this surface are large with respect to the layer thickness: the surface is decomposed into elements which are approximately plane and whose dimensions whilst being large compared to the layer width are small compared to the radius of curvature. The results of the preceding sections are applied to determine the height of the closure prisms and the orientation of their

---

[8] This discussion concerns of course the dimensions of the layers in the plane of the figure; in the plane perpendicular to the plane of the figure the quantities W_a and W_p hold for a thickness of 1 cm.

spontaneous magnetization at each point on the surface.

The layers, which have the form of elliptic discs will be limited in this way by prism-like rings whose triangular sections will deform slowly and continuously from one point to another around the disc whilst the spontaneous magnetization slowly changes direction[9].

Figure 13 represents the back half of two such rings. Their thickness reduces to zero on the generator kk' which corresponds to a plane tangent to a cylinder which is equally inclined with respect to $\vec{J}_1, \vec{J}_2$ and $\vec{J}$ so that each ring is cut into two half rings, one of type A the other of type P.

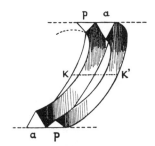

Fig. 13.
Perspective view of closure rings for a cylinder (back half). The thickness of the rings reduces to zero along the generator KK'.

The energy of formation of the two prismatic rings which close the flux of a pair of layers is thus completely determined. Everything else being equal it is proportional to the square d^2 of the thickness of the pair. Furthermore, for the same relative orientation of the crystal and the field but with two limiting homothetic surfaces it is proportional to two arbitrary homologous distances, for example to the long axis L of the ellipse which limits the layers; this energy could therefore be written in the form CLd^2 in which C is independent of the absolute dimensions of the crystal and the layers. But the number of layers contained in a cylinder length of 1 cm is inversely proportional to d: the energy which must be spent to close the lateral flux for a cylinder segment of 1 cm length is thus equal to CLd.

It would seem that the layers should multiply indefinitely to reduce this energy but in reality the surface energy of the layer walls must be considered[10],

---

[9] These annular domains are rather special since their magnetization does not have the same direction everywhere: their charge density $\rho = -\operatorname{div} J$ is however in the conditions indicated completely negligible as far as the energy is concerned.
[10] The wall energies of the annular prismatic domains can be neglected.

this is proportional on the one hand to the surface of an isolated wall, that is L^2, and on the other hand to the number of these walls that is to $1/d$; the energy can therefore be written as $C'L^2/d$. The existence of this energy tends to reduce the number of layers. The total energy of the structure W_t can be written

$$W_t = CLd + \frac{C'L^2}{d}. \qquad (42)$$

This energy is minimized for

$$d = \sqrt{\frac{LC'}{C}} \qquad (43)$$

and has the value $2\sqrt{L^3CC'}$, it should be noticed that in this situation the energy of the prismatic closure rings is equal to the wall energy.

The equation (43) shows that, all else being equal, the periodicity of the domain structures of a crystal limited by a cylindrical surface is proportional to the square root of its diameter.

(2) *The case of an ellipsoid* — When the crystal has the form of an elongated ellipsoid the layers no longer have a uniform thickness from one end to the other of the specimen since the dimensions of the ellipse which limits the layers vary: the layer thickness, maximum in the centre, diminishes towards the two extremities of the ellipsoid.

It becomes apparent that in the general case the notion of a periodic layer structure should be understood in a broad sense: the period is not in fact constant but varies slowly from one end of the crystal to the other.

(23) The secondary surface structure and the powder pattern lines: Unfortunately it seems to be extremely difficult to study experimentally this secondary structure which as we have just seen must exist according to theoretical arguements. Powder patterns constitute one of the rare experimental means for such a study so it is important to examine the relationship of these patterns to the secondary structure, but first of all it is necessary to understand the mechanism of the deposition of ferromagnetic powders onto an iron surface.

Now, in the surface structures described above, the walls within the closure prisms do not carry any magnetic charges whilst the charge density is everywhere equal to the perpendicular component of the mean magnetization on the external crystal surface: there is therefore no reason why the powders which are left to settle on the surface should agglomerate in any particular region. Nonetheless we have already noticed at the end of section 18 that in the real equilibrium structure small changes in the directions of the spontaneous magnetizations of the closure domains are necessarily produced and that these changes are *different* for a prism of type A to those for type B. Different magnetic charges will therefore appear on the surface regions corresponding to the prisms A and P.[11] From the point of view of the local external field it is as if these regions are either positive or negative: the external lines of force near the surface will have the form shown in Fig. 14. In principle the magnetic powder should deposit in the regions where the field gradient is largest that is at b, b' and c, c': *two* lines should thus be observed for each elementary motif that is one line per leaflet.

Nevertheless it must be remembered that the applied field always has a tangential component H, directed for example as in Fig. 14, which superposes on the local field created by the magnetized regions. The powder grains, under the influence of this field, will agglomerate as beads magnetized in the direction of H: it suffices to compare the position of their poles (Fig. 14) with the sign of the magnetic charges carried by the regions to see that these oriented beads will straddle the lines b, b' and will avoid the lines c, c'. Consequently equidistant powder deposits will be observed at positions b, b' separated from each other by the distance d equal to the periodicity of the structure: there will be one powder line for two layers[12].

However, it may well be that in certain particular situations it is possible to observe simultaneously both the b and c lines: this is the probable interpretation of certain 'doublets' observed by Kaya[13] on facets perpendicular to a binary axis.

(24) Application of the theory of the secondary surface structure to the case of a thin strip oriented and magnetized along a binary axis: In order to clarify our ideas and to study the effect of the field on the separation of the lines we will now consider rather thoroughly a simple example which roughly corresponds to an experiment carried out by Kaya.

---

[11] Kennard (Phys. Rev., 1939, 55, p 312), realising that the cobalt domain structure proposed by Landau and Lifschitz has no poles and so cannot give rise to powder pattern lines, Kennard proposes that the surface tension of the prism walls gives rise to a deformation of these walls and to the charges on the *internal* walls. These charges are in fact much weaker than those arising by the mechanism which we describe in §18.

[12] The disappearance of one line out of two is connected to certain weak field observations by McKeehan and Elmore (*Phys. Rev.* 1934, **46**, p. 226) on structures related to the polish of the surface layers: depending on the sense of the field one or other of the lines is observed. Elmore (*Phys. Rev.* 1937, **51**, p. 982) has given a good analysis of the mechanism.

[13] cf. for example Figs. 6 and 14 in the article by Kaya (*Z. für Physik* 1934, **90**, p. 551).

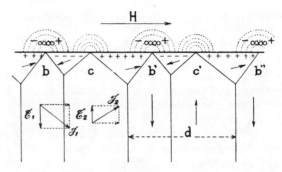

Fig. 14.
To illustrate the mechanism of powder deposition on the surface of a ferromagnetic material and the disappearance alternate lines.

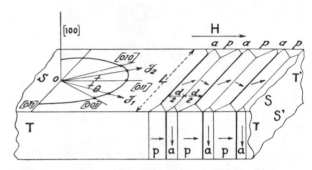

Fig. 15.
Elementary domains and secondary surface structure of a monocrystalline strip cut along a binary axis.

Consider a long thin strip (Fig. 15) in the form of a rectangular parallelipsed taken from the basal plane (100) of the cube: the two largest lateral faces S and S' will therefore be parallel to this plane. The long axis of the strip will be parallel to one of the binary axes of the (100) plane, for example the axis [011]. The magnetic field will always be applied along this direction and the corresponding demagnetizing factors will be supposed negligible. The two smaller lateral facets of the band T, T' will then be parallel to the plane (0$\bar{1}$1).

The magnetization process in question follows mode III with two phases whose spontaneous magnetization directions are in the (100) plane and make the same angle θ with the binary axis [011] related to the field H by the equation (20) of section 8

$$H\mathcal{J}_s = 2K \cos\theta (2\cos^2\theta - 1)$$

By symmetry the resultant magnetization $\vec{\mathcal{J}}$ is always directed along the binary axis [011] and the two phases occupy equal fractions of the total volume: the two layers which constitute the basic motif have the same thickness $d/2$, whilst the walls separating them are perpendicular to the [011] axis.

The spontaneous magnetization of the two phases is always parallel to the top and bottom surfaces S and S′: there will therefore be no closure domains on these faces. On the contrary, the other lateral faces T, T′ carry charges, of density $\pm \vec{J}_s \sin\theta$ depending on the nature of the phase underneath the point considered, these charges must be suppressed by suitable closure domains (Fig. 15).

Using the notation of section 19 we decompose the spontaneous magnetization of layer 1 into a vector $\vec{\mathcal{J}}$ of modulus $\vec{J}_s \cos\theta$, parallel to the [011] axis and equal to the mean magnetization, and into a vector $\vec{\mathcal{F}}_1$, of modulus $\vec{J}_s \sin\theta$ parallel to the direction [0$\bar{1}$1], the intercept of the wall plane with the plane S. The spontaneous magnetization of the layer 2 is decomposed in the same way into a vector $\vec{\mathcal{J}}$ and a vector $\vec{\mathcal{F}}_2 = -\vec{\mathcal{F}}_1$ (the vectors $\vec{\mathcal{G}}$ are zero). It is the flux $\vec{\mathcal{F}}_1 d/2$ of this vector $\vec{\mathcal{F}}$ in the layer 1 which must be closed back into the layer 2 without creating any charges on the surfaces T, T′ which are parallel to the mean magnetization.

Closure in position A — The spontaneous magnetization of the closure prisms is therefore parallel to the surface T and its orientation is defined by the angle α which it makes with the binary axis [011]. The three components into which we decompose this magnetization are then: the first, $\vec{\mathcal{J}}_s \cos\theta$ equal and parallel to the mean magnetization, the second $\vec{\mathcal{G}} = \vec{\mathcal{J}}_s \sin\theta$ parallel to the [100] axis, the third, $\vec{\mathcal{F}}$, equal to $\vec{\mathcal{J}}_s(\cos\theta - \cos\alpha)$ and directed in the opposite sense to the mean magnetization, it is this last component which closes the flux. The height h of the closure prism is given by the relationship (38) which becomes

$$d \sin\theta = 2h(\cos\theta - \cos\alpha).$$

The angle α must be such that $\cos\alpha$ lies between -1 and $\cos\theta$ and it must correspond to the minimum energy.

$W_{(\alpha,\theta)}$ denotes the energy required (per cm³) to bring the magnetization from the position it occupies within the layers 1 or 2 into its position in the closure prism defined by the angle α in the plane T: since, all else being equal, the volume of the closure prisms is proportional to h, the energy needed to produce the closure prism is proportional to $hW_{\alpha,\theta}$ that is to $\sin\theta W_{\alpha,\theta}/(\cos\theta - \cos\alpha)$. We need therefore to find the minimum of this expression as a function of α.

The total energy W_θ as a function of θ, for any direction in the plane (100) is readily found from the sum of the magnetic energy and the magnetocrystalline energy given by an appropriate transformation of the relationship (2).

$$W_\theta = \frac{K}{4}(2\cos^2\theta - 1)^2 - H\mathcal{J}_s \cos\theta.$$

At equilibrium θ is related to H by equation (20). Similarly the energy W_α corresponding to different possible positions in the plane T of the closure prism spontaneous magnetization is given by

$$W_\alpha = K\left(\cos^2\alpha - \frac{3}{4}\cos^4\alpha\right) - H\mathcal{J}_s \cos\alpha.$$

By definition we have

$$W_{(\alpha,\theta)} = W_\alpha - W_\theta$$

and an examination of the variation of the expression

$$\frac{\sin\theta}{\cos\theta - \cos\alpha} W_{(\theta,\alpha)},$$

shows it to be minimum for a value of α slightly smaller than $\pi/2$. The value of the minimum differs only slightly from the value corresponding to $\alpha = \pi/2$, that is

$$\frac{K}{4}\tan\theta(2\cos^2\theta - 1)(6\cos^2\theta + 1),$$

taking account of the value of $H\mathcal{J}_s$: one can suppose without any notable error[14] that, in the closure position A, *the spontaneous magnetization of the closure prisms is perpendicular to the plane S*, that is to say that it is *directed along* the fourfold magnetization axis in the plane of the surface T.

The closure of the lateral magnetization flux of a basic motif made up of two layers of total thickness d within the surface T requires an energy $d^2 W_a$ given by the expression

$$d^2 W_a = \frac{Kd^2}{16}\tan\theta(2\cos^2\theta - 1)(6\cos^2\theta + 1).$$

per cm length of the motif.

Closure in position P — In this position the component $\mathcal{F}$ instead of being directed in the opposite direction to the field is in the same sense and is equal to $\mathcal{J}_s (\cos\alpha - \cos\theta)$ where $\cos\alpha$ obeys the inequality $\cos\theta < \cos\alpha < 1$. The height h of the closure prism is given by

$$d \sin\theta = 2h(\cos\alpha - \cos\theta).$$

As above we must minimize $hW_{(\alpha,\theta)}$ but this minimum

---

[14] This error becomes smaller as the field decreases.

now corresponds exactly to $\alpha=0$, that is to the case where the *spontaneous magnetization of the closure prism is parallel to the applied field*. The closure energy $d^2 W_p$ is

$$d^2 W_p = \frac{K d^2}{4} \sin\theta \cos\theta (1 - \cos\theta)(2 + 3\cos\theta).$$

Closure of the lateral flux corresponding to a given element of the strip — According to section 21 the flux closure of an unlimited structure is produced by a combination of positions A and P in which the energy of formation of the closure prisms is equal to $d^2 W_a W_p/(W_a + W_p)$ for an element of the motif of thickness d and 1 cm wide. To relate this energy to a square centimetre of the surface T it is necessary to multiply by $1/d$, the number of motifs within this thickness. Furthermore the total surface of the separating walls is equal to $2/d$ cm² per cubic centimeter of material, the corresponding energy is therefore

$$2\gamma/d \text{ ergs cm}^{-3}.$$

Let L be the width of the strip in the plane of the surface S; consider a right angled prism of height L whose bases, each one centimeter square, are situated in the planes T and T'. The total energy W_t corresponding to this prism is then

$$W_t = \frac{2\gamma L}{d} + 2d \frac{W_a W_p}{W_a + W_p}.$$

This is minimum for

$$d = \sqrt{\frac{\gamma L (W_a + W_p)}{W_a W_p}}.$$

$d/\sqrt{L}$ is thus a function of $\cos\theta$. The periodicity d only depends on the width of the strip it is independent of its thickness and of its length.

Numerical example for the case of a strip 1 cm wide — We already know the expression for W_a and W_p; as for the surface energy values γ we have shown in a previous article (A) that they are given by

$$\gamma = \sin\theta \left\{ \sin\theta \sqrt{6 - 4\sin^2\theta} + \frac{6 - 7\sin^2\theta}{\sqrt{3}} \sinh^{-1} \sqrt{\frac{3\sin^2\theta}{6 - 7\sin^2\theta}} \right\} \times 0.70 \text{ ergs cm}^{-2}.$$

With these data and for different values of $\cos\theta$ and of the corresponding field H we have calculated the values of W_a, W_p and d for a strip of width $L=1$ cm the results are given in Table III.

In the absence of an applied field, that is at the position of remanent magnetization the periodicity is initially infinite. This is due to the fact that, at the position A, the magnetization direction of the closure domains coincides with an easy magnetization direction and so consequently there is no need to dissipate any energy to bring the spontaneous magnetization into this direction: the closure prisms can therefore be very big so as to reduce the number of layers and the corresponding wall energy.

The periodicity diminishes as the field increases and finally attains a more or less constant value in the range of 65–70 μm when the field reaches 150 gauss. The corresponding curve is shown in Fig. 16.

Comparison with Kaya's observations [18] — These observations were made on a disc of width $L'=2$ cm, cut from the plane (100) and magnetized along the [011] axis. We have used Kaya's photographs to measure the distances separating two powder pattern lines for different values of the magnetization, and

TABLE III.

$\cos\theta = \dfrac{\mathcal{J}}{\mathcal{J}_s}$	$W_a \cdot 10^{-4}$ (ergs cm^{-3}).	$W_p \cdot 10^{-4}$ (ergs cm^{-3}).	$\dfrac{W_a W_p}{W_a + W_p} \cdot 10^{-4}$ (ergs cm^{-3}).	γ (ergs cm^{-2})	d (microns)	H (gauss).
0,707	0	6,47	0	1,21	∞	0
0,72	0,39	6,25	0,37	1,19	179	13
0,73	0,69	6,06	0,62	1,17	137	24
0,75	1,30	5,67	1,06	1,12	103	47
0,80	2,73	4,54	1,71	0,99	76	113
0,85	3,96	3,29	1,79	0,81	67	190
0,90	4,73	1,98	1,40	0,59	65	281
0,95	4,56	0,77	0,66	0,32	70	385

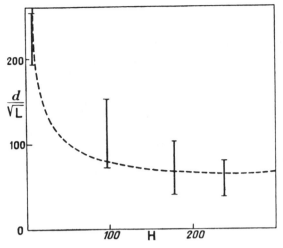

Fig. 16.
The dotted curve gives the theoretical separation of the powder pattern lines as a function of field whilst the vertical lines correspond to the experiments of Kaya: the height of the lines corresponds to the dispersion of the results.

consequently of the field according to equation (20). The measured distances are assumed to give the periodicity and are plotted against H in Fig. 16 after dividing each value by $\sqrt{L'} = 1.41$. The values obtained are considerably dispersed but show an overall agreement with the theoretical curve.

(25) The diagrams obtained by Sixtus [17]: The patterns obtained by Kaya are much less regular than those of Sixtus for strips of silicon–iron. Amongst the strips investigated by Sixtus there are two (the strips 1 and 7N) of width $L \simeq 1.55$ cm which are cut approximately from the (100) plane but the long axis along which the field is applied makes an angle of 30° with the nearest fourfold axis. The mean distances separating two adjacent powder pattern lines vary with the field as shown in Fig. 17. The methods indicated in the first part of this article were used to determine the spontaneous magnetization directions and the relative proportions of the two phases, the period d of the structure was then calculated following the method given above. The calculation is not difficult but is rather tedious; therefore we only give the results in the form of the dotted curve in Fig. 17. This is very similar to the previous curve for a strip oriented along the binary axis (for comparison the ordinates for that case must be multiplied by $\sqrt{L} = \sqrt{1.55} = 1.24$).

The calculated curve lies above the experimental curves for the two specimens but has the same general aspect: in particular the limiting value in strong fields is not far from the experimental values.

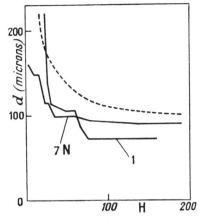

Fig. 17.
Separation of powder pattern lines as a function of field. The full curves correspond to the two specimens studied by Sixtus whilst the theoretical curve is shown dashed.

The rather large discrepancies which are observed in medium strength fields come, no doubt, from the fact that the strip is not exactly oriented in the plane (100); thus for example, in the case of the specimen 7N the normal to the strip instead of corresponding to a fourfold axis makes an angle of 20° with it. As a result the spontaneous magnetizations of the two phases are not parallel to the two largest lateral surfaces of the strip and closure domains must therefore form on these faces leading to a decrease in the spatial periodicity of the powder pattern lines.

(26) Influence of imperfections in the crystal: It should be remarked that the real structure detected by the powder deposit can differ considerably from the equilibrium structure: these false equilibria are shown very clearly in the experiments of Sixtus: the *mean* distance between the lines changes from 200 μm to 100 μm depending on whether the final field of 10 gauss is reached by increasing or by decreasing the field. These differences arise no doubt from the irreversibility of the wall generation process. Apart from this effect, which changes the mean separation of the lines, considerable variations are often observed in the individual separations: these are due to the passive resistance which the real crystal offers to wall movement which means that the walls cannot reach their minimum energy positions, equidistant one from the other, as readily as in the ideal crystal.

These passive resistances are responsible for the usual hysteresis phenomena: the measurement of the area of the magnetization cycle or of the coercive force gives the order of magnitude of possible irregularities agreeing with the observed values.

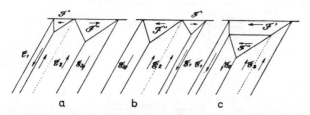

Fig. 18.
Different types of lateral flux closure corresponding to a three layer system. In the case of an infinite structure these different types should be combined in an appropriate manner.

(27) Structures with three phases: So far in this third section only two-phase structures have been treated. The theory of surface domains for three phases can be developed in an analogous way but the closure domains have forms and relative dispositions which are more complicated as is illustrated in Fig. 18. A large number of different arrangements allow the lateral flux to be closed and the energies of these systems are probably rather close to each other: the perturbing effect of lattice irregularities should therefore be more important than in the case of two-phase structures: the corresponding powder pattern lines are observed to be short and irregular.

Moreover large variations of periodicity should not be expected to occur as a function of the applied field as was the case with two phases. These variations were in fact related to the possibility of closure of the lateral flux by means of a third privileged direction not requiring any energy expenditure. However, with three phases all the prigileged directions are already occupied, the flux must now be closed in some other way requiring more energy, so the influence of the applied field is much smaller. We have not thought it useful therefore, to weigh this article down with calculations which are both complicated and without any real interest since there are at present no sufficiently complete experimental data which would allow the conclusions of the theoretical analysis to be verified.

REFERENCES

[1] *Z.f. Physik*, 1932, **74**, 295.
[2] *Phil. Mag.*, 1927, **4**, 641.
[3] *Physik. Z. der Sowjetunion*, 1935, **8**, 153.
[4] *Nature*, 1933, **132**, 517.
[5] *Cf.* Honda and Kaya, *Sc. Rep. Tohoku Imp. Univ.* 1926, **15**, pp 721 and 756; Kaya, *ibid.* 1928, **17**, 639.
[6] *Z.f. Physik*, 1933, **84**, 705.
[7] *Nature*, 1933, **132**, 517.
[8] *Z.f. Physik*, 1931, **69**, 78; *cf.* also Becker, *Ferromagnetism*, 117.
[9] *Z.f. Physik*, 1933, **84**, 705.
[10] *Z.f. Physik*, 1929, **56**, 649.
[11] Akulov, *Z.f. Physik*, 1931, **69**, 78; *cf.* also Becker, *Ferromagnetism*, 117. We have repeated this calculation in paragraph 9.
[12] *Cf.* Becker, *Ferromagnetism*, 122.
[13] *Proc. Roy. Soc. London*, 1925, **107**, 496.
[14] *Sc. Rep. Tohoku Imp. Univ.*, 1926, **15**, 721.
[15] *Z.f. Physik*, 1932, **71**, 442.
[16] *Phys. Rev.*, 1931, **38**, 1903; 1932, **41**, 507.
[17] *Phys. Rev.*, 1937, **51**, 870.
[18] *Z.f. Physik*, 1934, **89**, 796; 1934, **90**, 551.

C29 (1957)

CONCERNING THE INTERPRETATION OF THE MAGNETIZATION LAWS OF PYRRHOTITE

Note (1977): This article is drawn from the last part of a report presented at the National Colloquium held to commemorate the work of P. Weiss (Strasbourg, 8–10 July, 1957).

It is shown how the properties of pyrrhotite crystals are in fact explained by the phase theory whilst historically, and by a curious combination of circumstances, it was these same properties which lead to the development of the molecular field theory.

For a recapitulation of the magnetic properties of pyrrhotite §4 of the first part of C29 given in Chapter 1.

(1) The particularities of pyrrhotite: It is really very strange that P. Weiss was led to formulate the molecular field hypothesis as a result of his experiments on the magnetic properties of pyrrhotite. This is in fact one of the most complicated materials known: its chemical formula is close to Fe_7S_8 and it contains both ferric and ferrous ions. It has a vacancy lattice such that if we denote the unoccupied sites by the symbol T the formula would be more correctly Fe_7S_8T. At low temperature the vacancies are distributed in an ordered manner [1], this order disappears at a temperature twenty degrees above the Curie temperature [2]. Pyrrhotite has without doubt a ferrimagnetic nature which is related to the vacancy ordering.

Below the Curie point the evolution of the magnetic properties as a function of temperature is still imperfectly understood [3]. At room temperature this material is ferromagnetic in a plane, called the basal plane, and paramagnetic in the perpendicular direction (c axis). When the temperature is reduced to the liquid hydrogen region a ferromagnetism appears along the axial direction and becomes stronger and stronger. Finally the crystals are nearly always twinned and consist of an assembly of crystals possessing a common c axis and rotated through an angle of $\pi/3$ with respect to each other.

The magnetization of pyrrhotite was investigated several times by P. Weiss and his collaborators [4]. The most important conclusions of their research have been summarized above; another important result concerns the magnetization J created by a field H of constant magnitude rotating in the ferromagnetic plane XOY of a disc cut from this plane. When H turns from OX to OM the magnetization J turns from OX to OP maintaining a constant magnitude; then when H turns from OM to OY the magnetization turns from OP to OY and its magnitude varies so that its projection onto OY remains constant (Fig. 1). Let $\beta = Y\hat{O}P$, $\alpha = Y\hat{O}M$; the ratio α/β is usually very small.

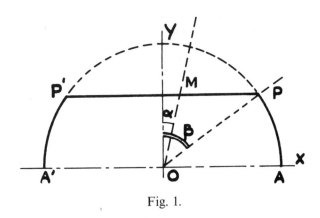

Fig. 1.

(2) The notion of phases in a ferromagnetic crystal: The complete interpretation of the magnetization curves of a single crystal of pyrrhotite involves the concept of *phases* which we have introduced previously [5] to interpret the properties of iron crystals. We showed that, for an ideal monocrystal cut into the form of an ellipsoid, the internal magnetic field is uniform and equal to the sum of a uniform applied field and of the demagnetizing field; and in addition that the elementary domains could be grouped into a finite number of distinct categories. The spontaneous magnetization directions of all the domains in the same category are mutually parallel and deviate in a continuous way from the easy magnetization directions with which they coincide when the internal field is zero. The domains of the same category constitute what we chose to call a *phase*. The maximum number of coexisting phases is equal to the number of easy magnetization directions: six in the case of iron counting two for each easy magnetization axis (the fourfold crystal axes); two in the case of uniaxial crystals when the axis is itself the easy magnetization direction.

The magnetization of the crystal increases from zero up to saturation following a succession of *modes* each characterized by the number of phases present at equilibrium: in this manner in the case of iron, the modes I, II, III and IV succeed one another as the field increases, these modes correspond respectively to 6, 3, 2 and 1 coexisting phases.

(3) The magnetization of pyrrhotite in its basal plane: We limit ourselves here to a study of the case of a crystal cut into the form of a flattened ellipsoid of revolution whose equatorial plane coincides with the basal plane XOY. Let OX be the easy magnetization axis and OY the perpendicular direction: we suppose that the field $\vec{H}$ is always parallel to the plane XOY.

Let $\vec{H}$ increase from zero upwards whilst remaining along the direction making an angle α with OY. The system contains two phases initially and the internal field $\vec{H}_i = \vec{H} - n\vec{J}$ (n = the demagnetizing factor of the disc) is necessarily parallel to OY so that the two phases can have the same thermodynamical potential and thereby coexist in equilibrium. The spontaneous magnetizations of these two phases are then symmetrical with respect to OY, make an angle β with this axis and have the same projection J_y onto OY.

As we have remarked above [6], the magnetocrystalline coupling can be represented by a 'structural' field $-NJ_y$ parallel to OY and the equilibrium equations for the magnetization can be written:

$$H_x - nJ_x = 0 \tag{1}$$

$$H_y - nJ_y = NJ_y . \tag{2}$$

These equations show that the mode I is equivalent to an anisotropic paramagnetic material whose susceptibility is equal to $1/n$ in the direction OX and $1/(N+n)$ in the direction OY.

It must be emphasized that J_x represents only the mean value of the magnetization component along OY: there are in reality two phases whose relative proportions are λ and μ and whose spontaneous magnetizations make the angles $+\beta$ and $-\beta$ with the OY axis. This is expressed by the two equations

$$(\lambda - \mu) J_s \sin \beta = J_x \tag{3}$$

$$(\lambda + \mu) J_s \cos \beta = J_y \tag{4}$$

which together with the condition $(\lambda + \mu) = 1$ allow λ, μ and β to be calculated. In ideal conditions the elementary domains should take on the form of thin planar layers piled up perpendicularly to OY: the magnetostatic energy is then zero.

When H is small, J_x and J_y are small according to equations (1) and (2), whilst $\cos \beta$ is small according to equation (9); equation (1) then shows that λ and μ should have similar values. So initially the two phases occupy equal volumes, then as H increases λ increases (if α lies between 0 and π) and finally when $\lambda = 1$ and $\mu = 0$ the transition from mode I to mode II occurs, only one phase remains in mode II. This transition corresponds to the equations

$$H_x = nJ_s \sin \beta ,$$
$$H_y = (N + n) J_s \cos \beta ,$$

which are equivalent to the parametric equations of an ellipse E centred on the origin with its short axis of length nJ_s parallel to OX and its long axis of length $(N+n)J_s$ parallel to OY as shown in Fig. 2. If we

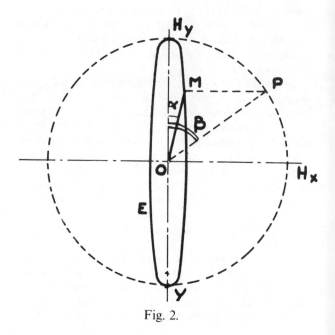

Fig. 2.

represent the field H by a vector OM we remark that the system has two phases when OM lies inside E and one phase when it is outside.

It is easy to verify that since M lies on an ellipse the field OM corresponds to a magnetization parallel to OP where P is the intersection of a circle of centre O and radius $(N+n)J_s$ with a line containing M and parallel to OX. When the applied field decreases but remains directed along OM, the mean magnetization decreases proportionally and itself remains parallel to OP.

In mode II only one elementary domain remains, its spontaneous magnetization is $\vec{J_s}$: the equilibrium position of the magnetization is obtained by setting $\vec{J_s}$ parallel to the resultant of the applied field and the 'structural' field: the demagnetizing field is parallel to $\vec{J_s}$ and so no longer plays any rôle.

(4) The interpretation of the observations made by Weiss: The theoretical approach which we have sketched above allows the observations of Weiss to be interpreted in detail. Let us examine, for example, what happens when a magnetic field of magnitude H = OQ (Fig. 3) rotates in the plane XOY starting from OX. When $\vec{H}$ rotates from OQ to OM the magnetization turns from OA to OP whilst remaining equal to J_s. This is mode II. Since the ratio N/n is of the order 100, as is the ratio of the axes of the ellipse, the angle α, the angle made by OM with OY, becomes very small as soon as the angle AOP becomes appreciable.

When H rotates further from OM to OM′ (symmetrical to M with respect to OY) we pass into mode I and the extremity of the vector $(N+n)\vec{J_s}$ describes a curve PR′P′ which is very close to the straight line PRP′ (R and R′ are the intersections with OY). OR = Hy = H cos α and according to equation (2) OR′ is equal to H; the ratio OR/OR′ is therefore equal to cos α. This is very close to unity since according to equations (6) and (7)

$$\tan \alpha = \frac{n}{N+n} \tan \beta,$$

and tan β is of the order unity.

As we recalled earlier P. Weiss did in fact notice that the extremity of the magnetization vector described a curve very close to the trajectory APRP′A′.

We note finally that if the crystal is twinned and the twins form planar layers across the complete width of the disc and lie perpendicular to the c axis then the demagnetizing factor n which enters in the equations must be smaller than the theoretical factor, it seems that this is what happened in the specimens examined by Weiss.

REFERENCES

[1] F. Bertaut, *C.R. Acad. Sci.* 1952, **234**, 295.
[2] F. K. Lotgering, Thesis, Utretch, 1956.
[3] R. Pauthenet, *C.R. Acad. Sci.*, 1952, **234**, 2261; Thesis, Grenoble, 1957.
[4] P. Weiss, *J. de Phys.*, 1905, **4**, pp469 and 829; P. Weiss and J. Kunz, *J. de Phys.*, 1905, **4**, 847; Ziegler, Thesis, Zurich, 1915, pp 1332, 1532 and 1587; 1905, **141**, pp 182 and 245; see also P. Weiss and G. Foex, Le Magnetisme (Armand Colin, Paris, 1926).
[5] L. Neel, *J. Phys. Rad.*, 1944, **5**, pp 241 and 265.
[6] *Cf.* C29, Chap. I, § 4.

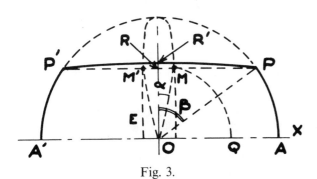

Fig. 3.

A58 (1947)

THE PROPERTIES OF FINE GRAINED CUBIC FERROMAGNETIC MATERIALS

Note (1977): In this article, whose publication was delayed by the war in order not to reveal the essential elements of a patent, we consider the problem of calculating the critical diameter below which a spherical bead of a cubic ferromagnetic material should be a *single domain*.
For associated properties see also articles B1, A59, A60.

In materials which are magnetically soft in the bulk, as is the case for iron, the variations of magnetization in the hysteresis cycle take place by movement of the walls separating the elementary domains. If the iron is reduced into fine particles whose dimensions are inferior to the wall thickness, which is around 500 Å, this mechanism is no longer valid since walls cannot exist in such small volumes of material. The magnetization variations must now proceed by the much more difficult process of simultaneous rotations of the spin ensemble. These ideas lead L. Weil, J. Aubry and the author to make high coercive force iron powders in 1942 [1]. Short accounts of the theory of this type of effect have already been made. It would seem useful to present a more precise description.

The critical dimension below which the magnetization of a particle always remains uniform must first be determined.

C. Kittel [3] compares the energy of a cube, with sides of length L, in a state of uniform spontaneous magnetization to the energy of the same cube in the state of a pole-free structure obtained by division into four equal elementary domains separated by infinitely thin 90° walls of surface energy γ. He shows that the uniform magnetization distribution is the more stable structure below the critical dimension $L_0 = (3\sqrt{2}\cdot\gamma/\pi\ _s^2)$. With $\gamma = 2$ ergs/cm^2 and $= 1700$ we find $L_0 = 100$ Å [4]. This result should be regarded with caution since the wall thickness, far from being negligible compared to the particle dimensions, is in reality five times bigger. In fact the concept of a wall has neither sense nor interest in this situation and it is best to approach the problem in another way.

Consider a spherical grain of radius R and volume V in two magnetization states. In the first state the magnetization is uniform and the energy W_1 simply reduces to the term $(\frac{2}{3}\pi \mathcal{J}^2 V)$, neglecting the magnetocrystalline energy which is an order of magnitude smaller. In the second state, the simplest pole free situation, the magnetization flux lines are circular and perpendicular to planes passing through a diameter Δ of the sphere; the energy W_2 reduces to the Weiss–Heisenberg term and is calculated from the known formula [5]

$$w = \frac{Na^2}{12}\left[(\nabla \mathcal{J}_x)^2 + (\nabla \mathcal{J}_y)^2 + (\nabla \mathcal{J}_z)^2\right],$$

where N is the molecular field coefficient, a the distance between magnetically active neighbours [2.86 Å for iron [6]] and ($\mathcal{J}_x$, $\mathcal{J}_y$, $\mathcal{J}_z$) the cartesian coordinates of the spontaneous magnetization. The energy density at a distance ρ from the axis Δ is equal to

$$(N\mathcal{J}^2 a^2/12\rho^2)$$

and by integration over the volume of the sphere

$$W_2 = \frac{N\mathcal{J}^2 a^2 V}{4R^2}\left(\log_e \frac{2R}{a} - 0{,}307\right).$$

The term $\log_e(2R/a)$ arises because the integral must be limited to the distance $a/2$ from the axis. With $N = 5890$ [4] we find that W_1 and W_2 are equal for $2R = 320$ Å. Below this critical diameter the magnetization of an iron sphere is therefore uniform.

Below the critical diameter the magnetization variations are due only to simultaneous rotations of the spin ensemble. If we now take account of the magnetocrystalline energy of a cubic material.

$$W_c = K(\alpha^2\beta^2 + \beta^2\gamma^2 + \gamma^2\alpha^2),$$

where (α, β, γ) are the direction cosines of the spontaneous magnetization with respect to the fourfold axes, hysteresis effects appear according to the mechanism studied by Becker [7] and Akulov [8]. When the magnetic field direction is along a fourfold axis the hysteresis cycle is rectangular with a coercive field $H_c = (2K/\mathcal{J})$. For an ensemble of non-interacting randomly oriented spheres a simple, but long, graphical calculation shows that if K is positive the mean hysteresis cycle has a coercive field $H_{cm} = 0.64\ K/\mathcal{J}$; for iron this corresponds to about 160 gauss.

It has been shown experimentally that it is possible to make iron powders with high coercive fields, and according to the experiments of F. Bertaut, the grain dimensions of these powders are in the region 200–300 Å, in excellent agreement with the critical diameter values given by our calculations. However, the coercive fields of 1000 gauss obtained by L. Weil cannot be explained by the magnetocrystalline anisotropy alone. It is most probable that the anisotropic form of the particles is responsible for the high coercive field. This will be the subject of a forthcoming article.

REFERENCES

[1] Brev. Fr., dep. Chambery, no. pv., 323, 7 April 1942.
[2] L. Neel, Magnetism in France between 1939 and 1945 [Rapport Congres Assoc. Fr. Avancement des Sci., Paris, Oct. 1945 (sous presse)]; The effect of internal demagnetising fields on the properties of ferromagnetic materials (Conf. Soc. Fr. de Phys., Paris, 6 and 8 Mars 1946); New applications of theoretical research to permanent magnets (Congres Elec. Grenoble, Jillet 1946, Ann. Univ. Grenoble, 22, 1946, p. 71).
[3] *Phys. Rev.*, 1946, **70**, 965.
[4] L. Neel, Cahiers de Physique, no. 25, 1944, p. 1.
[5] L. Landau and E. Lifshitz, *Phys. Z. des Sowjetunion*, 1935, **8**, 153.
[6] L. Neel, *Ann. Physique*, 1936, **5**, 232.
[7] *Z.f. Phys.*, 1930, **62**, 253.
[8] *Z.f. Phys.*, 1933, **81**, 790.

A127 (1963)

THE MAGNETOCRYSTALLINE ENERGY OF A MACROCRYSTAL SUBDIVIDED INTO QUADRATIC CRYSTALLITES

Abstract: A calculation is made of the apparent magnetocrystalline energy of an initially cubic FeNi monocrystal decomposed by irradiation into three categories of quadratic crystallites whose axes coincide with one of the three fourfold axes of the original crystal. An expression of the simplest form $K\alpha^2$ is taken to describe the anisotropy of each crystallite and this leads to a fourth order expression in the direction cosines (α, β, γ) of the magnetization.

Pauleve and Dautreppe have shown [1] that fast neutron irradiation of a cubic macrocrystal of equiatomic Fe–Ni in a magnetic field parallel to a fourfold axis produces a strong magnetic anisotropy due to the formation of quadratic crystallites coherent with the initial matrix. These crystallites, whose dimensions are determined by X-rays to be of the order of 200 Å, belong to three categories depending on which of the three fourfold axes of the monocrystal has become the unique fourfold axis of a crystallite. Pauthenet has shown [2] that the magnetocrystalline energy has the form $K_1\alpha^2 + K_2\alpha^4$, where the constant K_2, of the same order of magnitude as K_1, has the large value of 2.3×10^6 ergs/cm^3 which is difficult to explain *a priori*.

Furthermore the same author has found that a weak magnetic field is sufficient to magnetize the specimen to saturation along the direction of the magnetic field applied during the irradiation; this shows that the crystallites cannot be considered as being independent since a certain number of them have a difficult magnetization direction which coincides with the easy magnetization direction of the macrocrystal. Since the crystallite dimensions are probably smaller than the dimensions of the elementary domains it seems probable that there is close correlation between the directions of the spontaneous magnetizations in the different domains. Given a certain direction for the macroscopic magnetization, defined by the direction cosines (α, β, γ) taken with respect to the fourfold crystal axes, the local deviations of the spontaneous magnetization from this direction at all points within the different crystallites must always be small due to the smoothing effect of the exchange forces and internal magnetic fields due to the dispersion of the magnetization direction. In what follows we will develop this idea.

Suppose then that three categories of crystallites exist (1, 2, 3), and that these occupy respectively the relative volumes p, q, r

$$p + q + r = 1. \qquad (1)$$

The simplest expressions which we can choose for the magnetocrystalline energy densities are respectively for the three categories: $-P\alpha_0^2$, $-Q\beta_0^2$, $-R\gamma_0^2$; $(\alpha_0, \beta_0, \gamma_0)$ are the direction cosines of the spontaneous magnetization.

The crystallites are supposed on the average to be cubes of side L. Each is surrounded by other crystallites randomly distributed among the three categories and whose total number, around twelve, is sufficiently large to allow the real adjacent milieu to be replaced by a homogeneous medium with a uniform magnetization J_a, essentially equal to the spontaneous magnetization J_s, aligned along the direction of the macroscopic magnetization (α, β, γ).

As a result all the crystallites of the same category in the same external conditions have the same magnetization defined by the direction cosines $(\alpha + \lambda_i, \beta + \mu_i, \gamma + \nu_i)$ at the crystallite centre. At the crystallite surface this magnetization must be continuous with the magnetization of the fictitious ambient milieu which has the direction (α, β, γ). Supposing the crystallites to be cubic, this condition can be expressed by taking a cube corner as origin in each crystallite and setting.

$$F = \sin\frac{\pi x}{L} \sin\frac{\pi y}{L} \sin\frac{\pi z}{L}; \qquad (2)$$

F is essentially positive. It is then sufficient to suppose that, at the different points within each crystallite, the direction cosines which describe the local spontaneous magnetization direction are respectively equal to $(\alpha+\lambda_i F, \beta+\mu_i F, \gamma+\nu_i F)$, $(i=1, 2, 3)$. The λ_i, μ_i, ν_i, supposed to be small compared to unity, obey the three relations

$$\alpha\lambda_i + \beta\mu_i + \gamma\nu_i = 0, \quad (3)$$

which express the fact that the sum of the square of the direction cosines is equal to unity ignoring the second order terms. The following three relationships are also obeyed

$$p\lambda_1 + q\lambda_2 + r\lambda_3 = 0 \quad (4)$$

as are two analogous relationships obtained by replacing λ and μ then by ν, which express the fact that the mean magnetization direction is indeed the direction (α, β, γ).

Now that we have defined the magnetization direction at all points of the macrocrystal we can calculate the different energy terms involved.

Magnetocrystalline energy density E_c: This density is the mean density over the different points of the crystallite, the mean value of F is $8/\pi^3$ so that we find:

$$\left.\begin{aligned} -E_{c1} &= P\left(\alpha^2 + \frac{16}{\pi^3}\alpha\lambda_1\right); \\ -E_{c2} &= Q\left(\beta^2 + \frac{16}{\pi^3}\beta\mu_2\right); \\ -E_{c3} &= R\left(\gamma^2 + \frac{16}{\pi^3}\gamma\nu_3\right). \end{aligned}\right\} \quad (5)$$

We have neglected the second order terms $\lambda_1^2, \mu_2^2, \nu_3^2$.

Magnetostatic energy density E_m in the external field: Since the angle φ between the local magnetization and the mean magnetization is given by $\varphi^2 = (\lambda^2 + \mu^2 + \nu^2)F^2$ and the mean value of F is $1/8$ the required energy density is

$$E_{mi} = -H_T J_s \left[1 - \frac{1}{16}(\lambda_i^2 + \mu_i^2 + \nu_i^2)\right], \quad (6)$$

where $H_T = \alpha H_x + \beta H_y + \gamma H_z$ is the projection of the external field $\vec{H}$ onto J_s.

Exchange energy density E_w: This term takes account of the increased exchange energy due to the non-uniformity of the magnetization and is calculated using the Landau–Lifshitz formula

$$E_w = \frac{1}{12} N a_0^2 \left[(\text{grad } J_x)^2 + (\text{grad } J_y)^2 + (\text{grad } J_z)^2\right], \quad (7)$$

where N is the molecular field coefficient, a_0 is the distance between close neighbours in the crystalline lattice and J_x, J_y, J_z are the projections of J_s onto the three coordinate axes. The required energy density is

$$E_{wi} = \frac{\pi^2}{32} N J_s^2 \frac{a_0^2}{L^2} (\lambda_i^2 + \mu_i^2 + \nu_i^2). \quad (8)$$

Internal dispersion field energy density E_d: This term occurs because each crystallite is placed in a uniform magnetic field whose value is close to $4\pi J_s/3$ and at each point the local magnetization within the crystallite makes the angle φ with J_s as calculated above. The corresponding energy increase is given by

$$E_{di} = \frac{4}{3}\pi J_s^2 \frac{\overline{\varphi^2}}{2} = \frac{\pi J_s^2}{12}(\lambda_i^2 + \mu_i^2 + \nu_i^2). \quad (9)$$

It is not necessary to take account of the non-uniform magnetization within each crystallite: each crystallite in fact is equivalent to a series of concentric spherical shells. Each shell has a uniform magnetization so that the internal field is zero and there is no magnetostatic interaction between two shells.

The total mean energy density W_i of a crystallite is thus equal to $E_{ci} + E_{mi} + E_{wi} + E_{di}$ and the total energy density W_M of the macrocrystal equal to

$$pW_1 + qW_2 + rW_3$$

can finally be written as

$$\begin{aligned} W_M = &-(pP\alpha^2 + qQ\beta^2 + rR\gamma^2) \\ &- H_T J_s - \frac{16}{\pi^3}(pP\alpha\lambda_1 + qQ\beta\mu_2 + rR\gamma\nu_3) \\ &+ pC(\lambda_1^2 + \mu_1^2 + \nu_1^2) + qC(\lambda_2^2 + \mu_2^2 + \nu_2^2) \\ &+ rC(\lambda_3^2 + \mu_3^2 + \nu_3^2); \end{aligned} \quad (10)$$

with

$$C = \frac{H_T J_s}{16} + \frac{\pi^2 N J_s^2 a_0^2}{32 L^2} + \frac{\pi J_s^2}{12}. \quad (11)$$

By considering (α, β, γ) as constant we determine the nine unknowns λ_i, μ_i, ν_i which must satisfy the six equations (3) and (4) in such a way that the energy W_M is minimum. The calculation is rather long but can be carried through without difficulty by using the method of Lagrangian multipliers to yield finally

$$W_C = -(\alpha^2 p P + \beta^2 q Q + \gamma^2 r R) - \frac{64}{\pi^6}\frac{D}{C} \quad (12)$$

where

$$D = \alpha^2(1-\alpha^2)p(1-p)P^2 + \beta^2(1-\beta^2)q(1-q)Q^2$$
$$+ \gamma^2(1-\gamma^2)r(1-r)R^2 + 2\beta^2\gamma^2 qrQR \quad (13)$$
$$+ 2\gamma^2\alpha^2 rp RP + 2\alpha^2\beta^2 pq PQ.$$

We have obtained then the expression W_C for the energy density of the macrocrystal as a function of the direction cosines of the mean magnetization, that is to say the magnetization that can be detected macroscopically. It is as if we had a homogeneous specimen with a uniform magnetization equal to J_a and possessing an apparent magnetocrystalline energy W_C: this is why we did not include in W_C the term $-H_T J_s$ which occurs in W_M; it represents the energy of the macrocrystal magnetized in the external field so it must be separated from W_C. We note that the energy W_C is a fourth order function of the direction cosines even though the magnetocrystalline energy of the crystallites only contains second order-terms: the difficulty raised at the beginning of this note is thus dissipated.

There is however an important difference between the present case and that of a homogeneous material, in that the apparent spontaneous magnetization J_a is not constant: it is related to the real spontaneous magnetization J_s by the relationship

$$J_a = J_s\left(1 - \frac{\overline{\varphi^2}}{2}\right). \quad (14)$$

One easily finds that

$$\overline{\varphi^2} = \frac{8}{\pi^6}\frac{D}{C^2}. \quad (15)$$

As a result J_a is found to be a function of H_T and of (α, β, γ).

In the case initially cited of an irradiation carried out in a magnetic field parallel to a fourfold axis of the macrocrystal we must have $q=r$ and $Q=R$ by symmetry. The expression for D then reduces to

$$D' = \alpha^2(1-\alpha^2)[p(1-p)P^2 + q(1-q)Q^2 + 2pq PQ]$$
$$+ 2\beta^2\gamma^2 qQ^2. \quad (16)$$

When the irradiation is carried out in the absence of a magnetic field or when the field is parallel to a threefold axis of the macrocrystal W_C simplifies considerably and becomes

$$W'_C = \frac{64}{3\pi^6}\frac{P^2}{C}(\alpha^4 + \beta^4 + \gamma^4). \quad (17)$$

The expression for the apparent magnetocrystalline energy now contains only terms of the fourth order in (α, β, γ).

These results appear to agree with the experimental results [2].

REFERENCES

[1] J. Paulevé, D. Dautreppe, J. Laugier and L. Neel, *J. Phys. Rad.*, 1962, **23**, 841.
[2] J. Laugier, J. Paulevé and R. Pauthenet, *Comptes rendus*, 1963, **257**, 3340.

A85 (1951)
THE INFLUENCE OF THE SUBDIVISION INTO ELEMENTARY DOMAINS ON THE HIGH FREQUENCY PERMEABILITY OF CONDUCTING FERROMAGNETIC MATERIALS

Abstract: After critically assessing the previous work on the subject the author considers the theory of the influence of the subdivision into elementary domains on the high frequency permeability of conducting ferromagnetic materials. The domains near the surfaces are supposed to consist of plane parallel layers, of thickness d, perpendicular to the surface, and a rigorous solution of the problem is given in the case where the Bloch walls are infinitely deformable and extendable without any energy expenditure. The values of the real and imaginary components of the permeability are calculated as a function of a reduced frequency a which depends on d.

A theory taking account of the surface tension of the walls is developped for the case of very small d values. It is shown that when d is greater than a few microns the influence of the surface tension is negligible; as a result it is generally justifiable to neglect the wall energy.

(1) Introduction Experiment shows that in very weak fields and at high frequency the permeability μ of a magnetic material is a complex quantity

$$\mu = \mu_1 - j\mu_2 , \qquad (1)$$

which has a strong frequency dispersion. For a conducting material, such as soft iron, the real part μ_1 is equal at low frequency to the initial static permeability μ_0 then decreases noticeably above 10 Mc/sec and is reduced to $\mu_0/5$ at around 500 Mc/sec and finally tends towards unity. The imaginary part μ_2 is practically zero at low frequencies, it increases with the frequency and passes through a maximum which is of the order of $\mu_0/3$ at around 500 Mc/sec and then tends towards zero.

This phenomenon is in part simply due to the inhomogeneity of a magnetic material at the microscopic level because of its subdivision into elementary domains. In weak fields, magnetization variations are principally due to movement of the walls separating the elementary domains and are thus localized in a fraction of the order of a hundredth or a thousandth of the total volume. These movements are thus accompanied by induced micro-currents which tend to slow them down. But these micro-currents do not completely explain the dispersion of the permeability since it is known that insulating ferromagnetic materials such as the ferrites also show dispersion. There must therefore be inertial and quasi-frictional terms, probably of atomic origin, which oppose the movement of the walls. This is a manifestation of the fact that the atomic moments cannot instantaneously follow the variations of the applied magnetic field. The study of these relaxation phenomena is of considerable theoretical interest but we will not discuss them further here and we refer the reader to an excellent study by Kittel which was presented at the Colloque International de Ferromagnetism et d'Antiferromagnetism de Grenoble, in 1950 [1].

It would be very useful to have some precise information concerning the effect of the induced micro-currents so as to determine the part played by the wall dynamics in the variation of μ. The rôle of these micro-currents was first described by R. Becker [2] who calculated the damping due to the movement within the material of plane circular walls oriented perpendicular to the spontaneous magnetization. The objections which can be made to this theory are not so much based on the form of the proposed elementary domains, which show little agreement with our present day knowledge of their form and relative positions, but rather on the following fact: the interpretation of the permeability measurements at high frequencies is based on the theory of the skin effect. It would appear then indispensable to examine the theory of the skin effect of an inhomogeneous material composed of elementary domains and not to limit oneself to studying the micro-currents created within a bulk material by idealized magnetization variations to which no possible experiment corresponds. C. Kittel

[3] has examined this skin effect when the surface domains, of thickness h, are distributed as shown in Fig. 1. He supposes essentially that the walls aa', bb', ..., separating these domains are rigid and undeformable and that they move together parallel to themselves over the complete thickness h during the magnetization variations. This hypothesis is hardly

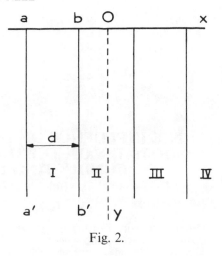

Fig. 2.

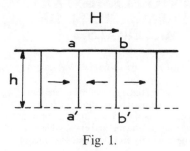

Fig. 1.

justifiable and we will show later (§ 12) that on the contrary it is more correct to suppose that the walls are *completely deformable*.

We will reconsider the theory therefore by supposing that we are dealing with elementary domains made up of planar layers of thickness d, piled one on top of the other like the leaves of a book and magnetized alternatively in opposite directions. The phenomena depend on the orientation of the layers with respect to the surface of the material: two main positions should be distinguished, the first has the plane of the layers perpendicular to the surface and the second has them parallel to the surface. We limit ourselves in this article to an examination of the first position.

(2) Defining the problem: Take three rectangular axes $Oxyz$: Oxz is the surface of the material, Oyz is parallel to the plane of the domain walls. The intersections of the walls on the plane Ozy are represented by aa', bb', etc in Fig. 2. The odd domains I, III, etc are magnetized along the positive direction of Oz, the even domains II, IV, etc are magnetized in the opposite direction. The system is in an external field H_e parallel to Oz

$$H_e = H_0 e^{j\omega t} \qquad (2)$$

whose circular frequency ω is large enough for the thickness of the skin layer to be small compared to both the thickness of the specimen under investigation and to the height of the domains along the direction Oy.

We will neglect the inertia of the walls and the forces related to the relaxation of the atomic moments. We will suppose for the moment that the walls are infinitely deformable and that the only pressures which determine their movement are, on the one hand, the magnetic pressure $2HJ_s$, where J_s is the spontaneous magnetization and H the magnetic field component along Oz, and on the other hand a return pressure towards the equilibrium position. The initial static susceptibility along Oz is χ_0, so that in the case of a slow magnetization variation the walls would appear to have a surface susceptibility of $\chi_0 d$, whilst the rest of the material has zero susceptibility, in other words unit permeability.

In the case where only the action of very weak fields is considered the wall movements are small compared to the distance d separating two neighbouring walls, so that without any noticeable error the walls can be supposed *fixed* and to have a surface susceptibility equal to $\chi_0 d$ per cm^2.

(3) Equations and boundary conditions of the problem: The problem has cylindrical symmetry. We use the following notation; ρ is the resistivity i_x and i_y are the current densities, H_t is the magnetic field parallel to Oz. Maxwell's equations in c.g.s. reduce to

$$4\pi i_x = -\frac{\partial H_t}{\partial y},$$

$$4\pi i_y = \frac{\partial H_t}{\partial x}, \qquad (3)$$

$$\rho\left(\frac{\partial i_x}{\partial y} - \frac{\partial i_y}{\partial x}\right) = -\frac{\partial H_t}{\partial t},$$

since the permeability is unity. We seek a solution of the form

$$H_t = H e^{j\omega t}, \qquad (4)$$

which substituted in the system of equations (3) shows that H is a function of x and y independent of time satisfying the partial differential equation:

$$\frac{\partial^2 H}{\partial x^2} + \frac{\partial^2 H}{\partial y^2} = \frac{4\pi j \omega}{\rho} H. \qquad (5)$$

The required solution is periodic

$$H(x + d, y) = H(x, y),$$

equal to H_0 for $y=0$ is equal to zero for infinite y and on the walls, situated for example at

$$x = \frac{1}{2}d \quad \text{and} \quad x = -\frac{1}{2}d,$$

satisfies a boundary condition which we will now determine.

Consider a wall P (Fig. 3) in a variable field H_t perpendicular to the plane of the figure and with the

Fig. 3.

positive direction going from front to back through the plane of the figure. The flux corresppponding to 1 cm length of wall is equal to $4\pi\chi_0 d H_t$ so that a small circuit element $abcd$, of area δs, is exposed to an electromotive force of $-4\pi\chi_0(\partial H_t/\partial t)$ per cm and this gives rise to a current of intensity

$$i = \frac{\delta s}{2\rho} \cdot 4\pi\chi_0 \, d \, \frac{\partial H_t}{\partial t}, \qquad (6)$$

directed in a positive sense along the arrows. If i_{t_1} and i_{t_2} are the tangential components of the current density on either side of P one should then write

$$i_{t_1} - i_{t_2} = \frac{4\pi\chi_0 d}{\rho} \frac{\partial H_t}{\partial t}. \qquad (7)$$

In the case of interest here, taking account of equations (3) and (4) and using the symmetry $\overline{i_y}(x) = -i_y(-x)$, this condition (7) can be written for $x = d/2$ as

$$-\frac{1}{2\pi}\frac{\partial H}{\partial x} = \frac{4\pi\chi_0 j \omega d}{\rho} H. \qquad (8)$$

In the following we simplify the nomenclature by setting

$$a = \frac{4\pi^2 \chi_0 d^2 \omega}{\rho}, \qquad (9)$$

and

$$r^2 = ja. \qquad (10)$$

The quantity a is a real number proportional to the frequency; r^2 is a pure imaginary quantity.

(4) Solution of the problem: We are looking for a solution of the form

$$H = H_0 \sum_{n=1}^{\infty} c_n \cos p_n x \, e^{-q_n y}, \qquad (11)$$

where each term in the summation should satisfy the equations (5) and (8). This gives the relationship

$$q_n^2 - p_n^2 = \frac{r^2}{\pi \chi_0 d^2}, \qquad (12)$$

which defines q_n as a function of p_n, and also the relationship

$$p_n \sin \frac{p_n d}{2} = \frac{2r^2}{d} \cos \frac{p_n d}{2}. \qquad (13)$$

If we let $\varphi_1, \varphi_2, \ldots, \varphi_n, \ldots$ be the successive roots of the complex transcental equation

$$\psi \tan \varphi = r^2, \qquad (14)$$

the values of p_n are given by

$$p_n = \frac{2\varphi_n}{d}. \qquad (15)$$

The condition that H is equal to H_0 for $y=0$ must now be used. This requires that the Fourier series $\Sigma c_n \cos p_n x$ be equal to unity in the interval $0 < x < d/2$. We will suppose that the functions $\cos p_n x$ form a complete orthogonal system: this is known to be true when r^2 is real. The orthogonality can in fact be verified. We have

$$Q_{nm} = \int_0^{\frac{d}{2}} \cos p_m x \, \cos p_n x \, dx \qquad (16)$$

$$= \frac{1}{2}\int_0^{\frac{d}{2}} [\cos(p_n + p_m)x + \cos(p_n - p_m)x]\, dx,$$

whence

$$2Q_{nm} = \frac{p_n \sin\frac{p_n d}{2} \cos\frac{p_m d}{2} - p_m \sin\frac{p_m d}{2} \cos\frac{p_n d}{2}}{p_n^2 - p_m^2} \quad (17)$$

But since p_n and p_m satisfy the equation (13) Q_{nm} must be zero when n is different to m. When n is equal to m,

$$Q_{nm} = \int_0^{\frac{d}{2}} \cos^2 p_n x\, dx = \frac{d}{4} + \frac{\sin p_n d}{4 p_n}, \quad (18)$$

so that taking account of equations (13) and (16)

$$Q_{nm} = \frac{d}{4}\left(1 + \frac{\sin 2\varphi_n}{2\varphi_n}\right) = \frac{d}{4}\left(1 + \frac{\sin^2 \varphi_n}{r^2}\right) \quad (19)$$

The Fourier coefficients c_n of the Fourier series development of a function $Y \equiv \Sigma c_n \cos p_n x$ are obtained in the usual way by multiplying both sides by $\cos p_n x$ and then integrating between 0 and $d/2$. For $Y=1$ this gives

$$c_n = \frac{2 \sin \varphi_n}{\varphi_n \left(1 + \dfrac{\sin^2 \varphi_n}{r^2}\right)}. \quad (20)$$

The function H is now completely defined.

(5) Calculation of the flux: From the physical point of view the quantity which interests us the most is the total magnetic flux Φ which passes through the material per unit parallel to the external field. This flux is composed of two terms Φ_1 and Φ_2 corresponding respectively to the flux of the field H within the elementary domains and to the magnetization due to the surface susceptibility attributed to the walls.

The first term is obtained by a double integration of H with respect to x and to y which gives

$$\Phi_1 = \frac{2H_0}{d} \sum_{n=1}^{\infty} \int_0^{\frac{d}{2}} \int_0^{\infty} c_n \cos p_n x\, e^{-q_n y}\, dx dy, \quad (21)$$

or

$$\Phi_1 = \sum_{n=1}^{\infty} \frac{H_0 c_n \sin \varphi_n}{q_n \varphi_n} \quad (22)$$

The wall flux can be written

$$\Phi_2 = 4\pi\chi_0 d \sum_{n=1}^{\infty} \frac{H_0}{d} \int_0^{\infty} c_n \cos\frac{p_n d}{2} e^{-q_n y}\, dy, \quad (23)$$

or again

$$\Phi_2 = 4\pi H_0 \chi_0 \sum_{n=1}^{\infty} \frac{c_n \cos \varphi_n}{q_n} \quad (24)$$

The sum gives

$$\Phi = \Phi_1 + \Phi_2 = \frac{\pi H_0 \chi_0 d^2}{r^2} \sum_{n=1}^{\infty} \frac{c_n q_n \sin \varphi_n}{\varphi_n} \quad (25)$$

This equation can be simplified by taking account of equations (12), (13), (14) and (15) to give

$$\Phi = \frac{2\pi \chi_0 d H_0 A}{r} \quad (26)$$

after setting

$$A = \sum_{n=1}^{\infty} \frac{2r^3 \left(1 + \dfrac{r^2}{4\pi\chi_0 \varphi_n^2}\right)^{1/2}}{\varphi_n (\varphi_n^2 + r^2 + r^4)} \quad (27)$$

The problem is thereby reduced to the determination of the roots of the transcendental equation (14) where r^2 is a pure imaginary number.

(6) Calculation of the complex permeability: The classical calculation for the skin effect in the case of a homogeneous ferromagnetic material gives the surface flux Φ as a function of the permeability μ and the amplitude H_0 of the applied alternating field. This gives

$$\Phi = \left(\frac{\rho\mu}{4\pi j \omega}\right)^{1/2} H_0, \quad (28)$$

which using the relations (9) and (10) can be rewritten as:

$$\Phi = (4\pi\chi_0 \mu)^{1/2} \frac{H_0 d}{2r}. \quad (29)$$

If we now compare equations (26) and (29) we find that for a material with surface domains the macroscopic behaviour is that of a homogeneous material with a complex permeability given by:

$$\mu = \mu_1 - j\mu_2 = 4\pi\chi_0 A^2. \quad (30)$$

This is, of course, the permeability given by applying the classical theory of the skin effect to the interpretation of the experimental data.

(7) Calculation of μ_L and μ_R: Some authors use another method to interpret their results. The part of the flux in phase with the field, that is the real part, can be considered as that of a material having a real fictitious permeability μ_L. Similarly the part of the magnetic flux with a $\pi/2$ phase shift wrt the field, the imaginart part, can be considered as belonging to a material with a fictitious permeability μ_R which is also real but different from the preceding value.

Since r is given by the product of a real quantity by $(1-j)$, μ_L and μ_R are defined by:

$$\sqrt{\mu_L} - j\sqrt{\mu_R} = (1-j)\, A\sqrt{4\pi\chi_0}, \qquad (31)$$

and comparison of this equation and equation (30) yields the following relationships

$$\mu_1 = \sqrt{\mu_R \mu_L}\,;\quad \mu_2 = \frac{1}{2}(\mu_R - \mu_L). \qquad (32)$$

(8) Calculation of A at low frequencies ($a<1$): The roots of equation (14) can be developed in increasing powers of r. The first root is found to be

$$\varphi_1 = r - \frac{r^3}{6} + \frac{r^5}{7^2} + \ldots, \qquad (33)$$

and the following roots are given by:

$$\varphi_{n+1} = n\pi + \frac{r_2}{n\pi} - \frac{r_4}{n^3\pi^3} + \ldots \quad n=1,2,3,\ldots \qquad (34)$$

Introducing the values of these roots into equation (27) which defines A and developing again in series of increasing powers of r we finally find

$$A = (1 - 0{,}167\, r^2 + 0{,}078\, r^3 + \ldots)\sqrt{1 + \frac{1}{4\pi\chi_0}} \qquad (35)$$

This expression combined with equation (30) can be used to verify that μ tends to $\mu_0 = 1 + 4\pi\chi_0$ as r tends to zero.

(9) Calculation of A at medium frequencies ($a<10$): In this case we calculate separately the first three terms of the summation A which correspond to the first three roots, calculated exactly, of equation (14) and we replace the sum of the following terms by an integral in which the expression for the corresponding roots is taken to be the expansion (34) limited to its first two terms. These calculations are extremely longwinded.

The Figs. 4 and 5 give the real and imaginary parts x_n, y_n of the roots ψ_n for $n=1,2$ and 3, as a function of

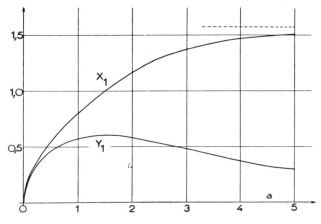

Fig. 4.
Values of the real and imaginary components X_1 and Y_1 respectively of the first root of the transcendental equation (14), as a function of a.

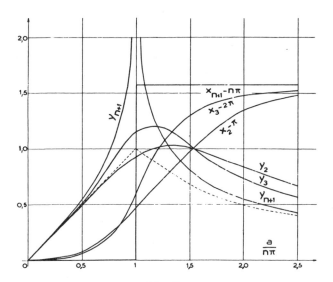

Fig. 5.
Values of the real and imaginary components of the second, third and $(n+1)$th roots of the transcendental equation (14) for a very large, as a function of $a/n\pi$.

the parameter a defined by equation (9). The asymptotic values of x_n and y_n when n tends to infinity are also indicated on Fig. 5. These values were all calculated by successive approximations using Kennelly's tables for the complex hyperbolic functions.

Finally, we remark that when a is large compared to $n\pi$ the asymptotic value of the root ψ_{n+1} can be written

$$\varphi_{n+1} = u - \frac{u}{1+r^2} + \frac{r^2 u^3}{3(1+r^2)^4}$$

with

$$u = \frac{\pi}{2} + n\pi \quad (36)$$

provided that $n \geq 2$. The asymptotic expression for the first root when $a > 1$ is given by

$$\varphi_1 = \frac{\pi r^2}{2(1+r^2)} \left[1 + \frac{\pi^2}{12(1+r^2)^2} + \ldots \right] \quad (37)$$

(10) Calculation of A at very high frequencies ($a > 10$): One possible method is to replace the summation A by two integrals taking n as the variable, the first integral corresponding to the interval $\frac{1}{2} < n < a/\pi$, and the second to the interval $a/\pi < n < \infty$. In the first integral the approximate value of the roots is taken to be $\psi_{n+1} = (\pi/2) + n\pi(1 + j/a)$; whilst in the second $\psi_{n+1} = n\pi + (ja/n\pi)$ and $1/a$ is neglected with respect to unity.

This approach gives

$$A = \frac{1}{\pi r} \sqrt{1 + \frac{1}{4\pi \chi_0}} \{\text{Log}(4.90 \, a^2) + 3.95 j\}, \quad (38)$$

but it is not possible to be sure of the validity of this approximation nor to fix the value of a above which it is acceptable: indeed when n is very large Fig. 5 shows that the roots φ_n are poorly represented by the asymptotic formulae given above in the region $a = n\pi$.

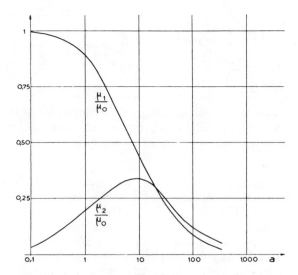

Fig. 6.
Values of the real and imaginary components of the apparent complex permeability as a function of the parameter a defined by equation (9).

(11) Results of the calculation: Fig. 6 summarizes the results of the calculation when μ_0 is supposed to be large compared to unity, greater than 50 at least, using the methods described in the three preceding paragraphs. μ_1/μ_0 and μ_2/μ_0 are shown as a function of a.

As far as the calculation of μ_1 is concerned the segments of the curve calculated by the different methods fit quite satisfactorily. This is not the case for the calculation of μ_2: the approximation of section 10 does not seem to converge satisfactorily. We have therefore found it necessary to refine the values of μ_2 when a is in the range 10 to 50 by using the values of μ_1, which seem to be more exact, in the Kramers–Kronig equations [4] which relate the components of the complex magnetic susceptibility $\chi = \chi_1 - j\chi_2$ and which appear to be applicable here. In particular we must have

$$\chi_2(\omega) = -\frac{2\omega}{\pi} \int_0^\infty \frac{\chi_1(\omega') - \chi_1(\infty)}{\omega'^2 - \omega^2} d\omega'. \quad (39)$$

(12) Influence of the surface tension of the walls: In the preceding calculations we have completely neglected the surface tension of the walls. But since the amplitude of the wall oscillations decreases with depth the wall cannot remain plane; the resulting deformation increases its surface. In order to estimate the corresponding error we will suppose that the structure of the surface domains is the same as that described at the beginning of section 2 but that the thickness of the elementary domains d is sufficiently small to allow the magnetic field H_t, which exists within the ferromagnetic material along OZ, to be practically independent of x, so that the propagation equations of the electromagnetic field reduce to

$$\frac{\partial^2 H_t}{\partial y^2} = \frac{4\pi}{\rho} \frac{\partial B_t}{\partial t}, \quad (40)$$

where B_t denotes a mean induction, independent of x.

In the absence of a field the walls take up an equilibrium position parallel to the plane OYZ. A field H_t, varying with t, causes the walls to deviate from their equilibrium position by a distance X_t which depends on t and y. At large depths $X_t = 0$. The magnetic moment variation, due to the movement of a single wall, between the ordinates y and $y + dy$ is equal to $2J_s X_t dy$ so that the mean induction B_t at depth y is:

$$B_t = H_t + \frac{8\pi J_s X_t}{d} \quad (41)$$

We now examine the equilibrium conditions for a

wall element of height dy. It is acted upon by the following forces

(a) a magnetic force, equal to $2H_t J_s dy$ and to first approximation parallel to Ox;

(b) a restoring force $-kX_t dy$ which tends to bring the wall back to its equilibrium position in the absence of a field. This force is parallel to Ox and its intensity is found by setting the susceptibility of the system in a constant field equal to the initial static susceptibility. This gives

$$k = \frac{4J_s^2}{d\chi_0}; \quad (42)$$

(c) the surface tension forces of the wall, equal to γ, tangential to the wall section through the plane OXY and applied to the two extremities of the element dy. Their projections along OX are respectively equal to

$$-\gamma \frac{\partial X_t}{\partial y} \quad \text{and} \quad \gamma\left(\frac{\partial X_t}{\partial y} + \frac{\partial^2 X_t}{\partial y^2} dy\right).$$

For the considered wall segment the equilibrium equation becomes

$$2H_t J_s = \frac{4J_s^2}{d\chi_0} X_t - \gamma \frac{\partial^2 X_t}{\partial y^2}. \quad (43)$$

We need to resolve the system of equations (40), (41) and (43) which give the general solution of the problem. If we suppose that the external field H_e is given by equation (2) the required solution must have

$$B_t = Be^{j\omega t}; \quad H_t = He^{j\omega t}; \quad X_t = Xe^{j\omega t}, \quad (44)$$

where B, H and X are functions independent of t, which satisfy the equation

$$\frac{\partial^2 H}{\partial y^2} = \frac{4\pi j\omega}{\rho} B. \quad (45)$$

and equations (44) and (43), ignoring the subscript t. Elimination of B and H in these equations shows that X must be a solution of the following fourth order differential equation

$$\frac{\partial^4 X}{\partial y^4} - S \frac{\partial^2 X}{\partial y^2} + P = 0 \quad (46)$$

where the quantities S and P are defined as follows:

$$S = \frac{4J_s^2}{\gamma d \chi_0} + \frac{4\pi j\omega}{\rho},$$

$$P = \frac{64\pi^2 J_s^2 j\omega}{\gamma \rho d}\left(\frac{1}{4\pi\chi_0} + 1\right).$$

The solution has the form:

$$X = C'e^{-m'y} + C''e^{-m''y} \quad (48)$$

where m' and m'' are two roots (with positive real parts) of the characteristic equation

$$m^4 - Sm^2 + P = 0. \quad (49)$$

The two coefficients C' and C'' are determined from the boundary conditions.

We know that, in the first place, the walls must always meet the surface of the body perpendicularly as is shown in Fig. 7 at (a). If this condition is not satisfied

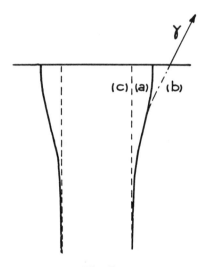

Fig. 7.

and if, for example, the wall was in position (b), the tangential component of the surface tension $\vec{\gamma}$ applied to the extremity of the wall at the surface of the body could not be in equilibrium. It is therefore necessary to set the derivative $\partial X/\partial y$ to zero at the surface ($y=0$), this gives

$$m'C' + m''C'' = 0. \quad (50)$$

It is also necessary to express the fact that, for $y=0$,

the magnetic field, given as a function of X by equation (43), is equal to the external field H_0. This gives a second relationship between C' and C'':

$$2H_0 J_s = \frac{4J_s^2}{\chi_0 d}(C' + C'') - \gamma(C'm'^2 + C''m''^2). \quad (51)$$

(13) Calculation of the magnetic flux: From the physical point of view the important quantity is the magnetic flux Φ which passes through the body, reduced to a length of 1 cm

$$\Phi = \int_0^\infty B dy. \quad (52)$$

B can be obtained from equation (45), the integration gives

$$\Phi = -\frac{\gamma \rho}{8\pi J_s j\omega}(C'm'^3 + C''m''^3). \quad (53)$$

By eliminating C' and C'' from the three equations (50), (51) and (53) and by replacing $m'^2 + m''^2$ and $m'^2 m''^2$ by their values S and P, deduced from the relationships between the roots of equation (49) it is finally found that

$$\Phi^2 = \frac{\rho \mu_0 H_0^2}{4\pi j \omega} A', \quad (54)$$

where

$$A' = \frac{1 + \dfrac{ja'}{\mu_0} + 2\sqrt{ja'}}{(1 + \sqrt{ja'})^2}, \quad (56)$$

and

$$a' = \frac{\gamma d \mu_0^2 \omega}{4\rho J_s^2}. \quad (56)$$

The initial static permeability $\mu_0 = 1 + 4\pi\chi_0$ appears to be multiplied by A': the apparent permeability is therefore $\mu_0 A'$. The coefficient A' depends only on the parameter a' which is proportional to the frequency. The following table gives the value of the real part $Re(A')$ and the imaginary part $Im(A')$ of A' for several values of a'.

TABLE

a'	0,1	0,3	1,0	3	10	30	100	300	1 000
$Re(A')$	0,98	0,92	0,78	0,60	0,40	0,24	0,14	0,08	0,04
$Im(A')$	0,19	0,29	0,36	0,35	0,27	0,19	0,12	0,07	0,04

(14) Discussion: We now have two theories, one which applies to thick domains, of width d, in which the wall energy γ is neglected and in which the flux is given by the expression (26), the other which applies to narrow domains where the flux is given by (54).

In the two cases, the form of the real permeability component as a function of $\log a$ or $\log a'$ are rather similar and by and large are related by a simple translation parallel to the frequency axis such that $a = 1.35 a'$. The two theories therefore give roughly the same values for μ_1 as a function of ω when a is equal to $1.35\, a'$ that is when γ is equal to 1.4 ergs/cm^2, $\mu_0 = 100$ and $J_s = 1\,700$ for a domain thickness $d = 5.2 \times 10^{-6}$ cm. This very small value is of the order of magnitude of the Bloch wall thickness: the domain thickness is always much greater, by a factor of at least ten for example. *It is therefore justifiable to neglect the wall energy in a first approximation* although it may be necessary to estimate the error involved.

As an example, we suppose that the frequency dependence of the apparent permeability decrease of the soft annealed iron used by R. Millership and F. V. Webster in their experiments [5] is due only to the subdivision into elementary domains. The real part of the permeability is then divided by 2 at 340 Mc/sec, corresponding to a value of a equal to 7.2, consequently the domain thickness is found to have the very reasonable value of $d = 3.3 \times 10^{-4}$ cm (for the same numerical values used above and $\rho = 10^4$ e.m.u.).

Assuming this value for d we find $a' = 0.085$. Referring to the preceding table it is seen that the corrections which must be made to μ_1, calculated according to equation (26), to take account of the wall energy, are less than 2% in the region where the permeability is already reduced by half due to the existence of the domains themselves. It is therefore legitimate to suppose the Bloch walls to be infinitely deformable.

It would be interesting to test the validity of equation (26) by experiments on soft iron treated so as to have large elementary domains of thickness greater than 10 microns for example. It is almost certain that in this case the decrease of the apparent permeability would be exclusively due to the subdivision into elementary domains and would correspond to the tenets of the theory proposed here.

I thank M. Pauthenet for his help in drawing the figures in this article.

REFERENCES

[1] C. Kittel, *J. de Phys.*, 1951, **12**, 291.
[2] R. Becker, *Phys. Z.*, 1938, **39**, 856; *Ann. Phys. Lpz.*, 1939, **36**, 340.
[3] C. Kittel, *Phys. Rev.*, 1946, **70**, 281.
[4] R. de Kronig, *Z. Techn. Phys.*, 1938, **19**, 509; *Phys. Z.*, 1938, **39**, 823.
[5] R. Millership and F. V. Webster, *Proc. Phys. Soc. B*, 1950, **63**, 783.

Chapter VII
BLOCK AND NÈEL WALLS
A50, A97, A118

A50 (1944)
SOME PROPERTIES OF FERROMAGNETIC DOMAIN WALLS

Abstract: The present article is concerned with a detailed study of the walls separating elementary domains in an idealized *iron* crystal (without any deformation or inclusions) with the aim of laying the foundations of the theory of the magnetization process in pure ferromagnetic materials.

It is shown first that the magnetization components perpendicular to the wall in adjacent domains are equal and that this perpendicular component remains constant across the wall transition layer: the spontaneous magnetization direction undergoes a sort of precession about the normal to the wall. The general equation for the wall energy γ is then expressed as a function of the magnetocrystalline energy and numerical values of γ are obtained for some important special cases. The most stable 180° walls are found to be those perpendicular to a fourfold axis: their energy is 1.4 ergs/cm².

The creation of 90° walls generally involves magnetostrictive constraints which increase the free energy. The most stable 90° walls have a surface energy of 1.2 ergs/cm² and are perpendicular to a twofold axis. This is verified by Bitter pattern observations.

90° walls are about 500 Å (50 nm) thick whilst the thickness of 180° walls, which can only be calculated by taking account of magnetostriction, is of the order 1800–2000 Å (180–200 nm).

Finally the effect of a magnetic field on 90° walls perpendicular to a twofold axis is examined and their energy as a function of the applied field is calculated. Bitter pattern observations should in principle allow these results to be experimentally tested.

(1) Introduction:

It is well known that an unsaturated ferromagnetic material is subdivided into what are called *elementary domains*. The magnetization in each domain, the so-called *spontaneous magnetization*, is uniform: its *intensity* $\mathfrak{J}_s$ is equal to the saturation magnetization intensity at the given temperature but its direction varies from one domain to another. These directions are not arbitrary: thus in the case of pure annealed iron in a magnetic field of up to ten gauss there are in practice only three possible magnetization directions, called *privileged directions*, which correspond to the three fourfold crystal axes. In weak fields, the observable magnetization variations, given by the resultant of the magnetizations of the different domains, are due almost exclusively to changes in the relative dimensions of the different domains: in each domain the spontaneous magnetization retains a constant intensity and direction, but the walls which separate the neighbouring domains move. The magnetization processes in weak fields are principally wall movement phenomena, but before this dynamic situation can be dealt with, it is essential to understand the nature and properties of stationary domain walls.

The first important step in this direction is due to Bloch [1]: he showed that walls are not just geometrical surfaces on either side of which the magnetization lies along different privileged directions but that a wall is in fact a transition layer whose width can reach several hundreds of Angstrom units and within which the magnetization direction rotates progressively from one direction to the other.

The explanation of this phenomenon is the following: the juxtaposition of two phases magnetized along different directions produces neighbouring atom pairs with non-parallel atomic moments and consequently with a non-minimal exchange energy. Calculation shows that this supplementary energy decreases as the transition region increases in thickness. But as the width increases there are more and more atoms whose magnetic moments are not aligned along an easy magnetization direction, and as a result the magnetocrystalline energy is increased. The wall finally assumes a thickness which minimizes the sum of the exchange and magnetocrystalline energies. All walls are therefore characterized by a certain surface energy γ which plays an essential role in low field magnetization processes. This energy is of the order of 1 erg/cm². Bloch, Becker and others have already calculated this energy in certain special cases.

In this article we intend to take up such calculations in the case of ideal, perfect, undeformed, crystals free of any inclusions. The case of cubic crystals, and of iron in particular, will be examined in detail. The results bring to light various aspects of wall properties which have been neglected up to the present: in particular the determination of the wall energy as a function of the magnetization direction orientation with respect to the wall; the exact way in which the magnetization rotates within the wall; the role of magnetostriction and the demagnetizing field, etc. This present article is purely theoretical and forms the basis for future publications concerning the properties of ferromagnetic materials in weak and medium fields. In particular the first of this future series will appear shortly and is referred to here by the letter B [2]. For the sake of concision we will often refer to the recent excellent book by Becker [3] designated by the letter F.

(2) Wall orientation with respect to the spontaneous magnetization directions of adjacent domains: Intuitively it seems most unlikely that a wall can take on any orientation with respect to the spontaneous magnetization directions of the two elementary domains which it separates. Consider the case of a polycrystalline, globally isotropic, material in the form of a cylinder sufficiently long that the longitudinal demagnetizing field be zero and that the perturbating effect of the ends be negligible. Place this cylinder in a field H parallel to its axis; as a result the mean magnetization $\vec{J}$ is also parallel to the axis. The magnetic energy W_M of the system is given by the volume integral throughout space

$$W_M = - \int \vec{H} \vec{J} dv + \int \frac{H_d^2}{8\pi} dv \qquad (1)$$

in which H_d represents the variable magnetic field due to the internal free charges. The energy is minimum when H_d is zero at all points, that is when there are no internal free charges. It is therefore probable that the structures in which the domains are so disposed as to satisfy this condition will play a role of particular importance in domain theory.

Consequently we limit our investigation to charge free walls. Their orientation, defined by the unit vector $\vec{n}$ normal to the wall, satisfies the relation

$$\vec{h}_1 = \vec{h}_2 \qquad (2)$$

in which $\vec{J}_1$ and $\vec{J}_2$ are the spontaneous magnetizations of the adjacent domains.

(3) The magnetization direction within a wall: Consider a plane wall satisfying condition (2) and take φ_0 as the angle between the spontaneous magnetization vectors $\vec{J}_1$ and $\vec{J}_2$ and the wall normal which is directed towards the domain of magnetization $\vec{J}_2$. Let θ_1 be the angle between a reference direction in the wall plane and the projection of $\vec{J}_1$ onto the wall plane; similarly θ_2 is the angle between the projection of $\vec{J}_2$ and the same reference direction. The magnetization direction at any point within the wall is defined completely by the angle φ which it makes with the normal and by the angle θ between its projection onto the wall plane and the reference direction.

Within the wall the magnetization turns progressively from its initial direction (φ_0, θ_1) to its final direction (φ_0, θ_2): but there are of course an infinity of ways by which this rotation can be affected. It will be shown later (§ 10) *that the wall energy is minimum when the normal magnetization component remains constant and equal to $J_s \cos\varphi$ across the transition layer*. All other cases involve the creation of a double layer of magnetic charge with a relatively large supplementary magnetic energy.

We will suppose then that within the wall the angle θ changes continuously from θ_1 to θ_2 whilst φ remains constant and equal to φ_0: since the subscript no longer serves any useful purpose from hereon we write φ for φ_0. The magnetization direction is seen to undergo a sort of precession about the wall normal.

Consider then a *plane* wall and choose three rectangular coordinate axes with the abscissa OX normal to the wall. The angle θ is then a function of x alone, we aim to determine this function by minimizing the total energy. As we have already pointed out this energy is due to the increase both of the magnetocrystalline energy and of the exchange energy.

(4) Magnetocrystalline energy: The magnetocrystalline energy w_c represents the coupling energy between the spontaneous magnetization and the crystal lattice: it depends only on the direction of the spontaneous magnetization, that is to say, in the present case, only on the parameter θ. This energy can be represented quite generally by a function $f(\theta)$ (per cubic centimeter). We take the zero level of this energy to be the energy corresponding to the initial or the final orientations

$$f(\theta_1) = f(\theta_2) = 0. \qquad (3)$$

In substances with cubic symmetry, the magnetocrystalline energy is generally expressed as a function of the direction cosines (α, β, γ) of the spontaneous magnetization with respect to the cube axes; it is sufficient to limit the expression to the 4th order terms:

$$w_c = K(\alpha^2\beta^2 + \beta^2\gamma^2 + \gamma^2\alpha^2). \qquad (4)$$

In the case of iron the different experimental values of the constant K are in quite good agreement (F, p. 123) and give $K = 4.3 \times 10^5$ ergs/cm^3. The expression (4) is transformed into polar coordinates (φ, polar angle; θ, azimuth) by taking the wall normal as the polar axis as was set out in the previous paragraph. Equation (4) then reduces to a function $f(\theta)$ of the azimuthal angle θ, since the polar angle φ remains constant across the wall.

Those cases in which the polar axis, which is normal to the wall, coincides with a symmetry axis of the crystal lattice find particularly important applications. Expressions for the magnetocrystalline energy as a function of the corresponding new coordinates are given below.

the polar axis is a fourfold axis,

$$w_c = K \left[\sin^2 \varphi - \frac{7}{8} \sin^4 \varphi - \frac{1}{8} \sin^4 \varphi \cos 4\theta \right] ; \quad (5)$$

the polar axis is a threefold axis,

$$w_c = K \left[\frac{\cos^4 \varphi}{3} + \frac{\sin^4 \varphi}{4} - \frac{\sqrt{2}}{3} \cos \varphi \sin^3 \varphi \cos 3\theta \right] ; \quad (6)$$

the polar axis is a twofold axis,

$$w_c = \frac{K}{4} \left[(1 - 4 \sin^2 \varphi + 4 \sin^4 \varphi) + (6 \sin^2 \varphi - 4 \sin^4 \varphi) \sin^2 \theta - 3 \sin^4 \varphi \sin^4 \theta \right]. \quad (7)$$

In these three cases the origin plane, corresponding to $\theta = 0$, passes through the polar axis and one of the other fourfold axes.

In magnetically uniaxial material, such as cobalt or strongly stretched or compressed nickel, the magnetocrystalline energy can be written in the form

$$w_c = C \sin^2 \theta \quad (8)$$

where θ represents the angle between the axis and the spontaneous magnetization direction.

The general expression for the *overall* magnetocrystalline energy W_c of a 1 cm² wall segment corresponds to the energy of the volume enclosed by a cylinder of base 1 cm² and whose axis is perpendicular to the wall

$$W_c = \int_{-\infty}^{+\infty} f(\theta) \, dx. \quad (9)$$

(5) Exchange energy: The Heisenberg exchange interactions which are responsible for ferromagnetism act only over short distances and are almost uniquely localized at an atom A and its $2p$ nearest neighbours at distance a; these $2p$ neighbours are the only ones of consequence as far as the magnetization is concerned. These interactions can be expressed in the convenient terminology of Weiss as being equivalent to a fictitious magnetic field, the molecular field, equal to $N\vec{\mathcal{J}}$ where the coefficient N is a constant and $\vec{\mathcal{J}}$ is the value of the spontaneous magnetization at the position of atom A. It is as if the atom A exerts a field $N\vec{\mathcal{J}}/2p$ on any one of its active neighbours B. The coupling energy is minimum when the magnetic moment μ of the atom B is parallel to $\vec{\mathcal{J}}$, that is to say parallel to the moment of the atom A.

When the moment μ deviates from the direction $\vec{\mathcal{J}}$ by the angle $\delta\psi$, the potential energy of the atom pair A and B increases by w_{ab} which, putting $_s = |\quad|$ gives

$$w_{ab} = \frac{N\mathcal{J}_s}{2p} \mu (1 - \cos \delta \psi) = \frac{N\mathcal{J}_s \mu}{4p} (\delta \psi)^2. \quad (10)$$

In the present case $\varphi\psi = \sin \delta \; \delta\theta$ where $\delta\theta$ is the difference in the values of θ corresponding to the two atoms A and B; if λ is taken as the angle between the wall normal (x axis) and the line joining the centres of A and B the difference between the abscissa of A and B is $a \cos \lambda$ giving:

$$\delta\theta = \frac{d\theta}{dx} \delta x = a \frac{d\theta}{dx} \cos \lambda.$$

Finally, we obtain

$$w_{ab} = \frac{N\mathcal{J}_s \mu}{4p} a^2 \sin^2 \varphi \left(\frac{d\theta}{dx} \right)^2 \cos^2 \lambda, \quad (11)$$

The total coupling energy of the atom A with its $2p$ active neighbours is obtained by multiplying the expression (11) by $2p$ and by replacing $\cos^2 \lambda$ by its mean value of $1/2$ (the distribution of the active neighbours about A obeys cubic symmetry). If there are v atoms contained in 1 cubic centimeter the exchange energy for a volume dv is given by $w_e dv$ where the total coupling energy of atom A is multiplied by $1/2 \, vdv$, the factor $1/2$ arises because pairs such as AB must be counted only once. Since $v\mu = \mathcal{J}_s$ one obtains

$$w_e = \frac{N\mathcal{J}_s^2}{12} a^2 \sin^2 \varphi \left(\frac{d\theta}{dx} \right)^2. \quad (12)$$

This expression is valid for all cubic systems. It can be simplified by introducing the *demagnetizing energy* E required to pass from the ferromagnetic to the disordered state. As Weiss has shown

$$E = \frac{1}{2} N \mathcal{J}_s^2 . \qquad (13)$$

In the following, for the sake of brevity we set

$$A = \frac{Ea^2}{6} \sin^2 \varphi . \qquad (14)$$

With this notation the exchange energy W_e of a wall element of surface 1 cm^2 is written as

$$W_e = \int_{-\infty}^{+\infty} w_e \, dx = \int_{-\infty}^{+\infty} A \left(\frac{d\theta}{dx}\right)^2 dx . \qquad (15)$$

(6) Determination of the azimuth θ as a function of x[1]: We require that the total wall energy $\gamma = w_c + w_e$ be minimum. The variation $\delta\gamma$ of γ corresponding to an arbitrary variation $\delta\theta$ of θ should then be zero. Writing $\theta' = d\theta/dx$ we obtain the expression

$$\delta\gamma = \int_{-\infty}^{+\infty} \delta(w_c + w_e) \, dx =$$
$$= \int_{-\infty}^{+\infty} \left[2A\theta' \frac{d}{dx} \delta\theta + f'(\theta) \delta\theta \right] dx = 0 .$$

Integrating by parts and using the fact that:

$$\theta' \frac{d}{dx} \delta\theta = \frac{d}{dx} (\theta' \delta\theta) - \delta\theta \frac{d\theta'}{dx}$$

and that $\theta' \delta\theta$ is zero at the two limits we obtain

$$\delta\gamma = \int_{-\infty}^{+\infty} \left[f'(\theta) - 2A \frac{d\theta'}{dx} \right] \delta\theta \, dx = 0 .$$

$\delta\gamma$ will be zero for any $\delta\theta$ if the term in brackets is zero

$$f'(\theta) - 2A \frac{d\theta'}{dx} = 0 . \qquad (16)$$

Multiplying by θ' and integrating we find

$$f(\theta) = A \left(\frac{d\theta}{dx}\right)^2 \qquad (17)$$

which satisfies the limiting conditions [$\theta' = 0$ for $\theta = \theta_1$ and $\theta = \theta_2$]. This relationship shows that at each point in the wall the increases in the exchange energy and magnetocrystalline energy are equal, so that $w_e = w_c$.

[1] The classical method due to Becker (F, p. 191) is used in this paragraph.

Equation (17) gives:

$$dx = \sqrt{A} \frac{d\theta}{\sqrt{f(\theta)}} . \qquad (18)$$

Substituting for this value of dx in the expression (9) integrating and multiplying the result by two to take account of the exchange energy we find the following expression for the surface energy of a wall in a cubic material

$$\gamma = a \sin\varphi \sqrt{\frac{2E}{3}} \int_{\theta_1}^{\theta_2} \sqrt{f(\theta)} \, d\theta . \qquad (19)$$

Replacing $f(\theta)$ by the value of the magnetocrystalline energy for a uniaxial substance (eq. (8)) we obtain the values for the wall energy given by Becker (F, p. 192). It should be remarked that E represents here the total demagnetizing energy per cubic centimetre whilst the I of formula (13) in Becker is equal to $2E/p$. Moreover, the a in our formula (19) represents the distance between magnetically active atoms whilst in formula (18) in Becker a represents the lattice parameter.

We now apply equation (19) to some special cases of importance for iron.

(7) 90° walls: The 90° walls considered are walls separating domains whose spontaneous magnetization $\vec{\mathcal{J}}_1$ and $\vec{\mathcal{J}}_2$ lie along the two different fourfold axes OA and OB (Fig. 1). The basic condition (2) concerning the equality of the normal magnetization components indicates simply that the wall normal lies in the plane OQD, perpendicular to the vector $\vec{\mathcal{J}}_1 - \vec{\mathcal{J}}_2$. Among this infinity of possible orientations we will investigate three particular cases where the wall normal coincides either with the fourfold axis OQ, or with the twofold axis OD or with the threefold axis OT.

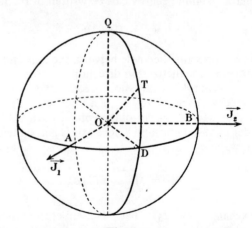

Fig. 1.

(a) *90° walls perpendicular to a fourfold axis* — The wall plane is perpendicular to the polar axis OQ and therefore coincides with the plane AOB (Fig. 1). Within the wall the magnetization rotates in this plane maintaining the angle $\varphi = \pi/2$ with respect to the normal OQ. Since $\varphi = 1$ the expression (5) reduces to $f(\theta) = K \sin^2 \theta \cos \theta$. Using the fact that the initial and final magnetization directions are respectively $\theta_1 = 0$ and $\theta_2 = \pi/2$ and that

$$\int_{\theta_1}^{\theta_2} \sqrt{f(\theta)} \, d\theta = \int_0^{\frac{\pi}{2}} \sqrt{K \sin^2 \theta \cos^2 \theta} \, d\theta = \frac{1}{2} \sqrt{K},$$

the wall energy according to equation (19) is

$$\gamma_Q = \frac{a}{2} \sqrt{\frac{2EK}{3}}. \tag{20}$$

(b) *90° walls perpendicular to a twofold axis* — The spontaneous magnetization directions make an angle $\varphi = \pi/4$ with the wall normal OD. Introducing this value of φ into the expression for the magnetocrystalline energy (7) we obtain

$$f(\theta) = K \left(\frac{1}{2} \sin^2 \theta - \frac{3}{16} \sin^4 \theta \right)$$

and noting that the initial and final magnetization directions correspond to $\theta_1 = 0$ and $\theta_2 = \pi$

$$\int_{\theta_1}^{\theta_2} \sqrt{f(\theta)} \, d\theta = \sqrt{K} \int_0^\pi \sqrt{\frac{1}{2} \sin^2 \theta - \frac{3}{16} \sin^4 \theta} \, d\theta =$$

$$= \sqrt{\frac{K}{2}} \left(1 + \frac{5}{2\sqrt{6}} \sinh^{-1} \sqrt{\frac{3}{5}} \right),$$

whence the wall energy γ

$$\gamma_D = \frac{a}{2} \sqrt{\frac{2EK}{3}} \left(1 + \frac{5}{2\sqrt{6}} \sinh^{-1} \sqrt{\frac{3}{5}} \right) =$$

$$= 1{,}727 \, \frac{a}{2} \sqrt{\frac{2EK}{3}}. \tag{21}$$

(c) *90° walls perpendicular to a threefold axis* — In this type of wall the magnetization rotates whilst maintaining a fixed angle φ with the threefold axis equal to the angle between a three and a fourfold axis. This gives $\cos \varphi = 1/\sqrt{3}$ and substitution of this value in the expression for the magnetocrystalline energy, equation (6), gives $f(\theta) = \frac{4K}{27}(1 - \cos 3\theta)$, then using the fact that θ varies across the wall from $\theta_1 = 0$ to $\theta_2 = 2\pi/3$ we obtain

$$\int_{\theta_1}^{\theta_2} \sqrt{f(\theta)} \, d\theta = \int_0^{\frac{2\pi}{3}} \sqrt{\frac{4K}{27}(1 - \cos 3\theta)} \, d\theta =$$

$$= \frac{8}{9} \sqrt{\frac{2}{3}} K,$$

yielding the wall energy γ_T

$$\gamma_T = a \sin \varphi \sqrt{\frac{2E}{3}} \cdot \frac{8}{9} \sqrt{\frac{2}{3} K} = \frac{16}{27} a \sqrt{\frac{2EK}{3}}. \tag{22}$$

A comparison of these three results shows that it is the 90° walls perpendicular to a fourfold axis which have the lowest surface energy: however, we will show later that because of magnetostrictive effects these are not the most stable type.

(8) 180° walls: We consider now 180° walls separating adjacent domains whose magnetizations $\vec{J}_1$ and $\vec{J}_2$ lie along the same fourfold axis OQ (Fig. 2)

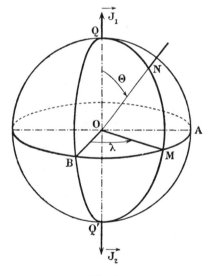

Fig. 2.

but in opposing directions. According to the fundamental conditions (2) the wall normal is perpendicular to the vector $\vec{J}_1 - \vec{J}_2$, i.e. to OQ: the wall plane can therefore be any plane passing through QQ', for example the plane QMQ'. Let λ be the angle between this plane and the meridion origin plane passing through the fourfold axis OB; within the wall the magnetization direction ON rotates from OQ to OQ' within the plane QMQ'. Let θ be the angle between ON and OQ; expressed in terms of λ and θ the magnetocrystalline energy is obtained by replacing φ by θ and θ by λ in equation (5)

$$f(\theta) = K \left[\sin^2 \theta - \left(\frac{7}{8} + \frac{1}{8} \cos 4\lambda \right) \sin^4 \theta \right].$$

In a given wall λ is constant whilst θ varies from 0 to π; putting $\mu = \frac{7}{8} + \frac{1}{8} \cos 4\lambda$ we find

$$\int_{\theta_1}^{\theta_2} \sqrt{f(\theta)} \, d\theta = \sqrt{K} \int_0^\pi \sqrt{\sin^2 \theta - \mu \sin^4 \theta} \, d\theta =$$

$$= \sqrt{K} \left[1 + \frac{1-\mu}{\sqrt{\mu}} \sinh^{-1} \frac{\mu}{\sqrt{1-\mu}} \right].$$

Applying the general formula (19) we obtain the wall energy

$$\gamma = a \sqrt{\frac{2 \, EK}{3}} \left[1 + \frac{1-\mu}{\sqrt{\mu}} \sinh^{-1} \sqrt{\frac{\mu}{1-\mu}} \right]. \tag{23}$$

This is a function of μ which is minimum for $\mu = 1$, that is for $\lambda = 0$; a 180° wall has minimum energy γ when it lies perpendicular to a fourfold axis; the energy is given by

$$\gamma = a \sqrt{\frac{2 \, EK}{3}}. \tag{24}$$

This value is just twice that given by equation (20): in this position a 180° wall is equivalent to two juxtaposed 90° walls perpendicular to the same fourfold axis.

The maximum energy of a 180° wall is 1.38 times greater than the minimum energy given by (24). It corresponds to $\mu = 3/4$ that is $\lambda = \pi/4$: in this position the wall is perpendicular to a twofold axis.

To summarize, in the case of iron the wall energy is always of the order of $a\sqrt{2EK/3}$ no matter the orientation.

(9) Numerical value of the wall energy: K is equal to 4.3×10^5 ergs/cm^3 in iron. Moreover we know that the active neighbours from the magnetic point of view are not the 8 nearest neighbours placed on the threefold axes but the 6 next nearest neighbours on the fourfold axes at a distance a equal to the lattice parameter $a = 2.86$ Å. The value of the demagnetization energy E which it would be possible to obtain from specific heat measurements would not be very precise. It is preferable to use equation (13) and to determine the molecular field coefficient N by using the Weiss relationship $N = \Theta/C$, where Θ is the Curie temperature and C the Curie constant per unit volume. The best measurements of the spontaneous magnetization for iron at room temperature give $\mathcal{J}_s = 1710$, the atomic Curie constant is equal to 1.265, the Curie temperature is close to 1040°K, and the density is 7.8 giving $N = 5890$ whence $E = 8.6 \times 10^9$ ergs/cm^3 and finally

$$a \sqrt{\frac{2 \, EK}{3}} = 1{,}4 \text{ ergs/cm}^2 ; \tag{25}$$

this gives the order of magnitude of the surface energy of a wall.

(10) Effect of a magnetic double layer: Throughout this article we have supposed that the magnetization must rotate within a wall so as to maintain its perpendicular component constant. We justify this supposition by showing that when the magnetization rotates in a plane *perpendicular* to the wall the energy of the wall is much larger. Suppose, then, that within the wall the spontaneous magnetization rotates in a plane perpendicular to the wall from an initial position OA parallel to the wall into the antiparallel position OA' and let θ be the angle made by the spontaneous magnetization with OA at the point x.

This magnetization distribution is no longer one of zero divergence and gives rise to a field H, perpendicular to the wall, obeying the Poisson equation

$$\frac{\partial H}{\partial x} = 4\pi \rho = -4\pi \operatorname{div} \vec{\mathcal{J}} = -4\pi \frac{\partial \mathcal{J}_x}{\partial x},$$

integration gives $-H = 4\pi \mathcal{J}_x = 4\pi \mathcal{J}_s \sin \theta$ since this field is zero outside the wall. The corresponding energy density W_m is

$$W_m = \frac{H^2}{8\pi} = 2\pi \mathcal{J}_s^2 \sin^2 \theta. \tag{26}$$

This magnetic energy term must be added to the usual magnetocrystalline energy term in the calculation of the wall energy. This term is much the bigger since $2\pi \mathcal{J}_s^2 = 1.83 \times 10^7$ ergs/cm^3 whilst $K = 4.3 \times 10^5$ ergs/cm^3. Neglecting the latter

$$\gamma = a \sqrt{\frac{2 \, E}{3}} \int_0^\pi \sqrt{2\pi \mathcal{J}_s^2 \sin^2 \theta} \, d\theta =$$

$$= 4 a \mathcal{J}_s \sqrt{\frac{\pi E}{3}} = 17{,}5 \text{ ergs/cm}^2. \tag{27}$$

This energy is 12 times greater than the usual 180° wall energy. This considerable increase in the free energy can be avoided if there are no free charges; this is why div $\vec{\mathcal{J}} = 0$ and the angle between the magnetization and wall normal remains constant.

(11) 90° wall thickness: We now consider the angular variation of θ within the wall taking as an example a 90° wall perpendicular to a fourfold axis. The application of equation (18) to this case yields the differential equation

$$dx = a \sqrt{\frac{E}{6K}} \frac{d\theta}{\cos\theta \sin\theta},$$

which after integration gives the following relationship between x and θ, taking the origin at the wall centre

$$x = a \sqrt{\frac{E}{6K}} \log \mathrm{tg}\,\theta. \quad (28)$$

The continuous line in Fig. 3 shows the variation of θ as a function of x where the abscissa is plotted in units of the reduced length $l_0 = a\sqrt{E/6K}$ which according to the numerical values calculated previously is equal to 165 Å. In principle the notion of wall thickness has no meaning since θ takes on the values 0 or $\pi/2$ only at an infinite distance from the wall. In practice three quarters of the total rotation are contained in a length $3l_0$, that is roughly 500 Å. We will consider this to be the practical value of the 90° wall thickness.

(12) 180° wall thickness: Suppose for simplicity that the wall lies perpendicular to a fourfold axis OX. Take two perpendicular axes OY, OZ along the fourfold axes in a plane P parallel to the wall. The magnetization rotates across the wall in the plane YOZ from the initial direction OY into the antiparallel direction; the magnetization direction passes through the intermediate position OZ so that a 180° wall can be considered as the juxtaposition of two 90° walls perpendicular to the same fourfold axis. The wall energy is therefore easily calculated but a difficulty arises concerning the wall thickness: the two 90° wall elements which constitute the 180° wall should be separated by an infinite distance since in a 90° wall $\theta = \pi/2$ only at an infinite distance from the wall. In other words the two antiparallel domains should always be separated by a domain magnetized along a perpendicular direction, this comes down to accepting that 180° walls do not exist.

We show now that magnetostrictive effects remove this difficulty and lead to a profound modification of the preceding conclusions.

Consider first an infinite crystal uniformly magnetized along the direction OM lying in the plane YOZ (OX, OY, OZ are easy directions), this direction makes an angle θ with OY. The relative variations in length along a direction having direction cosines (β_1, β_2, β_3) are known (F, p. 280) to be given as a function of the spontaneous magnetization direction (α_1, α_2, α_3) by the formula

$$\frac{\delta l}{l} = \frac{3}{2} \lambda_{100} \left(\alpha_1^2 \beta_1^2 + \alpha_2^2 \beta_2^2 + \alpha_3^2 \beta_3^2 - \frac{1}{3} \right) +$$
$$+ 3 \lambda_{111} (\alpha_1 \alpha_2 \beta_1 \beta_2 + \alpha_2 \alpha_3 \beta_2 \beta_3 + \alpha_3 \alpha_1 \beta_3 \beta_1), \quad (29)$$

where λ_{100} and λ_{111} are the values of the saturation longitudinal magnetostriction along the [100] and [111] directions. The formula above shows that there is a difference in magnetostriction between the position where the magnetization makes an angle θ with OY and that where it lies along OY. Equation (29) shows this difference to be given by the following relationship for a direction in the plane YOZ making an angle φ with OY

$$\left(\frac{\delta l}{l}\right) - \left(\frac{\delta l}{l}\right)_0 = \frac{3}{2} \lambda_{100} (\sin^2 \varphi - \cos^2 \varphi) +$$
$$+ 3 \lambda_{111} \cos\theta \sin\theta \cos\varphi \sin\varphi. \quad (30)$$

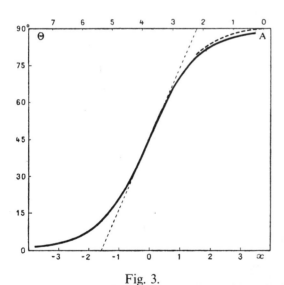

Fig. 3.
Showing the angular variation of the spontaneous magnetization direction within a wall perpendicular to a fourfold axis: 90° wall — full line; 180° wall — dashed line. The widths are expressed in terms of the reduced length $l_0 = a\sqrt{E/6K} = 165$ Å for iron.

As the magnetization rotates from the direction OY into the direction OM the material undergoes a deformation in the plane YOZ which can be characterized by the corresponding components of the deformation tensor A_{ij}. It is readily shown that

$$A_{22} = -A_{33} = \frac{3}{2}\lambda_{100}\sin^2\theta$$

$$A_{23} = A_{32} = \frac{3}{2}\lambda_{111}\cos\theta\sin\theta.$$

Suppose now that in this same infinite crystal the negative x region contains a domain magnetized along OY and the positive x region an antiparallel domain. These two domains are separated by a wall within which the magnetization rotates: a wall element between x and $x+dx$, were it isolated, would undergo a θ dependent deformation parallel to the plane YOZ, this deformation is defined by the components A_{22}, A_{23}, A_{33} of the tensor A_{ij}. However since the wall is thin and is enclosed between two semi-infinite uniformly magnetized domains the wall elements cannot deform in their own plane. A constraint is therefore exerted on each element. Were the element isolated, this constraint would produce a deformation $-A_{22}, -A_{33}, -A_{23}$ equal and opposite to the magnetostrictive deformation. The elastic energy associated with a small deformation of a cubic crystal is given by

$$W_{él} = \frac{1}{2}C_1(A_{11} + A_{22} + A_{33})^2 +$$
$$+ C_2(A_{11}^2 + A_{22}^2 + A_{33}^2) + 2C_3(A_{12}^2 + A_{23}^2 + A_{13}^2)$$

and in the present case this becomes

$$W_{él} = \frac{9}{2}C_2\lambda_{100}^2\sin^4\theta + \frac{9}{2}C_3\lambda_{111}^2\sin^2\theta\cos^2\theta. \quad (31)$$

Magnetostriction, therefore, produces an additional term which must be added to the magnetocrystalline energy $f(\theta)$. This magnetostrictive term is given by the above expression which is also a function of θ having the form

$$W_{él} = K'\sin^2\theta\cos^2\theta + D\sin^4\theta. \quad (32)$$

The values of K' and D can be calculated according to Becker (F, pp. 143 and 280) by taking the values of the elastic constants as

$$C_1 = 1{,}46 \cdot 10^{12} \qquad C_2 = 0{,}48 \cdot 10^{12}$$
$$C_3 = 1{,}12 \cdot 10^{12}$$

and the magnetostrictive constants as

$$\lambda_{100} = 19{,}5 \cdot 10^{-6} \qquad \lambda_{111} = -18{,}8 \cdot 10^{-6}.$$

We obtain

$$K' = 1780 \text{ ergs/cm}^3 \qquad D = 820 \text{ ergs/cm}^3.$$

The value of $f(\theta)$ to be used in the calculation of the wall energy can therefore be written as

$$f(\theta) = (K + K')\sin^2\theta\cos^2\theta + D\sin^4\theta.$$

Since K' and D are much smaller than K they have a negligible influence on the wall energy. But a finite value is now obtained for the wall thickness. The differential equation (18) describing the relationship between θ and x is written as

$$dx = a\sqrt{\frac{E}{6}}\frac{d\theta}{\sqrt{(K+K')\sin^2\theta\cos^2\theta + D\sin^4\theta}}.$$

Integrating and taking the origin of the coordinates at the wall centre

$$x = a\sqrt{\frac{E}{6(K+K')}}\sinh^{-1}\left(\sqrt{\frac{K+K'}{D}}\cotg\theta\right). \quad (33)$$

The variation of θ as a function of x for the left hand half of the wall is shown as the dashed curve in Fig. 3. To the left of the figure the dashed and full curves superpose. The scale corresponding to the dashed curve is shown along the top. The complete curve is obtained by continuing the curve symmetrically about A. The curve corresponding to half a 180° wall is seen to be almost identical to that for a 90° wall except in the region near $\theta = \pi/2$. The total width of a 180° wall is of the order of 11–12 l_0, i.e. 1800–2000 Å. The difficulties raised at the end of the previous section are therefore dissipated.

(13) The influence of magnetostriction on 90° walls: The results obtained in the previous section lead to other important conclusions. Consider an iron single crystal, with fourfold axes OXYZ, which we suppose to be subdivided into three elementary domains A, B, C separated by two walls S, S' parallel to the plane YOZ (Fig. 4). We will suppose that the two

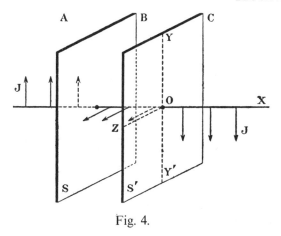

Fig. 4.

external domains A, C are magnetized along OY and the antiparallel direction OY' whilst the central domain B is magnetized along OZ, perpendicular to OY. These three domains will obviously produce a distortion within the material due to the magnetostriction: in particular if the central domain is much thinner than the total thickness the external domains impose their own parameters on the central domain which will thus be in a state of elastic strain. According to equation (31) the energy of the domain B is found to be 820 ergs/cm^3 for $\theta = \pi/2$. It is as if the walls S, S' separating B from A and C were submitted to a perpendicular pressure of 820 bars directed from A and from C into B. If the walls S and S' remain mobile they will come together so that the domain B will shrink and finally disappear leaving behind a single 180° wall whose structure was described in the previous section. This shows that in the absence of an applied field the walls S and S', and all 90° walls perpendicular to a fourfold axis, cannot exist in a state of stable equilibrium: they are always subjected to a perpendicular pressure of up to 820 bars.

But 90° walls perpendicular to a fourfold axis only represent a special case amongst all possible 90° walls: it is necessary therefore to determine whether other orientations exist for which there are no magnetostrictive constraints. Consider therefore two domains magnetized respectively along $\alpha_1 = 1$, $\alpha_2 = \alpha_3 = 0$ and $\alpha'_1 = 0$, $\alpha'_2 = 0$, $\alpha'_3 = 0$. Equation (29) allows the difference between the magnetostrictions of these domains to be calculated along the direction defined by the direction cosines $(\beta_1, \beta_2, \beta_3)$:

$$\left(\frac{\delta l}{l}\right)_{\alpha'} - \left(\frac{\delta l}{l}\right)_{\alpha} = -\lambda_{100}(\beta_2^2 - \beta_1^2). \quad (34)$$

This difference is zero for the two planes $\beta_2 = \beta_1$ and $\beta_2 = -\beta_1$. This second plane is perpendicular to the internal bisector of the two magnetization directions and is a possible 90° wall: that which is perpendicular to a twofold axis. *So amongst all possible plane walls separating two domains magnetized at 90° to each other and which respect the principle of equal perpendicular magnetization components there is only one which produces no magnetostrictive constraints: this wall is perpendicular to a twofold axis. In the absence of other perturbations*, especially those due to demagnetizing field effects, this then is the only 90° wall which can exist in an ideal single crystal of iron.

It is possible moreover to define more exactly what we mean by the notion of an ideal single crystal. Walls can no longer propagate freely in deformed crystals containing impurities but are acted upon by a quasi-frictional force. The perpendicular pressures which must be exerted to move a 90° wall are of the order of $H_c \mathcal{J}$, where H_c is the coercive field and $\mathcal{J}$ the spontaneous magnetization of the material. Since $\mathcal{J} \simeq 1700$ the magnetostrictive pressures required to move unstable 90° walls are of the same order of magnitude as those necessary to overcome the quasi-frictional effects which produce coercive fields H_c of the order of 0.5 gauss. As a result when the coercive field of an iron specimen is considerably greater than 0.5 gauss all possible types of 90° wall may exist within the specimen whilst if the coercive field is well below this limit only those walls perpendicular to a binary axis can exist.

Experiment confirms these predictions. When only two categories of elementary domains with non-antiparallel spontaneous magnetizations exist in a pure iron single crystal whose coercive field is below 0.5 gauss then these domains take the form of parallel stacked sheets whose plane is perpendicular to a binary axis. This is shown in the powder patterns (Bitter method) produced by Sixtus [5] and by Kaya [6] (cf. B, first and second part).

However, even in an ideal crystal 90° walls with other orientations may be observed, for example perpendicular to a threefold axis. This situation arises when demagnetization field effects *impose* the existence of three different types of mutually parallel elementary plane sheet domains, the sheet planes can in this case only be perpendicular to a threefold axis (cf. B, first and second parts).

Finally, we remark that the selection between different 90° wall orientations which is affected by magnetostriction no longer functions in the case of 180° walls: this is because the magnetostriction is independent of the *sense* of the spontaneous magnetization direction. Two domains magnetized antiparallel therefore exert no mechanical constraint on each other whatever the orientation of the separating wall.

(14) The influence of the magnetic field on the wall energy: It remains to say a few words about the influence of the magnetic field since in the preceding sections we supposed the field to be zero. This problem appears to be complicated but can in fact be greatly simplified since in practice the internal magnetic field can only assume certain particular orientations with respect to a wall in equilibrium. Consider the case of a circular iron disc which is randomly oriented except that its plane is not perpendicular to a fourfold axis. Experiment and theory both show (B, part one) that just after the first bend in the magnetization curve the walls are always perpendicular to a threefold axis whilst the internal field is parallel to this same axis. We will not consider this example further but rather the following one which is better known from an experimental point of view: namely a disc cut perpendicular to a fourfold axis placed in a field parallel to its plane. For a weak internal field H_i the crystal subdivides into domains having the form of plane sheets magnetized along the two fourfold axes OX and OY in the plane of the disc. The domain walls are perpendicular to the bisector OD of the angle XOY (Fig. 5). *These walls are therefore 90° walls perpendicular to a binary axis*, this wall type was

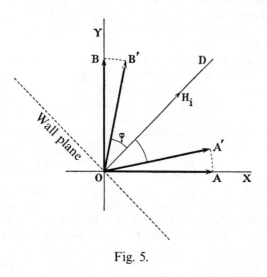

Fig. 5.

examined in section 7b. When the applied field increases (whilst remaining in the quadrant XOY) the internal field increases also and remains parallel to the bisector OD as long as the two sets of elementary domains coexist in the crystal. However as H_i increases the two spontaneous magnetization directions deviate from their initial positions OA and OB, aligned with the OX and OY axes, and turn symmetrically towards OD following OA′ and OB′. The angle between these directions and OD is such that the sum of the magnetic field energy $-H_i \mathcal{J}_s \cos \Phi$ and the magnetocrystalline energy $(K \cos^2 2\varphi)/4$ be minimum. It is easily shown [7] that φ is related to H_i according to

$$H_i \mathcal{J}_s = 2K \cos \varphi (2 \cos^2 \varphi - 1).$$

When H_i reaches the value $2K/\mathcal{J}_s \simeq 500$ gauss the magnetization directions on both sides of the wall become parallel to the field and so the wall disappears in a continuous and reversible way.

Wall energy — It is easy to obtain the corresponding evolution in the value of the wall energy. As we have shown previously (§ 3 and § 10) the magnetization direction rotates within the wall so as to maintain a fixed angle φ with the polar axis OD: the magnetic field energy therefore remains constant as the azimuthal angle θ varies from 0 to π. The increase in the magnetocrystalline energy is given by the following expression obtained from equation (6)

$$f(\theta) = w_c(\theta, \varphi) - w_c(0, \varphi) =$$
$$= \frac{K}{4}[(6 \sin^2 \varphi - 4 \sin^4 \varphi) \sin^2 \theta - 3 \sin^4 \varphi \sin^4 \theta].$$

According to (19) the wall energy γ is then given by

$$\gamma = a \sin \varphi \sqrt{\frac{2E}{3}} \int_0^\pi \sqrt{f(\theta)}\, d\theta,$$

which finally yields

$$\gamma = \frac{a}{2}\sqrt{\frac{2EK}{3}} \sin \varphi \left\{ \sin \varphi \sqrt{6 - 4 \sin^2 \varphi} + \frac{6 - 7 \sin^2 \varphi}{\sqrt{3}} \sinh^{-1} \sqrt{\frac{3 \sin^2 \varphi}{6 - 7 \sin^2 \varphi}} \right\} \quad (35)$$

This wall energy decreases quite rapidly with φ as shown by the following table.

TABLE

$\cos \varphi = \mathcal{J}:\mathcal{J}_s$	0,707	0,72	0,73	0,75	0,80	0,85	0,90	0,95	1,00
γ [ergs/cm^2]	1,21	1,19	1,17	1,12	0,99	0,81	0,59	0,32	0,00

As we have shown elsewhere (B, 3rd part) a study of the variations of periodic powder pattern lines as a function of applied field gives an experimental measure of the variation of γ with φ. The few available experimental results do not allow formula (35) to be verified precisely: they do show nonetheless a variation of the predicted order of magnitude.

REFERENCES

[1] F. Bloch, *Z.f. Physik*, 1932, **74**, 295.
[2] L. Néel, Les lois de l'aimantation et de la subdivision d'un monocristal de fer en domaines élémentaires, *J. de Phys.*, 1944, **5**, pp 241 and 265 (see A52, Chapter VII)
[3] Becker, Ferromagnetismus (J. Springer, Berlin, 1939).
[4] L. Néel, *Ann. de Phys.*, 1936, **5**, 232; cf. also the articles C5 and C7.
[5] Sixtus, *Phys. Rev.*, 1937, **51**, 870.
[6] Kaya, *Z.f. Physik*, 1934, **89**, pp 796 and 90; 1934, **89**, 551.
[7] Akulov, *Z.f. Physik*, 1931, **69**, 78.

A97 (1955)
BLOCH WALL ENERGY IN THIN FILMS

Note (1977): This type of wall is now known as a Néel wall.

It is shown that the Bloch wall energy in a continuous thin film first increases as the film thickness decreases then passes through a maximum and finally returns to its usual value when the film becomes extremely thin.

F. Bloch has shown [1] that between two elementary ferromagnetic domains, with spontaneous magnetizations J_1 and J_2 there is a wall of finite width $2a_0$ within which the spontaneous magnetization direction turns progressively from the direction of J_1 into the direction of J_2. Let the wall width have the arbitrary value $2a$. The wall energy is the sum of two terms the first of which E_w represents the Weiss–Heisenberg term (exchange) and is inversely proportional to a; the second E_c is the magnetocrystalline energy and is directly proportional to a. The resulting total energy E_t can be written as

$$E_t = E_w + E_c = \frac{1}{2}\gamma_0\left(\frac{a_0}{a} + \frac{a}{a_0}\right) \quad (1)$$

where γ_0 and a_0 are two constants. The equation shows that E_t has a minimum value of γ_0 when $a = a_0$. In the case of a wall separating two antiparallel domains in iron the numerical values of the constants [2] are $\gamma_0 = 1.4$ ergs/cm^2 and $a_0 = 10^{-5}$ cm.

It has become customary to consider γ_0 and a_0 as two constants whose value depends only on the nature of the material. In reality this is true only in the case of walls whose lateral extent is very big with respect to a_0. This condition is not always respected since both elementary domains of a few microns in size and much thinner continuous films can also exist in practice.

Consider for example two domains magnetized in antiparallel directions in the plane of a thin film of thickness D and separated by a plane wall which is perpendicular to the film plane and parallel to the spontaneous magnetization of the adjacent domains. Within the wall the spontaneous magnetization rotates smoothly through 180° whilst remaining parallel to the wall plane. The volume magnetic charge density is zero everywhere but a surface charge appears at the intersection of the wall and the film surface: within a strip of width $2a$ the surface charge density varies between 0 to $+J_s$ or 0 to $-J_s$ (J_s = spontaneous magnetization). A magnetostatic term appears in the overall energy expression, this increases the wall energy and tends to reduce the wall width.

To evaluate these effects in terms of D we suppose that the wall is modified but conserves the same width across the entire thickness of the thin film. The magnetostatic energy is roughly the same as for two strips of width a and of charge densities $+J_s$ and $-J_s$ separated by a distance D. The corresponding energy is approximately that of the demagnetizing field of a straight elliptical cylinder of cross-section aD, of axial ratio D/a and of magnetization J_s parallel to the longer axis D. The magnetostatic energy E_m per cm^2 of wall becomes

$$E_m = \frac{2\pi a^2 J_s^2}{a + D}.$$

As we show later a must be supposed small with respect to D. In these conditions the total energy $E'_t = E_w + E_c + E_m$ can be written to a first approximation in the form

$$\frac{2E'_t}{\gamma_0} = x + \frac{1}{x} + \frac{1}{2}p^3 x^2$$

with

$$x = \frac{a}{a_0} \quad ; \quad p^3 = \frac{8\pi J_s^2 a_0^2}{\gamma_0 D}.$$

The minimum of E'_t is found very easily. Let a_m and γ_m be the thickness and energy of the wall for E'_t minimum, the figure shows the values of a_0/a_m (curve F) and γ_m/γ_0 (curve A) as a function of the parameter p.

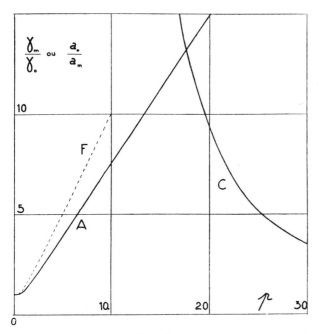

Fig. 1.

Consider the example of an iron film ($J_s = 1\,700$) of thickness $D = 5 \times 10^{-5}$ cm; then $p^3 = 105$ and the curves yield the values $a_m = 0.21\, a_0$ and $\gamma_m = 3.6\, \gamma_0$. In a thin iron film 0.5 µm thick the Bloch walls are thus five times thinner and have an energy four times greater than a conventional wall: the effect is therefore an important one.

The form of the curve A, which has as asymptote the straight line $\gamma_m/\gamma_0 = 3p/4$, shows that γ_m should increase indefinitely as D decreases. What happens in reality is that when a becomes large compared to D the spontaneous magnetization turns across the wall from the direction J_1 to the direction J_2 by a rotation within the film plane rather than in the wall plane: this reduces the minimum of the overall energy. The magnetostatic energy in this case is approximately that of two parallel strips of width D carrying surface charge densities of $+J_s$ and $-J_s$ and separated by a distance a. This energy is to a first approximation equal to that of a straight elliptical cylinder of cross-section aD, of axial ratio D/a and magnetized in the direction of the longer axis which is now of length a. The magnetostatic energy of a 1 cm^2 wall element is

$$E'_m = \frac{2\pi a D J_s^2}{a + D}$$

When D is small compared to a this energy tends towards the value $2\pi D J_s^2$ independently of the wall thickness a. In these conditions the wall conserves its usual thickness a_0 whilst its energy γ_m is a function of D given by

$$\gamma_m = \gamma_0 + 2\pi D J_s^2 .$$

For the same numerical values as those used earlier for iron the values of γ_m/γ_0 as a function of p are shown as the curve C in the figure. The curve C intersects the curve A near $p = 17.5$ and $\gamma_m/\gamma_0 = 13.5$. This represents the transition point at which the spontaneous magnetization changes from one rotational mode to the other: the regions of the curves A and C close to this point are not very precise due to the approximations made in the denominator of E_m or E'_m.

In conclusion it follows that, as the thickness of a thin continous film diminishes, the energy of a wall perpendicular to the film plane goes through a maximum and then for very thin films tends towards its usual bulk value. In the case of iron this maximum occurs close to $D = 100$ Å and has a value at least ten times the usual energy.

This process may give rise to at least part of the increase in coercive field which is observed as the thickness of thin continuous ferromagnetic films decreases [3].

REFERENCES
[1] *Z. Physik*, 1932, **74**, 295.
[2] L. Néel, *Cahiers de Phys.*, 1944, **25**, 1.
[3] L. Néel, Remarques sur la théorie des propriétés magnétiques des couches minces et des grains fins (Colloque sur les propriétés magnétiques des lames minces, Alger, 1955).

A118 (1962)
A NEW METHOD FOR MEASURING BLOCH WALL ENERGIES

Abstract: A Bloch wall, initially perpendicular to the surfaces of a thin ferromagnetic film, is deformed by the action of the magnetic field due to an electric current within the film. The intersections of the wall with the two opposite film surfaces are displaced in opposite directions. This displacement is calculated as a function of the wall energy and the current density.

The only direct method of measuring the energy of a Bloch wall which has been used successfully up to the present seems to be that of C. P. Bean and R. N. de Blois [1]: this method consists in magnetizing longitudinally a cylindrical wire then, by a sudden inversion of the magnetic field, in creating a cylindrical wall which, starting from the wire surface tends to decrease in diameter under the influence of its surface tension and to finally disappear on the wire axis. A steady longitudinal magnetic field is used to create a magnetic pressure which maintains the wall in equilibrium. The surface tension energy of the wall can be deduced from the value of this field. This method can only be applied to isotropic ferromagnetic materials and requires wires which may sometimes be difficult to make.

We propose here a different and novel method which uses thin ferromagnetic single crystal films of uniform thickness (surfaces parallel) a few tenths of a millimeter thick. 180° Bloch walls aligned perpendicular to the film surfaces are created in the film and the intersections of the wall with the two surfaces are observed using the Bitter method. An electric current is then passed through the film, which must be a good conductor, along the direction perpendicular to the spontaneous magnetization directions of the domains adjacent to the wall under study. This current produces a magnetic field having reversed values on the two surfaces of the film. The two wall/surface intersections tend to be displaced in opposite directions: an equilibrium is established between the magnetic pressure and the surface tension forces which oppose the increase in the total surface of the wall. The surface energy is deduced from the values of the electric current density and the displacement of the intersections of the wall with the film surfaces.

The theory of this experiment is established in this article.

Equilibrium state of a wall in zero applied field: Take three rectangular coordinate axes OXYZ with an origin O situated midway between the two film surfaces, and with the plane XOZ parallel to the surface planes. We suppose that we are concerned only with cylindrical walls with axes parallel to OZ separating elementary domains magnetized along OZ in one sense or the other.

Consider first a plane wall making an angle φ with the normal OY to the surface planes, the angle is taken as positive going from OY to OX. The wall energy per unit surface (γ) is in general a function of φ.

Consider now the equilibrium of a wall traversing the film in the absence of an applied field. In this wall take a small rectangular element ABCD with the sides AB and CD of length dS and the sides BC and DA of unit length and parallel to OZ. If we consider only the forces parallel to the plane XOY we find that the element ABCD is acted upon by surface tension forces in its plane and by a couple $-(\partial\gamma/\partial\varphi)$. Since the element is in equilibrium the other two wall segments situated on either side of it must exert the following forces:

(1) on DA, a force $-\gamma$ in the wall plane and a force $-(\partial\gamma/\partial\varphi)$ perpendicular to the wall plane;

(2) on BC, forces equal and opposite to those above.

When the side BC of the surface element is situated on the upper surface of the crystal the forces γ and $(\partial\gamma/\partial\varphi)$ are no longer compensated by the reaction of the rest of the Bloch wall: their resultant is necessarily perpendicular to the crystal surface otherwise the wall would move. The angle φ must therefore take a value φ_s which is a solution of the equation

$$\operatorname{tg}\varphi = -\frac{1}{\gamma}\frac{d\gamma}{d\varphi}, \qquad (1)$$

this value φ_s will be given the name 'fitting angle'.

Wall equilibrium in the presence of a magnetic field: If we now apply parallel to OZ a magnetic field varying with Y and of intensity H sufficiently small that the spontaneous magnetization directions of the elementary domains remain essentially in the direction of easy magnetization OZ then the wall no longer remains plane: φ is no longer constant and takes for example the value φ on BC and $\varphi + d\varphi$ on DA.

In these conditions the external forces, parallel to the plane XOY which act on the element ABCD are the following:

(1) a force PdS normal to the wall and applied to the centre M of the element. This force is due to a difference in magnetic pressure $P = -2HJ_s$ between the two elementary domains adjacent to the wall (spontaneous magnetization J_s);

(2) forces $-\gamma$ and $-(d\gamma/d\varphi)$ applied on DA respectively parallel and perpendicular to the plane tangent to the wall at DA.

(3) forces

$$\gamma + (d\gamma/d\varphi)\,d\varphi \quad \text{and} \quad (d\gamma/d\varphi) + (d^2\gamma/d\varphi^2)\,d\varphi$$

applied on BC, respectively parallel and perpendicular to the plane tangent to the wall at BC.

The resultant of the projection of these forces onto the plane tangent to the wall at M is zero when the terms in $d\varphi^2$ are neglected. To the same approximation the resultant of the projection of these forces onto the wall normal at M should be zero, hence

$$P\,dS = \left(\gamma + \frac{d^2\gamma}{d\varphi^2}\right) d\varphi. \qquad (2)$$

Introducing the radius of curvature of the wall $R = dS/d\varphi$ the preceding relation can also be written

$$PR = \gamma + \frac{d^2\gamma}{d\varphi^2}, \qquad (3)$$

which can be considered as a generalization of Laplace's law for the equilibrium of thin membranes with a finite surface tension.

Case of a ferromagnetic material with a wall energy independent of its orientation: We first consider the case where the surface tension has a value γ_0 independent of ϕ. Consider a wall passing through OZ. In the absence of any current the wall is plane and since the 'fitting angle' defined by (1) is zero the wall is normal to the crystal surfaces and lies along the plane YOZ.

The wall is now influenced by the magnetic field obtained by passing an electric current of uniform density I through the film parallel to OX. The magnetic field acting on a point Y is parallel to OZ and equal to $4\pi IY$. The magnetic pressure acting on a point Y in the wall is therefore $-P = 8\pi IJ_s Y$. Under the effect of this pressure the wall deforms symmetrically with respect to the axis OZ. The wall coordinates X, Y satisfy the differential equation deduced from (2) by replacing dS by $dY/\cos\varphi$

$$-8\pi IJ_s Y\,dY = \gamma_0 \cos\varphi\,d\varphi. \qquad (4)$$

Integrating this equation and giving the integration constant a value such that the value of φ at the film surface, for $Y = L$, be equal to the fitting angle value of zero

$$4\pi IJ_s(L^2 - Y^2) = \gamma_0 \sin\varphi. \qquad (5)$$

The value of φ_c of φ at the film centre, for $Y = 0$, is given by

$$\sin\varphi_c = \frac{4\pi IJ_s L^2}{\gamma_0}. \qquad (6)$$

This shows that φ_c increases with I and attains the value $\pi/2$ for a value I_0 of the current density given by

$$I_0 = \frac{\gamma_0}{4\pi J_s L^2}. \qquad (7)$$

When I is smaller than I_0, putting $i = I/I_0$ and $y = Y/L$ the equation $X = f(Y)$ for the wall follows from (5) by using the relation $dX = \mathrm{tg}\varphi\,dY$.

After integration we find

$$\frac{X}{L} = \int_0^y \frac{i(1 - y^2)\,dy}{\sqrt{1 - i^2(1 - y^2)^2}} = g(y). \qquad (8)$$

The current i provokes the displacement of the intercept of the wall with the upper surface of the crystal by an amount X_s given by

$$X_s = L \cdot g(1). \qquad (9)$$

The displacement on the lower surface is $-X_s$. When i is small X_s can be developed in the form

$$\frac{X_s}{L} = \frac{2}{3} i + \frac{8}{35} i^3 + \frac{32}{231} i^3 + \cdots \qquad (10)$$

When I approaches I_0, i tends to unity and X_s tends to infinity. When I is greater than I_0 the wall coincides with the plane XOZ.

The determination of the initial slope of the experimental curve giving X_s as a function of I, or the determination of I_0, allows γ_0 to be deduced.

Case of a ferromagnetic material with a wall energy dependent on the orientation: As an example take the case of a cubic crystal whose fourfold axes, supposed to be easy magnetization directions, coincide with the three rectangular coordinate axes defined previously. The wall energy has a minimum value φ_0 for $\varphi = 0$ and $\varphi = \pi/2$ and a maximum value $\gamma_0 + \gamma_1$ for $\varphi = \pi/4$ and $3\pi/4$. Suppose that the variation of γ with φ can be expressed in the form

$$\gamma = \gamma_0 + \frac{1}{2}\gamma_1 - \frac{1}{2}\gamma_1 \cos 4\varphi = \gamma_0 + 4\gamma_1 \sin^2\varphi - 4\gamma_1 \sin^4\varphi. \quad (11)$$

The equilibrium equation for the wall is obtained from equation (3). Note that in a region like that of positive Y where the magnetic pressure keeps the same sign and does not fall to zero the expression $\gamma + d^2\gamma/d\varphi^2$ cannot be zero either since this would require that the radius of curvature of the wall be zero as well; this is not physically admissible. As a result φ must be smaller than a certain value φ_m, the smallest root between 0 and $\pi/2$ of the equation

$$\gamma + \frac{d^2\gamma}{d\varphi^2} = \gamma_0 + 8\gamma_1 - 60\gamma_1 \sin^2\varphi + 60\gamma_1 \sin^4\varphi. \quad (12)$$

This supposes of course that equation (12) has roots, that is to say that γ_1 is not too small. We suppose that this situation holds. The other case is not of much interest since the solution would be similar to that corresponding to a fixed wall energy.

Replacing R in equation (3) by $dY/(\cos\varphi\, d\varphi)$ gives the differential equation

$$-8\pi J_s IY\, dY = (\gamma_0 + 8\gamma_1 - 60\gamma_1 \sin^2\varphi + 60\gamma_1 \sin^4\varphi) \cos\varphi\, d\varphi, \quad (13)$$

which is immediately integrable. The integration constant is found by setting φ, the 'fitting angle' given by (1), equal to zero for $Y = L$. This gives

$$4\pi J_s I(L^2 - Y^2) = (\gamma_0 + 8\gamma_1)\sin\varphi - 20\gamma_1 \sin^3\varphi + 12\gamma_1 \sin^5\varphi. \quad (14)$$

According to the previous discussion the second member of this equation increases uniformly as Y decreases from L to O. The largest value φ_c of φ corresponds then to $Y = 0$. Putting

$$j = \frac{4\pi J_s IL^2}{\gamma_0 + 8\gamma_1} \quad (15)$$

where j is the reduced current density this value φ_c is given by

$$\sin\varphi_c - \frac{20\gamma_1}{\gamma_0 + 8\gamma_1}\sin^3\varphi_c + \frac{12\gamma_1}{\gamma_0 + 8\gamma_1}\sin^5\varphi_c = j. \quad (16)$$

The angle φ_c is at most equal to the limiting angle φ_m defined above as the smallest root of the equation (12). Let j_m be the limiting current density obtained by replacing φ_c by φ_m in equation (16). The wall is seen to deform more and more as j increases from 0 to j_m. When j becomes greater than j_m the wall should suddenly break free and rotate into the plane XOZ: its imprint on the surface plane should suddenly disappear. The phenomenon is thus quite different to what happens in a material with a fixed wall energy: indeed in this last case as the current density tends to its maximum value I_0 the displacement X_s of the wall imprint on the crystal surfaces tends continuously to infinity.

The wall equation is given by the integral

$$X = \int_0^Y \text{tg}\,\varphi\, dY, \quad (17)$$

where φ is given as a function of Y by the equation (14). The displacement X_s of the wall trace on the crystal surface is therefore given by equation (17) with L as the upper limit of the integral.

Equations (14), (17) and (18) can be resolved numerically but series expansions can also be used. For example putting

$$B = \frac{4\pi J_s I(L^2 - Y^2)}{\gamma_0 + 8\gamma_1}, \quad (18)$$

The expansion of $\sin\varphi$ as a function of powers of B is deduced from (14):

$$\sin\varphi = B + \frac{20\gamma_1}{\gamma_0 + 8\gamma_1}B^3 + \frac{1104\gamma_1 - 12\gamma_0}{(\gamma_0 + 8\gamma_0)^2}B^5 + \cdots \quad (19)$$

and the expansion of $\tan\varphi$

$$\text{tg}\,\varphi = B + \frac{\gamma_0 + 48\gamma_1}{2(\gamma_0 + 8\gamma_1)}B^3 + \frac{3\gamma_0^2 + 192\gamma_0\gamma_1 + 10944\gamma_1^2}{8(\gamma_0 + 8\gamma_1)^2}B^5 + \cdots \quad (20)$$

Putting this value into equation (16) we come down to the integration of a polynomial in Y and obtain finally

$$\frac{X_s}{L} = \frac{2}{3} j + \frac{8}{35} \frac{\gamma_0 + 48\gamma_1}{\gamma_0 + 8\gamma_1} j^3 +$$
$$+ \frac{32}{231} \frac{\gamma_0^2 + 64\gamma_0\gamma_1 + 3648\gamma_1^2}{\gamma_0^2 + 16\gamma_0\gamma_1 + 64\gamma_1^2} j^5 + \cdots \quad (21)$$

Note that for $\gamma_1 = 0$, equation (21) reduces to equation (10).

As for the maximum displacement X_{sm} beyond which the wall suddenly breaks free, this is calculated by determining the smallest root, $\sin\varphi_m$, of equation (16) in which φ_c is replaced by φ_m and by inserting this value of j_m into equation (21).

We have shown previously [2] that in a cubic crystal $\gamma_1 = 0.38\,\gamma_0$

$$\sin\varphi_m = 0{,}48 \quad ; \quad j_m = 0{,}301 \quad ; \quad j = \frac{0{,}311\, J_s\, IL^2}{\gamma_0}.$$

The displacement X_s is then given by

$$\frac{X_s}{L} = 0{,}667\, j + 1{,}088\, j^3 + 4{,}69\, j^5,$$

and the maximum displacement for $X_{ms} = 0.243\, L$.

Note however that the equation (11) which we have taken initially as representing the angular variation of the wall energy is only rather roughly valid. If more precise results than those obtained here are required it will be necessary to use the correct value of γ [2].

This method of measuring Bloch wall energies has been successfully used in experiments by R. Aléonard and P. Brissonneau [3].

REFERENCES

[1] Magnetic properties of metals and alloys, published by the American Society for Metals, Cleveland, Ohio, 1958.
[2] L. Néel, Cahiers de Physique, no. 25, 1944, p. 1–20.
[3] *Comptes rendus*, 1962, **254**, 2934.

Chapter VIII

WEAK FIELDS: ANHYSTERETIC MAGNETIZATION
A45, A44, A47, A48, A144

Ferromagnetic hysteresis can be described in a particularly simple way and by rigorous laws in the Rayleigh region, i.e. the region where applied magnetic fields are small compared with the coercivity. These laws, established by Lord Rayleigh, are discussed here (in A45) for the case of changes in magnetization which arise from the displacement of boundaries separating elementary domains. However, as will be discussed in A91 in Chapter IX, the laws are equally valid for an assembly of ferromagnetically coupled grains. The Rayleigh laws therefore have a general scope and constitute a fundamental property of hysteresis.

Alongside their theoretical interest, the laws enjoy an important rôle in applications concerned with weak fields such as magnetic recording. The protection of ships against magnetic mines can also be cited in this context. Mines are detonated by the action of the field created by the ship's magnetization which is itself induced by the terrestrial magnetic field. Néel has proposed a method whereby the induced magnetization can be annulled by an opposing permanent magnetization obtained by the action of a transient field H_A: this magnetization is equal to $\frac{1}{2}b\,H_A^2$ where b is the irreversible term in the Rayleigh law. This procedure, known as 'degaussing' described in article A41 (not reproduced in this volume) was successfully used in 1940 at the onset of the 'mine war'.

A45 (1942)

THEORY OF RAYLEIGH'S LAWS OF MAGNETIZATION

I. DISPLACEMENTS OF AN ISOLATED WALL

Abstract: This article is devoted to the theoretical interpretation of the laws discovered in 1887 by Lord Rayleigh and which have remained without explanation since. These laws give a complete description, in terms of two coefficients a and b, of the hysteresis loops of a ferromagnet when subjected to a magnetic field H small compared with the coercive field; a and b relate to reversible and irreversible phenomena respectively. The initial magnetization curve can be expressed as $\mathcal{J} = a\mathrm{H} + b\mathrm{H}^2$.

This theory rests on an exact analysis of the displacements of domain walls in a non-uniform medium. The point of departure is a flexible analytic expression for the free energy as a function of the wall displacement. This expression depends on two parameters only, α and P_0. Thereafter, by means of an artifice, the total path of the wall is divided amongst various simply structured domains. Throughout all possible loops, the wall remains enclosed within the domain in which it was originally contained.

The first part of the paper sets out the laws which govern the mean displacement of an isolated wall: they possess a structure similar to Rayleigh's laws. A second part will be devoted to the case of multidomain materials and the interpretation of coercive force.

INTRODUCTION

(1) The Rayleigh laws: The properties, after demagnetization, of a ferromagnetic body placed in magnetic fields small compared with the coercive field, are described by the Rayleigh laws [1]: when a *decreasing magnetic field* reaches a value H_1, with a corresponding magnetization of $\mathcal{J}_1$, and the field is now *increased* from H_1 to H, the resulting magnetization $\mathcal{J}$ is given by

$$\mathcal{J} - \mathcal{J}_1 = a(\mathrm{H} - \mathrm{H}_1) + \frac{b}{2}(\mathrm{H} - \mathrm{H}_1)^2 \quad \mathrm{H} > \mathrm{H}_1 \quad (1)$$

Similarly when an *increasing magnetic field* reaches H_2 with a magnetization value of $\mathcal{J}_2$ and is now *decreased* the resulting magnetization can be expressed algebraically as

$$\mathcal{J} - \mathcal{J}_2 = a(\mathrm{H} - \mathrm{H}_2) - \frac{b}{2}(\mathrm{H} - \mathrm{H}_2)^2 \quad \mathrm{H} < \mathrm{H}_2. \quad (2)$$

The coefficient a retains the same value regardless of the sense of variation of the field: it corresponds to the *reversible* magnetization whereas b corresponds to the *irreversible* magnetization. Using equations (1) and (2) the location of the peaks of the symmetrical loops described about the origin between $-\mathrm{H}_1$ and $+\mathrm{H}_1$ can be deduced as

$$\mathcal{J}_1 = a\mathrm{H}_1 + b\mathrm{H}_1^2. \quad (3)$$

The area of the loop is

$$\mathrm{W}_1 = \frac{4}{3} b \mathrm{H}_1^3. \quad (4)$$

Since 1887 these laws have received no satisfactory explanation but Weiss and de Freudenreich [2] showed that it was possible to envisage a *fictitious material* endowed with similar properties. This substance comprised an assembly of elementary domains each possessing a symmetrical rectangular hysteresis loop characterized by a coercive field H_c; let $dv = \varphi d\mathrm{H}_c$ be the volume fraction occupied by the domains in the coercivity range H_c to $\mathrm{H}_c + d\mathrm{H}_c$. In addition the authors assume that each domain is subject to a constant uniform field H_a arising from the neighbouring domains: let $\psi d\mathrm{H}_a$ be the probability of finding this field in the range H_a to $\mathrm{H}_a + d\mathrm{H}_a$. If the probabilities φ and ψ are independent and possess constant non-zero values in the region of $\mathrm{H}_c = 0$ and $\mathrm{H}_a = 0$ respectively, then the characteristics of the Rayleigh laws are recovered. The same idea was used by Preisach [3] who specified the values of φ and ψ required to describe the overall hysteretic properties of certain alloys. He also used an ingenious graphical representation to provide a clear explanation of certain characteristic features associated with magnetic viscosity[1].

Unfortunately, although it is possible to attribute a physical meaning to H_c and H_a, it is not possible to justify, *a priori*, the choice of φ and ψ values that must be given to obtain Rayleigh's laws. In point of fact it is not a theory but, rather, a purely formal representation difficult to reconcile with our present knowledge of the magnetization processes in small fields which mainly involve reversible and irreversible domain wall movements. Co-operative reversals of magnetization, domain by domain, do not take place.

(2) The different processes of magnetization: According to these concepts which are supported by a wide range of experimental data, a ferromagnetic body consists of a multitude of elementary domains (Weiss domains), generally small compared with the grain size and magnetized *uniformly* to saturation. Changes of magnetization provoked by an external field involve either the *rotation* of magnetization within a domain or the *growth of certain domains* at the expense of their neighbours, i.e. wall motion.

Rotation of magnetization is linked to magnetocrystalline anisotropy. Thus for a pure unstrained iron crystal in zero applied field, the magnetization in each domain aligns with one of the cube edges. These are the so-called easy axes of magnetization. In order to produce an appreciable rotation of the magnetization a field of about 10 Oe is required. Parallelism of the external field with the different domain magnetizations of a polycrystal is only effectively reached in a field of 500–600 Oe. This anisotropy is characterized by an energy constant, the anisotropy constant K, equal to 4.4×10^5 erg cm^{-3} for iron (cf. A, p 112).

It is easy to reduce the coercive force of iron well below 1 Oe. Under these conditions all the hysteresis phenomena of special interest here take place without significant deviation of the magnetization of the various domains from their initial easy axes. By a similar token the energy dissipated in the hysteresis cycle can be as low as 100 erg cm^{-3}, a value 1000 times smaller than the energy difference produced by rotation. This reveals the possibility of experimentally *separating* these two processes. *We shall limit our study hereafter to wall displacements between domains in which the magnetization directions are assumed to remain fixed along one of the easy axes.*

In iron, the domain magnetization can adopt six possible orientations. This leads to two types of walls: *180° walls* which separate two domains with magnetization directed antiparallel and *90° walls* which separate two domains with magnetization directions which make an angle of 90° with each other.

In an ideal crystal the three $\langle 100 \rangle$ axes are strictly equivalent energetically but in fact, owing to impurities or other causes, certain regions depart *locally* from cubic symmetry. One of the axes corresponds to a slightly smaller energy than the other two[2] but the two possible senses along this axis are obviously equivalent. We characterize the three possible directions by an index i which takes on values 1, 2 and 3, and the two possible senses for each direction i, by the notation $+i$ and $-i$. The magnetocrystalline energy of a volume element $dxdydx$ with co-ordinates x, y, z and in which the magnetization is directed along i can be written as $u_i(x, y, z) \, dxdydz$. By symmetry $u_i = u_{-i}$. We assume that u *varies haphazardly from one point in the crystal to another* and that deviations from the average value remain small compared with the anisotropy energy K_1.

At first sight it appears that in the absence of a magnetic field, the magnetization at each point x, y, z, always follows the i axis which corresponds to the lowest energy u_i. This is certainly the case when u_i remains smaller than u_j and u_k in a domain *whose extent is very large vis à vis interatomic distances*. However, when the perturbations are on a very small scale, too frequent changes in magnetization direction are hindered by the concomitant appearance of new walls, accompanied, as we shall see later, by an energy expenditure greater than that which corresponds to the difference $u_k - u_i$ or $u_j - u_i$.

(3) Wall energy: Imagine, to begin with, a wall in the form of a geometric plane on one side of which the spins are in the i direction and on the other side in the j direction. The spins of two nearest neighbour atoms, situated on either side of the wall are not parallel and, as a consequence of exchange effects, the energy is considerable: of the order of 10^2 erg cm^{-2} of wall area when the spins are antiparallel. This energy can be very much reduced by the formation of a transition layer across which the spins change progressively from one direction to the other. If the exchange energy were the only consideration this layer would tend to be as wide as possible, a situation opposed by the magnetocrystalline energy. This is because a thicker layer implies a greater region over which the magnetization is not along an easy axis. In fact, as first shown by Bloch [4], there exists an optimum thickness for which the surface energy, γ, is a minimum. Becker (A, p 189) has given an elementary calculation of γ.

---

[1] See also Becker and Doring, Ferromagnetismus, Berlin 1939, 221. This work will be hereafter designated as reference A.

[2] The axes, or rather the directions of minimum energy, are subject in addition to fluctuations in direction but generally this effect is negligible. It will not be discussed further.

Direct experiments [5] have confirmed the value of γ predicted by theory.

In the case of interest here, it is found for a 90° wall that

$$\gamma = a \sqrt{\frac{I K_1}{2}}, \qquad (5)$$

where a is the lattice parameter, I is the exchange energy per cm^3 and K_1 is the anisotropy constant. The 180° wall has twice this energy value. The calculation yields $\gamma \sim 1.5$ erg cm^{-2} for iron.

(4) The rôle of small perturbations: Consider a large domain in which u_i is everywhere smaller than u_j and u_k with the exception of a small cubic domain of side l for which this condition is not fulfilled. The average value, $\bar{u}_j$, of u_j inside the cube could for example be smaller than the other mean values $\bar{u}_i$ and $\bar{u}_k$. Outside the cube the magnetization lies along i. If inside it lies along j, then a wall of surface area $6l^2$ and energy $6l^2\gamma$ appears. If on the other hand, it keeps to the i direction, then there is no longer a wall energy — but an energy of $(\bar{u}_i - \bar{u}_j)l^3$ must be provided to overcome magnetocrystalline forces. These two energies are equal when the side of the cube has a length l_0 defined by

$$l_0 = \frac{6\gamma}{\bar{u}_j - \bar{u}_i}.$$

The magnetization changes direction only if l is greater than l_0. In well-annealed pure iron, where $\bar{u}_j - \bar{u}_i$ certainly does not exceed 1000 erg cm^{-3}, l_0 is of the order of 10^{-2} cm.

This is a sizeable value which corresponds to several thousand interatomic distances. Now it is certainly true that local defects on a much smaller scale exist in the crystal lattice of iron, e.g. those associated with the insertion of carbon atoms. Thus for a carbon content of 0.01% a cube of side 10 interatomic distances will contain, on average, one atom of carbon.

The local inhomogeneities of the iron lattice can, therefore, be classified under two categories depending upon their scale with respect to the critical distance l_0.

In *large scale* perturbations the magnetization is oriented along the i axis which corresponds to the smallest u_i value. On the contrary, *for small scale perturbations the magnetization direction is unaffected by local variations of u_i.* The actual direction of the magnetization corresponds to the smallest of the three mean values $\bar{u}_i$, taken over a volume whose linear dimensions are large compared with l_0. *If the three averages are equal, the three directions are equiprobable.*

In what follows we examine especially the *small-scale effects* and assume, for simplicity, *that no perturbations exist on the large scale. A priori*, the six possible directions and senses for the domain magnetization are equally probable.

(5) Energy related to wall displacement: We now study the displacement of a wall under the action of a magnetic field. To simplify matters, we imagine an unlimited cylindrical domain split into two by a plane wall Σ of surface S perpendicular to its generators. The position of the wall is determined by its abscissa x on an axis X'OX parallel to the generators. This wall separates two regions: on the negative x side the magnetization remains constant and equal to $\bar{I}$: on the positive x side, the magnetization is equal to $\bar{J}$.

The magnetocrystalline and exchange forces are assumed everywhere to be subject to local variations such that the surface energy γ at a point y, z of the wall situated at x is a function of x, y, z. When the wall, originally at x_0, suffers a displacement dx, the corresponding change in the surface energy is,

$$dw_p = dx \int_\Sigma \left(\frac{\partial \gamma}{\partial x}\right)_{x_0} dy\, dz.$$

Across the region between x_0 and $x_0 + dx$, the magnetization changes from direction j to direction i, involving an expenditure of magnetocrystalline energy of

$$dw_u = dx \int_\Sigma [u_i(x_0, y, z) - u_j(x_0, y, z)]\, dy\, dz.$$

It should be noted that this last term is zero for a 180° wall where $j = -i$ and $u_i = u_{-i}$. The sum $dw_p + dw_u$ is the differential with respect to x of a function W, defined apart from a constant, which, when x increases, *fluctuates* about a mean *constant* value. This is because, by hypothesis, large scale perturbations are presumed not to exist.

Let a magnetic field denoted by the vector H be applied. The change is magnetic energy now associated with the same wall displacement is

$$dw_H = dx \int_\Sigma (\bar{J} - \bar{I})\bar{H}\, dy\, dz = (\bar{J} - \bar{I})\bar{H} S\, dx.$$

The action of the magnetic field is equivalent to a *hydrostatic pressure* V given by,

$$V = (\bar{J} - \bar{I})\bar{H}. \qquad (6)$$

The wall is *in equilibrium* at point x if

$$V = \frac{1}{S} \frac{dW}{dx}, \quad (7)$$

but stability is assured only if

$$\frac{d^2 W}{dx^2} > 0.$$

In order to simplify the notation and until stated otherwise, magnetic fields will be measured in terms of equivalent pressure V.

Displacements of the wall Σ depend only on V and the *characteristic function* W/S.

PART ONE
STUDY OF THE DISPLACEMENTS OF AN ISOLATED WALL

(6) The form of the characteristic function: *A. First approximation* — Kersten [6] chose a sinusoidal function of x for W/S. There is no justification for this choice which leads to nothing significant in weak fields and provides no explanation of Rayleigh's laws.

To provide a better representation of the non-uniformity of the substance, we assume, in a first approximation, that W/S is a function of x represented by a *polygonal contour* A_n, A_n+1, A_n+2, etc each side of which corresponds to a length $2l$ along the x-axis (Fig. 1). *We shall assume that the slopes of the sides are distributed at random with no correlation between adjacent slopes.* Let $p = \frac{d}{dx}\frac{W}{S}$ be the slope of a side and p the *reduced slope* defined by $p = P/P_0$. The probability that an arbitrarily chosen side has a *reduced slope* between p and $p+dp$ is defined as $\omega(p)dp$. The calculation will be developed for the particular case of the Gaussian law.

$$\omega(p) = \frac{1}{\sqrt{\pi}} e^{-p^2}. \quad (8)$$

The characteristic function is then completely defined by the two parameters l and P_0.

B. Second approximation — The sharp angles of the polygon have no physical reality so in order to eliminate discontinuities in the first derivative of the characteristic function, we proceed as follows. Let AO and OC be two adjacent sides of the contour with M_1 and M_2 their midpoints (Fig. 2). *We replace the contour*

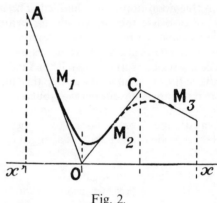

Fig. 2.

$M_1 O M_2$ *by a parabolic arc which meets the sides AO and OC tangentially at M_1 and M_2 respectively.* This procedure is repeated for successive intervals. Finally, we obtain a continuous curve with a continuous first derivative consisting of a succession of parabolic arcs each corresponding to the same interval $2l$ along X'OX. Taking O as the origin of co-ordinates the equation of the arc $M_1 M_2$ can be written in terms of the slopes $-P$ and Q of the sides AO and OC as,

$$\frac{W}{S} = (P + Q) \frac{l}{4} - (P - Q) \frac{x}{2} + (P + Q) \frac{x^2}{4l} \quad -l < x < l. \quad (9)$$

The model of the characteristic function so obtained is defined in terms of two constants l and P_0. To what extent this gives a faithful representation of an actual characteristic function is a question reserved until later.

In the following sections we investigate the displacement in a magnetic field of a plane wall of constant area having a characteristic function of the

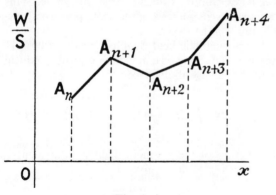

Fig. 1.

form envisaged above. Since this function consists of a succession of parabolic arcs the movement of a wall along an elementary parabolic arc is first studied: the arc is completely defined as shown above, by the slopes $-P$ and Q of the two adjacent sides from which it is formed.

(7) The motion of a wall on the arc of the parabola $(-P, Q)$: Stable equilibrium depends on the value of the second derivative of the characteristic function which, from equation (9), is

$$\frac{1}{S}\frac{d^2w}{dx^2} = \frac{1}{2l}(P + Q).$$

If $P+Q$ is negative, the wall is never in stable equilibrium regardless of the applied field value.

If $P+Q$ is positive, several cases occur which are now examined.

(A) $P>0$; $Q>0$. Here the arc (M_1M_2 in Fig. 2) has the form of a valley. In zero external field the lowest point of the valley, i.e. the vertex of the parabola, is a position of stable equilibrium for the wall at x_0 where

$$x_0 = l\frac{P - Q}{P + Q}. \qquad (10)$$

The effect of a magnetic field, measured by the equivalent hydrostatic pressure V (cf. § 5), moves the wall to a point x which satisfies relation (7) and, taking account of relations (9) and (10), can be written as

$$x - x_0 = \frac{2lV}{P + Q}.$$

The corresponding change in the magnetization is given by the vector $S(x-x_0)(\bar{J}-\bar{I})$, proportional to the displacement $(x-x_0)$.

To simplify the discussion that follows a *reduced displacement* $i=x/2l$ will be employed such that actual displacements are obtained by multiplying the reduced quantities by $2l$. The quantity i can also be considered as a *reduced magnetization* (a scalar quantity) from which real changes in magnetization can be obtained on multiplying by the vector quantity $2lS(\bar{J}-\bar{I})$. In a similar way, the slopes are measured by reduced slopes $p=P/P_0$, $q=Q/P_0$ etc. and the magnetic fields by a reduced field $h=V/P_0$.

With these new units we get

$$i - i_0 = \frac{h}{p + q}. \qquad (11)$$

It can easily be verified that the wall attains the right-hand extremity of the arc, M_2, for $h=q$ and the left hand extremity for a reduced negative field of $h=-p$. *In short, when h varies from $-p$ to $+q$ the wall moves reversibly from M_1 to M_2, proportional to h: it is as if the wall possessed a constant susceptibility of $(p+q)^{-1}$.*

(B) The preceeding position remains true for those cases not included in A but for which $P+Q$ is still positive; the only difference is that the limiting values of the field, $-p$ and q, no longer embrace the zero value.

Let us now see what happens when the field reaches, for $h=q$, the extremity M_2 of the arc, and is now increased infinitesimally to a value of $h+\varepsilon$. Two possibilities are open. Let q and q' be the slopes of the two sides which correspond to the next parabola M_2M_3 (Fig. 2). If $q'>q$ this arc is *stable* and $h+\varepsilon$ lies between q and q'. The wall therefore continues to move *reversibly* beyond M_2 without discontinuity at M_2.

If $q'<q$, the arc is *unstable* and the wall displaces *irreversibly* until it arrives at a point on some further parabola where the slope of the characteristic function equals $h+\varepsilon$.

(8) The movement of a wall along a series of parabolic arcs: Let us now suppose that *a single plane wall of constant area moves along a series of N parabolic arcs*, defined by the $N+1$ slopes of $N+1$ successive sides of a polygonal contour. The slopes are given by the probability law (8). An infinite number of possible polygonal contours exists with $N+1$ sides. We shall calculate the *average* displacements of the wall when the magnetic field, always *weak*, is allowed to vary in an arbitrary manner. The reduced field is referred to as weak if its modulus is small compared with unity[3].

To specify the problem completely the initial position of the wall must still be defined. In fact, *a priori*, the wall in zero field can occupy initially any of a number of stable positions on the contour[4]. However if the wall is previously subjected to a slowly decreasing a.c. field it reaches a well-defined position in zero field. This position will be taken as the initial one. Experiments involving measurements in small fields are usually preceded by a *demagnetization* according to the mechanism just described.

(9) Categorization of the sides of the polygonal contour: We study the displacements of a wall in a field h whose modulus is always less than a positive quantity ε which is less than one, i.e.

$$|h| < \varepsilon, \qquad (12)$$

To do this the $N+1$ sides of the contour must be classified according to the values of their reduced slope

---
[3] This condition applied to the actual field H implies that the scalar product $\bar{H}(\bar{J}-\bar{I})$ is small compared with P_0.
[4] We see later that there are $N/4$ possible equilibrium positions.

compared with $-\varepsilon$ and ε. Essentially, a side with slope greater than ε cannot be crossed, from left to right, as long as the field h obeys condition (12). This is not the case if the slope is less than ε. Three categories of sides can therefore be distinguished, A, B, C. Category A includes sides with slopes greater than ε; category B, those with slopes between $-\varepsilon$ and $+\varepsilon$; the final category C, has sides with slopes less than $-\varepsilon$.

(10) Division of the polygonal contour into domains: The contour can be decomposed, completely and uniquely, into a series of *domains* of greater or lesser length, which belong to the following four types; AB^nA, AB^nC, CB^nA and CB^nC, where the index n is an arbitrary positive integer (or zero). The notation, AB^nC for example, signifies a succession of $n+2$ sides with the first one (reading from the left) belonging to category A, the next n to category B and the last one to category C. Sides in categories A or C are always common to two domains: one starting and the other finishing there. Domains of type ABC start in the middle of an A side and finish in the middle of a C side. As a result a domain with index n occupies a length of the polygon contour equal to $n+1$ sides, and is formed of $(n+1)$ parabola arcs.

Following demagnetization, the wall must be inside a domain of the type CB^nA.

Let us assume that during a demagnetization the wall is inside an AB^nA type domain when the modulus of the field reaches ε. If the field is now decreased to a value of $-\varepsilon$ the wall, moving leftwards, will certainly by this time have reached the right hand extremity of the arc AB. (The left-hand extremity of AB marks the end of this domain.) In point of fact this wall has traversed the B sides and is not found on the arc AB which is unstable. This is the case because the slope of side A is positive and has a greater modulus than that of the second side B. The wall has thus crossed irreversibly the arc AB and continued leftwards down the slope, outside the domain. Moreover, it is impossible to re-climb this slope since the maximum positive value of the demagnetizing field, ε, is less than the slope of A. The wall therefore, cannot return to the AB^nA domain. Analogous reasoning shows the impossibility of finding the wall in domains of types AB^nC and CB^nC.

(11) The number of CB^nA domains contained within a polygonal contour of N sides: According to equation (8) the probability η of finding a slope in the interval $-\varepsilon$ to ε, is

$$\eta = \frac{2}{\sqrt{\pi}} \int_0^\varepsilon e^{-p^2} dp.$$

The probability that a side, picked at random, belongs to categories, A, B, C is therefore, respectively: $\left(\frac{1}{2} - \frac{\eta}{2}\right)$, η, and $\left(\frac{1}{2} - \frac{\eta}{2}\right)$. It may be deduced immediately that the number ν_n of CB^nA domains contained within a polygon of N sides is

$$\nu_n = N \left(\frac{1}{2} - \frac{\eta}{2}\right)^2 \eta^n. \quad (13)$$

In the case of small ε, then, to third order:

$$\eta = \frac{2\varepsilon}{\sqrt{\pi}}. \quad (14)$$

The significance of the domain concept may be stated thus: *Regardless of the changes in the magnetic field, provided its modulus remains less than ε, the wall stays in the domain originally occupied.* The problem of the mean displacement of a wall is solved if we know (a) the average initial partition of a wall amongst domains and (b) the laws governing the displacement inside a given type of domain.

(12) The probability of wall capture by a given domain: Consider *one* domain of type CB^nA whose terminal sides, C and A, have slopes p and p' ($p'<0$; $p>0$). It can be shown (see Appendix I at the end of the second part of this article) that the probability of a wall occupying this domain after demagnetization is given by,

$$\bar{\rho}_{pp'} = \frac{1}{2N} (\theta_p \theta_{p'})^{-1/2}, \quad (15)$$

where θ_x is defined through,

$$\theta_x = \frac{1}{\sqrt{\pi}} \int_{|x|}^{+\infty} e^{-x^2} dx. \quad (16)$$

(13) The probability of wall capture by all the domains of a given type: The total number of CB^nA type domains for all possible values of n whose extreme sides C and A have characteristic slopes p and p', lying between p and $p+dp$ and $p'+dp'$, is according to equation (8) equal to

$$\frac{N}{\pi} e^{-(p^2 + p'^2)} dp\, dp'.$$

From equation (15), the probability of finding a wall amongst such domains is therefore $\mathfrak{J}_p \mathfrak{J}_{p'} dp dp'$ with,

$$\mathcal{J}_p = \frac{1}{\sqrt{2\pi}} \theta_p^{-1/2} e^{-p^2}. \qquad (17)$$

The total number of domains capable of sustaining walls in the initial state can be obtained by letting p vary from $-\infty$ to 0 and p' from 0 to $+\infty$. It is easy to show that

$$\int_{-\infty}^{0} \int_{0}^{+\infty} \mathcal{J}_p \, \mathcal{J}_{p'} \, dp \, dp' = 1,$$

which agrees with the sense of probability attributed to $\mathcal{J}_p \mathcal{J}_{p'} dp dp'$.

Furthermore, from relation (13) we see that *the probability $\varphi_n dp dp'$ of finding the wall in a domain CB^nA of a given n and slopes p,p' lying between p and $p+dp$ and p' and $p'+dp'$ is*

$$\varphi_n \, dp \, dp' = \eta^n \, \mathcal{J}_p \, \mathcal{J}_{p'} \, dp \, dp'. \qquad (18)$$

It is easy to verify, using (13) and (14) that

$$\sum_{n=0}^{\infty} \eta^n \int_{\varepsilon}^{\infty} \int_{\varepsilon}^{\infty} \mathcal{J}_p \mathcal{J}_{p'} dp dp' = 1.$$

We now examine the laws of magnetization for a wall enclosed within each type of domain.

Henceforth, the characteristic slopes of the C and A sides for each domain studied will be referred to as $-p$ and q respectively.

(14) Magnetization from CA domains: The characteristics p and q vary from ε to $+\infty$. Since $|h| < \varepsilon$, the susceptibility of a domain $(-p, q)$ remains constant at $(p+q)^{-1}$.

The average magnetization i_0 due to these domains can therefore be written as

$$i_0 = h \int_{\varepsilon}^{\infty} \int_{\varepsilon}^{\infty} \frac{\mathcal{J}_p \mathcal{J}_q}{p+q} \, dp \, dq. \qquad (19)$$

This expression can be written in a slightly different form on putting

$$\lambda = \int_{0}^{\infty} \int_{0}^{\infty} \frac{\mathcal{J}_p \mathcal{J}_q \, dp \, dq}{p+q}, \qquad (20)$$

λ is a numerical constant equal to 0.81. Noting that when p is small $\mathcal{J}_p = 1/\sqrt{\pi}$, we get

$$i_0 = h \left[\lambda - \frac{2}{\sqrt{\pi}} \int_{\varepsilon}^{+\infty} \mathcal{J}_q \log \frac{q+\varepsilon}{q} dq - \frac{2\varepsilon}{\pi} \log 2 \right]. \qquad (21)$$

This magnetization is always reversible and proportional to h.

(15) Magnetization due to CBA domains: Consider a domain with characteristics $-p, r, q$ $(p > \varepsilon; -\varepsilon < r < \varepsilon; q > \varepsilon)$. This magnetization i_1 will be the sum of a term i'_1 corresponding to negative r, and a term i''_1, when r is positive (each possibility is equally likely).

(a) r is *negative*. Here the wall, initially on the arc BA, remains on this arc when h is positive; the associated constant susceptibility is $(q-r)^{-1}$. The magnetization arising from these domains is therefore, using (18),

$$i'_1 = \frac{h\eta}{2\varepsilon} \int_{\varepsilon}^{\infty} dp \int_{\varepsilon}^{\infty} dq \int_{-\varepsilon}^{0} \frac{\mathcal{J}_p \mathcal{J}_q}{(q-r)} dr, \qquad (22)$$

which becomes in the event of small ε,

$$i'_1 = +\frac{h}{2} \frac{2}{\sqrt{\pi}} \int_{\varepsilon}^{\infty} \mathcal{J}_q \log \frac{q+\varepsilon}{q} dq. \qquad (23)$$

(b) r is *positive*. The wall, initially on the arc CB with a susceptibility $(p+r)^{-1}$, transfers to the arc BA when h reaches the value of r; the susceptibility then becomes $(q-r)^{-1}$. When $h<r$, the magnetization is $h/(p+r)$. When $h>r$, it is $\dfrac{h}{p+r} + \varphi$, with

$$\varphi = (r-h)\left(\frac{1}{p+r} - \frac{1}{q-r}\right).$$

Taking due regard of relations (18) and (22), the average magnetization from these domains is given by

$$i''_1 = i'_1 + \frac{h\eta}{2\varepsilon} \int_{\varepsilon}^{+\infty} dp \int_{\varepsilon}^{+\infty} dp \int_{0}^{h} \varphi \, \mathcal{J}_p \, \mathcal{J}_q \, dq \qquad (24)$$

Assuming that ε is small and restricting the calculation to terms in h^3 and the main part of the integral, we obtain,

$$i''_1 = i'_1 + \frac{h^3}{3\pi\varepsilon} + \cdots \qquad (25)$$

As in the previous section, the magnetization is reversible, but this time there is a term in h^3 which depends on ε.

(16) Domains of type CB^2A: Let $-p, a, b, q$, be the slopes of the four corresponding sides. It is convenient to distinguish between two extremely important cases which depend upon the relative values of a and b.

(1) $a<b$. The domain comprises 3 parabolic arcs with susceptibilities from left to right respectively of: $(p+a)^{-1}$, $(b-a)^{-1}$, $(-b+q)^{-1}$. The wall passes from the first to the second arc when $h=a$, from the second to the last when $h=b$. The phenomena are completely reversible like those encountered for the domains discussed above.

(2) $a>b$. Let AB, BB', B'C, be the three constitutive arcs of the domain (Fig. 3). The middle arc BB' is convex upwards since $a>b$. It is unstable. Fig. 4 represents the magnetization (i.e. the abscissa x of the wall in Fig. 3), plotted along the Oi axis as a function of h (along the Oh axis).

Consider an initial negative value of field: the wall at m in Fig. 3 is indicated at α in Fig. 4. When the field is increased the wall moves from m towards B and arrives there when $h=a$. During this process the magnetization is proportional to the field, the susceptibility is $(p+a)^{-1}$ and the straight line $\alpha\beta$ in Fig. 4 is described. When h exceeds the value a by an infinitesimal quantity ζ the wall passes across the unstable arc BB' and moves irreversibly to a point n on arc B'C where the slope is $a+\zeta$. Upon further increase of the field, the magnetization grows proportionally but now with a different proportionality constant equal to $(q-b)^{-1}$: the straight line section $\gamma\delta$ is described. Similar behaviour takes place on retracing the field except that the discontinuity occurs at point B' for $h=b$. The hysteresis loop shown in Fig. 4 is thus described.

Our task is to determine the average magnetic properties of these domains when a and b take all possible values between $-\varepsilon$ and $+\varepsilon$.

We make use of an artifice. Each irreversible domain $(-p, a, b, q)$ can be uniquely associated with a reversible domain $(-p, b, a, q)$ and vice versa.

It is easy to show that the total magnetization of the two domains is identical to the sum of the magnetizations from two CBA type domains characterized by $(-p, a, q)$ and $(9p, b, q)$ together with the three cycles shown in Fig. 5.

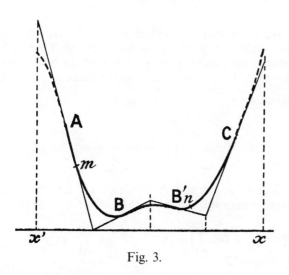

Fig. 3.

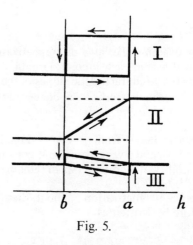

Fig. 5.

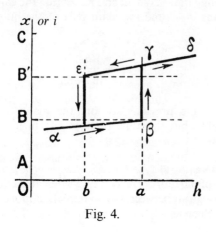

Fig. 4.

1. A rectangular loop of height 1 and width $a-b$;
2. Two horizontal branches joined by the diagonal of the preceding loop;
3. A fictitious cycle, as shown in III, with horizontal side branches *which form extensions of each other*. The slope of the lower branch is $\dfrac{-1}{p-b}$, while that of the upper branch is $\dfrac{-1}{p+a}$. The area of the loop is

$$\frac{(a-h)^2}{2}\left[\frac{1}{p+a}+\frac{1}{q-b}\right];$$

which is always small compared with the area $(a-b)$ of the rectangular loop I because p and q are large compared with $(a-b)$. The relative error introduced by neglecting loop III bears only on the irreversible part of the magnetization; it is of the order of $\dfrac{a}{p} < \dfrac{\varepsilon}{p}$.

Since the average magnetic properties of CBA type domains have already been calculated, it now suffices to study the average properties of loops I and II. We note that the preceding calculation has effectively eliminated the influence of p and q which leaves the necessity of dealing with an average area relative to a and b only. Since ε is small, all values of a and b between $-\varepsilon$ and $+\varepsilon$ are equally likely.

Obviously the loop I arises from an irreversible domain $(-p, a, b, q)$ whereas loop II comes from a reversible domain $(-p, b, a, q)$. The others play only the rôle of correction terms.

Every cycle, I or II, can be represented by a point x, y in the (x, y) plane such that x is the slope of the first B-type side and y is the slope of the second B side. The pair of domains coupled together above corresponds to points with co-ordinates (a, b) and (b, a). All possible points are contained within a square limited by the lines $x = \pm \varepsilon$ and $y = \pm \varepsilon$, of area $4\varepsilon^2$. This representation is quite symmetric. The probability of finding the representative point of a loop within a rectangle of area $dx\,dy$ is clearly[5] $dx\,dy/4\varepsilon^2$. The points representing reversible (type II) loops are found within the triangle $\alpha\beta\gamma$ whereas those representing irreversible (type I) loops are within triangle $\alpha\delta\gamma$ (Fig. 6).

(17) The average magnetic properties of reversible (type II) loops: Let us say that a field h is applied. We now draw the lines Hy' and Hx' corresponding to the equations $x = h$ and $y = h$ which cut $\alpha\gamma$ at H. These two lines divide the triangle $\alpha\beta\gamma$ into three regions (u), (v) and (w) (Fig. 6).

Taking the average magnetization as origin we see that the magnetization in region (u) corresponds to the upper branch of the cycle and equals $\tfrac{1}{2}$; in region (w) it is $-\tfrac{1}{2}$. The resultant magnetization from both these regions is

$$\bar{i}_u + \bar{i}_w = \frac{1}{4\varepsilon^2} \times \frac{1}{2} (\text{area } ax'H - \text{area } \gamma y'H) =$$

$$= \frac{1}{8\varepsilon^2} \left[\frac{1}{2}(\varepsilon + b)^2 - \frac{1}{2}(\varepsilon - b)^2 \right] = \frac{h}{4\varepsilon}.$$

In the rectangular region (v) it is easily shown that the magnetization of a loop represented by the point m

---

[5] It is assumed that ε is small compared with unity: if not the probability should be multiplied by $e^{-(a^2+b^2)}$.

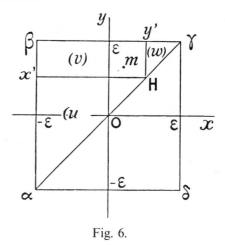

Fig. 6.

with co-ordinates x, y, (in the $x'Hy'$ system of axes) is given by

$$i = \frac{1}{2} \frac{x-y}{x+y},$$

from which, after putting $Hx' = x_0$ and $Hy' = y_0$, we obtain the mean magnetization as

$$\bar{i}_v = \frac{1}{4\varepsilon^2} \cdot \frac{1}{2} \int_0^{x_0} \int_0^{y_0} \frac{(x-y)\,dx\,dy}{x+y} = \frac{1}{8\varepsilon^2}$$

$$\{(x_0^2 - y_0^2) \log(x_0 + y_0) + y_0^2 \log y_0 - x_0^2 \log x_0\};$$

But $x_0 + y_0 = 2\varepsilon$ and $x_0 - y_0 = 2h$ which leads to, assuming h is small and neglecting terms in h^5,

$$\bar{i}_v = \frac{1}{4\varepsilon}\left[h(2\log 2 - 1) - \frac{h^3}{3\varepsilon^2}\right].$$

The average magnetization of reversible type II loops is therefore

$$\bar{i}_p = \bar{i}_u + \bar{i}_v + \bar{i}_w = \frac{h}{2\varepsilon}\log 2 - \frac{h^3}{12\varepsilon^3}. \qquad (26)$$

(18) The average magnetic properties of irreversible (type I) loops: This assembly of loops is fairly close to that investigated formally by Weiss and Freundenreich and by Preisach (cf. § 1). However, these authors assumed, *a priori* and without the least justification, that a ferromagnetic body contains an internal assembly of *elementary domains*, each corresponding to one of these loops. In particular, the need to ascribe a constant value to the density of domains near the origin constitutes an unrealistic hypothesis when irreversible domains only are considered — this is tantamount to ignoring half of the representative plane to the left of the diagonal αγ. *This difficulty disappears when account is taken of this half of the plane, which corresponds to reversible phenomena.* Moreover, the constant value of the density arises logically and necessarily from the analysis which has just been developed.

Before studying the average magnetic properties of the group of loops represented by points within the triangle αδγ, we shall determine the initial states of the magnetization within these cycles when $h=0$. For this purpose the triangle is split into different regions.

In the upper triangle OAγ, the magnetization equals $-\frac{1}{2}$ because it corresponds to the lower branch of the type I cycles: this will be denoted schematically (Fig. 7) by the minus sign $(-)$. By a similar token the

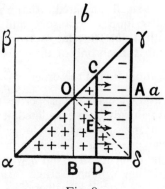

Fig. 8.

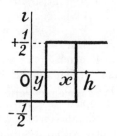

Fig. 9.

The positive value h_0 having been reached, let the field be decreased to h. All cycles situated above the line $b=h$, indicated by C'D' in Fig. 10, will take a

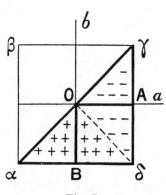

Fig. 7.

magnetization in the triangle OBα corresponds to the upper branch. This is equal to $+\frac{1}{2}$ and will be denoted by the plus sign $(+)$. In principle the magnetization in the square OBδA should be arbitrary, i.e. the domain magnetization can be equally positive or negative.

In fact, however, the material is demagnetized in an a.c. field reducing to zero. It is easy to demonstrate, by means of methods analogous to those we are going to use, that the square OBδA must be divided into two by the diagonal Oδ. The magnetization is $+\frac{1}{2}$ in triangle OBδ and $-\frac{1}{2}$ in triangle OδA.

Starting from this initial state, let a positive field, h, be applied. All domains with $x<h$ (cf. Fig. 9) should have magnetization of $+\frac{1}{2}$. These are situated to the left of the line $a=h$ shown as CD in Fig. 8. The magnetization of cycles located in triangle OEC is reversed. The magnetization acquired, i, is proportional to the area of the triangle, i.e. to h^2. For $h=h_0$ we have

$$i_0 = \frac{h_0^2}{4\epsilon^2}. \tag{27}$$

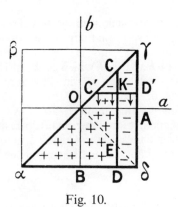

Fig. 10.

magnetization value of $-\frac{1}{2}$. This time it is the domains in triangle CC'K which reverse and we have

$$i - i_0 = -\frac{1}{2}\frac{(h-h_0)^2}{4\epsilon^2} \tag{28}$$

(H decreasing from $+h_0$ to $-h_0$). This magnetization law will hold until K reaches E when $h=-h_0$, and we have $i = -i_0$. If at this point, the field is increased, the following magnetization law will hold.

$$i = -i_0 + \frac{1}{2}\frac{(h+h_0)^2}{4\epsilon^2} \qquad (29)$$

(H increasing from $-h_0$ to h_0).

These are the characteristic features of the phenomenon studied by Rayleigh: an irreversible term of second order is obtained[6].

Finally we note from equations (13) and (14) that, to third order in ϵ, the wall has a probability of $4\epsilon^2/\pi$ of occupying a domain of type CB^2A. The contribution, i_2, of these domains to the total magnetization is —*for the case of a first magnetization* —

$$i_2 = 2\log 2\,\frac{\epsilon h}{\pi} + \frac{h^2}{\pi} - \frac{h^3}{3\pi\epsilon} + \frac{2\epsilon}{\sqrt{\pi}}(i_1'' + i_1'). \qquad (30)$$

(19) The average total magnetization: We now take ϵ and h as being the same order of magnitude and calculate the total magnetization neglecting 3rd order terms. As a result, domains of the type CB^3A can be discounted: their contribution is effectively third order because the magnetization of a wall occupying such a domain is of order unity and the probability of occupation by the wall is of order ϵ^3. It is sufficient, therefore, to combine the contributions i_0, i_1 and i_2 arising from domain types CA, CBA and CB^2A respectively and which are given by equations (21), (23), (25) and (30).

The term i_0 is of order ϵ; $i_1 = i_1' + i''$ is of order ϵ^2. Consequently, the last term in (30) can be neglected

---

[6] Equations (27), (28) and (29) should be compared with equations (3), (1) and (2) respectively of § 1.

since it is of the order of i_1 multiplied by ϵ. Thus, for the initial magnetization we have

$$i = i_0 + i_1 + i_2 = \lambda h + \frac{h^2}{\pi}. \qquad (31)$$

All the second order terms containing ϵ have vanished. Whereas the λh term is reversible, the term h^2/π is not and must be replaced by $+h^2/2\pi$ or $-h^2/2\pi$ depending upon whether the ascending or descending branch of the hysteresis loop is being traced.

The importance of this result lies less in the particular numerical values[7] of the coefficients in equation (31) than in the remarkable nature of the series expansion in increasing powers of h, the first two of which appear in the equation. Obviously any theory must eventually lead to a series but here *we see a specifically new feature: the odd terms are reversible and the even terms are irreversible*. The generality of this result goes beyond the case for which it has just been demonstrated and applies to any probability law which governs the distribution of sides on the polygonal contour.

We have just shown, therefore, that the elementary displacement of an isolated wall obeys a mode of magnetization which conforms to Rayleigh's laws. It remains to be seen whether the total magnetization given by the sum of the elementary phenomena conforms in a similar way.

In the second part we shall study the case of a material containing many domains separated by variously oriented walls and, using similar methods, calculate the coercive field.

---

[7] NB. λ is a numerical constant $=0.81$.

II. MULTIPLE DOMAINS AND COERCIVE FIELD

Abstract: In this second part the results relating to the displacement of an isolated wall (obtained in the first part) are applied to the study of the properties of a material consisting of a large number of domains. In addition the study of the wall movements in strong fields allows the coercive field H_c to be determined without the introduction of any new parameters. The final result for the first magnetization curve is

$$J = \alpha J_s (0{,}38 h + 0{,}16 h^2),$$

where J_s is the saturation magnetization, α is a number which depends on the number of domains and $h = rH/H_c$. The coefficient r is a slowly varying function of α which is specified. The expressions obtained for a, b, H_c can be checked experimentally and the predicted values for the dimensionless quantity bH_c/a are found to be fairly close to the observed values. This verification is limited to cases for which rotational processes are negligible. The relevant values of the initial rotational susceptibility a' can be specified as a function of the anisotropy constants K_1 and K_2. It is found that a' is

$$-- J_s^2 / \left(2K_1 + \frac{3}{2} K_2 \right)$$

or $J_s^2 / 3K_1$, depending upon whether the easy axis be a cube diagonal or cube edge.

(20) Laws of magnetization for a multidomain material: The results obtained above relating to the laws controlling the displacement of an isolated plane wall are now applied to the case of wall movement in a material containing many walls separating elementary domains.

We assume first, with Landau, that the domains adopt a configuration which eliminates free poles inside the body: when the shape-dependent demagnetizing field is zero, *the internal field is then everywhere equal to the external field*. We shall neglect variations in elastic deformation due to magnetostriction. We shall also take the elementary domains as given, without prejudging the mechanism which controls their number and assume that they are independent.

A) Single crystal uniaxial material

The simplest case is that of a *uniaxial* material in which the *c-axis coincides with the easy direction of magnetization*. Assume that the magnetic field acts parallel to the axis. The domains can be imagined as a stack of sheets alternately magnetized in opposite senses and separated by a series of 180° walls. We suppose that the walls possess unique values l and P_0 of the characteristic constants (cf. § 6) throughout the material. When the state of the material changes from saturation in one sense to saturation in the opposite sense the walls *sweep* out the whole volume V of the body. The following relation therefore exists between the total surface wall area, S, and the average total excursion, L, of a wall during magnetization reversal,

$$V = LS. \qquad (32)$$

If the reduced quantities of magnetization and field in equation (31) are replaced by their real values and noting that $\bar{J} - \bar{I} = 2 J_s$, then the magnetization per unit volume J on the initial magnetization curve is

$$J = \alpha J_s \left(\frac{\lambda J_s}{P_0} H + \frac{2 J_s^2}{\pi P_0^2} H^2 \right), \qquad (33)$$

where J_s is the saturation magnetization and $\alpha = \dfrac{8 l S}{V}$.

From equation (32) it follows that $\alpha = \dfrac{8 l}{L}$ which represents the *reciprocal of the number of minima of the characteristic function contained within the average total path of a wall*.

B) Polycrystalline uniaxial material

In this case also the magnetization is given by equation (33) but not J_s^2 and J_s^3 must be replaced by the values of $J_s^2 \cos^2 \alpha$ and $J_s^3 \cos^3 \alpha$ averaged over a hemisphere. These averages are $J_s^2/3$ and $J_s^3/4$ respectively and so

$$J = \alpha J_s \left(\frac{\lambda J_s}{3 P_0} H + \frac{J_s^2}{2 \pi P_0^2} H^2 \right). \qquad (34)$$

C) Cubic material

The problem is much more complicated for a cubic material because there are two types of walls, viz 90° and 180° walls. Not surprisingly each possesses different characteristic values[8], l and P_0. The fractions, V_1 and V_2 of the total volume swept out by both types of walls must also be introduced, making 5 parameters in all. Such a complication at this stage is premature and we confine ourselves to the use of equation (34), in which α and P_0 represent average values for all possible walls.

(21) The movement of walls in strong fields: The theoretical formula (34) is only valid for weak fields, i.e. small compared with $P_0/\mathcal{J}_s$. We now consider the inverse problem of a field greater[9] than $P_0/\mathcal{J}_s$. We assume that in the initial zero field state the wall is situated at one of the extreme points of its total free path and, furthermore, that this position is arbitrary vis-à-vis the structure of the characteristic function. A fairly strong field is now applied and we calculate the mean displacement of the wall — but limited to the case where the displacement is of the order of L which follows if we assume that l/L, i.e. α, is very small. In this case the reversible part of the displacement, which is of the order of l, is negligible compared with the irreversible part. This is in contrast to the Rayleigh region. It is sufficient here to use the polygon contour as the characteristic function.

In an applied field h the wall will move until it meets an obstacle, i.e. a side with slope greater than h. If N be the number of sides of the polygonal contour contained within the path length L, the number of obstacles n, is

$$n = N \int_h^{+\infty} \frac{1}{\sqrt{\pi}} e^{-p^2} dp = N\theta_h \qquad (35)$$

in the notation of equation (16), (§ 12).

It can easily be shown (equation (50) of Appendix 2) that the mean free path, t, of the wall is given by

$$t = \frac{L}{n}(1 - e^{-n}). \qquad (36)$$

(22) Application to the coercive field: Let the initial state defined in the preceding section correspond to the ideal remanent magnetization. To obtain the state of zero magnetization the wall must move a distance of L/2 since a displacement of L produces saturation in the opposite sense. The average number of obstacles corresponding to this free path of L/2 is obtained by

---
[8] Furthermore, for 90° walls, $\mathcal{J}_s$ in equation (34) must be replaced by $\mathcal{J}_s/\sqrt{2}$.
[9] Rotational processes are still assumed to be negligible.

letting $t = L/2$. A solution of this transcendental equation furnishes $n = 1.6$. The corresponding magnetic field is the *coercive field*.

On substituting this value of n into equation (35) and noting that $N = L/2l = 4/\alpha$, we obtain

$$\theta_h = 0.4\,\alpha, \qquad (37)$$

with a unique solution $h = r$ which is a number equal to the reduced coercive field. It is a slowly varying function of α: when α varies from 0.2 to 0.005, the quantity only changes from 1 to 2.5.

A) Single crystal uniaxial material

In this case the reduced value of the coercive force r equals $2\,\mathcal{J}_s H_c/P_0$, where H_c is the actual value of the coercivity. Equation (33), valid for weak fields, can then be written in the form

$$\mathcal{J} = \alpha \mathcal{J}_s \left(\frac{\lambda r}{2} \frac{H}{H_c} + \frac{r^2}{2\pi} \frac{H^2}{H_c^2} \right). \qquad (38)$$

B) Polycrystalline uniaxial material

To take account of the isotropic distribution of crystallographic axes over the unit sphere the quantity $\mathcal{J}_s$ in the reduced coercive field $\mathcal{J}_s H_c/P_0$, must be replaced by $\mathcal{J}_s \cos\alpha$ averaged over a hemisphere. We have therefore $\mathcal{J}_s H_c = rP_0$ and equation (34) becomes

$$\mathcal{J} = \alpha \mathcal{J}_s \left(\frac{\lambda r}{3} \frac{H}{H_c} + \frac{r^2}{2\pi} \frac{H^2}{H_c^2} \right) \qquad (39)$$

This formula would apply equally well to a cubic material like iron (where the cube edges are easy directions) if the domain structure consisted of groups of plane parallel slabs separated by 180° walls — as in Fig. 11.

Fig. 11.

For the analogous slab structure separated by 90° walls, a formula similar to (39) is obtained with $\mathcal{J}_s$ simply replaced by $\mathcal{J}_s/\sqrt{2}$.

(23) Initial rotational susceptibility: The best experimental check on the preceding formulae and one which corresponds closest to theory requires a single crystal with a unique easy axis. This would be the case for a cobalt single crystal along the c-axis. In the absence of such measurements we may examine the conditions under which the formulae could be applied to cubic crystals.

A first condition demands that in the field regime studied, the change in magnetization produced by reversible rotation inside the domains must be small compared with changes produced by wall motion. In the case of a cubic material we express this rotation as a function of the anisotropy constants K_1 and K_2. By definition, the magnetocrystalline energy density, W, can be written in terms of the direction cosines of the magnetization (with respect to the cube edges) as

$$W = K_1(\alpha^2\beta^2 + \beta^2\gamma^2 + \gamma^2\alpha^2) + K_2\alpha^2\beta^2\gamma^2,$$

where terms of 8th degree and above are ignored. We assume now that, in the initial state, the magnetization directions in the various domains which constitute the polycrystal are distributed isotropically over a unit sphere. The resultant magnetization is zero but an applied field provokes a slight rotation of the magnetization vectors which gives rise to a magnetization component parallel and proportional to the field,

$$\mathcal{J} = a'H.$$

The calculation of a' presents no difficulty. The result depends upon whether [111] is an easy axis (the case of nickel) or [100] is an easy axis (the case of iron)

$$[111]\ a' = -\frac{\mathcal{J}_s^2}{2K_1 + \frac{2}{3}K_2}\ ;\quad [100]\ a' = \frac{\mathcal{J}_s^2}{3K_1}. \tag{40}$$

For iron, with $K_1 = 4.4 \times 10^5$, we find that $a' = 2.2$. In the case of nickel for which $K_1 = -5 \times 10^4$ and $K_2 = 0$ we have $a' = 3$. The experiments of Radovanovic [7] and Kahan [8] on pure nickel give initial susceptibilities at room temperature in the range 2 to 11. Thus there are instances where the initial susceptibility of nickel seems due entirely to rotation. As a result nickel can hardly be recommended for a study of all motion. The same can be said of the Fe_2Ni and Fe_2Co alloys studied by Weiss and Freudenreich [9]. They obtained room temperature values of $a = 9.4$ and $a = 7.55$ respectively for the two materials. Using the appropriate anisotropy constants the calculated rotational susceptibilities are (approximately) $a' = 10$ and $a' = 11$. Once again the rotation masks wall displacement.

The situation is much more satisfactory with regard to iron because the initial susceptibility of ordinary soft iron is about 30, whereas the rotational susceptibility is 2.2. For this reason iron was chosen to verify the theory. Ideally, a suitable correction to the observed values should be made for the rotational susceptibility. Unfortunately this is not possible because the latter depends on the values of the anisotropy constant K_1 which in turn depend upon deformation and, in steels, the percentage of carbon. In principle K_1 values obtained from the $1/H^2$ term in the law of approach to saturation could be used (cf A, p. 167) but the data is unavailable. We shall not, therefore, attempt to make a correction which might be spurious.

(24) Power relation between a and b, independent of H_c: A comparison of equations (3) and (39) leads to the following relation for a polycrystalline material:

$$\frac{a^2}{b\mathcal{J}_s} = \frac{2\alpha\pi}{9}\lambda^2 = 0{,}45\,\alpha. \tag{41}$$

This allows α to be deduced from the known experimental values of a, b and $\mathcal{J}_s$: in fact α varies from 0.001 for very well annealed soft iron to 0.2 for quenched steels and magnet steels. In very soft iron therefore one wall exists for 1000 minima of the characteristic function. However, it does not necessarily follow that the number and total surface area of walls are smaller for soft iron than for steel because the scale of deformation could be smaller in the former and the minima closer together.

Experiment shows that a and b increase with temperature especially in the vicinity of the Curie point. Weiss and his co-workers [8] have demonstrated in certain cases, for a given material, power relations between a and b, independent of temperature. Thus for iron, Renger [10] found

$$b = 0{,}068\,a^{5/3}.$$

A macroscopic interpretation of such a relation must assume, to begin with, that the number of elementary domains is independent of temperature. This supposition appears valid: furthermore, when the temperature is raised, strains must change the amplitude of variations in the characteristic function without appreciable alteration of the number of minima. As a result α is independent of temperature, or nearly so at least. In addition $\mathcal{J}_s$ varies much less, relatively, than a and b, except near the Curie point. Under these conditions equation (41) predicts b to be proportional to a^2. Having regard to the number of approximations that have been made the experimental result of a 5/3 power law gives satisfactory agreement with the theoretical result of 2.

TABLE I.

$\dfrac{1}{\alpha}$	r	$\dfrac{3r}{2\pi\lambda} = \dfrac{bH_c}{a}$
5	0,99	0,58
10	1,24	0,74
20	1,45	0,86
50	1,70	1,00
100	1,87	1,12
200	2,04	1,22
500	2,23	1,33
1 000	2,37	1,41
2 000	2,50	1,49

(25) Relation between a, b, and H_c: According to equations (3) and (39) we have

$$\frac{a}{b} = \frac{2\pi\lambda}{3r} H_c \, . \qquad (42)$$

The constant r again depends on α but not markedly so — see Table I. Equation (42) therefore constitutes a base for checking the theory. Unfortunately, a, b and H_c have rarely been measured together.

Table II, drawn up using the results of Gumlich and Rogowski [11] allows some relative comparisons between carbon and silicon steels. Generally speaking, the experimental value of bH_c/a is between 2 and 5 times greater than the calculated value, but varies within relatively constricted limits as predicted by the theory. On the contrary, the value of b varies a thousand fold between specimens.

However, results from experiments being carried out in our laboratory by Glinski indicate that the theory works much better for modern magnet steels with higher α values.

A close study shows that the discrepancies between theory and experiment cannot be attributed to either the particular and schematic form of the domains used or to uncertainties in the average orientation of the walls with respect to the magnetization or to a mixture of several categories of walls with different values of P_0. These various effects, even if they were to modify α by a factor of 3, would only change bH_c/a by 10–20%. This inconsistency between theory and experiment is more likely to be related to defects in the schematic representation of the characteristic function.

Better agreement between theory and experiment can be arrived at even if the concept embodied in the first approximation is retained, i.e. that of the polygonal contour. This is achieved by using a distribution law for the slopes which differs slightly from that of Gauss. Hence if law (8) is replaced by

$$\pi(p) = \frac{1}{1 + |p|^n}$$

the relative proportion of slopes with a higher modulus is greatly increased vis-à-vis the Gaussian law. This leads to a marked increase of H_c and consequently, of the ratio bH_c/a. Excellent results are obtained for $n = 3$. Obviously it is impossible to establish, *a priori*, whether a Gaussian law gives the best possible description of the complex inhomogeneities in the internal stress distribution.

Furthermore, even if a Gaussian law is adopted it is possible to imagine a correlation between the slopes of two adjacent sides which would tend to make them

TABLE II.

Composition		Treatment	a	b	H_c	$\dfrac{bH_c}{a}$	$\dfrac{1}{\alpha}$
% C	% Si						
0,044	0,004	well annealed	25	240	0,37	3,5	600
0,29	4,45	annealed	40	450	0,66	7,3	470
		unannealed	36	190	1,25	6,6	260
0,56	0,18	annealed	10,3	3,6	7,1	2,5	60
		quenched	4,6	0,28	44,3	2,7	20
0,99	0,10	annealed	5,8	0,20	16,7	0,6	10
		quenched	3,4	0,27	52,4	4	40
3,1	3,27	annealed	14	5,7	4,6	1,9	50

neighbours of one another. Under these conditions the relative proportion of CN²A type domains is increased and hence the value of bH_c/a.

It is difficult to appreciate the respective role played by these two effects. Until more information is forthcoming it is best to treat r as an *empirical numerical coefficient*. Since, a priori, r varies little with the nature of the material and above all with temperature there would appear to be scope for a study of widely different materials to determine whether (i) the values of r are coherent and (ii) they are a function of a single variable α. Similarly for a given material with negligible rotation, it would be interesting to investigate the temperature variation of a, b and H_c.

(26) Variation of the reversible susceptibility with magnetization: With a homogeneous distribution of small deformations the initial reversible susceptibility is proportional to the surface area of the wall. For simplicity we have adopted a layered domain structure such that the movement of a wall involves no change in area. This hypothesis is certainly not correct. Consider, for example, the case of a single crystal with one easy axis; when saturation in one sense is attained, domains magnetized in the opposite sense will not exist and the surface area of wall is zero. In fact this is not strictly true because, in all probability, very small reverse *nuclei* still remain which will serve as seeds for the demagnetization. The wall area of these domains is negligible. Therefore the area of the wall, essentially zero at $\mathfrak{J} = \pm \mathfrak{J}_s$, must pass through a maximum at $= 0$. Current theory finds it difficult to describe the laws governing this variation. In work on the reversible magnetization of ferromagnetic crystals in weak fields Kondorsky [12] assumes that the area of walls between two phases i and k which occupy volume fractions v_i and v_k of the material is proportional to $v_i v_k$. This rather arbitrary (despite the justification that Brown has tried to give it [13]) hypothesis does, nevertheless, give the general gist of the phenomena. The theory developed above assumed a constant surface area of walls: *it is therefore valid only in the vicinity of $\mathfrak{J} = 0$*, i.e. when the area becomes a maximum and can be considered as constant. Moreover, the maximum is fairly flat, since, according to Kondorsky's hypothesis, the area diminishes by only a quarter when the magnetization has reached half the remanence value.

The calculation of the coercive field is not changed noticeably on account of this new view point since it depends (logarithmically) on the free path of the walls and not on the wall area. Now, for the same volume swept out and the same initial area at zero magnetization, the free linear path of a wall is only 50% larger for a variable wall area than for one that is fixed.

This makes only a few percent difference to the coercive field and is therefore negligible.

A calculation to determine the variation in initial susceptibility along the initial magnetization curve (and hence to recover the semi-empirical result of Gans) must take changes of the total wall area into account. A route similar to that opened by the work above should be followed.

APPENDIX I
THE PROBABABILITY THAT A GIVEN DOMAIN WILL CAPTURE A WALL DURING DEMAGNETIZATION

Consider a domain of the type CB^nA where p' and p are the slopes of the sides C and A ($p' < -\varepsilon$; $p > \varepsilon$). Assume that this domain is sited at a point m on the segment DE which represents the polygonal contour and within which the wall can move. Without loss of generality we can assume that the slopes of the N sides of DE are all different and denote by 2η a positive number smaller than the modulus of all the differences between the slopes of two arbitrary sides. The demagnetizing process is defined as follows: let H be a positive number with a modulus greater than all the slopes. The demagnetizing field, h, is ascribed the following alternating positive and negative values:

$$H, -(H-\eta), H-2\eta, -(H-3\eta), H-4\eta, \text{etc.}$$

until $|h|$ is smaller than ε.

Between D and m there are n' sides having slopes a' with a modulus greater than $|p'|$; similarly between m and E there are n sides having slopes a with a modulus greater than $|p|$. We now write the slopes of all sides between D and E in the order in which they are met.

$$a'_{n'} \ldots a'_3, a'_2, a'_1, p', p, a_1, a_2 \ldots a_n \quad (43)$$

with the condition $|a'_j| > |p'|$ for all j and $|a_k| > |p|$ for all k. All values of a and a' in the series (43) are deleted which do not fulfil the following conditions: $|a'_j| > |a'_k|$ and $|a_j| > |a_k|$ for all $k < j$. A new series is obtained,

$$b'_{k'} \ldots b'_2, b'_1, p', p, b_1, b_2, \ldots, b_k, \quad (44)$$

in which the b and b' values satisfy the following condition

$$|b'_k| > \cdots > |b'_2| > |b'_1| > |p'|$$
$$|b_k| > \cdots > |b_2| > |b_1| > |p|. \quad (45)$$

Theorem

In order that, after demagnetization, the wall shall be found in CB"A i.e. between p and p', it is a necessary and sufficient condition that all b' values be negative and all b values be positive.

The condition is sufficient. If the final position were not between p' and p it would be elsewhere, for example between b_j and b_{j+1}. Now let $H - 2r\eta$ and $H - 2(r+1)\eta$ be the two positive values of the demagnetizing field which include b_j. When $h = H - 2r\eta$, let us suppose that the wall lies between b_j and b_{j+1}: it will leave there and move leftwards beyond p' for a value of $-[H - (2r+1)\eta]$ and when $h' = H - 2(r+1)\eta$ it will no longer return to the interval $b_j b_{j+1}$ since $b' < b_j$. Regardless of j, at the end of the demagnetization the wall cannot be situated between b_j and b_{j+1}: it must therefore lie between p and p'.

The condition is necessary. In fact, let it be supposed that the condition is not fulfilled. From the positive b' slopes and the negative b slopes choose the one with the greatest modulus, e.g. b_j. Enclose b_j by the values:

$$H - (2r+1)\eta < |b_j| < H - (2r-1)\eta < |b_{j+1}|.$$

For $h = H - 2r\eta$ the wall lies between b_j and b_{j+1}; for $h = -[H - (2r+1)\eta]$ the wall cannot cross b_j since

$$b_j < -[H - (2r+1)\eta] \ ;$$

Thereafter it always remains between b_j and b_{j+1}.

A series a' will be referred to as *good* when the series of b' values extracted from it includes no positive term: a series a is good when the corresponding series b includes no negative term. *In order that the wall lies ultimately between C and A, it is necessary therefore that the two series a and a' be good* — as indicated by the theorem.

Three obvious lemmas may be stated.

Lemma I.
If a series a is good, it remains so after the addition of an arbitrary positive term in arbitrary position.

Lemma II.
If a series a is *good*, it remains so after the addition of a negative term c with modulus less than a_1, in an arbitrary location, provided that it is not to the left, between p and a_1, in which case the series becomes bad if $|c| > p$ but remains good if $|c| < p$.

Lemma III.
A good series remains so if terms are removed from it.

The first and second lemmas justify the fact of not having included terms in the series a of modulus less than p.

The series a comprises n terms. The $n!$ possible permutations are equiprobable. We shall determine the number of good permutations. We first write the slopes in the order of increasing modulus

$$c_1, c_2 \ldots \ldots c_n \quad \text{with:} \quad |c_1| < |c_2| < \cdots < |c_n|$$

Let us assume that we have determined the number Q_p of good permutations corresponding to the last p terms and evaluated Q_{p+1} as a function Q_p. If the term c_{n-p} is removed from a good permutation of the last $p+1$ terms, then according to lemma III, a good permutation of the last p terms is obtained. All good permutations of the last $p+1$ terms will therefore be obtained by addition of c_{n-p} to the good permutations of the last p terms. Dependent upon the sign of c_{n-p}, two cases (equiprobable, *a priori*) may occur because the probability function is even. If $c_{n-p} > 0$, then c_{n-p} can be added in $(p+1)$ different positions: if $c_{n-p} < 0$, only p positions are possible because according to lemma II, the position to the left is not good. *On average*, therefore, we have

$$Q_{p+1} = \left(p + \frac{1}{2}\right) Q_p,$$

from which

$$Q_n = \frac{1}{2} \cdot \frac{3}{2} \cdots \cdots \frac{2n-1}{2}.$$

The ratio of the number of good permutations to the total number is hence

$$\frac{Q_n}{n!} = \frac{1 \cdot 3 \cdots \cdots 2n-1}{2 \cdot 4 \cdots \cdots 2n}.$$

According to a formula of Wallis, we have,

$$\frac{1 \cdot 2 \cdots \cdots 2n-1}{2 \cdot 4 \cdots \cdots 2n} = (1 + \epsilon) \sqrt{\frac{2}{(2n+1)\pi}}$$

where ϵ tends to 0 as n tends to ∞.

When n and n' are large, the probability ρ that the domain CB"A contains a wall is hence

$$\rho_{nn'} = \frac{Q_n}{n!} \cdot \frac{Q_{n'}}{n'!} = \frac{1}{\pi} \frac{1}{\sqrt{nn'}}. \tag{46}$$

We denote by N_1 and N_2 the number of sides of the polygonal contour DE whose moduli are greater than $|p|$ and $|p'|$ respectively. Putting $x = mE/DE$ we have on average

$$n = N_1 \cdot \frac{mE}{DE} = N_1 x \quad \text{et} \quad n' = N_2 \frac{Dm}{DE} = N_2(1-x).$$

The relation (46) can then be written as

$$\rho_{nn'} = \frac{1}{\pi\sqrt{N_1 N_2}} \cdot \frac{1}{\sqrt{x(1-x)}}. \quad (47)$$

Furthermore, since *a priori* m can occupy any position along DE, the mean value of $\rho_{nn'}$ is obtained by taking the average value of $1/\sqrt{x(1-x)}$ in this interval. This average is π so that

$$\bar{\rho}_{nn'} = \frac{1}{\sqrt{N_1 N_2}}. \quad (48)$$

For the case of a Gaussian law the number, N_1, of obstacles along the contour DE with modulus greater than $|p|$ is (according to the notation[10] of section 12) $2N\theta_p$. Similarly $N_2 = 2N\theta'_p$. Equation (48) therefore becomes

$$\bar{\rho}_{pp'} = \frac{1}{2N}(\theta_p \theta_{p'})^{-\frac{1}{2}}. \quad (49)$$

Expression (48) is general. Its validity is independent of the probability function $\pi(p)$ which governs the distribution of slopes, provided that $\pi(p)$ is an even function of p. It shows that for two domains situated on DE the one which is enclosed between the steepest sides i.e. has the biggest values of $|p|$ and $|p'|$, has the greater chance of capturing a wall during the demagnetization.

APPENDIX II
CALCUALTION OF THE FREE PATH

Let us assume that the obstacles are distributed randomly along an infinite line such that *on average* there are n in a length L. Imagine now an arbitrary segment AB of length L, chosen from the infinite line, and let C be the obstacle closest to A situated between A and B. In the absence of any obstacle between A and B then B takes on the role. It is proposed to calculate the mean value of the path AC, equal to the free path.

[10] N.B.

$$\theta_x = \int_x^\infty \frac{1}{\sqrt{\pi}} e^{-x^2} dx.$$

Let A be the origin and p the probability of finding no obstacle between A and a point D with abscissa l. The probability $p + dp$ of finding no obstacle between A and a point D' with abscissa $l + dl$ is the product of the probability p of not finding an obstacle between A and D and the probability of finding none along DD'. Obviously the latter is $(1 - (n/L)dl)$. Therefore, we have finally $dp = -(n/L)p \, dl$ which after integration yields $p = Ce^{-n/L}$.

As p must tend to 1 as l tends to zero we may deduce that $C = 1$. Now take a finite segment AB of length L. The probability of finding no obstacle between A and B is e^{-n}. The corresponding mean path is Le^{-n}.

When obstacles exist between A and B, the probability of a path between l and $l + dl$ is equal to the product of the probability of finding no obstacle between A and D, i.e. $e^{-n/L}$, and the probability of finding an obstacle between l and $l + dl$, i.e. $(n/L)dl$. The corresponding path is therefore

$$\int_0^L l e^{-\frac{n}{L}l} \frac{n}{L} dl = \frac{L}{n} - Le^{-n} - \frac{L}{n}e^{-n}.$$

After adding to the mean path above that which corresponds to the case where there are no obstacles between A and B, the total average path, t, is obtained as

$$t = \frac{L}{n}(1 - e^{-n}). \quad (50)$$

As was to be expected, this excursion is equal to L for $n = 0$ and tends to L/n when n is large.

REFERENCES
[1] Lord Rayleigh, *Phil. Mag.*, 1887, **23**, 225.
[2] P. Weiss and J. de Freudenreich, *Arch. Sc. Phys. Nat.*, Genève, 1916, **42**, 449.
[3] F. Preisach, *Z.f. Physik*, 1935, **94**, 277.
[4] F. Bloch, *Z.f. Physik*, 1932, **74**, 295.
[5] W. Doring, *Z.f. Physik*, 1938, **108**, 137; W. Doring and H. Haake, *Phys. Z.*, 1938, **39**, 865.
[6] Kersten, *Phys. Z.*, 1938, **9**, 860.
[7] D. Radovanovic, *Arch. Sc. Phys. Nat.*, Genève, 1911, **32**, 315.
[8] T. Kahan, *Ann. de Phys.*, 1938, **9**, 105.
[9] P. Weiss and J. de Freudenreich, *Arch. Sc. Phys. Nat.*, Genève, 1915, **39**, 125; 1916, **42**, 5.
[10] K. Renger, Thèse, Zurich, 1913; P. Weiss and K. Renger, *Archiv. f. Elektrotechnik*, 1914, **2**, 406.
[11] E. Gumlich and W. Rogowski, *Ann. der Physik*, 1911, **34**, 235.
[12] E. Kondorsky, *Phys. Rev.*, 1938, **53**, 319 and 1022.
[13] W. F. Brown, *Phys. Rev.*, 1949, **75**, 568.

A44 (1942)
THE RAYLEIGH LAWS FOR SOME MAGNET STEELS
with
G. GLINSKI

The analysis of the displacement of walls which separate elementary Weiss domains in a medium characterized by irregular perturbations has provided one of the authors [1] with a theoretical interpretation of the Rayleigh laws [2]. These are concerned with a description of the magnetization behaviour in weak fields. The properties of a ferromagnetic subjected to a field H small compared with the coercive field H_c, are completely defined by the coefficients a and b which appear in the initial magnetization curve: $\mathcal{J} = aH + bH^2$. By means of various approximations the theory shows that on putting $bH_c/a = 0.42r$ and $a^2/b\,\mathcal{J}_s = 0.91\alpha$, where $\mathcal{J}_s$ is the saturation magnetization, the two *numbers* r and α are linked by the relation

$$\int_r^{+\infty} \frac{1}{\sqrt{\pi}} e^{-p^2}\, dp = 0{,}4\,\alpha. \tag{1}$$

This equation must be tested experimentally. A great deal of experimental data shows that the value of α ranges between 1/1000 for soft materials and several tenths for the best magnet steels. According to equation (1) the corresponding variation in r is much smaller: from 2 to 1. Unfortunately, apart from some already ancient work of Gumlich and Rogowski [3] there is a lack of simultaneous measurements of a, b and H_c. In order to fill this gap one of the authors (Glinski) has made a study (with a magneto-galvanometer) of a certain number of ferromagnetic alloys in the form of fairly elongated ellipsoids. The results of the study are summarized in the table where a and b are represented per gram of each material.

	H_c	a	b	α	bH_c/a
unannealed electrolytic iron	2,9	2,65	0,9	0,024	0,98
tungsten steel	52	0,44	0,003 8	0,32	0,45
cobalt steel	116	0,169	0,000 68	0,31	0,47
Ni–Al	410	0,048	0,000 059	0,41	0,50

The values of α are not very accurate because the saturation values were not determined with precision but this has negligible repercussion on the evaluation of r. With α values of 0.4, 0.2 and 0.02 respectively, the theory shows that the ratio bH_c/a ($=0.42r$) must be 0.37, 0.42 and 0.71. The experimental values, shown in the last column of the table, are very close to their theoretical counterparts and vary in the same sense. This result is all the more remarkable because the theory is very schematic and the coefficients α, b and H_c vary between extremely wide limits: thus b varies by a factor of 20,000.

These preliminary but satisfactory results indicate that the proposed theory constitutes a useful guide to the experimental study of ferromagnetic materials in weak fields. In the case of the magnet steels so far studied even numerical agreement has been obtained. However, the agreement is less good for the very soft steels investigated by Gumlich and Rogowski. But it now seems possible to unravel with success the vast experimental domain of weak fields.

REFERENCES

[1] L. Néel, *Cah. Phys.*, 1942, **12**, 1; 1943, **13**, 18.
[2] *Phil. Mag.*, 1887, **23**, 225.
[3] *Ann der Phys.*, 1911, **34**, 235.

A THEORETICAL TREATMENT OF THE EFFECT OF THE DEMAGNETIZING FIELD ON ANHYSTERETIC MAGNETIZATION

Abstract: When a magnetic material with constant susceptibility a_0 is subjected to a demagnetizing field characterized by a coefficient N, the effective susceptibility a is given by

$$\frac{1}{a} = \frac{1}{a_0} + N.$$

The author shows that this formula is no longer valid when irreversible phenomena are involved. In the particular case of anhysteretic magnetization, the new anhysteretic susceptibility c, in the presence of a demagnetizing field, is given by the following formula

$$c = \frac{1}{N}(1 - e^{-c_0 N}).$$

where c_0 is the anhysteretic susceptibility of a closed circuit. As soon as $c_0 N$ exceeds a value of several units then, to all intents and purposes, the effective anhysteretic susceptibility becomes equal to $1/N$.

(1) Ideal or anhysteretic magnetization: Regardless of its initial state a ferromagnetic body can be reduced to a final state of zero magnetization by exposure to an a.c. field. Initially the field strength must be greater than the coercive force but is then slowly reduced to zero. If a constant d.c. field H_0 is superimposed on the a.c. field the body retains at the end of the process a magnetization $\mathcal{J}_0$ which depends only on H_0. This magnetization, termed *ideal* or *anhysteretic*, depends neither on the initial state nor on the way the amplitude of the a.c. field is decreased as a function of time — as long as the decrease is slow. If $\mathcal{J}_0$ is plotted as a function of H_0 the anhysteretic magnetization curve (*b*) is obtained (Fig. 1).

In the vicinity of the origin $\mathcal{J}_0$ is proportional to H_0 and we can write

$$\mathcal{J}_0 = c_0 H_0. \qquad (1)$$

We shall call c_0 the *anhysteretic susceptibility*. It is generally 20 to 50 times greater than the initial reversible susceptibility: in effect when the process is completed and the field H_0 removed, the magnetization decreases insignificantly from $\mathcal{J}_0$ to $\mathcal{J}_0'$ (Fig. 1).

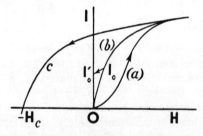

Fig. 1.
(a) Initial magnetization curve. (b) Anhysteretic curve.

(2) The effect of a demagnetizing field: It has been assumed that the procedures above were effected in a closed magnetic circuit. Let us now suppose the existence of a demagnetizing field H_d, proportional to the magnetization $\mathcal{J}$ and directed in the opposite sense

$$H_d = -N\mathcal{J}. \qquad (2)$$

For a material whose magnetization is given *reversibly* by the formula

$$\mathcal{J} = a_0 H, \qquad (3)$$

260

the effect of the demagnetizing field is obtained by replacing H in equation (3) by $H - N\mathcal{J}$. On resolving with respect to $\mathcal{J}$ we obtain

$$\mathcal{J} = aH, \quad (4)$$

where the new apparent susceptibility a is related to the actual value a_0 by the formula

$$\frac{1}{a} = \frac{1}{a_0} + N. \quad (5)$$

It is tempting to apply the same reasoning to the anhysteretic susceptibility and write a relation analogous to (5) between the true susceptibility c_0 and its apparent value c in the presence of a demagnetizing field. This would not be correct, as is proved by experiment. In order to be exact about this point we examine more clearly the nature of anhysteretic magnetization.

(3) The process by which anhysteretic magnetization is acquired: The anhysteretic magnetization $\mathcal{J}_0$ is acquired *progressively* as the alternating field is gradually reduced. To begin with all the elementary domains participate in magnetization reversals. Then, inside some domains, the magnetization becomes fixed in a definite direction: these domains are, as it were, *frozen* whilst the magnetization in the remainder continues to alternate. As the a.c. field is gradually diminished the number of frozen domains increases until they are all frozen. This property is particularly obvious for an elementary domain possessing a symmetric rectangular loop. The magnetization of such a domain can only change sign for as long as the a.c. field is greater than the coercive field. When it becomes smaller the magnetization remains constant at either $+\mathcal{J}$ or $-\mathcal{J}$; thereafter it is *frozen*.

As the a.c. field decreases the total magnetization of the frozen domains increases from zero to the final value of $\mathcal{J}_0$, whereas the magnetization of the remaining domains adopts alternating values of $+\mathcal{J}'$ and $-\mathcal{J}'$, where $\mathcal{J}'$ decreases slowly with time from saturation to zero.

We denote by $gd\lambda$ the fraction of the magnetization which freezes when the absolute value of the a.c. field decreases from λ to $\lambda - d\lambda$: g is a function of λ and H_0. By definition

$$\mathcal{J}_0 = \int_\infty^0 g\, d\lambda. \quad (6)$$

If H_0 is sufficiently small, we may put $g = kH_0$ where k is simply a function of λ. Taking note of (1) we then have

$$c_0 = \int_\infty^0 k\, d\lambda. \quad (7)$$

(4) Introduction of the demagnetizing field: The effect of a demagnetizing field is now examined. We denote by $\mathcal{J}$ and $\mathcal{J}'$ the fractions of the magnetization which are frozen and reversing respectively at a certain moment in the demagnetization when the external a.c. field is changing from $+h$ to $-h$. The magnetization therefore varies between $\mathcal{J} + \mathcal{J}'$ and $\mathcal{J} - \mathcal{J}'$ and the *total field* which acts on the body varies between $H_0 + h - N(\mathcal{J} + \mathcal{J}')$ and $H_0 - h - N(\mathcal{J} - \mathcal{J}')$; it is the sum of an a.c. field with amplitude λ equal to $h - N\mathcal{J}'$ and a constant field of $H_0 - N\mathcal{J}$. As far as the a.c. field is concerned, it is as if the amplitude were passing from h to $h - N\mathcal{J}'$, but this has little importance since the initial amplitude is always very large and the final amplitude is zero, $\mathcal{J}'$ being zero at the end of the process. In fact we have already noted that modification to the mode of decrease of the a.c. field does not change the result of the experiment.

On the other hand since the actual field acting is $H_0 - N\mathcal{J}$ the fraction $d\mathcal{J}$ of frozen magnetization acquired while the amplitude of the a.c. field decreases from λ to $\lambda - d\lambda$ is

$$d\mathcal{J} = k(H_0 - N\mathcal{J})\, d\lambda \quad (8)$$

When λ changes from $+\infty$ to zero the frozen magnetization grows from zero to the final, desired, value of $\mathcal{J}_0^d$: from which we find after integrating and taking account of relation (7)

$$\int_0^{\mathcal{J}_0^d} \frac{d\mathcal{J}}{H_0 - N\mathcal{J}} = \int_\infty^0 k\, d\lambda = c_0. \quad (9)$$

Carrying out the integration and re-arranging we obtain

$$\mathcal{J}_0^d = (1 - e^{c_0 N})\frac{H_0}{N}. \quad (10)$$

The new apparent anhysteretic susceptibility c, defined by

$$\mathcal{J}_0^d = cH_0, \quad (11)$$

is thus linked to the true susceptibility c_0 by the relation

$$c = \frac{1}{N}(1 - e^{-c_0 N}). \quad (12)$$

This relation is completely different from that of (5) pertinent to reversible susceptibilities. Fig. 2 is a graph

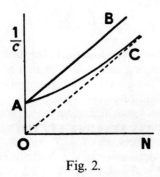

Fig. 2.

in which the variation of $1/c$, the reciprocal of the apparent anhysteretic susceptibility, is plotted as a function of N. Let OA be the initial value $1/c_0$ corresponding to $N=0$. If the phenomenon were reversible we would obtain the straight line AB parallel to the bisector OC. In fact the relation (12) is represented by AC which is asymptotic to OC: as soon as $c_0 N$ exceeds a value of several units, the apparent anhysteretic susceptibility becomes practically equal to $1/N$ and *no longer depends on the nature of the material*.

These theoretical predictions have been confirmed by experiment[1].

[1] Néel, Forrer, Janet and Baffie. See paper A48.

A48 (1943)
ANHYSTERETIC MAGNETIZATION AND THE DEMAGNETIZING FIELD.
EXPERIMENT AND THEORY
with R. FORRER, Mlle N. JANET and R. BAFFIE

Abstract: The authors bring into relief the discrepancies between the various experimental determinations of the initial anhysteretic susceptibility c_0. Whereas closed magnetic circuit techniques using toroid or yoke furnish finite values, methods using rods or ellipsoids give, after correction for the demagnetizing field, infinite values. This is because the correction in question is not exact in its usual form when, as in the case here, irreversible phenomena are involved.

A theory of Néel explains the discrepancies: the initial anhysteretic susceptibility c, without corrections, is given as a function of the demagnetizing factor by the formula $c = [1 - \exp(-c_0 N)]/N$. As soon as N becomes moderately large, c is practically equal to $1/N$, which gives an effective value of infinity according to the classic correction. In addition measurements of the anhysteretic susceptibility as a function of the demagnetizing field have been made on two types of magnet steels: these confirm the proposed formula.

Definition: The *anhysteretic magnetization* is defined as the magnetization, $\mathcal{J}$, acquired by a body placed in a constant field H and subjected to an a.c. field whose amplitude, initially large compared with the coercive field, is slowly decreased to zero. The magnetization depends only on H and is zero when $H = 0$. We shall denote by *anhysteretic susceptibility*, the ratio $\mathcal{J}/H$.

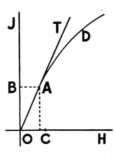

Fig. 1.

Experiments of Steinhaus and Gumlich: The first work on initial anhysteretic susceptibility in weak fields, H, was done by Steinhaus and Gumlich [1]. They made a magnetometer study of moderately elongated ellipsoids with an aspect ratio of about 20 to 1. The initial tangent OT to the anhysteretic curve D (Fig. 1) had a slope of $1/N$, N being the demagnetization factor, so that at point A (Fig. 1) the demagnetizing field $-N\mathcal{J}$ was equal to $-N \times CA = \overline{AB}$, equal and opposite to the applied field $\overline{OC}$: *the internal field was zero.* After correction for the demagnetizing field the anhysteretic curve was therefore tangential to the magnetization axis. In other words the anhysteretic susceptibility was *infinite* and even a theoretical explanation of this curious phenomenon had been given.

Experiments of Kahan and of Forrer and Baffie: However, after a magnetometer study of drawn nickel and cobalt wires (100:1 aspect ratio), Kahan [2] working at the Weiss laboratory in Strasbourg demonstrated a deviation between the line OT with slope $1/N$ and the initial tangent to the anhysteretic curve. According to the nomenclature of Forrer [3] there is thus, in certain cases, an internal "structural" field which adds to the demagnetizing field.

In 1942 Forrer and Baffie [4] using a magnetogalvanometer determined the anhysteretic curve of iron–nickel–aluminium ellipsoids having a coercive field of 400 Gauss. They confirmed that the initial tangent was practically identical with the line OT thereby showing that the initial anhysteretic susceptibility of good modern magnet steels was infinite. The demagnetizing factor had been determined in two complementary ways: from the geometrical dimensions and the "permalloy" method. The same workers confirmed their results by a ballistic galvanometer technique applied to a rod.

Experiments of Marquaire and of Néel and Janet: In the meantime M. Marquaire, chief engineer at the Aciéries d'Ugine, had made closed circuit measurements on magnet steels using the toroid or yoke method. He found that the initial anhysteretic susceptibility far from being infinite possessed relatively low values. Similarly, an abundance of data, not yet published, collected over several months by Néel and Janet on a wide variety of ferromagnetic materials also shows that the initial anhysteretic susceptibility is always finite: it is only 20 to 50 times greater than the initial reversible susceptibility.

The source of disagreement between the experimental results: We were therefore confronted with essentially contradictory results which might be summarized as follows. In a closed circuit where there is no demagnetizing field the initial anhysteretic susceptibility is finite: in measurements done on ellipsoids and after correction for the demagnetizing field, it is infinite.

Néel [5] has observed that the usual argument which allows deductions to be made about what would happen in closed circuit from measurements actually done in the presence of a demagnetizing field really applied only to *reversible phenomena;* anhysteretic magnetization is essentially an *irreversible phenomenon*. The development of this idea allows the apparent anhysteretic susceptibility, c, to be calculated as a function of the demagnetizing field and the value, c_0, in closed circuit. It is found that

$$\frac{1}{c} = \frac{N}{1 - e^{-c_0 N}} \quad (1)$$

whereas the usual classic argument gives

$$\frac{1}{c} = \frac{1}{c_0} + N . \quad (2)$$

Formula (1) shows that the apparent anhysteretic susceptibility becomes practically equal to $1/N$ as soon as $c_0 N$ exceeds a value of several units. *The discrepancies enunciated above are thus explained qualitatively.*

It appeared necessary therefore to make a quantitative check of equation (1). To this end we used a yoke method but replaced the usual single rod by two identical half-rods separated by a "pole-gap" of width e. The two opposite ends of the rods fit into the branches of the yoke (Fig. 2).

Determination of the coefficient N of the demagnetizing field: We know that a pole-gap whose width e is small compared with the length L of the rod is equivalent to

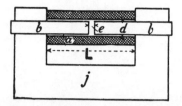

Fig. 2.
a, Magnetizing coils; b, rods under investigation; j, yoke = 10 cm; d = 1 cm.

an *average* demagnetizing field with demagnetizing factor $N = 4\pi e/L$, the reluctance of the rest of the circuit being assumed negligible. Moreover this equality is only valid if the transverse dimensions of the rod are large compared with e. So, for a rod with square section of side 1 cm and a pole-gap of 0.2 mm, the calculated value is already too small by 7%.

The calculation is made easier if a rigid magnetization model is used. Thus for a rod of square section of side d,

$$N = \frac{4\pi e}{L} - \frac{8e^2}{Ld}\left(1 + \text{Log}\frac{2d}{e}\right). \quad (3)$$

We have also *measured* N by tracing the major hysteresis loops of two rods of soft iron, for different pole-gap values e, and comparing the tangents at points of inflexion on these loops with that for the loop $e = 0$. N is deduced from their slopes by means of an equation of the type (2). The values so obtained are represented by the circles in Fig. 3, whereas the curve calculated from equation (3) is traced as the dash–dot line. The experimental curve, the solid line, is clearly below the theoretical curve but has a similar shape.

This difference derives in part from the approximations that have been made to simplify the theory and also from the fact that L which appears in equation (3) must be somewhat larger than the distance between the two internal faces of the yoke. It should also be noted that, especially for large pole-gaps, the concept of the average demagnetizing field makes little sense: the factor N in particular must depend on the average magnetization. In this respect it would have been preferable to experiment with ellipsoids of revolution of various lengths. Only temporary practical reasons led us to adopt the set-up described above.

Be that as it may, we have used the experimental values of N determined in this way in what follows below.

Results of the measurements: Two rods each of a Fe–Ni–Al magnet steel and a 2% cobalt steel were studied. For different values of e the rods were demagnetized in

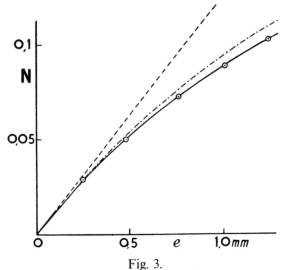

Fig. 3.
The demagnetizing field as a function of pole-gap of a rod 10 cm in length and of square section of side 1 cm. The extremities are short-circuited magnetically.
 --- Elementary formula $n = 4\pi e/L$;
 —·—·— Formula (3) in the text;
 · Experimental points.

a reducing a.c. field, a weak d.c. field being present also of magnitude H. When the field is fairly weak the magnetization obtained, $\mathcal{J}$, is proportional to it: the corresponding anhysteretic susceptibility $c = \mathcal{J}/H$ in c.g.s. units per cm³ is given in the table.

Rod material	e in mm	c in c.g.s.	$\dfrac{c_0}{c}$	N	$c_0 N$
Fe–Ni–Al magnet $H_c \sim 400$ gauss $c_0 = 10,1$	0,	10,1	1	0	0
	0,14	9,25	1,09	0,017	0,17
	0,27	8,6	1,18	0,030	0,30
	0,52	7,55	1,34	0,053	0,53
	0,76	6,85	1,47	0,072	0,73
	1,04	6,6	1,53	0,090	0,91
2% Co steel $H_c \sim 120$ gauss $c_0 = 30,4$	0	30,4	1	0	0
	0,07	25,4	1,20	0,008	0,24
	0,15	22,4	1,36	0,018	0,55
	0,29	18,7	1,63	0,032	0,97
	0,58	13,9	2,18	0,058	1,76
	0,76	11,8	2,58	0,072	2,19

Comparison with theoretical results: In order to compare these results with theory we make use of the reduced co-ordinates: $y = c_0/c$ and $x = c_0 N$ where c_0 is the true anhysteretic susceptibility, without demagnetizing field. Under these conditions, relation (1) is represented by the curve

$$y = \frac{x}{1 - e^{-x}}, \qquad (4)$$

shown as Γ in Fig. 4, whereas the classic relation (2) gives the straight line AB with equation $y = 1 + x$. The

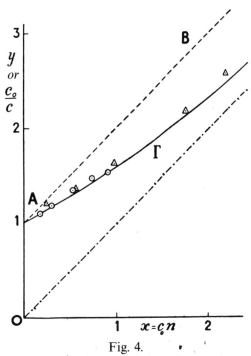

Fig. 4.
Variation of the reciprocal of the anhysteretic susceptibility as a function of the demagnetizing field. The line AB is given by the classic relation whereas the new theory proposed here is represented by the curve Γ.
 ⊙ experimental points for an Fe–Ni–Al alloy;
 △ experimental points for an Fe–Co alloy.

experimental results, circles and triangles, *are clearly in favour of the proposed theory* which would therefore appear to be correct for magnetically hard materials. Thus far it has not been possible to complete this study on softer materials because the yoke is not suitable for such measurements.

Conclusion: It follows from this article that the effect of the demagnetizing field is not simply to shear the hysteresis loop as has been supposed hitherto. In particular as soon as the demagnetizing field attains a modest value the anhysteretic curve rapidly becomes tangential to the line $\mathcal{J} = H/N$. This property could be exploited for the experimental determination of the demagnetizing field

REFERENCES

[1] *Verh. d. D. Phys. Ges.*, 1915, **17**, 369.
[2] *J. de Physique*, 1934, **5**, 9 and 463.
[3] *C.f.* Forrer and Martak, *J. de Physique*, 1931, **2**, 198.
[4] *Cahiers de Physique*, 1943, **15**, 57.
[5] Néel, Théorie de l'effet du champ démagnétisant sur l'aimantation anhystérétique, *Cah. Phys.*, 1943, **17**, 47 (Chap. VIII, A-47).

A144 (1969)

ON THE INCREASE OF THE APPARENT SUSCEPTIBILITY OF A FERROMAGNETIC BODY PRODUCED BY THE SUPERPOSITION OF AN ALTERNATING FIELD

Abstract: The action of an additional field of intensity H_a and angular frequency ω_a is to increase the apparent susceptibility of a ferromagnetic body relative to that in a small original field of angular frequency ω_b. The effect varies in a regular way when ω_b/ω_a increases from 0 to $+\infty$ without any particular singularity at $\omega_b = \omega_a$. It becomes important when H_a approaches a value close to that of the coercive field.

It has been known [1] for a long time that the apparent susceptibility of a ferromagnetic specimen placed in a magnetic field of low frequency and intensity small compared with the coercive field, is considerably increased by the superposition of an additional field of higher frequency and a magnitude comparable with that of the coercive field. Intuitively the phenomenon is explained by saying that it is no longer the initial susceptibility which is an effective factor but rather, the anhysteretic susceptibility, which is considerably higher: indeed, it can be argued by analogy from the fact that the magnetization acquired in a small static field is very much increased by the superposition of an a.c. field decreasing to zero from an initial value large compared with the coercive field. By the same token it was thought [2] that this increase in susceptibility would not occur when the frequency of the additional field was less than that of the coercive field. However, a recent experimental study by Cohen [3] has shown, without any valid explanation, that the susceptibility increases in this case also.

It is therefore of interest to analyze more precisely the process and calculate the change in magnetization as a function of time. We assume that the specimen is subject to a total field H_s equal to the sum of two alternating fields $H = H_a \cos \omega_a t$ and $h = h_b \cos \Phi$, with $\Phi = \omega_b t + \Phi_b$, sufficiently small so that the magnetization always remains in the Rayleigh region. This limitation is not essential but simply convenient in that it allows of simplification in the calculations. In this region the magnetization J is defined for the initial magnetization curve by

$$J = aH_s + bH_s^2, \qquad (1)$$

and for the recoil curves by

$$\Delta J = a\, \Delta H_s \pm \frac{1}{2} b\, (\Delta H_s)^2 . \qquad (2)$$

The term in a is reversible and depends only on the actual value of the total field H_s: as far as interest here is concerned it may be temporarily neglected and attention devoted solely to the irreversible term in b.

The results are particularly simple when the values of h_b and $\omega_b h_b$ relative to the original field are very small compared with the values H_a and $\omega_a H_a$ of the additional field. Let

$$h_b \ll H_a ; \quad \omega_b h_b \ll \omega_a H_a \qquad (3)$$

We shall assume from now on that these conditions are fulfilled.

At any given moment the value of the magnetization J depends upon the preceding extremum values of H_s which therefore must be known. As a result of the conditions (3) the times at which H_s reaches the extremum values coincide with the times $t_n = n\pi/\omega_a$ (n being an arbitrary integer) at which H attains extrema. The extrema of H_s are therefore $(-1)^n H_a + h_b \cos \Phi_n$ where Φ_n is the value of Φ corresponding to $t = t_n$. Let us show now that the extrema, J_n, of the magnetization J are given by the expression

$$J_n = b\,[(-1)^n\, H_a^2 + 2\, H_a h_b \cos \Phi_n] + C , \qquad (4)$$

where C is a constant.

To do this it is sufficient to show that if the relation (4) be true for n it is also true for $n+1$. In fact let us

267

suppose that n is even: when t increases from t_n to t_{n+1}, the amplitude of the change in H_s is $-(2H_a + h_b \cos \Phi_n - h_b \cos \Phi_{n+1})$. The corresponding variation in the magnetization $J_{n+1} - J_n$ is hence equal to

$$- (b/2) (2 H_a + h_b \cos \Phi_n - h_b \cos \Phi_{n+1})^2$$

since a recoil curve is described. By neglecting terms in h_b^2 one notes that J_{n+1} is obtained if n is replaced by $n+1$ in formula (4): this provides the required proof. The same reasoning is valid when n is odd.

Throughout the whole time interval

$$n\pi/\omega_a < t < (n + 1) \pi/\omega_a ,$$

and taking account of the conditions (3), the specimen magnetization J always changes in the same sense as the magnetization J_0 which the specimen would acquire if h_b were zero. It always corresponds to one and the same recoil curve. We find then

$$J = J_0 + 2b\, H_a h_b \cos \Phi_n - (-1)^n b\, H_a h_b$$
$$(\cos \Phi - \cos \Phi_n) [\cos \omega_a t - (- 1)^n] . \quad (5)$$

We are interested here in the component, j, of J with angular frequency ω_b. We can write

$$j = j_\parallel \cos \Phi + j_\perp \sin \Phi ,$$

where $j_\parallel$ and $j_\perp$ denote the terms which are in phase and in quadrature with h respectively.

If k and $k+N$ are two integers, then

$$j_\parallel = \lim_{N=\infty} \frac{2\omega_a}{N\pi} b H_a h_b \sum_{n=k}^{k+N} \int_{\frac{n\pi}{\omega_a}}^{\frac{(n+1)\pi}{\omega_a}} \{2 \cos \Phi_n - (-1)^n (\cos \Phi - \cos \Phi_n) \times [\cos \omega_a t - (- 1)^n]\} \cos \Phi\, dt, \quad (6)$$

as well as an analogous expression for $j_\perp$ obtained after replacing the factor $\cos \Phi dt$ by the factor $\sin \Phi dt$ in the expression (6).

We can then write

$$j_\parallel = 2b\, H_a h_b k_\parallel ; \quad j_\perp = 2b\, H_a h_b k_\perp , \quad (7)$$

where the two numerical factors $k_\parallel$ and $k_\perp$ are given by the expressions

$$k_\parallel = \frac{1}{2} \left[\frac{\omega_a}{\pi\omega_b} \sin \frac{\pi\omega_b}{\omega_a} + 1 + \frac{1}{\pi} \frac{\omega_a \omega_b}{\omega_a^2 - \omega_b^2} \sin \frac{\pi\omega_b}{\omega_a} \right] ; \quad (8)$$

$$k_\perp = \frac{1}{2} \left[\frac{\omega_a}{\pi\omega_b} \left(1 - \cos \frac{\pi\omega_b}{\omega_a}\right) - \frac{1}{\pi} \frac{\omega_a \omega_b}{\omega_a^2 - \omega_b^2} \left(1 + \cos \frac{\pi\omega_b}{\omega_a}\right) \right]. \quad (9)$$

TABLE

$\dfrac{\omega_b}{\omega_a}$	$k_\parallel$	$k_\perp$	$\sqrt{k_\parallel^2 + k_\perp^2}$
0,00	1,000	0,000	1,000
0,25	0,980	0,104	0,986
0,50	0,924	0,212	0,948
0,75	0,843	0,282	0,889
1,00	0,750	0,318	0,815
1,25	0,660	0,321	0,734
1,50	0,585	0,297	0,656
1,75	0,531	0,257	0,590
2,00	0,500	0,212	0,543
2,25	0,488	0,171	0,517
2,50	0,488	0,139	0,507
2,75	0,495	0,118	0,509
3,00	0,500	0,106	0,511
3,25	0,503	0,099	0,513
3,50	0,504	0,095	0,513
3,75	0,502	0,090	0,510
4,00	0,500	0,085	0,507
∞	0,500	0,000	0,500

The table gives several values of $k_\parallel$, $k_\perp$ and $\sqrt{k_\parallel^2 + k_\perp^2}$. It can be confirmed for example that $\sqrt{k_\parallel^2 + k_\perp^2}$ varies fairly regularly from 1000 to 0.500 when ω_b/ω_a increases from 0 to $+\infty$ without any singularity when $\omega_a = \omega_b$.

We now re-introduce the reversible term in ah and note that under the action of an additional field H, the susceptibility in the field h, equal to a for $H = 0$, is now characterized by the two values $s_\parallel$ and $s_\perp$:

$$s_\parallel = a + 2b H_a k_\parallel \quad \text{and} \quad s_\perp = 2b H_a k_\perp , \quad (10)$$

corresponding to the alternating components of the magnetization, of frequency ω_b, in phase and in quadrature respectively, with the field h. The relative increase of the parallel susceptibility $s_\parallel$, or what may be termed the "gain factor of the field H", is thus equal to $1 + (2b/a) H_a k_\parallel$.

The parallel susceptibility $s_\parallel$ therefore changes from $a + 2b H_a$ to $a + b H_a$ when ω_b/ω_a grows from 0 to $+\infty$. These two extreme limits may be interpreted in a simple way.

Let us note first of all that $a + 2b H_a$ is the differential susceptibility s_a along the initial magnetization curve in the field H_a; this curve is also the locus of the peaks of symmetrical loops with increasing amplitude. When $\omega_a \gg \omega_b$, the increase in the size of the change of magnetization during the course of a cycle in the field h is exactly equal to $s_a h_b$. This is because H experiences a very large number of maxima at the moment when h is close to one of its own maxima h_b.

On the contrary, for $\omega_a \ll \omega_b$, the apparent susceptibility in field h must be the average differential susceptibility s_m along the symmetrical loop of amplitude H_a. Indeed by definition and on referring to the second Rayleigh law given by equation (2) we obtain

$$s_m = \frac{\int_0^{\frac{\pi}{a}} [a + bH_a(1 - \cos \omega_a t)] \, dt}{\int_0^{\frac{\pi}{a}} dt} = a + bH_a.$$

(11)

The close parallel between $s_{\parallel}$ and the differential susceptibility along and at the limits of the loop with amplitude H_a helps to predict the nature of the variation of $s_{\parallel}$ beyond the Rayleigh region. Just as the differential susceptibility, so $s_{\parallel}$ must pass through a pronounced maximum when H_a attains the coercive field value H_c and must decrease thereafter.

Taken together, these results provide a qualitative explanation of all Cohen's experiments. A quantitative interpretation is not possible for several reasons. Firstly, the exact values of a and b for the Permalloy used are not known. Secondly, Cohen did not measure $s_{\parallel}$ but the attenuation of the field h produced by a screen in the form of Permalloy sheet rolled into a circular cylinder and subjected to a field H. If the cylinder had possessed a constant permeability μ and had been infinite the interior field h_i would be given as a function of the external field h by the formula

$$h_i = \frac{2R}{\mu e} h,$$

where R and r denote the radius and thickness of the cylinder respectively; but none of these conditions was realized. Thirdly, the fields h and H are not generally parallel. Finally the sheet was not so thin that induced currents could be ignored — and correction is difficult.

We can only check that the right order of magnitude is obtained. The average values of commercial molybdenum Permalloy 4-79, the material used in the experiments under consideration, are: $a = 2500$; $b = 125,000$ (in emu). Using these values it is found that for $H_a = 0.04$, close to the coercive field, the maximum gain factor of the field H must go from 5 to 3 when ω_b/ω_a increases from 0 to $+\infty$. This is certainly the order of magnitude that may be deduced from the experimental data.

REFERENCES
[1] W. Steinhaus and E. Gumlich, *Verhandl. deut. phys. Gesell.*, 1915, **17**, 369.
[2] Von W. Albach and G. A. Voss, *Z. angew. Phys.*, 1957, **9**, 111.
[3] D. Cohen, *Appl. Phys. Lett.*, 1967, **10**, 67.

Chapter IX
COERCIVE FORCE: MAGNETS
B1, A51, A57, A59, A60, A91

B1 (1942)
IMPROVEMENT IN THE MANUFACTURE OF PERMANENT MAGNETS

Note (1977): This concerns a short extract of a patent taken out in April 1942, but not published until 9 years later because of the war. The patent was taken out by the Société d'Electrochimie et d'Electrométallurgie d'Ugine and names as inventor Louis Neel. More exactly the experimental part rests on the work of Louis Neel and Jacque Aubry, inspired and guided by the theoretical concepts of Louis Neel.

Contrary to the known methods, the object of the present invention, which is also a process to manufacture permanent magnets from powders, allows the use of certain metals and their alloys which have only weak coercivities in bulk form and for this reason have never been used as materials for good permanent magnets. As will be seen below, one can, in this way, obtain not only excellent permanent magnets, but this aim is achieved at low net cost because of the nature of the metals used.

The object of the invention is a process for making permanent magnets by agglomeration of powders. It is characterised by the fact that one uses powders consisting mainly of one of the ferromagnetic metals, iron, nickel or cobalt, or of ferromagnetic alloys or, more generally, of a mixture of metals or elements capable of forming a ferromagnetic alloy or compound, where the individual grains have extremely small size, of the order of colloidal particles. One compacts these powders with or without a binder at temperatures low enough to prevent sintering or coalescence of the grains.

The applicant has in effect discovered the extraordinary fact that powders of metals, alloys or combinations of the above-mentioned metals and elements, when their particles are of sufficiently small size, of the order of colloidal particles, e.g. a hundredth of a micrometer to one micrometer, possess by themselves a very high coercive force which can reach 2000 oersted and more, and they are thus eminently suitable for the manufacture of high quality permanent magnets.

A51 (1944)
EFFECT OF CAVITIES AND INCLUSIONS ON THE COERCIVE FORCE

Abstract: This report takes up an idea proposed by Kersten, according to which holes and non magnetic inclusions play the role of obstacles to the movement of walls separating individual domains and thus cause a coercive force to appear; according to this author, the coercive force of tempered steels arises from inclusions (cementite) of a mean radius of 0.6 micrometers. It is shown that the theory of Kersten, based only on the variations of domain wall area is unacceptable since it neglects the demagnetizing field which causes 100 times greater energy variations.

It is shown that holes can be divided into two categories — the large holes (dimensions over 0.1 μm in iron) which have not the effect of obstacles but that of nuclei which expand indefinitely as the applied field exceeds a critical value which is inversely proportional to their diameter; the small holes (smaller than 0.1 μ) which do act as obstacles: the coercive force is calculated as a function of the diameter d of these holes, taking account of the demagnetizing field. The coercive force H_c grows with d and is proportional to the $\frac{2}{3}$rd power of the total volume of the holes.

When a given volume of cavities or inclusions is distributed throughout the ferromagnetic, the coercive force starting from zero, grows with the mean size of the holes, passes through a maximum as they attain a size of 0.1 microns, and then decrease. For iron the maximum coercive force, corresponding to an inclusion volume of 10%, is of the order of 100 oe. It is proposed to interpret in this way the variation of the coercive force during heat treatment in alloys where a non-magnetic phase is precipitated during low temperature annealing.

Thus inclusions, together with irregular strains due to internal stresses, cause the existence of coercive force: in iron the former are more important; in nickel, on the other hand, the latter.

INTRODUCTION

(1) Origin of hysteresis: Nowadays one attributes ferromagnetic hysteresis to the *irreversible movement of the walls between domains*, the direction of the spontaneous magnetization of the two adjacent domains remaining constant.

In an ideal crystal without cracks, impurities or deformations, an infinitely small field can move a plane domain wall. Let us, however, assume that there are irregularly distributed internal stresses, arising from whatever cause. A movement dx of a wall element ds is now associated with a variation dw of the potential energy, arising partly from the surface energy variations connected with the deformation, and partly because the magnetization of the volume $dsdx$ has changed from one to another, energetically not equivalent direction.

Thus the wall is moved by the force exerted by the magnetic field in surroundings which are the more uneven the greater the irregularities in the crystal lattice. Such a movement consumes energy: It can be compared with that of a vehicle on a horizontal plane which has irregular humps. The 'rises' can occur strictly speaking in a fully reversible manner, but this is not the case for the 'falls' where the wall velocity is limited by the retarding field due to eddy currents induced by the changes in spontaneous magnetization. The energy released as Joule heat represents the energy content of a hysteresis cycle, and we can see how the coercive force depends on the irregularities of the wall movement.

To develop a formal theory of the domain wall movements and hysteresis requires only a definite description of the energy changes due to the obstacles to wall movement, without discussing their origin. This we have done recently for the case of the Rayleigh laws [1]. The fact remains that from the point of view of the internal structure of the metal it is essential to know the true nature of these irregularities. Becker and his coworkers [2] have postulated that they are due to irregularities of internal stresses and this hypothesis is the basis of their explanation of the hysteresis and of the coercive force.

However, Kersten [3] after a discussion of the stress theory, has recognized its limits and inadequacy and he proposes to replace it by what he calls the inclusion theory, (Fremdkörper theorie).

(2) Kersten's theory of inclusions: This theory, based on the notion of surface tension of the walls, introduced by Bloch [4], consists essentially of the recognition that a *wall is in equilibrium when it occupies a position where its surface is minimized.* In order to displace it, one has to supply the energy which corresponds to the increase in surface area due to the displacement.

Let us assume, with Kersten, that *holes* or *non-magnetic inclusions* are distributed over the volume of the ferromagnetic body: for simplicity let us further assume that they are all spherical of equal radius ρ, the centres occupying the lattice points of a simple cubic lattice of spacing s (Fig. 1). Now take a plane domain wall BB′, parallel to an elementary cube face, which separates two domain magnetized anti-parallel, and with a surface wall energy γ.

It is evident that without an external magnetic field, the stable equilibrium positions of the wall coincide with the lattice planes like AA′, CC′, ... These are the positions where the wall surface is a minimum. Let us now move the wall by an amount x, smaller than ρ, from its equilibrium position. The surface which cuts each sphere in the wall is now reduced by $\pi(\rho^2 - x^2)$, and hence the total energy W by a term proportional to $-\pi(\rho^2 - x^2)\gamma$. A wall element of wall of surface area s^2, which contains on average one sphere, experiences after this movement, a restoring force:

$$d = -\frac{\partial W}{\partial x} - 2\pi x \gamma.$$

Now let us apply a magnetic field H in a direction as shown in Fig. 1. This exerts a force directed to the right on the elements s^2, of value $2\,\mathfrak{I}_s H_s^2$ ($\mathfrak{I}_s$ = spontaneous magnetization). The element moves until the restoring force is equal to the magnetic force, hence the displacement is given by equation (1).

$$2\mathfrak{I}_s H s^2 = 2\pi x \gamma. \qquad (1)$$

For small fields, the wall movement is proportional to the field and the process is reversible. The maximum restoring force corresponds to $x = \rho$ and is $-f_m = 2\pi\rho\gamma$. At this point the wall is tangential to the spheres and the magnetic field H_c which balances this force is given by

$$H_c = \frac{f_m}{2\,\mathfrak{I}_s s^2}. \qquad (2)$$

If H exceeds this critical value by an infinitely small amount the wall separates from the sphere and moves *irreversibly* through the whole lattice. The field H_c behaves like a coercive force.

Expressing this critical field as a function of the radius ρ and the volume α of the inclusion per cubic centimeter of material ($\alpha = \tfrac{4}{3}\pi\rho^3 s^{-3}$), we obtain

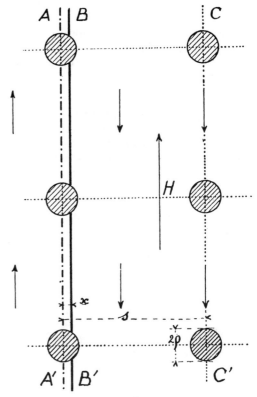

Fig. 1.

$$H_c = \left(\frac{4\pi}{3}\right)^{\frac{2}{3}} \frac{f_m \alpha^{\frac{2}{3}}}{2\rho^2 \mathfrak{I}_s} \qquad (3)$$

and if one expresses this maximum restoring force for the case of iron, it becomes

$$H_c = \left(\frac{3}{4\pi}\right)^{\frac{2}{3}} \frac{\pi \gamma \alpha^{\frac{2}{3}}}{\mathfrak{I}_s \rho} = 0{,}99\,\frac{\alpha^{\frac{2}{3}}}{\rho} \cdot 10^{-3}. \qquad (4)$$

At constant radius, the critical field is proportional to the 2/3rd power of the total volume of inclusions or holes.

Kersten applies this theory to annealed carbon steels and assumes that all the carbon, in the form of cementite, causes the non-magnetic inclusions. According to experiments quoted, H_c varies as the 2/3rd power of the percentage of carbon for concentrations between 0 and 2%C, which seems to indicate that the mean volume of the inclusions remains constant and one deduces from the numerical values that the mean radius of the inclusions is 0.6 microns.

This theory by Kersten, briefly described, appears even more plausible, as the author develops it further, successfully interpreting the values of the initial permeability and the temperature variation of the coercive force in these steels. But, before going further, it is important to note that Kersten does not mention the role of the demagnetising field of the inclusions. Is this factor actually negligible?

(3) The role of the demagnetizing field: Let us suppose, as above, that the magnetization is uniform and let us put $OD = x$, the distance of the wall P to the centre O of the cavity ($x < \rho$). Let us put $x = \rho \cos \delta$. Concerning the calculation of the energy, it appears as though the sphere carries magnetic surface charges of density equal to the normal component of the magnetization. These charges are positive in the region AF and A'F', negative in regions AF' and A'F (Fig. 2). For the corresponding magnetic energy W_d one obtains without difficulty [1]:

$$W_d = 4\pi^2 \mathcal{J}_s^2 \rho^3 \sum_{n=1}^{\infty} \frac{n(n+1) c_n^2}{(2n+1)^2}, \quad (5)$$

where the coefficients c_m are defined as functions of the Legendre polynomials $P_n(\cos \delta)$ as

$$c_n = \frac{P_{n-2} - P_n}{2n-1} - \frac{P_n - P_{n+2}}{2n+3}.$$

When $x = \rho$, $\cos \delta$ equals 1 and the wall is tangent to the cavity; the coefficient $c_1 = 1$ whilst the subsequent ones are 0, and one finds

$$W_0 = \frac{1}{2} \cdot \frac{4}{3} \pi \mathcal{J}_s \cdot \frac{4}{3} \pi \rho^3 \mathcal{J}_s.$$

One recognises the demagnetization energy of a uniformly magnetized sphere of magnetization $\mathcal{J}_s$ and radius ρ.

A study of the variation of the expression in equation (5) shows that W_d diminishes with x and goes through a minimum for $x = 0$, when the wall cuts the sphere into two equal halves. This minimum is equal to $0.46 W_0$. The difference in energy between the 2 extreme positions is hence equal to $0.548/9\pi^2 \mathcal{J}^2 \rho^3$. For a sphere of radius 0.6 microns and $\mathcal{J}_s = 1710$, one obtains 2.97×10^{-6} ergs. In Kersten's theory the difference arising from surface energy between the two same positions is

$$\pi \rho^2 \gamma = 1{,}56 \cdot 10^{-8} \text{ ergs}$$

taking γ as 1.4 ergs/cm². *The demagnetizing energy is 200 times greater than the surface tension energy.*

It is therefore absolutely impossible to follow

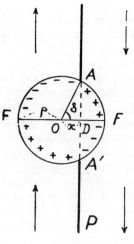

Fig. 2.

Kersten in the development of a theory based on variations of surface energy of the walls, which neglects energetically much more important contributions. It is necessary to take up completely anew the problem of the cavities and non-magnetic inclusions and their influence on domain wall movement.

(4) Complexity of the problem: At first one might be tempted simply to transpose Kersten's theory by replacing the surface energy by the magnetostatic energy W_d. The restoring force is $f_m = (\partial W_d / \partial x)$ and using equation (5) for W_d one obtains graphically, as maximum restoring for f_m:

$$f_m = 0{,}86 \cdot \frac{8}{9} \pi^2 \mathcal{J}_s^2 \rho^2, \quad (6)$$

and substituting this into equation (3) one obtains in the case of iron [7]

$$H_c = 0{,}86 \left(\frac{3}{4\pi}\right)^{2/3} \frac{4}{9} \pi^2 \mathcal{J}_s \alpha^{2/3} = 2480 \alpha^{2/3}. \quad (7)$$

H_c depends only on the total volume of the cavities and not on their radius. This formula[1] is totally in conflict with experiment, since it gives a coercive force of about 1000 Gauss for a steel with 1% carbon in the form of cementite.

In reality the problem is not so simple, because our original assumptions are not acceptable, especially for large holes. *The magnetization cannot be considered as uniformly rigid in their vicinity.* We must first study

---

[1] It would be more rigorous to take account of the true statistical distribution of the holes, as we have done in the case of the Rayleigh laws, but for reasons of simplicity we will here use the simpler method of Kersten (cavities at the nodes of a cubic lattice).

what occurs near to large holes. Subsequently we will study the case of small holes for which the assumption of a uniform magnetization is acceptable.

STUDY OF LARGE INCLUSIONS

In a recent study concerned with the laws of magnetization and the domain structure of an iron single crystal [5], we have shown that the presence of free, internal magnetic charges, greatly increases the free energy and therefore the elementary domains take up shapes and relative positions such that the free charges are minimized. The same will be the case here in the neighbourhood of large cavities. We will therefore investigate what is the structure of the elementary domains which reduces and even completely eliminates these free charges. These structures depend very sensitively on what happens on the surface of the cavity, and they must therefore present aspects which vary greatly depending on its shape. There is no question of studying all the possibilities, but we will take the limiting condition for the simplest cases, corresponding to a cube shaped cavity with faces perpendicular to the four-fold crystal axes. We take two cases, one where the cavity is isolated inside a uniformly magnetized domain and one where it is cut into two equal havles by a domain wall parallel to a cube face.

(5) The wall cuts the cavity in two equal halves: Let C be a cubic hole of side d, cut in half by a wall P. With uniform magnetization, the free poles are arranged as shown in Fig. 3a, and the magnetic energy is of the same order of magnitude as calculated above for a sphere of equal volume cut in half by a wall, i.e. $0.46 . 2/3 . \pi . \mathscr{T}_s^2$ per unit volume. or 2.8×10^6 ergs/cm^3.

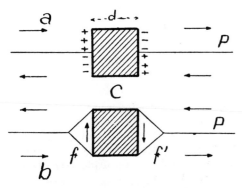

Fig. 3.
a. Domain wall P bisecting a large cube-shaped cavity;
b. Formation of closure domains f and f'.

Since we have assumed that the size of the cavity is large compared to the domain wall thickness, which is of the order of tenth of a micron [6 and 11] one can regard the walls as simple planes with a certain surface energy γ. Furthermore we assume that the applied magnetic field acting on the system is sufficiently weak for the spontaneous magnetization to remain directed along a four-fold axis (the preferred axis for iron in zero field).

The problem consists in finding a system of elementary domains with spontaneous magnetization, as far as possible directed along the four-fold axes with the domain walls having no free poles. This latter condition in effect implies that the magnetostatic energy term in the expression of the total energy becomes zero. Fig. 3b shows one solution of the problem: it suffices to imagine two domains f and f', called *closure domains*, having the shape of triangular prisms magnetized parallel to the cube faces of the cavity. The energy of this structure becomes simply the surface energy of the walls dividing the domain which were created in this process, i.e. $(2\sqrt{2}-1)d^2\gamma$. No term of magnetocrystalline energy appears because the magnetization in the prisms follows a four-fold axis. Thus, finally, for a cavity of side 1 micron, the energy of the structure in Fig. 3b becomes 2.5×10^{-8} ergs, about *100 times smaller* than that corresponding to the structure 3a (2.8×10^6 ergs).

Hence it is certain that firstly *the arrangement 3a is impossible*, and secondly that *the energy of the real structure is equal or less than that of 3b*. Furthermore, in this particular case it seems difficult to find a structure with smaller energy than 3b; we will hence consider that this is a good approximation to the real structure.

In this example we have been able to make the free poles disappear because the domain wall cut the cube into two equal parts: *the magnetic flux which passes through a cylinder circumscribing the cavity* and with axis parallel to the magnetization is therefore zero. It is evident that whatever the shape of the cavity, one can always put the wall in a position to realize this condition: hence one can always find structures without free poles, but they are often more complex and their energy slightly larger, because the magnetization of the closure domains is often not exactly collinear with the direction of easy magnetization. Nevertheless, in general the same conclusion prevails: whatever the shape of the cavity, provided its dimensions are of the order of 1 micron or more, it is possible to find a structure with an energy some ten times smaller than that which corresponds to uniform magnetization and the factor grows with the dimension of the cavity.

Inversely, when the size of the cavity becomes small, of the order of a hundredth of a micron, the energy of the structure with rigid uniform magnetization becomes smaller than that with closure domains. There is then no sense in imagining the formation of closure domains since the domain wall thickness is larger than the cavity.

(6) Isolated hole in a uniformly magnetized domain: We take as example the simplest case of a cubic cavity ABCD (Fig. 4a) with faces perpendicular to the four-fold crystal lattice directions. The face AB carries a positive charge amounting to $Q = d^2 \mathcal{T}_s$ (d ... cube edge), whereas the face CD has a total charge $-Q$. It is no longer possible to arrange for the free charge to disappear by a suitable arrangement of new domains, as was done in the preceeding paragraph. In fact it can easily be shown that, whatever arrangement we adopt (provided it is symmetrical about a plane Y–Y'), the algebraic sum of the free charges to the left of the plane Y–Y', remains constant and equal to Q, whilst the sum of the charges to the right remains equally constant at $-Q$.

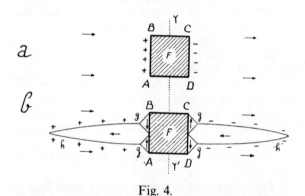

Fig. 4.
a. Large isolated cube-shaped cavity immersed inside a domain; b. Formation of a spike with closure domains resulting in a considerable reduction in magnetic energy.

Nevertheless, it is impossible that the structure shown in Fig. 4a should exist in a real substance because it implies the existence of magnetic fields of several thousand gauss near AB or CD, sufficient to turn the magnetization at these points into a direction parallel to these two faces. Hence the spontaneous magnetization is profoundly perturbed in the vicinity of the cavity, but it is extremely difficult to develop a rigorous theory of the situation.

Still, from a qualitative point of view it appears certain that the charges Q have to be distributed over a large area in order to reduce the magnetic energy: to do that, new domains with new walls must be established. The stretching out of the charges has to stop, therefore, at the point where the magnetostatic energy gain is compensated by the increase in wall energy.

Below we describe two such structures, but it must be well understood that we only deal with examples solely thought up to illustrate the possibility of reducing in this manner the magnetic energy due to the presence of the cavity. The real structures could well be quite different: we hope only to estimate the order of the magnitude of the new energy after the perturbation.

The three directions parallel to the cube (ABCD) sides of the cavity F (Figs. 4 and 5) are directions of easy magnetization. In a first example, we make the charges, arising from the magnetization parallel to AD, disappear, as is shown in Fig. 5a, by means of domains such as hhhh, which are magnetized along AB. If the walls of these new domains are inclined at 45° to AB, the new structure has no free magnetic charges. In reality, the charges are pushed out to infinity and one has introduced surface energy of infinite walls, because the domains h are infinitely long. The structure will therefore take the shape of Fig. 5b with spike shaped secondary domains h'h' of finite length. The density of the surface charges becomes smaller as the length, and with it the surface energy, grows. Hence an equilibrium defined by minimum energy must establish itself.

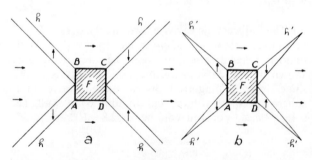

Fig. 5.
Another schematic arrangement illustrating the possibility of reducing the magnetic energy by the formation of secondary domains around a large cube-shaped cavity

In a second example shown in Fig. 4b, we disperse the charges lengthwise by means of supplementary domains h and h' with the shape of cigar like spikes, magnetized antiparallel to the principal domain. These two domains are joined to the walls AB and CD of the cavity by four intermediary closure domains gggg, magnetized in a perpendicular direction. The bases of these prisms are right angled isosceles triangles. The only free charges in this structure are sited along the walls of the two half-wedges, their density becoming smaller as their length increases.

In the two examples mentioned one can easily estimate the energy at least when the spikes are sufficiently long. It is found that the energy of the structure in equilibrium is about the same in the two examples, and we will only calculate it here for the case of Fig. 4b.

In this case the total energy reduces practically to the demagnetizing energy of the spikes and the surface energy of the walls[2]. In effect, if the spike is sufficiently long, the surface of the closure domains g is negligible compared with that of the walls of the spikes.

Let us estimate *the order of magnitude* of the total energy of a cigar shaped spike of length kd. To do this it is good enough to liken the spike in question both as regards its surface and its mean demagnetizing factor to an ellipsoid of revolution[3] of the same length, kd, and the same equatorial cross section $d^2/2$. The volume V, the surface S and the demagnetizing coefficient N of this ellipsoid are given by the following formulae, which become more accurate as k increases.

$$S = \sqrt{\frac{\pi^3}{8}} kd^2 \qquad V = \frac{1}{3} d^2 k \qquad (8)$$

$$N = \frac{8}{k^2} (\log k \sqrt{2\pi} - 1),$$

from which the energy becomes[4]:

$$W_t = S\gamma + 2N\mathcal{J}_s^2 V, \qquad (9)$$

Considering the parameter of elongation k as variable, this energy will be minimum for

$$\frac{\partial W_t}{\partial K} = 0,$$

which gives us **k** as a function of **d**:

$$\frac{3}{32} \sqrt{\frac{\pi^3}{2} \frac{\gamma}{d\mathcal{J}_s^2}} = \frac{1}{k^2} (\log k \sqrt{2\pi} - 2). \qquad (10)$$

The values of d so determined for different values of the parameter k, are collected in Table I, where we have taken $\gamma = 1.4$ ergs/cm^2 and $\mathcal{J}_s = 1710$ cgs units.

TABLE I.

k	d (microns)	W_t (ergs)	$\dfrac{W_t}{W_0}$
10	0,145	1,62.10^{-8}	0,88
20	0,37	1,90.10^{-7}	0,61
50	1,56	7,9 .10^{-6}	0,34
100	5,0	1,58.10^{-4}	0,205
200	16,8	3,5 .10^{-3}	0,120
400	57	8,2 .10^{-2}	0,069

The values of W_t are also shown in the table. Finally, in order to allow comparison of the energies corresponding to the two structures a and b of Fig. 4, the ratio of the energy W_t of the structure 4b to W_0 of the structure 4a assumed equal to the demagnetization energy of a spherical hole of same volume.

An inspection of the table shows that for holes of a diameter of some tenths of microns, the difference in energy between the two structures is very small, but this is no longer so as the diameter becomes a few microns: the large holes are therefore surrounded by a secondary structure whereas near small holes the magnetization retains its original direction.

(7) Effect of the magnetic field: The secondary domain structures which establish themselves spontaneously around large cavities, are very sensitive to the action of a magnetic field. Thus in the example illustrated in Fig. 4b, a magnetic field in the same direction as a spike, tends to elongate this spike. If it attains a critical value H_s, the *spike grows indefinitely first in length and then in thickness*.

It is easy to obtain an upper limit of this critical field

---

[2] We suppose for the moment that the system is not subjected to any magnetic field.

[3] In a spherical cavity the spike corresponds to that approaching a paraboloid of revolution with its apex in the equatorial plane. This form is characterized by the points situated at the extremities of the major axes. Let us remember, in this context, how we have shown elsewhere (Cahiers de Physicque 1941, No. 4, p. 57) that an isolated domain, completely immersed in a phase uniformly magnetized in the opposite sense, cannot take up the shape of a prolate ellipsoid of revolution: in effect the magnetic charges distributed in the surface create outside the ellipsoid a sufficient magnetic field to suppress the rounded end caps and to transform them into points. Briefly this means again that the walls separating two antiparallel domains cannot carry but a feeble density of charges and therefore must be everywhere almost parallel to the magnetization. The paraboloid, to which reference was made above, fulfils this condition the better as k grows. Nevertheless, in the same paper we have also shown that, regarding the surface energy as well as the magnetic energy, the results are on the whole equivalent if one replaces the paraboloid shaped spikes by ellipsoids of revolution of equal section and elongation, and this is how we have proceeded here for simplicity's sake.

[4] If the spike of uniform magnetization $\mathcal{J}_s$ is isolated in a non-magnetic medium, its demagnetizing energy, as is known, becomes $\frac{1}{2} N \mathcal{J}_s^2 V$. But here it is immersed in a phase of magnetization $-\mathcal{J}_s$, which doubles the charges and quadruples the energy.

H_s if one neglects the effect of the demagnetizing field. For a growth δk in length of the spike, the surface increases by $\sqrt{\pi^3/8}.d^2\delta k$ and the volume by $1/3 d^3 \delta k$. Hence the surface energy grows by $\sqrt{\pi^3/8}d^2\gamma\delta k$ whilst the magnetic energy in the external field decreases by $\frac{2}{3}d^3 H \mathcal{J}_s \delta k$. The total increase equals $(\sqrt{\pi^3/8}.\gamma - 2/3 dH \mathcal{J}_s)d\delta k$. If H exceeds a certain critical value H_s, called starting field, given by $H_s = 3/2\sqrt{\pi^3/8}.\gamma/d\mathcal{J}$, one can see that the energy is diminished by an increase in the length of the spike: the spike ought to grow indefinitely. A more complete discussion regarding the state of the demagnetizing field has already been given by Döring [7 and 2] for a similar problem: the correct value for H_s is sensibly equal to the one derived above, except for a factor 5/6; it is for iron

$$H_s = \frac{5}{4}\sqrt{\frac{\pi^3}{8}}\frac{\gamma}{d\mathcal{J}_s} = \frac{1}{d}\cdot 2{,}02.10^{-3}. \quad (13)$$

This critical field is inversely proportional to the size of the cavity, it is 20 Gauss for a cavity of 1μ diameter. The corresponding curve representing H_s is shown by GCD in Fig. 7 in logarithmic coordinates for both axes.

Results of the same order are obtained for the scheme of Fig. 5b: the critical field is given by an equation similar to (13) with a slightly different numerical coefficient.

(8) The role of large cavities as nuclei of demagnetization: We will now see that a large, isolated hole in a uniformly magnetized domain, constitutes what one calls a nucleus of demagnetization. When the opposing field is sufficiently high, new domains bounded by new domain walls develop indefinitely around the cavity. These new domains are magnetized antiparallel to the principal domain for the case of Fig. 4b and at 90° to the principal domain for the structure 5b.

The nucleation of these new domains and these new walls is the easier the larger the dimensions of the cavity.

One knows already that the presence of nuclei is indispensible to explain the presence of domain walls whose movements are called for in the phenomena of hysteresis. Let us consider a small, very elongated bar magnetized to saturation lying along a four-fold axis in a single crystal of iron. In the absence of any inversely magnetized nucleus. To demagnetize such a bar in the absence of any inversely magnetized nucleus one must begin by rotating the spontaneous magnetization at any point and for this to happen the magnetocrystalline anisotropy forces must be overcome. If α is the angle of rotation of the sponeaneous magnetization with respect to the axis of the bar (assumed small) the magnetocrystalline energy is $K\alpha^2$ (K being the anisotropy constant), whereas the magnetic energy in the applied field H equals $-H\mathcal{J}_s \cos\alpha \approx H\mathcal{J}_s[1-(\alpha^2/2)]$. It is clear from the expression for the total energy that the initial equilibrium remains stable as long as the negative applied field is below $2K/\mathcal{J}_s$, i.e. about 500 Gauss[5], whereas in these conditions the coercive force of a single crystal of iron can be smaller than 0.1 Gauss.

One must therefore accept the existence of nuclei which cannot be attributed to thermal fluctuations. Their nature is very little known[6]. The role played in this context by cavities or inclusions is hence particularly interesting.

Naturally, it is very probable that besides nuclei formed by large cavities, there also exist others of different origin, but of whose nature we have as yet no information.

(9) The large holes as obstacles to domain wall propagation: It now remains to examine the role which the large holes can play as obstacles to domain wall movement. First we must note that this problem only makes sense as long as the applied field H which pushes the wall is below the critical field H_s given by equation (13). In fact, if the applied field exceeds H_s each cavity, far from being an obstacle, gives rise to new walls.

Suppose therefore $H < H_s$: to solve the above problem one must find in which way the wall moves from the position of Fig. 3b where it bisects the cavity, to that of Fig. 4b or 5b, where it is completely detached from the domain system associated with the cavity. After calculating the energy corresponding to each position of the domain wall one can deduce the restoring force which the cavity exerts on the wall. Unfortunately the problem posed here has very probably no sense: it is in effect not at all certain that by forcing a wall to leave the energetically very favourable position of Fig. 3b, one ends up with a structure of the type 4b or 5b, comprising a cavity and its wall system, isolated in a uniformly magnetized domain. One can equally well imagine a phenomenon of the kind illustrated in Fig. 6a to 6d in which the cavity becomes the starting point of new elementary domains which develop gradually as the wall recedes from its original position.

---

[5] Considerations of this type are the basis of an old theory of the coercive force due to Becker (Z. für Physik 1930, **62**, 253), later abandoned by its author in favour of the theory of wall displacements. See also Akulov (Z. für Physik 1933, **81**, 790).

[6] See with respect of this subject the beautiful experiments by Sixtus (Phys. Rev. 1935, **48**, 425) and by Döring and Haake (Phys, Z. 1938, **39**, 865) on artificially created nuclei.

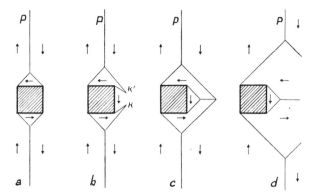

Fig. 6.
Sketch showing how a wall P which initially bisects a cube-shaped cavity can give rise to new domains as it separates from the cavity. The energy of the demagnetising field remains zero except in (b) where it is very small.

The difficulties of the possible situations are such that it appears completely hopeless to outline even a scanty theory. Summing up, one can only state that in a pure non deformed substance which contains a number of inclusions of mean diameter d, distributed in any manner, the coercive force is smaller or equal to the limit H_s, given in equation (13). This limit is, moreover, quite small since it hardly exceeds 2 Gauss for holes of 10 micron diameter.

STUDY OF SMALL HOLES

It follows from paragraph 5 and especially Table 1 (para. 7), that a cavity of size below 0.1 micron cannot acquire any secondary structure because this would increase the free energy. Let us note, besides, that the thickness of the walls separating elementary domains in iron is also about a tenth of a micron, and it is not surprising that in these circumstances new walls cannot be formed. Meanwhile, near the cavity the magnetization does not, for all that, retain the direction it had without the hole present, because it is deflected by the magnetic field due to the opposing charges on the cavity surface. A difficult and approximate calculation can show that the reduction of the free energy associated with this re-orientation becomes quickly quite negligible as the hole size decreases. Since we are here only interested in the energy aspect and the order of magnitude of the effect, we will neglect this re-orientation: we will henceforth assume that the presence of a small hole does not modify the direction of spontaneous magnetization in its vicinity.

We intend now to calculate the energy of a cavity as a function of its position relative to the wall. For simplicity we assume a spherical cavity. Further, we now assume that the cavity is small compared to the wall thickness. The results will therefore not be absolutely accurate except for very small holes, e.g. perhaps of the order of hundredths of microns, but they will nevertheless give an order of magnitude acceptable for holes of tenths of micron sizes.

(10) Exchange energy and magnetocrystalline energy of the small cavities: If a point M passes through a wall separating two regions magnetized antiparallel[7], its spontaneous magnetization turns in a continuous manner from one preferred direction to the other, remaining parallel to the plane of the wall [6, para 3]. The angle θ of the magnetization with one of the preferred axes varies therefore from 0 to π and only depends on the abscissa of the point M taken normal to the wall.

Consider a cavity of volume dv, small compared with the wall thickness, penetrating into the interior of the wall to a point where the direction of magnetization is marked by the angle θ. In this same volume of substance the magnetization goes from this direction θ to the preferred direction and hence the magnetocrystalline energy decreases by $K \sin^2 \theta \cos^2 \theta dv$.[8] At the same time the exchange energy decreases by an equal amount so that the total energy decrease is $2K \sin^2 \theta \cos^2 \theta dv$. It is a maximum for $\theta = \pi/4$ when the cavity lies in the middle of one of the two half 90°-walls which constitute a 180° wall. The energy therefore equals $K/2 = 2.15 \times 10^5$ ergs per cm^3 of wall, in iron.

(11) Magnetic energy of a small cavity: Take a small spherical hole diameter ρ, which is assumed not to perturb the direction of the magnetizatuon around it. Let θ be the angle of the direction of the spontaneous magnetization in that plane of the wall which passes through the centre of the cavity. At neighbouring points situated at a distance x from this plane, the direction of magnetization is therefore defined by $\theta + x(\partial \theta/\partial x)$, and this we take as valid for $|x| \leqslant \rho$. In these circumstances one can show (see Note II) that the magnetic energy due to the charges arising on the wall, becomes

$$W_d = \frac{8\pi^2 \mathcal{J}_s^2 \rho^3}{9} \left[1 - \frac{2}{25} \rho^2 \left(\frac{\partial \theta}{\partial x} \right)^2 \right]. \quad (14)$$

Outside the wall $(\partial \theta/\partial x) = 0$, the energy becomes

---

[7] The two antiparallel directions lie necessarily in the plane of the wall.

[8] K, magnetocrystalline constant for iron is about equal to 4.3×10^5 ergs per cm^3.

equal to the first factor of (14), equal to the magnetic energy of a uniformly magnetized sphere. On the other hand we have found that inside a wall (A, para 6)

$$\frac{\partial \theta}{\partial x} = \frac{1}{a} \sqrt{\frac{6K}{E}} \cos \theta \sin \theta . \qquad (15)$$

In this equation a is the lattice parameter (2.86 Å) and E the energy of demagnetization (8.6×10^9 ergs per cm^3 for iron). That part of the magnetic energy which depends on θ, becomes finally per unit volume:

$$-\frac{8\pi \mathfrak{J}_s^2 K}{25 E a^2} \rho^2 \cos^2 \theta \sin^2 \theta =$$
$$= - 1,8.10^{17} \rho^2 \cos^2 \theta \sin^2 \theta . \qquad (16)$$

It varies with θ in the same way as the magnetocrystalline energy derived in the previous paragraph. But whilst the magnetocrystalline energy (per unit volume of cavity) was independent of ρ, the magnetostatic energy is proportional to ρ^2. The two energies are equal for $\rho = 220$ Å.a One finds therefore that for cavities smaller than some tens of angstroms, the variations of the energy of the demagnetizing field are negligible compared with those of the magnetocrystalline energy.

(12) Calculation of the coercive force: We have at first to determine the maximum restoring force exerted by the walls on the cavity. Now, following the two preceding paragraphs, the *total* energy W of a small cavity is of the form

$$W = - C \cos^2 \theta \sin^2 \theta \qquad (17)$$

with

$$C = \frac{4}{3} \pi \rho^3 \left[2K + \frac{8\pi \mathfrak{J}_s^2 K}{25 E a^2} \rho^2 \right] . \qquad (18)$$

From equation (17) we deduce the restoring force f, and, taking account of equation (15) it becomes finally

$$f = - \frac{\partial W}{\partial x} = \frac{C}{2a} \sqrt{\frac{6K}{E}} \sin^2 2\theta \cos 2\theta . \qquad (19)$$

This expression has a maximum for $\cos^2 2\theta = 1/3$, i.e. the maximum restoring force becomes

$$f_m = \frac{C}{3a\sqrt{3}} \sqrt{\frac{6K}{E}} . \qquad (20)$$

In order to get the coercive force produced by a regular distribution of small holes of radius ρ situated at the nodes of a simple cubic lattice we have to insert this value of f_m simply into equation (3), and we obtain

$$H_c = \left(\frac{3}{4\pi}\right)^{\frac{2}{3}} \sqrt{\frac{6K}{E}} \frac{C}{6\sqrt{3}\rho^2 \mathfrak{J}_s a} \alpha^{\frac{2}{3}} , \qquad (21)$$

where α is the relative volume of the cavities. Taking the value of C from equation (18), H_c for iron becomes

$$H_c = [4,73 . 10^7 \rho + 0,99 . 10^{19} \rho^3] \alpha^{\frac{2}{3}} . \qquad (22)$$

The coercive force is again proportional to $\alpha^{2/3}$. The values of the proportionality constant are given in Fig. 7 in logarithmic coordinates for both axes, as a function of the hole diameter $d = 2\rho$ (branch ABG of the curve).

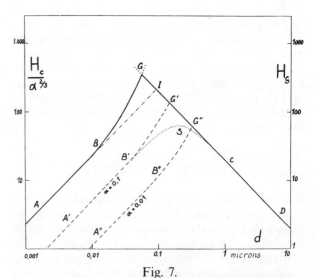

Fig. 7.
Case of iron: plot in logarithmic coordinates as a function of cavity diameter d; GCD: the field which causes nucleation indefinitely and the coercive force respectively A'B'G': cavities 10% of total volume, A"B"G" cavities 1% of total volume (R.H. scale in Gauss) and ABG: value of $H_c/\alpha^{2/3}$ (L.H. scale).

GENERAL DISCUSSION

In order to obtain the preceding results we have given the cavities particular shapes, cubes or spheres, but it is evident that we have arrived at results which are very near to those for any shaped holes, provided they are sufficiently close and on the whole isotropic. Only the numerical coefficients will be slightly different.

(13) Nature of the cavities or inclusions: Before entering into a discussion which will be based on the graphs of Fig. 7, one must first examine the real nature of the inclusions and holes which are the subject of this theory. The numerical values cited refer to inclusions of cementite in iron. Because of its high coercive force we suppose that cementite behaves on the whole like a non-magnetic substance. But the concept of inclusions is obviously more general, we can have impurities of any kind, voids inside or between the crystallites. We can also have some more or less amorphous cement which binds the crystallites together, and which may play the role of inclusions because its magnetic properties are very different from those of the crystallites. Moreover, all these inclusions are generally of sufficiently large size, of the order of microns or more, and finally, apart from the cementite they are of accidental nature.

On the other hand, in the case of segregation by precipitation of a supersaturated phase, we have inclusions of variable dimensions, often very small, the presence of which is inevitable. This is the case of the precipitation of a supersaturated solution of beryllium in nickel, where the nuclei of precipitated beryllium, form inclusions. Similarly, a solid solution of gold and nickel decomposes at low temperatures into two phases, of which one is very gold-rich and non-magnetic and hence constitutes inclusions. The presence of inclusions in a ferromagnetic medium is quite general.

(14) Influence of the subdivision on the coercive force: Let us examine as an example the base of iron containing 10% inclusions in the form of cavities of equal diameter **d** regularly distributed.

When d is greater than 0.1 microns the very numerous cavities act as nuclei and the maximum coercive force is given by the straight line CD.

If d is less than 0.1 micron we must use equation (22) with the corresponding value of $\alpha^{2/3}$ and we get the curve A'B'G' (Fig. 7).

So the coercive force, at first very small for very great dispersion of the inclusions, grows with the size of the particles and passes through a maximum which is of the order of a hundred gauss for inclusions of 0.1 micron diameter[9]. Then as the particles continue to grow, the coercive force diminishes.

If the volume of inclusions had been 1 percent, we would have obtained curve A"B"G"CD. Let us observe in passing that the orders of magnitude so obtained agree satisfactorily with those observed in steels.

---

[9] The maximum coercive force is given by point G', but it is clear that the angular section B'G'C doesn't represent a physical reality and arises from the schematic character of the theory. A rounding off as shown by B'SC is more likely.

Moreover, one can see that the maximum coercive force which can be obtained with any proportion of inclusion, is given by the point I, intersection of two straight lines AB and CD. The coordinates d_m and H_{cm} of point I for iron are given by

$$d_m = 2{,}3\, a\, \sqrt{\frac{E}{K}} = 0{,}93 \cdot 10^{-5} \text{ cm} \qquad (23)$$

$$H_m = 0{,}87\, \frac{K}{\mathcal{J}_s} = 218 \text{ gauss.} \qquad (24)$$

These two formulae show in a general way that the most effective inclusion diameter from the point of view of coercive force, is of the order of the wall-thickness (about $3a\sqrt{E/6K}$ for a 90° wall) and that the maximum coercive force is about $K/\mathcal{J}_s$.

Kersten has already remarked that in his theory the cavities have maximum effectiveness when their diameter is equal to the wall thickness, but the significance of his result is totally different from ours since in the descending branch of the curve $H_c = f(d)$ in his theory, the cavities act as sources of variation of surface energy, whereas here we deal with a limiting field caused by the effect of nuclei.

One cannot refrain from relating the variation of the coercive force with the diameter of the cavities, characterised by a maximum for a well-defined diameter to the phenomena observed in the study of the coercive force of alloys where segregation by precipitation occurs and *the coercive force always passes through a maximum as a function of the tempering time*. In the inclusion theory this is immediately evident because it is generally admitted that the precipitate particles, at first excessively fine, grow progressively in the course of tempering. The magnitude of the optimum diameter, 0.1 microns, tallies with what we know of the processes of precipitation.

Does this mean that we have to abandon the old theory of internal stresses which has hitherto been exclusively adopted to explain these phenomena? This would indeed be tempting because one would have to assume considerable internal stresses withstanding a tempering in an abnormal manner. It appears rather more probable to think of the two causes as occurring simultaneously and that the coercive force of alloys arises partly from irregular internal stresses, and partly from non-magnetic inclusions finely dispersed in the magnetic phase: In materials with large magnetostriction and weak anisotropy, like nickel, the internal stress will have the more important effect, whereas in substances with small magnetostriction and large anisotropy, like iron, the effect of inclusions is predominant.

NOTE I

Magnetic energy of a sphere which cuts through the plane separating two anti-parallel magnetized regions: Consider a sphere (Fig. 8) centre 0, radius ρ, with 3 rectangular axes OX, OY, OZ, immersed in a medium with spontaneous magnetization directed along OZ for $x > x_0$, and directed antiparallel for $x < x_0$ ($|x_0| < \rho$). Let us take OX as the polar axis and define by r, δ and φ

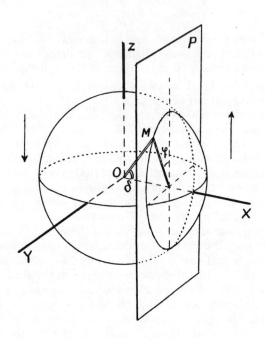

Fig. 8.

respectively the radius vector, polar angle and azimuth of a point M. The surface density of charges on the surface of the shell is:

$$\sigma = \epsilon \mathcal{J} \sin \delta \cos \varphi, \quad (25)$$

where $\varepsilon = 1$ for $x > x_0$ and -1 for $x < x_0$.

We put (P_n ... Legendre polynomial of order n)

$$P'_n \doteq \frac{dP_n(\cos \delta)}{d(\cos \delta)}.$$

The surface energy σ can now be written in the form:

$$\sigma = [c_1 P'_1 + c_2 P'_2 + \ldots + c_n P'_n + \ldots] \sin \delta \cos \varphi, \quad (26)$$

because $P'_n \sin \delta \cos \varphi$ is a particular solution of the Laplace equation. To calculate the coefficient C_n one need only equate equations (25) and (26), multiply term by term by $P'_n \sin \delta$ and put $\cos \delta = t$. One then multiplies by dt and integrates between -1 and $+1$. But from the known properties of the Legendre polynomials (7) one deduces that

$$\int_{-1}^{+1} (1 - t^2) P'_n P'_m \, dt$$

is zero for $n \neq m$ and equal to

$$\frac{2n(n+1)}{2n+1}$$

for $m = n$.

But we have, putting $\rho t_0 = x_0$

$$\int_{-1}^{+1} \epsilon (1 - t^2) P'_n \, dt = - \int_{-1}^{t_0} (1 - t^2) P'_n \, dt + \int_{t_0}^{1} (1 - t^2) P'_n \, dt;$$

but also

$$(1 - t^2) P'_n = \frac{n(n+1)}{2n+1}(P_{n-1} - P_{n+1}) =$$

$$= \frac{n(n+1)}{2n+1} \left[\frac{P'_n - P'_{n-2}}{2n-1} - \frac{P'_{n+2} - P'_n}{2n+3} \right],$$

from which finally, calculation gives

$$c_n = \frac{P_{n-2} - P_n}{2n-1} - \frac{P_n - P_{n+2}}{2n+3}, \quad \text{for } t = t_0, \quad (28)$$

Calling V_e and V_i respectively, the potential outside and inside the sphere arising from the distribution of the charges, we can write

$$V_e = \left[a_1 \frac{P'_1}{r^2} + \ldots + a_n \frac{P'_n}{r^{n+1}} + \ldots \right] \sin \delta \cos \varphi,$$

$$V_i = \left[b_1 P'_1 r + \ldots + b_n P'_n r^n + \ldots \right] \sin \delta \cos \varphi. \quad (29)$$

Since the potential must be continuous as we traverse the sphere and the normal components $-\partial V / \partial r$ of the field differ by $4\pi\sigma$, we obtain the following relations:

$$\frac{a_n}{\rho^{n+1}} = b_n \rho^n \; ; \quad 4\pi \mathcal{J} c_n = (n+1)\frac{a_n}{\rho^{n+2}} + nb\,\rho^{n-1}, \tag{31}$$

These allow us to calculate the coefficients a_n and b_n. Putting $r = \rho$ we get the potential at the surface of the sphere, V_s, and by surface integration over the sphere

$$\int \tfrac{1}{2}\sigma V_s ds$$

the energy W, where $ds = \rho^2 \sin\delta d\delta d\varphi$. Finally this gives

$$W = 4\pi^2 \mathcal{J}^2 \rho^3 \sum_{n=1}^{\infty} \frac{n(n+1)^2}{(2n+1)^2} C_n^2 \tag{32}$$

Taking only the first five terms of the series, we get the following values for the factor following the summation sign Σ.

$\cos\delta = t$	0	0,1	0,2	0,3	0,4	0,5	0,6	0,7	0,8	0,9	1,0
$\frac{W}{4\pi^2 \mathcal{J}^2 \rho^3} \times 10^3$	102	105	112	123	136	152	170	188	204	217	222

NOTE II

Magnetic energy of a sphere immersed in a medium with slowly varying direction of spontaneous magnetization: Using the same notation as in Note I, we now immerse the sphere in a medium in which the direction of magnetization rotates as a function of x, remaining parallel to the plane YOZ. Suppose that the angle θ of the magnetization with OX varies slowly so that, in the interval $-\rho < x < \rho$, we can put with sufficient approximation:

$$\theta = \theta_0 + x \cdot \frac{\partial \theta}{\partial x} = \theta_0 + ct, \quad \text{with } c = \rho \cdot \frac{\partial \theta}{\partial x}.$$

If we orientate the axis OX so that $\theta_0 = 0$, the direction cosines of the magnetization become: O, $\sin ct$ and $\cos ct$ and the density σ of charge distribution on the surface of the sphere becomes (if we neglect terms of third order in t):

$$\sigma = \left[c \cos\delta \sin\delta \sin\varphi + \left(1 - \frac{c^2}{2}\cos^2\delta\right)\sin\delta\cos\varphi \right] \mathcal{J}. \tag{33}$$

But

$$\frac{\sin\delta\cos\varphi}{r^2}, \quad \frac{\cos\delta\sin\delta\sin\varphi}{r^3}, \quad \frac{\left(\cos^2\delta - \frac{1}{5}\right)\sin\delta\cos\varphi}{r^4}$$

are particular solutions of the Laplace equation and one can write the potentials outside and inside the sphere V_e and V_i in the form:

$$V_e = \frac{A}{r^3}\cos\delta\sin\delta\sin\varphi + \frac{B}{r^2}\sin\delta\cos\varphi +$$
$$+ \frac{C}{r^4}\left(\cos^2\delta - \frac{1}{5}\right)\sin\delta\cos\varphi, \tag{34}$$

$$+ C' r^3 \left(\cos^2\delta - \frac{1}{5}\right)\sin\delta\cos\varphi. \tag{35}$$

The coefficients A, B, C, A', B', C', are again obtained by the condition that the potential is continuous as we go through the sphere, and the normal component of the field $-\partial V/\partial r$ is equal to $4\pi\sigma$. Replacing r by ρ in the expression for V_e we obtain the potential V_s on the surface as

$$V_s = 4\pi \mathcal{J}\rho \left[\frac{c}{5}\cos\delta\sin\delta\sin\varphi + \right.$$
$$\left. + \frac{1}{3}\left(1 - \frac{2c^2}{35}\right)\sin\delta\cos\varphi - \frac{c^2}{14}\cos^2\delta\sin\delta\cos\varphi \right]. \tag{46}$$

The energy W of the charge distribution is hence given by

$$W = \frac{1}{2} \int_0^\pi \int_0^{2\pi} \sigma V_s \rho^2 \sin\delta\, d\delta\, d\varphi, \tag{37}$$

and on integrating, we obtain

$$W = \frac{8\pi^2 \mathcal{J}^2 \rho^3}{9}\left(1 - \frac{2c^2}{25}\right). \tag{38}$$

A57
FOUNDATIONS OF A NEW GENERAL THEORY OF THE COERCIVE FORCE

Summary: In the introduction (§ 1 to 4) the principles of the theory of the coercive force as acknowledged at present, based on the motion of movement of walls separating domains, are recalled. The wall is restrained by *local variations* of its surface energy which are due either to elastic lattice perturbations (Becker, Döring etc.), or to the presence of non-magnetic inclusions (Kersten). The calculations are always of the form that changes occur in a medium with plane, rigid walls with lattice deformations or faults distributed on a regular simple cubic lattic.

The author shows (§ 7) that, preserving the hypothesis of a plane, rigid wall, but with an essentially irregular arrangement of lattice faults, which is *a priori* much more likely, the greatest possible variations in surface tension can in reality only cause negligible coercivity, of the order of tenths of gauss. If now one assumes that the walls are deformable, the resulting coercive force increases but remains nevertheless of the order of one gauss (§ 8–12). Hence the mechanisms called in by the existing theories cannot account for the experimentally observed coercive forces of several hundred gauss.

The author proposes a new mechanism (§ 12 and 13) based on the existence of fluctuations in the *direction* or *magnitude* of the spontaneous magnetization inside the domains. These fluctuations in direction are caused by elastic lattice deformations and those in magnitude by inequalities of concentration or the presence of inclusions or cavities. Because of the existence of these fluctuations, the divergence of the vector of magnetization is no longer zero and the resulting fictitious magnetic charges create a magnetic energy which varies according to the position of the wall, independent of the surface tension of the wall which only plays a secondary role.

Based on these considerations, an approximate expression for the coercive force is calculated (§ 20–25), fluctuations either as arising from variations in internal lattice stresses (equations 84, 85, 86) or as arising from cavities and inclusions (equations 98 and 99). It is shown (§ 26) that the magnitude of the coercive force so obtained is indeed similar to that observed: the inclusions play a preponderant role in the case of iron, and the strains in the case of nickel. Finally it is shown (§ 27) that equation (99) accounts very satisfactorily for a group of experimental results collected by Kersten in relation to the effect of inclusions on the coercivity of iron. The proposed theory is sufficiently general and flexible to allow the interpretation of a large number of experimental facts.

In the course of this work the author has been led to establish and to use an approximate mathematical representation of variations in surface tension of domain walls (§ 5–6) as well as an approximate expression for fluctuations in magnetic energy caused by irregularly distributed internal stresses (§ 14–19).

INTRODUCTION

(1) Coercive force and variations of the surface energy of domain walls: The works of Becker [1, pp 203–217], Kondorski [2], Kersten [3, 5] etc. agree in attributing hysteresis phenomena to local variations of surface energy of the walls separating elementary domains. Let us imagine a *plane* wall P, separating 2 domains magnetized antiparallel, which can be displaced remaining parallel, and denote by the abcissa x the displacement of the mid-plane of the wall along the axis OX, perpendicular to the wall. Suppose that the surface energy γ of the wall P per unit area varies with x in an irregular manner, (curve Γ, Fig. 1). In the absence of a magnetic field the wall occupies a minimum energy position such as A, A'. If now we apply a field H parallel to the spontaneous magnetization $\mathcal{J}_s$ of the domain to the left of A, the wall experiences a force $2H\mathcal{J}_s$ per unit area in the direction OX and moves along OX until that force is balanced by the force $-\partial\gamma/\partial x$. Thus, if H is small, the wall moves *reversibly* and returns to A after removal of the field. The reversibility is maintained until the wall reaches at B, point of inflexion of the curve Γ. Let h_c be the corresponding critical field. If the field exceeds, by however small an amount, this value h_c, the wall moves to a point such as C where the slope $d\gamma/dx$ of the curve Γ is equal to $-2H\mathcal{J}_s$, and this displacement can be quite large. The movement is *irreversible* since in removing the field the wall returns only to the nearest minimum at A' and not to its original position at A. One sees that one can establish a formal theory of hysteresis on these ideas. In particular the coercive force is closely related to the critical field corresponding to a maximum value of $(d\gamma/\partial x)_m$ the slope of curve Γ [4]. We have

---

[1] This is tantamount to saying that the wall cannot be regarded as a geometrical plane but has an appreciable thickness of several hundred angstroms.

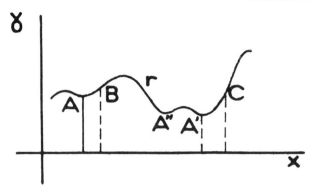

Fig. 1.

$$H_c \sim \frac{1}{2\mathcal{J}_s}\left(\frac{\partial \gamma}{\partial x}\right)_m . \qquad (1)$$

For a physical understanding we now have to specify the origin of these surface tension variations. Two phenomena have been considered: Variations in internal stress (Becker, Döring [1], Kondorski [2]) and the presence of non-magnetic inclusions (Kersten [3]).

(2) Effect of internal stresses: Becker has shown [1, pp 132–146] that if a ferromagnet, which for simplicity's sake we suppose to be isotropic with a saturation magnetostriction λ_s, is subjected to a simple tension σ making an angle θ with the direction of spontaneous magnetization, the energy is increased by a term of elastic deformation energy equal to $3/2\,\lambda_s\sigma\sin^2\theta$. The constant $C=\frac{3}{2}\lambda_s\sigma$ thus plays a role analogous to that of the anisotropy constant K so that the wall energy which is $2a\sqrt{JK}$ in the undeformed substance, becomes $2a\sqrt{J(K+bC)}$ as a result of the tension σ. In these expressions a is the lattice parameter, J the exchange energy per cm^3 and b a numerical constant of the order of 1, which depends on the orientation of the wall with respect to the crystal axes.

Any change in the direction or magnitude of the stress and strain is thus accompanied by a change in the surface energy. With strong deformation forces, C becomes of the order of magnitude K and one can already see that the amplitude $\delta\gamma$ of the varuations of wall energy can become of the order of γ itself, i.e. some ergs per cm^2.

Supposing that the deformations only depend on x, the wall surface energy which was discussed above, can vary by a quantity $\delta\gamma$ equal to some ergs per cm^2 in an interval $d=x_1x_2$. This interval can be very small if the variations in stress are on a very small spacial scale. Yet this interval cannot become smaller than the wall thickness e, because of the averaging effects which arise with a smaller scale. Since e is of the order of some hundreds of angstroms, one can see that the critical field

$$H_c = \frac{\delta\gamma}{2d\mathcal{J}_s} \qquad (2)$$

can reach some hundreds of gauss for very high strains. This is about the magnitude of the coercive forces of good permanent magnets.

(3) Effect of inclusions (Fremdkörper theory of Kersten): For a long time the workers in the field of permanent magnets have observed that impurities and non-soluble additions increase the coercive force of iron, solely by their presence and not by the action of strains. In order to explain this, Kersten has recently remarked that in the presence of non-magnetic inclusions, the mean wall energy would depend on the position of the wall. Let us have spherical cavities distributed at the lattice points of a simple cubic lattice (Fig. 2). It is clear that in the position P where the wall cuts a certain number of spheres, all those parts of the wall which lie inside the spheres have zero surface energy, the mean wall energy per unit surface area is thus smaller than in position P' where the wall passes between the spheres. If the holes occupy a volume comparable to that of the magnetic material the variations in wall energy can be of the order of γ. The magnitude of the corresponding critical field is again comparable with that observed experimentally.

(4) Critique of the preceeding theories: One can raise serious objections to the theories which have just been described. They assume in effect that the walls are planes and that the stresses which from now on we will

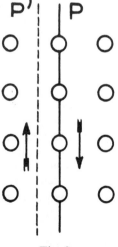

Fig. 2.

call more generally "accidents", are also planes because they are independent of y and of z.[2] An accident of a given thickness thus extends laterally parallel to the wall for distances which are at least equal to the lateral dimensions of the wall. For example, an accident of a thickness of $0.1\,\mu$, corresponding to the maximum effectiveness, extends laterally for at least $10\,\mu$ since this is the order of magnitude of the domains and their enclosing walls. These theoretical accidents differ enormously from the real ones which evidently have on average equal dimensions in all directions. But now, if this is so, a plane domain wall of lateral dimensions, large compared to those of the accidents, must average all the effects, the increases as well as the reductions of surface energy and the resultant total must be much smaller than that calculated above.

It follows that in order to establish definitely whether the hypothesis that the coercive force resulting from variations in wall surface energy can satisfactorily account for experimental results, it is essential to revert to more precise methods of calculation. This will be the object of the first part of this paper.

PART ONE

General theory of the propagation of a domain wall with variable surface energy

(5) Mathematical description of the propagation of a domain wall with varying surface energy: Let us first consider a substance with irregular distortions without otherwise defining the origin of the faults which can be of any kind, internal strains, inclusions, variations in composition etc. Let us simply assume that the scale of the irregularities is small compared to the domain dimensions and that their magnitude is not large enough to destroy the notion of conventional domains. These perturbations cause local variations in the surface tension of the walls and will at the outset establish a manageable mathematical representation of these conditions.

We refer our substance to a 3 dimensional rectangular coordinate system OXYZ. The surface energy γ_T of an element of area ds central about the point $M(x, y, z)$, is an irregular function of the coordinates x, y, z. We put γ_0 as the mean value and $\gamma = \gamma_T - \gamma_0$ the irregularly varying part of this function. Otherwise the accidents are mainly characterised by their *intensity* and *extension* in space.

The intensity is connected with γ^2, the mean square of γ. In a very strongly disturbed substance $\overline{\gamma_2}$ is of the same order of magnitude as γ_0^2.

The extension of size of the accidents is connected with η, *a correlation distance* such that in two points M and M' which are far apart with respect to η, the values of γ can be considered to be independent, but if M and M' are close compared to η the local values of γ are equal. We assume essentially that the correlation distance η is small relative to the domain dimensions.

Consider now in the substance an original volume B, e.g. a cube of sides $2\pi L$ which is of the order of the domain size. Inside this cube the purely varying part of the wall energy can be developed as a Fourier series

$$\gamma = \sum_{pqr} \gamma_{pqr} \begin{matrix}\sin\\ \cos\end{matrix}\frac{px}{L} \begin{matrix}\sin\\ \cos\end{matrix}\frac{qy}{L} \begin{matrix}\sin\\ \cos\end{matrix}\frac{rz}{L}, \qquad (3)$$

with the summation over all positive and zero values of p, q, r (except $p=q=r=0$) and let us note that to each set of p, q, r belong 8 coefficients $\gamma_{p,q,r}$ corresponding to the 8 possible combinations of sine and cosine. We suppose that the *fictitious* substance is obtained by repeating the original volume periodically along the 3 coordinate directions with a period of $2\pi L$, and this substance possesses properties practically identical to those of a real material.

The values of the coefficients $\gamma_{p,q,r}$ are related to the values of γ in the region B by the expression

$$\gamma_{pqr} = \frac{1}{\pi^3 L^3} \int_B \gamma \begin{matrix}\sin\\ \cos\end{matrix}\frac{px}{L} \begin{matrix}\sin\\ \cos\end{matrix}\frac{qy}{L} \begin{matrix}\sin\\ \cos\end{matrix}\frac{rz}{L}\, dx dy dz, \quad (4)$$

where the integration extends over the whole interior of B.[3] Changing the position of B in the real substance results in changing the values in the equation such that the coefficients $\gamma_{p,q,r}$ have no longer a definite value but must rather be considered as arbitrarily changing variables which we furthermore suppose to be independent. One can therefore derive the following expressions where a bar above a symbol signifies the mean value of the quantity with regard to all possible positions of the reference volume element B.

$$\overline{\gamma_{pqr}} = 0, \qquad \overline{\gamma_{pqr}\gamma_{p'q'r'}} = 0. \qquad (5)$$

In fact, only the mean squares $\overline{\gamma^2_{p,q,r}}$ of the coefficients enter the calculation. It is therefore a question of calculating the mean squares as functions of $\overline{\gamma^2}$ and of η, the constants which characterise the perturbations.

(6) Calculation of the mean squares of the coefficients of the Fourier series development of an irregular function: If we put $d\tau = dx dy dz$ and $d\tau' = dx' dy' dz'$ and square equation (4) we obtain, for instance for a

---

[2] In reality in Kersten's theory the accidents are not planes but since he deals with identical accidents situated on the lattice points of a simple cubic lattice, with walls parallel to one of the cube faces, this comes mathematically to the same point.

[3] This equation only applies when the product $p.q.r$ differs from zero. When one of the quantities p, q, r is zero one must multiply the second term by $\frac{1}{2}$ and if two are zero by $\frac{1}{4}$.

term with 3 cosines:

$$\gamma^2_{pqr} = \frac{1}{\pi^6 L^6} \int_B \int_B \gamma\gamma' \cos\frac{px}{L} \cos\frac{px'}{L} \cos\frac{qy}{L} \cos\frac{qy'}{L} \cos\frac{rz}{L} \cos\frac{rz'}{L} d\tau d\tau' \,. \quad (6)$$

Take a cube C of side twice the correlation length η, centred about the point $M(xyz)$. Because of our approximation we suppose that if the point $M'(x', y', z')$ is outside the cube, the mean value of $\gamma\gamma'$ for all positions relative to the cube is zero, whereas if M' is inside the cube this mean value is $\overline{\gamma^2}$. Under these conditions, if p, q and r are such that $p\eta/L$, $q\eta/L$ and $r\eta/L$ are small compared to $\pi/2$, the integral (6) reduces to

$$\overline{\gamma^2_{pqr}} = \frac{8\eta^3 \overline{\gamma^2}}{\pi^6 L^6} \int_B \cos^2\frac{px}{L} \cos^2\frac{qy}{L} \cos^2\frac{rz}{L} d\tau, \quad (7)$$

and finally,

$$\overline{\gamma^2_{pqr}} = \frac{8\eta^3 \overline{\gamma^2}}{\pi^3 L^3} \quad (8)$$

If on the contrary $p\eta/L$, $q\eta/L$, and $r\eta/L$ are large compared to $\pi/2$, the mean value of $\cos px'/L \cdot \cos qy'/L \cdot \cos rz'/L$ in the cube C tends to zero and so does $\overline{\gamma^2_{pqr}}$. Now let us put $N = \pi L/2\eta$. We can then reconsider and describe the results by saying that since p, q and r are all smaller than N, the mean square $\overline{\gamma^2_{qpr}}$ is given by

$$\overline{\gamma^2_{pqr}} = \frac{\overline{\gamma^2}}{N^3} ; \quad (9)$$

and is independent of p, q and r, but it is zero if any one of the quantities p, q and r is larger than N. These results are very simple and we can accept them a priori as defining the simplest ideally irregular perturbation.

1st remark: The preceding calculations do not apply except when the product pqr is zero. If one of the quantities p, q or r is zero, one has, for instance:

$$\overline{\gamma^2_{oqr}} = \frac{1}{2} \frac{\overline{\gamma^2}}{N^3} \quad (10)$$

and when two are zero

$$\overline{\gamma^2_{oor}} = \frac{1}{4} \frac{\overline{\gamma^2}}{N^3} \,. \quad (11)$$

2nd remark: In a sense the approximations of this paragraph mean that we divide the base cube B into $8N^3$ cubes equal to cube C and that in each of the latter, the value of γ is constant and independent from that in all other cubes C. One finds these $8N^3$ independent irregular variables in the form of $8N^3$ independent coefficients γ_{pqr} of the development of equation (3), which is logically satisfactory.

3rd remark: Such a representation of the irregularities of the domain wall surface energy is undoubtedly not strictly mathematically rigorous, but at any rate it appears to approach much closer to reality than the method of plane accidents. Moreover, let us recall that similar methods used to calculate the approach to saturation, yield satisfactory results which agree with experimental facts [6][7].

4th remark: All these results only apply in a substance where the correlation length η is larger than the domain wall thickness e. If η is smaller than e one must limit the summation in (3) to values of p, q and r, each smaller than the limit $N' = \pi L/2e$. In fact, as we have already said above (para 2) the wall integrates the accidents of small wavelength and, all other things being equal, their influence becomes less and less as the wavelength decreases below e. For simplicity we finally suppose that the γ_{pqr} are zero if any one of the values p, q and r is larger than the smaller of the two numbers N and N'.

(7) Study of the movement of a plane wall: Now let us study the movement of a *plane* wall, parallel to XOY. Its surface energy per cm^2 as a function of z is obtained by averaging the expression (3) over x and y and hence reduces to

$$\gamma = \sum_r \gamma_{oor} \frac{\sin}{\cos} \frac{rz}{L}, \quad (12)$$

where the summation does not extend further than 2N terms corresponding to the zero values of p and q. One deduces

$$\frac{\partial \gamma}{\partial z} = \sum_r \gamma_{oor} \frac{r}{L} \frac{+\cos}{-\sin} \frac{rz}{L}. \quad (13)$$

To calculate the mean coercive force we must know the mean value of the maximum of $\partial\gamma/\partial z$ for all possible positions of the cube B, $(\partial\gamma/\partial z)_m$. To find this we first calculate the mean square of $\partial\gamma/\partial z$, which, using (13) is

$$\overline{\left(\frac{\partial \gamma}{\partial z}\right)^2} = \sum_r \frac{1}{2} \frac{r^2}{L^2} \overline{\gamma^2_{oor}}, \quad (14)$$

and using equation (11) it becomes

$$\overline{\left(\frac{\partial \gamma}{\partial z}\right)^2} = \sum_{r=1}^{N} \frac{r^2}{4L^2} \frac{\overline{\gamma^2}}{N^3} = \frac{1}{12} \frac{\overline{\gamma^2}}{L^2}. \quad (15)$$

To establish a relation between $\overline{(\partial \gamma/\partial z)^2}$ and $\overline{(\partial \gamma/\partial z)^2}$ we use a method previously applied to the interpretation of the Rayleigh laws in weak fields. We approximate the function $\gamma = f(z)$ by a polygonal contour of 2N sides, each having a length L/N and with slopes given by a Gauss probability law. In each case thus we conserve the same number 2N of irregular independent variables. It is then quite easy to see that the fraction $\overline{(\partial \gamma/\partial z)^2_m}:\overline{(\partial \gamma/\partial z)^2}$ is of the order of magnitude log 2N. Now using expression (1) of the coercive force, we get

$$H_c \sim \sqrt{\frac{\log 2N}{12}} \frac{\sqrt{\overline{\gamma^2}}}{2\mathfrak{J}_s L}. \quad (16)$$

As we have already observed, $\sqrt{\overline{\gamma^2}}$ is maximally of the order of γ_0 i.e. a few units of c.g.s. L is of the order of 10^{-3} cm, N (or N') is at most of the order of one hundred. The coercive force given by (16) is hence about 1 gauss for the strongest possible perturbations of the wall energy, well below the values of several hundred gauss given by the methods of Döring, Kornetzki, Kersten etc.

(8) Study of movements of a deformable wall: The preceding result does, however, not prove that the coercive force does not originate from inequalities of the surface tension of the wall; this is because another hypothesis of the above mentioned authors is unacceptable: that of a plane wall. The wall must deform and bulge irregularly as it moves. We will now drop the restriction of planeness and study the consequences of this fact. For this we will use the expression for γ given in (3), but to simplify the formulation and later calculations we will assume that the terms corresponding to both p and q being simultaneously zero, vanish. This causes no difficulty since we have shown that their influence is negligible and that besides these terms, independent of x and y, do not tend to deform the wall.

We assume that the coefficients γ_{pqr} are sufficiently small compared to γ_0 so that the wall is not much deformed and that any position of the wall, on average parallel to the plane XOY and of mean position h_0, can be represented by the equation

$$z = h_0 + \sum_{pq} h_{pq} \genfrac{}{}{0pt}{}{\sin}{\cos} \frac{px}{L} \genfrac{}{}{0pt}{}{\sin}{\cos} \frac{qy}{L}. \quad (17)$$

In principle this wall divides two regions: on the side of negative z a region where the spontaneous magnetization $\mathfrak{J}_s$ is everywhere parallel to the direction OX, assumed a preferred direction; on the side of positive z a region with antiparallel spontaneous magnetization. In reality the spontaneous magnetization undergoes local deviations in the vicinity of the wall, and these we will also have to take into account.

We propose to determine the possible position (or positions) of equilibrium of the wall subjected to a field parallel to OX. To do this we will have to determine the corresponding values of the coefficients h_0 and h_{pq}. Therefore we will determine the minimum of the free energy F as a function of these coefficients.

(9) Calculation of the different energy terms: This energy F, which we will give per unit surface area, is the sum of three terms: the magnetic energy F_H in the externally applied field H, the energy of the wall F_P and the energy F_D of the so called magnetic field of *dispersion* which arises from the magnetic charges distributed along the wall because it makes an angle with the spontaneous magnetization.

Energy in the applied field: This depends only on the field H and the mean position h_0 of the wall; except for a constant it is:

$$F_H = -2h_0 \mathfrak{J}_s H. \quad (18)$$

Wall energy: This consists of two terms. The first F'_P arises from the increase in surface area of the wall due to the deformation. To calculate this, taking only terms to the second order, it is sufficient to assume a constant surface tension γ_0, and this gives

$$F'_P = \frac{\gamma_0}{8L^2} \sum_{pq} (p^2 + q^2) h_{pq}^2. \quad (19)$$

The second term F''_P corresponds to the local variations of the surface energy. It is obtained by developing γ according to equation (3) into a Taylor series in z about the value $z = h_0$ to refer to the value of z given in equation (17). One then forms the mean with respect to x and y and replaces the true element of wall area by its projection on the plane XOY. Thus one obtains

$$F''_P = \gamma_0 + \sum_{pq} \left\{ \frac{h_{pq}}{4} \sum_r \frac{r}{L} \gamma_{pqr} \genfrac{}{}{0pt}{}{+\cos}{-\sin} \frac{rh_0}{L} \right\}. \quad (20)$$

To write this equation (20) we use the development into a Taylor series of the Fourier series (3). This process is only valid for the case where the values of the distances $z - h_0$ of the real wall from the mean plane of the wall, remain small versus the smallest of the quantities L/p,

L/q. This is quite realistic because on the one hand, from the physical point of view, the series (3) can easily be considered to be limited and on the other hand one can study perturbations of sufficiently small amplitude for the distance $z - h_0$ to be smaller than any previously fixed limit.

Finally, the wall energy F_P is given by

$$F_P = F'_P + F''_P. \quad (21)$$

Magnetic energy: If the two domains which are separated by the wall retain a rigid magnetization, there appears on that wall a surface density of magnetic charges σ, given by:

$$\sigma = \frac{2\mathcal{J}_s}{L} \sum_{pq} ph_{pq} \begin{array}{c} -\cos \\ +\sin \end{array} \frac{px}{L} \begin{array}{c} \sin \\ \cos \end{array} \frac{qy}{L}. \quad (22)$$

The effect of these charges, in the first order, is as if the density σ were a priori sitting on the mean wall of the position h_0, instead of being on the real wall (17). These charges create a magnetic field with components perpendicular to OX, which deflect the spontaneous magnetization from its original direction parallel to OX. For instance, in the lower domain with negative z, let β and γ be the angles, assumed to be small, which the projection of the magnetization on the planes XOY and XOZ makes respectively with OX. At each point the spontaneous magnetization is in equilibrium under the action of the magnetocrystalline forces which arise from a potential $K(\beta^2 + \gamma^2)$ and of a magnetic field arising from a potential V. The equilibrium conditions are:

$$2K\beta + \mathcal{J}_s \frac{\partial V}{\partial y} = 0 \; ; \quad 2K\gamma + \mathcal{J}_s \frac{\partial V}{\partial z} = 0. \quad (23)$$

In the upper domain the equations are similar; one only has to replace $-\mathcal{J}_s$ by $+\mathcal{J}_s$.

In these equations we neglect the Weiss–Heisenberg interactions at short distances. This is justified as long as the radius of curvature of the real deformed wall is large compared with the wall thickness (0.1 μ for iron). We assume that this condition applies.

Since the magnetization is no longer uniform, a volume density m of magnetic charges appears, and this is connected with the potential V by Poisson's equation $\Delta V + 4\pi m = 0$ and is given to the second order by:

$$m = -\operatorname{div} \vec{\mathcal{J}} = -\mathcal{J}_s \left(\frac{\partial \beta}{\partial y} + \frac{\partial \gamma}{\partial z} \right)$$

Using the 4 preceding equations to fix the extent of β, γ and m one finds finally that the potential partial derivatives of V satisfy the equation

$$\frac{\partial^2 V}{\partial x^2} + Q \left(\frac{\partial^2 V}{\partial y^2} + \frac{\partial^2 V}{\partial z^2} \right) = 0 \quad (24)$$

after using the abbreviation

$$Q = 1 + \frac{2\mu \mathcal{J}_s^2}{K}. \quad (25)$$

We must now find two analytical expressions for the potential which are solutions of (24), the one valid in the region of negative $z - h_0$ giving V, the other for $z - h_0$ positive, giving V'. V = 0 for $z =$ infinity and on the wall itself satisfying the condition:

$$\left(\frac{\partial V}{\partial z} \right)_0 - \left(\frac{\partial V'}{\partial z} \right)_0 = 4\pi (\sigma + \gamma_0 \mathcal{J}_s + \gamma'_0 \mathcal{J}_s), \quad (26)$$

Here the index 0 signifies the coordinates of the plane of the wall, and we take account of the charges σ and the charges arising from the deviation of the spontaneous magnetization. Using (23) and (25), equation (26) can also be written as

$$\left(\frac{\partial V}{\partial z} \right)_0 - \left(\frac{\partial V'}{\partial z} \right)_0 = \frac{4\pi\sigma}{Q}. \quad (27)$$

So that one finds for V:

$$V = \sum_{pq} V_{pq} \begin{array}{c} -\cos \\ +\sin \end{array} \frac{px}{L} \begin{array}{c} \sin \\ \cos \end{array} \frac{qy}{L} e^{m(z-h_0)} \quad (28)$$

with the notations

$$m = \frac{1}{L} \sqrt{q^2 + \frac{p^2}{Q}} \quad (29)$$

$$V_{pq} = \frac{4\pi p \mathcal{J}_s}{mLQ} h_{pq}. \quad (30)$$

To get an equation for V' one only has to replace V by V' and m by $-m$ in the one equation, (28).

Knowing V and V' it is easy to deduce β, β' and β', γ'.

In addition it is important to find the energy of the total arrangement, which is obtained by adding to the magnetic energy at every point the magnetocrystalline energy and to integrate over the total volume. Taking account of equation (23) one gets:

$$W = \int \left[\frac{H^2}{8\pi} + K(\beta^2 + \gamma^2) \right] d\tau$$

$$= \frac{1}{8\pi} \int \left[\left(\frac{\partial V}{\partial x} \right)^2 + Q \left(\frac{\partial V}{\partial y} \right)^2 + Q \left(\frac{\partial V}{\partial z} \right)^2 \right] d\tau. \quad (31)$$

Having carried out all calculations, the energy corresponding to one cm^2 of the wall is:

$$F_D = \frac{1}{16\pi} \sum_{pq} \frac{p^2 + Qq^2}{mL^2} V_{pq}^2 = \quad (32)$$

$$\frac{\pi \mathcal{J}_s^2}{L\sqrt{Q}} \sum_{pq} \frac{p^2}{\sqrt{p^2 + Qq^2}} h_{pq}^2 .$$

If we had assumed a rigid magnetization, which is equivalent to taking K as infinite, the corresponding energy would be obtained by putting Q=1 in the above equation. One can see that the deformation of the direction of magnetization which occurs in reality divides the energy by a factor of at least $\sqrt{Q}$, i.e. by at least 6.6 for iron and 5.7 for nickel at ordinary temperature, and by more at high temperatures. By neglecting this deformation one would incur a great error!

(10) Equilibrium of the wall: The total energy which is obtained by inserting the values of F_H, F'_P, F''_P and F_D, given by equations (18), (19), (20) and (32) is a function of h_0 and h_{pq}. At the wall equilibrium position, the partial differential quotients of the total energy with respect to any one of these variables, become zero which gives the equations

$$-2\mathcal{J}_s H - \sum_{pq} \left\{ h_{pq} \sum_r \frac{r^2}{4L^2} \gamma_{pqr} \frac{\sin}{\cos} \frac{rh_0}{L} \right\} = 0. \quad (33)$$

$$2h_{pq} A_{pq} + \sum_r \frac{r}{4L} \gamma_{pqr} \frac{+\cos}{-\sin} \frac{rh_0}{L} = 0 \quad (34)$$

in which, to simplify the writing, we have put:

$$A_{pq} = \frac{\pi \mathcal{J}_s^2}{L\sqrt{Q}} \frac{p^2}{\sqrt{p^2 + Qq^2}} + \frac{\gamma_0}{8L^2}(p^2 + q^2) \quad (35)$$

Eliminating the $h_{p,q}$ parameters, (33) takes the form:

$$-2\mathcal{J}_s H + \frac{\partial \Gamma}{\partial h_0} = 0, \quad (36)$$

with the notation

$$\Gamma = \sum_{pq} \left\{ \frac{1}{4A_{pq}} \sum_r \frac{r}{4L} \gamma_{pqr} \frac{-\cos}{+\sin} \frac{rh_0}{L} \right\} . \quad (37)$$

Equation (36) represents the equilibrium position of a fictitious plane and rigid wall separating the two elementary domains in question, which has an energy which is an irregular function of the position h_0 given by equation (37). The complex problem of the displacement of a deformable wall is here reduced to the much simpler problem of the displacement of a rigid wall.

(11) Determination of the coercive force: In order to calculate the coercive force, we first determine $d\Gamma/dh_0$ using equation (37).

$$\frac{\partial \Gamma}{\partial h_0} = \sum_{pq} \left\{ \frac{1}{32 A_{pq}} \sum_{rr'} \frac{rr'^2}{L^3} \gamma_{pqr} \gamma_{pqr'} \frac{-\cos}{+\sin} \frac{rh_0}{L} \frac{\sin}{\cos} \frac{r'h_0}{L} \right\} .$$

(38)

Then we calculate the mean square of the quantity (38) for all values of h_0 and all possible positions of the base B, which gives, if we remember that the right angle terms vanish in the mean;

$$\overline{\left(\frac{\partial \Gamma}{\partial h_0}\right)^2} = \sum_{pqrr'} \left(\frac{1}{32}\right)^2 \frac{1}{4 A_{pq}^2 L^6} \overline{\gamma_{pqr}^2} \overline{\gamma_{pqr'}^2} r^2 r'^4 . \quad (39)$$

Carrying out the summation with respect to r and r' and replacing $\overline{\gamma_{pqr}^2}$, $\overline{\gamma_{pqr'}^2}$ by their values given in equation (9), we get simply

$$\overline{\left(\frac{\partial \Gamma}{\partial h_0}\right)^2} = \left(\frac{1}{32}\right)^2 \frac{N^2}{15 L^6} (\overline{\gamma^2})^2 \sum_{pq} \frac{1}{A_{pq}^2} \quad (40)$$

What really interests us is the mean square of the maximum $(d\Gamma/dh_0)_m$ of $d\Gamma/dh_0$. Using arguments similar to those used in paragraph 7, one can deduce that the ratio

$$\overline{\left(\frac{\partial \Gamma}{\partial h_0}\right)^2_m} : \overline{\left(\frac{\partial \Gamma}{\partial h_0}\right)^2}$$

must be of the order of magnitude of log 4N because the expression of the energy given in equation (37) contains only 4N truly independent terms in sine and cosine. According to equations (1) and (4) one obtains finally the following approximate value of the coercive force

$$2H_c \mathcal{J}_s \sim \frac{N\overline{\gamma^2}}{32 L^3} \sqrt{\frac{\log 4N}{15} \sum_{pq} \frac{1}{A_{pq}^2}} . \quad (41)$$

Calculation of $\Sigma(1/A_{pq}^2)$:- The most important terms of this sum correspond to small values of A_{pq}, amongst which we will begin to calculate those corresponding to $p=0$. One sees easily from equation (35) that

$$\sum_q \frac{1}{A_{0q}^2} = \left(\frac{8L^2}{\gamma_0}\right)^2 \sum_{q=1}^N \frac{1}{q^4} \sim \left(\frac{8L^2}{\gamma_0}\right)^2 \frac{\pi^4}{90} . \quad (42)$$

Furthermore, we must multiply this expression by 2 to take account of the terms in sine and cosine, and divide it by 4, because in replacing $\overline{\gamma_{pqr}^2}$ and $\overline{\gamma_{pqr'}^2}$ by their value we have consistently used equation (9) whereas we should have used (10) because p is zero.

Now let us calculate the terms of the sum for which p is not zero. The preponderant terms correspond to small values of p for which p/Q is negligible compared with q. Also, by neglecting $p^2\gamma_0/8L^2$ versus $q^2\gamma_0/8L^2$ we commit an error leading to an overestimate in the summation, which, as we will see is not important. Having said all this, one has:

$$A_{pq} = \frac{K}{2L}\frac{p^2}{q} + \frac{\gamma_0}{8L^2}q^2 \tag{43}$$

and

$$\frac{1}{A_{pq}^2} = \left(\frac{8L^2}{\gamma}\right)^2 \frac{q^2}{\left(q^3 + \frac{4KL}{\gamma_0}p^2\right)^2}. \tag{44}$$

The sum with respect to q is simply evaluated by transforming it into an integral:

$$\sum_q \frac{1}{A_{pq}^2} \sim \frac{1}{3}\left(\frac{8L^2}{\gamma_0}\right)^2 \left[\frac{1}{\frac{4KL}{\gamma_0}p^2} - \frac{1}{N^3 + \frac{4KL}{\gamma_0}p^2}\right]. \tag{45}$$

The second term in the bracket is negligible compared to the first. It now remains to sum with respect to p, which becomes

$$\sum_{p=1}^\infty \frac{1}{p^2} = \frac{\pi^2}{6} \tag{46}$$

and finally

$$\sum_{pq} \frac{1}{A_{pq}^2} \sim \frac{\pi^2}{18}\left(\frac{8L^2}{\gamma_0}\right)^2 \frac{\gamma_0}{4KL} \quad (p=0 \text{ excluded}). \tag{47}$$

But, $\gamma_0/4KL$ is always small compared to unity since γ_0 is of the order of one, K of the order of 10^5 and L of the order of 10^{-3}. It follows that the sum (47) is negligible compared to that in (42). The physical significance of this result is however clear: The only deformations of the wall which have a significant effect are cylindrical with axes parallel to the spontaneous magnetization. The wall on the whole resembles an undulating sheet with undulations parallel to the magnetization; the demagnetizing field opposes other deformations.

Finally, according to (41) and (42), the coercive force is:

$$H_c \sim \frac{1}{30}\frac{N\overline{\gamma^2}}{L\mathcal{J}_s\gamma_0}\frac{\pi^2}{8}\sqrt{\frac{\log 4N}{3}}. \tag{48}$$

It becomes larger in the same measure as L/N becomes smaller, but the smallest value that that quantity can take is of the order of e, the thickness of the wall, i.e. about 10^{-5} cm. As $\mathcal{J}_s$ is of the order of 10^3 it follows that for the greatest possible deformations, equation (48) gives a coercive force which does not exceed a few Gauss and which is hence considerably below the experimentally observed values.

Thus, if one takes account of wall deformations one obtains a coercive force (equation (48)) quite a lot larger than with the hypothesis of a plane wall (equation (16)), but which nevertheless does not exceed 10% of the experimental values. The variations of the surface tension of the wall do not on the whole play anything but a quite secondary part in the origin of the coercive force.

Hence, the hypotheses which are actually assumed do not permit an interpretation of the experimental facts. It is therefore necessary to search for a real mechanism which prevents wall displacement.

SECOND PART

New theory of the coercive force

(12) On the certain role of the internal stresses and cavities: Even if the variations of the wall surface tension play only a subordinate role in the explanation of the coercive force, it remains quite certain that local anisotropies due to the presence of internal stresses must play an extremely important part in the phenomena of hysteresis. Of this one can give two convincing proofs: Firstly, one knows that an anneal which causes a relaxation of the stresses of deformation also results in a considerable reduction of the coercive force. Secondly, since in substances with zero magnetostriction internal stresses do not cause any magnetic anisotropy, one must expect that such materials, e.g. permalloys, have very weak coercive forces. This is precisely what is observed.

Besides, as we have already said (para. 3), the cavities and non-magnetic inclusions equally have an influence.

We will show that in these two cases, the magnetization of the elementary domain is no longer uniform and that the magnetic disperse fields which result, hinder the free movement of the domain walls.

(13) The magnetic disperse fields and the coercive force: Let us have initially two neighbouring elementary domains, magnetized in antiparallel directions and separated by a plane wall parallel to the direction of magnetization. When the crystal contains no deformations or inclusions, the magnetization is perfectly uniform and equal to the theoretical spontaneous magnetization inside each domain and the wall of separation is able to move freely without expenditure of energy.

Now imagine in the interior of the crystal irregularly distributed stresses. Except in some rare spots, these disturbances tend to deflect the spontaneous magnetization from the ideal original direction of easy magnetization. A new distribution of the spontaneous magnetization tends to be established: But this new distribution, which need not have zero divergence, is accompanied by the appearance of internal free magnetic charges, therefore we have *magnetic disperse fields*. This requires energy and hence opposes the deflections of the spontaneous magnetization and limits their magnitude. We are dealing here, on the whole, with *magnetic interactions* between different regions of the same elementary domain.

Hence an equilibrium establishes itself, characterised by a certain density m of magnetic charges: $m = -\text{div } \mathcal{J}$, which varies from point to point inside each domain. If the spontaneous magnetization reverses by 180°, as happens after the passage of a wall, the new equilibrium is identical to that before, except that the sign of the charges is reversed.

Now consider a domain A (Fig. 3a) magnetized to the right, inside which we have drawn some islands of magnetic charges, some positive, some negative; the domain A is separated by a wall P from a domain B, magnetized in the opposite sense. One can see that if the wall takes the position of Fig. 3b, in which it bisects a certain number of these islands, the potential energy of the system becomes smaller. The passage of the wall in fact changes the sign of the charges: the bisection of an island thus causes opposite charges to be in the vicinity of each other, which decreases the potential energy. The passage of a wall through a domain in this way is hence accompanied by continuous fluctuations of the potential energy, which causes hysteresis phenomena according to the mechanism described in paragraph 1. The variations of the surface energy of the wall play only a minor role and as a first approximation, can be neglected.

The appearance of hysteresis and coercive force is thus tied to variations of the direction of the spontaneous magnetization inside the domains. Let us note straightaway that irregular variations of the magnitude of the spontaneous magnetization also give rise to internal magnetic charges and hysteresis effects by a mechanism very similar to the one above. Such variations in magnitude of the spontaneous magnetization will be produced by the presence of a non-magnetic phase finely dispersed inside the principal phase.

(14) Mathematical expression of the perturbations of the energy due to irregular internal stresses: The mathematical working out of the theory of which the principles have been outlined, is unfortunately very difficult. We shall have to establish and use very crude methods, but we hope they will suffice to give at least an order of magnitude estimate of the phenomena.

Take first an elementary domain referred to three rectangular axes OXYZ magnetized along OX, the preferred direction, and let the magnetization $\mathcal{J}_s$ be confined along OX by magnetocrystalline forces and eventually by a field H. If in a volume dv the spontaneous magnetization deviates a little from OX in a direction defined by its direction cosines β and γ with respect to OY and OZ, the energy of the system increases by $(K + (H \mathcal{J}_s/2))(\beta^2 + \gamma^2)$ where K is the anisotropy constant. If we now assume that the substance is perturbed by internal stresses, the preferred direction will deviate more or less and the magnetocrystalline energy will be more or less modified; one must hence add to the expression of the energy of the system, which should be a minimum for the equilibrium state, a perturbation term $F(\beta, \gamma)$ which is a function of x, y and z, and the energy W_t of the system becomes

$$W_t = \int \left[F(\beta, \gamma) + \left(K + \frac{H \mathcal{J}_s}{2} \right) (\beta^2 + \gamma^2) \right] dv.$$

(49)

This expression of the energy does not yet take account of the magnetic interactions in the sense discussed above; we have still to study the manner in which we will introduce above; we have still to study the manner in which we will introduce them. For the time being the difficulty consists in giving F a form permitting to develop the final calculations, notably to conserve the linear character of the equation of the

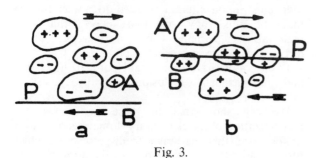

Fig. 3.

second order partial derivatives which determine the magnetic potential. The most general form which we can adopt is the following:

$$F = C[r(\beta^2 + \gamma^2) - 2(\beta\beta' + \gamma\gamma')], \qquad (50)$$

where C is a constant independent of x, y, z, with dimensions of an energy, r is a number and β' and γ' functions of x, y, z, which can be developed into a Fourier series in the interior of the base:

$$\beta' = \sum \beta'_\lambda \sin\frac{p'x}{L} \cos\frac{q'y}{L} \sin\frac{r'z}{L},$$
$$\gamma' = \sum \gamma'_\lambda \sin\frac{p'x}{L} \sin\frac{q'y}{L} \cos\frac{r'z}{L}. \qquad (51)$$

The summation is extended over all possible values of the index λ, which for simplicity has replaced the quadruple indices $p'q'r't$ where $p'q'$ and r' can have all integer positive values and t can take values from 1 to 8, corresponding to the 8 possible combinations of sine and cosine.

The functions β', γ' are irregularly varying functions in the sense indicated in paragraph 5. The coefficients β'_λ, γ'_λ of the development are irregular variables of which the mean square is constant as long as p', q', r' remain smaller than a certain limit N and zero from there onwards. This limit becomes proportionally greater as the correlation distance of the disturbances becomes smaller. In what follows we put, for simplification

$$\lambda = \frac{p'}{L}, \quad \mu = \frac{q'}{L}, \quad \nu = \frac{r'}{L}, \quad l = \frac{N}{L}. \qquad (52)$$

(15) Principle of the determination of C, r, β' and γ', from the characteristics of the disturbance: The following manner of the irregular internal stresses which perturb the substance will be pictured: we will assume that they are pure tensions σ_i of constant magnitude but with irregular distribution of the orientation along all directions in space, such that on average they are isotropic. Further, in order to make the scheme a little more general, we will assume that these tensions affect only a fraction of the total volume of the substance, leaving a part $(1-v)$ of the body undisturbed.

To determine the value of F which corresponds to the distribution just described, we employ an expedient: We will assume that K, the magnetocrystalline energy constant, is zero, we neglect the magnetic interactions and calculate the law of approach to saturation of the substance by two different methods, the rigorous classical one and a method where we put in an energy F of the type given by equation (50). *We will assign values to C, r, β' and γ' so that this second method gives the same results as the first one.* The energy is thus completely determined and the expression so obtained will allow us to study the influence of the magnetic interactions.

(16) Determination of the law of approach by the classical method: To obtain the law of approach of the perturbed substance we consider that is composed of small elements $d\tau$ subjected to uniform tension σ_i, constant in magnitude but randomly distributed in direction. Since, by hypothesis, we neglect the magnetic interaction, we obtain the law of approach of the assembly by averaging the laws of approach of the elements $d\tau$. In reality the set of small domains here considered, has hysteresis: to be precise the law of approach in this case represents the magnetization $\mathcal{J}'$ obtained after magnetization to saturation, and subsequent reduction of the field to zero. Calling the angle of the spontaneous magnetization with the direction of tension, θ, the magnetoelastic energy becomes, as we have recalled in paragraph 2:

$$W = \frac{3}{2}\lambda_s \sigma_i \sin^2\theta = C \sin^2\theta. \qquad (53)$$

An identical calculation has already been done by Gans [8] for the case of the law of approach of polycrystalline cobalt, where the magnetocrystalline energy can take the form of equation (53). The law of magnetization depends only on the parameter $\xi = (H\mathcal{J}_s/C)$; hence it has the form:

$$\mathcal{J}' = \mathcal{J}_s[1 - vf(\xi)], \qquad (54)$$

since the perturbation only applied to a fraction v of the substance, the rest being saturated for all values of the field. When the field is weak, one obtains the following series:

$$f(\xi) = \frac{1}{2} - \frac{1}{3}\xi + \frac{3}{32}\xi^2 - \frac{1}{30}\xi^3 + \ldots \qquad (55)$$

When it is very strong, one has, near saturation:

$$f(\xi) = \frac{4}{15}\frac{1}{\xi^2} + \frac{32}{105}\frac{1}{\xi^3} + \ldots \qquad (56)$$

This law of approach is shown as A on Figure 4 for the case $v=1$.

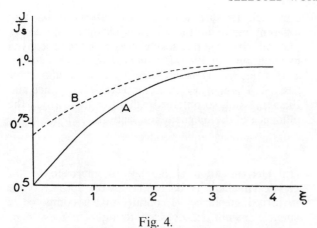

Fig. 4.
Curve A: approach to saturation of a polycrystalline uniaxial substance, calculated neglecting the magnetic interactions. Curve B: the same but taking the interactions into account.

(17) Determination of the law of approach, using the expression F for the energy of perturbation: Since K is zero and magnetic interactions are neglected the local equilibrium conditions for the magnetization under the action of the field H, are simply

$$-\frac{\partial F}{\partial \beta} = H\mathcal{J}_s \beta, \quad -\frac{\partial F}{\partial \gamma} = H\mathcal{J}_s \gamma, \quad (57)$$

From this one finds β and γ. With the approximation we use, the local magnetization in the direction of the field H is equal to:

$$\mathcal{J} = \mathcal{J}_s \left(1 - \frac{\beta^2 + \gamma^2}{2}\right).$$

Averaging this over all values of x, y, z, one finally gets for the law of approach

$$\mathcal{J} = \mathcal{J}_s \left[1 - \frac{P}{\left(\frac{\xi}{2} + r\right)^2}\right], \quad (58)$$

where, for simplicity, we have put

$$P = \frac{1}{16} \sum_\lambda (\beta'^2_\lambda + \gamma^2_\lambda) \quad (59)$$

and

$$\xi = \frac{H\mathcal{J}_s}{C}$$

(18) Determination of P and r: First we will assume that the constants C defined in equations (50) and (53) are identical. Then, in order to express that the approximate law (58) coincides best with the exact law (54) near a certain magnetization $\mathcal{J}$, we determine P and r so that in that point the two curves of magnetization have the same tangent. This gives

$$\frac{P}{\left(\frac{\xi}{2} + r\right)^2} = vf(\xi); \quad \frac{-P}{\left(\frac{\xi}{2} + r\right)^3} = v\frac{\partial f(\xi)}{\partial \xi} \quad (61)$$

The values of P/v and r so determined are shown by two curves in Fig. 5 as functions of $F(\xi)$ as parameter, i.e. against $(\mathcal{J}_s - \mathcal{J})/v\mathcal{J}_s$.

The very small perturbations are particularly interesting; the corresponding values are: $P/v = 1/15$ and $r = -2/7$.

(19) Application of the proposed method to the study of the influence of the magnetic interactions on the law of approach to saturation: In order to find the modifications to the law of approach (54) which result from consideration of the magnetic interactions, we consider first P and r as two constants to which we attribute, once the calculation is completed and for each value of the magnetization obtained, values corresponding to the final value at each point of the parameter $f(\xi) = (\mathcal{J}_s - \mathcal{J})/v\mathcal{J}_s$.

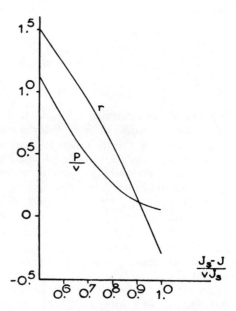

Fig. 5.
Values of the coefficients r and P/v as a function of the parameter $(\mathcal{J}_s - \mathcal{J})/v\mathcal{J}_s$.

Calling V the magnetic interaction potential, the two components along OY and OZ of the corresponding magnetic field are $-\partial V/\partial y$ and $-\partial V/\partial z$, and the local equilibrium conditions given in (57) must be replaced by:

$$-\frac{\partial F}{\partial \beta} = H\mathcal{I}_s\beta + \mathcal{I}_s\frac{\partial V}{\partial y}; \quad -\frac{\partial F}{\partial \gamma} = H\mathcal{I}_s\gamma + \mathcal{I}_s\frac{\partial V}{\partial z}, \quad (62)$$

On the other hand, the components of the local magnetization along the three axes are:

$$\mathcal{I}_x = \mathcal{I}_s, \quad \mathcal{I}_y = \beta\mathcal{I}_s, \quad \mathcal{I}_z = \gamma\mathcal{I}_s, \quad (63)$$

From this the density of magnetic charges becomes:

$$m = -\operatorname{div}\mathcal{I} = -\mathcal{I}_s\left(\frac{\partial \beta}{\partial y} + \frac{\partial \gamma}{\partial z}\right), \quad (64)$$

and, with abbreviations:

$$Q = 1 + \frac{4\pi\mathcal{I}_s^2}{H\mathcal{I}_s + 2Cr}; \quad R = \frac{2C}{H\mathcal{I}_s + 2Cr} \quad (65)$$

we get for the Poisson equation $\Delta V + 4\pi m = 0$, using equations (51) and (52):

$$\frac{\partial^2 V}{\partial^2 x} + Q\left(\frac{\partial^2 V}{\partial y^2} + \frac{\partial^2 V}{\partial z^2}\right) \quad (66)$$

$$= -4\pi\mathcal{I}_s R \sum_\lambda (\mu\beta'_\lambda + \nu\gamma'_\lambda)\sin\lambda x \sin\mu y \sin\nu z.$$

This is satisfied by a potential V of the following form

$$V = 4\pi\mathcal{I}_s R \sum_\lambda \frac{\mu\beta'_\lambda + \nu\gamma'_\lambda}{\lambda^2 + Q(\mu^2 + \nu^2)} \sin\lambda x \sin\mu y \sin\nu z. \quad (67)$$

Inserting this into equation (62), one deduces the values of β and γ and from there the mean magnetization in the OX direction, by averaging the local magnetization for all possible values of x, y, z, as

$$\mathcal{I} = \mathcal{I}_s\left(1 - \frac{\beta^2 + \gamma^2}{2}\right)$$

This expression obtained is of zero degree in λ, μ, ν. We replace λ^2, μ^2 and ν^2 by their mean values which are all equal, and noting that the mean value of $\beta'_\lambda\gamma'_\lambda$ is zero, the magnetization becomes finally:

$$\mathcal{I} = \mathcal{I}_s\left[1 - \frac{R^2}{8}\frac{Q^2 + Q + 2{,}5}{(1+2Q)^2}\sum(\beta'^2_\lambda + \gamma'^2_\lambda)\right]. \quad (68)$$

In general, even if the internal disturbances are large, Q is large compared with 1. Under these conditions the expression (68) is further simplified, and, with the notation already used, reduces to:

$$\mathcal{I} = \mathcal{I}_s\left[1 - \frac{P}{2\left(\frac{\xi}{2} + r\right)^2}\right] \quad (69)$$

Now let us calculate the values of P and r corresponding to a particular previously given value of in a substance of which a fraction v is perturbed by internal stresses. One knows the value a of the parameter $(\mathcal{I} - \mathcal{I}_s)/v\mathcal{I}_s$ and from the curves of Fig. 5 one finds the corresponding value of P/v and of r.

But, from equation (69) one has:

$$\frac{P}{2v\left(\frac{\xi}{2} + r\right)^2} = \frac{\mathcal{I}_s - \mathcal{I}}{v\mathcal{I}_s} = a.$$

from which one gets the value of the parameter ξ, and finally $\mathcal{I}$ as a function of H. In this way the values of Table 1 are determined.

In the case where $v=1$, the curve for the magnetization as a function of the parameter ξ has been drawn in Fig. 4, curve B. Comparing this with curve A, one sees that the magnetic interactions facilitate the magnetization. Let us recall incidentally that we had previously [7] calculated the law of approach to saturation for nickel by a similar method

TABLE 1

ξ	$\dfrac{\mathcal{I}_s - \mathcal{I}}{v\mathcal{I}_s}$	$\dfrac{P}{v}$	r
0,02	0,30	0,500	0,925
0,20	0,25	0,375	0,755
0,51	0,20	0,270	0,565
0,86	0,15	0,190	0,365
1,29	0,10	0,130	0,160
1,59	0,075	0,107	0,050
2,02	0,050	0,090	−0,06
2,81	0,025	0,075	−0,18

and obtained results which correspond with experiment. The preceding calculation is considerably more general.

(20) Propagation of a wall in a substance with irregularly distributed internal stresses: We will now study the movement of a wall parallel to the plane XOY which divides two domains magnetized antiparallel, one along positive z, the other in the opposite direction along negative z. We will use expression (50) for the energy of perturbation. The calculation of the equilibrium of the spontaneous magnetization inside an elementary domain is analogous to the previous case (para. 19), except that now we will assume that it is the magnetocrystalline anisotropy, and not the magnetic field which tends to keep the magnetization in the preferred direction OX. In equation (49) we therefore put $H=0$ and $K \neq 0$, and we have to replace the product $H \mathcal{J}_s$ by $2K$ each time it recurs in the formulae. Equation (65) defining Q and R must be replaced by the following

$$Q = 1 + \frac{2\pi \mathcal{J}_s^2}{K + Cr}; \quad R = \frac{C}{K + Cr}. \quad (70)$$

Taking account of this change, equations (62), (63), (64) and (66) remain valid, but the course of the calculation will change because we must now consider the two domains which are divided by the wall. In the upper domain, the equations quoted above remain unchanged, but in the lower domain the sign of $\mathcal{J}_s$ must be reversed.

(21) Calculation of the energy of the system as a function of the wall position h: We will neglect the surface energy of the wall and moreover assume that the wall deforms in a manner which does not generate any magnetic surface charges. We will later specify the limits of validity of these hypotheses.

Before calculating the energy of the system one must determine the equilibrium state of the magnetization on either side of the wall. Since the magnetization changes sign when the wall is traversed, one must find two expressions for the potential V, the one V_1 with respect to the upper domain, the other V_2, for the lower domain, which must on the one hand satisfy the partial differential equations (66), and on the other hand obey the boundary conditions on the wall surface. Since the wall carries no charges, these conditions are:

$$V_1 = V_2, \quad \frac{\partial V_1}{\partial z} = \frac{\partial V_2}{\partial z} \quad \text{for } z = h. \quad (71)$$

To simplify the calculations we assume that the summations Σ, which figure in the preceding equations, do not extend beyond a single term. In that case, the expressions for the potential V_1 and V_2 become:

$$V_1 = v \sin \lambda x \sin \mu y \sin \nu z + A e^{-n(z-h)} \sin \lambda x \sin \mu y,$$

$$V_2 = -v \sin \lambda x \sin \mu y \sin \nu z + B e^{n(z-h)} \sin \lambda x \sin \mu y,$$

(72)

with the following notations:

$$v = \frac{4\pi R \mathcal{J}_s (\mu \beta'_\lambda + \nu \gamma'_\lambda)}{\lambda^2 + Q(\mu^2 + \nu^2)}, \quad (73)$$

$$n^2 = \mu^2 + \frac{\lambda^2}{Q}. \quad (74)$$

The constants A and B are found by satisfying the boundary conditions (71), that means

$$B + A = 2v \frac{\nu}{n} \cos \nu h,$$

$$B - A = 2v \sin \nu h. \quad (75)$$

Having thus determined the equilibrium of the magnetization, we shall calculate the corresponding energy. The local energy density is the sum of three terms: first the classical magnetic energy term

$$\frac{1}{8\pi} \left[\left(\frac{\partial V}{\partial x}\right)^2 + \left(\frac{\partial V}{\partial y}\right)^2 + \left(\frac{\partial V}{\partial z}\right)^2 \right]$$

then the magnetocrystalline energy $K(\beta^2 + \gamma^2)$ and the magnetoelastic energy F. One verifies easily that apart from a term in β'^2 and one in γ'^2, the sum of the two last terms is

$$\frac{\mathcal{J}_s^2}{4(Cr + K)} \left[\left(\frac{\partial V}{\partial y}\right)^2 + \left(\frac{\partial V}{\partial z}\right)^2 \right].$$

Now, the terms in β'^2 and β'^2 are of no interest to us because in the integration they contribute only terms independent of h. Finally, the total energy is:

$$W = \frac{1}{8\pi} \int \left[\left(\frac{\partial V}{\partial x}\right)^2 + Q\left(\frac{\partial V}{\partial y}\right)^2 + Q\left(\frac{\partial V}{\partial z}\right)^2 \right] d\tau. \quad (76)$$

The calculation of the energy in the two domains presents no problem. One finds that in a cylinder with axis parallel to OZ and a base area of 1 cm², W has the value:

$$W = \frac{v^2}{16\pi} Q\left(\frac{v^2}{n} + n\right) \cos^2 vh. \qquad (77)$$

(22) Calculation of the critical field: This energy is a sinusoidal function of the position h of the wall. For the wall to be in equilibrium under a pressure $2H\,\mathcal{J}_s$ of a field along OX, one must have:

$$2H\mathcal{J}_s = -\frac{\partial W}{\partial h} = \frac{v^2 Q v}{16\pi}\left(\frac{v^2}{n} + n\right) 2 \sin vh \cos vh. \qquad (78)$$

The critical field for which the wall moves suddenly through the whole domain, corresponds to the maximum of $\partial W/\partial h$, which becomes, inserting the expression for v:

$$H_c = \frac{\pi R^2 \mathcal{J}_s}{2} \frac{v(\mu\beta'_\lambda + v\gamma'_\lambda)^2}{n[\lambda^2 + Q(\mu^2 + v^2)]}. \qquad (79)$$

This calculation corresponds to a particular value of the index λ. We will allow that one obtains an approximate value of the critical field brought on by a perturbation which is characterised by a certain value of P given by equation (59), corresponding to the set of possible values of the index λ, by taking the mean of the critical fields due to elementary perturbations each corresponding to a single value of λ, but so that for each the value of P is the same as above. Equation (59) is therefore to be replaced by:

$$P = \frac{1}{16}(\beta'^2 + \gamma'^2). \qquad (80)$$

This means we must take the mean of equation (79) with respect to all possible values of λ, μ, v. Since this expression is homogeneous and of zero order in λ, μ, v, one can, besides, impose the condition: $\lambda^2 + \mu^2 + v^2 = 1$. Moreover we limit ourselves to the case where Q is large, which is always true in practice even with strong deformations. Remembering that the mean value of β'^2_λ is equal to the mean value of γ'^2_λ, one can write, as far as mean values are concerned and since $\beta'_\lambda \cdot \gamma'_\lambda = 0$:

$$\overline{(\mu\beta'_\lambda + v\gamma'_\lambda)^2} = \frac{1}{2}(\beta'^2_\lambda + \gamma'^2_\lambda)(\mu^2 + v^2). \qquad (81)$$

Hence the mean critical field becomes:

$$\overline{H}_c \sim \frac{4\pi \mathcal{J}_s P R^2}{Q} \overline{\left(\frac{\mu^2 + v^2}{\mu^2 + v^2 + \frac{\lambda^2}{Q}} \cdot \frac{v}{n}\right)}. \qquad (82)$$

Since Q is large, one can replace

$$\frac{\mu^2 + v^2}{\mu^2 + v^2 + \lambda^2 : Q}.$$

by one. The last bracketed term of (82) is thus reduced to the mean value which v/n takes when λ, μ, v, are the coordinates of any point taken at random on the unit sphere. One finds easily for large Q

$$\overline{\left(\frac{v}{n}\right)} = \frac{2}{\pi}(1{,}386 + \log\sqrt{Q}). \qquad (83)$$

One can see that this term varies very slowly with Q. Taking the values of Q and R given in (70), one gets finally:

$$\overline{H}_c \sim \frac{4}{\pi} \frac{P}{\mathcal{J}_s} \frac{C^2}{K + Cr}\left[1{,}386 + \log\sqrt{\frac{2\pi\mathcal{J}_s^2}{K + Cr}}\right]. \qquad (84)$$

As the constant 2K plays now a role analogous to $H\,\mathcal{J}_s$, the parameter ξ of paragraphs 16, 17, and 18, is here given by

$$\xi = \frac{2K}{C}, \qquad (84\text{ bis})$$

and the values of P and r in equation (84) are functions of ξ, i.e. of 2K/C, given in Table 1.

Two cases are of particular interest.

(1) *The energy of elastic deformation C affecting a portion v of the substance, is small compared to the magnetocrystalline energy K*, ξ is large and P/v is hence 1/15. Under these conditions one has:

$$\overline{H}_c \sim \frac{4}{15\pi} \frac{vC^2}{K\mathcal{J}_s}\left[1{,}386 + \log\sqrt{\frac{2\pi\mathcal{J}_s^2}{K}}\right]. \qquad (85)$$

(2) The contrary case, *where the energy C is large compared to K*; ξ is small, P/v = 0.925. One obtains

$$\overline{H}_c \sim 0{,}69\, \frac{vC}{\mathcal{J}_s}\left[1{,}386 + \log\sqrt{\frac{6{,}8\mathcal{J}_s^2}{C}}\right]. \qquad (86)$$

When the substance is very strongly perturbed, v is nearly 1 and the expression approaches closely to:

$$H_c \sim \frac{C}{\mathcal{J}_s} \qquad (87)$$

which corresponds to the mechanism in which pure

rotations take place simultaneously in each crystallite of the polycrystalline substance. The substance is supposed to consist of an assembly of many crystallites, with negligible mutual interactions, which are deformed homogeneously in all possible directions [1, pp 216–217].

This agreement is certainly very comforting because the approximations on which the preceding calculations are based appear less justified as the perturbations become stronger. Nevertheless, we arrive for strong perturbations at the same result as with the theory of pure rotations which, in its turn, becomes more applicable as the perturbations become stronger, since the magnetic interaction between domains, which the rotation theory neglects, become comparatively weaker and weaker. It seems therefore that the methods of calculation which have been used, lead to correct orders of magnitudes, irrespective of the relative strength of the perturbation.

(23) Influence of the wall surface energy: In the preceding calculations the energy arising from the increase of the wall surface due to its being not plane, have been neglected: in fact it has been assumed that the wall deforms in a manner which does not cause any magnetic surface charges. An elementary but lengthy calculation which is not of enough interest to be given here, shows that this energy term doesn't become comparable to the others which have been considered, until the moment when the wavelength $2\pi/\lambda$, $2\pi/\mu$, $2\pi/\nu$ of the perturbations becomes small enough to be comparable with the wall thickness.

For a given deviation β, γ of the magnetization the term here neglected is positive: it is thus equivalent to a reduction of the coercive force. We are here dealing with an effect similar to the following which have also been neglected.

(1) The effect of the integration of small perturbations by the wall thickness when the latter becomes large compared with the correlation distance.

(2) The Weiss–Heisenberg energy which is no longer negligible when the changes in direction of the spontaneous magnetization are on a sufficiently small scale to be comparable to the wall thickness.

The set of the aforementioned facts therefore tempts us to suppress in the Fourier series development of β' and γ', which define the perturbations, terms (if they exist) for which $2\pi/\lambda$, $2\pi/\mu$, $2\pi/\nu$ are smaller than the wall thickness. We will take these complementary perturbing effects into account in an approximate manner.

(24) Propagation of a wall when the spontaneous magnetization varies irregularly from point to point: Before discussing equations (85) and (86) we will study the other cause of the internal disperse magnetic fields: the existence, inside the elementary domains of non-magnetic particles or inclusions (or less magnetic than the principal phase). In an even more general way this means that the spontaneous magnetization $\mathcal{J}_s$ varies locally from point to point. Before tackling the problem analytically we will assume that the spontaneous magnetization can be developed into a Fourier series

$$\mathcal{J}_s = \mathcal{J}_m + \sum_\lambda i_\lambda \cos \lambda x \sin \mu y \sin \nu z , \qquad (88)$$

in which we assume the coefficients i_λ to be small with respect to the mean spontaneous magnetization $\mathcal{J}_m$. All the calculations of the preceding paragraphs apply without great modification. In particular, the equations (57) or (62), of the equilibrium of the spontaneous magnetization remain identical to a first order of approximation, which we consider here, since the i_λ are small compared with $\mathcal{J}_m$. The only difference arises from equation (63) defining the three components of the spontaneous magnetization, which must now be written to an accuracy of first order:

$$\mathcal{J}_x = \mathcal{J}_m + \sum_\lambda i_\lambda \cos \lambda x \sin \mu y \sin \nu z ,$$

$$\mathcal{J}_y = \beta \mathcal{J}_m , \quad \mathcal{J}_z = \gamma \mathcal{J}_m . \qquad (89)$$

The value m of the density of the interior magnetic charges becomes

$$m = -\operatorname{div} \vec{\mathcal{J}} = \lambda i_\lambda \sin \lambda x \sin \mu y \sin \nu z -$$
$$- \mathcal{J}_m \left(\frac{\partial \beta}{\partial y} + \frac{\partial \gamma}{\partial z} \right) . \qquad (90)$$

Here, as before, we have assumed for simplicity that the summation extends no further than a single value of the index λ.

The Poisson equation (66) must be replaced by

$$\frac{\partial^2 V}{\partial x^2} + Q \left(\frac{\partial^2 V}{\partial y^2} + \frac{\partial^2 V}{\partial z^2} \right) =$$
$$= -4\pi [\lambda i_\lambda + R \mathcal{J}_m (\mu \beta'_\lambda + \nu \gamma'_\lambda)] \sin \lambda x \sin \mu y \sin \nu z . \qquad (91)$$

The solutions away from the wall are always of the type (72) but equation (73), which gives v, is now changed to the following:

$$v = 4\pi \frac{\lambda i_\lambda + R \mathcal{J}_m (\mu \beta'_\lambda + \nu \gamma'_\lambda)}{\lambda^2 + Q(\mu^2 + \nu^2)} . \qquad (92)$$

The rest of the calculation is identical, and one gets finally for the coercive force, instead of (79) the equation below:

$$H_c = \frac{\pi\nu}{2\mathcal{J}_m n} \frac{[\lambda i_\lambda + R\mathcal{J}_m(\mu\beta'_\lambda + \nu\gamma'_\lambda)]^2}{\lambda^2 + Q(\mu^2 + \nu^2)}. \quad (93)$$

As before we are specially interested in the mean value of the field of equation (93), with respect to all values of λ, μ, ν. As it is clear that the mean values of $i_\lambda \beta'_\lambda$ and $i_\lambda \gamma'_\lambda$ are zero, we find that the total mean coercive force is the sum of two terms: the one $\overline{H}_c$ which we have already found in (82) and (84) arising from irregular internal stresses, and the other $\overline{H}'_c$ corresponding to characteristic effects of irregularities in the spontaneous magnetization, which is given by the following equation:

$$\overline{H}'_c = \frac{\pi i_\lambda^2}{2Q\mathcal{J}_m} \left[\overline{\lambda^2 \nu \left(\mu^2 + \nu^2 + \frac{\lambda^2}{Q}\right)^{-1} \left(\mu^2 + \frac{\lambda^2}{Q}\right)^{-\frac{1}{2}}} \right]. \quad (94)$$

To simplify the expression in brackets and since we are only concerned with orders of magnitude of the phenomena, we replace by its average value the value of the factors of the product which differ only little from that average value, i.e. we replace $\mu^2 + \gamma^2 + \lambda^2/Q$ by $2/3 + 1/3Q$, which is about 2/3. The average of the product $\lambda^2 \nu(\mu^2 + \lambda^2/Q)^{-1/2}$ can now be calculated without difficulty if Q is large. One finds: $(2/3\pi)(0.386 + \log\sqrt{Q})$. The equation for the coercive force thus becomes

$$\overline{H}'_c = \frac{K + Cr}{4\pi\mathcal{J}_m} \frac{i_\lambda^2}{\mathcal{J}_m^2} \left(0,386 + \log\sqrt{\frac{2\pi\mathcal{J}_m^2}{K + Cr}}\right). \quad (95)$$

We recall that for small internal stresses (C small compared to K) $r = 0.18$ and for large ones $r = 0.925$.

We allow, in analogy with the results of the rigorous treatment made above in paragraphs 8 to 11 of the influence of the variations of the wall surface energy, that if a perturbation of the spontaneous magnetization is given by a series such as (88), the total coercive force is represented by a sum of a series of individual terms like (95). One obtains hence the total coercive force by replacing i_λ^2 in (95) by the sum $\Sigma_\lambda i_\lambda^2$.

(25) Calculation of $\Sigma\ i_\lambda^2$ for some simple cases: Suppose for instance that the substance consists of two finely dispersed elements, one with spontaneous magnetization i_1 occupying a volume v_1, the other with spontaneous magnetization i_2 occupying a volume v_2. Of course $v_1 + v_2 = 1$. The mean spontaneous magnetization is $v_1 i_1 + v_2 i_2$ and the mean square of the spontaneous magnetization is $v_1 i_1^2 + v_2 i_2^2$. But, using equation (88), one gets

$$\overline{\mathcal{J}_s^2} = \mathcal{J}_m^2 + \frac{1}{8} \sum i_\lambda^2, \quad (96)$$

and thence

$$\sum i_\lambda^2 = 8v_1 v_2 (i_1 - i_2)^2. \quad (97)$$

Putting this value into the expression for the coercive force we obtain, by limiting ourselves to small perturbations of the internal stresses:

$$\overline{H}'_c = \frac{K}{4\pi\mathcal{J}_m} \frac{8v_1 v_2 (i_1 - i_2)^2}{\mathcal{J}_m^2} \left(0,386 + \log\sqrt{\frac{2\pi\mathcal{J}_m^2}{K}}\right). \quad (98)$$

Let us use this formula for the case where one of the constituents consists of non-magnetic inclusions ($i_2 = 0$) and calling $v = v_2$ the volume of these inclusions, we have finally[4]

$$\overline{H}'_c = \frac{2Kv'}{\pi\mathcal{J}_m} \left(0,386 + \log\sqrt{\frac{2\pi\mathcal{J}_m^2}{K}}\right). \quad (99)$$

(26) General remarks and relative order of magnitude of the coercive force due to deformations and to inclusions: Before using the preceding equations to interpret experimental results, we must remember that they refer essentially to the schematic model of a single, plane wall, which is one average parallel to the direction Δ of the applied field, and which divides two elementary domains of some tens of microns thickness magnetized on the whole, one in the direction Δ, the other antiparallel to it. Moreover, the various disturbances of the lattice are assumed to be small enough and not extending far enough to give rise to new elementary domains[5]. The coercive force found is, in short, the mean field necessary to move the wall by a distance of the order of magnitude of half the domain thickness.

---

[4] This equation differs profoundly from the one given in a previous paper (Cahiers de Physique 1944, **25**, p. 21, A57) devoted to the effect of cavities and inclusions on the coercive force. The reason is that in simplifying the treatment we were wrong, firstly by placing the cavities on the nodes of a simple cubic lattice and secondly by assuming that the magnetization is rigid outside the wall. The equation (99) given here is thus much preferable to equation (21) of the quoted paper.

[5] Thus one excludes the large perturbations, equal or larger than e.g. 10 μ. In addition, refer to the paper by Néel already cited in paragraph 4, ref. [4].

In reality one has many domains separated by walls of different orientation and very complicated perturbations of the lattice. The formulae can hence not pretend to give more than a first rough approximation. They neglect particularly a phenomenon which is sometimes encountered and relates to the nuclei which give rise to the birth of walls.

However that may be, the coercive force for wall movements presents itself by two terms, a first one $\overline{H_c}$ given by equation (84) relating to deformation caused by internal stresses, and a second $\overline{H_c'}$ given by equation (98) relating to cavities and inclusions. Depending on circumstances it may be advantageous to use equation (84) in one of the forms (85) or (86) and equation (98) in the form (99).

It is now important to specify the order of magnitude of the coercive force relating to these two causes in order to establish whether the mechanisms treated are in fact able to account for the observed phenomena. For this we apply the theory to iron and to nickel in which we assume that a fraction v of the volume is subjected to irregularly oriented internal stresses of intensity $\sigma_1 = 30 \, \text{kg/mm}^2$ and that another fraction v' is taken up by inclusions.

Iron: For iron we have $\mathfrak{J}_s = 1700$, $\lambda_s = 20 \times 10^{-6}$, $C = (3/2)\lambda_s \sigma_i = 9.0 \times 10^4$, $K = 4.3 \times 10^5 \, \text{ergs/cm}^3$. The parameter ξ defined in (84b) is equal to 9.5, i.e. large and equation (85) is applicable. For the rest one uses (99) to get the total:

$$H_c = 2{,}1 \, v + 360 \, v' \qquad \text{(iron)} \qquad (100)$$

Nickel: For nickel we have $\mathfrak{J}_s = 500$, $\lambda_s = 30 \times 10^6$, $C = (3/2)\lambda_s \sigma_i = 13.5 \times 10^4$, $K' = -5 \times 10^4 \, \text{ergs/cm}^3$. But the constant K which appears in our formulae refers to the energy which keeps the spontaneous magnetization in the easy direction; here the ternary axis. We see easily that with the notations of paragraph 14, the magnetocrystalline energy W_m takes, near this direction, the form

$$W_m = \frac{K'}{3}\left[1 - 2(\beta^2 + \gamma^2)\right] \qquad (101)$$

and hence $K = -2K'/3 = 3.3 \times 10^4$. This gives therefore $\xi = 2K/C = 0.244$. One must apply equation (84) and, using Table I one finds $P/v = 0.355$ and $r = 0.73$. Thus finally we get:

$$H_c = 330 \, v + 97 \, v' \qquad \text{(nickel).} \qquad (102)$$

Comparing equations (100) and (102) one sees that in iron the inclusions have the preponderant effect whilst in nickel the internal stresses are important. One sees also that with reasonable values of v and v' one arrives at coercive forces which are well of the order of magnitude of the fields observed.

(27) The difficulties of verification of the theory of internal stresses:

In the theory just expounded, the internal stresses which perturb a magnetic substance are characterised by two parameters: their intensity σ_i, i.e. the parameter $C = (3/2)\lambda \cdot \sigma_i$ and their extent defined by the fraction v of the total volume in which they are active. The measure of the coercive force represents simply a relation between these two parameters. In order to be able to determine these parameters, one can study the law of approach in strong fields which are treated in paragraph 19 of this theory. In this law of approach one must naturally take into account the influence of the polycrystallinity and magnetocrystalline energy. A corresponding theory of the law of approach has already been published [7]. Moreover, one must be certain, on the one hand that the substance is not subject to homogeneous elastic deformation which cannot be represented by equation (50), and on the other hand that there are sufficiently few cavities and inclusions present so as not to give rise to a large coercive force.

To verify this theory one needs an independent method to evaluate the internal stresses; unfortunately this doesn't exist apart from X-rays which have insufficient sensitivity and offer difficulties of interpretation.

Kersten [8][1, p. 157] has proposed a certain number of magnetic methods to estimate the internal stresses but they do not take the magnetic interactions between elementary domains into account, nor does it seem possible without detailed discussion to compare their results with those found in the present theory. This will be the subject of a later paper.

(28) Verification of the formula which relates the coercive force to the volume of the inclusions:

The verification of equation (99) relating to inclusions meets with better conditions than equation (84) relating to internal stresses: it is in fact much easier to know the inclusion volume than the strength of internal stresses.

To this purpose we use the experimental results collected by Kersten [7] relating to inclusions of Cu, Fe_3C, Al_2O_3, etc. in iron. We are dealing with annealed alloys where it is supposed that the coercive force is plotted on a log–log scale against the inclusion volumes which have been calculated by Kersten on the basis of the limits of solid solubility; for instance iron only dissolves 3% of titanium and only the excess contributes to form inclusions of Fe_3Ti. Only for the case of cementite have we modified Kersten's calculation of v by taking account of the substance's own magnetization. In effect we have here equation

(98) with $i_1 = 1700$ and $i_2 = 1052$ and the fictitious volume which has to be used in (99) to maintain its validity in this case, and this means that the weight p of carbon per gram of alloy gives a volume $v = 2.3p$.

On this graph one can also see the theoretical straight line given in equation (100). The general agreement between theory and experiment is impressive if one considers the imperfections and approximations of the theory and also the phenomena which have been neglected. For instance we have taken the values of pure iron for K and $\mathcal{J}_m$ whereas we should have inserted the values for the limiting solid solutions which are unfortunately not known. Moreover it is very possible that some inclusions cause elastic deformations of the lattice in spite of the anneal, and the calculated coercive force is too small. This will occur for Fe_3W_2. Inversely, if an important proportion of the inclusions is smaller than the domain wall thickness ($0.1\ \mu$) or on the other hand, consists of very large particles (greater than $10\ \mu$) the coercive force given by equation (99) is too large. This will be so for copper and certain forms of cementite. A good example of this latter case cited by Kersten [7, p. 68] is sulphur, which even in the smallest concentration comes out in large inclusions visible in the microscope. Correspondingly, the coercive force of iron containing 10% by volume of sulphur is only 2.2 Gauss, fifteen times less than the value given by (99).

In brief, the set of results cited by Kersten seem to favour more the theory which we have developed, than with his own theory which is subject to severe objections in its mathematical treatment and which gives a $v^{2/3}$ dependence of the coercivity, in visible disagreement with the points of Fig. 6.

(29) Thermal variation of the coercive force: It appears that equations (84) and (99) permit immediate calculation of the thermal variation of the coercive force if we know that of K and $\mathcal{J}_m$, provided that C, v and v' remain constant: this is in fact true for v' and it is reasonable to take the temperature variation of C and v to be small up to the Curie point, around which H_c varies rapidly. One has thus an indirect test for the validity of the theory[6]. In reality the matter is alas much more complicated because as one approaches the Curie point to within about $100°$, K becomes extremely small causing a strong increase in the wall thickness, which ends up by having the magnitude of the elementary domains: the notion of an elementary domain loses its meaning and so does the preceding theory. In a sense one could take account of this by varying the limit of cut-off of the development in (51) and (88) as has been indicated before (Remark 4, para. 6). We do not yet possess the basis for making such calculations specially as concerns the thermal variations of K.

---

[6] In his book [5], Kersten has made calculations of the thermal variation of H_c which appear to agree with experiment, but he assumes that the exchange energy per cm^3, A, is independent of temperature. In reality, A varies approximately as the square of the spontaneous magnetization as is shown by experiment (specific heats) and also by a stricter study of the theory. The agreement found is thus fortuitous.

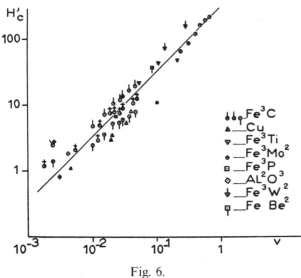

Fig. 6.
Coercive force of a series of iron alloys according to the results collected by Kersten as a function of the relative volume v of the non-magnetic phase (logarithmic coordinates). The straight line corresponds to the theory established in this paper.

REFERENCES

[1] R. Becker and W. Doring, *Ferromagnetismus*, Springer (Berlin, 1939).
[2] Kondorski, *Sowjet Phys.*, **11**, 1937, p. 597.
[3] Kersten, *Physik. Z.*, **44**, 1943, p. 53.
[4] L. Néel, *Cah. de Phys.*, no. 12, 1942, p. 1 and no. 13, 1943, p. 18.
[5] Kersten, *Grundlagen einer Theorie der ferromagnetischen Hysterese und der Koerzitivkraft* (Hirzel, Leipzig, 1943).
[6] L. Néel, *C.R. Acad. Sc.*, **220**, 1945, p. 738; *J. Phys. Rad.*, **9**, 1948, p. 184. (A62)
[7] L. Néel, *C.R. Acad. Sc.*, **220**, 1945, p. 814; *J. Phys. Rad.*, **9**, 1948, p. 193. (A63)
[8] Gans, *Ann. der Physik*, **15**, 1932, p. 28; *Physik. Z.*, **33**, 1932, p. 15.
[9] M. Kersten, *Z.f. Physik*, **71**, 1931, p. 553; **76**, 1932, p. 505; **82**, 1933, p. 723; **85**, 1933, p. 708.

A59 (1947)
THE COERCIVE FORCE OF A POWDER WITH ANISOTROPIC GRAINS OF A CUBIC FERROMAGNETIC SUBSTANCE

Remark (1977: The publication of this note, related to Note A-58 (Chapter VI) was delayed for 5 years because of the war and the necessity not to divulge to the occupying force the elements of a patent.

Summary: Calculation of the coercive force of a powder consisting of single domain ferromagnetic grains of negligible magnetocrystalline anisotropy but possessing shape anisotropy. Evaluation of the influence of the magnetic interactions between the grains as a function of the bulk density.

When the dimensions of the grains of a powder of a ferromagnetic substance are less than a critical value, which for iron is about 300 Å, the spontaneous magnetization in each grain is uniform and changes in magnetization arise only from rotations of the spontaneous magnetization [1].

If the grain has the shape of a prolate ellipsoid of revolution of volume V, and M and N are the demagnetising coefficients along the long and short axis respectively, and if one neglects the magnetocrystalline energy, the total energy is reduced to $CV\alpha^2$, where α is the cosine of the angle of the spontaneous magnetization with the major axis, and C a coefficient equal to $\frac{1}{2}(M-N)\mathcal{J}^2$. Everything behaves as if we dealt with a uniaxial substance, the major axis of the ellipsoid playing the role of the direction of easy magnetization. This grain exhibits hysteresis according to the mechanism studied by R. Becker [2] and N. Akulov [3]. If the direction of the applied field is along the major axis, the hysteresis loop is rectangular with a coercivity $H_c = 2C/\mathcal{J}$. For an assembly of similar ellipsoids, without interactions, randomly oriented, a graphical method shows that the mean loop has a coercive force $H_{cm} = 0.96 C/\mathcal{J}$. If the ratio of the ellipsoids axes a/b is near one: $a/b = 1 + \varepsilon$, one has $N - M = 8\pi\varepsilon/5$ and $H_{cm} = 0.96(4\pi\mathcal{J}\varepsilon/5)$, which for iron becomes $H_{cm} = 4100\varepsilon$.

One gets a relatively true picture of a real powder by considering it as a mixture of ellipsoids of different eccentricities. The simplest assumption is to regard ε as a parameter which can vary between 0 and its maximum value 5/4 corresponding to $N - M = 2\pi$, i.e. a cylinder, and to suppose that the relative volumes of ellipsoids with ε between ε and $\varepsilon + d\varepsilon$ are proportional to $d\varepsilon$. A graphical method shows that the mean loop so obtained has a coercive force of $0.26\pi\mathcal{J}$, 1400 Gauss for iron.

In compressed powders one must take the mutual interactions between the grains into account. The simplest way of doing this consists of considering that in any system of cavities distributed isotropically in a ferromagnetic substance of uniform spontaneous magnetization $\mathcal{J}$, the magnetic energy of the system per cm^3 is $(2\pi/3)v(1-v)\mathcal{J}^2$, where v is the relative volume of the cavities. Referring to unit volume of ferromagnetic substance, the energy is $(2\pi/3)(1-v)\mathcal{J}^2$. For powder particles, the presence of neighbouring particles multiplies the energy by $1-v$. As in the calculation indicated above, the coercivitiy is proportional to the energy, one obtains as a value for the coercive force $H_{cm} = 0.26\pi\mathcal{J}(1-v)$. Introducing d, the apparent powder density and d_0 the density of the solid metal, this becomes

$$H_{cm} = 0{,}26\,\pi\mathcal{J}\left(1 - \frac{d}{d_0}\right) \qquad (1)$$

For an iron powder of density 4.5 the calculation gives 500 gauss. As L. Weil [4] has shown, equation (1) is in agreement with experimental findings; it also gives the upper limit of coercive force which can be attained by a powder of given apparent density if all grains are below the critical size.

It seems thus that shape anisotropy of particles is the origin of the hysteresis of ferromagnetic powders of cubic materials. As has already been noted [1] the magnetocrystalline energy plays only a secondary role. This is naturally not the case for non-cubic and strongly anisotropic substances like manganese-bismuth where, as Guillaud [5] has shown, the magnetocrystalline anisotropy is the true cause of the coercive force.

REFERENCES

[1] L. Néel, *Comptes rendus*, **224**, 1947, p. 1488. (A58)
[2] *Zeits. f. Physik*, **62**, 1930, p. 253.
[3] *Zeits. f. Physik*, **81**, 1933, p. 790.
[4] *Comptes rendus*, **224**, 1947, p. 923.
[5] *Thèse*, Strasbourg, 1943.

A60 (1947)

THEORY OF THE ANISOTROPY OF CERTAIN MAGNET STEELS HEAT TREATED IN A MAGNETIC FIELD

Summary (1977): A single domain ferromagnetic particle isolated in a non-magnetic matrix, tends to take spherical shape under the influence of surface tension, but tends to elongate in the direction of magnetization to minimise the demagnetizing field. Calculation of the equilibrium shape. Consequences for magnetic properties.

Consider an alloy having a magnetic phase consisting of small grains of constant volume v, immersed in a non-magnetic phase. If the temperature is high enough for atomic diffusion, the grains of the magnetic phase must become more spherical under the action of the interfacial tension γ between the two phases. But if we apply a magnetic field sufficient to give the grains their saturation magnetization $\mathcal{J}_s$, the grains take the form of ellipsoids, elongated along the field direction. In effect, calling k the eccentricity of the ellipse, the particle energy is the sum of the surface tension energy S_γ and the demagnetizing field energy $\frac{1}{2}(N\mathcal{J}^2 v)$ where S and N are given by

$$S = \frac{2\pi}{k}\left(\frac{3v}{4\pi}\right)^{\frac{2}{3}}(1-k^2)^{-\frac{1}{6}}(\arcsin k + \sqrt{1-k^2}),$$

$$N(k) = 4\pi\frac{1-k^2}{2k^3}\left(\ln\frac{1+k}{1-k} - 2k\right).$$

This energy is a minimum for a non-zero value of k. If $\mathcal{J}^2/\gamma$ is small, the eccentricity k_m corresponding to the minimum is given by

$$k_m^2 = \frac{3}{4}\left(\frac{4\pi}{3}\right)^{\frac{2}{3}}\frac{\mathcal{J}^2 v^{\frac{1}{3}}}{\gamma}.$$

If we cool the system in a saturating magnetic field to room temperature, where diffusion is negligible, the grains stabilise in an elongated shape which they subsequently keep whatever the applied magnetic field. Thus the total system shows magnetic anisotropy. When it is magnetized to saturation along the common major axes of the ellipsoids, the demagnetizing field energy per cm³ is in fact equal to $(\frac{1}{2})V(1-V)N(k_m)\mathcal{J}^2$, if V is the relative total volume of the magnetic phase. When magnetized to saturation in a perpendicular direction, the demagnetizing field energy is $(\frac{1}{2})V(1-V)N'(k_m)\mathcal{J}^2$, calling $N'(k_m)$ the demagnetizing coefficient along the minor ellipsoid axis, related to $N(k_m)$ by $N(k_m)+2N'(k_m)=4\pi$. When k is small, the difference W_a of the two energies, the anisotropy energy, is given by

$$W_a = \frac{3\pi}{10}\left(\frac{4\pi}{3}\right)^{\frac{2}{3}}V(1-V)\frac{\mathcal{J}^4 v^{\frac{1}{3}}}{\gamma}.$$

It is probably a mechanism of this nature which is responsible for the magnetic anisotropy of certain magnet steels based on *iron–nickel–aluminium* [1], especially *Alnico V* [2] *or Ticonal*, after cooling from 1300°C in a magnetic field of some thousands of gauss. At ordinary temperatures, and without cobalt, these alloys consist of two phases [3], both bodycentred cubic lattices but with lattice spacings which differ by about 0.4%. One phase is composed almost exclusively of iron and strongly magnetic, the other is relatively weakly magnetic. The structure is probably the same if cobalt is present.

In order to obtain an estimate of γ one can assume that the two phases have crystal lattices with parallel axes and fit together by a system of G. I. Taylor type dislocations [4] analogous to that which one supposes for the fit of the elements having mosaic structure [5]: the calculation shows that the corresponding deformation energy is of the order of 100 ergs/cm². One can also calculate the deformation energy which corresponds to the coercive force of about 400 Gauss arising from the internal stress, by using the recent theories of the coercive force [6]. One finds about 10^8 ergs/cm^3. Relating this figure to the surface of the grains of which one knows the approximate size, (about 10^{-5} cm) since they are of the same order as the thickness of the walls of the elementary domains, if the alloy was treated to obtain maximum coercivitiy, one gets a value of 300 ergs/cm^2. This value is evidently a maximum so that an approximate value of $\gamma = 200$ ergs/cm^2 appears reasonable.

Taking $V = \frac{1}{2}$; $\mathfrak{J} = 1700$; $v^{1/3} = 10^{-5}$ cm, one finds $W_a = 2.5 \times 10^5$ ergs/cm^3. Now, Lliboutry has measured the anisotropy of a sample of Alnico V and found 2.2×10^5 ergs/cm^3. Naturally, in order for the mechanism to operate, the Curie point θ must be sufficiently high so that at temperatures a little below θ the diffusion be rapid enough. That is why ordinary iron–nickel–aluminium alloys, which have too low a Curie point, orient very little.

The proposed mechanism gives a right order of magnitude for the anisotropy energy, while the earlier explanations based on magnetostriction were unsatisfactory since the energies which they gave were only 10^3 ergs/cm^3.

REFERENCES

[1] D. A. Oliver and J. W. Shedden, *Nature*, **142**, 1938, p. 209.
[2] B. M. Smith, *Gen. El. Rev.*, **45**, 1942, p. 210.
[3] A. J. Bradley and A. Taylor, *Proc. Roy. Soc.*, A, **166**, 1938, p. 353.
[4] *Proc. Roy. Soc.*, A, **145**, 1934, p. 362.
[5] J. M. Burgers, *Proc. Phys. Soc.*, **52**, 1940, p. 23.
[6] L. Néel, *Comptes rendus*, **223**, 1946, pp. 141 and 198; *Annales de l'Université de Grenoble*, **22**, 1946, p. 299. (A57)

A91 (1954)

OBSERVATIONS ON THE THEORY OF THE MAGNETIC PROPERTIES OF HARD MATERIALS

Remark (1977): Whether we deal with fine grains which interact or movements of Bloch walls, we find for weak fields the same Rayleigh laws.

Summary: The author shows that it is possible to improve the theory of magnetic hysteresis of fine, single domain particles in a very simple way by taking account on the one hand of the dispersion of the coercive force of the individual particles and on the other hand of the interaction field between the grains and of the dispersion of this field about its mean value. One succeeds in this way to account for the main characteristics of the hysteresis cycle and to interpret especially the Rayleigh laws relating to weak fields. One obtains without any arbitrary initial conditions numerical values which agree remarkably well with experimental results relating to the known good permanent magnets.

(1) The theory of the single domain particles: The processes of magnetization of ferromagnetic materials consist of two elementary processes: The movements of the walls which separate the elementary Weiss domains and the rotation of the magnetic moments of these domains. It is this latter mechanism which undoubtedly plays a preponderant role when we deal with hard magnetic materials, i.e. those which are used for good permanent magnets. These phenomena are particularly simple to interpret when the substance contains two phases of which one, non-magnetic, constitutes a matrix which surrounds magnetic particles of the second phase. If the grains are sufficiently small, at least of the order of tenths of a micron, they contain only a single Weiss domain. The resulting magnetic moment of the particle keeps a constant value, only its direction can change. This is how in the magnets of compressed iron powder the magnetic phase consists of the iron grains whereas the matrix is formed by the voids which subsist between the grains after compression. Equally in magnets of the Fe–Ni–Al type, one deals with a magnetic phase with properties similar to iron immersed in a non-magnetic phase consisting mainly of nickel and aluminium.

The laws of magnetization of an isolated single domain particle are well known [1–4]. In general one assumes for simplicity a grain of cylindrical symmetry where the total energy of the grain in a field H is expressed by the equation

$$W = - MH - MH_c \cos^2 \theta , \qquad (1)$$

where θ is the angle of the moment M with the axis of revolution and H_c is a constant, called the critical field, which characterises the magnetic anisotropy of the grain: Shape anisotropy (elongated particle), magnetocrystalline anisotropy (coupling of the magnetic moment with the crystal lattice) or still magnetoelastic anisotropy (internal stresses). It suffices to say that W is a minimum.

When the applied field is parallel to the grains axis, the magnetization as a function of the field can be represented by a rectangular hysteresis loop, symmetrical about the origin and characterised by two discontinuities of magnetization corresponding to fields $-H_c$ and $+H_c$. When the applied field is perpendicular to the axis, the component m of the moment in the field direction varies reversibly with H in the interval $-H_c < H < +H_c$

$$m = \frac{MH}{H_c} . \qquad (2)$$

Outside this interval, m is equal to $-M$ or $+M$ depending on whether H is below $-H_c$ or above $+H_c$. Finally, in intermediate cases where the grain axis makes an angle with the field direction, one obtains a hysteresis loop of complex shape with a coercivity and with discontinuities of magnetisation which became smaller as the axis approaches the direction perpendicular to the field.

We now consider a mass of identical independent grains with randomly directed axes. The mean hysteresis loop is shown by curve A in Fig. 1. It is characterised by a remanent magnetic moment equal to half the saturation moment and a coercive force equal to 0.48 H_c, i.e. about half the maximum coercive force, which is that of the grain whose axis is parallel to the field. This curve has evidently some points of similarity with the

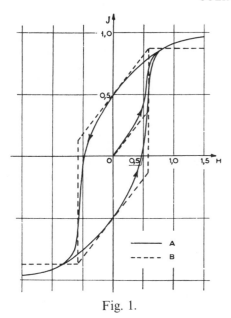

Fig. 1.

hysteresis loop of hard materials. In particular, it gives a sensible value for the remanent magnetization as well as the right order of magnitude for the coercivity.

One could, furthermore, improve the theory by introducing a mass of grains with different degrees of elongation suitably distributed from spheres to infinite cylinders. But this is not enough to avoid the weaknesses of the theory. In practice, it is certain that the ferromagnetic grains are far too close to one another to be considered as independent. In spite of the complications which we encounter, we must discard the assumption of independence of grains and study in what degree the introduction of interactions modifies the results of the simple theory to allow us to explain certain experimental results which are incompatible with that latter theory.

Amongst the incompatibilities we have the Rayleigh laws relating to small fields. Indeed, according to the simple theory, if after demagnetization one applies a field H_m, and then reduces the field to O, the magnetization ought to remain independent of the field, whereas experiment gives a parabolic variation. Note also that the simple theory gives an infinite anhysteretic susceptibility while experiment yields a finite one.

Proximity effect: It is possible to calculate the influence of the interactions on the coercive force when the anisotropy of the grains is solely shape anisotropy. In fact, for isolated grains of any shape which are randomly oriented with respect to the spontaneous magnetization J_0, the magnetostatic energy per cm³ of grains is equal to $(2/3)\pi J_0^2$. On the other hand it can be shown that any given system of cavities or particles of any shape, provided only that it is on average, isotropic and magnetized to saturation, has a magnetostatic energy equal to $(2/3)\pi J_0^2 v(1-v)$, where v is the fraction of the total volume occupied by the magnetic constituent. The magnetic energy per unit volume of this constituent is therefore $(2/3)\pi J_0^2(1-v)$: as a consequence of the proximity of neighbouring grains, the energy is multiplied by $(1-v)$, and the coercive force, proportional to the energy, is also multiplied by $(1-v)$. Hence one expects that the coercive force of magnets of compressed iron powder must be proportional to the mean density of the compacted substance. This conclusion was very satisfactorily verified by Weil [5]. The highest quality magnet corresponds to the maximum of the average energy, i.e. to the maximum of $v(1-v)$ at $v=\frac{1}{2}$, and this is confirmed by experiment.

The Lorentz field: The proximity effect represents just one of the many aspects of the role of the interactions between the grains. In a mass of grains of mean magnetization J_m each grain is, because of the neighbouring grains, exposed to a kind of Lorentz field of mean strength $(4/3)\pi J_m$. Since this field is zero at the coercive point, it does not influence the value of coercivity but distorts profoundly the other parts of the hysteresis loop. In fact, in the materials which are here of interest, it corresponds near remanence to a considerable magnetic field of about 1800 e.m.u.

The fluctuations of the Lorentz field: From grain to grain this field fluctuates strongly about the mean, and it is possible to calculate the fluctuations under the assumptions that there is no correlation between the direction of the magnetic moments of neighbouring grains and also that the grains are disposed on the lattice points of a face centred cubic lattice of spacing a. The magnetic moment m of a grain is therefore $a^3 J_s/4$, where J_s is the mean saturation magnetization of the system. To simplify further, let us suppose that the magnetic moment of each grain can lie only along one of the two directions parallel and antiparallel to the mean magnetization J_s. If we put $p = J_m/J_s$, the respective probabilities of these 2 orientations are $(1+p)/2$ and $(1-p)/2$.

Let us consider two neighbouring domains A and B and assume that the line joining their centre, makes an angle θ with the resultant magnetization direction of J_m. The mean component h_m along the direction of J_m of the field h exerted by B on A is

$$h_m = \frac{m}{r^3}(3\cos^2\theta - 1)p \quad \text{avec} \quad r = \frac{a}{\sqrt{2}} \quad (3)$$

The mean square of the excess of h over the average h_m is therefore:

$$\overline{(h - h_m)^2} = \frac{m^2}{r^6}(3\cos^2\theta - 1)^2(1 - p^2), \quad (4)$$

and finally the mean square H_j^2 of the difference of the interaction field with respect to the mean field, due to the 12 neighbours of A is:

$$H_j^2 = \frac{24}{5}(J_s^2 - J_m^2). \qquad (5)$$

An analogous argument allows the calculation of the mean square $H_j'^2$ of the field of fluctuation along a direction perpendicular to J_m. One finds

$$H_j'^2 = \frac{18}{5}(J_s^2 - J_m^2). \qquad (6)$$

These two last equations show that the fluctuations of the Lorentz field are greatest near the condition where the mean magnetization $J_m = 0$. The amplitude of the fluctuations, for $J_s = 850$ is about 1700 emu. This is a considerable field, of the order of the critical field of the most anisotropic grains. These fluctuations must therefore play an important role.

The probability of finding a fluctuation field between certain limits follows a Gaussian distribution. However, with the approximations which we will make below, we will replace this distribution by a simple rectangular one corresponding to the same mean square. We will thus assume that the probable values of the fluctuation field are uniformly distributed between the limits $\pm H_f \sqrt{3}$.

(2) The fundaments of the theory of the ideal magnet: To get a better theory of hysteresis, one must take the following into account:
(A) The different orientations of the grains.
(B) The proximity effect.
(C) The distribution of the coercivities of the individual grains.
(D) The Lorentz field.
(E) The fluctuations of the Lorentz field.

Up till now, the different theories proposed have only taken points A and B into account. We propose to outline below a schematic theory of the ideal permanent magnet which takes the other factors into account. We suppose that this magnet consists of a magnetic phase (e.g. iron) of spontaneous magnetization $J_0 = 1700$ emu, which occupies a fraction $v = \frac{1}{2}$ of the total volume, in the form of very small grains dispersed in a non-magnetic phase. The average saturation magnetization is thus 850 emu, i.e. an order of magnitude found both in magnets of compressed iron-powder and in magnets of the Fe–Ni–Al type.

We assume that the grains are approximately prolate ellipsoids of revolution with all possible eccentricities, so that the coercivities of the individual grains, if isolated, are distributed between O (sphere) and $2\pi J_0$ (cylinder). However, with $v = 0.5$ the proximity effect halves this variation to between O and πJ_0. To avoid great mathematical difficulties which arise from the distribution of orientations, we replace the grains having a critical field between h_c and $h_c + dh_c$, randomly oriented and corresponding to a fraction of dJ_s of the saturation magnetization, by the following much simpler distribution.

(1) Grains with axes parallel to the applied field, with critical field $h_c' = 0.54\, h_c$, corresponding to the fraction $0.5\, dJ_s$ of the saturation magnetization. These grains contribute the irreversible part of the variations in magnetization.

(2) Grains with axis perpendicular to the applied field, with the same critical field h_c' as the ones above, corresponding to a fraction of $0.37\, dJ_s$ of saturation. These grains contribute the reversible part of the changes in magnetization.

These proportions and coercivities have been chosen so that the resulting hysteresis loop and curve of initial magnetization (curve B, Fig. 1) due to these 2 categories of grains approach as closely as possible to those corresponding to grains with critical field h_c oriented at random (curve A, Fig. 1). After what was said above, the critical field greater than H_c' is equal to $0.54\, \pi J_0$, i.e. 2880 emu. We will assume in what follows that the values of h_c' are uniformly distributed between O and H_c'.

We will now calculate the two Rayleigh coefficients a and b and the coercive force of the assumed model.

Initial reversible susceptibility (Rayleigh's a-term): The initial susceptibility comes from the grains with axes perpendicular to the applied field, subject, in addition to the latter, to the fluctuation field h_j.

Suppose that the total variation of the macroscopic magnetization is so small that the topography of the fluctuation field can be assumed to be invariable and consider all grains between h_c' and $h_c' + dh_c'$: they are exposed to fluctuation fields uniformly distributed between $-H_f\sqrt{3}$ and $+H_f\sqrt{3}$, and, as we will show below, h_c' in absolute value is always inferior to $H_f\sqrt{3}$. The grains with h_f greater than h_c' are positively saturated whilst those with h_f less than h_c' are negatively saturated. The resultant moment of grains with h_f between $-h_c'$ and $+h_c'$ is zero. When a small field H is applied, these limits $-h_c'$ and $+h_c'$ are replaced by $-h_c' + H$ and $+h_c' + H$; so that the application of a field H has simply the effect of reversing the magnetization of grains from $-J_0$ to $+J_0$ for which the fluctuation field is in an interval of magnitude H. The contribution to the total magnetization of these grains is

$$\frac{0{,}37\, J_s H\, dh_c'}{H_f\sqrt{3}\ \ H_c'}.$$

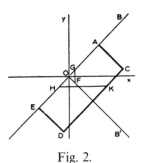

Fig. 2.

Adding the contribution of the grains with different values of h'_c, one arrives at the expression for the initial reversible susceptibility:

$$a' = 0{,}37 \, J_s/H_f \sqrt{3} \, ; \tag{7}$$

but we have not yet taken account of the Lorentz field which corresponds to a demagnetizing field NJ_m with $N = 4\pi/3$. Including this last effect we obtain finally the definitive value of the initial susceptibility coefficient a as:

$$a = \frac{a'}{1 - Na'}. \tag{8}$$

We must still calculate H_f, which is the square root of the mean square of the fluctuation field for $J_m = 0$. For this we use equations (5) and (6). If we assume that of the 12 neighbours of a given grain, 12×0.5 are magnetized in the field direction and 12×0.37 perpendicular to the field, we have

$$H_f = 1{,}932 \, J_s \, , \tag{10}$$

$$a' = 0{,}11 \, , \quad Na' = 0{,}46 \, , \tag{11}$$

and finally

$$a = 0{,}20 \, . \tag{12}$$

Irreversible initial magnetization (Rayleigh's b-term): This results from the grains with axes parallel to the applied field. A grain with critical field h'_c and exposed to a fluctuating field h_f describes under the influence of an externally applied field an assymmetrical hysteresis loop characterised by a lower critical field $-h'_c - h_f$ and an upper critical field $h'_c - h_f$. Each of these grains can be represented on a plane diagram by a point of abscissa $x = h'_c - h_f$ and ordinate $y = -h'_c - h_f$. With the initial assumptions, the points representing the model are uniformly distributed inside a rectangle ACDE (Fig. 2) with sides parallel to the bisectors OB and OB' of the coordinate axes, and the abscissae of the corners C and D are respectively $H'_c + H_f\sqrt{3}$ and $G'_c - H_f\sqrt{3}$. Weiss and Freudenreich [6] have observed long ago that such an arrangement of assymetric loops obeys the Rayleigh laws for weak fields, for which

$$J = a'H + b'H^2 \tag{13}$$

is the curve of initial magnetization and the returning branch from the maximum excursion J_m, H_m is

$$J - J_M = a'(H - H_M) - \frac{b'}{2}(H - H_M)^2 \, . \tag{14}$$

The a' term corresponds naturally to grains with perpendicular axes, studied in the preceding paragraph.

After demagnetization in a decreasing alternating field, the grains which are represented by points below OB' are positively magnetized, those above OB' are negatively magnetized. During the first magnetization, the application of a field H reverses the magnetic moment of the grains whose representative points are in triangle OFG with a side OF parallel to OY and abscissa equal to H (see Neel [7]). This results in a term $b'H^2$ with

$$b' = \frac{0{,}50 \, J_s}{2\sqrt{3} \, H'_c H_f}. \tag{15}$$

We must introduce the Lorentz field, neglected till now, which gives finally a value of the coefficient b of the irreversible term in H^2 of the law of initial magnetization

$$b = \frac{b'}{(1 - Na')^3}. \tag{16}$$

Introducing the numerical values adopted above, one gets: $b' = 2{,}6 \times 10^{-5}$; $b = 1{,}7 \times 10^{-4}$.

Coercive force: When, after having magnetized to saturation, one applies a negative magnetic field, one reaches the coercive force $-H_c$ when the resultant magnetization is zero. As far as the grains with axis parallel to the field are concerned, it is easily shown [7] that if in Fig. 2 one draws a straight line HK parallel to the axis OX of ordinate $-H_c$, the grains corresponding to the trapezoidal region HKCA are negativelgnetized, whereas those of HEDK are positively magnetized. The corresponding resulting magnetization is

$$\frac{0{,}50 \, J_s}{2\sqrt{3} \, H'_c H_f} [H_c^2 - (H'_c - H_c)^2] \, .$$

On the other hand, the reversible magnetization with axes perpendicular to the field, is

$$\frac{0.37 J_s H_c}{H_f \sqrt{3}},$$

and, by making the total magnetization which is equal to the sum of these two terms zero, one gets the relation

$$H_c = 0.287 H_c'. \qquad (18)$$

and with the numerical values adopted

$$H_c = 830. \qquad (19)$$

(3) Comparison with experimental results: Let us now compare the values so obtained with typical experimental results for compressed iron powder magnets or Fe–Ni–Al magnets. We get the following table

	a	b	H_c
theory	0.20	1.7×10^{-4}	830
iron powder	0.25	2.7×10^{-4}	473
Fe–Ni–Al	0.26	1.8×10^{-4}	557

Considering the crude approximations which have been made above, the ideal magnet of our theory has properties surprisingly similar to those of real materials.

One need not be surprised, that the theory gives a greater coercive force than any which has ever been observed for these types of magnet. In fact, the theory would give 550 instead of 830 if one assumed that the particle shape distribution, instead of being between sphere and infinite cylinder were between sphere and ellipsoid of an axial ratio 3:1 which is certainly closer to reality.

It now remains to complete the study of the model for the domain of high magnetizations and to calculate for example the remanent magnetization and the hysteresisless susceptibility: this would be very complicated because one would have to consider the fact that the mean square of the fluctuation field is proportional to $J_s^2 - J_m^2$, and, since we deal with large amplitudes of the variation of the magnetization, one would have to introduce a Gaussian distribution of the fluctuating fields. Provisional, fairly laborious calculations appear to give results of correct orders of magnitude. Let us simply remark in passing that with this new way of looking at the problem, there is no longer any reason why one should obtain a remanent magnetization exactly half the saturation magnetization.

(4) Conclusions: By introducing and putting into effect, in a very simple manner, the interaction between domains in the form of a kind of Lorentz field and its fluctuations, one obtains a model which has magnetic properties very near to those of the hard magnetic materials.

In particular, it enables us to interpret the value of the reversible susceptibility quantitatively and to explain why the hard materials obey the Rayleigh laws just as well as the soft materials for which the interpretation is entirely different and is based on wall movements. It is very interesting to note that the mechanism which is studied here gives a value of $bH_c/a = 0.70$, very similar to 0.58 given in the theory of the Rayleigh laws for wall movements, when walls are close to each other [7]. We have already noted elsewhere [8][9] that, as far as the coercive force is concerned, the theory of wall movements and that of rotations gives, under these conditions, results which approach each other. The same is the case here. This statement is very reassuring because the two theories approach the problem of very strongly disturbed substances in a completely different way. In the one, one starts from a large, almost perfect crystal, in the other from very fine grains, isolated from each other. In the end one obtains, nevertheless, similar results.

REFERENCES

[1] N. Akulov, *A. Phys.* **81**, 1933, p. 790.
[2] L. Néel, *Compt. Rend.* **224**, 1947, pp. 1488, 1550. (A58, A59)
[3] E. C. Stoner and E. P. Wohlfarth, *Phil. Trans.* **240**, 1948, p. 599.
[4] L. Néel, *Ann. Géophys.* **5**, 1949, p. 99. (A69)
[5] L. Weil, *Compt. Rend.* **225**, 1947, p. 229; **227**, 1948, p. 1347.
[6] P. Weiss and J. P. de Freudenreich, *Arch. Aci. Natur.* Genève, **42**, 1916, p. 449.
[7] L. Néel, *Cahiers de Physique*, 1942, no. 12, p. 1; 1943, no. 13, p. 18. (A45)
[8] L. Néel, *Ann. Univ. Grenoble*, **22**, 1946, p. 299. (A57)
[9] L. Néel, *Physica* **15**, 1949, p. 225.

Chapter X
RANDOM FIELDS: CREEP AND TILTING
A101, A102, A106, A107, A109, A108

A101 (1970)

ACTION OF SUCCESSIVE RANDOM MAGNETIC FIELDS ON THE MAGNETIZATION OF FERROMAGNETIC SUBSTANCES

Summary: After a review of the notion of irreversible susceptibility, the variation in magnetization produced by successive applications of a random magnetic field is calculated. The results of this investigation are applied to the theory of thermal fluctuation after-effect.

In investigations of magnetic hysteresis, much interest is paid to the local magnetic properties, in the neighbourhood of a point. There are two reasons for this. Firstly, from a knowledge of the magnetic history of the sample, i.e. the way in which the given point has been reached, it is easy to predict the effects of subsequent small variations of the magnetic field. Secondly, the effects on the magnetization of various treatments or physical processes, magnetic after effect, shocks etc., can frequently be described very simply by introducing small fictitious fields superimposed on the applied magnetic field.

Let us therefore suppose that after a certain treatment the purpose of which is to define the initial state (consisting for example of demagnetization by an alternating field which slowly decreases to zero), by increasing the magnetic field along the curve CA (Fig. 1), we reach a point A of magnetization J_A in the field H_A. The slope S_D of the curve CA at the point A is called the *differential susceptibility*. Now let the applied field change from H_A to $H_A + \delta H$, where δH is assumed to be small compared with the coercive field. Two cases arise.

If δH is negative, the representative point describes a backward line AD, the slope of which, S_R, is smaller than the slope S_D of AC at the point A. The change in magnetization obtained in this way is essentially reversible: if, after such a reduction in the field, we return to the initial value H_A, the magnetization likewise recovers its initial value J_A. Hence S_R is known as the *reversible susceptibility*.

The situation is very different if δH is positive. The representative point describes the segment AA′, the continuation of AC, of slope S_D; the corresponding

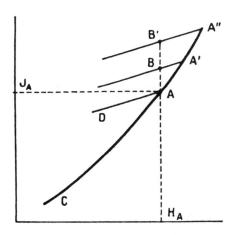

Fig. 1.

increase in the magnetization is

$$\delta J = S_D \delta H.$$

When the magnetic field is subsequently reduced to its initial value H_A, the representative point describes a new backward line A′B, parallel to AD. The initial magnetization has thus increased by an amount $\delta J' = AB$ given by

$$\delta J' = S_D \delta H - S_R \delta H = S_A \delta H$$

The quantity $S_A = S_D - S_R$ is called the *irreversible susceptibility*. Clearly the quantities S_A, S_D, S_R depend not only on the current values J_A and H_A of the

315

magnetization and the field, but also on the whole magnetic history of the sample.

Let us now assume that after a treatment defining its initial state as above, the sample is successively subjected to fields $H_A + h_1$, $H_A + h_2$, ..., $H_A + h_{n-1}$, $H_A + h_n$, where the additional fields h_j are small in absolute value compared with the coercive field. It follows from the above analysis that after the final field $H_A + h_n$ is applied, the representative point J_n of the final magnetization must lie on a backward line $A''B'$ of slope S_R which passes through a point A'', located on the extension of CA with abscissa equal to $H_A + h_m$, where h_m is the largest in algebraic value of the fields h_1, h_2, ..., h_n.

The final magnetization J_n in the field $H_A + h_n$ is then given by

$$J_n = J_A + S_D h_m - S_R(h_m - h_n)$$

and can be written

$$J_n = J_A + S_A h_m + S_R h_m \quad (1)$$

The magnetization has thus undergone an irreversible change $S_A h_m$ proportional to the irreversible susceptibility S_A and to the greatest value h_m reached by the fields h_j, and a reversible change $S_R h_n$ proportional to the reversible susceptibility S_R and to the final value of the fields h_j.

A very interesting case is when the h_j are random in nature. Let us suppose for example that this random field is defined by the probability $p(x)dx$ (independent of j) that h_j lies between $h_r x$ and $h_r(x+dx)$, such that the average value of h_j is zero and the average value of h_j^2 is equal to h_r^2.

As will be seen from the examples which we investigate below, the physically interesting quantity is not the magnetization J_n obtained after applying one particular series of n fields $H_A + h_j$, but rather the mean value J_{nm} of the values J_n obtained by reapplying a very large number of times a series of n successive fields $H_A + h_j$, where each series is separated from the preceding one by a return to the initial state by demagnetizing in a decreasing alternating field. In these conditions, defining $x_n h_r$ as the mean value of h_m, where h_m is the greatest value reached by the random fields h_j over the n successive changes, and noting that the average value of h_n is zero, the following relation can be deduced from (1)

$$J_{nm} = J_A + S_A x_n h_r \quad (2)$$

Calculation of x_n We call P the probability that the random field h_j is smaller than $x h_r$, that is

$$P = \int_{-\infty}^{x} p \, dx . \quad (3)$$

The probability that h_j is greater than $x h_r$ is therefore $1 - P$. The probability that h_j is greater than $x h_r$ at least once in two trials is equal to the probability $1 - P$ that it is larger than $x h_r$ during the first trial plus the probability $P(1 - P)$ that it is greater than $x h_r$ during the second trial, i.e. altogether $1 - P^2$. Similarly, the probability that h_j is greater than $x h_r$ after n trials is $1 - P^n$.

It follows that in n trials the probability that the largest value reached by h_j lies between $x h_r$ and $(x + dx) h_r$ is equal to

$$1 - P^n(x) - [1 - P^n(x + dx)] = \frac{dP^n}{dx} dx .$$

The mean value $x_n h_r$ of the largest random field h_j obtained in a large number of successive series each containing n trials is equal to

$$x_n = \int_{-\infty}^{+\infty} x \, \frac{dP^n}{dx} dx . \quad (4)$$

Numerical values of x_n The value of x_n given by (4) is general for any p. The case in which p is Gaussian

$$p = \frac{1}{\sqrt{2\pi}} e^{-\frac{x^2}{2}},$$

and where P is consequently simply related to the Galton function, is of particular importance. We propose to calculate the corresponding numerical values.

Integration by parts twice reduces expression (4) to the equivalent form

$$x_n = \int_0^\infty (1 - P^n) dx - \int_{-\infty}^0 P^n dx . \quad (5)$$

Case of n equal to or less than 10 The integral

$$I_n = \int_0^\infty G^n dx ,$$

is tabulated, where G is given by

$$G = \sqrt{\frac{2}{\pi}} \int_x^\infty e^{-\frac{x^2}{2}} dx .$$

The integrals I_n have been calculated by H. J. Godwin [1]. When x is positive, P is equal to $1 - G/2$; when x is negative, on changing x to $-x$, P can be replaced by $G/2$, so that one can write

$$x_n = \int_0^\infty \left[1 - \left(1 - \frac{G}{2}\right)^n\right] dx - \int_0^\infty \left(\frac{G}{2}\right)^n dx,$$

or alternatively,

$$x_n = \left[n\frac{I_1}{2} - \frac{n(n-1)}{1.2}\frac{I_2}{4} + \cdots + (-1)^{n-1}\frac{I_n}{2^n}\right] - \frac{I_n}{2^n}.$$

The values calculated in this way are listed in column (a) of the table.

Case of n greater than 10 The second integral of the second term in relation (5) is negligible in this case. The first integral can then be evaluated by numerical integration using tables of the Galton function. The results of column (a) of the table have been determined in this way.

An approximate value for this first integral can be obtained by noting that it must be equal to the value of x at which P^n equals $1/2$: the tables give the result immediately. These values of x_n are shown in column (b) of the table. As can be seen, they differ from the exact values by less than 1%.

$n-1$	x_n (a)	$n-1$	x_n (a)	x_n (b)	x_n (c)
0	0	20	1.889	—	—
1	0.5642	50	2.257	—	—
2	0.8463	100	2.511	—	—
3	1.0293	200	2.747	—	—
4	1.1629	500	3.037	—	—
5	1.2671	1 000	3.241	3.198	3.219
6	1.3521	10 000	3.852	3.812	3.824
8	1.4850	100 000	4.384	4.347	4.355
10	1.5860	1 000 000	4.861	4.827	4.833

Instead of calculating the value of x corresponding to $P^n = (1/2)$ from the table, the asymptotic expansion can be used,

$$P = 1 - \frac{1}{x\sqrt{2\pi}} e^{-\frac{x^2}{2}} \left[1 - \frac{1}{x^2} + \frac{1.3}{x^4} - \frac{1.3.5}{x^6} + \cdots\right]$$

which, limited to its first terms, gives

$$x_n^2 = 2\left[\log n - \log(x_n\sqrt{2\pi \log 2})\right]. \tag{6}$$

The corresponding values are listed in column (c) of the table. As a result of a fortunate cancellation of errors, the accuracy is even better than with the exact value of P.

Application to thermal fluctuation after-effect The random field h_j may for example be caused by thermal fluctuations. We have shown that the mean square field of thermal fluctuations was given by [2]

$$h_f^2 = h_r^2 = \frac{4\pi kT}{3v},$$

where v is the volume affected by a Barkhausen discontinuity, k is the Boltzmann constant and T the absolute temperature. Let us now introduce a so-called reorganization time constant θ, of the order of 10^{-10} s, which is the smallest time interval that must separate two observations of the field h_j at the same point so that the two values obtained may be considered to be independent. In these conditions, if t is the interval separating the time at which the observation of the magnetization J_t is made from the time that the field H_A is established, then it is as if the substance had been subjected, in addition to the field H_A, to n successive applications of a random field where n is given by

$$n = t/\theta.$$

Since n is very large (of the order of 10^{10}), the value of x_n given by relation (6) can be used, and setting this value in equation (2), one finds

$$J_t = J_A + S_A h_f \sqrt{2(Q' + \log t)} \tag{7}$$

where

$$Q' = -\log \theta - \log(x_n \sqrt{2\pi \log 2})$$

equation (7) is identical to one that can be deduced from equation (86) in reference [2]. When t is about a second, x_n is close to 6.5 and Q' is approximately 20, in agreement with the experimental results of the same article.

In a later Note we shall describe a much more novel application of the notions described here.

REFERENCES

[1] *Quart. J. Appl. Math.*, 1952, **5**, part I, 109; this work was brought to my attention by M. Kuntzmann.
[2] L. Néel, *J. Phys. Rad.*, 1950, **11**, 49; see also *J. Phys. Rad.*, 1951, **12**, 339.

A102 (1957)

AN ATTEMPT TO INTERPRET THE CREEP IN HYSTERESIS LOOPS

Summary: The principal experimental features of creep phenomena are reviewed. It is shown that they can be interpreted by supposing that successive hysteresis loops, described between the same limits and macroscopically identical, in fact differ microscopically, giving rise to random coupling fields between elementary domains.

Let us take a sample of ferromagnetic material exhibiting hysteresis, and define its initial state by demagnetization in an alternating field which decreases slowly to zero, starting from an initial value which is large compared with the coercive field. If the sample is then subjected to a succession of alternate magnetic fields $+H_A$, $-H_A$, $+H_A$, ..., symmetrical about the origin, the successive loops are practically identical to each other after the second or third; in other words the loop becomes fixed. The same is not true for asymmetrical loops, obtained by subjecting the sample to successive fields H_A, H_B, H_A, H_B ... such that H_B is different from $-H_A$. In these conditions the magnetization-field loop is no longer closed and a continual shift of the whole loop is observed. This phenomenon, which was noticed first by J. A. Ewing [1], has been investigated much more recently by L. Lliboutry [2], who gave it the name of *creep*[1], after a metaphor used by H. Bouasse [3] in an investigation of magnetization-torsion loops.

Suppose that H_B is slightly smaller than H_A and of the same sign. Designating by J_n the magnetization in the field H_B after the nth application of the field H_A, Lliboutry found that the difference $J_{n+1} - J_n$, for n up to 40, is given by the empirical formula

$$J_{n+1} - J_n = \frac{B}{1 + Kn^{0.9}}$$

A formula of this kind had already been used by M. Ascoli [4] to represent the effects of successive shocks on the value of the remanent magnetization. For higher values of n, very recent work by Nguyen Van Dang [5] indicates that $J_n - J_1$ is proportional to $(\log n)^a$, where the exponent a lies between 0.5 and 1. According to [5] the amplitude of the phenomenon, which is measured

[1] *reptation* in French

in the example given above by the coefficient B, goes through a maximum when H_B is equal to zero for a value of H_A close to the coercive field H_c.

It should be pointed out that creep seems to be a general property of all polycrystalline ferromagnetic substances, with values which, when expressed in the reduced coordinates H/H_c and J/J_s (where J_s is the saturation magnetization), are always of the same order of magnitude.

Creep cannot be explained on the basis of the conventional models normally used in simple theories of hysteresis; another mechanism must be found. Let us assume as a working hypothesis that there is a small random magnetic field superimposed on the applied fields H_A and H_B between which the successive hysteresis loops are described. We shall call this field the *creep field*, and attribute to it the following properties.

We assume firstly that this creep field varies from one point to another in the sample, and that its mean value is zero in each direction in space. The creep field however is taken to be constant at any given point and independent of time for as long as the applied field H_A (or H_B) remains constant. There is therefore no relation between this creep field and the thermal fluctuation field which we proposed previously to explain fluctuation after-effect [6], and which is a random function of time. We shall picture the creep field as being purely and simply superimposed on the thermal fluctuation field. Finally, let us designate by h_r (mean square creep field) the root mean square of the creep field.

After the field H_A is first applied, thus giving a certain spatial distribution of the creep field, let us apply the field H_B, then again H_A. We assume that the new distribution of creep field is entirely independent of the initial distribution. In general we shall assume that each new application of the field H_A after a return to the field H_B gives rise to a creep field distribution which is independent of the previous ones, but is characterized by the same value h_r of the mean square creep field.

It is important to note that this creep field acting at a point with coordinates (H_A, J_A) in the hysteresis loop is in fact related to the whole magnetization loop lying between the extremum points (H_A, J_A) and (H_B, J_B). If we consider another series of loops lying between (H_A, J_A) and (H_C, J_C), the creep field at the same point (H_A, J_A) of the loop is characterized by a value of the mean square creep field different to the previous one.

We call $p(x)dx$ the probability that the creep field lies between the limits $h_r x$ and $h_r(x + dx)$ at a given point M in the sample and after the nth application of the field H_A. We assume a Gaussian form for p

$$p = \frac{1}{\sqrt{2\pi}} e^{-\frac{x^2}{2}}.$$

We thus assume that at any point M in the sample, the mean values over a very large number of loops of the creep field and of its square are respectively given by 0 and h_r^2.

To complete the description of the properties of the creep field, a correlation distance d_c must be introduced, such that for two points M and M' in the sample separated by a distance greater than d_c the creep fields at the same nth application of the field H_A are entirely independent of each other; for two points separated by a distance less than d_c, the creep fields have the same value.

In order to determine the effects of the creep field on the magnetization, we shall use the results of an earlier investigation [7] of the effects of successive random fields. It was shown there that if S_A is the irreversible susceptibility, the application of n successive random fields caused an increase in magnetization equal to $S_A h_m$, where h_m is algebraically the largest of these n fields. This is precisely what happens in the neighbourhood of any point M, within a domain centred on M, and of dimensions smaller than the correlation distance d_c. The average effect over all the domains in the sample, each one being similar to the above, is the same as the average effect that would be obtained by repeating a very large number of times the application of a series of n random fields in the neighbourhood of the same point M, starting from the same initial state. In reference [7] we showed that the corresponding magnetization J_{nm} was given by the expression

$$J_{nm} = J_A + S_A x_n h_r$$

where x_n is a number which depends on n and is given by

$$x_n = \int_{-\infty}^{+\infty} x \frac{dP^n}{dx} dx,$$

and P denotes the integral

$$P = \int_{-\infty}^{x} p\, dx;$$

The numerical values of x_n were also calculated [7].

There is however an important difference between the present problem and the one which we investigated earlier. In the latter, the irreversible susceptibility S_A corresponded to the straightforward succession of random fields which occurred. In the present case, however, the application of two consecutive random fields is always separated by a return to the field H_B. The corresponding irreversible susceptibility must therefore be different from the ordinary irreversible susceptibility S_A and we shall denote it by S_{AB}.

For the sake of brevity let us in the following statement of definitions, denote by $j(H_1, H_2, \ldots, H_k)$ the magnetization obtained in the ferromagnetic sample by subjecting it successively to the fields $H_1, H_2, \ldots$, the final field being H_k. The sample has a well defined initial state, symbolized by the letter j, obtained for example through a demagnetization in a decreasing alternating field. In these conditions, if H_A is positive and δH an increase in the field which is small compared with the coercive field, then the classical irreversible susceptibility is defined by the relation

$$S_A \delta H = j(H_A + \delta H, H_A) - j(H_A)$$

By analogy, in the problems of creep of interest here, we could define an irreversible susceptibility S'_{AB} through the relation

$$S'_{AB} \delta H = j(H_A + \delta H, H_B, H_A) - j(H_A)$$

It should be pointed out that in this way of increasing the magnetization, $S'_{AB} \delta H$ is defined as the difference in magnetization of two states, the first corresponding to two successive applications of the field H_A (in order to apply the field $H_A + \delta H$, the field H_A must first have been attained), and the second corresponding to a single application of the field H_A. Now the effects of the creep fields, whose existence we assume, are different in the two cases. They are equivalent in the first state to an additional field $x_2 h_r$ (according to reference [7], $x_2 = 0.5642$), and in the second state to an additional field $x_1 h_r$ ($x_1 = 0.0000$), so that the final fields in these two states are not the same.

The correct definition of the irreversible susceptibility required is therefore

$$S_{AB} \delta H = j(H_A + \delta H, H_B, H_A) - j(H_A, H_B, H_A) ; \quad (1)$$

where the final fields H_A, taking into account the creep fields, are really equal in both states.

The susceptibility S_{AB} can be used to define the effects of creep at the upper extremity of the cycle, of coordinates J_A, H_A. To calculate the effects of creep at the lower extremity of the cycle, of magnetization J_B in the field H_B, a second irreversible susceptibility can be analogously defined

$$S_{BA} \delta H = j(H_A + \delta H, H_B) - j(H_A, H_B), \quad (2)$$

where H_A is still taken to be positive and H_B is of smaller modulus than H_A.

In these conditions, the magnetizations J_{An} and J_{Bn} obtained after the nth application of the fields H_A and H_B can be respectively written

$$J_{An} = J_A + S_{AB} x_n h_r, \quad (3)$$

$$J_{Bn} = J_B + S_{BA} x_n h_r. \quad (4)$$

We call the term in h_r the *creep term*.

The susceptibilities S_{AB} and S_{BA} are in general quite different from S_A: thus in the Rayleigh domain (very weak fields), with a relation describing the initial magnetization,

$$J = aH + bH^2$$

the following values are easily found:

$$S_A = 2bH_A; \quad S_{AB} = S_{BA} = b(H_A + H_B)$$

In symmetric cycles $H_B = -H_A$; for small amplitudes the susceptibilities S_{AB} and S_{BA} are zero, but this result appears to be valid for all amplitudes, at least with the simple models usually employed to account for the phenomena of hysteresis. It follows that for symmetric cycles, there should be no creep, in complete agreement with the experimental results.

The creep terms appearing in equations (3) and (4) are the result of superimposing the random creep field on the field H_A, i.e. on whichever of the two fields H_A or H_B is greater in absolute value. In principle there should also be creep terms corresponding to the superposition of the creep field on H_B. However it is easy to see that the cumulative effect of the random creep fields which would be felt when these fields were added to the field H_A do not occur for the field H_B. Experiment indeed shows that the magnetization $j(H_A, H_B, H_A)$ is virtually independent of H_B and does not notably differ from $j(H_A)$ as long as H_B lies between $-H_A$ and $+H_A$. The effects of small variations of H_B caused by the addition of a random creep field which is small compared with H_B are thus completely wiped out by the subsequent application of the field H_A.

Up to this point we have assumed that the initial state was defined by the application of an alternating field which decreased slowly to zero. The sequence of events would, however, be quite different if the initial state was defined in another way: for example by saturating the sample in a very large positive magnetic field and then bringing the field to zero, or again by superimposing a constant magnetic field on the alternating demagnetizing field. The values of S_{AB} and S_{BA} would still be defined by equations (1) and (2), but would take different values, while equations (3) and (4) would still be valid.

It remains now to discuss the physical significance of the randomness of the creep field. We believe that after two successive circuits round a magnetization loop between the limits J_B and J_A, although the final magnetizations J_A are equal macroscopically, they must differ microscopically and correspond to totally different distributions of the elementary domains, or at least of a fraction of them. Similarly, the distribution of magnetic poles and internal dispersion fields must also differ in the two cases. In short, one can classify the dispersion fields into two categories. In the first, the dispersion fields readopt the same values at each successive application of H_A: these do not interest us here. In the second, for each new application of H_A the distribution of internal dispersion fields is entirely independent of the previous distributions. The total volume occupied by the elementary domains reorganized in this way during each cycle must be proportional to $J_A - J_B$. If we allow that the dispersion field acting at a point M in the sample is caused by the sum of a very large number of elementary terms of different magnitude and orientation, due to the action of a statistical distribution of poles at different distances, we are forced to conclude firstly that the probability law which governs the creep field is Gaussian, and secondly that its mean square h_r^2 is proportional to $J_A - J_B$.

We suspect however that this proportionality is valid only for a sufficiently large variation of the magnetization $J_A - J_B$: for small amplitudes, the distributions of the elementary domains must be identical in successive loops. A minimum amplitude must be required so that small initial differences, due for example to the random nature of the thermal fluctuation field, may be amplified by a chain of interactions which finally gives distributions of the elementary domains which are independent of the previous ones at the end of each loop.

A comparison of the proposed theory with experiment will be treated in later publications. For the moment let us simply state that for large values of n the theory gives a creep term proportional to $(\log n)^{1/2}$, which appears to be compatible with the experimental values of Nguyen Van Dang [5].

REFERENCES

[1] *Phil. Trans. Roy. Soc. London*, 1885, **176**, 523.
[2] *Ann. Phys.*, 1951, **6**, 731.
[3] *Ann. Chim. Phys.*, 1907, **10**, 199.
[4] *Nuovo Cimento*, 1902, **3**, 6.
[5] Comm. Soc. Franç. Phys., Grenoble, 20 February 1957.
[6] L. Nèel, *J. Phys. Rad.*, 1950, **11**, 49; 1951, **12**, 339.
[7] L. Nèel, *Comptes rendus*, 1957, **245**, 2442.

A106 (1958)

ON THE EFFECTS OF A COUPLING BETWEEN FERROMAGNETIC GRAINS EXHIBITING HYSTERESIS

Summary: The properties of a set of pairs of interacting ferromagnetic grains are examined. It is shown that, depending on the kind of hysteresis loops described, the magnetization resulting from the second application of a field H is not, as is the case in the usual models, necessarily equal to the magnetization produced by the first application: it may be smaller or larger.

Ferromagnetic hysteresis can be explained using a model consisting of a set of mutually independent grains, each having a rectangular symmetric hysteresis loop. Each grain is thus characterized by a saturation magnetic moment a and a critical field c. The ensemble is defined by a certain distribution function M giving the saturation moment Mdc of the grains whose critical field lies between c and $c+dc$. Physically, the elementary grains correspond for example to anisotropic grains, sufficiently small to be single domain, whose direction of minimum energy is parallel to the applied magnetic field.

Such a model, whose simplicity is tempting, takes account quite well of the phenomena in the high magnetization region, if a suitable distribution function is taken. However it does not conform to the two Rayleigh laws describing the properties in fields much smaller than the coercive field.

To account for these two laws, one must use the model consisting of a set of grains with asymmetrical loops and different increasing and decreasing critical fields c_1 and c_2, which was originally proposed by P. Weiss and J. de Freudenreich [1], then taken up by F. Preisach [2], and subsequently by L. Néel [3]. This model is a great improvement on the previous one in its explanation of the magnetic properties over the whole range of magnetization and field. This is to be expected anyway, since the distribution function $M(c_1, c_2)$ is now a function of two variables instead of one.

Physically, in order to prove the existence of grains with asymmetrical loops, it is necessary to introduce interactions between the grains [4]. When a grain whose critical field is c is subjected by its surroundings to actions which are equivalent to a magnetic field h, it is the same as if the given grain had an asymmetric loop with critical fields $c-h$ and $-c-h$. But clearly the coupling field h must depend on the state of magnetization of the whole set of grains so that the critical fields of a given grain cannot be considered to be constants. The set of grains with asymmetric loops assumed above, therefore, does not correspond to any physical model. Only in the cases limited to the effects of weak variations of the magnetic field about the origin, and where h can be considered to be constant as a first approximation, can a proper correspondence be established between an asymmetrical grain model and a set of real interacting grains.

Since the properties of a set of interacting grains seem *a priori* extremely difficult to treat rigorously, we propose as a first stage to examine the properties of a set of mutually independent *pairs*, each consisting of two interacting grains. For two given grains of a pair, of moments a and a', the simplest way to introduce an interaction is to assume that the first grain exerts a magnetic field $+na$ or $-na$ on the second, and that the second exerts a field $+na'$ or $-na'$ on the first, according to the sign + or − of the magnetic moment of the grain in consideration.

The simplest physical picture of such a coupling is that of magnetic dipolar interactions. The sign and the value of the coupling coefficient n depends on the angle between the line joining the centres of the two grains and the direction of minimal energy, which is assumed to be the same for the two grains. With this model it is expected that the coercive fields c and c' of the two grains will be of the order of a and a' and that the absolute values of n will be of the order of unity for grains in contact. Moreover, the average value of n must be close to zero.

In a subsequent Note we shall describe a simple and

general method for plotting the hysteresis loops of these pairs. The results are as follows. The loops are symmetrical about the origin and belong to one of the six types A, A', B, C, C', K shown on Fig. 1. The abscissae of the points P, Q, R, S, T have the values of $OP = c - na'$; $OQ = c' - na$; $OR = -c - na'$; $OS = -c' - na$; $OT = -OS$ for type A' and $OT = -OR$ for type A. The ordinates of the points M and N are $OM = a + a'$ and $ON = a' - a$. The vertical sections of the loops correspond to irreversible paths in the direction of the arrows. In reality, the pairs can also occur in unstable magnetic states which vanish after a first saturation magnetization: we have not represented them here.

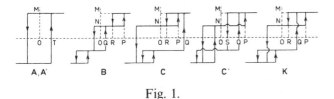

Fig. 1.

For given values of a and a', taking $a < a'$, the types of loops depend on the values of c and c'. The types corresponding to different regions of the plane (c, c') are shown on Fig. 2a for positive coupling $(n > 0)$ and on Fig. 2b for negative coupling $(n < 0)$. In both figures $OF = OF' = |n|(a' - a)$; $OG = OG' = |n|(a' + a)$; the straight lines passing through F, F', G, G', are parallel to the first bisector. Apart from the loops A and A' of standard shape, the other loops have complex shapes which are mainly symmetrical, but containing asymmetrical subloops. The properties of an ensemble of pairs share at the same time the properties of a set of symmetric grains and those of a set of asymmetrical grains.

In this way the anhysteretic remanent magnetization, obtained by superimposing a decreasing alternating field on a very small positive steady field, is equal to $a + a'$ for the loops A and A', to $a' - a$ for the loops B, C, K and to $a - a'$ for the loops C.

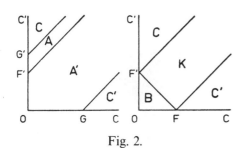

Fig. 2.

The initial anhysteretic susceptibility of an ensemble of pairs is thus infinite, as is that of an ensemble of grains with symmetric loops of the standard shape.

Let us now consider the properties of a set of pairs in weak fields. For example, let us take the type C' loops corresponding to very small values of n and c', with negative coupling $(n < 0)$. After demagnetization in an alternating field which decreases to zero, the remanent magnetization, whose absolute value is $a' - a$, has an equal probability of being positive or negative. In the former case, the type-C' loop behaves, for small variations of the magnetic field about zero, like a symmetric loop of coercive field c' and of moment a' subjected to an additional steady field $+na$. In the latter case, the same type-C' loop behaves under the same conditions like a symmetric loop identical to the previous case but subjected to an additional steady field $-na$.

It follows that a set of type-C' loops, such that the values of c' and n are uniformly distributed in the neighbourhood of $c' = 0$ and about $n = 0$, will have for small variations of the magnetic field about zero properties similar to those of a set of asymmetrical loops [3]: in particular, it obeys both Rayleigh laws. The same argument applies to loops of type C. As for the other loops of type A, A', B or K, their contribution to the variations in magnetization in weak fields is negligible compared to the contribution from loops of type C and C'.

These various remarks suggest that it must be possible to give an excellent representation of a real ferromagnetic substance by means of an ensemble of interacting pairs. To fit the properties of the model to those observed, the distribution function M at one's disposal now depends on at least three independent variables, for example c, n, and $a' - a$. This is an interesting but complex problem which will be kept for future discussion.

Another point which appears much more important will now be examined: the properties of certain K-type pairs subjected to magnetization loops between the field limits H and H'. We start from the initial state obtained by demagnetization in an alternating magnetic field decreasing to zero. The remanent magnetization of a pair in zero field is zero on average, because it has equal probabilities of having the value $ON = a' - a$ or the opposite value $-ON$. The external conditions, corresponding to the different possible magnetization loops, correspond in each case to a point P on a plane XOY (Fig. 3) with cartesian coordinates $OX = H$ and $OY = H'$. In order to simplify this discussion, we shall restrict our attention to the case where H' lies between $+H$ and $-H$, where H is taken to be positive and represents the first field applied after demagnetization. The representative points P are then located inside the quadrant bounded

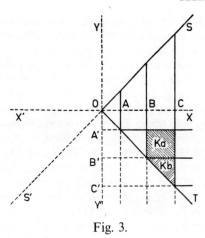

Fig. 3.

by the first bisector OS and the second bisector OT.

Furthermore, the K-type pair under consideration has three increasing critical fields corresponding to the three straight lines parallel to OY passing through the points A, B, C (Fig. 3), with equations $X = c + na'$, $X = c' - na$, $X = c - na'$ (where n is negative), and three decreasing critical fields corresponding to the three straight lines parallel to OX passing through A', B', C', and symmetrical to the first about the second bisector OT.

Using the notation of a previous Note [5], we denote by $j(H_1, H_2, \ldots, H_k)$ the magnetization of a pair obtained by applying successive fields of values $H_1, H_2, \ldots, H_k$, starting from the average initial state as defined above. It is known that conventional magnetization loops are *closed*, i.e. they follow the relations

$$j(H) = j(H, H', H) = \ldots$$

and

$$j(H, H') = j(H, H', H, H') = \ldots$$

The same is true here, provided that the point P lies inside the quadrant SOT and outside the two hatched regions Ka and Kb (Fig. 3). When the point P falls inside these two regions, new properties appear.

The region Ka is bounded by the four straight lines going through B, C, B' and A'. When the point P lies inside it, it is found that $j(H) = a'$, $j(H, H') = a' - a$ and $j(H, H', H) = a' - a$. For later loops, the same values are obtained. Thus for loops in the region Ka, $j(H, H', H)$ is smaller than $j(H)$. This is a new and quite remarkable property with no equivalent in the models used hitherto. We note that the loops in the region Ka are essentially asymmetrical.

The region Kb is bounded by the two straight lines passing through C and through B' as well as by OT. When the point P lies inside it is found that $j(H) = a'$, $j(H, H') = -a' - a$ and $j(H, H', H) = a' - a$. This time the second magnetization in the field H, $j(H, H', H)$, is greater than the first magnetization $j(H)$. This is also a new and remarkable property, and is foreign to the models used previously. We note that the loops in the region Kb are symmetric or almost so.

When one deals with an ensemble containing a large number of pairs, the resulting loop is generally not closed. The resultant magnetization of the ensemble (denoted by the capital J) is generally such that $J(H, H', H)$ is different from $J(H)$, but the sign of the difference depends on the values of H and H', as well as on the form of the distribution function M. Nevertheless, in the case of symmetric loops traced between +H and −H, corresponding to the points of the bisector OT, the loop of a given pair *can* belong to Kb but never to Ka. For the ensemble of pairs one must therefore have $J(H, -H, H) > J(H)$. Similarly, for asymmetrical loops traced between H and O and corresponding to points on the OX axis, the elementary loops are never of the Kb type (since $-c' + na$ is essentially negative) but can belong to Ka (when $c + na'$ is negative). For the ensemble of pairs it therefore follows that $J(H, O, H) < J(H)$.

To be specific, let us consider an ensemble of pairs with negative coupling corresponding to given values of a, a' and n, where c and c' can take all possible values. Each pair corresponds to a point in the (c, c') plane of Fig. 4. Pairs of type K will then correspond to the points located inside the rectangle DFF'D' ($OF = OF' = na - na'$). It is easy to show that the points of this rectangle which satisfy the condition $J(H, -H, H) > J(H)$ relevant to symmetric loops are the points lying to the right of the straight line EG, whose equation is $c = H + na'$, and above the straight line. ET, defined by $c' = H + na$ (Fig. 4). These are the points located inside the hatched quadrilateral EGFT. The existence of unclosed loops requires the presence of pairs made up of grains with different volumes ($a \neq a'$).

It can also be shown that the points which satisfy the relation $J(H, O, H) < J(H)$, pertaining to asymmetrical loops are the points lying in the quadrilateral EGFT

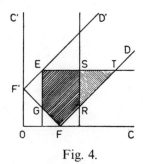

Fig. 4.

and satisfying the additional condition of being to the left of the straight line SR given by $c = -na'$: these are therefore points located inside the pentagon EGFRS. Since this pentagon forms only a part of the quadrilateral EGFT, it follows that the resultant loops satisfy the inequality

$$J(H, -H, H) - J(H) > J(H) - J(H, O, H).$$

It is of interest to estimate the order of magnitude of the difference $J(H, -H, H) - J(H)$ relative to the saturation magnetization. For this it would be necessary to know the distribution function, the form of which is completely unknown at present: suffice it to say that values of 2 to 5% seem reasonable.

The question now is whether these phenomena, peculiar to interacting pairs of grains, correspond to observable facts in the hysteresis loops of real substances. We shall not discuss this point here, but it would seem already that several anomalies in the initial loops, observed by Nguyen Van Dang [6] in the course of his investigations into creep phenomena, must be identified with the anomalies which we have just examined.

We note that the type C' loops are capable of taking on a *negative* thermoremanent magnetization, thus adding a new mechanism to those already known [7].

REFERENCES

[1] *Arch. Sc. Phys. et Nat.*, 1916, **42**, 449.
[2] *Z. Physik*, 1935, **94**, 277.
[3] *Cah. Phys.*, 1942, **12**, 2; 1943, **13**, 18.
[4] L. Néel, *Appl. Sc. Res.*, 1955, **4**, 13.
[5] L. Néel, *Comptes rendus*, 1957, **244**, 2668.
[6] *Comptes rendus*, 1958, **246**, 2357.
[7] L. Néel, *Phil. Mag. Suppl.*, 1955, **4**, 191.

A107 (1958)
COUPLING BETWEEN ELEMENTARY FERROMAGNETIC DOMAINS: TILTING EFFECT

Summary: The author resumes, using a different method, the investigation he undertook previously into couplings between domains, and confirms the earlier results. It is shown that deviations from loop closure are not restricted to the first cycle of the applied field, but also extend to the succeeding ones, though with rapidly decreasing amplitude.

In a previous Note [1] it was shown that starting with a model consisting of an ensemble of mutually independent ferromagnetic grains with symmetric rectangular hysteresis loops, and by introducing interactions between the grains, the resultant hysteresis loop of the ensemble was not always closed. By executing, for example, after demagnetization, a hysteresis loop between the fields $+H$ and $-H$, the magnetization obtained after the second application of the field H is larger than the magnetization produced by the first. In contrast, for asymmetrical loops between O and H, the magnetization given by the second application of the field H is smaller than that given by the first.

These properties are not specific to the model adopted. Similar results are obtained by introducing interactions between components other than the grains: for example Bloch walls moving in an inhomogeneously perturbed medium.

We shall restrict ourselves to showing this for the case of asymmetrical loops traced between O and H, after demagnetization in an alternating field which slowly decreases to zero. We shall simply assume that the component has an initial magnetization curve OA and a closed reverse loop APCDA (Fig. 1a). We shall simplify these properties by replacing the initial magnetization curve by the straight line OA of slope a (Fig. 1b) and the reverse loop by the straight line AB of slope b which is capable of being followed reversibly for as long as the field remains within the limits O and H. If the field takes on a value greater than H, the representative point goes to a point such as E, on the continuation of OA; the corresponding reversal line remains parallel to AB.

In this scheme, b and more particularly a depend on the field H; a has a maximum in the neighbourhood of

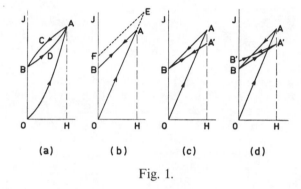

Fig. 1.

the coercive field H_c. The element we have just described will be specified as being of first order.

Let there then be two elements I and II of first order with susceptibilities a_1 and a_2 for the initial magnetization curve, and b_1 and b_2 for the reversal line. For variations J_1, J_2 the magnetization of the two elements, we associate a coupling energy $-nJ_1J_2$: n is a kind of molecular field interaction coefficient. It follows that element I is subjected to the field $H + nJ_2$ and element II to the field $H + nJ_1$.

Determination of the magnetization of the two elements when the interaction is positive $(n>0)$ — This problem can be solved graphically by drawing the curves J_1 and J_2 of the magnetization of the two elements as a function of the reduced field H/n in two rectangular axis systems respectively, $J_1O_1h_1$ and $J_2O_2h_2$. The axes $J_2O_2h_2$ are so arranged that the coordinates of O_2 with respect to $J_1O_1h_1$ are equal to $-H/n$ and $+H/n$, that O_2h_2 is parallel to O_1J_1 and that O_2J_2 is parallel to O_1h_1 (Fig. 2a). The points of intersection P of the magnetization curves of the two elements correspond to the equilibrium state of the

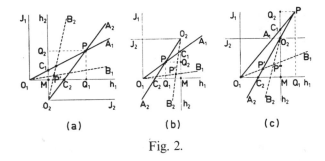

Fig. 2.

ensemble in the field H. The equilibrium is stable when the slope of the tangent at P to the magnetization curve of element I is smaller than that of the tangent at the same point on the magnetization curve of element II, where the two slopes are taken in the same system of axes $J_1 O_1 h_1$; in the opposite case the equilibrium is unstable.

Thus let $O_1 A_1$ and $O_2 A_2$ (Fig. 2a) be the initial magnetization lines, with slopes $a_1 n$ and $a_2 n$, for the two elements. The intersection P corresponds to the magnetization adopted by the two interacting elements, when after demagnetization the field has been changed from O to H: the coordinates PQ_1 and PQ_2 of the point P in the system of axes $h_2 M h_1$, where M is the intersection of $O_1 h_1$ and $O_2 h_2$, respectively represent the magnetizations of the elements I and II. They are larger than the magnetizations MC_1 and MC_2 which the two elements would have taken in the same field H had they been isolated (C_1 and C_2 are the respective intersections of $O_1 A_1$ and of $O_2 A_2$ with $O_2 h_2$ and $O_1 h_1$).

The two straight lines $O_1 A_1$ and $O_2 A_2$ become parallel when $a_1 a_2 n^2$ reaches unity. The point P is pushed to infinity: the two elements become magnetized spontaneously parallel to each other in zero field.

If after magnetizing in the field H, the field is brought back to O, the system behaves as if a field $-H$ were applied to elements of susceptibility b_1 and b_2, corresponding to the reversal lines. The corresponding variations in magnetization are the coordinates with their signs changed, within the system of axes $h_2 M h_1$, of the point of intersection P′ between the two reversal lines $O_1 B_1$ and $O_2 B_2$ of slopes $b_1 n$ and $b_2 n$. When the field is brought back to the value H, the new changes in magnetization correspond to the same point P′ and are of opposite sign to the preceding ones. It follows that the second application of the field H produces the same magnetization as the first. The interactions modify the resulting susceptibility, but do not alter the general shape of the magnetization curves, which remain similar to those of a first order element (Fig. 1b).

Negative interaction ($n<0$) — In this — the most interesting — case, the axes $J_2 O_2 h_2$ are arranged so that the coordinates of O_2 with respect to $J_1 O_1 h_1$ are both equal to $-H/n$ and that $O_2 J_2$ and $O_2 h_2$ are respectively antiparallel to $O_1 h_1$ and $O_1 J_1$ (Fig. 2b): the notation is identical with that of Fig. 2a.

The changes in magnetization induced by the application of the field H and then its removal are represented by the coordinates of the points P and P′: they are smaller than the changes in magnetization of the elements taken in isolation, but have the same general form. The resultant magnetization of the two interacting elements thus has the form shown in Fig. 1b, as in the preceding case.

The case we have just analysed is that of Fig. 2b where the points C_1 and C_2 lie respectively between O_2 and M and between O_1 and M: it corresponds to fairly weak interactions. When the interactions become strong and $a_1 a_2 n^2$ reaches unity, the point P is pushed to infinity and the two elements become magnetized spontaneously antiparallel to each other in zero field. An intermediate case is that in which only $a_1 n$ is less than -1, while the product $a_2 n$ lies between 0 and -1; it is shown in Fig. 2c, where the notation of Fig. 2a has become maintained. The point C_1 lies beyond O_2.

The magnetization $J_1 = Q_1 P$, taken upon the application of the field H, is positive, but the magnetization $J_2 = Q_2 P$ of the element II is negative. When the field H is removed, the changes in magnetization of the two elements are both negative, i.e. for the element II, in the same direction as the previous change. It follows that the representative point of these variations is no longer at the intersection of the two reversal lines $O_1 B_1$ and $O_2 B_2$, but at the intersection P′ of $O_1 B_1$ with the straight line $O_2 A_2$ of the initial magnetization of the element II.

It is only at the second application of the field H that the change in magnetization of the element II at last is of opposite sign to the previous variation: the representative point is then at P″. The magnetization loops are shown in Fig. 1c. The point A′ obtained at the second application of the field H lies below A: indeed, the resultant change in magnetization in this second application, equal to the sum of the coordinates of P″, is smaller than the sum of the coordinates of P′, since the slope of P′P″, namely $-b_1 n$, is less than unity. During the subsequent variations of the field between the limits O and H, the resultant magnetization reversibly traces out the straight line BA′. We shall say that loops of type 1c characterize an element of second order. The magnetic properties of such an element are essentially defined by three values a, b, c of the susceptibility ($a>b>c$), corresponding to the first growth of the field, its removal, and then to its restoration. Only c is a truly reversible susceptibility.

This process can be pursued further by conceiving a first order element of susceptibilities a_1 and b_1, interacting with a second order element of

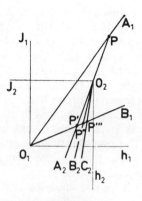

Fig. 3.

susceptibilities a_2, b_2, c_2. Let us assume, furthermore, that of the products of the type $-an$ or $-bn$, only the product $-a_1 n$ is greater than unity. The representative diagram is that of Fig. 3 where the slopes of the straight lines $O_1 A_1, O_1 B_1, O_2 A_2, O_2 B_2, O_2 C_2$ are equal to $-a_1 n, -b_1 n, -a_2 n, -b_2 n, -c_2 n$, in their respective system of axes. It is easy to verify that the representative points corresponding to the successive applied field values H, O, H, O are the points P, P', P'', P'''. The loops are shown in Fig. 1d, where the slopes of the successive lines of magnetization OA, AB, BA', A'B' become steadily smaller since the sums of the coordinates of the points P, P', P'', P''' become steadily smaller. We point out that the last line A'B' is reversible. We have just described the third order element.

By introducing more and more complex interactions, elements of arbitrary order n can be generated whose magnetization diagram consists of $n+1$ straight lines of decreasing slope. The lower and upper extremities, B_j and A_j, of these lines must tend to the limiting points B_∞ and A_∞.

The quantity $-an$ can rarely be larger than unity. Indeed it can be shown that since the largest value of a is of the order of J_s/H_c (where J_s is the remanent magnetization and H_c the coercive field) this situation corresponds to the case where the maximum interaction energy nJ_s^2 is of the order of the internal hysteresis energy $H_c J_s$ of an element. The probability $1/r$ that a given element should satisfy the inequality $-an > 1$ must therefore be small. Furthermore, it is certainly impossible to satisfy the relation $-bn > 1$: it is known that in materials like iron the value of b is well under a tenth of the largest value of a.

Moreover, the interactions in an nth order element are the result of interactions between n first order elements of which $n-1$ satisfy the inequality $-an > 1$. The probability of finding an nth order element is therefore of the order $1/r^n$. In reality it must be much smaller yet because the steric hindrances must limit strictly the number of first order elements capable of interacting in pairs with appreciable values of n. Finally, considering the series of straight line segments AA', BB', A'A'', B'B'', ..., each term can only be a small fraction of the preceding one, so that the limiting points A_∞ and B_∞ will in practice be reached very quickly.

The results obtained by the method expounded in this Note thus confirm those previously obtained [1], but with great precision. The original method considered only interactions between two grains: thus it was found that the points A', A'', ..., on the one hand and B, B', B'', ..., on the other, merged. Strictly speaking, this is not so and the deviations from closure of the loops must appear not only at the first field cycle, but also at the succeeding ones.

Arguments analogous to those used above, albeit more complicated, can be advanced for symmetric loops described between $-H$ and $+H$. It can be shown that the deviations from loop closure occur not only at the first half cycle of the field but also at succeeding ones, although with rapidly decreasing amplitude. This has to do, in accordance with the previous results, with a slight recovery of the loop whose average slope increases: this is again a kind of tilting but in the positive direction. We recall that tilting was negative for asymmetrical loops traced between O and H. As symmetric loops are free of creep, positive tilting will be easier to investigate than negative tilting.

The experimental observations, in particular the as yet unpublished results of Nguyen Van Dang, seem to confirm the existence of positive tilting in symmetric loops, and negative tilting combined with creep in asymmetrical loops. They also show that the effect of tilting is not confined to the first cycle, but also extends into the following ones, decreasing very rapidly.

REFERENCE
[1] L. Néel, *Comptes Rendus*, 1958, **246**, 2313.

A109 (1959)
THE EFFECTS OF INTERACTIONS BETWEEN ELEMENTARY FERROMAGNETIC DOMAINS: TILTING AND CREEP

Abstract: The author expounds an elementary theory of coupled ferromagnetic square-loop domains, based on a detailed investigation of the properties of two coupled domains. Some new phenomena are theoretically predicted, in particular "tilting". This is a positive or negative change of the magnetiaation in a field H_A, induced by varying the field a number of times between H_A and H_B. If the coupling of the part of the sample whose magnetization changes during the loop between H_A and H_B and the rest of the sample is represented by a statistically fluctuating field, another new phenomenon, "reptation" (creep), occurs: the hysteresis loop between H_A and H_B undergoes a gradual displacement, depending on the number of alternations of the field. These predicted effects have been experimentally observed.

(1) Introduction: In order to collate, interpret and predict the extremely complicated phenomena which make up ferromagnetic hysteresis, one has to turn to the simplest possible models and to specify the correspondence between the properties of the model and those of the real object. The two main models used until now start either with a plane Bloch wall undergoing translations in an irregularly perturbed medium, or with a single domain grain. The first model is more suitable for soft magnetic substances and the second for hard substances, but in the limit of highly perturbed materials, both give essentially equivalent results. Here we shall make more frequent use of the second model.

When the applied magnetic field H is parallel to the minimum energy direction of an isolated single domain, the magnetization processes are purely irreversible: the hysteresis loop is symmetric and rectangular and is completely defined by a saturation magnetic moment a and a critical or coercive field c. On the contrary, in the perpendicular direction, the processes are reversible: the component m of the magnetic moment along H is proportional to H in the interval

$$-a < m < +a.$$

Considering now an ensemble of randomly oriented, mutually independent grains whose critical fields are suitably distributed, one has quite a good picture of ferromagnetic hysteresis, for the irreversible as well as for the reversible effects. When one is interested more in the irreversible than in the reversible effects, the model can be satisfactorily simplified by giving a common orientation to all the grains: we shall in fact do just that in what follows. In spite of their simplicity, such models allow one to treat fruitfully questions as complex as that of thermoremanence, the magnetic memory of rocks or of magnetic fluctuation after effect. On the other hand, they fail for the detailed analysis of certain properties of hysteresis loops, such as for example the two Rayleigh laws for weak fields.

At this stage it is necessary to improve the model and to elaborate its properties. Clearly it is desirable to take account of the interactions between grains and to investigate their effects. These are very varied and very complicated, and are subject to an extremely difficult mathematical treatment which requires severe approximations to be introduced.

A fairly simple effect is one to which we have given the name *proximity effect* [1]. It appears in a particularly simple form when the anisotropy of the grains reduces to anisotropy of shape. For any system of grains of arbitrary shape which are magnetized to saturation and have geometric shapes which are on average isotropic, it can be shown that the magnetostatic energy per unit volume is equal to $\frac{2}{3}\pi J_s^2 v(1-v)$, where J_s is the saturation magnetization and v the fraction of the total volume occupied by the magnetic phase. The energy per unit volume of this phase is therefore proportional to $(1-v)$, and hence the coercive field, which is proportional to the energy, is also proportional to $(1-v)$. The experiments of L. Weil [2] have confirmed this conclusion.

It can also be remarked that the ensemble of grains must exert on any one given grain a coupling field, which is analogous to the Lorentz field in dielectric theory and is proportional to the mean field; but this approach does not introduce any interesting or novel feature. It is much more fruitful to take account of the spatial fluctuations of this field, i.e. of the fact that it differs from one grain to another as a result of irregularities in the spatial distribution of the grains and in the orientation of their magnetic moments [1]. Thus let us suppose that each grain is subjected to a certain coupling field h coming from its neighbours in addition to any external magnetic field which may be applied.

Furthermore, let us suppose that h is *constant*, and that the grain under consideration, assumed to be isolated, is characterized by a symmetric hysteresis loop of critical field c. On account of h, the grain now has an *asymmetrical* hysteresis loop defined by a lower critical field $-c-h$ and an upper critical field $+c-h$. The *model of grains with asymmetrical loops* is composed of an assemblage of such grains and is defined by a certain function of two variables $m(h, c)$ which gives the total magnetic moment dm of the grains in a coupling field, whose value lies between h and $h+dh$ and whose coercive field lies between c and $c+dc$.

This model has been in use for a long time. P. Weiss and J. de Freudenreich [3] used it first in 1916 and showed that it allowed a very simple interpretation to be given to the Rayleigh laws for the magnetization in weak fields, on the sole condition that $m(0, 0)$ be different from zero. F. Preisach used it [4] to show in the case of magnetic after effect, how the principle of superposition could fail. More generally, it seems that this model is necessary to interpret all the properties associated with ferromagnetic hysteresis in weak fields.

The interest and the simplicity of this model are such that it was natural to try to extend its use to high magnetizations by determining a distribution relation $m(h, c)$ capable of giving satisfactory results in the whole range of hysteresis. These attempts have not given very convincing results. In particular, for magnetically hard substances such as those used for the manufacture of permanent magnets, we concluded that it was impossible to account for the relatively high value of the initial anhysteretic susceptibility by using for the values of the distribution function $m(h, c)$ those that fitted best the interpretation of the other magnetic properties inside the hysteresis loop: the anhysteric susceptibility obtained is always three or four times smaller than the actual value [5].

Conversely, the model of grains with asymmetric loops must not be considered only as a simple extension and improvement of the model of grains with symmetric loops. In reality, as we showed a long time ago [6], for the region of weak magnetization it gives a formally rigorous representation of the movement of a Bloch wall in an irregularly perturbed medium. But as soon as the magnetization becomes large, it is no longer possible to establish a correspondence between the model and reality.

Returning now to the investigation and the representation of the interactions between the grains with symmetric cycles, one cannot avoid the fact that the coupling field h must depend on the state of magnetization of the assemblage of grains: we have shown elsewhere [1] that the mean square $\langle h^2 \rangle$ should be proportional to $J_s^2 - J_m^2$, where J_s and J_m are respectively the saturation magnetization and the average magnetization of the assemblage of grains. Thus the field h is a function of the magnetization J_m, as are likewise the critical fields $-c-h$ and $+c-h$ of a given grain: these critical fields cannot be considered to be constants. To take them as constant is meaningful only in the Rayleigh region, i.e. in a sufficiently narrow region for there never to be two neighbouring grains whose magnetic moments both change direction during the variations of the macroscopic magnetization under investigation. Indeed, with this restriction, the grains whose magnetic moment may change are surrounded by grains whose moments stay fixed and whose effects can therefore be represented by a coupling field of constant intensity h.

In summary, neither for the movements of Bloch walls nor for a set of interacting single domain grains, is it possible outside the Rayleigh range to give a physical meaning to the magnetic properties of an assemblage of grains with asymmetrical loops and to make a real model fit it.

It therefore seems necessary to go back to the direct investigation of interactions between grains of a three dimensional assemblage. Unfortunately, this problem appears *a priori* to be inextricable.

(2) Pairs of interacting grains: As we shall see below, however, interesting results can already be obtained by examining the properties of a set of mutually independent *pairs*, each composed of two interacting grains [7]. We suppose that these grains, whose loops are symmetric and rectangular, have moments a and a' and critical fields c and c'. We define the interactions by assuming that the first grain exerts a mean interaction magnetic field $+na$ or $-na$ on the second, while the second exerts a field $+na'$ or $-na'$ on the second, the sign $+$ or $-$ depending on the magnetic moment of the grain in consideration.

The magnetization loops of such pairs belong to one of the six types A, A', B, C, C', K shown on Fig. 1. For given values of a and $a'(a' > a)$, the types obtained depend on the values of c, c' and n, and are indicated in the (c, c') plane for positive coupling $(n > 0)$ in Fig. 2a and for negative coupling $(n < 0)$ in Fig. 2b. Apart from the types A and A' which are conventional, the other

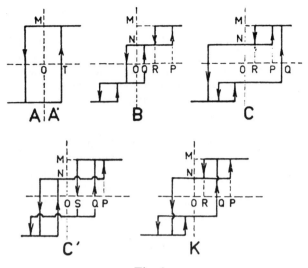

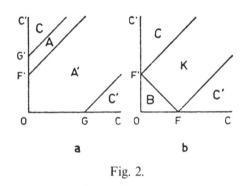

Fig. 1.

Fig. 2.

types are characterized by complex forms which are generally symmetrical, but in which one can discern secondary asymmetrical loops. The properties of an assemblage of such pairs relate simultaneously to those of an assemblage of grains with symmetric loops and one with asymmetric loops.

Thus the anhysteretic remanent magnetization, obtained by superimposing a small steady positive field upon an alternating field which decreases to zero, is equal to $a+a'$ for the types A and A', to $a'-a$ for the loops B, C, K, and to $a-a'$ for C' loops. The *average* anhysteretic remanent magnetization of an assemblage of pairs is therefore positive and finite (the negative remanent magnetization of the C' loops is cancelled by that of the C loops). It follows that the initial anhysteretic susceptibility of the assemblage is infinite, as is that of an assemblage of grains with symmetric loops, and in quite good agreement with the experimental results.

In contrast, taking for example pairs of type C, as long as the applied fields do not exceed the value OQ, they behave like grains with asymmetrical loops: the critical fields of these grains are $+OR$ and $+OP$, when the random effects of demagnetization give a positive remanent magnetization $+ON$, and $-OR$ and $-OP$ when the remanent magnetization is negative. Type C' pairs behave analogously. A set of pairs of type C and C' such that the values c, c' and n are uniformly distributed in the neighbourhood of $c=c'=0$ and around $n=0$, thus has, for small variations of the magnetic field around zero, properties which are similar to those of a group of grains with asymmetrical loops. In particular it obeys the two Rayleigh laws (in weak fields the contribution from the types A, A', B and K is negligible compared with that of types C and C').

These various remarks suggest that it is possible to give an excellent description of the hysteresis of a real ferromagnetic substance using an assemblage of pairs of the above types, providing one has available a suitable distribution for the values of a, a', c, c' and n. The value of this new model is unquestionable, since on the one hand it improves the agreement with experiment, and on the other furnishes a more satisfactory physical picture; it has, however, the drawback of increased complication.

(3) Tilting phenomena: Nonetheless the real interest of the model consisting of an assemblage of pairs lies elsewhere: this model has peculiar properties which are foreign to those of models used previously, and which, as it happens, correspond to properties of the real material [7].

To state these properties, we shall conform to a notation proposed previously [8] whereby $j(H_1, H_2, \ldots, H_k)$ denotes the magnetic moment of a grain or of a pair obtained by subjecting it, after demagnetization in a decreasing magnetic field, to successive magnetic fields H_1, H_2, ..., and lastly H_k (in general this is an average value, since the demagnetization generates remanent moments $+m_r$ or $-m_r$ with equal probability). The corresponding quantities for an assemblage of pairs will be represented by notations of the same kind but where j is simply replaced by J.

Having stated this, we note that in all the models used until now to represent hysteresis, whether with symmetric or asymmetric loops, the magnetization taken on at the second application of a field H_A is the same as that produced by the first application. More specifically, given a positive field H_A and a field H_B lying between $-H_A$ and $+H_A$, the relation

$$j[H_A, H_B, H_A] = [H_A]$$

is always satisfied: the return loop (H_B, H_A) closes on the second application of the field H_A.

Loops of type A, A'C and C' (Fig. 1) also have the same property, but the K type loop behaves quite differently.

We consider firstly the case of partial loops traced symmetrically about the origin ($H_B = -H_A$).

When H_A is smaller than OQ, one finds indeed

$$j[H_A, -H_A, H_A] = j[H_A] = 0$$

and similarly when H_A is larger than OP one also gets

$$j[H_A, -H_A, H_A] = j[H_A] = OM.$$

On the other hand, when H_A lies between OQ and OP, it is easy to see that

$$j[H_A, -H_A, H_A] = OM$$

and that

$$j[H_A] = (OM + ON)/2$$

The magnetization taken on at the second application of the field H_A is greater than that produced at the first, i.e.

$$j[H_A, -H_A, H_A] > j[H_A]$$

The succeeding loops do not modify further the value of j.

Using a notation whose meaning is obvious, we can also say that

$$j[H_A, (-H_A, H_A)^n]$$

is independent of n when n is greater than or equal to 1. To this effect we give the name *positive tilting*.

Now let us consider the asymmetrical partial loops described between the field H_A and the field $H_B = 0$. When H_A is smaller than OQ, one obtains

$$j[H_A, O, H_A] = j[H_A] = 0$$

and similarly when H_A is larger than OP

$$j[H_A, O, H_A] = j[H_A] = OM,$$

while in the interval $OQ < H_A < OP$ one finds

$$j[H_A, O. H_A] = ON$$

and

$$j[H_A] = (OM + ON)/2$$

This time, the magnetization taken on at the second application of the field H_A is smaller than the magnetization produced by the first:

$$j[H_A, O, H_A] < j[H_A];$$

this is *negative tilting*. As above, $j[H_A, (O, H_A)^n]$ is independent of n when n is larger than or equal to 1.

In fact this last property is related to the fact of having reduced the interactions to those between elements of the same group, each group being composed of only two elements. It is established in a subsequent article [9] that if one considers groups composed of an increasing number of elements, the magnetization $j[H_A, (H_B, H_A)^n]$ become independent of n only for increasingly large values of n. In the limit, $j[(H_A, (-H_A, H_A)]^n$ increases indefinitely with n while $j[H_A, (O, H_A)^n]$ decreases indefinitely, towards limiting values which in all likelihood are very quickly reached, in an exponential manner.

In the whole preceding discussion we have chosen for the initial state of the type K pairs the state obtained by demagnetization in a decreasing alternating magnetic field, but similar processes occur with initial states obtained by other methods. For example, one may choose for the initial state that obtained by the application of a very high negative magnetic field. Then it is found that in the interval

$$OQ < H_A < OP$$

the magnetizations take on the values

$$j'[H_A] = OM$$

and

$$j'[H_A, O, H_A] = ON.$$

This displays a negative tilting effect whose magnitude is twice as large as that obtained after demagnetization in an alternating field (here we have replaced j by j' to emphasize the difference between the initial states chosen in the two cases).

The same phenomena will occur, although the smaller amplitude, in an assemblage of pairs made up of a mixture of types A, A', C, C' and K.

It is important to realize that the results we have just obtained are the direct consequence of the existence of couplings between two systems (here two grains) endowed with hysteresis. The fact of having chosen grains with rectangular hysteresis loops as the elements of each system is quite irrelevant. Indeed, we have shown [9] that tilting effects of the same kind were liable to occur as a consequence of allowing interactions between any two systems which are characterised only by their reversible susceptibility and their irreversible susceptibility, without having to call into play the mechanism of the hysteresis phenomena.

The investigation of the interactions between the

two elements of a pair of grains thus leads to novel theoretical predictions which appear to correspond to incontrovertible experimental results. For many years competent experimenters have observed that successive hysteresis loops described between the same limits do not close, but the smallness of the effects, the apparent irregularity of the observed deviations, sometimes positive, sometimes negative depending on the experimental conditions, have indubitably discouraged observers from undertaking a detailed investigation. The recent systematic experiments of Nguyen Van Dang [10] qualitatively confirm the existence of positive or negative tilting effects in the conditions as predicted in the above account, but also show that other more complex effects must be superimposed on them.

(4) Creep phenomena: When a succession of hysteresis loops are described between two fixed limits, it is by no means assured that the two extremum states A and B, which by definition correspond to macroscopic magnetizations that are always the same, should have microscopic structures which also are identical. One can comprehend that two successive states of magnetization having the same resultant magnetization correspond to different distributions, if not of all the elementary domains, then at least of a significant fraction of them. We shall thus adopt as a working hypothesis [8] that in the course of the successive loops described between the same limits the magnetic moment distribution of those domains belonging to a certain fraction F of the sample is entirely independent of the preceding distributions: the successive distributions are subject only to the condition that they give the same resultant magnetic moment. Let us examine the consequences of such an assumption.

We start for example in the state A, with magnetization J_A, and consider the coupling field which represents the influence of the other domains (magnetic dipolar interaction amongst others) on any given domain G. This field is the sum of a term which at every cycle takes on the same value and is therefore invariant, plus a random term. The fixed term is due in part to the influence of the domains which do not belong to the fraction F, and in part to those of the fraction F, as a result of the fact that their resultant magnetic moment stays fixed throughout the successive loops. The random part comes from the action of the domains of F associated with the random nature of their magnetic moment distribution. We give this random part the name *creep field* and we denote its mean square value by h_r^2. This creep field possesses the following properties. In a given state A it varies from one point to another in the sample with a mean value of zero. At a given point in the sample its value averaged over the successive loops is also zero.

If in the state A in consideration the differential susceptibility of the sample were purely reversible, the effects of the random fields acting at the different points would cancel out on average and, on a macroscopic scale, these fields would not be observable. In fact, part of the differential susceptibility which we denote by S_{AB} is irreversible: the double suffix simply indicates that the value of this susceptibility depends on both states A and B which bound the successive loops. The irreversibility of this susceptibility entails the following consequences [11]: the magnetization produced by the random field h in the neighbourhood of a given point P is maintained and remains unchanged as long as the random field at P subsequently continues to be smaller than h; however this magnetization alters, taking the value $S_{AB}h'$ whenever the random field takes on a value h' larger than h. The magnetization near P produced by the creep field at the nth passage through the loop is thus equal to $S_{AB}h_n$, where h_n is the largest value taken by the creep field in the n successive occupations of the state A, or, which comes to the same thing, after n drawing lots.

We shall denote $x_n h_r$ the average value of h_n obtained by repeating these n draws a very large number of times: x_n is a numerical coefficient the calculated values of which are listed in the following Table [11].

For values of n greater than 1000, x_n is given with an error of less than one per cent by the relation

$$x_n^2 = 2[\log n - \log(x_n\sqrt{2\pi} \log 2)]$$

and varies asymptotically as $(\log n)$.

TABLE OF VALUES OF x_n

$n-1$	x_n	$n-1$	x_n
0	0	10	1.586
1	0.564	100	2.511
2	0.846	1 000	3.241
3	1.029	10 000	3.852
4	1.163	100 000	4.384
5	1.267	1 000 000	4.861

The irreversible susceptibility S_{AB} now remains to be defined precisely, bearing in mind that the changes in magnetization depend not only on the applied fields but also on the number of partial loops described. We shall therefore start with a sample, demagnetized by a decreasing alternating field, which is subjected to a series of successive loops between a positive field H_A and a field H_B of value between $-H_A$ and $+H_A$: the field H_A is applied first. The susceptibility S_{AB} appropriate to the extremity A of the partial loop is defined [8] by the relation

$$S_{AB}\delta H = J[H_A + \delta H, H_B, H_A] - J[H_A, H_B, H_A],$$

where δH is a very small positive field. Similarly for the extremity B of the partial loop an irreversible susceptibility is defined.

$$S_{BA}\delta H = J[H_A + \delta H, J_B] - J[H_A, H_B]$$

The susceptibilities S_{AB} and S_{BA} are in general quite different from the conventional irreversible differential susceptibility S_A pertaining to the initial magnetization curve. Thus, in the Rayleigh region (fields much smaller than the coercive field) and with an initial magnetization relation of the form

$$J = aH + bH^2$$

one finds the following values

$$S_A = 2bH_A; \quad S_{AB} = S_{BA} = b(H_A + H_B)$$

One can also see that for symmetric loops ($H_B = -H_A$), the susceptibilities S_{AB} and S_{BA} vanish. This important result, established here for the case of small amplitudes, is also valid for large amplitudes, as shown by experiment; moreover, this is in agreement with the properties of the classical models used to explain hysteresis.

If we now combine the above results, we conclude that the value of the magnetization J_{An} corresponding to the extremity A of the partial loop traced out between the fields H_A and H_B, must depend on the number of previous loops n performed after the demagnetization, and is given by the relation

$$J_{An} = J_A + S_{AB} x_n h_r$$

Similarly, the magnetization J_{Bn} which corresponds to the lower limit of the partial loop, is given by

$$J_{Bn} = J_B + S_{BA} x_n h_r$$

As S_{AB} is approximately equal to S_{BA}, it follows that the partial hysteresis loop traced out between H_A and H_B must undergo some kind of overall displacement, with a progressive translation proportional to x_n, i.e. proportional to $(\log n)^{1/2}$, when the number of loops performed becomes very large: this effect has been given the name *creep* (*reptation*). For this effect to occur, it is essential that the loops be asymmetrical: indeed it was shown above that the irreversible susceptibilities S_{AB} and S_{BA} are zero when the loops are symmetrical.

The creep terms J_{An} and J_{Bn} given by the above equations are the consequence of the sum of the random field and H_A, i.e. whichever of H_A or H_B has the larger absolute value, as is shown by the relations used above to define S_{AB} and S_{BA}. At first sight one might think that creep terms exist due to the superposition of h on H_B: but the cumulative effect of the random fields which reveals their effects when these fields are added to H_A, does not occur for the field H_B. In other words, reapplication of the field H_A wipes out practically all the effects of the field H_B: this is why the magnetization $J[H_A, H_B, H_A]$ is almost independent of H_B and is roughly equal to $J[H_A]$ as long as H_B lies between $-H_A$ and $+H_A$. The effects of small variations of H_B due to the addition of a small random field are thus completely obliterated by the subsequent application of the field H_A. More accurately, this means that the irreversible susceptibility S'_{BA}, which defines the effect of the superposition of a random field on H_A and which is given by

$$S'_{BA}\delta H = J[H_A, H_B + \delta H, H_A, H_B] - \\ - J[H_A, H_B, H_A, H_B],$$

where δH is a very small field, is zero. It is easy to show this for the grain model with asymmetric loops.

In the effects we have just investigated, the initial state of the sample was obtained by demagnetization in a decreasing alternating magnetic field, but it is clear that analogous phenomena must occur when the initial states are obtained differently. The theory is the same: only the values S_{AB} and S_{BA} of the irreversible differential susceptibility change.

The effects of creep which we have just described have actually been known experimentally for a long time. Observed first in 1885 by J. A. Ewing [12], they were rediscovered by L. Lliboutry [13] and by J. C. Barbier [14] in their investigations on the effects of shocks and on magnetic fluctuation after effect, and are now the subject of a systematic and extensive examination by Nguyen Van Dang [10].

The principal difficulty of this investigation is the impossibility of isolating pure creep: creep always occurs in conjunction with tilting, and there are interfering perturbations caused by the induced currents and especially the different kinds of magnetic after-effect: diffusion after-effect and fluctuation after-effect. Limiting ourselves to creep and tilting, and representing by a straight line AB the return loop described between the extremum fields H_A and H_B, we observe the following double process as the number n of circuit cycles increases: translation of the straight line AB proportional to $(\log n)^{1/2}$ for large values of n — this is creep; rotation of AB which tends either to approach the field axis (negative tilting of the asymmetrical loops), or to move away from the field axis (positive tilting of symmetric loops). In both cases there is a limiting asymptotic position reached exponentially.

There is no doubt that the theories sketched out here, and particularly the theory of tilting effects, are extremely rough and simplified. It would be worthwhile to give them a more extensive and rigorous mathematical foundation. They seem at least to have the merit of attracting attention to the existence and to the interest of a whole series of very little known phenomena. Admittedly, until now the complexity of hysteresis phenomena has always discouraged experimenters from making systematic investigations of it and from turning it into proper scientific knowledge.

REFERENCES

[1] L. Néel, *Appl. Sc. Research*, 1954, **B4**, 13.
[2] L. Weil, *C.R.Acad. Sci.*, 1947, **225**, 229; 1948, **227**, 1347.
[3] P. Weiss and J. de Freudenreich, *Arch. Sc. Nat.*, Geneve, 1916, **42**, 449.
[4] F. Preisach, *Z. Physik*, 1935, **94**, 227.
[5] L. Néel, unpublished.
[6] L. Néel, *Cahiers de Physique*, 1942, **12**, 1; 1943, **13**, 18.
[7] L. Néel, *C.R. Acad. Sci.*, 1958, **246**, 2313.
[8] L. Néel, *C.R. Acad. Sci.*, 1957, **244**, 2668.
[9] L. Néel, *C.R. Acad. Sci.*, 1958, **246**, 2963.
[10] Nguyen Van Dang, *Soc. Fr. Physique*, Grenoble, 20 fèvrier 1957; *C.R. Acad. Sci.*, 1958, **246**, 2357 and 3034.
[11] L. Néel, *C.R. Acad. Sci.*, 1957, **244**, 2441.
[12] J. A. Ewing, *Phil. Trans. Roy. Soc. London*, 1885, **176**, 523.
[13] L. Lliboutry, *Ann. Physique*, 1951, **6**, 731.
[14] J. C. Barbier, *Ann. Physique*, 1954, **9**, 84.

A108 (1959)
COMBINED EFFECT OF RANDOM FIELDS OF CREEP AND OF THERMAL FLUCTUATIONS

Summary: The author investigates the influence of the superposition of these two random fields on the hysteresis of ferromagnetic bodies and then shows how measurements of the differential susceptibility, made subsequently to the execution of the creep loops, allow the determination of the various characteristic constants of the creep.

When a ferromagnetic sample which has been demagnetized in an alternating magnetic field which slowly decreases to zero, is then subjected to a positive field H_A in which it takes on the magnetization J_A, it is known that a small additional field h_a, applied for a short time, produces an irreversible increase $S_A h_a$ in the magnetization; S_A is the conventional irreversible susceptibility. If, subsequently, the field is reduced from H_A to a value H_B intermediate between $-H_A$ and $+H_A$ and then brought back again to H_A, part of the irreversible increase $S_A h_a$ is destroyed and only a fraction $S_{AB} h_a$ survives: S_{AB} is the compound irreversible susceptibility [1].

The description of the magnetic properties of the sample in the neighbourhood of the point (H_A, J_A), and particularly the effects of the random fields of creep and of thermal fluctuations, requires a knowledge of these irreversible susceptibilities S_A and S_{AB}.

The random thermal fluctuation field gives rise to a magnetic after-effect whose effects are the same as that of a very high frequency alternating field, of more than 10^8 Hz, whose amplitude h_t increases with time [2]:

$$h_t = S(Q + \log t) \quad (1)$$

where S is of the order of several thousandths of the coercive field and Q is a numerical constant of the order of several tens [3].

The magnetization J_t in the field H_A therefore increases with the time of application of the field H_A according to the relation

$$J_t = J_A + S_A h_t \quad (2)$$

Furthermore, the existence of a random creep field due to coupling between elementary domains and to the random redistribution of a part of these domains during the execution of the partial loops between H_A and H_B has the following consequences [4]: after n successive applications of the field H_A, separated by $n-1$ returns to the value H_B, the final magnetization J_n is an increasing function of n given by the expression

$$J_n = J_A + S_{AB} x_n h_r \quad (3)$$

Here, h_r is the mean square creep field (related to the amplitude of the partial loop executed between H_A and H_B), and x_n is the mean of the greatest value taken in n trials by a reduced laplacian (or gaussian) variable. Thus, for example, $x_1 = 0$, $x_2 = 0.564$, $x_3 = 0.846$, etc.

Actually, the two kinds of random fields whose effects we have just examined act simultaneously. Let us determine for example the magnetization J_{nt} produced by a sequence of n applications of the field H_A, each of length t_0 and separated from the preceding application by a reduction of the field to the value H_B. One can write

$$J_{nt} = J_A + (S_A - S_{AB})S(Q + \log t_0) + S_{AB} h_f \quad (4)$$

The second term of the right hand side of (4) refers to the fraction of irreversible increases in magnetization which is destroyed at each return of the field to the value H_B, and which depends only on the length of the final application t_0 of the field H_A. The third term in equation (4) concerns the undestroyed fraction, the only one to give rise to creep, and for which the total length of application of the field H_A is equal to nt_0.

The value of h_f remains to be determined. The simplest assumption is to take h_f to be equal to the sum of the random thermal field and the creep field, i.e.,

$$h_f = S(Q + \log nt_0) + x_n h_r. \quad (5)$$

In fact, the process is not quite so simple. On the ith application of the field H_A, a given point in the sample is actually subject to the sum of the thermal fluctuation field and the random field $x_i h_r$. At each new application of H_A, the random field takes on a new value, so that the real problem is that of the superposition of the random thermal field with a field which varies appropriately as a function of time. We have shown previously [2] that the effect of such a process was the same as that of a fictitious field h_f given by the expression

$$h_f = S \left(Q + \log \int_0^t \exp\left[\frac{h}{S}\right] dt \right). \quad (6)$$

Here, h is taken to be equal to $x_i h_r$ during the ith period of application of the field H_A, and, when the field takes the value H_B, equal to a sufficiently large negative value so that the corresponding value of the exponential is negligible. Under these conditions the integral reduces to a sum and h_f can then be written in the form

$$h_f = S \left(Q + \log t_0 + \log \sum_{i=1}^n \exp\left[\frac{x_i h_r}{S}\right] \right), \quad (7)$$

where the x_i are n independent reduced gaussian variables.

The calculation of the sum appearing on the right hand side of equation (7) is quite difficult, but one can obtain an approximate result using the following method [5].

Firstly we set

$$\log \sum_{i=1}^n \exp\left[\frac{x_i h_r}{S}\right] = \varphi(c) \log n, \quad (8)$$

with

$$c = \frac{h_r}{S} (2 \log n)^{-\frac{1}{2}} \quad (9)$$

We note that it is easy to calculate $\varphi(c)$ when c is very small or very large. (a) $c \ll 1$. The exponentials of the left hand side of equation (8) are expanded in series, neglecting all but the first three terms; it is found that since the sum of the x_i is zero on average,

$$\varphi(c) = 1 + c^2 \quad (10)$$

(b) $c \gg 1$. The sum occurring in the first term of equation (8) now reduces to approximately the largest term, which we shall denote $\exp(x_n h_r/S)$. But the average value of x_n is approximately equal (4) to $(2 \log n)^{1/2}$, and so

$$\varphi(c) = 2c \quad (10')$$

One then notes that for $c = 1$ the values of $\varphi(c)$ and of its derivative, either from the expression (10) or (10') are the same. It is, therefore, probable that by adopting expression (10) for $c < 1$ and expression (10') for $c > 1$, a reasonably satisfactory approximation will be obtained in the whole interval of c. Finally, the values of h_f are then given by the expressions

$$h_f = S \left[Q + \log n t_0 + \frac{h_r^2}{2S^2} \right]; \; h_r < S(2 \log n)^{1/2}; \quad (11)$$

$$h_f = S(Q + \log t_0) + x_n h_r \; ; \; h_r > S(2 \log n)^{1/2}; \quad (12)$$

Expression (11), which pertains to the case where the random creep field is small compared to the thermal fluctuation field, shows that the usual creep term in $x_n h_r$ which occurs in Expression (5) has disappeared: there is suppression of ordinary creep. However, a thermal creep term in $S \log n$ remains. In contrast, expression (12) pertaining to the case where the random creep field is much larger than the thermal fluctuation field, shows that the thermal creep term in $S \log n$ vanishes and only the ordinary creep term remains. The presence of the various random fields modifies the magnetic properties of the sample in the neighbourhood of the point (H_A, J_A) and in particular the values of the irreversible increase in magnetization produced by a temporary additional field h_a.

In regard firstly to the random fields of thermal fluctuations, let us for example consider the effect of an additional field h_a superimposed for a time t' on a main field H_A which has already been acting alone for a time t. The magnetization obtained immediately after the field is returned to the value H_A is found by applying equation (6), and can be written

$$J = J_A + S_A S [Q + \log(t + t' e^{h_a/S})]. \quad (13)$$

In particular, when h_a is sufficiently large compared with S, this relation becomes

$$J = J_A + S_A S(Q + \log t') + S_A h_a \quad (14)$$

Thus the increase in magnetization produced by the temporary action of the field h_a is equal to $h_a S_A + (\log t' - \log t) S S_A$, and not the conventional value $h_a S_A$: this gives an indirect means of investigating the effects of thermal fluctuations, which has already been exploited by J. C. Barbier [3].

We now consider the effect of the random creep fields

on the irreversible increase of magnetization produced by an additional field h_a acting temporarily during the nth application of the field H_A. We first want to know the fraction of the sample for which the return to the field H_B destroys the irreversible increase in magnetization produced by the additional field. The last application of the field is thus the only one that counts in this case, so that the irreversible increase in magnetization is simply equal to $(S_A - S_{AB})h_a$. Thermal fluctuations are neglected for the moment.

In the remaining fraction of the sample in which the magnetization is not wiped out, the phenomenon is more complicated. Indeed, the different points in the sample have already been subjected to the temporary action of the successive creep fields, with mean maximum value equal to $x_n h_r$. Roughly speaking, the additional field should produce no effect as long as it is smaller than $x_n h_r$, while the effect must be proportional to $(x_a - x_n)h_r$ for $x_a > x_n$ (for simplicity we set $h_a = x_a h_r$). The final total increase in magnetization will thus be

$$\Delta J = (S_A - S_{AB}) x_a h_r, \qquad x_a < x_n; \quad (15)$$

$$\Delta J = S_A x_a h_r - S_{AB} x_n h_r, \qquad x_a > x_n. \quad (16)$$

Actually the situation is slightly more complicated, since at each point of the sample the additional field $x_a h_r$ is superimposed on the random creep field $x h_r$ which characterizes the nth application of the field H_A. It is thus the resultant field $(x + x_a)h_r$ which, when it is larger than $x_n h_r$, produces an irreversible increase in magnetization. The mean value h_m of the effective field is then given by the relation

$$h_m = h_r \int_{-z}^{+\infty} (x + z) y\, dx, \quad (17)$$

with $z = x_a - x_n$, where $y\,dx$ is the probability that the random field lies between xh_r and $(x + dx)h_r$.

Since x is a reduced gaussian variable, one can set

$$y_z = (2\pi)^{-\frac{1}{2}} \exp\left(-\frac{z^2}{2}\right);$$

$$Y_z = (2\pi)^{-\frac{1}{2}} \int_0^z \exp\left(-\frac{z^2}{2}\right) dz. \quad (18)$$

An elementary calculation gives the final increase in magnetization required

$$\Delta J = (S_A - S_{AB}) x_a h_r + S_{AB} h_r F \quad (19)$$

where F is a function of x_a given by one of the two following expressions

$$F = y_z + z\left(\frac{1}{2} + Y_z\right), \qquad z > 0; \quad (20)$$

$$F = y_{-z} + z\left(\frac{1}{2} - Y_{-z}\right), \qquad z < 0. \quad (21)$$

When h_a is small, z is roughly equal to $-x_a$, so that if n is sufficiently large, bigger than 100 for example, y_{-z} is practically zero and Y_{-z} is equal to 1/2, so that F is very small. The increase ΔJ is then given by relation (15).

When, however, h_a is large, y_z tends to zero and Y_z to 1/2, so that ΔJ is given by relation (16). In the intermediate region a knowledge of F gives the shape of the curve. The following table gives a few values of F calculated for $n = 1000$ (in this case $x_n = 3.24$).

TABLE OF VALUES OF F ($n = 1000$)

x_a	F	x_a	F	x_a	F
0	0.000	2.5	0.133	4.48	1.292
0.5	0,001	3.0	0.291	4.98	1.757
1.0	0.003	3.24	0.399	5.48	2.243
1.5	0.017	3.48	0.531	5.98	2.741
2.0	0.052	3.98	0.873	6.48	3.246

For the last four points, F is approximately equal to $x_a - x_n$: the curve coincides with its asymptote given by equation (16).

One can thus see the importance of determining the curve (19): by a set of measurements carried out following a series of n creep loops, one can determine both the composite susceptibility S_{AB} and the mean creep field $x_n h_r$.

Of course, account has to be taken of the after-effect of thermal fluctuations, using equation (6). We call t_0 the duration of application of the field H_A at each of the n creep loops, t the time interval separating the start of the nth application of the field H_A from the first magnetization measurement before application of the additional field, t' the total duration of application of the additional field h_a. The second magnetization measurement is performed after the field is returned to the value H_A. The required increase in magnetization, which is equal to the difference between the preceding magnetization measurements, is given by the following relation, which replaces equation (16)

$$\Delta J = (S_A x_a - S_{AB} x_n) h_r +$$
$$+ S \left(S_A \log \frac{t'}{t} + S_{AB} \log \frac{t}{t_0} \right), \tag{22}$$

Relation (22) is valid when x_a is much larger than x_n and when h_r is larger than $S(2 \log n)^{1/2}$.

REFERENCES

[1] L. Nèel, *Comptes rendus*, 1957, **244**, 2668.
[2] L. Nèel, *J. Phys. Rad.*, 1950, **11**, 49 ; 1951, **12**, 339.
[3] J. C. Barbier, *Ann. Phys.*, 1954, **9**, 84.
[4] L. Nèel, *Comptes rendus*, 1957, **244**, 2441.
[5] This method was pointed out to us by Paul Lèvy.

Chapter XI

MAGNETIC AFTER-EFFECT
A73, A78, A84

These three articles deal with the effect of time on the magnetization, i.e. magnetic after-effect, which is sometimes called magnetic viscosity (in French, le traînage magnétique).

Two types can be distinguished: irreversible after-effect, or more correctly *fluctuation after-effect*, of a thermodynamic character and present in all ferromagnetic substances, and reversible after-effect, or more correctly *diffusion after-effect* arising from diffusion inside the crystalline lattice.

A73 (1950)
THEORY OF MAGNETIC AFTER-EFFECT IN BULK-MATERIALS IN THE RAYLEIGH REGION

Summary: This paper is devoted to the theoretical study of magnetic after-effect of bulk substances in magnetic fields weak compared to the coercive force, i.e. in the Rayleigh region. We suppose that thermal fluctuations enable the walls separating elementary domains to get over obstacles normally characterized by a critical field H, by the action of a field h less than H.

In the first part, taking as a starting point a very general expression for the probability as a function of $H-h$ of clearing obstacles, we show that the effect of fluctuations at the end of a time interval t is roughly equivalent to that of a decreasing fictitious alternating field of initial amplitude $S(0.577+C+\log t)$, where S and C are constants depending on the nature of the susbtance and largely independent of the magnetic field and time. We deduce an expression for the initial magnetization, remanent magnetization, etc., as a function of h and t as well as the effect of tempering at temperature T. We also deduce the reversible susceptibility and the losses in a weak alternating magnetic field and we show the existence of a loss angle δ, which is proportional to S and, as a first approximation independent of the field and the frequency. We find a relation between δ and the variation of the reversible susceptibility with frequency.

In the second part, fluctuations of the magnetic field of dispersion created by fluctuations of the spontaneous magnetization direction are attributed to thermal fluctuations: we find that the average fluctuation field squared is equal to $\dfrac{4\pi kT}{v}$ where v is the average volume affected by the magnetization discontinuities. We likewise evaluate the order of magnitude of the decay time of a given distribution of the dispersion field and finally we calculate the probability of a wall crossing an obstacle, as a function of $H-h$ as well as the values of S and Q. For magnets of the Al–Ni–Co type, we find that S is close to 2 Oe and that Q is close to 20.

(1) Introduction: The idea of explaining magnetic after-effect by thermal fluctuations is not new: in particular it was expressed in a very interesting work by Preisach [1]. In a recent paper [2] we tackled the quantitative study for a ferromagnetic material consisting of very fine and independent grains. However, although such a model is well adapted to the study of baked clays and lavas it is not suited to ordinary ferromagnetic materials such as iron and steels where a large part of the magnetization arises from displacement of domain walls: in particular it does not explain the so characteristic laws of Lord Rayleigh relating to weak fields. Therefore we will now take up this question with a more suitable model.

We thus propose a general theory of *after-effect due to fluctuations*. Naturally this after-effect can coexist with after-effects of different origins, for example with that due to the very easy diffusion of carbon and nitrogen atoms in the crystal lattice of α-iron, on which there are fine theoretical and experimental studies by Snoek [3]. This *diffusion after-effect* is especially evident in very pure iron treated in an appropriate way.

The theory of ferromagnetic hysteresis based on analysis of the displacement of walls separating elementary domains [4] shows that getting over a given obstacle requires application of a magnetic field h at least equal to a certain critical field H characterizing the obstacle in question. However, it is reasonable to imagine that thermal fluctuations help the wall surmount the obstacle, in which case h is a little less than H, although this is classically impossible. We will describe this effect by the probability $\dfrac{dt}{\tau}$ of seeing a wall which is stopped in front of an obstacle at time t get over it between t and $t+dt$. The time constant τ is clearly larger for larger $H-h$ and smaller for higher temperatures.

Let us put, in a general way

$$H - h = F(\tau, T), \qquad (1)$$

where F is a function increasing with τ and temperature T.

Expanding F as a series close to some value t of τ, we can write

$$H - h = S(Q + \log \tau), \qquad (2)$$

where S and Q are functions of t and possibly temperature, defined by

$$S = \left(\frac{\partial F}{\partial \log \tau}\right)_{\tau=t}, \quad Q = \left(\frac{F}{S} - \log \tau\right)_{\tau=t}. \quad (3)$$

We will assume that S and Q vary *slowly* with $\log t$.

In the first part of this paper we develop a formal and general theory for the effect of time on magnetization phenomena, based upon formula (2), especially applicable to fields weak compared to the coercive force: in short, a generalization of Rayeigh's laws.

In the second part of the paper we study a mechanism capable of leading to a relation like (1) and calculate the magnitude of the constants S and Q for Fe–Ni–Al–Co type magnets.

I.
EFFECT OF TIME ON THE MAGNETIZATION IN THE RAYLEIGH REGION

(2) Explanation of the conventional Rayleigh laws: In order to determine the rôle of time in magnetization we must firstly set out a general theory of hysteresis which is satisfactory in the absence of this additional factor. Such a theory exists only in the Rayleigh region, i.e. when the magnetization and field are small compared to the saturation magnetization and coercive force respectively. The corresponding laws, which are particularly simple and general, had been stated by Lord Rayleigh and explained by means of wall displacements by L. Néel [5].

Let us briefly recall the principle of this explanation. A wall P, which on average is flat, separates two elementary domains magnetized antiparallel to one another. Let the position of the wall on the axis Ox normal to the average plane of the wall be x. As a result of the various factors disturbing the crystal lattice–internal forces, inclusions etc., the energy W of the system in the absence of an external field is an unknown function of x which can be represented by an irregular line with each element $A_n, A_{n+1}, A_{n+2}, \ldots$, corresponding to a given length of the x-axis. We assume that the slopes of the elements are distributed at random with no correlation between the slopes of two consecutive elements. Then, *in weak fields compared to the coercive force* and for a large number of similar walls we show that from the point of view of magnetization, on average everything happens as if we had dealt with an assembly E of imaginary, independent ferromagnetic grains characterized by two critical fields a and b relating to increasing and decreasing applied fields, respectively. Let point M with rectangular co-ordinates a and b represent each grain in a plane T. The assembly E is defined by the resultant saturation magnetic moment $\frac{1}{2}\beta ds$ of the grains situated in the surface element ds[1]; β is in theory a function of a and b. The Rayleigh region corresponds to that part of the plane T centred around the origin O inside which the *density β can be considered as constant*.

In the theory, very different aspects of magnetization cycles are attributed to these imaginary grains, depending on the sign of the difference $a-b$:

(1) $a > b$. Asymmetric rectangular hysteresis cycles result (Fig. 1A) with upper and lower discontinuities at respectively $h = b$ and $h = a$.

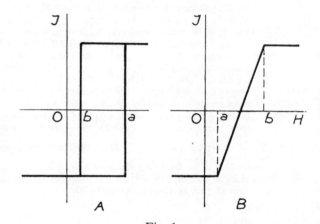

Fig. 1.
Hypothetical cycles equivalent to wall displacements.

(2) $a < b$. There is no hysteresis — the variation of magnetization with field is reversible as indicated by the line of Fig. 1B.

Consider grains of the second category corresponding to points of the plane T situated above the first bisector B'B (Fig. 2). Considering reversible phenomena which are not dependent on time, at least to the approximation here, their existence in the expression for the magnetization appears simply as a reversible term which in the Rayleigh region is proportional to the applied field h. We will say nothing further about these grains.

Grains of the first category which give rise to irreversible phenomena are situated below the first bisector BB'.

(3) Magnetization in a field h: The magnetization state of the assembly E in a field h is obtained by tracing the two equations $b = h$ and $a = h$, i.e. the lines AC and AD, in the half-plane T which interests us (Fig. 2). The

---
[1] Magnetic moment reversal of grains contained in the element ds therefore produces a total change equal to βds.

MAGNETIC AFTER-EFFECT

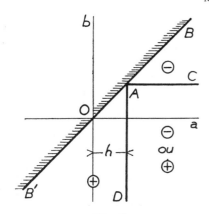

Fig. 2.
The half plane T and quadrant Q bounded by AC and AD.

two lines define a quadrant DAC whose significance is fundamental — outside this quadrant the magnetization of corresponding grains takes a clearly defined value which depends only on the current value of the field; it is negative in the triangle BAC and positive in the triangle B'AD. On the contrary, inside the quadrant DAC the direction of magnetization of the grains does not depend on the current value of the field, but on previous states, i.e. on the *magnetic history* of the body.

For example, let the magnetic field have a value $g+dh$, close to h. The quadrant moves to D'A'C' (Fig. 3). Let I be the point of intersection of A'C' with AC.

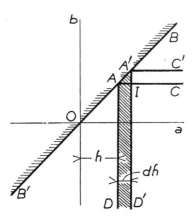

Fig. 3.
The quadrants Q and Q' represented by DAC and C'A'C' relating to field values h and $h+dh$. When the field increases from h to $h+dh$ the only grains whose moments change are situated in the hatched region DAA'ID.

The magnetization changes only for grains situated in the band DAA'ID': they all become magnetized in the positive sense, whereas grains in the band CIA'C' keep the negative magnetization they had when the field was h. This example illustrates how to determine, step by step, the final magnetic state of a system, knowing the initial state.

To define the initial magnetic state of the assembly E and make it independent of its *magnetic history* we can, for example, apply a very strong magnetic field and then restore zero field. We thus obtain the remanent magnetization after saturation and the magnetic state of the grains is represented by Fig. 4α. We can also demagnetize E by means of a slowly decreasing alternating magnetic field. The second bisector OB" (Fig. 4β) then divides the half-plane T into two regions,

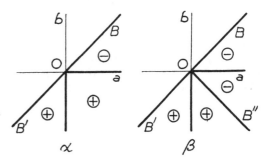

Fig. 4.
α, state of grains G after saturation in a positive field; β, state of the same grains after application of an alternating field slowly decreasing to zero.

an upper negatively magnetized region and a lower positively magnetized region. The resultant magnetization of E is zero. This state has the advantage over the previous one of being thermodynamically the most stable, and henceforth we will call it simply the *demagnetized* state.

By way of example, Fig. 5α shows the magnetic state of the grains after applying a magnetic field h to the demagnetized assembly E. Leaving out reversible terms, the magnetization $\mathfrak{J}$ of E is due to grains in the triangle OCD and is therefore proportional to the square of the field. We have

$$\mathfrak{J} = \beta h^2. \qquad (4)$$

Fig. 5β shows the magnetic state of the grains after removing h. The remanent magnetization $\mathfrak{J}_r$ is given by

$$\mathfrak{J}_r = \frac{1}{2}\beta h^2. \qquad (5)$$

345

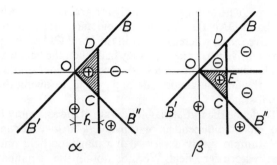

Fig. 5.
When a field h is applied to the previously demagnetized assembly E the state of the grains is represented by α. In bringing the applied field to zero the state of grains is represented by β.

(4) Rôle of time in the case of a steady magnetic field: As stated in the introduction, we suppose that it is possible in the absence of thermal fluctuations, to reverse the magnetization of an initially negatively magnetized grain by applying a field h less than the critical field a, this reversal being controlled by a time constant τ given by

$$a - h = S(Q + \log \tau), \quad (6)$$

where S and Q relate to values of $\log t$ close to $\log \tau$ and are not dependent on a and b.

Magnetization of the sub-assembly E_b — Consider now a sub-assembly E_b composed of grains from E in which the lower critical field is between b and $b + \delta b$. The possible values of a are thus spread out between b and $+\infty$. In the absence of fluctuations, the change $\delta \mathcal{J}$ in the magnetization of E_b results from grains with a value of a less than h. It is written

$$\delta \mathcal{J} = \beta (h - b) \delta b. \quad (7)$$

Thermal fluctuations allow the magnetic moment of certain grains, with a larger than h to reverse into the positive sense, on condition that the position occupied after reversal is always thermodynamically more stable than the initial position. For this, it is necessary that $\frac{a+b}{2}$ be less than h, i.e.

$$a < 2h - b. \quad (8)$$

We will assume that for reasonable experimental times, the value of τ deduced from (6) is very small for $a = h$ and very large for $a = 2h - b$, and that h is not too close to b.

For a value τ of the relaxation time the proportion P of reversed grains at the end of a time t is equal to

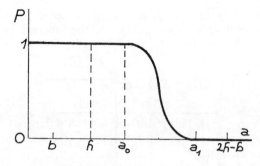

Fig. 6.
Proportion P of reversed grains at time t as a function of the upper critical field a.

$1 - e^{-t/\tau}$, it is a function of a represented by the curve of Fig. 6: a_0 and a_1 are values of a corresponding to time constants τ much smaller or much larger than t. It follows that P is practically equal to 1 for $a = a_0$ and to 0 for $a = a_1$. The change in magnetization $\delta \mathcal{J}$ of the grains E_b at the end of time t caused by the application of the field h can be written

$$\delta \mathcal{J} = \beta \delta b (a_0 - b) + \beta \delta b \, R, \quad (9)$$

where we put

$$R = \int_{a_0}^{a_1} \left(1 - e^{-\frac{t}{\tau}}\right) da. \quad (10)$$

By changing the variable $x = \frac{t}{\tau}$, R transforms into an exponential integral,

$$R = S \int_0^{x_0} (1 - e^{-x}) \frac{dx}{x} = S(C + \log x_0), \quad (11)$$

where C is the Euler constant, equal to 0.577, and where

$$x_0 = t \exp \left(Q + \frac{h - a_0}{S} \right). \quad (12)$$

Substituting these values in expression (9) the required change in magnetization becomes

$$\delta \mathcal{J} = \beta \delta b \, [h - b + S(Q + C + \log t)]. \quad (13)$$

The effect of thermal fluctuations is therefore equivalent to the effect which would be produced by a fictitious supplementary magnetic field, equal to $S(Q + C + \log t)$, added to the applied field h.

This supplementary thermal field which we will

denote by G(t, T) is a function of time t and of temperature T closely related to the function F(t, T) defined in the introduction. We easily find

$$G(t, T) = S(Q + C + \log t) = F + C \frac{\partial F}{\partial \log t}. \quad (14)$$

Notes: (1) This result still holds if β depends on a and b provided that the variations are weak in the interval $a_0 a_1$.

(2) In the applications which are the object of the second part of this paper, the second term of the last member of relation (14) is negligible (several percent) compared with the first term. $G(t, \tau)$ is therefore practically equal to F(t, T).

(3) The value of G(t, T) does not depend on the value of the lower critical field b characterizing the chosen sub-assembly. Moreover, we can follow analogous reasoning to that preceding for a sub-assembly E_a of grains with upper critical fields between a and $a + \delta a$ and initially magnetized in the positive sense. By subjecting E_a to a field h one shows that the grains whose moments orient negatively are those which correspond to values of b greater than $h - G(t, T)$.

Magnetization of the assembly E — Finally, given some magnetic state of the assembly E at time $t = 0$, let us apply a magnetic field h which remains constant: the state of E at a time t is obtained by tracing in the half-plane T a quadrant Q(h, t, T) bounded by the two lines A'D' and A'C' of equations

$$\left. \begin{array}{l} a = h + G(t, T), \\ b = h - G(t, T); \end{array} \right\} \quad (15)$$

a and b representing also the co-ordinate of the corner A' of the quadrant Q (Fig. 7).

Inside the quadrant Q grains keep their initial magnetic state, positive or negative. Outside the quadrant Q they take their equilibrium state in the field h — those for which $\frac{a+b}{2}$ is greater than h are magnetized negatively, the others are magnetized positively. The boundary between the two regions is formed by the line P'A' parallel to the second bisector OB''.

In principle the same method can be used to determine the magnetic state of E after a field h_1 has been applied between $t = 0$ and $t = t_1$, and a field h_2 between $t = t_1$ and $t = t_2$, etc. It is sufficient to consider the successive quadrants $Q(h_1, t - 0, T)$, $Q(h_2, t_2 - t_1, T)$, etc. and to determine step by step the successive magnetic states.

However the method fails when h_1 and h_2 are too close to one another as we are going to show. We

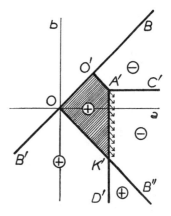

Fig. 7.
State of the grains in the initially demagnetized assembly E after applying a field h for a time t (cf. Fig. 5 in which time has not been taken into consideration).
Quadrant Q (h, t, T) is represented by D'A'C'.

calculate G(t, T) rigorously when h varies as a function of time. To simplify, we will limit ourselves to the case where S and Q are constants.

(5) Role of time t in the case of a magnetic field dependent on t: If the grains of the sub-assembly E_b are all initially negatively magnetized, the probability P of finding a grain with a positive magnetization, for a given value of a, satisfies the differential equation

$$dP = \frac{1 - P}{\tau} dt, \quad (16)$$

giving,

$$P = 1 - \exp \left\{ - \int_0^t \frac{dt}{\tau} \right\}. \quad (17)$$

The change in magnetization $\delta \mathcal{J}$ of E_b is still given by relation (9) but with a value for R which becomes

$$R = \int_{a_0}^{a_1} \left[1 - \exp \left\{ - \int_0^t \frac{dt}{\tau} \right\} \right] da. \quad (18)$$

Let us change the variable,

$$x = \int_0^t \frac{dt}{\tau}. \quad (19)$$

At time t constant, we find from equation (6)

$$da = S \frac{d\tau}{\tau} = - S\tau d\left(\frac{1}{\tau}\right), \quad (20)$$

347

and hence,

$$dx = -\frac{x}{S} da \qquad (21)$$

and

$$R = S \int_0^{x_0} (1 - e^{-x}) \frac{dx}{x} = S(C + \log x_0), \qquad (22)$$

with

$$x_0 = e^{\left(Q - \frac{a_0}{S}\right)} \int_0^t e^{\frac{h}{S}} dt. \qquad (23)$$

After calculation, the change in magnetization of E_b can be written

$$\delta \mathcal{J} = \beta \delta b (h_e - b), \qquad (24)$$

where

$$h_e = S \left[Q + C + \log \int_0^t e^{\frac{h}{S}} dt \right]. \qquad (25)$$

It looks as if, neglecting the fluctuations, we have simply applied a magnetic field h_e to the sub-assembly.

Let us apply these results to an experiment in two stages: the first of duration t with a constant field h, the second of duration t' with a field h', which is also constant. Thus we obtain

$$h_e = S(Q + C) + S \log \left\{ t e^{\frac{h}{S}} + t' e^{\frac{h'}{S}} \right\}. \qquad (26)$$

This equation shows that the result of the experiment is independent of the sequence of the stages: also it shows that when the absolute value of $h - h'$ is at least equal to 4S or 5S the result of the experiment is practically determined by the more intense field, e.g. h. Everything happens as if the weaker field h' had no effect. Then equation (26) reduces to

$$h_e = h + S(Q + C + \log t), \qquad (27)$$

in agreement with equations (13) and (24).

In the same way one could show that subjecting E_b to a magnetic field of intensity h during time intervals t and t' separated by a period during which the field has a value at least 4S or 5S less than h, the result is the same as that obtained with a single experiment of duration $t + t'$ and field h.

All these results can be extended without much modification to the more general case where S and Q are slowly varying functions of $\log t$.

These results particularly demonstrate the fact that the boundaries of the quadrants $Q(h, t, T)$ only result from a mathematical model: physically the boundaries form the centre of a transition region whose thickness is the order of 4S to 5S, inside which the probable value of the magnetic moment of a grain changes gradually. It follows that the method described at the end of section 4 can only give correct results when the boundaries of successive quadrants do not fall in regions where the boundaries of previous quadrants would have produced rapid changes, as a function of a and b, of the probable magnetization of the grains. These restrictions do not relate to crossing the boundaries at right angles where the effects which are of second order, are generally negligible.

(6) Applications of the theory to several particular cases: Let the initial state still be that obtained after demagnetization in an alternating field slowly decreasing to zero, and suppose that the temperature T_0 is constant.

Initial magnetization curve — Apply a field h at time $t = 0$. Let C'A'D' (Fig. 7) represent the quadrant $Q(h, t, T)$. Grains whose magnetization is reversed at time t are the grains situated inside the trapezium OK'A'O' where K' defines the intersection of A'D' with the second bisector OB". The area of this trapezium is equal to $h^2 + 2hG(t, T)$ and disregarding reversible terms proportional to the field, the magnetization is written

$$\mathcal{J} = \beta [h^2 + 2h G(t, T_0)]. \qquad (28)$$

The magnetization is a function increasing with time connected with the reversal of the magnetic moment of grains situated close to the boundary A'K'. Substituting the value of G from equation (14) we obtain

$$\mathcal{J} = \beta h^2 + 2\beta S(Q + C + \log t) h. \qquad (28b)$$

Remanent magnetization — Apply a field at time $t = 0$, remove it at time t and observe the magnetization at time $t + t'$. Let the quadrants $Q(0, t', T)$ and $Q(h, t, T)$ be represented by CAD and C'A'D' and let K and K' be the intersections of A'D' with AC and with the second bisector OB" (Fig. 8). Grains whose moments reverse after those operations are those which are situated inside the triangle AKK' of side equal to $h + G(t, T) - G(t', T)$. The remanent magnetization is then written

$$\mathcal{J}_r = \frac{\beta}{2} [h + G(t, T_0) - G(t', T_0)]^2. \qquad (29)$$

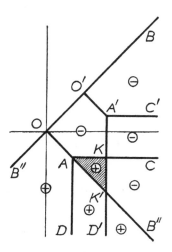

Fig. 8.
State of the grains in the initially demagnetized assembly E after applying a field h for a time t, then removing the field for a time t' (cf. Fig. 5β in which time has not been taken into consideration). Quadrants Q(0, t', T) and Q(h, t, T) are represented by DAC and D'A'C' respectively.

Naturally, this formula ceases to be useful when A approaches K', i.e. when $\mathcal{J}_r$ is zero.

Effect on the remanent magnetization of tempering at temperature T_1 — Apply a field h at temperature T for a time t, then heat the body in a zero field to a temperature T_1 above T for a time t'. Finally, bring the body back to T and observe the magnetization after time t''. Reasoning as above shows that the remanent magnetization $\mathcal{J}'_r$ which remains after this treatment is

$$\mathcal{J}'_r = \frac{\beta}{2}\left[(h + G(t, T_0) - G(t', T_1)\right]^2, \quad (30)$$

as long as G(t'', T) remains less than G(t', T_1).

Magnetic after-effect — Formulae (28) and (29) show that the magnetization is a function of time, i.e. it shows *magnetic after-effect*. If S and Q [cf. equation (2)] are constant and if h is large compared to G, the magnetization varies linearly with $\log t$. Formula (30) shows that the remanent magnetization decreases following tempering and becomes independent of the time t'' — the magnetic after-effect decreases and there is *stabilization*. In the same way, it is possible to stabilize the remanent magnetization by the action of a weak alternating field $h \sin \omega t$, slowly decreasing to zero from an initial value h_0. The remanent magnetization $\mathcal{J}''_r$ which remains after this treatment is given by

$$\mathcal{J}''_r = \frac{\beta}{2}\left[h - h_0 + G(t, T_0) - G(\tau, T_0)\right]^2 \quad (31)$$

and does not depend on the time of the measurement. The time τ is the order of $\frac{1}{\omega}$.

(7) Properties of the assembly E in a very weak alternating field: Let an alternating field act on E given by

$$h = h_0 \sin \omega t, \quad (32)$$

where h_0 is small compared to S. Let the probabilities of finding the magnetization of a grain directed in the positive or negative sense be respectively $x_a + x$ and $x_b - x$ where a and b are the critical fields of the grain and x_a and x_b two constants such that

$$x_a + x_b = 1 \quad (33)$$

where x is a sinusoidal function of time t. During the time interval dt, x increases by dx and satisfies the differential equation

$$dx = (x_b - x)\frac{dt}{\tau_a(h)} - (x_a + x)\frac{dt}{\tau_b(h)}, \quad (34)$$

where $\tau_a(h)$ and $\tau_b(h)$ are probabilities of changing from the negative to the positive position and the reverse, given approximately by the following formulae derived from the fundamental equation (2):

$$\left. \begin{array}{l} S[Q + \log \tau_a(h)] = a - h, \\ S[Q + \log \tau_b(h)] = h - b; \end{array} \right\} \quad (35)$$

S and Q are slowly varying functions of ω obtained by replacing t by $\frac{1}{\omega}$ in relations (3). In zero field we obtain an equilibrium defined by

$$\frac{x_b}{\tau_a(0)} = \frac{x_a}{\tau_b(0)} \quad (36)$$

and, using relation (33) we have

$$x_b = \frac{e^{\frac{a}{S}}}{e^{\frac{a}{S}} + e^{-\frac{b}{S}}}, \quad x_a = \frac{e^{-\frac{b}{S}}}{e^{\frac{a}{S}} + e^{-\frac{b}{S}}}. \quad (37)$$

Now substitute for $\tau_a(h)$, $\tau_b(h)$, x_a and x_b in differential equation (34) and change the variable:

$$\left. \begin{array}{l} a = z + u, \\ b = z - u. \end{array} \right\} \quad (38)$$

Then expand $\exp(h/S)$ and $\exp(-h/S)$ up to terms of the first degree in h. We then obtain, neglecting terms of second order in xh, the equation:

$$\frac{dx}{dt} + 2xe^{Q-\frac{u}{S}}\cosh\frac{z}{S} = \frac{he^{Q-\frac{u}{S}}}{S\cosh\frac{z}{S}}, \quad (39)$$

whose stable solution is written:

$$x = \frac{2e^{2Q-\frac{2u}{S}}}{\omega^2 + 4e^{2Q-\frac{2u}{S}}\cosh^2\frac{z}{S}} \cdot \frac{h_0 \sin\omega t}{S}$$

$$- \frac{e^{Q-\frac{u}{S}}}{\omega^2 + 4e^{2Q-\frac{2u}{S}}\cosh^2\frac{z}{S}} \cdot \frac{\omega h_0 \cos\omega t}{S\cosh\frac{z}{S}}, \quad (40)$$

On the other hand the saturation magnetic moment of the grains contained in the element $dadb$ is equal to $\frac{\beta}{2}dadb$. But $dadb = 2dudz$ so the resultant moment dM of grains corresponding to the element $dudz$ is a function of time given by

$$dM = 2\beta x\, du\, dz. \quad (41)$$

(8) Power lost by after-effect: The power d^2W lost by after-effect by grains of the element $dudz$ is, by definition,

$$d^2W = 2\beta\, du\, dz \int_{t=0}^{t=1} x\, dh, \quad (42)$$

with

$$dh = h_0 \omega \cos\omega t\, dt. \quad (43)$$

Integrating with respect to t, we obtain

$$d^2W = \frac{\beta\omega^2 h_0^2}{S\cosh\frac{z}{S}} \cdot \frac{e^{-\frac{Q}{S}}}{\omega^2 + 4e^{2Q-\frac{2u}{S}}\cosh^2\frac{z}{S}} du\, dz. \quad (44)$$

To find the power dissipated by the assembly E this expression is integrated firstly with respect to u between zero and $+\infty$, then with respect to Z between $-\infty$ and $+\infty$, which finally gives

$$W = \beta\omega h_0^2 \int_0^\infty \operatorname{arctg}\left\{\frac{2}{\omega}e^Q \cosh\frac{z}{S}\right\} \frac{dz}{\cosh^2\frac{z}{S}}. \quad (45)$$

If Q is large compared to unity, the arctangent is practically always equal to $\frac{\pi}{2}$ unless z is very small, so that

$$W = \frac{\pi}{2}\beta\omega h_0^2 \int_0^\infty \frac{dz}{\cosh^2\frac{z}{S}} = \frac{\pi}{2}\omega\beta S h_0^2. \quad (46)$$

The losses w per cycle are obtained by dividing W by $\frac{\omega}{2\pi}$, i.e.

$$w = \pi^2 \beta S h_0^2. \quad (47)$$

They are practically independent of the frequency, which is a very surprising result.

(9) Reversible susceptibility of the assembly E in a very weak alternating field: The magnetic moment of the element $dudz$ possesses a component in phase with the field h, to which corresponds a reversible susceptibility χ_r of the assembly E which is written:

$$\chi_r = \frac{4\beta}{S}\int_{-\infty}^{+\infty} dz \int_0^{+\infty} \frac{e^{2Q-\frac{2u}{S}}\, du}{\omega^2 + 4e^{2Q-\frac{2u}{S}}\cosh^2\frac{z}{S}}, \quad (48)$$

Integrating firstly with respect to u, then changing the variable:

$$\operatorname{tgh}\frac{z}{S} = x \quad (49)$$

we find

$$\chi_r = \beta S \int_0^1 \operatorname{Log}\frac{1 + A - x^2}{1 - x^2} dx, \quad (50)$$

where to simplify we have put

$$A = \frac{4e^{2Q}}{\omega^2}, \quad (51)$$

A is always very large, so neglecting terms in $\frac{1}{A}$ we find, finally,

$$\chi_r = 2\beta S(Q + 1 - \log \omega). \tag{52}$$

This expression is to be compared with the coefficient of h in expression (28b) for the magnetization of the assembly E placed in a field h for a time t, where h is this time large compared to S. This coefficient χ'_r is:

$$\chi'_r = 2\beta S(Q + C + \log t). \tag{53}$$

We deduce an approximate expression for the susceptibility χ''_r in a *strongly alternating field* by replacing t by a time equal to a quarter of a period, i.e. $t = \frac{\pi}{2\omega}$. Then we find

$$\chi''_r = 2\beta S(Q + 1,029 - \log \omega). \tag{54}$$

This formula is practically identical to (52). It follows that the reversible susceptibility, i.e. the coefficient of h in the expression for the magnetization of E keeps a constant value independent of the relative values of h and S.

(10) Formulae involving the loss angle: There is a loss angle δ, i.e. a phase lag δ of the induction with respect to the magnetic field, corresponding to the losses W given by equation (46). As a function of δ, these losses are

$$W = \frac{\omega \mu h_0^2 \operatorname{tg} \delta}{8\pi}, \tag{55}$$

where μ is the permeability. Comparison with (46) gives

$$\mu \operatorname{tg} \delta = 4\pi^2 \beta S. \tag{56}$$

As a first approximation, this loss angle is independent of the frequency and of the intensity of the magnetic field. It corresponds to the losses called residuals obtained after subtraction of the losses by hysteresis and by Foucault currents and was demonstrated experimentally for the first time by Jordan. The theory given here directly links it to the after-effect of semi-static experiments.

Moreover, given the magnetization $\mathcal{J}$ in a field h corresponding to the initial magnetization curve given by equation (28), the corresponding induction B_i is

$$B_i = h + 4\pi \mathcal{J} = \mu h. \tag{57}$$

Using relations (3) and (14) which relate G, F, S and Q, we derive

$$\frac{\partial B_i}{\partial \log t} = 8\pi \beta h \frac{\partial G}{\partial \log t} = 8\pi \beta h \left(S + C \frac{\partial S}{\partial \log t} \right). \tag{58}$$

If S varies slowly with $\log t$, we can neglect the last term of (58) and then comparing (56), (57) and (58) we obtain

$$\frac{1}{B_i} \frac{\partial B_i}{\partial \log t} = \frac{2}{\pi} \operatorname{tg} \delta, \tag{59}$$

which relates the loss angle to the slope of the curve $B = f(\log t)$. Note that we have already obtained [2] an identical formula in the study of an assembly of grains with *symmetrical* loops not satisfying the Rayleigh laws: here by contrast, it concerns an assembly of grains with *asymmetrical* loops.

Now let $\mu_r = 1 + 4\pi \chi_r$ be the permeability in a weak alternating field, defined by the ratio of the component of the induction parallel to the field to the value of this field. From equation (52):

$$\frac{\partial \chi_r}{\partial \log \omega} = -2\beta S, \tag{60}$$

from where, comparing with equations (56) and (57) we derive

$$-\frac{1}{\mu_r} \frac{\partial \mu_r}{\partial \log \omega} = \frac{2}{\pi} \operatorname{tg} \delta. \tag{61}$$

As above, this formula is identical to that obtained in the study of grains with symmetrical loops.

It is important to note that formulae (59) and (60) remain valid when the magnetization of the body arises from the superposition of the magnetization due to the system E of grains G and another magnetization, *independent of time*, of some other origin.

Note: The hysteresis losses, w_r for a loop between $-h$ and $+h$ are given by

$$w_r = \frac{4}{3} \beta h^3. \tag{62}$$

From equations (47) and (62) the magnetic field h_e for which the after-effect losses equal the hysteresis losses is given by

$$h_e = \frac{3\pi^2}{4} S. \tag{63}$$

below h_e the after-effect losses are higher than the hysteresis losses; above h_e it is the inverse.

II.
THE ROLE OF THERMAL FLUCTUATIONS OF THE INTERNAL DISPERSION FIELD

(11) Fluctuations of the internal field: Thermal fluctuations help the walls to get over obstacles by several different mechanisms. One of the most effective seems to be due to an internal magnetic field produced by changes in the direction of the spontaneous magnetization under the influences of thermal agitation: in effect the spontaneous magnetizations corresponding to different points of the same elementary domain are not exactly parallel to the overall direction, as is shown, moreover, by the decrease of the macroscopic spontaneous magnetization with increasing temperature. This non-uniformity gives rise to an internal magnetic field, or fluctuation field which varies irregularly with time, with or against the external applied field and so makes passing over the obstacles easier.

To define and calculate this field, it is first necessary to note that the irreversible magnetization variations arise in the final analysis from discontinuous Barkhausen jumps, each affecting a volume v of the ferromagnetic substance. Therefore it is the mean value of the fluctuation field in this volume v which is of importance to us here. In the volume v affected by a Barkhausen jump, two elementary domains exist separated by a wall which sweeps through all this volume in the course of the jump. Strictly speaking, it is necessary to calculate the average fluctuation field under these complex conditions. However, as this field is probably little different to the average fluctuation field which prevails in an element of the same volume v inside a single elementary domain, we will restrict ourselves to this latter calculation for an ideal substance, devoid of internal stresses and inclusions.

(12) Study of the thermal variations of the spontaneous magnetization: Refer the substance to three rectangular axes OXYZ, OZ being parallel to the direction of magnetization at absolute zero. At temperature T variations appear defined by the direction cosines α and β of the spontaneous magnetization with OX and OY. We can represent these by Fourier series,

$$\left. \begin{array}{l} \alpha = \sum \alpha_{pqr} \exp[2\pi i(px + qy + rz)], \\ \beta = \sum \beta_{pqr} \exp[2\pi i(px + qy + rz)], \end{array} \right\} \quad (64)$$

where the Σ signs are extended here, as in the following, to the integer values, positive and negative, of p, q, r subject to the condition $p^2 + q^2 + r^2 < R^2$, R being chosen such that the number of summed terms is equal to the number of degrees of freedom of the atomic magnetic moments contained in 1 cm³. As α and β are real, the co-efficients α_{pqr}, β_{pqr} must, furthermore, satisfy the relations:

$$\alpha^*_{pqr} = \alpha_{-p,-q,-r} \quad \text{et} \quad \beta^*_{pqr} = \beta_{-p,-q,-r},$$

where the asterisk indicates the imaginary conjugate.

To the magnetic state characterized by equations (64) corresponds an energy larger than that at complete saturation ($\alpha \equiv \beta \equiv 0$), which we will choose henceforth as the zero energy. This increase in energy is the sum of the three following terms; first the exchange energy E_w, second the magnetocrystalline energy E_c, and third the energy of internal dispersion fields.

The exchange energy E_w is given by the following expression, first established by Landau:

$$E_w = \frac{1}{12} Na^2 [(\text{grad } \mathcal{J}_x)^2 + (\text{grad } \mathcal{J}_y)^2 + (\text{grad } \mathcal{J}_z)^2], \quad (65)$$

in which $\mathcal{J}_x$, $\mathcal{J}_y$, $\mathcal{J}_z$ are the three components of the spontaneous magnetization, N is the molecular field coefficient and a the distance between neighbouring magnetically active atoms. By applying (65) to the distribution $\mathcal{J}_x = \alpha \mathcal{J}_0$, $\mathcal{J}_y = \beta \mathcal{J}_0$, $\mathcal{J}_z = \mathcal{J}_0$ we find

$$E_w = \frac{\pi^2 N a^2 \mathcal{J}_0^2}{3} \\ \times \sum (p^2 + q^2 + r^2)(\alpha_{pqr}\alpha^*_{pqr} + \beta_{pqr}\beta^*_{pqr}). \quad (66)$$

If OZ is the easy direction and K the anisotropy constant, E can be written

$$E_c = K \sum (\alpha_{pqr}\alpha^*_{pqr} + \beta_{pqr}\beta^*_{pqr}). \quad (67)$$

Finally, the internal field energy can be found from Poisson's equation,

$$\Delta V = 4\pi \text{ div } \vec{\mathcal{J}}, \quad (68)$$

where V is the magnetic potential and which integrates directly. We find finally

$$E_m = 2\pi \mathcal{J}_0^2 \sum \frac{(p\alpha_{pqr} + q\beta_{pqr})(p\alpha^*_{pqr} + q\beta^*_{pqr})}{p^2 + q^2 + r^2}. \quad (69)$$

In the expression for the total energy,

$$E = E_w + E_c + E_m,$$

the term in $\alpha_{pqr}\alpha^*_{pqr}$ is affected by a coefficient C given by

$$C = \frac{\pi^2 N a^2 \mathcal{J}_0^2}{3}(p^2 + q^2 + r^2) + K + \frac{2\pi \mathcal{J}_0^2 p^2}{p^2 + q^2 + r^2}. \quad (70)$$

For iron, where $K = 10^5$ ergs cm^{-3} and $\mathcal{J}_0^2 = 3 \times 10^6$, the second term is generally negligible compared with the third. As regards the first and third terms their relative value depend on p, q, r: they are on average equal for

$$p^2 + q^2 + r^2 = \frac{2}{\pi N a^2}, \quad (71)$$

that is to say for an average wave length

$$\lambda = \frac{1}{\sqrt{p^2 + q^2 + r^2}} = a\sqrt{\frac{\pi N}{2}}. \quad (72)$$

For iron, with $N = 5890$ and $a = 2.86$Å, λ is close to 300 Å. It results that for terms of the series (64) whose wavelengths are larger than this, the internal dispersion magnetic field constitutes the main part of the total energy.

In purely classical statistics, the values of α_{pqr} and of β_{pqr} are found by making a linear substitution for these quantities, so as to change the expression for the energy into a sum of squares and to assign to each an average energy of $\frac{kT}{2}$. We can moreover quantify the problem by saying that the component of the kinetic moment in the OZ direction changes only by multiples of $\frac{h}{2\pi}$ and then, at low temperatures, we find that the Bloch law of approach to saturation as $T^{3/2}$ follows in a very intuitive manner.

In a quite similar way, the same method of calculation applied to a polycrystalline substance [6], gives a magnetization law consistent with the experimental results and in agreement with the theoretical results of Holstein and Primakoff [7] obtained by another method. Therefore the basis of our calculation appears to be sound.

(13) Calculation of the mean value in volume v of the component of the dispersion field parallel to the spontaneous magnetization: For wavelengths large compared to 300 Å, the energy reduces to magnetic terms which, substituting

$$\gamma_{pqr} = p\alpha_{pqr} + q\beta_{pqr}, \quad \gamma'_{pqr} = p\alpha_{pqr} - q\beta_{pqr}$$

(73)

are written

$$E_m = 2\pi \mathcal{J}_0^2 \sum \frac{\gamma_{pqr} \gamma^*_{pqr}}{p^2 + q^2 + r^2}. \quad (74)$$

There are also $\gamma'_{pqr}\gamma_{pqr}$ terms to which correspond oscillations of large amplitude to zero divergence not giving rise to a magnetic field and of no interest to us here. To a term in $\gamma_{pqr}\gamma^*_{pqr}$, corresponds an average energy equal to $\frac{kT}{2}$, divided equally between the three rectangular components of the field. The component H_z in particular, is written:

$$H_z = \sum h_{pqr} \exp[2\pi i(px + qy + rz)], \quad (75)$$

and the average energy corresponding to each term is $\frac{1}{3}\frac{kT}{2}$, or

$$\overline{h_{pqr} h^*_{pqr}} = \frac{4}{3}\pi kT. \quad (76)$$

The average value H_m of H_z in a rectangular parallelepiped of sides a, b, c, of volume $v = abc$ and centre x_0, y_0, z_0 is written

$$H_m = \frac{1}{v}\int_v H_z \, dv$$

$$= \sum \frac{h_{pqr}}{v} \frac{\sin \pi p a}{\pi p} \frac{\sin \pi q b}{\pi q} \frac{\sin \pi r c}{\pi r}$$

$$\times \exp[2\pi i(px_0 + qy_0 + rz_0)]. \quad (77)$$

whose mean square is, by factorizing $\overline{h_{pqr} h^*_{pqr}}$,

$$\overline{H_m^2} = \frac{4\pi kT}{3v^2} \prod_{x=a,b,c}\left[\sum_{p=-\infty}^{\infty}\frac{\sin^2 \pi p x}{\pi^2 p^2}\right]. \quad (78)$$

Replacing the sums by integrals we obtain,

$$\overline{H_m^2} = \frac{4\pi kT}{3v^2}\left[\int_0^\infty \frac{\sin^2 x}{x^2} dx\right]^3 abc, \quad (79)$$

and finally, as the integral between brackets is equal to 1,

$$\overline{H_m^2} = \frac{4\pi kT}{3v}. \quad (80)$$

This method of calculation is justified by the fact that

when a, b, c are large compared to λ, the terms from which it would be necessary to account for the non-magnetic parts of the energy form only a negligible part of the sums or integrals. To simplify the notation we will put

$$H_f^2 = \overline{H_m^2} \; ; \qquad (81)$$

H_f is inversely proportional to the square root of the volume v and independent of its shape. It is close to 0.4 Oe for a cube of side 1 μ at 300 K. The values of the mean fields are distributed following a Gaussian law, so that at a given time the probability $\bar{\omega}(H_m)dH_m$ of finding this field between H_m and $H_m + dH_m$ is

$$\bar{\omega}(H_m) = \frac{1}{H_f \sqrt{2\pi}} \exp\left\{-\frac{H_m^2}{2H_f^2}\right\}. \qquad (82)$$

(14) The reorganization time θ: Now let us suppose that to get over an obstacle which would require a field H in the absence of fluctuations, there is only available a field h which is less than H. The probability $\frac{dt}{\tau}$ of passing over the obstacle during the time interval dt, mentioned in the Introduction, is equal to the probability of the field H_m exceeding H$-h$ at least once during the same time interval. For this, we must know how the fluctuation field H_m evolves with time. This can be represented by splitting the time into successive equal intervals of suitable duration θ, and supposing that in each interval the average field is constant and given at random by the law (82) without any correlation between the values of H_m corresponding to two consecutive intervals. Under these conditions the probability is

$$\frac{1}{\tau} = \frac{1}{\theta} \int_{H-h}^{\infty} \bar{\omega}(H_m) dH_m . \qquad (83)$$

If (H$-h$) is large compared to H_f we can write, approximately,

$$\frac{1}{\tau} = \frac{H_f}{(H-h)\theta\sqrt{2\pi}} \exp\left\{-\frac{(H-h)^2}{2H_f^2}\right\}. \qquad (84)$$

Therefore, it would be useful to know how to determine, *a priori*, the reorganization time θ which can also be defined as the minimum time between two observations of the field H_m such that the values obtained can be considered independent.

The calculation of Q unfortunately appears to be too difficult. The most we can state, as shown by an elementary calculation which need not be reproduced here, is that there is very strong coupling between the magnetic field "wave" defined for example by relation (75), and the thermo-elastic waves from the Debye theory of the specific heat of solids: in effect the elastic forces give rise to magnetic couples which impart a precessional motion to the spontaneous magnetization, with transfer of elastic energy to magnetic energy. The reorganization time θ of the magnetic field "waves" must therefore be the same order of magnitude as the reorganization time θ' of the magnetoelastic waves, which is defined in a similar way. The calculation of θ' is not easier than θ, but we can compare it to the thermal fluctuation decay time θ'' when the amplitude of a sinsuoidal thermal fluctuation of wavelength L is divided by e. An elementary calculation gives

$$\theta'' = \frac{L^2}{4\pi^2 h}, \qquad (85)$$

where h is the diffusivity which is the order of 0.1 for substances which interest us. For $\frac{L}{2} = 10^{-5}$ cm, we thus obtain $\theta'' = 10^{-10}$ s, which should also give the order of magnitude of θ for domains of volume $v = 10^{-15}$ cm³; θ therefore varies as $v^{2/3}$. However the fragility of such reasoning is obvious.

(15) Calculation of F(τ, T): From the definition of F(τ, T) given in (1) we can write, using relation (84):

$$H - h = F(\tau, T) = H_f \sqrt{2(Q' + \log \tau)} \qquad (86)$$

where

$$Q' = \log \frac{H_f}{H - h} - \log \theta - \frac{1}{2}\log 2\pi \qquad (87)$$

and

$$H_f = \sqrt{\frac{4\pi kT}{3v}}. \qquad (88)$$

It appears difficult to determine *a priori* values of H_f and Q' for a given substance: H_f depends in effect on v and Q' depends principally on θ. In the best permanent magnets like Fe–Ni–Al–Co, the elementary domains are certainly very small, of the order of the wall thickness, i.e. 10^{-5} cm, which would make v the order of 10^{-15} cm³ with corresponding values of H_f the order of 13 Oe at 300 K. Accepting the reasoning of section 14, values of θ for the same magnets are the order of 10^{-10} s. Then from relations (86) and (87), with H the order of 500 Oe, it can be seen that for values of h corresponding to relaxation times τ of 1 s, Q' is in the region of 20.

From the logarithmic nature of Q', it can be seen that Q' will vary relatively little from one substance to another. When at H_f it must decrease rapidly as the substance becomes softer as the number of Barkhausen discontinuities increases greatly. Thus, for a soft iron with coercive force $H_c = 0.5$ Oe, Bozarth and Dillinger [8] find an average value for v of 10^{-9} cm^3, corresponding to $H_f = 1.3 \times 10^{-2}$ Oe. As the coercive force of Fe–Ni–Al–Co is around 500 Oe we can see that for the two substances considered the ratio $\dfrac{H_f}{H_c}$ is the same and equal to about $\dfrac{1}{40}$.

If we expand F as a series of the region of some value of τ, we obtain (cf. § 1),

$$H - h = S(Q + \log \tau), \qquad (89)$$

with

$$S = \frac{\partial F}{\partial \log \tau} = \frac{H_f}{\sqrt{2(Q' + \log \tau)}} \qquad (90)$$

and

$$Q = \left(\frac{F}{S} - \log \tau\right) = 2Q' + \log \tau. \qquad (91)$$

With the preceding example of Fe–Ni–Al–Co materials for times τ the order of a second, we obtain a value for S very close to 2 Oe. Now, we have shown above that for the initial magnetization curve it is as if we add a term depending on time like $S \log t$, to the applied field: this would appear easy to show experimentally since for these materials it varies by 8 Oe between 1 mn and 1 h.

(16) Brownian motion of the wall: In a recent note [9] we have shown that the spins inside a wall can undergo a series of precessions under the influence of different perturbing couples created by thermoelastic Debye waves: it results in a characteristic motion agitating the wall capable of helping it pass over obstacles. This effect is expressed by a time constant τ' given by

$$\frac{1}{\tau'} = C \exp\left\{-\frac{v \mathcal{J}(H-h)^{\frac{3}{2}}}{3kTH^{\frac{1}{2}}}\right\} \qquad (92)$$

which is to be compared with relation (84). A quite long discussion shows that log C is the same order of magnitude as $-\log \theta$, within a factor of two for example. To compare the efficiency of the two mechanisms in question it suffices to compare the argument of the exponentials. If ρ is the ratio of the argument of the exponential of relation (92) to that of relation (84) we find

$$\rho = \frac{8\pi}{9} \frac{\mathcal{J}}{H^{\frac{1}{2}}(H-h)^{\frac{1}{2}}}. \qquad (93)$$

With $\mathcal{J} = 1700$, $H = 500$, $H-h = 80$ as above, we obtain $\rho = 24$ and therefore the role of brownian motion of the wall is negligible compared to that of the internal fluctuation field.

(17) Conclusions: Knowing the expression for the characteristic function $F(\tau, T)$ given by relation (86) and dependent on the two parameters v and Q', the laws of the magnetic after-effect in weak fields are completely determined. The particularly simple cases of the initial magnetization curve and remanent magnetization are given by equations (28), (29) and (30), whereas equation (56) gives the loss angle in an alternating field.

It remains now to see how well the experimental results compare to the formulae established by the present work, and in the event that they are verified, to see to what extent the experimental values of v and Q' agree with the rough estimates made above. Unfortunately experimental material which could be used for this is very meagre. Several workers in the Laboratoire d'Electrostatique et de Physique du Métal have been working to fill this gap for two years and their results will be shortly published and discussed. Without prejudging them, it seems likely that the theory proposed here may explain and link together many experimental facts.

REFERENCES

[1] F. Preisach, *Z.f. Physik*, 1935, **94**, 277.
[2] L. Néel, *Ann. de Géographique*, 1949, **5**, 99.
[3] J. L. Snoek, *Physica*, 1938, **5**, 663; 1939, **6**, 161, 591, 797; 1941, **8**, 711, 734, 745; 1942, **9**, 862; *Ned. T. Natuurk.*, 1942, **9**, 417; New developments in ferromagnetic materials, Elsevier, Amsterdam, 1947.
[4] Becker and Doring, Ferromagnetismus, Springer Berlin, 1939.
[5] L. Néel, *Cahiers de Physique*, 1942, **8**, 65; **12**, 1; 1943, **13**, 18.
[6] L. Néel, *C.R. Acad. Sci.*, 1945, **220**, 814; *J. de Physique*, 1948, **9**, 193.
[7] *Phys. Rev.*, 1940, **58**, 1098; 1941, **59**, 388.
[8] *Phys. Rev.*, 1932, **41**, 345.
[9] *C.R. Acad. Sci.*, 1949, **228**, 1210.

A78 (1951)
THE MAGNETIC AFTER-EFFECT

Summary: It is shown that there are at least two types of magnetic after-effect–reversible after-effect and irreversible after-effect — and the simplest way of formally describing their properties is by introducing two fictitious after-effect magnetic fields dependent on time and superimposed upon the applied magnetic field.

The fictitious field for reversible after-effect originates from the progressive stabilization with time of the spontaneous magnetization along the direction it occupies. This stabilization is associated with physical diffusion in the interior of the ferromagnetic substance. It is noted that strains due to magnetostriction are not sufficient to explain the stabilization of the diffusion and a mechanism involving magnetocrystalline coupling is proposed.

With regard to the irreversible after-effect, it results from various effects of thermal fluctuations, the most important of which appears to be due to fluctuations of the internal dispersion field arising from fluctuations of the spontaneous magnetization direction.

Finally, we have assembled various numerical results which define the magnitude of the two types of after-effect.

INTRODUCTION

(1) Categories of magnetic after-effects: Here under the name of magnetic after-effect we include everything connected with the influence of time on magnetization phenomena in ferromagnetic bodies, with the following exceptions:

(a) the influence of induced currents.

(b) the effects of alternating magnetic fields with frequencies higher than about 10^5 Hz.

(c) effects of physical or chemical alterations or transformations of the ferromagnetic substance.

Leaving aside very fine grains whose properties have been dealt with elsewhere [1], the experimental evidence distinguishes at least two categories of magnetic after-effect, which are possibly superposable, and which we will call *reversible after-effect* and *irreversible after-effect*:

(a) Reversible after-effect — (1) The variation in magnetization following a given change in field tends towards a limiting value as the time t tends to infinity. (2) This effect is only found within a certain temperature range. (3) It obeys the superposition principle. (4) The loss angle in an alternating field is very dependent on the frequency and temperature. (5) The initial permeability decreases with increasing time since demagnetization. (6) This after-effect is only observed in certain substances.

(b) Irreversible after-effect — (1) The change in magnetization is approximately proportional to $\log t$. (2) This after-effect hardly varies with temperature. (3) It does not obey the superposition principle. (4) The loss angle varies little with the frequency and temperature. (5) The initial permeability is not affected. (6) It affects all ferromagnetic substances.

Reversible after-effect has been known since the work of J. A. Ewing [2] and since then has been frequently studied, for example by H. Atorf [3], G. Richter [4], C. E. Webb and L. H. Ford [5], J. L. Snoek [6], H. Falenbrach [7], and H. Wilde [8]. Irreversible after-effect which was more recently discovered by F. Preisach has been studied by P. Courvoisier [10], R. Street and J. C. Woolly [11], L. Néel [12], J. C. Barbier [13] and L. Lliboutry [14].

In this paper, these two types of after-effect are studied by introducing the new idea of a fictitious after-effect magnetic field. This allows a simplification and generalization of the laws relating to the phenomena whilst directly distinguishing between the two after-effects. In addition, we propose a new mechanism to interpret reversible after-effect and briefly explain a new theory for it, a more detailed account of which will appear in another publication.

Likewise we will summarize the theory of irreversible after-effect, referring to a previous publication [12] for further details.

(2) Definition of the fictitious after-effect field: Experiment shows that when the magnetic field applied to a ferromagnetic substance is varied by H, the corresponding change in magnetization is the sum of two terms:

$$J = J_0 + J_1(t) \tag{1}$$

The first, J_0, corresponds to a quasi-instantaneous change in magnetization, whereas the second, $J_1(t)$ is a function of time t.

We can represent the phenomenon as due to an additional fictitious field $H(t)$, which we will call the *after-effect field*, superimposed upon the applied field and which varies in a suitable way with time. Obviously such an idea is only of interest if $H(t)$ can be expressed in a simple way as a function of the initial state and of H. We will show later that this is actually the case.

(3) Definition of the various differential susceptibilities: Since the after-effect field is generally very small, it is relevant first to examine the behaviour of a ferromagnetic substance in a very small field superimposed upon the main field. After magnetizing a body to a value J by changing the field, a further very small change dH of the field *in the same sense* produces a change in magnetization dJ, whereas a change *in the opposite sense* $-dH$ produces a change $-dJ'$, whose absolute value is smaller than dJ. We will put

$$\frac{dJ}{dH} = a + c, \quad \frac{dJ'}{dH} = a \quad (1b)$$

and call $a+c$ the total differential susceptibility, a the *differential reversible susceptibility* and c the *differential irreversible susceptibility*. Determination of c depends upon the course followed to reach the value J, whereas the experiments of Gans [15] show that to a first approximation a is only a function of J.

When the substance has been demagnetized by a decreasing alternating field, the initial magnetization curve according to Lord Rayleigh [16] is,

$$J = a_0 H + b H^2 \quad (2)$$

provided that H is small compared to the coercive force H_c. It follows that at the point (j, H) the differential susceptibilities are

$$a = a_0, \quad c = 2bH. \quad (3)$$

On the branch of the hysteresis loop obtained by decreasing the field from its maximum value H_m to a value H, we have,

$$a = a_0, \quad c = b(H_m - H) \quad (4)$$

provided that $|H| < |H|$. In particular, at remanence $H = 0$ and we have $c = bH_m$.

(4) Reversible after-effect and irreversible after-effect: Two categories of after-effect fields with quite distinct properties can be distinguished.

The first, $H_r(t)$ is a field opposed to the imposed magnetization change and whose absolute value decreases, whilst keeping the same sign and tends to zero as t tends to infinity (Fig. 1).

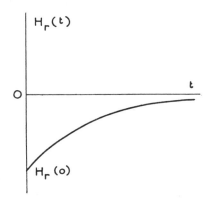

Fig. 1.
Variation of the reversible after-effect field $H_r(t)$ with time.

The initial value $H_r(0)$ of this field is finite. The instantaneous change in magnetization J_0 therefore occurs under the action of a field $H + H_r(0)$, with H and $H_r(0)$ being of opposite signs, whereas the subsequent change $J_1(t)$ is related to the after-effect field by the formula

$$J_1(t) = (a + c)[H_r(t) - H_r(0)], \quad (5)$$

where $a+c$ is the total differential susceptibility. Therefore we call this *reversible after-effect*.

We can also think of the after-effect field as a fluctuating field, sometimes positive and sometimes negative, whose maximum absolute value $H_i(t)$ increases with time (Fig. 2).

It follows from section 3 that the change in magnetization associated with such a field constantly changing its sign, is written:

$$J_1(t) = c H_i(t), \quad (6)$$

where c is the differential irreversible susceptibility. Therefore we call this *irreversible after-effect*.

We propose to show that using very simple expressions for $H_r(t)$ and $H_i(t)$ the magnetic after-effect of real substances can be described in all regions of the hysteresis loop.

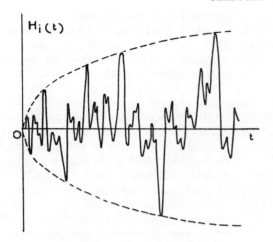

Fig. 2.
Variation of the irreversible after-effect field $H_i(t)$ with time.

I. REVERSIBLE AFTER-EFFECT

(5) The stabilization of the spontaneous magnetization: We will assume that the spontaneous magnetization is stabilized when it has been in a certain direction for a long time. Its direction is defined by the three direction cosines α', β', γ' with respect to three rectangular axes (the three easy directions of magnetization). Now if the direction α, β, γ of J_s is suddenly changed a supplementary term F_s must be added to the usual expression for the energy. This stabilization energy F_s per cm³ is given by:

$$F_s = -W(\alpha^2\alpha'^2 + \beta^2\beta'^2 + \gamma^2\gamma'^2), \quad (7)$$

where W is essentially positive.

Now let the magnetization which was previously stabilized in the α', β', γ' direction suddenly be changed at time $t=0$ to another direction α'', β'', γ''. Let it stay in this position for a time t. The energy that is now required to bring it rapidly into any direction α, β, γ is found from the expression:

$$F_s = -W(\alpha^2\alpha'^2 + \beta^2\beta'^2 + \gamma^2\gamma'^2)G(t) - W(\alpha^2\alpha''^2 + \beta^2\beta''^2 + \gamma^2\gamma''^2)[1-G(t)], \quad (8)$$

where $G(t)$ is a suitable function of time such that $G(0)=1$ and $G(\infty)=0$.

Following, among others, G. Richter [4] and R. Becker [17], we put

$$G(t) = \int_0^\infty g(\tau) e^{-\frac{t}{\tau}} d\tau \quad \text{avec} \quad \int_0^\infty g(\tau) d\tau = 1. \quad (9)$$

A good representation of the phenomena is obtained by putting

$$g(\tau) = \frac{1}{\tau \log \dfrac{\tau_{max}}{\tau_{,in}}}$$

in the interval $\tau_{min} < \tau < \tau_{max}$ and $g(\tau)=0$ outside this interval.

(6) Calculation of the reversible after-effect field: Consider now a *90° wall* which has been in a certain position $x=0$ for a long time. Consider moving it suddenly at time $t=0$ to a position x. As a consequence of this, in certain regions the magnetization will move from its stabilized direction. Therefore work must be done. It is as if a fictitious field $h_r(0)$ directed in such a way so as to restore the wall to its initial position, acts on the system. If, for instance, a wall parallel to the yOz plane is moving in the Ox direction and separates two domains whose spontaneous magnetizations, of intensity J_s, are in the Oy and Oz directions respectively, then using a theory of walls given previously [18] the fictitious field directed along Ox is a function of x such that:

$$h_r(0) = -\frac{Wd_0}{J_s} \frac{\partial}{\partial x}\left(\frac{x}{d_0} \coth \frac{x}{d_0} - 1\right). \quad (10)$$

where d_0 is a length of the order of a third the wall thickness, equal to 165 Å for pure iron, given by

$$d_0 = a\sqrt{\frac{E}{6K}}, \quad (11)$$

where a is the distance between two neighbouring atoms, E is the Weiss–Heisenberg magnetization energy and K the anisotropy constant.

When x is small compared with d_0 relation (10) reduces to

$$h_r(0) = -\frac{2Wx}{3J_s d_0}, \quad (12)$$

where for large values of x, $h_r(0)$ tends towards a constant value,

$$h_r(0) = -\frac{W}{J_s} \quad (13)$$

which in practice is attained for $x=4d_0$. The curve $h_r(0)=f(x)$ is shown as a solid line in Fig. 3.

With regard to *180° walls* we can consider these as

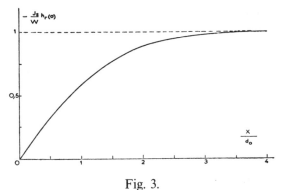

Fig. 3.
The variation with displacement x of the reversible after-effect field opposed to the movement of a 90° wall at time $t=0$.

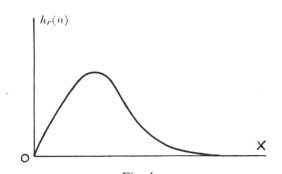

Fig. 4.
Schematic variation with displacement x of the reversible after-effect field opposed to the movement of a 180° wall at time $t=0$.

two juxtaposed 90° walls. The corresponding $h_r(0)$ curve is represented schematically in Fig. 4 which is seen to be very different from that of Fig. 3. In this case, $h_r(0)$ starts by increasing, passes through a maximum, decreases and then tends towards zero which is reached in practice for x equal to 5 or $6d_0$, that is for the thickness of the 180° wall. This fact is easy to understand as the stabilization can only have an effect inside the wall because the two oppositely magnetized domains which separate the wall always have the same energy whatever the stabilization state.

When the wall is kept at the position x for a time interval t, a new stabilization is produced and the fictitious field $h_r(t)$ which acts on the wall tends towards zero following the law:

$$h_r(t) = h_r(0)\, G(t). \qquad (14)$$

It must be noted that $h_r(t)$ only has this value at this position: if the wall is displaced even by the action of this field, $h_r(t)$ will change because of this. Nevertheless, to a first approximation when the displacement is small compared with the wall thickness $h_r(t)$ is still given by equation (14).

The fictitious field that we have just defined is no other than the reversible after-effect field relating to an isolated wall, which is why we have denoted it by $h_r(t)$. To obtain the macroscopic after-effect (defined in section 4), it is necessary to take the average for all the different walls. To simplify the following we will suppose that all the 90° walls in the substance are identical to those described above and we will denote their equivalent total surface area per cm³ by S_0. Therefore we can interchange $h_r(t)$ and $H_r(t)$.

For further simplification we will suppose that the substance only contains 90° walls. Indeed, we consider that this type of wall alone can explain the phenomena related to changes in magnetization greater than several c.g.s. units: the role of 180° walls is usually negligible.

(7) **Disaccommodation or reversible aging:** Let us demagnetize a ferromagnetic substance at time $t=0$ so that the effects of previous stabilizations are removed. We can then perform two main types of experiments. In the first type magnetic measurements are made, for example of permeability, under conditions such that the measurement time is small compared to the relaxation time appearing in the expression for $G(t)$. Therefore the measurement time t is taken as a parameter. In the second type, after demagnetization we wait for a time sufficient for stabilization to be complete. We then apply a constant magnetic field and study the change in the magnetization as a function of time. In this section, we will first consider the theory relating to experiments of the first type.

When a sudden change in magnetization J is brought about at time t, the walls move on average by x given by the expression

$$J = x\, S_0\, J_s, \qquad (15)$$

in which S_0 denotes the surface area of the walls per cm³ of substance. It is useful to distinguish the two cases where x is small or large compared to the wall thickness, that is, compared to $3d_0$.

When x is small compared to d_0, the after-effect field found from equations (12) and (14) is written:

$$H_r(t) = -\frac{2\, W\, [1 - G(t)]}{3\, J_s^2\, d_0\, S_0}\, J. \qquad (16)$$

This field, which is proportioned to the magnetization, can be considered like a demagnetizing field. Consequently the initial susceptibility a' of the substance at time t is related to the value at $t=0$ by the relation

$$\frac{1}{a'} = \frac{1}{a_0} + \frac{2W[1 - G(t)]}{3 J_s^2 d_0 S_0}. \qquad (17)$$

Thus the initial susceptibility a' continuously decreases with time from the value a_0 to an asymptotic value a_1, corresponding to the value of a' given by equation (17) with $G(t)=0$. This is the well-known phenomena of disaccommodation of permeability (cf. H. Atorf [3], C. E. Webb and L. H. Ford [5], E. Becker [17], J. L. Snoek [6]).

When the applied field is strong enough for x to be larger than the wall thickness, the after-effect field reduces to $\dfrac{W[1-G(t)]}{J_s}$ which is independent of the field H. Neglecting bH^2 in Rayleigh's law, the average susceptibility a'' between $H=0$ and H, which is equal to a_0 at $t=0$, is given at time t by

$$a'' = a_0 - \frac{W[1 - G(t)]}{J_s H} a_0. \qquad (18)$$

This shows that in weak fields the average susceptibility between 0 and H deviates hyperbolically from the straight line of Rayleigh's law $J/H = a_0 + bH$. In total, taking account of the term in bH^2, the field variation of the average susceptibility is represented by a curve like that of Fig. 5, where the asymptote, indicated by a dashed line, also represents the Rayleigh straight line which would have been obtained in the absence of stabilization.

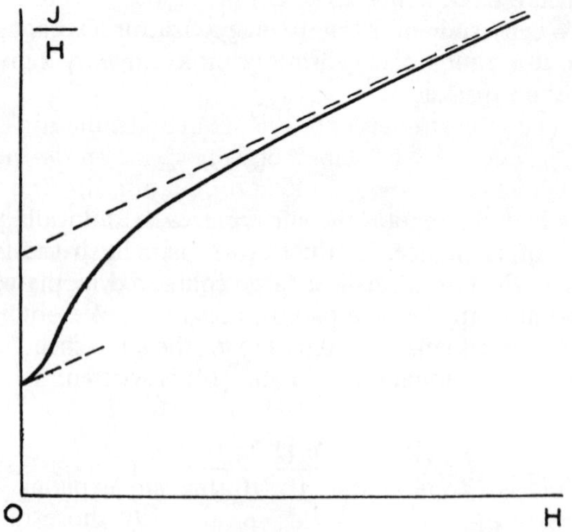

Experiments by Pawlek [32] on silicon–iron with 3% Si saturated with nitrogen in atmospheres containing from 1% to 10% nitrogen by volume, illustrate the phenomena very clearly. Whereas pure silicon–iron obeys the Rayleigh law, the permeability curves of silicon–iron saturated with nitrogen measured at 50 c/s after stabilization show anomalies in weak fields like that indicated in Fig. 5, which become more accentuated for higher nitrogen contents. H. Wilde and G. Bosse [19] found a similar curve for a dynamo lamination containing carbon (see Fig. 2 of their paper, p. 215).

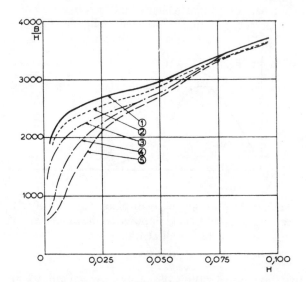

Fig. 6.
Variation of the permeability as a function of field for samples of silicon–iron with various nitrogen contents (after Pawlek [32]).

In the same sort of way, the divergence from the Rayleigh law for some substances of very high permeability as found by Sixtus [21] probably arises from this phenomenon. Using the results of these workers we can calculate the reversible after-effect $-H_r(0) = \dfrac{W}{J_s}$ for various substances from the difference between the curve and its asymptote. Moreover, knowing a_1 and a_0 it is possible to deduce the value of $3d_0 S_0$, which represents the volume occupied by the walls in 1 cm³ of substance. The results shown in Table I indicate the order of magnitude of the after-effect field for some quite different cases.

The results expressed by formula (18) can also be

Fig. 5.
Variation of J/H as a function of H for rapid measurements made after stabilization.

TABLE I.
Reversible after-effect field values.

Substance	$-H_r(0)$	$3S_0d_0$	Reference
	(Oersted)	(cm^3)	
Silicon–iron + X%N$_2$ by volume	0.006	?	Pawlek
Dynamo lamination + X%C	0.0075	0.45×10^{-3}	Wilde and Bosse
Silicon–iron + X%C	0.0003	?	Sixtus
Nickel–iron + ?	0.00036	1.2×10^{-2}	Sixtus
Iron + 0.006%C	0.035	0.8×10^{-3}	Snoek
Ex-carbonyl iron + X%C	0.008	?	Richter

looked at in another way. A long time after demagnetization, in order to obtain the same magnetization as that which would be obtained soon after demagnetization a stronger magnetic field is required. The difference is precisely equal to the reversible after-effect field $-H_r(0) = \dfrac{W}{J_s}$. In Fig. 7, two such curves are plotted for iron with 0.006% carbon from

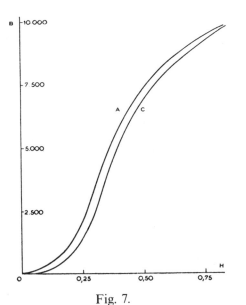

Fig. 7.
Magnetization curves for 0.006%C iron measured (A) soon after demagnetization and (C) after stabilization (after the results of Snoek [20]).

the results of Snoek [20, Figs. 3 and 4, p. 801], and can be approximately superimposed by a translation $-H_r(0) = 0.035$ Oe. This value corresponds to a stabilization energy of about 60 ergs cm^{-3}. Snoek finds $a_0 = 110$ and $a_1 = 16$ and therefore from equation (17) for $t = \infty$, $3S_0d_0 = 0.8 \times 10^{-3}$, which is the correct order of magnitude, close to the figure obtained for the dynamo lamination (cf. Table I).

(8) Change of the magnetization with time: Now let us consider the second type of experiment in which a certain field H is applied for example at the time origin $t = 0$ after the substance has been allowed to stabilize after demagnetization. Here also it is convenient to distinguish two cases where the total wall displacement x produced by the field H is either small or large compared to the wall thickness.

(a) *The case of the small wall displacements* — We suppose x is small compared to d_0. The problem is easy to solve when the after-effect field is still small compared to H, that is to say when a_1 (defined in section 7) is close to a_0. Then we find

$$\frac{J}{H} = a_0 + (a_1 - a_0) \, G(t). \qquad (19)$$

The magnetization J increases with time from the initial value $a_1 H$ to the value $a_0 H$.

When $a_0 - a$ is the order of a_1, or larger, the problem becomes difficult unless there is only a single time constant, in which case G(t) reduces to $e^{-t/\tau}$. Then, as already shown by J. L. Snoek [6],

$$\frac{J}{H} = a_0 + (a_1 - a_0) \, e^{-\dfrac{a_1 t}{a_0 \tau}}. \qquad (20)$$

The new time constant is equal to $\dfrac{a_0 \tau}{a_1}$. It depends on a_0, that is to say on the force which keeps the wall in its equilibrium position. Because of the variety of walls which exist in a substance it may be quite possible that the magnetization after-effect depends on several time constants even though the stabilization may depend only on one.

When the after-effect field is small and equation (19) applicable, the principle of superposition is valid because of the linear character of equation (19) and the differential equations of which G(t) is the solution. We can thus calculate the value of the remanent magnetization, J_r, at time $t + t'$ obtained by applying a field H from the time $t = 0$ to the time t. We obtain

$$\frac{J_r}{H} = (a_0 - a_1) \, [G(t') - G(t + t')]. \qquad (21)$$

It indeed appears that this formalism explains the experiments in weak fields, particularly those of G. Richter [4].

(b) *The case of large wall displacements* — Now let us consider displacements for which x is larger than the wall thickness. In this case, $H_r(0)$ is practically constant and equal to $\dfrac{-W}{J_s}$. The part of the magnetization which depends on time has a value after an infinite time given by equation (5) as follows:

$$J_1(\infty) = -(a+c)H_r(0). \qquad (22)$$

Therefore it is simply proportional to the total differential susceptibility at the point (J, H).

In the paper of G. Richter [4, Fig. 23, p. 626], given as a function of ΔB, that is of $4\pi J_0$ we find the value of β or $4\pi J_1(\infty)$ corresponding to that total change with time of the remanent magnetization produced by the prolonged action of a field H_m; β divided by the corresponding total differential susceptibility, $a_0 + bH_m$, should therefore be a constant. In Fig. 8, $\dfrac{\beta}{a_0 + bH_m}$ is plotted against J_0 and shows that this is indeed the case. As soon as the change in

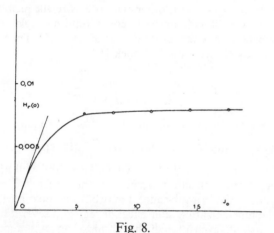

Fig. 8.
Reversible after-effect field as a function of the instantaneous change in magnetization, for an ex-carbonyl iron sample studied by Richter [4].

magnetization exceeds 5 e.m.u. the after-effect field $-H_r(0)$ stays practically constant, its value 0.008 Oe corresponding to a stabilization energy of 13.5 ergs cm^3.

Naturally we ought to compare the experimental curve of Fig. 8 with the theoretical one of Fig. 3. This example, as well as that for iron of Snoek referred to in section 7, shows that it is the 90° walls which are responsible for the reversible after-effect because 180° walls always give a zero value for $H_r(0)$ for large wall displacements (Fig. 4). They also illustrate the advantage of the formalism proposed in this paper,

because the same value of $H_r(0)$ accounts for the after-effect in the major part the hysteresis cycle.

Even the change in magnetization as a function of time can be derived from equations (5) and (14), but these are only valid when the amplitude $J_1(\infty)$ of the after-effect corresponds to a displacement of the wall distinctly smaller than the wall thickness (the amplitude of the after-effect must not be confused with the total amplitude of the wall displacement which also includes the instantaneous displacement). When the amplitude of the after-effect is larger the calculations are more difficult.

(9) Anomalous after-effect: Various experimental details of reversible after-effect can easily be interpreted using the ideas proposed here. Take for example the anomalous after-effect, good examples of which are given by A. Mitkewitch [22]. Suppose a wall is stabilized in a zero field at position $x=0$. At time $t=0$ a positive field H is applied which brings the wall to x_0, then at time t the field is reduced to H', still in the positive sense, and the wall returns to x'_0. If x'_0 and $x_0 - x'_0$ are both larger than the wall thickness, the after-effect field which acts on the wall at x'_0 at a time $t+t'$ is obtained by applying the principle of superposition; and is equal to $\dfrac{W}{J_s}[2G(t+t') - G(t')]$.

This field is positive and tends towards zero when t is very large: the after-effect therefore occurs in the same sense as the change in magnetization which immediately preceded it. In contrast, when t is small, the after-effect field tends towards zero from a negative value: therefore the after-effect occurs in the opposite sense to the change of magnetization which had immediately preceded it. This is anomalous after-effect.

(10) The effect of the temperature: The results established by G. Richter [4] of course remain valid in the presentation here. To change the temperature from T_0 to T_1, all the elementary time constants τ appearing in the expression for $G(t)$ must be multiplied by the same factor,

$$f = e^{\theta\left(\dfrac{1}{T_1} - \dfrac{1}{T_0}\right)}. \qquad (23)$$

It follows that for the two temperatures T_0 and T_1 the magnetization curves as a function of temperature will coincide after a simple translation along the time axis. The activation temperature θ of carbonyl–iron studied by G. Richter is for example about 11,000 K.

Because of this temperature dependence reversible after-effect can only be observed in the temperature interval within which the elementary time constants are in the range of the measurement times for example from several seconds to several hours in the case of a magnetometer. Below this temperature interval stabilization never occurs and the substance keeps its initial state. Above it, stabilization is so rapid that it is achieved before the first possible measurement. Experiment shows that reversible after-effect is only observable in a temperature interval of about ten degrees.

Experiments of H. Fahlenbrach [7] show that certain substances possess several after-effect "bands". For the type of observation between 1 and 30 minutes, silicon–iron shows two "bands", one centred around 50°C and the other around 400°C, whereas a nickel–iron which also possesses two "bands", has then at 150°C and 400°C.

The ratio of the relaxation times $\frac{\tau_{max}}{\tau_{min}}$ used to account for the experimental results appear to be very variable. For ex-carbonyl iron, G. Richter [4] finds $\tau_{max}/\tau_{min} = 30$ whereas H. Wilde [8] finds ratios of the order of 500 for the dynamo lamina and Heraeus "trafoperm 35 M²", and 3000 to 5000 for Krupp "hyperm I".

(11) The loss angle in reversible after-effect: With alternating current, reversible after-effect is manifested by the existence of a loss angle δ which measures the phase lag between the induction and the magnetic field. In principle such experiments show nothing more than the quasi-static experiments described above, but in practice they allow us to examine more easily the case of small time constants. In this way G. Richter [4], J. L. Snoek [6, 23], R. Becker [17] and H. Wilde [8] have calculated the loss angle as a function of the quasi-static after-effect characteristics in the case of very weak fields.

Note that at a given temperature the loss angle is significant only in a well defined frequency range which is broader for greater τ_{max}/τ_{min} ratios. When this ratio is very large, the loss angle can stay constant within a wide range of frequencies (cf. for example the calculations of H. Wilde [8]).

H. Jordon showed a long time ago that after deducting the losses from induced currents and from hysteresis, certain substances showed residual losses corresponding to a loss angle independent of the field and the frequency. In fact more recent research [33, 34] has shown it is more complicated and that the loss angle often depends on the frequency and temperature and that in a number of cases there is also superposition of two phenomena [8]. Therefore it is probably a case of reversible and irreversible after-effect losses (cf. § 17) being superimposed. Present experimental results do not allow their separation and are still too fragmentary to be usefully discussed, so we will not discuss this subject any further.

(12) Physical origin of the stabilization: J. L. Snoek [24] discovered that diffusion of carbon or nitrogen atoms in the crystalline lattice gave rise to the reversible after-effect in iron. He showed that the after-effect disappears on complete purification of the iron and reappeared when carbon or nitrogen was reintroduced. More recent work of Falhenbrach [7] and of Wilde [8] have confirmed the conclusions of Snoek, which in addition are supported by the close relationship, shown by G. Richter [25], between magnetic after-effect and mechanical after-effect which is also due to the diffusion of carbon and whose theory is quantitatively verified. For details we refer the reader to a recent account by Snoek [23].

Although the presence of carbon or of nitrogen seems to be definitely established as the cause of after-effect, the mechanism of its action is less well known. Snoek starts with the fact that the diffusion of interstitial carbon atoms occurs in such a way as to relieve the stress which the lattice is under and so attributes after-effect to relaxation of the magnetostrictive stresses prevailing in 180° walls. We know in fact, that for these walls stresses due to magnetostriction exist corresponding to a supplementary energy density P of the order of

$$P = \frac{9}{4} G \lambda_{100}^2, \qquad (24)$$

where G is a shear modulus $\left(G = \frac{c_{11} - c_{12}}{2}\right)$ and λ_{100} the saturation longitudinal magnetostriction. When the deformation is slow compared to the time taken by the carbon atoms to change their place then after the carbon atoms have been redistributed among all the available interstitial sites, it appears as if the shear modules assumes a weaker value, say $G - g$, such that

$$g = \frac{2 V C G^2 k^2}{3 R T}, \qquad (25)$$

where V is the atomic volume, C the number of carbon atoms per iron atom, R the gas constant and k a constant equal to 0.86. The supplementary energy density P then decreases by an amount W, which is actually the stabilization energy, given by

$$W = \frac{3 V C G^2 k^2 \lambda_{100}^2}{2 R T}. \qquad (26)$$

In the example given by Snoek, with $C = 2.8 \times 10^{-4}$, it is therefore 8 erg cm^{-3}.

Unfortunately, as evident from the analysis of the phenomena given above, stabilization of 180° walls only gives an after-effect when the wall displacements are less than the wall thickness, that is to say when the magnetization changes do not exceed several e.m.u., whereas the experiments by Snoek, as well as those by Richter, show that a considerable after-effect exists for large changes of magnetization. Therefore the stabilization must be produced in the interior of the elementary domains and be shown during 90° wall displacements. Therefore Snoek's explanation is not sufficient.

(13) Generalization of Snoek's interpretation: We can generalize Snoek's explanation by noting that magnetostriction gives rise to elastic stresses inside elementary domains and not only inside the walls. For example, consider two categories of domains composed of flat planes stacked one upon another alternatively magnetized along the two directions of easy magnetization, assumed to be situated in the plane of the domains. The magnetostriction energy is therefore of the order of P/2, where P is given by equation (24), giving a stabilization energy equal to a half of that given by equation (26), i.e. 4 ergs cm^{-3} in the iron studied by Snoek. However, this energy appears to be too weak to explain the experimental results because in this example the difference between the magnetic energies before and after stabilization, as deduced graphically from Fig. 7, is of the order of 29 ergs cm^{-3}. If we accept that 90° wall displacements contribute a half of the total magnetization change, which is a rather optimistic estimate, then the theoretically calculated value scarcely attains 2 ergs cm^{-3}. Therefore, the mechanism invoked by Snoek, even after some improvement, remains far from adequate.

(14) Diffusion stabilization by means of magnetocrystalline coupling: We therefore should investigate whether the carbon is capable of causing stabilization by a mechanism other than magnetostriction. Indeed it seems that *magnetocrystalline coupling* is capable of playing a much more important role than magnetostriction.

Consider the α-iron lattice. The A–B pairs of two neighbouring atoms, 2.86 Å apart, are formed by an atom A and its six neighbours placed on the quaternary axes passing through A: the line of centres A–B can therefore be parallel to one of the three quaternary axes Oxy. We know that the carbon atoms enter the lattice in the centre of the AB pairs. Three different positions exist according to the orientation of the AB pair of the two iron atoms adjacent to the carbon atom being considered. Suppose that all the carbon atoms are similarly situated, the A–B pairs of the neighbours being, for example, parallel to Ox, then, even supposing that the insertion of the carbon atoms does not change the lattice parameters we obtain a substance, which is a martensite, crystallographically and magnetically anisotropic. The Ox direction is no longer magnetically equivalent to the Oy or Oz directions and so to change the spontaneous magnetization from the Ox direction to the Oy direction requires a certain energy, which to a first approximation is proportional to the ratio C of the number of carbon atoms to the number of iron atoms, or, CK per gram atom of iron where K is thus the magnetocrystalline anisotropy constant of an ideal martensite saturated with carbon.

It follows that the three positions for insertion of the carbon are not equivalent when a spontaneous magnetization exists. The carbon atoms, initially supposed to be distributed at random, tend to occupy sites such that the energy of the system will be least. Once placed in this way, they stabilize the spontaneous magnetization in the direction that it occupies, creating a magnetic anisotropy. An elementary calculation shows that the stabilization energy W arising from this process is

$$W = \frac{K^2 C}{3 RTV}. \quad (27)$$

With a concentration of carbon atoms C equal to 2.8×10^{-4}, a value for W equal to 60 ergs cm^{-3} requires a value of K equal to 3.4×10^8 ergs for a gram molecule of the ideal martensite FeC. This is a very reasonable result, comparable to the molecular anisotropy constant of MnBi, which according to Guillard [27] is 3.8×10^8 ergs.

This mechanism is therefore capable of giving a stabilization energy of the correct magnitude. It must apply equally well to the action of nitrogen as to that of carbon. Unfortunately however it appears to be difficult to formally prove its existence. Furthermore we must note that stabilization by magnetocrystalline coupling coexists with stabilization by magnetostriction, but that, in relation to magnetic after-effect, the latter is negligible compared to the former. In contrast, for mechanical after-effect, stabilization by magnetostriction is the sole cause.

An interpretation of the same sort can be applied to the mixed manganese–zinc ferrite containing excess Fe$_2$O$_3$, which according to Snoek shows a large reversible after-effect. Here we are concerned with diffusion of vacancies, or *holes*, in the spinel lattice and the mechanism should be analogous to that just described. It could then be explained why according to Snoek [23], this magnetic after-effect is not associated with a mechanical after-effect, because there is no necessary connection between stabilization due to magnetocrystalline coupling and stabilization by magnetostriction: the latter can be zero without the former being so.

II.
IRREVERSIBLE AFTER-EFFECT

(15) Expression for the after-effect field: We attribute irreversible after-effect to the action of thermal fluctuations which help the walls separating elementary domains to cross over barriers opposing their movement. The collection of experimental results are interpreted satisfactorily by letting the irreversible after-effect field defined above have the value

$$H_i(t) = S(Q + \log t), \qquad (28)$$

where S and Q vary only slowly as a function of $\log t$ and are sensibly the same in all parts of the hysteresis loop. Naturally we suppose that t is never small enough for $(Q + \log t)$ to be zero. We will justify expression (28) later.

We have shown (cf. § 4) that the after-effect $J_1(t)$ associated with $H_i(t)$ is written

$$J_1(t) = c H_i(t), \qquad (29)$$

where c denotes the irreversible differential susceptibility.

We know that c is very large in the vicinity of the coercive force, on the steepest part of the limiting hysteresis loop. Here, P. Courvoisier [10], for the first time it seems, observed irreversible after-effect in a hard carbon steel and showed that $J_1(t)$ was proportional to c. Also in this region R. Street and J. C. Woolley [11] again observed irreversible after-effect in an alnico with a coercive force of 350 Oe and verified that $J_1(t)$ and c varied in parallel along the whole length of the descending branch of the limiting hysteresis loop. Values of S, calculated from the results of various authors are given in Table II.

(16) Irreversible after-effect in the Rayleigh region: Let us suppose that we apply a field H at time $t = 0$ to a substance previously demagnetized by a decreasing alternating field. The magnetization J at time t can be written from equations (1), (2), (3), (6) and considerations of section 4 as

$$J = J_0 + J_1(t) = a_0 H + b H^2 + 2 b H H_i(t) \qquad (30)$$

which is a linear function of $\log t$.

We are also able to calculate the remanent magnetization J_r obtained at time $t + t'$ after applying the field H up to time t. This is equal to the magnetization obtained by applying the field $H + H_i(t)$, removing this field and taking the average magnetization due to a fluctuating field $\pm H_i(t')$. As the reversible term in a_0 disappears, it becomes

$$J_r = b [H + H_i(t)]^2 - \frac{b}{2} [H + H_i(t) + H_i(t')]^2 \\ + \frac{b}{4} [2 H_i(t')]^2, \qquad (31)$$

or, after simplification,

$$J_r = \frac{1}{2} b [H + H_i(t) - H_i(t')]^2. \qquad (32)$$

Irreversible after-effect in the Rayleigh region was discovered in 1947 at the Laboratory in Grenoble, independently of the work cited above (§ 15), and has been systematically studied since by L. Lliboutry [14] and especially J. C. Barbier [13] who have, in particular, verified the validity of formulae [30] and [32].

When $H_i(t)$ and $H_i(t')$ are small compared to H, we can write

$$\left. \begin{array}{l} J = \text{const.} + 2 b \, \text{SH} \log t. \\ J_r = \text{const.} + b \, \text{SH} \log t - b \, \text{SH} \log t'. \end{array} \right\} \qquad (33)$$

We see that the after-effect, measured by the coefficient of the $\log t$ term is twice as large in the field H than at the corresponding remanent state. The values for S so obtained for various samples are indicated in Table II together with the coercive force

TABLE II

Substance	S (Oe)	H_c (Oe)	Reference	Magnetization-region
Ni–Zn ferrite	0.002	0.29	J. C. Barbier	R
Ni–Zn ferrite	0.0009	0.35	”	R
Pure iron	0.0005	0.5(?)	”	
Soft steel	0.0009	1.7	L. Lliboutry	R
Medium-hard steel	0.0014	5.3	”	R
Hard steel	0.116	15	P. Courvoiser	H_c
Ni (powder)	0.52	125	C. Barbier	R
Ferro-cobalt	0.15	155	”	R
Annealed alnico	0.31	200	”	R
Treated alnico	1.7	350	R. Street and	H_c
	1.5	350	J. C. Woolley	R

H_c. The letters H_c or R indicate whether the measurements have been made close to the coercive force or within the Rayleigh region. We note that for alnico, the same values of S are obtained in the two regions. The laws of irreversible after-effect may therefore be explained in a particularly simple way by the introduction of an irreversible after-effect field in the form of equation (28).

We should point out that in order to detect the after-effect L. Lliboutry used either a small additional magnetic field h [14] or a small shock which plays the same rôle as a field [28]. When h is of the same order of magnitude as S, the effects become complicated. For example, it is possible to show [12] that the successive action of a field H for a time t, and of field $H+h$ for a time t', where H and $H+h$ are close, is equivalent to the action of a fictitious field H_f given by the relation

$$H_f = SQ + S \log \left[t e^{\frac{H}{S}} + t' e^{\frac{H+h}{S}} \right] \quad (34)$$

(17) The loss angle in irreversible after-effect for the Rayleigh region: In an alternating field, additional energy losses are associated with irreversible after-effect. It can be expressed by a loss angle δ, which is independent of the magnetic field intensity and also, to a first approximation, independent of the frequency. It can be easily shown [12] that

$$\text{tg}\, \delta = \frac{\pi b S}{a_0}. \quad (35)$$

As $\frac{b H_c}{a_0}$ is of the order of unity [31], the loss angle is of the order of $\frac{S}{H_c}$. These losses by after-effect superpose purely and simply on the hysteresis losses which themselves remain equal to $\frac{4 b H^3}{3}$ ergs per cycle and per cubic centimetre, whatever the frequency. In fact it is easy to verify that the terms in bH^2 in the expression of the magnetization laws do not depend on time and keep the same value as in the classical theory; only terms of the first degree in H depend on time and so give rise to a loss angle independent of field. Furthermore if S is independent of $\log t$, this loss angle is independent of the frequency.

(18) Invalidity of the superposition principle: Formula (33) which gives the time variation of the remanent magnetization and which is readily verified by experiment, illustrates that the principle of superposition does not apply here, since the term in $\log t$ is independent of the term in $\log t'$. Besides, it would not be applicable since the after-effect is only related to the irreversible terms in the magnetization law. F. Preisach [9] found the physical significance of the failure of the superposition principle in the fact that the elementary domains which play a part in the after-effect at remanence are not the same domains as those which come into play in the after-effect under field H.

(19) The mechanism of thermal fluctuations [12]: Several mechanisms certainly exist in which thermal fluctuations help walls cross barriers opposing their displacement but they all pose difficult problems. The most effective mechanism appears to be the effect of internal dispersion magnetic fields produced by thermal fluctuations of the spontaneous magnetization about its mean direction [12].

We let the fluctuation field which helps a wall cross a barrier corresponding to a Barkhausen discontinuity of volume v be equal to the average value H_m of the component of the dispersion field in a given direction in the same volume v taken inside an elementary domain. This average value H_m is itself a complex function of time t whose squared value, averaged with respect to t, is obtained by saying that the average energy is the order of $kT/2$. More exactly, we find

$$\frac{1}{8\pi} \overline{v H_m^2} = \frac{kT}{6}. \quad (36)$$

Also, we suppose that the possible values of H_m are distributed following a Gaussian law and we denote by θ the smallest time interval which must separate two observations so that the corresponding values of H_m can be considered as practically independent. Given this, we let the fluctuation field h relating to a time interval t possess a value such that there is a 50% probability of H_m exceeding this value at least once during this time interval. Then we obtain, approximately,

$$\frac{\theta}{t} = \sqrt{\frac{\overline{H_m^2}}{2\pi h^2}} \exp\left\{ -\frac{h^2}{2 \overline{H_m^2}} \right\}. \quad (37)$$

This shows that h is a function of $\log t$ which close to $t = \tau$ has the form

$$h = S(Q + \log t) \quad (38)$$

with

$$S^2 = \frac{4\pi k T}{6 v (Q' + \log \tau)}, \quad (39)$$

and

$$Q' = \frac{1}{2}(Q - \log \tau) = -\log \theta + \frac{1}{2} \log \frac{2kT}{3vh^2}. \quad (40)$$

It is difficult to calculate the reorganization time θ, but it can be estimated as the order of 10^{-10} s corresponding to values of $-\log \theta$ the order of 20 to 25. Then comparison of the preceding three equations shows that $\frac{2kT}{3vh^2}$ is the order of $\frac{1}{4\pi Q'}$ and so Q' is close to 20. If t only varies by several orders of magnitude around a geometric mean τ, then to a first approximation the field is a linear function of $\log t$, in agreement with experiment.

As regards the order of magnitude of S, given that the average volume of Barkhausen discontinuities range from 10^{-9} cm^3 for mild steel to 10^{-15} cm^3 for the best magnet steels (estimated volume of a cube of side equal to the wall thickness), we must expect a range of values, at room temperatures, from 0.002 Oe for soft materials to 2 Oe for hard materials. This agrees well with the experimental order of magnitude.

It is certain that v is also a function of temperature, probably decreasing as T increases. If, nevertheless, we suppose that v is constant, formula (39) shows that S varies as $T^{1/2}$. Experiments by P. Courvoisier [10] show that for steel S varies as T, but J. C. Barbier [30] finds a variation rather as $T^{3/4}$ for alnico. For the same material, R. Street and J. C. Wooley [11] find that the product of S and the irreversible differential susceptibility C varies as T between 86 and 523 K. Therefore this point is not yet clarified.

Formula (39) also shows that S is not independent of time but must, in reality, increase as t decreases. The time interval during which measurements of after-effect have been made is not sufficiently long to say with certainty whether or not this prediction is in agreement with experiment. Note that from formula (39) the loss angle δ, being proportional to S, must become infinite for $\log \tau = -20$, t being the order of a quarter period. Perhaps the enormous increase of losses in ferrites at high frequencies the order of a megacycle, can be attributed to this phenomenon.

Remark of Dr Snoek: We must distinguish between phenomena affecting the value of the initial permeability and those which affect the permeability at high inductions. It seems beyond doubt that the experiments of Webb and Ford can be explained solely by the magnetostriction energy in the wall. The mechanism proposed by M. Néel probably takes place in high inductions, but, without definite proof, new experiments appear to be necessary.

Reply of Prof. Néel: The experiments of Richter, like those of Dr Snoek, favour the same mechanism at both high and low inductions, which energetically cannot be attributed to magnetostriction.

Remark of Dr Roberts: I would like to suggest tentatively a mechanism for weak-field after-effect which does not require the presence of impurity atoms. If the change of magnetization is small enough, then, for a given average size of domain, the average number of spins that must reverse will be less than the average number contained in the domain wall. This would seem to result in the appearance of irregularities in the domain wall and to require an additional field H to supply the extra exchange energy involved.

Question of Prof. Becker: How can one experimentally distinguish between $H_r(t)$ and $H_i(t)$?

Reply of Prof. Néel: The two fields have very different effects on the magnetic properties. For example, the initial permeability is affected by $H_r(t)$ but not by $H_i(t)$. Likewise, the ratio between the after-effect in field H and the after-effect at remanence (acquired in the same field) is very different in the two cases.

Question of Dr Kurti: What is the order of magnitude of the time constants of magnetic after-effect, and is the theory applicable to all time constants?

Reply of Prof. Néel: In reversible, or diffusion after-effect all orders of magnitude are possible depending on the temperature, but it is unlikely that the simple theory proposed here is applicable when the time constant is less than 10^{-6} to 10^{-7} s. Other phenomena must then be taken into account.

Remark of Prof. Forrer: Experiments in progress on the change of Curie point in samples of iron containing occluded gases can be interpreted by a decrease in the interaction strength along one axis only. Therefore in some parts, the iron becomes tetragonal.

Remark of Dr Michel: It would be interesting to confirm the results obtained in a body centred cubic lattice (α-Fe with C or N) with experiments relating to a face centred cubic lattice. In this respect Bernier has shown that nickel allows carbon or nitrogen into octahedral vacancies.

Remark of Dr Epelboin: It seems to me that the idea of a fictitious reversible after-effect field proposed by M. Néel could be demonstrated by comparison of the permeability as a function of the alternating magnetic field obtained in the presence of this after-effect, with that measured in a time interval short enough to eliminate this effect. Recently such curves have been obtained in our laboratory by A. Marais using a sheet of silicon–iron (cf. Fig. 9).

Remark of Prof. Gorter: In thermodynamics, I distinguish between reversible and irreversible phenomena. After-effects are always irreversible and so it seems to me that the name reversible after-effect is unfortunate — we should find another name.

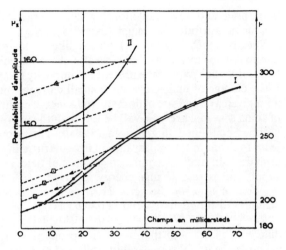

Fig. 9.
Permeability–magnetic field characteristics at 800 c/s of a laminated silicon–iron sheet, heated (sample I) or annealed (sample II). The continuous curves are for measurements made in the presence of after-effect, whereas the dashed curves have been taken in its absence by making the measurements in a time which is negligible compared to the after-effect time constant.

Reply of Prof. Néel: I agree with Prof. Gorter. We could call reversible after-effect by the name diffusion after-effect and irreversible after-effect by fluctuation after-effect. The only disadvantage is that it presupposes the explanation of these phenomena.

REFERENCES

[1] L. Néel, *Ann. Geophys.*, 1949, **5**, 99.
[2] J. A. Ewing, *Proc. Roy. Soc.*, 1889, **46**, 269.
[3] H. Atorf, *Z. Physik*, 1932, **76**, 513.
[4] G. Richter, *Ann. Physik*, 1937, **29**, 605.
[5] C. E. Webb and L. H. Ford, *J. Inst. El. Eng.*, 1934, **75**, 787.
[6] J. L. Snoek, *Physica*, 1938, **5**, 663.
[7] H. Fahlenbrach, *Ann. Physik*, 1948, **2**, 355.
[8] H. Wilde, *Frequenz*, 1949, **3**, 309.
[9] F. Preisach, *Z. Physik*, 1935, **94**, 277.
[10] P. Courvoisier, *Sitz. ber. Bayer. Ak. Wiss.*, 1945–1946, **10**, 89.
[11] R. Street and J. C. Woolley, *Proc. Phys. Soc.*, A, 1949, **62**, 562.
[12] L. Néel, *J. Phys. Rad.*, 1959, **11**, 49.
[13] J. C. Barbier, *C.R. Acad. Sci.*, 1950, **230**, 1040.
[14] L. Lliboutry, *C.R. Acad. Sci.*, 1950, **230**, 1042.
[15] R. Gans, *Ann. Physik*, 1910, **33**, 1065.
[16] Lord Rayleigh, *Phil. Mag.*, 1887, **23**, 225.
[17] R. Becker and W. Doring, Ferromagnetismus (Springer, Berlin, 1939).
[18] L. Néel, *Cahiers de Physique*, 1944, **25**, 1.
[19] H. Wilde and G. Bosse, *Frequenz*, 1948, **2**, 214.
[20] J. L. Snoek, *Physica*, 1939, **6**, 797.
[21] K. Sixtus, *Z.f. Physik*, 1943, **121**, 100.
[22] A. Mitkewitch, *J. Phys. Rad.*, 1936, **7**, 133.
[23] J. L. Snoek, New developments in ferromagnetic materials, Amsterdam, Elsevier, 1947.
[24] J. L. Snoek, *Physica*, 1939, **6**, 161; 1941, **8**, 711 and 734.
[25] G. Richter, *Ann. Physik*, 1938, **32**, 683.
[26] D. Polder, *Philips. Res. Rep.*, 1945–1946, **1**, 5.
[27] C. Guillaud, *Thèse*, Strasbourg, 1943.
[28] L. Lliboutry, *C.R. Acad. Sci.*, 1950, **230**, 1586.
[29] H. Jordan, *Elecktr. Nachr.-Techn.*, 1924, **1**, 7.
[30] J. C. Barbier, *Ann. de Physique*, 1954, **9**, 84.
[31] L. Néel, *Cahiers de Physique*, 1942, **12**, 1; 1943, **13**, 18.
[32] F. Pawlek, *Arch. f. Eisenhuttenwesen*, 1943, **9**, 363.
[33] W. B. Ellwood, and V. E. Legg, *Bell Syst. Techn. J.*, 1937, **16**, 212.
[34] K. H. Stewart, *Proc. Inst. Electr. Eng.*, 1950, **97**, 121.

A84 (1952)
THEORY OF THE DIFFUSION AFTER-EFFECT

Summary: After having recalled the existence and theory of fluctuation after-effect (or irreversible after-effect), the author develops a theory for diffusion after-effect (or reversible after-effect) in which magnetocrystalline coupling acts between the positions of interstitial impurity atoms, which are at the origin of the phenomenon, and the spontaneous magnetization direction. He shows that it is as if a so-called stabilization energy, which can have values up to the order of 100 ergs cm^{-3}, must be added to the ordinary magnetic anisotropy. He deduces a general expression for the pressure experienced at a given time by a 90° or 180° wall arising from the diffusion of impurity atoms, knowing the previous positions of the wall. He gives an approximate method for the calculation of the diffusion effects when these can be compared to a small perturbation of the ordinary magnetization process.

Several simple examples of the application of the theory are then given and allow explanation of the more classical experiments on disaccommodation and after-effect, for sinusoidal alternating fields and discontinuous changes in the applied field, respectively. The author then studies more complicated examples and explains in particular, the anomalous after-effect of Mitkewitch as well as experiments of Webb and Ford on the alternating current permeability of silicon–iron sheets. He shows how one can explain changes in this permeability with the time since demagnetization and since application of the measuring field, as well as its variation with the magnetic field.

The theory appears to give a satisfactory qualitative explanation of the presently known characteristics of the diffusion after-effect.

(1) Introduction: We have already shown in a Report presented at the International Conference on Ferromagnetism and Antiferromagnetism in Grenoble in 1950 [1] that we could conveniently distinguish two quite different types of magnetic after-effect which we called by the names of *irreversible or fluctuation after-effect* and *reversible or diffusion after-effect*.

Fluctuation after-effect affects all ferromagnetic substances and obeys particularly simple and well studied laws [2, 3, 4], which we will not consider here. It seems to result from internal fluctuations of dispersion fields.

Diffusion after-effect, as shown by Snoek in a remarkable series of publications [5], is associated with diffusion of impurity atoms within the ferromagnetic lattice. In the case of iron, for example, we are concerned with carbon or nitrogen atoms. According to Snoek, internal stresses play an essential role in the phenomenon but we have shown that, rather, it is necessary to consider the effects of magnetocrystalline energy. We have already given the basic essentials of this theory in the paper cited above, treated simple examples and indicated experimental verifications of the formulae obtained.

Unfortunately lack of space prevented us from treating the theory adequately, or studying the behaviour of 180° walls and the after-effect in alternating fields. We will therefore now give a longer and more rigorous presentation, taking the opportunity to explain the quite old but very careful experimental work of C. E. Webb and L. H. Ford [6] on the effect of time on the alternating current permeability. However, we will not repeat the experimental verifications already explained in the Report cited, to which we refer the reader.

We will limit ourselves to the study of the diffusion of carbon atoms, but all our results are applicable, *mutatis mutandis*, to the diffusion of hydrogen or nitrogen atoms. Also diffusion of holes corresponding to possible vacant positions in the lattice must be accompanied by analogous behaviour.

(2) Carbon in the α-iron lattice: Consider a body centred cubic crystal of iron (α-iron) its quaternary axes parallel to the rectangular axes $Oxyz$. We know that such a crystal can dissolve a small amount of carbon whose atoms occupy interstitial sites such as c (Fig. 1) in the crystal lattice. Each carbon atom therefore has two Fe neighbours diametrically opposite —the line D which joins the centres of the two neighbours is parallel to Ox, Oy, or Oz. Depending on the orientation of D we can distinguish three categories of interstitial sites which we will call x, y or z. With perfectly cubic symmetry the three sites will be exactly equivalent. In reality, the existence of a spontaneous magnetization introduces an asymmetry and differentiates between the x, y and z sites.

We can take this asymmetry into account by

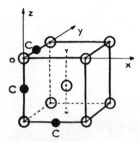

Fig. 1.
Position of the carbon atoms in the α-iron crystal lattice.

attributing to *each interstitial site occupied by a carbon atom*, an energy equal to $w \cos \psi^2$, where ψ is the angle of the corresponding line D with the spontaneous magnetization direction. It follows that the occupied x, y or z sites are characterized by energies $w\alpha^2$, $w\beta^2$ or $w\gamma^2$ respectively, where α, β and γ are the direction cosines defining the spontaneous magnetization relative to $Oxyz$.

In order to theoretically determine this energy it is necessary to know the effect of the presence of carbon atoms on the magnetocrystalline interactions. With the present state of the theory this is not possible and at most we can only estimate by comparison the order of magnitude of w. Imagine an ideal substance of formula FeC, that is a sort of martensite in which the iron atoms are placed exactly as in α-iron but where all the interstitial sites of the same sort, say x, and only these, are occupied by carbon atoms. This substance will certainly possess a very strong anisotropy, of a tetragonal character, with an anisotropy constant which ought to be comparable to those of known strongly anisotropic bodies, for example MnBi, studied by C. Guillard [7] with a value of 3.8×10^8 ergs/mole. Dividing this value by Avagadro's number we can see that it is not unreasonable to expect w to be the order of 0.6×10^{-15} erg.

We must note that w is of a purely magnetocrystalline origin and does not show any necessary relationship with possible changes of the lattice parameters in the Ox, Oy or Oz directions. In contrast, Snoek [5] studied the redistribution of carbon atoms induced by an elastic deformation of the lattice; for example he found that a dilation of the lattice in the Ox direction favoured occupation of the x sites. This is therefore a very different problem, and we will return to it later.

(3) Equilibrium distribution of the carbon atoms: Let an α-iron crystal contain c carbon atoms per cubic centimetre. We assume that c is small compared to the number of iron atoms, which is, nevertheless in agreement with the low solubility of carbon in α-iron. Let α, β, γ define the direction of spontaneous magnetization.

When equilibrium is reached at a temperature T, the number of carbon atoms situated on the x, y or z sites is given by l_0, m_0, or n_0 respectively:

$$\left. \begin{aligned} l_0 &= C_0 \exp\left\{-\frac{w\alpha^2}{kT}\right\}, \\ m_0 &= C_0 \exp\left\{-\frac{w\beta^2}{kT}\right\}, \\ n_0 &= C_0 \exp\left\{-\frac{w\gamma^2}{kT}\right\}, \end{aligned} \right\} \quad (1)$$

where the constant C_0 is determined by the condition

$$l_0 + m_0 + n_0 = c. \quad (2)$$

At ordinary temperatures kT is the order of 10^{-14} erg. Furthermore, as the magnitude of w indicated above, 0.6×10^{-15} erg, can be considered as a maximum, then it follows that as a reasonable approximation, w/kT can be considered small and the exponentials can be expanded neglecting terms in $\frac{w^2}{k^2 T^2}$:

$$\left. \begin{aligned} l_0 &= \frac{c}{3}\left[1 + \left(\frac{1}{3} - \alpha^2\right)\frac{w}{kT}\right], \\ m_0 &= \frac{c}{3}\left[1 + \left(\frac{1}{3} - \beta^2\right)\frac{w}{kT}\right], \\ n_0 &= \frac{c}{3}\left[1 + \left(\frac{1}{3} - \gamma^2\right)\frac{w}{kT}\right], \end{aligned} \right\} \quad (3)$$

This equilibrium distribution requires a certain time to be established. However, once established there is not time for it to be modified during very rapid changes in the orientation of the spontaneous magnetization, so that the part of the energy of the system due to the presence of carbon atoms is written, per cubic centimetre,

$$E_d = w(l_0 \alpha'^2 + m_0 \beta'^2 + n_0 \gamma'^2). \quad (4)$$

where α', β', and γ' are the direction cosines of the spontaneous magnetization during such a process. With values of l_0, m_0, and n_0 given by equations (3), and neglecting two terms which do not depend on the orientation of the spontaneous magnetization, we find

$$E_d = -W_0(\alpha^2 \alpha'^2 + \beta^2 \beta'^2 + \gamma^2 \gamma'^2), \quad (5)$$

where

$$W_0 = \frac{cw^2}{3kT}. \quad (6)$$

Also, the fact of having let the spontaneous magnetization remain for a long time in a given direction α, β, γ introduces an anisotropy term into the expression for the energy. We will call this the *stabilization energy*. It is important to determine the possible magnitude of this energy. In samples of iron rich in carbon, like those studied by Snoek [5] the atomic ratio carbon/iron is 2.8×10^{-4} which corresponds to $c = 2.37 \times 10^{19}$. For $T = 300$ K and $w = 0.6 \times 10^{-15}$ erg we therefore find $W_0 = 70$ ergs cm^{-3}.

In Snoek's theory the elastic deformations of the lattice which give rise to the redistribution of carbon arise from the magnetostriction. We have shown previously [1] that, for the same carbon concentration, the energy thus brought into play only attains a value of 8 ergs cm^{-3}, and that, only inside the Bloch walls, i.e. in about one thousandth of the total volume for the substances in question. On the contrary, the stabilization energy W_0 considered here takes the value given by relation (6) in the entire volume of the substance. Therefore by comparison, it seems legitimate to neglect Snoek's energy.

(4) Rate of acquiring stabilization: This depends particularly upon the height of the potential barriers which separate two stable positions next to a carbon atom. For example let l_0, m_0 and n_0 be the number of carbon atoms situated on the x, y, and z sites in the equilibrium distribution corresponding to a given fixed direction α, β, γ of the spontaneous magnetization. Then let $l_0 + l$, $m_0 + m$ and $n_0 + n$ be their number at time t. The number $-dl$ of carbon atoms which leave x sites during the time interval dt is proportional to dt, to $l_0 + l$ and to some coefficient v_α which depends principally on the height of the potential barrier and to a lesser degree on α^2:

$$dl = -(l_0 + l) v_\alpha dt. \quad (7)$$

Assuming the shape of the "ascent" of the potential barrier does not depend upon the shape of the "descent" on the opposite side, then the atoms which leave x sites must distribute themselves equally between y and z sites (directly passing from x to x is not possible). The balance between atoms gained and lost by x sites is therefore written:

$$dl = \left[-(l + l_0) v_\alpha + \frac{1}{2}(m + m_0) v_\beta + \frac{1}{2}(n + n_0) v_\gamma \right] dt. \quad (8)$$

with two analogous equations for dm and dn. In the equilibrium state l, m and n are zero by definition; likewise for dl, dm and dn. Therefore v_α, v_β and v_γ are necessarily connected by the relations

$$v_\alpha = \frac{v}{l_0}, \quad v_\beta = \frac{v}{m_0}, \quad v_\gamma = \frac{v}{n_0}, \quad (9)$$

where v is some parameter. We will give the physical significance of these relations (9) later. Taking account of relations (9), equation (8) can be written:

$$dl = \frac{v}{2}\left(\frac{m}{m_0} + \frac{n}{n_0} - 2\frac{l}{l_0}\right) dt: \quad (10)$$

with two analogous expressions for dm and dn.

We have however assumed that $\frac{w}{kT}$ is small compared to unity, and so the departures from equilibrium remain small and l_0, m_0 and n_0 remain close to $\frac{c}{3}$. As a first approximation we can then consider these three quantities equal in equation (10), which, combined with the fact that $l + m + n = 0$, transforms this equation to

$$dl = -\frac{3}{2} vl \, dt, \quad (11)$$

with two analogous equations obtained by replacing l by m or n.

Now, the stabilization energy defined by equation (4) is a linear and homogeneous function of l_0, m_0 and n_0, and more generally of $l_0 + l$, $m_0 + m$ and $n_0 + n$: therefore it must obey an equation of the type similar to (11). If E is the stabilization energy at a time t we therefore have

$$dE = -\frac{3}{2}(E - E_d) v \, dt. \quad (12)$$

The stabilization energy therefore exponentially approaches its equilibrium value E_d with a time constant given by

$$\theta = \frac{2}{3\nu}. \qquad (13)$$

Now suppose that the spontaneous magnetization direction α, β, γ instead of being fixed as was previously the case, varies as some function of time. Let us now calculate the value of the stabilization energy $E(t)$ at time t. This will naturally be a linear and homogeneous function of $\alpha'^2, \beta'^2, \gamma'^2$ where α', β', γ' correspond to different instantaneous potential positions of the spontaneous magnetization at time t. At any time τ from between the time origin 0 and the final time t, the equilibrium value $E_d(\tau)$ of the stabilization energy is a function of τ through α^2, β^2 and γ^2 [cf. (5)]. The differential equation (12) then takes the form

$$\frac{dE}{d\tau} + \frac{E}{\theta} = \frac{E_d(\tau)}{\theta} \qquad (14)$$

This equation integrates easily. We will assume that at time $\tau = 0$ the anisotropy energy is initially zero, or to be more precise, that the carbon atoms are equally distributed between x, y and z sites. We will show later how this initial state can be reached in practice (cf. § 9). With these conditions, supposing that the second term of equation (14) is zero, then by changing the integration constant, we obtain

$$E(t) = \int_0^t E_d(\tau) e^{-\frac{t-\tau}{\theta}} \frac{d\tau}{\theta}. \qquad (15)$$

(5) Range of time constants: The time constant θ is certainly very sensitive to possible variations in the height of the potential barrier separating adjacent sites, i.e. to the activation energy w_a necessary to reach the top of the barrier. The relationship between θ and w_a is, in effect, of the type

$$\theta = C \exp\left\{\frac{w_a}{kT}\right\}, \qquad (16)$$

where the factor C only varies slowly with T and w_a.

Note here that by substituting in equation (9) the value of l_0 given by equation (1) we obtain

$$\nu_\alpha = \frac{2}{3\,CC_0} \exp\left\{-\frac{w_a - w\alpha^2}{kT}\right\}. \qquad (16b)$$

This equation shows us that it appears as if the activation energy is decreased by the interaction energy between carbon atoms and the spontaneous magnetization, as logically expected.

Now returning to equation (16), it can be seen that different lattice perturbations, such as inclusions, dislocations, impurity atoms etc. which locally modify values of w_a, will give rise to considerable variations in θ without necessarily modifying W_0 by very much. According to Snoek [5] $\frac{w_a}{k}$ is around 8900 K and so at ordinary temperatures it follows that θ changes by a ratio of 400 as w_a changes by only 20%. The change in W_0 must be of a similar order to that in w_a.

To account for this range of time constants in an approximate and sufficiently simple way, let us suppose that we can subdivide the crystal into very small regions of several angstroms in diameter, each corresponding to a certain time constant. Then the probability, $p(\theta)d\theta$, of finding a time constant between θ and $\theta + d\theta$ must obey the relation

$$\int_0^\infty p(\theta)\,d\theta = 1. \qquad (17)$$

With these conditions, the stabilization energy given by equation (15) is the sum of the energies of the different regions, such that in general

$$E(t) = \int_0^t E_d(\tau)\, g(t-\tau)\, d\tau, \qquad (18)$$

where the function $g(t-\tau)$ is defined by

$$g(t-\tau) = \int_0^\infty p(\theta)\, e^{-\frac{t-\tau}{\theta}}\, d\theta. \qquad (19)$$

Further, the most natural and simplest theory is to suppose that the activation energies w_a are uniformly distributed between a lower limit w' and an upper limit w''. The probability of finding w_a between w_a and $w_a + dw_a$ is therefore equal to $\frac{dw_a}{(w''-w')}$, which from equation (16) gives

$$p(\theta) = \frac{kT}{\theta(w''-w')} = \frac{1}{\theta(\log\theta_2 - \log\theta_1)}, \qquad (20)$$

where θ_2 and θ_1 are the time constants corresponding respectively to w'' and w'. This distribution law, which in the following we call the *logarithmic distribution* was proposed and used with success by G. Richter [8] and R. Becker [9] to explain the experimental results on ex-carbonyl iron.

(6) Energy associated with wall displacements: Now suppose that the changes in the spontaneous magnetization direction envisaged above arise from displacements of a flat Bloch wall parallel to the yOz plane separating two uniformly magnetized elementary domains. Let the position of the wall be given by the abscissa u in an appropriate plane fixed with respect to the wall. In general u is a function of time, $u(\tau)$.

We will assume that while the wall moves it keeps the same thickness and structure, i.e. that the spontaneous magnetization direction at a point $M(x, y, z)$ only depends on the distance $x - u$ of the point M from the reference plane of the wall. This theory is without doubt not rigorously correct but must be a good approximation, as the structure of a wall results from an equilibrium between Weiss–Heisenberg forces, corresponding to energies of the order of 10^{10} ergs cm^{-3} for iron, and magnetocrystalline forces which correspond to energies of the order of 5×10^5 ergs cm^{-3}, and therefore there seems to be little likelihood that the stabilization energy, which is less than 100 ergs cm^{-3} is capable of changing the structure.

The energy associated with a 1 cm^3 surface element of wall is obtained by integrating $E_d(\tau)$ within a cylinder of base 1 cm^2, whose axis is perpendicular to the plane of the wall, i.e. parallel to the Ox direction. Now in the expression for $E_d(\tau)$ given by equation (5) α, β, γ and α', β', γ' correspond respectively to directions assumed by the spontaneous magnetization when the wall is situated at $u(\tau)$ and when it occupies an arbitrary virtual position u' at time t. The integral of $E_d(\tau)$ from $x = -\infty$ to $x = +\infty$ is therefore a function of the only variable U:

$$U = u(\tau) - u' \qquad (21)$$

Thus we will put

$$F(U) = -\frac{1}{W_0} \int_{-\infty}^{+\infty} E_d(\tau) \, dx \qquad (22)$$

From equation (18) the total energy associated with a wall element of 1 cm^2 is therefore given by

$$E_s(t) = -W_0 \int_0^t F(U) g(t - \tau) \, d\tau \qquad (23)$$

This energy is a function of the virtual displacement u'. We find from the principle of virtual work, that at time t and at position u' the wall is subjected to a pressure P given by

$$P = -\frac{\partial E_s}{\partial u'} = \frac{\partial E_s}{\partial U}, \qquad (24)$$

and if we put

$$f(U) = \frac{\partial F(U)}{\partial U}, \qquad (25)$$

we can write

$$P = -W_0 \int_0^t f(U) g(t - \tau) \, d\tau . \qquad (26)$$

In reality the position of the wall for $\tau = t$ is equal to $u(t)$. The pressure P experienced by the wall from the after-effect created by diffusion of carbon atoms, is obtained by giving U the value

$$U = u(\tau) - u(t). \qquad (27)$$

in equation (26). This pressure P is a function $P(u, t)$ of $u(\tau)$ and time t.

(7) Calculation of F(U) and $f(U)$: Before proceeding any further, we must calculate the values of $f(U)$ for the two categories of walls found in iron, i.e. 90° walls and 180° walls.

(A) 90° walls — Consider a 90° wall parallel to the yOz plane separating two elementary domains, one situated on the negative x side with a spontaneous magnetization parallel to Oy, and the other on the positive x side with a spontaneous magnetization parallel to Oz. Inside the wall the magnetization turns progressively from the Oy direction into the Oz direction whilst remaining in the yOz plane. Let ψ be the angle between the spontaneous magnetization and Oy and u be the position of the wall, or more precisely the position of the plane corresponding to $\psi = \frac{\pi}{4}$. The angle ψ is a function of the only variable $x - u$ given [10] by

$$\mathrm{tg}\, \varphi = \exp\left\{\frac{x - u}{d}\right\}, \qquad (28)$$

with

$$d = d_0 \left(\frac{E}{6K}\right)^{1/2}, \qquad (29)$$

where K is the magnetocrystalline energy, E the molecular field energy and d_0 the distance between magnetically interacting atoms. For iron, d is the order of 165 Å.

For the two positions u and u' of the wall, we therefore have

$$Q = \alpha^2 \alpha'^2 + \beta^2 \beta'^2 + \gamma^2 \gamma'^2 = 2 \cos^2 \varphi \cos^2 \varphi'$$
$$- \cos^2 \varphi - \cos^2 \varphi' + 1. \qquad (30)$$

In the following, we will neglect the term $+1$ because it

amounts to adding a constant term to the energy. Then to abbreviate, let us put

$$a = \exp\left\{-\frac{2u}{d}\right\}, \quad a' = \exp\left\{-\frac{2u'}{d}\right\}, \quad (31)$$

and replacing $tg\psi$ and $tg\psi'$ by the values deduced from relation (28), we find

$$Q - 1 = \frac{a + a'}{a - a'}\left[\frac{1}{1 + a\exp\left(\frac{2x}{d}\right)} - \frac{1}{1 + a'\exp\left(\frac{2x}{d}\right)}\right] \quad (32)$$

This integrates easily from $x = -\infty$ to $x = +\infty$ by changing the variable,

$$v = 1 + a\exp\left(\frac{2x}{d}\right). \quad (33)$$

Finally we obtain

$$F(u - u') = \int_{-\infty}^{+\infty} (Q - 1)\,dx = \frac{1}{2}d\frac{a + a'}{a - a'}\log\frac{a'}{a}, \quad (34)$$

Reintroducing u and u' and replacing $u - u'$ by U it becomes simply

$$F'(U) = -U\coth\frac{U}{d}, \quad (35)$$

where the accent on F(U) simply indicates that it applies to a 90° wall. We find $f'(U)$ immediately from relation (25).

(B) *180° walls* — Suppose in this case that inside the wall the magnetization turns in the yOz plane from the negative to the positive Oz direction. For a wall whose centre is at $x = 0$, the relationship between ψ and x is [10, 11, 12],

$$tg\,\varphi = g\sinh\frac{x}{d}, \quad (36)$$

where g is given by

$$g = \left[\frac{P}{(1 + P)}\right]^{\frac{1}{2}}, \quad (37)$$

with

$$P = \frac{9}{4}(c_{11} - c_{12})\frac{\lambda_{100}^2}{K}; \quad (38)$$

where c_{11} and c_{12} are moduli of elasticity, K is the anisotropy constant and λ_{100} the saturation longitudinal magnetostriction along the quaternary axis. For iron P is close to 2×10^{-3}.

Consider then two walls situated at $\frac{U}{2}$ and $-\frac{U}{2}$. For these we have

$$tg\,\varphi = g\sinh\frac{2x - U}{2d}, \quad tg\,\varphi' = \sinh\frac{2x + U}{2d}. \quad (39)$$

As above, we integrate $Q - 1$ from $x = -\infty$ to $x = +\infty$. Terms in $\cos^2\psi$ and $\cos^2\psi'$ give integrals independent of U which are not of interest to us here. Using the values of ψ and ψ' given by expanding the hyperbolic sines in equation (39) and with

$$\exp\frac{x}{d} = z^{\frac{1}{2}}, \quad \exp\frac{U}{d} = a, \\ \frac{4}{g^2} = A', \quad \frac{4}{g^2} - 1 = A, \quad (40)$$

we can write, denoting differentials by ∂ instead of d,

$$2\cos^2\varphi\cos^2\varphi'\partial x = \\ = \frac{d\,A'^2\,a^2\,z\,\partial z}{(a^2z^2 + A\,az + 1)(z^2 + A\,az + a^2)}, \quad (41)$$

or,

$$2\cos^2\varphi\cos^2\varphi'\,\partial x = \frac{d}{P}\left[\frac{a(a^2 + 1)z + A}{a^2z^2 + az\,A + 1} - \frac{\frac{a^2 + 1}{a}z + A a^2}{z^2 + A az + a^2}\right]\partial z, \quad (42)$$

where

$$P = \frac{1}{A'^2}\left(a - \frac{1}{a}\right)\left[\left(a + \frac{1}{a}\right)^2 - A^2\right]. \quad (43)$$

Next it is necessary to integrate from $z = 0$ to $z = +\infty$ but the two terms of expression (42) are then infinite. Therefore we integrate each term just to N, take the difference and then make N tend towards infinity. These integrations only require changing the variables $az = t$ and $z = at$ to lead to classical integrals.

After some calculation we obtain

$$F''(U) = 2d \cosh^2 V \, \frac{\dfrac{U}{d} \coth \dfrac{U}{d} - V \coth V}{\cosh^2 \dfrac{U}{d} - \cosh^2 V}, \quad (44)$$

with

$$\cosh V = \frac{2}{g^2} \quad (45)$$

and where the double accent on the F(U) simply indicates that it applies to a 180° wall.

With the value of g indicated above, the ratio $\dfrac{\cosh^2 \dfrac{U}{d}}{\cosh^2 V}$ is negligible compared to unity when $\dfrac{U}{d}$ is less than 4. F''(U) then reduces to $-2U \coth \dfrac{U}{d}$, or double the energy of a 90° wall.

Finally, $f''(U)$ is calculated from equation (25).

(8) Variation of $f(U)$ with U: Note first of all that F(U) is an even function of U and that $f(U)$ is an odd function. When U is small compared to d, $f(U)$ can be expanded as a series from the preceding formulae which for a 90° wall gives,

$$f'(U) = -\frac{2U}{3d} + \frac{4U^3}{45 d^3} + \cdots \quad (46)$$

For a 180° wall, $f''(U)$ is nearly exactly double the preceding value.

When U tends towards positive infinity, $-f'(U)$ tends towards unity and $-f''(U)$ tends towards zero. Table I gives values of $f(U)$ in the two cases considered. For 180° walls we have used $g^2 = 2 \times 10^{-3}$.

Fig. 2 gives the variation of $-\tfrac{1}{2} f''(U)$ as a function of $\dfrac{U}{d}$ for a 180° wall. The curve shows a clear plateau around $\dfrac{U}{d} = 4$. For 90° walls, $-f'(U)$ is practically represented by the preceding curve up to $\dfrac{U}{d} = 4$, beyond which $-f'(U)$ is almost equal to unity.

(9) Initial state and movement of a wall: Formula (26) gives the pressure P(u, t) experienced by a wall as a result of the redistribution of carbon atoms at time t, knowing the position $u(\tau)$ of the wall at previous times and assuming that initially at $\tau = 0$ the carbon atoms are equally distributed among all the sites.

To bring about this initial distribution, demagnetization in a decreasing alternating field can be used. At every point during this process the spontaneous magnetization takes up very varied orientations so that the average values of α^2, β^2 and γ^2 are equal during the demagnetization and the carbon atoms must distribute equally on the x, y and z sites.

TABLE I.

$\dfrac{U}{d}$	$-f'(U)$ (90° wall)	$-\tfrac{1}{2}f''(U)$ (180° wall)
0.5	0.323	0.323
1	0.589	0.589
1.5	0.774	0.774
2	0.885	0.885
2.5	0.945	0.945
3	0.975	0.975
3.5	0.989	0.989
4	0.995	0.995
4.5	0.998	0.996
5	0.999	0.98
5.5	1.000	0.95
6		0.90
6.5		0.81
7		0.67
7.5		0.53
8		0.39
8.5		0.245
9		0.125
9.5		0.062
10		0.031

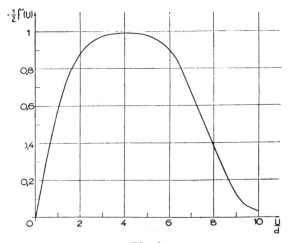

Fig. 2.
Variation of $-\tfrac{1}{2}f(U)$ as a function of U/d for a 180° wall.

However for this mechanism to be effective it seems that the total duration of the demagnetization must be comparable to the largest time constant θ. But this is not generally the case; demagnetization lasts only several minutes whereas θ often reaches several hours or several days. Nevertheless demagnetization by decreasing alternating current still seems to give good results from this point of view. Therefore it must act in a different way, such as the following, for example. The demagnetization gives rise to a complete redistribution of elementary Weiss domains inside the substance, which after demagnetization occupy entirely different positions bearing no relation to their previous positions. In these circumstances, the equal distribution of carbon atoms between the sites is realized *on average* inside each domain and the properties of the system will not be very different than if the equal distribution was in reality realized at each point.

Suppose then that this equal distribution of carbon atoms is realized at time $\tau = 0$. After this, in order to determine the movement of the wall under the influence of an applied magnetic field H, which is possibly a function of time, it is necessary to write down that at every instant in time the algebraic sum of the pressures experienced by the wall is zero. Apart from the pressure of the magnetic after-effect $P(u, t)$, we must also take into account, first, the magnetic pressure equal to the scalar product $\vec{H}(\vec{J}_1 - \vec{J}_2)$ of the applied magnetic field and the difference between the spontaneous magnetizations J_1 and J_2 of the two domains next to the wall, and second, the pressure $R(u)$ which attracts the wall back into its equilibrium position and which only depends on the current position u of the wall. This restoring pressure is related to various perturbations of the crystal lattice e.g. internal stresses, dislocations, etc. Finally, the equilibrium equation is written:

$$R(u) + \vec{H}(t)(\vec{J}_1 - \vec{J}_2) + P(u, t) = 0. \qquad (47)$$

This, in principle, resolves the problem of the displacement with time of *a* specific wall.

To extend the solution to the magnetization changes in a real substance we must account for in particular different possible wall orientations with respect to the exterior field, and also for the fact that different walls have different $R(u)$ functions. We know what an essential rôle this diversity of $R(u)$ functions plays in the interpretation of Rayleigh's laws [13]. But as it is quite evident that the general problem cannot be solved, we will consequently restrict ourselves to several schematic cases. We will assume for example that all the 90° and 180° walls are characterized by the same restoring forces $R'(u)$ and $R''(u)$. Later on we will examine the effect of a range of these functions in a very simple case.

We will also assume that all the walls of various orientations, contained in 1 cm³ of the sample being studied, can be replaced by two fictitious walls, one a 90° wall of surface area S' and the other a 180° wall of area S''. We will define the orientations of these walls as in section 7 and we will apply the external field in the Oy direction. The magnetic pressure is therefore equal to $J_s H(t)$ for the 90° wall and $2J_s H(t)$ for the 180° wall, where J_s is the spontaneous magnetization.

The change in the Oy component of the macroscopic magnetization J can therefore be considered as the sum of two terms J' and J'' related to the displacements u' and u'' of the walls by

$$J' = J_s u' S' \quad \text{and} \quad J'' = 2 J_s u'' S''. \qquad (48)$$

On the other hand, it is sometimes useful to represent the after-effect pressures by equivalent fictitious magnetic fields — the after-effect fields $H'(u', t)$ and $H''(u'', t)$ defined by

$$H'(u', t) = \frac{P'(u', t)}{J_s} \quad \text{and} \quad H''(u'', t) = \frac{P''(u'', t)}{2 J_s}. \quad (49)$$

(10) Variation of the susceptibility with time: After demagnetization at time $\tau = 0$ let the magnetic field be kept zero until time t when the susceptibility is measured by a ballistic method on applying a field H. First suppose that the response time of the measuring instrument is small compared to the after-effect time constant θ. Under these conditions $u(\tau)$ is zero in the time interval $0 < \tau < t$ and $u(t)$ simply represents the position of the wall soon after application of the field. The variable U therefore reduces to $-u(t)$ and $f(U)$ is independent of τ. Now let us put

$$\int_0^t g(t - \tau) d\tau = G(t). \qquad (50)$$

Using relation (19) which defines $g(t - \tau)$ we find that $G(t)$, which is equal to zero for $t = 0$, always increases with t and tends towards unity as t tends to infinity. For example with a single time constant

$$G(t) = 1 - e^{-\frac{t}{\theta}}. \qquad (51)$$

Relations (59) and (62) established later give $G(t)$ in the logarithmic case.

The magnetic pressure which acts on the wall is therefore

$$P(u, t) = W_0 f(u) G(t), \qquad (52)$$

so that the equilibrium equation relating to a 90° wall is written:

$$R'\left(\frac{J'}{S'J_s}\right) + HJ_s = -W_0 f'\left(\frac{J'}{S'J_s}\right) G(t), \quad (53)$$

taking account of relation (48) between u' and J', with an analogous expression for 180° walls.

It follows from this equation that the susceptibility $\frac{J'}{H}$ decreases with increasing time since the demagnetization, and therefore are we explaining the phenomenon known as *disaccommodation* of permeability. Equation (53) has a particularly simple form when H and consequently J' are small. In this case we can reduce R' and f' to the first term of their series expansion, which gives

$$-\frac{J'J_s}{\chi'_0} + J_s H = \frac{2 W_0 J'}{3 S' J_s d} G(t), \quad (54)$$

where χ'_0 is the initial susceptibility relating to 90° walls in the absence of after-effect. It follows from this equation that in weak fields it appears as if the magnetic after-effect is equivalent to an internal demagnetizing field proportional to the magnetization, i.e. $N'J'$, where

$$N' = \frac{2 W_0 G(t)}{3 S' J_s^2 d}, \quad (55)$$

for 90° walls. For 180° walls an analogous result is obtained with a demagnetizing field coefficient N'' given by formula (55) with S' simply replaced by $2S''$.

In higher fields where the displacement u of the walls exceeds $4d$ for 90° walls and $10d$ for 180° walls, $f(u)$ reduces to -1 and 0 respectively. For 90° walls it appears as if the substance were subjected to a supplementary fictitious field H' given by

$$H' = \frac{W_0 G(t)}{J_s}. \quad (56)$$

It is interesting to note that this supplementary field depends neither on the applied field nor on the form of the restoring pressure $R(u)$. For 180° walls the effects of diffusion of carbon on the susceptibility disappear when u exceeds $10d$. In principle therefore, the study of after-effect as a function of field allows separation of the parts played by 90° walls and 180° walls.

We have shown previously ([1], § 7 and 8, Figs. 7 and 8) that the experiments of Richter, as well as those of J. L. Snoek can be very satisfactorily explained by the theory just developed.

(11) General method of approximation: The example just studied is one of the special cases where the integral equation (47) is degenerate and can be solved in a simple and rigorous way. It already becomes more complicated when the response time of the measuring instrument is close to or greater than some of the after-effect time constants θ.

In this case, and several others which will be studied later, the following approximate method gives some simple results. We consider the after-effect as a *small* perturbation and in the expression for U [cf. (27)] we suppose that the difference between the actual position of the wall and the position it would occupy at the same time in the absence of after-effect is sufficiently small to be neglected. Also in this theory the values of $u(\tau)$ and $u(t)$ which appear in U refer to positions the wall would occupy at times τ and t, *in the absence of after-effect*, with fields of $H(\tau)$ and $H(t)$. The problem is therefore very much simplified because there is no longer an unknown function $u(\tau)$ within the summation.

This method gives a first approximation $u_1(\tau)$ to the position of the wall which is sufficient in most cases. If we want a better approximation, these values of $u_1(\tau)$ must be used to find a better value for U and then deduce a second approximation $u_2(\tau)$, and so on.

Let us apply this method to calculate the position of the wall at time $t = t_1 + \varepsilon$ where t_1 denotes the time when the constant field H is applied after being zero in the interval $0 < \tau < t_1$. This is the position calculated for example, from the maximum deflection of a ballistic magnetometer. In the absence of after-effect, the wall stays at rest at the origin until $\tau = t_1$ so that U is simply equal to $-u(t)$. Afterwards, the wall remains still in the interval $t_1 < \tau < t_1 + \varepsilon$ and U is therefore zero to a first approximation. The pressure P is therefore given by

$$P = -W_0 f(-u) \int_0^{t-\varepsilon} g(t-\tau) \, d\tau$$

$$= W_0 f(u) [G(t) - G(\varepsilon)], \quad (57)$$

using the definition of $G(t)$ given by relation (50). Equation (57) can be substituted in equation (52) and then by replacing $G(t)$ by $G(t) - G(\varepsilon)$ the problem can be solved in the case where ε is no longer negligible compared to all the θ values.

Let us now consider formula (57) for the particular case of a logarithmic distribution of time constants defined by formulae (19) and (20). By definition,

$$G(t) = \int_0^t g(t-\tau) \, d\tau$$

$$= \int_0^t \int_{\theta_1}^{\theta_2} \frac{e^{\frac{\tau-t}{\theta}}}{\log \theta_2 - \log \theta_1} \frac{d\tau \, d\theta}{\theta^2}, \quad (58)$$

and integrating first with respect to τ, then with respect to θ we obtain

$$G(t) = \int_{\theta_1}^{\theta_2} \frac{1 - e^{-\frac{t}{\theta}}}{\log \theta_2 - \log \theta_1} \frac{d\theta}{\theta} \tag{59}$$

$$= 1 + \frac{1}{\log \frac{\theta_2}{\theta_1}} \left[Ei\left(-\frac{\theta_2}{t}\right) - Ei\left(-\frac{\theta_1}{t}\right) \right]$$

When $\theta_1 < t < \theta_2$ we can use asymptotic formulae from expansion of the exponential integrals $Ei(-x)$, i.e.,

$$\left. \begin{array}{l} Ei(-x) = C + \log x - x \dfrac{1}{2} \dfrac{x^2}{2!} - \dfrac{1}{3} \dfrac{x^3}{3!} + \ldots, \\ \\ |x| < 17, \end{array} \right\} \tag{60}$$

$$\left. \begin{array}{l} Ei(-x) = -\dfrac{e^{-x}}{x}\left(1 - \dfrac{1!}{x} + \dfrac{2!}{x^2} - \dfrac{3!}{x^3} + \ldots \right), \\ \\ |x| > 17, \end{array} \right\} \tag{61}$$

Where C is Euler's constant, equal to 0.5772. Then we obtain

$$G(t) = 1 - \frac{C + \log \theta_1 - \log t}{\log \theta_2 - \log \theta_1}. \tag{62}$$

If $\theta_1 < \varepsilon < \theta_2$, we obtain $G(\varepsilon)$ by replacing t by ε in the preceding equation, which finally gives

$$G(t) - G(\varepsilon) = \frac{\log t - \log \epsilon}{\log \theta_2 - \log \theta_1}. \tag{63}$$

(12) Evolution of the magnetization with time: We can also imagine another type of experiment. After demagnetization at $\tau = 0$, the magnetic field is held at zero until time t_1 when a constant field H is applied and then the magnetization change with time measured. Such an experiment illustrates the properly called *magnetic after-effect*, whereas the experiment in section 10 refers to disaccommodation of the permeability, following the terminology used by Snoek.

In short, this problem has already been treated in the preceding section. All we need to do is to replace ε by $t - t_1$. Formulae (53) to (56) are applied by replacing $G(t)$ by $G(t) - G(t - t_1)$ which shows the close relationship between disaccommodation and after-effect. In the logarithmic case, the magnetization is a function of time, of the form $A + B \log t$, at least for $\theta_1 < t < \theta_2$. We have already met a relation of this type in the irreversible, or thermal fluctuation after-effect. Therefore when the magnetization shows this type of behaviour we are not able to conclude anything about the origin of the after-effect.

Using the method of the preceding paragraph we can treat more complicated cases. For example, demagnetize the substance at $\tau = 0$, then apply a field H_0 during the time interval $0 < \tau < t_1$, a field H_1 in the interval $t_1 < \tau < t_2$ and finally a field H_2 from time t_2 until the measurement time t. Let u_0, u_1 and u_2 denote successive positions of the wall, corresponding, in the absence of after-effect, to fields H_0, H_1 and H_2. Furthermore, to within the approximation used, u_2 is equal to the actual position of the wall at time t. It follows that the variable U is equal to $u_0 - u_2$ in the interval $0 < \tau < t_1$, to $u_1 - u_2$ in the interval $t_1 < \tau < t_2$, and to zero for $\tau > t_2$. Using relation (50) the magnetic pressure P at time t, defined by equation (26), is written

$$-\frac{P}{W_0} = f(u_0 - u_2) \int_0^{t_1} g(t - \tau) \, d\tau$$

$$+ f(u_1 - u_2) \int_{t_1}^{t_2} g(t - \tau) \, d\tau, \tag{64}$$

or,

$$-\frac{P}{W_0} = f(u_0 - u_2) \left[G(t) - G(t - t_1) \right]$$

$$+ f(u_1 - u_2) \left[G(t - t_1) - G(t - t_2) \right]. \tag{65}$$

This formula is easily generalized to more complicated cases. Moreover, important simplifications can be made both in the case where the differences $u_0 - u_2$ and $u_1 - u_2$ are small compared to d, and in the case where they are large. In the first case, $f'(u)$ or $f''(u)$ reduce to $-\dfrac{2U}{3d}$ or $-\dfrac{4U}{3d}$; in the second, $f''(u)$ reduces to zero and $f'(u)$ becomes -1 or $+1$ depending on whether U is positive or negative.

(13) Normal and anomalous after-effect: When a field H is applied at $\tau = t$, after demagnetization at $\tau = 0$ and the evolution of the magnetization in a constant field followed, the after-effect pressure stays equal to $W_a f(u)[G(t) - G(t - t_1)]$. When there is only one after-effect time constant θ the bracket becomes $e^{-t/\theta}(e^{-t_1/\theta}-1)$. This is always positive and decreases monotomically towards zero as the time t tends to infinity. It is obviously the same in the case where any number of time constants superimpose. As u and $W_o f(u)$ are of opposite sign, the magnetic pressure must always oppose the displacement of the wall and tend towards zero monotomically as t tends to infinity. The after-effect is therefore produced in the same sense as the change in magnetization which immediately precedes it. There is never a change of sign, and this we call *normal after-effect*.

Now take a more complicated case, such as that which resulted in formula (65). First let H_0 be zero, then give H_1 a positive value, and then return the field to a value H_2, less than H_1 but still positive. Let the relative values of these fields be such that u_2 and $u_1 - u_2$, which are both positive, be large compared to d. Finally assume the logarithmic hypothesis and that $t, t-t_1$ and $t-t_2$ are large compared to θ_1 and small compared to θ_2. None of these different assumptions are however essential and they are only introduced to simplify the presentation and the calculations: in other cases we would obtain analogous results to those which follow.

The magnetic pressure P determined from equations (64) and (65) is then

$$P = \frac{W_0}{\log \theta_2 - \log \theta_1} \log \frac{(t-t_1)^2}{t(t-t_2)}, \quad (66)$$

so that it varies with time in the same sense as

$$Y = \frac{(t-t_1)^2}{t(t-t_2)}. \quad (67)$$

From the derivative it can be seen that when t_2 is greater than $2t_1$, Y always decreases as t increases and tends to 1 for infinite t. Also the magnetic after-effect occurs in the same sense as the impressed change of magnetization immediately preceding it, on condition that the previous state, i.e. the application of the field H_1 had been maintained for longer than the field H_0.

When t_2 is less than $2t_1$, Y decreases until t reaches the value $\frac{t_1 t_2}{2t_1 - t_2}$ and then increases tending towards 1 as t tends to infinity. Hence when the previous state has not been held for a long time the sample preserves the memory of an older state and the after-effect finishes by being produced in the sense of the penultimate impressed change in the magnetization. However during a short time interval immediately following application of the final field, the after-effect is in the normal sense. Summarizing, the after-effect changes sign at time $\frac{t_1 t_2}{2t_1 - t_2}$. This is *anomalous after-effect* and the experiments of A. Mitkewitch [14] provide good examples of this. The essential characteristics of the phenomenon appear to be satisfactorily explained by the theory given above.

(14) After-effect in alternating fields. Case of small amplitudes: As there are many studies of disaccommodation of permeability in alternating fields, we will consider this from a theoretical point of view. Suppose that after demagnetization at time $\tau = 0$ the field is kept zero until $\tau = t_1$ when we apply a magnetic field

$$H = H_m \sin \omega t, \quad (68)$$

of constant amplitude, to which corresponds, in the absence of after-effect, a displacement u of the wall given by:

$$u = u_0 \sin \omega t \quad (69)$$

which we will assume is small compared to d. With the approximate method of section 11 the variable U becomes $-u_0 \sin \omega t$ in the interval $0 < \tau < t_1$ and $-u_0(\sin \omega t - \sin \omega \tau)$ in the interval $t_1 < \tau < t$. Limiting the expansion of $f'(U)$ to the first term, the magnetic pressure relating to a 90° wall in the case where there is only a single time constant, is given by

$$\frac{3Pd}{2W_0} = -\int_0^{t_1} u_0 \sin \omega t \, e^{-\frac{t-\tau}{\theta}} \frac{d\tau}{\theta}$$
$$+ \int_{t_1}^{t} u_0 (\sin \omega \tau - \sin \omega t) e^{-\frac{t-\tau}{\theta}} \frac{d\tau}{\theta}, \quad (70)$$

which after integration becomes

$$\frac{3Pd}{2W_0 u_0} = -\left(\frac{\omega^2 \theta^2}{1+\omega^2 \theta^2} - e^{-\frac{t}{\theta}}\right) \sin \omega t \quad (71)$$

$$- \frac{\omega \theta}{1+\omega^2 \theta^2} \cos \omega t$$

$$-(\sin \omega t_1 - \omega \theta \cos \omega t_1) \frac{e^{-\frac{t-t_1}{\theta}}}{1+\omega^2 \theta^2}$$

The last term is of no particular interest; it is a transient non-periodic term which depends on the phase of the current at the instant t_1 and which does not affect alternating current measurements. The sin ωt term is opposite in phase to the field H. When the period of the alternating current $\frac{2\pi}{\omega}$ is small compared to θ this term is practically equal to $-(1-e^{-t/\theta})\sin\omega t$ and shows that when the time elapsed since demagnetization increases from zero to infinity, it appears as if the system is subjected to a fictitious alternating magnetic field, of opposite phase to the applied field, with intensity increasing from zero to $\frac{2W_0 u_0}{3dJ_s}$. This fictitious field does not depend on the time t_1 at which the field $H_m \sin \omega t$ was applied.

The $\cos \omega t$ term is in quadrature with the applied field. It follows that u, i.e. in the end the magnetization, introduces a phase lag ε on the applied field, which gives rise to dissipation of energy. This phase angle is given by

$$\epsilon = \frac{\epsilon_0 \omega \theta}{1 + \omega^2 \theta^2}, \quad (72)$$

with

$$\epsilon_0 = \frac{2W_0 u_0}{3 d J_s H_0} = \frac{2 W_0 \chi_0'}{3 d S' J_s^2} \quad (73)$$

where χ_0 is the initial susceptibility of the system in the absence of after-effect.

In the logarithmic case, if the time constants vary between θ_1 and θ_2, we find

$$\begin{aligned}\epsilon &= \frac{\epsilon_0}{\log \theta_2 - \log \theta_1} \int_{\theta_1}^{\theta_2} \frac{\omega \theta}{1 + \omega^2 \theta^2} \frac{d\theta}{\theta} \\ &= \epsilon_0 \frac{\text{arctg}\,\omega\theta_2 - \text{arctg}\,\omega\theta_1}{\log \theta_2 - \log \theta_1}\end{aligned} \quad (74)$$

and if θ_1 and θ_2 are respectively very small and very large compared to $\frac{2\pi}{\omega}$, the loss angle becomes

$$\epsilon = \frac{\pi \epsilon_0}{2(\log \theta_2 - \log \theta_1)} = \frac{\pi k T \epsilon_0}{2(W'' - W')}, \quad (75)$$

from equation (20). This loss angle depends neither on the field nor on the frequency of the measuring alternating field. Also, it does not depend upon the time elapsed since demagnetization or since applying the measuring field. The energy losses therefore show the essential characteristics of residual losses as first demonstrated by Jordon [15]. These arise solely from the range of time constants and are also found in the after-effect due to thermal fluctuations [2], as we have shown previously. Therefore it is not so much on these characteristics than on the logarithmic variation of the magnetization with time that it is possible to distinguish between diffusion after-effect and fluctuation after-effect.

The phase angle ε depends on the temperature in a way which, though not being exponential, can nevertheless be quite sensitive: the temperature T appears explicitly in formula (75) whereas χ_0' depends on ε_0 and W_0 which, in their turn, particularly for the latter, vary with temperature. Therefore it is possible *a priori* to determine the sense of the resultant variation of ε.

(15) After-effect in alternating fields. Case of large amplitudes: When the amplitude u is large compared to d, using the same approximation as above, equation (26) giving the magnetic pressure for a 90° wall in the case of a single time constant θ, can be written as follows:

$$\begin{aligned}\frac{P}{W_0} = &\int_0^{t_1} \eta \exp\left(\frac{\tau - t}{\theta}\right)\frac{d\tau}{\theta} + \\ &+ \int_{t_1}^{t} \eta' \exp\left(\frac{\tau - t}{\theta}\right)\frac{d\tau}{\theta};\end{aligned} \quad (76)$$

where η is equal to -1 or $+1$ depending on whether $u_0 \sin \omega t$ is positive or negative, and where η' is equal to -1 or $+1$ depending on whether $u_0(\sin \omega t - \sin \omega \tau)$ is positive or negative.

The first of the integrals in (76), which we will denote by I_1, is equal to $\eta\left[\exp\left(-\frac{t-t_1}{\theta}\right) - \exp\left(-\frac{t}{\theta}\right)\right]$. Expanding η as a Fourier series it becomes

$$\begin{aligned}I_1 = -\frac{4}{\pi}\left(e^{-\frac{t-t_1}{\theta}} - e^{-\frac{t}{\theta}}\right) \times \\ \left(\sin \omega t + \frac{1}{3}\sin 3\omega t + \cdots\right).\end{aligned} \quad (77)$$

It is easy to calculate the second integral I_2 of equation (76) by parts. Let us put, in general:

$$\omega t = k\pi + \varphi,$$

where k is an integer and the modulus of ψ is less than $\frac{\pi}{2}$. Then it is convenient to distinguish two cases depending on whether k is even or odd. Let us put in addition,

$$Q = \exp\left(-\frac{\pi}{\omega\theta}\right), \quad Q' = \exp\left(-\frac{2\varphi}{\omega\theta}\right). \quad (78)$$

Then the portion p of I_2 relating to the integration interval $t - \frac{2\pi}{\omega} < \tau < t$ is

$$p = \eta''(1 - 2QQ' + Q^2), \quad (79)$$

where η'' is equal to -1 or $+1$ depending on whether k is even or odd. On the other hand, the portions of I_2 relating to integration intervals comprising between $t - \frac{4\pi}{\omega}$ and $t - \frac{2\pi}{\omega}$, between $t - \frac{6\pi}{\omega}$ and $t - \frac{4\pi}{\omega}$, etc. are successive terms of a geometric progression of ratio Q^2, first term p and number of terms equal to $\frac{(t-t_1)\omega}{2}$. In total that gives

$$I_2 = \frac{p}{1 - Q^2}\left(1 - e^{-\frac{t-t_1}{\theta}}\right). \quad (80)$$

But $\frac{p}{(1-Q^2)}$ is a periodic function of ψ (and not of t) which can be expanded as a Fourier series. Replacing ψ by ωt we have

$$\frac{p}{1-Q^2} = -\frac{4}{\pi}\coth\frac{\pi}{\omega\theta}\left[\frac{2\omega\theta}{4 + \omega^2\theta^2}\sin\omega t\right.$$

$$+\frac{4}{4 + \omega^2\theta^2}\cos\omega t - \cdots$$

$$-\frac{2\omega\theta}{4 + 9\omega^2\theta^2}\sin 3\omega t \quad (81)$$

$$\left. -\frac{4}{12 + 27\omega^2\theta^2}\cos 3\omega t + \cdots\right].$$

Finally, substituting this value into equation (81) and adding the value of I_2 thus obtained to I_1 given by relation (77) we obtain $\frac{P}{W_0}$ in the form of periodic terms of frequency $\omega, 3\omega, \ldots$, decaying exponentially. The most interesting case is that where the period $\frac{2\pi}{\omega}$ of the alternating field is small compared to θ. Limiting ourselves to terms in $\sin\omega t$ and $\cos\omega t$ and neglecting unity in comparison to $(\omega\theta)^2$ we obtain:

$$\frac{P}{W_0} = -\frac{4}{\pi}\left[\left(1 - \frac{2}{\pi}\right)e^{-\frac{t-t_1}{\theta}} - e^{-\frac{t}{\theta}} + \frac{2}{\pi}\right]\sin\omega t$$

$$-\frac{16}{\pi^2\omega\theta}\left(1 - e^{-\frac{t-t_1}{\theta}}\right)\cos\omega t. \quad (82)$$

Unfortunately, even with all our simplifications this formula has a much more complicated structure than the corresponding formula (71) relating to small amplitudes. Therefore, the repercussions of after-effect on alternating field measurements are more difficult to explain than in the case of discontinuous changes of the applied field.

Concerning, for example, the term in $\cos\omega t$ which represents the part of the after-effect pressure in quadrature with the applied field, we note that it now depends on $t - t_1$, i.e. on the time elapsed between the application of the field and the measurement. In particular, this term is zero at the instant the field is applied. It does not seem that the corresponding variation of the loss angle has been experimentally reported until now. We also note that the amplitude of the after-effect pressure does not depend on the amplitude H_m of the applied field and that the loss angle must vary inversely with H_m. This also has not been reported. Perhaps it is difficult to see in the presence of loss angles of different origins increasing with the field, as for hysteresis.

We can make similar observations about the part of the after-effect pressure which varies in opposite phase with the applied field. We will treat an interesting example.

(16) The variation of the permeability in alternating fields:

Formula (82) relating to a single time constant can be extended to the logarithmic case. We find

$$\frac{P}{W_0} = -\frac{4}{\pi}\int_{\theta_1}^{\theta_2}\left[\left(1 - \frac{2}{\pi}\right)e^{-\frac{t-t_1}{\theta}}\right.$$

$$\left. - e^{-\frac{t}{\theta}} + \frac{2}{\pi}\right]\frac{\sin\omega t\, d\theta}{\theta(\log\theta_2 - \log\theta_1)} \quad (83)$$

When the following two inequalities are satisfied:

$$\theta_1 \ll t \ll \theta_2 \quad \text{et} \quad \theta_1 \ll t - t_1 \ll \theta_2$$

and the period $\frac{2\pi}{\omega}$ is also small compared to θ_1 it follows from formulae (59), (60) and (61) that relation (83) takes the form:

$$\frac{P}{W_0} = -\frac{4\sin\omega t}{\pi}$$

$$\times\left\{\frac{2}{\pi}\left(1 - \frac{C + \log\theta_1}{\log\theta_2 - \log\theta_1}\right) + \frac{1}{\log\theta_2 - \log\theta_1}\right.$$

$$\left. \times\left[\log t - \left(1 - \frac{2}{\pi}\right)\log(t - t_1)\right]\right\}. \quad (84)$$

The first part of the expression in the brackets is independent of time and is of no interest to us. This is not so for the second part. If we suppost t_1 is zero, then the term in square brackets reduces to $\frac{2}{\pi}\log t$ and consequently the amplitude of the magnetic pressure always decreases with time.

It follows that if the alternating current permeability is measured soon after demagnetization and the measuring field left on, the permeability always decreases with time, as shown by the curve A of Fig. 3, plotted in arbitrary units.

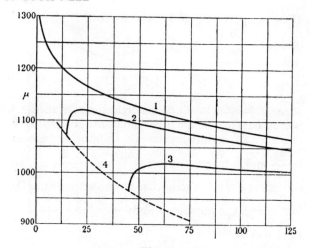

Fig. 4.
Experimental variation of the permeability with time, after Webb and Ford. The units of the time axis is minutes.

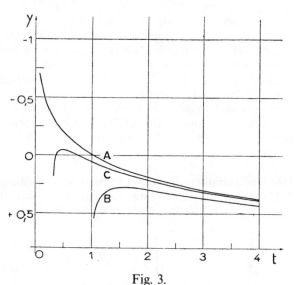

Fig. 3.
Theoretical variation of the permeability with time (arbitrary scales).

Now suppose on the contrary that the measuring field is not applied until time t after demagnetization, and is then kept on. The term between brackets in relation (48) starts by decreasing, passes a minimum for $t = \frac{\pi t_1}{2}$ and then increases continually. Therefore the permeability starts by increasing, passes a maximum, then decreases as shown by the curves B and C in Fig. 3, for $t_1 = 1$ and $t_1 = 0.3$ respectively.

This quite remarkable behaviour of the variation of permeability with time has been observed, under various experimental conditions, by C. E. Webb and L. H. Ford [6]. Here (Fig. 4) we have reproduced Fig. 1 of their paper. The similarity of curves 1, 2 and 3 with our curves A, B and C is evident, so it seems that we hold the key to these curious phenomena.

(17) Effect of the scatter of restoring forces: Among the phenomena connected with disaccommodation there is one which is particularly interesting and which was first observed by Webb and Ford [6] and later confirmed by G. Montalenti [16]. The permeability of 4% Si–iron sheet decreases as the time since demagnetization increases. We have just given the theory for this phenomenon. For example, let μ_1 be the permeability measured in alternating current 2 minutes after demagnetization and μ_2 be the permeability two weeks later. Experiment shows that the ratio $\frac{\mu_1 - \mu_2}{\mu_2}$ first increases with the amplitude of the measuring field H_m, passes a maximum around $H_m = 0.015$ Oe and then decreases rapidly, as shown by Figs. 5 and 6.

The simplified theory developed above, on the contrary, shows that the ratio $\frac{\mu_1 - \mu_2}{\mu_2}$ should decrease uniformly with the field. Without doubt, this disagreement originates in the over simplifications used, particularly in giving all the walls the same restoring force to the equilibrium position. We will show that properties are obtained which closely approximate to the experimental results when 180° walls are considered for which the *ordinary* restoring forces, for a displacement of about *d are negligible compared to forces associated with the diffusion stabilization*. We had implicitly excluded such walls from our preceding study as we limited ourselves to after-effect as a small perturbation. However even the existence of Rayleigh's laws relating to weak fields shows that such walls with very weak restoring forces must exist.

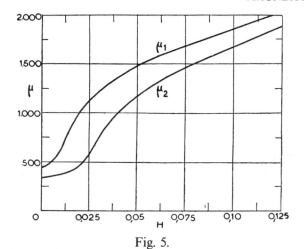

Fig. 5.
Curves of permeability against field: 2 minutes (μ_1) and two weeks (μ_2) after demagnetization (after Webb and Ford).

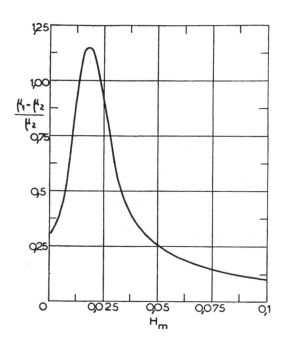

Fig. 6.
The ratio $\dfrac{\mu_1 - \mu_2}{\mu_2}$ as a function of field (after Webb and Ford).

Walls of this nature experience only small displacements less than $4d$ as long as the applied magnetic field remains less than a certain limit H_s corresponding to the position in which the applied magnetic pressure $2H_s J_s$ is equal to the maximum after-effect pressure, i.e. the maximum in Fig. 2. This gives, sensibly,

$$H_m = \frac{W_0 G(t)}{J_s}. \tag{85}$$

Beyond this limit, the wall "takes off" and experiences a very large displacement: if the restoring forces are negligible it sweeps over the entire elementary domain. Therefore as H surpasses the value H_m there is a discontinuity in the permeability. The field at which this discontinuity occurs depends on time: immediately after the demagnetization it is produced in a very small field H' but a long time after a much larger field H" is needed. We obtain variations of permeability such as those shown schematically in Fig. 7, with a maximum of the ratio $\dfrac{\mu_1 - \mu_2}{\mu_2}$ in the interval $H' < H < H''$.

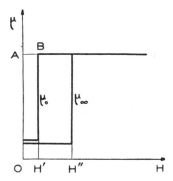

Fig. 7.
Schematic theoretical variation of permeability with field: soon after demagnetization (μ_0) or after a considerable time (μ_∞).

This schematic analysis shows that this maximum disappears if μ_1 is measured extremely soon after demagnetization, i.e. at less than the smallest time constant θ. Even in the weakest fields, therefore, we should obtain a high value of μ_1, represented by AB in Fig. 7 such that $\dfrac{\mu_1 - \mu_2}{\mu_2}$ no longer shows a maximum. In practice, experiments have shown that for 4% Si–iron sheet time constants considerably less than one second are present and so experimental conditions correspond well with the occurrence of a maximum.

For a precise calculation, apart from the distribution of time constants, it is necessary to know the distribution of restoring forces between the different walls. This is not well known at present, and in the absence of such results we have calculated the variations of the permeability with field for a substance in which the area dS of 90° walls with restoring pressures between a^2u and $(a+da)^2$ is $\dfrac{3S_0 a^2 da}{a_0^3}$ for a between 0 and a_0, and zero outside this interval. We suppose that the substance also contains 180° walls of the same total area with restoring forces twice as great as the preceding ones but distributed in the same way. To make the calculation easier, $f'(u)$ has been approximated to $-\dfrac{2u}{3d}$ in the interval $0 < U < \dfrac{3d}{2}$ and to -1 beyond; $f''(u)$ has been approximated to $-\dfrac{4U}{3d}$ for $0 < U < \dfrac{3d}{2}$ and to -2 for $\dfrac{3d}{2} < U < \dfrac{15d}{2}$ and to 0 beyond this. We have done the calculations for the following two values of $G(t)$:

$$G(t) = 0{,}15 \frac{da_0^2}{W_0}, \quad G(t) = 0{,}75 \frac{da_0^2}{W_0},$$

Fig. 8, plotted in arbitrary units, shows the corresponding values of the permeabilities μ_1 and μ_2 as well as $\dfrac{\mu_1 - \mu_2}{\mu_2}$. These curves show a striking likeness to those of Webb and Ford. Therefore, it seems quite reasonable to explain their results in this way.

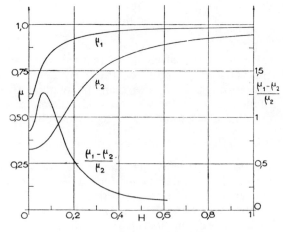

Fig. 8.
Theoretical variation, for a simple model, of μ_1, μ_2 and $\dfrac{\mu_1 - \mu_2}{\mu_2}$ as a function of field; μ_1 and μ_2 are permeabilities for measurements made at different times after demagnetization.

(18) Conclusions: It appears justifiable to conclude that the proposed theory qualitatively explains the known properties of disaccommodation and diffusion after-effect.

It remains to compare theory and experiment quantitatively. This is however not easy, because sufficiently good experimental accuracy can only be obtained by studying substances which possess a large after-effect for which the approximate method outlined above (§ 11) certainly does not apply, except in special cases. It will also be important to take account of the relative proportions of 90° and 180° walls as well as the range of restoring forces $R'(u')$ and $R''(u'')$.

If the theory finally proves correct, we would be able to use its results to determine the constant W_0 and the function $G(t)$, which suffice to characterize the after-effect. It would then be interesting to study the variations of W_0 as a function of temperature and concentration of carbon, nitrogen or hydrogen, as well as the relations between the function $G(t)$ and the physical state of the substance.

One could raise the objection od the difficulty of separating diffusion after-effect from fluctuation after-effect. We can distinguish them by the fact that fluctuation after-effect does not affect the permeability whereas diffusion after-effect does. In a most general way, when a large change in magnetization is produced by a discontinuous change of the applied field, the fraction of the magnetization change depending on time is of the form $\chi H_f(t)$ where $H_f(t)$ is a fictitious field dependent on time and independent of the magnetization. In the case of fluctuation after-effect, χ is the irreversible susceptibility corresponding to the experimental conditions, whereas for diffusion after-effect χ is the reversible susceptibility. As the ratio between these two susceptibilities varies a lot from point to point on the hysteresis curve, it is possible to separate these two types of after-effect.

I thank M. Pauthenet for help in making the figures.

REFERENCES

[1] L. Néel, *J. Physique Rad.*, 1951, **12**, 339.
[2] L. Néel, *J. Physique Rad.*, 1950, **11**, 49.
[3] J. C. Barbier, *C.R. Acad. Sci.*, 1950, **230**, 1040; 1952, **234**, 415.
[4] J. C. Barbier, *J. Physique Rad.*, 1951, **12**, 352.
[5] J. L. Snoek, *Physica*, 1938, **5**, 663; New developments in ferromagnetic materials, Elsevier, Amsterdam, 1947, § 16.
[6] C. E. Webb and L. H. Ford, *J. Inst. Electr. Eng.*, 1934, **75**, 787.
[7] C. Guillaud, *Thèsis*, Strasbourg, 1943.
[8] G. Richter, *Ann. Physik*, 1937, **29**, 605.
[9] R. Becker and W. Doring, Ferromagnetismus, Springer Berlin, 1939.
[10] L. Néel, *Cahiers de Physique*, 1944, **25**, 1.
[11] E. Lifshitz, *J. Phys, U.S.S.R.*, 1944, **8**, 337.
[12] C. Kittel, *Rev. Mod. Physics*, 1949, **21**, 541.
[13] L. Néel, *Cahiers de Physique*, 1942, **12**, 1; 1943, **13**, 18.
[14] A. Mitkewitch, *J. Physique Rad.*, 1936, **7**, 133.
[15] H. Jordan, *Electr. Nachr. Techn.*, 1934, **1**, 7.
[16] G. Montalenti, *Ric. Scientifica e Ricostruz.*, 1947, **17**, No. 7–8, 3.

Chapter XII

DIRECTIONAL ORDER: IRRADIATION
A92, A119

A92 (1954)
SURFACE MAGNETIC ANISOTROPY AND ORIENTATIONAL SUPERLATTICES

Note (1977): This paper is devoted to the investigation of the anisotropic distribution of the orientation of pairs of neighbouring atoms A–A, A–B, B–B in a solid solution AB, to which the name 'orientational superlattice' is given. More recently Chikazumi has examined the same problem and given it the name 'directional order' which is in current use and which seems preferable.

Abstract: In the first part of this article it is suggested that the magnetocrystalline and magnetoelastic energy of a ferromagnetic body be considered as the sum of elementary terms each related to a bond, i.e. to a pair of nearest neighbour atoms. On this basis, the theory of magnetostriction and of magnetocrystalline anisotropy is evolved, and on comparison with the experimental results, the values of the parameters which characterize the binding energy are obtained. From the same premises it is shown that in ferromagnetic bodies there must exist a *surface anisotropy energy*, which depends on the orientation of the spontaneous magnetization relative to the surface, and which, furthermore, is quite separate from the classical phenomenon of the shape demagnetization field. This surface energy, of the order of 0.1 to 1 erg/cm^2, is likely to play an important rôle in the properties of dispersed ferromagnetic materials whose elements have dimensions of less than 100 Å.

In the second part, it is shown using the above assumptions, that in ferromagnetic solid solutions with at least two constituents, if they are heat treated in a magnetic field, the nearest neighbouring atoms must distribute themselves anisotropically around a given atom and give rise to an *orientational structure*. On quenching, this superlattice may be maintained in pseudoequilibrium at low temperature, and is revealed by the appearance of a uniaxial magnetic anisotropy. The phenomena are made explicit by calculations, in particular the effect of the concentration, in the case of different simple cubic lattices and in the case of a substance which is isotropic by compensation. The calculated anisotropy is of the order of 10^3 to 10^5 ergs/cm^3, but in exceptional circumstances may be much bigger. This theory accounts for the properties of ferronickels heat treated in a magnetic field, and in particular, of single crystals of permalloy. The possible rôle these effects may have is suggested in the case of alnico V and in oriented cobalt ferrites.

In the last part it is shown that it is possible to generate an orientational superlattice in an arbitrary solid solution by means of elastic deformation at high temperature and preservation by quenching. If the solid solution is ferromagnetic, the superlattice thus generated gives rise to uniaxial magnetic anisotropy. On the basis of a rough theory of the elastic properties of solid solutions devised for this purpose, the phenomenon is tested mathematically: for ferro-nickels anisotropies of 10^4 ergs/cm^3 are found for stresses of 10 kg/mm^2. It is suggested that the uniaxial magnetic anisotropy of cold rolled, quasi single crystals of ferronickels be interpreted by the creation of such superstructures, where the plastic deformation allows the atoms to take up the equilibrium distribution corresponding to the system of applied strains. In rolled or stretched polycrystalline ferronickels the anisotropy is of opposite sign to the preceeding case; it is suggested that this may be explained by the same mechanism as the phenomena of regeneration following metallurgical creep.

INTRODUCTION

(1): In this article we shall be concerned with two new effects of the magnetocrystalline forces which are present in ferromagnetic substances: surface magnetocrystalline energy and orientational superlattices. The actual existence of these two phenomena seems beyond question, but it is important to estimate their order of magnitude as accurately as possible. To this purpose we have evolved a theory of magnetostriction and magnetic anisotropy which differs significantly from the classical theory. Our procedure does not have the same formal rigour as the classical method, but we wish to make clear that it is not our intention to investigate magnetostriction and anisotropy *per se*, but simply to estimate the magnitude of certain interatomic forces which occur in these phenomena.

We take for example a cubic crystal set in three rectangular coordinate axes which are parallel to the four fold axes; let $\beta_1, \beta_2, \beta_3$ be the direction cosines defining the orientation of the spontaneous magnetization J_s, and A_{ij} the components of the elastic deformation tensor. Simple symmetry considerations show that, if the expansion is limited to terms of second order in $\beta_1, \beta_2, \beta_3$, the magnetic energy density D_m of the system, which is related to the orientation of the spontaneous magnetization, must be of the form

$$D_m = K \Sigma' \beta_i^2 \beta_j^2 + B_1 \Sigma \beta_i^2 A_{ii} + 2 B_2 \Sigma' \beta_i \beta_j A_{ij}, \quad (1)$$

where K, B_1 and B_2 are three coefficients characteristic of the material. In this expression, as elsewhere below, the symbol Σ indicates the sum over the three values 1, 2 and 3 of the subscript i and over the three values 1, 2 and 3 of the subscript j, if it occurs: this is, therefore, a sum with either three or nine terms. In contrast, the symbol Σ' which always contains two subscripts contains only three terms corresponding to the three combinations of i and j 1–2, 2–3 and 3–1 respectively.

Equation 1 is the starting point of the classical theory, as stated for example in an article by Kittel [1].

Furthermore, by writing for the elastic energy density, D_{el},

$$D_{el} = \frac{C_1}{2} \Sigma^2 A_{ii} + C_2 \Sigma A_{ii}^2 + 2 C_3 \Sigma' A_{ij}^2, \quad (2)$$

where C_1, C_2 and C_3 are the elastic constants, one has a sufficient basis to form a consistent theory of magnetic anisotropy, of magnetostriction and of the effect of mechanical strains. This theory, which is formal in essence, is undoubtedly rigorous, but has the drawback of neglecting important aspects of these phenomena of magnetocrystalline couplings.

We shall follow another method whereby the magnetic energy of the system is considered to be the sum of a series of terms each related to a *bond*, i.e. to the forces acting between atoms taken pairwise. Clearly this method is open to the objection that these elementary energies are not likely to be additive: this means for example that with three atoms A, B, and C, that part of the energy of the pair AB which is associated with the orientation of J_s must also depend on the nature and position of the atom C. This is true, but similar objections can be raised for the theory of ferromagnetism itself or the theory of order–disorder in solid solutions both of which use methods of the same kind: this does not prevent these theories from being useful.

I
MAGNETOSTRICTION, BULK AND SURFACE MAGNETIC ANISOTROPY

(2) Expression for the bond energy: The elementary interaction energy $w(r, \varphi)$ must be a function of the separation r of the two atoms which form the bond and of the angle φ between the spontaneous magnetization J_s and the line connecting the centres of the two atoms. This energy may be expanded in a series of the form

$$w = g_1(r) P_2(\cos \varphi) + g_2(r) P_4(\cos \varphi) + \cdots, \quad (3)$$

where the P_n are the Legendre polynomials of order n and where the coefficients $g_n(r)$ are functions only of r.

Since both atoms of the bond possess a magnetic moment μ parallel to J_s, the coefficient $g_1(r)$ must contain the term $-3\mu^2/r^3$ corresponding to the dipolar magnetic coupling. It has been known for a long time that this magnetic dipolar coupling is much too weak to explain the observed anisotropy and magnetostriction [3]: $g_1(r)$ must therefore contain other terms, the origin of which is still very much an open question, but is probably connected with spin–orbit coupling. It is known only that these terms must decrease with distance faster than $1/r^3$ so that, in the first approximation introduced so far, only interactions between pairs of nearest neighbour atoms need be taken into account. We note incidentally that this approximation is not valid for magnetic dipolar coupling which corresponds to long range interactions and hence gives rise to the shape demagnetization field.

Lastly, denoting by r_0 the separation of two nearest neighbour atoms in the crystal and setting $r = r_0 + \delta r$, the elementary energy w can be written in the form

$$w = \left(-\frac{3\mu^2}{r^3} + l + m\delta r \right)\left(\cos^2 \varphi - \frac{1}{3} \right) \\ + (q + s \delta r)\left(\cos^4 \varphi - \frac{30}{35} \cos^2 \varphi + \frac{3}{35} \right), \quad (4)$$

where only the terms in first order in δr have been kept and l, m, q and s are four coefficients which are functions of r_0. Equation (4) is used as the starting point of the theory.

(3) Magnetic energy of the distorted crystal: In order to calculate the magnetostriction one must first determine the variation of magnetic energy produced by a stress defined by the tensor A_{ij}. We shall henceforth presume an initially cubic crystal. We select an arbitrary atom for origin with three rectangular coordinate axes $Oxyz$ parallel to the four-fold axes. The position of the next neighbour atom to the atom at the origin is defined by the direction cosines $\alpha_1, \alpha_2, \alpha_3$ of the straight line which connects the origin to the centre of this atom. After distortion, the position of the atom is defined by the new direction cosines

$$\begin{aligned} \alpha'_1 &= c(\alpha_1 + A_{11}\alpha_1 + A_{12}\alpha_2 + A_{13}\alpha_3), \\ \alpha'_2 &= c(\alpha_2 + A_{21}\alpha_1 + A_{22}\alpha_2 + A_{23}\alpha_3), \\ \alpha'_3 &= c(\alpha_3 + A_{31}\alpha_1 + A_{32}\alpha_2 + A_{33}\alpha_3), \end{aligned} \quad (5)$$

where the coefficient c, which is defined by the condition

$$\alpha_1'^2 + \alpha_2'^2 + \alpha_3'^2 = 1,$$

is given in a first approximation by the relation

$$c = 1 - \Sigma A_{ij}\alpha_i\alpha_j \qquad (6)$$

The change δr in the distance of the atom considered from the origin is

$$\delta r = r_0 \Sigma A_{ij}\alpha_i\alpha_j. \qquad (7)$$

Also, the variation $\delta \cos^2 \varphi$ in $\cos^2 \varphi$ is given by

$$\delta \cos^2 \varphi = \Sigma^2 \alpha_i' \beta_i - \Sigma^2 \alpha_i \beta_i$$
$$= 2 \Sigma \alpha_i \beta_i \Sigma' A_{ij}\beta_i\alpha_j - 2\Sigma^2 \alpha_i\beta_i \Sigma A_{ij}\alpha_i\alpha_j, \qquad (8)$$

where $\beta_1, \beta_2, \beta_3$ are the direction cosines of the spontaneous magnetization J_s.

Taking into account the above values, the change δw produced in the energy w can be written as follows, keeping only the main terms in l and m

$$\delta w = l\delta \cos^2 \varphi + m \left(\cos^2 \varphi - \frac{1}{3}\right)\delta r. \qquad (9)$$

What is actually of interest is the mean value δw_m for all possible positions of the neighbouring atom with respect to the atom at the origin. In this average, taken over the α_i, only remain the terms containing even powers of each of the α_i. Table I lists the mean values for the simplest cubic systems of the various products of the α_i which we shall need in what follows.

Table I.

Notation	Isotropic bond distribution	Simple cubic $n=6$	Body centred cubic $n=8$	Face centred cubic $n=12$
$s_2 = \overline{\alpha_i^2}$	1/3	1/3	1/3	1/3
$s_4 = \overline{\alpha_i^4}$	1/5	1/3	1/9	1/6
$s_{22} = \overline{\alpha_i^2 \alpha_j^2}$	1/15	0	1/9	1/12
$s_6 = \overline{\alpha_i^6}$	1/7	1/3	1/27	1/12
$s_{42} = \overline{\alpha_i^4 \alpha_j^2}$	1/35	0	1/27	1/24
$s_{222} = \overline{\alpha_1^2 \alpha_2^2 \alpha_3^2}$	1/105	0	1/27	0

From the mean value δw_m for one bond the change D_m' in magnetic energy density is obtained by multiplying by the number of bonds contained in 1 cm^3, $nN_a/2V$, taking care not to count them twice: n is the number of nearest neighbours, N_a Avagadro's number and V the atomic volume. When all the calculations are carried out one finds

$$D_m' = B_1 \Sigma A_{ii}\beta_i^2 + 2 B_2 \Sigma' A_{ij}\beta_i\beta_j + B_3 \Sigma A_{ii}, \qquad (10)$$

and the three coefficients B_1, B_2, and B_3 are given by the following relations

$$\left.\begin{array}{l} B_1 = \dfrac{nN_a}{2V}[2l(s_{22} + s_2 - s_4) + mr_0(s_4 - s_{22})], \\[6pt] B_2 = \dfrac{nN_a}{2V}[2l(s_2 - 2s_{22}) + 2mr_0 s_{22}], \\[6pt] B_3 = \dfrac{nN_a}{2V}(mr_0 - 2l)s_{22}. \end{array}\right\} \qquad (11)$$

Except for the term in $A_{11} + A_{22} + A_{33}$ which corresponds to a homogeneous expansion, independent of the orientation $\beta_1, \beta_2, \beta_3$ of J_s, and which is of no interest to us here, the last two terms of equation (1) reappear here, the form of which had been postulated *a priori* in order to define D_m.

(4) Calculation of the magnetostriction: The calculation is standard [1]. The total energy, which is the sum of the magnetic energy D_m' given in (10) and the elastic energy D_{el} given in (2), is minimised with respect to the components A_{ij}. Neglecting a term which is independent of the orientation of J_s, one finds

$$A_{ii} = -\frac{B_1 \beta_i^2}{2C_2}, \quad A_{ij} = -\frac{B_2 \beta_i \beta_j}{2C_3}. \qquad (12)$$

The expression for the magnetostriction in the direction $\gamma_1, \gamma_2, \gamma_3$ is given by

$$\frac{\delta \lambda}{\lambda} = \Sigma A_{ij}\gamma_i\gamma_j, \qquad (13)$$

so that finally

$$\frac{\delta \lambda}{\lambda} = \frac{3}{2}\lambda_{100}\left(\Sigma \beta_i^2 \gamma_i^2 - \frac{1}{3}\right) + 3\lambda_{111}\Sigma' \beta_i\beta_j\gamma_i\gamma_j. \qquad (14)$$

The constants λ_{100} and λ_{111}, equal respectively to the longitudinal magnetostriction along the fourfold and the threefold axes, have the values

$$\lambda_{100} = -\frac{B_1}{3C_2}, \quad \lambda_{111} = -\frac{B_2}{3C_3}. \quad (15)$$

Thus, for the face centred cubic system, one has

$$\left.\begin{array}{l}\lambda_{100} = -\dfrac{nN_a}{72\,VC_2}(6l + mr_0),\\[2mm]\lambda_{111} = -\dfrac{nN_a}{36\,VC_3}(2l + mr_0).\end{array}\right\} \quad (16)$$

Conversely, knowing the experimental values of the two principal longitudinal magnetostrictions λ_{100} and λ_{111}, one can deduce the values of the two basic constants $N_a l$ and $N_a mr_0$; thus for the face centred cube one finds

$$\left.\begin{array}{l}\dfrac{nN_a l}{2V} = \dfrac{9}{2}(C_3 \lambda_{111} - 2C_2 \lambda_{100}),\\[2mm]\dfrac{nN_a mr_0}{2V} = 9(2C_2 \lambda_{100} - 3C_3 \lambda_{111}).\end{array}\right\} \quad (17)$$

It must be added that the principal interest of this theory of magnetostriction lies in its ability to give the order of magnitude of the two constants l and m. Assuming the following values for nickel

$$\lambda_{100} = -54 \times 10^{-6}, \quad \lambda_{111} = -27 \times 10^{-6}$$

$$C_2 = 0.45 \times 10^{12}, \quad C_3 = 1.185 \times 10^{12};$$

$$V = 6.55 \text{ cm}^3$$

one obtains finally

$$N_a l = 0.8 \times 10^8 \text{ ergs}, \quad N_a mr_0 = 4.6 \times 10^8 \text{ ergs}.$$

(5) Extension of the theory to solid solutions: The above considerations can be extended easily to metallic solid solutions. If, for example, one has a solid solution with two constituents A and B, of atomic concentrations c_a and c_b respectively ($c_a + c_b = 1$), three categories of bonds A–A, A–B, B–B should be distinguished, to which correspond three sets of basic constants l and m: l_{aa}, l_{ab}, l_{bb} and m_{aa}, m_{ab} and m_{bb}. Moreover, if it is an ideal solid solution, i.e. with no correlation between the species of two nearest neighbours, the numbers of bonds A–A, A–B and B–B are respectively proportional to c_a^2, $2c_a c_b$, and c_b^2. The above relations remain valid provided l is given the value

$$l = c_a^2 l_{aa} + 2 c_a c_b l_{ab} + c_b^2 l_{bb}, \quad (18)$$

with a similar formula for m.

It follows that in a solid solution, magnetostriction must vary quadratically as a function of atomic concentration. Roughly, this is correct for ferronickels. Between 40% Ni and 100% Ni, according to the experiments of Bozorth [4], the two following formulae give an approximate description of the concentration dependence of the principal magnetostriction constants of this series of alloys:

$$\lambda_{111}.10^6 = -27 c_a^2 + 134 c_a c_b + 13 c_b^2,$$

$$\lambda_{100}.10^6 = -55 c_a^2 + 340 c_a c_b - 245 c_b^2,$$

where the suffix a indicates nickel and b indicates iron. Using the formulae (17) and taking the same elastic constants for iron as for nickel, one finds (in units of ergs)

$$N_a l_{ab} = -3.8 \times 10^8, \quad N_a m_{ab} r_0 = -7.3 \times 10^8;$$

$$N_a l_{bb} = +12 \times 10^8, \quad N_a m_{bb} r_0 = -28 \times 10^8.$$

These values are in fact compatible with those deduced from the magnetostriction of pure iron (body centred cubic lattice) if the interactions with the six second nearest neighbours are counted in addition to those of the eight first neighbours.

(6) Magnetocrystalline anisotropy: Investigating this question, one must distinguish between *free anisotropy*, in which the crystal lattice is free to distort, and *constant size anisotropy*, in which the crystal lattice is constrained to maintain a fixed size ($\delta r = 0$). The former corresponds to the normal experimental conditions, but as we shall see later, the relative difference between the two is generally small.

In order to determine the constant size anisotropy, one has just to calculate the mean value of the energy w given by expression (3), for $\delta r = 0$, with respect to all possible orientations of the bonds in the given crystal lattice. Here we are interested only in cubic lattices: in all of them the mean value of $\cos^2 \varphi$ is equal to 1/3. It follows that the term in $g_1(r)$ vanishes from the average and that it becomes necessary to take into account the terms in $g_2(r)$. The average value of $\cos^4 \varphi$ is given by

$$\overline{\cos^4 \varphi} = \overline{\Sigma^4 \alpha_i \beta_i} = s_4 + (6s_{22} - 2s_4) \Sigma' \beta_i^2 \beta_j^2. \quad (19)$$

The anisotropy energy density D_a is therefore of the type shown by the first term on the right hand side of equation (1). The anisotropy constant K is obtained by multiplying the mean value of w by the number of bonds contained in 1 cm³, $nN_a/2V$, and finally one finds the relation

$$D_a = K(\beta_1^2 \beta_2^2 + \beta_2^2 \beta_3^2 + \beta_3^2 \beta_1^2), \quad (20)$$

with
$$K = \frac{c N_a q}{V},$$

where the coefficient q is defined in equation (4) and where c is a numerical coefficient equal to 0 for an isotropic medium, to 1 for the face centred cubic lattice, to 16/9 for the body centred cube, and to -2 for the simple cube. As K is of the order of 10^5 ergs/cm^3 in cubic metallic ferromagnets, it follows that $N_a q$ is also of the same order of magnitude, while $N_a l$ is of the order of 10^9 to 10^8 ergs. It must, therefore, be concluded that in expression (3) the terms in $g_2(r)$ are at least a thousand times smaller than the terms in $g_1(r)$.

A few words should be said about free anisotropy, which occurs when the crystal is free to deform. To the energy density given by equation (20) must be added an energy density D'_a, obtained by giving the components A_{ij} in the expression $D_{el} + D'_m$ for the total energy, the values deduced from the magnetostriction, which are given by equation (12). Taking everything into account, one finds

$$D'_a = K' \Sigma' \beta_i^2 \beta_j^2$$
$$= \frac{9}{2}(C_2 \lambda_{100}^2 - C_3 \lambda_{111}^2)(\beta_1^2 \beta_2^2 + \beta_2^2 \beta_3^2 + \beta_3^2 \beta_1^2). \quad (21)$$

Examination of the numerical values shows that K' is small compared with the values of the anisotropy constant found experimentally: in practice, the error introduced by taking the free anisotropy to be equal to the constant size anisotropy is not large.

(7) Surface anisotropy: In the calculation of the anisotropy constant K, the terms in $g_1(r)$ vanish because the nearest neighbour atoms to a given atom are distributed around it with cubic symmetry. This is no longer the case when the atom under consideration lies at the surface of the crystal: the average value of $P_2(\cos \varphi)$ is not generally zero in this case.

For the magnetic dipolar term, on account of its associated long range interactions, the effects of *shape demagnetization field* recur. These effects are well known and it is unnecessary to expand further on them.

For the term in l in $g_1(r)$, the corresponding surface energy per unit surface atom is obtained by summing the values of $l \cos^2 \varphi$ for all the nearest neighbouring atoms and by multiplying the result by 1/2 to avoid counting the same binding energies twice. Hence an energy w_s is obtained which depends both on the orientation of the spontaneous magnetization with respect to the crystal surface and on the orientation of this surface with respect to the crystal axes. In Table II are listed the values of w_s for the most important cases. Multiplication of these values by the number of atoms present per unit surface area gives the values of the surface energy density D_s. Very often D_s takes the simple form

$$D_s = K_s \cos^2 \theta, \quad (22)$$

in particular in the important case of the (1 1 1) side of the face centred cube: θ is the angle that the spontaneous magnetization makes with the normal external to the surface. In other cases, the factor K_s takes on a more complicated form and it is expressed in terms of the direction cosines β_1, β_2, β_3 of the spontaneous magnetization with respect to the fourfold axes of the crystal. It is pointed out that D_s is zero for the (1 1 1) and (1 0 0) faces of the body centred cube, but this would not have been the case had the interactions with the second nearest neighbours been included. These interactions are particularly large in this type of lattice.

It was calculated above that the values of $N_a l$ were of the order of 10^8 to 10^9 ergs. With $r_0 = 2.5$ Å, K_s is found to range from 0.13 to 1.3 ergs/cm^2. These values appear to be too weak to affect notably the properties of bulk ferromagnetic materials.

It is conceivable that this surface energy becomes important at high frequency when the skin depth becomes very small, as in experiments on magnetic resonance. However, even at 2×10^{10} Hz, the skin depth in a metallic ferromagnetic is still of the order of 10^{-4} cm, so that from the ratio of the surface energy to the effective volume of the ferromagnetic material, energy densities of at most 10^4 ergs/cm^3 are obtained, corresponding to fictitious perturbing fields of 10 Oe. At first sight these fields seem to be too weak to have much influence on resonance experiments and the determination of the g factor. Nonetheless, it would be worthwhile to investigate this question more thoroughly.

(8) Surface anisotropy in ferromagnetic materials with finely dispersed phases: The surface anisotropy energy must play a larger part when the ferromagnet is subdivided into elements of very small size, of the order of 100 Å or less, either by fragmentation into very fine grains, or through a precipitation mechanism in an alloy with several phases.

Let us examine firstly the very simple case of a precipitate consisting of small flat platelets of thickness **d** with spontaneous magnetization J_s, separated by nonmagnetic platelets of the same thickness. The mean energy density D corresponding to the demagnetizing field is equal to

TABLE II.

Crystal lattice	surface plane	w_s	Energy density D_g
face centred cubic	(111)	$-\dfrac{l}{2}\cos^2\theta$	$-\dfrac{l}{r_0^2\sqrt{3}}\cos^2\theta$
	(100)	$-\dfrac{l}{2}\cos^2\theta$	$-\dfrac{l}{2r_0^2}\cos^2\theta$
	(011)	$+\dfrac{l}{8}(\beta_2-\beta_3)^2$	$+\dfrac{l}{4r_0^2\sqrt{2}}(\beta_2-\beta_3)^2$
body centred cubic	(111)	0	0
	(100)	0	0
	(011)	$-\dfrac{2}{3}l\beta_2\beta_3$	$-\dfrac{l}{r_0^2\sqrt{2}}\beta_2\beta_3$
simple cubic	(111)	0	0
	(100)	$-\dfrac{l}{2}\cos^2\theta$	$-\dfrac{l}{2r_0^2}\cos^2\theta$
	(011)	$+\dfrac{l}{2}\beta_1^2$	$+\dfrac{l}{2r_0^2\sqrt{2}}\beta_1^2$
12 isotropically distributed neighbours		$+\dfrac{4}{9}l\cos^2\theta$	$+\dfrac{8l}{9\sqrt{3}\,r_0^2}\cos^2\theta$

$$D = \frac{\pi}{2}J_s^2 \cos\theta, \qquad (23)$$

where θ denotes the angle between J_s and the plane of the platelets. The energy density corresponding to the surface magnetocrystalline energy can also be written

$$D' = \frac{K_s \cos^2\theta}{d} \qquad (24)$$

These two energies are equal for

$$d = \frac{2K_s}{\pi J_s^2} \qquad (25)$$

With $K_s = 1.3$ ergs/cm^3 and $J_s = 1000$, one obtains $d = 80$ Å. For this value of d, the real total anisotropy is twice that of the demagnetizing field. But if K_s is negative, for the same absolute value, the total anisotropy is zero. One can therefore appreciate that these surface anisotropies are likely to be an important factor in the size of precipitates whose order of magnitude is similar to that corresponding to the coercive field of good permanent bulk magnets, such as the alnicos: for instance, Kronenberg has shown recently [5] that the transverse dimensions of the precipitated grains in properly treated alnico V were close to 40 Å.

The results are similar for isolated fine grains. Let us take the example of a grain which is sufficiently small to contain a single elementary domain in the shape of an elongated ellipsoid of revolution with major axis $2a$ and equatorial diameter $2b$. An elementary calculation shows that the anisotropic surface magnetocrystalline energy W_s is given by

$$W_s = \frac{\pi ab\, K_s}{e}\left[\left(\frac{3}{e^2}-4\right)\arcsin e + \left(2-\frac{3}{e^2}\right)e\sqrt{1-e^2}\right]\cos^2\theta, \qquad (26)$$

where θ is the angle between J_s and the axis of revolution and e the eccentricity ($e^2 a^2 = a^2 - b^2$). When e is small, this energy reduces to

$$W_s = -\frac{16}{15} \pi a b e^2 K_s \cos^2 \theta. \tag{27}$$

Moreover, the energy W_d of the shape dependent demagnetizing field can be written as a function of the principal coefficients of the demagnetizing field N_a and N_b, in the form

$$W_d = -\frac{2\pi}{3} J_s^2 ab^2 (N_b - N_a) \cos^2 \theta, \tag{28}$$

which when e is small reduces to

$$W_d = -\frac{16}{30} \pi^2 ab^2 e^2 J_s^2 \cos^2 \theta. \tag{29}$$

The two energies W_s and W_d are equal for

$$b = \frac{2 K_s}{\pi J_s^2}, \tag{30}$$

i.e. exactly the same value as the critical thickness d given by relation (25). Expression (30) is probably valid as a first approximation for an arbitrary value of the eccentricity e. Indeed, one can confirm that it is exact for circular cylinders, when $e = 1$.

The critical equatorial radii are thus of the order of 80 Å for $K_s = 1.3$ ergs/cm^3, i.e. a value close to the size of the fine grains used in the manufacture of powder agglomerate magnets.

II
ORIENTATIONAL SUPERLATTICES CAUSED BY SPONTANEOUS MAGNETIZATION

(9) The notion of orientational superlattice: If it is assumed that the binding energy contains a magnetic term w given in a first approximation by a relation of the type

$$w = (l + m \delta r) \left(\cos^2 \varphi - \frac{1}{3} \right), \tag{31}$$

where φ is the angle between the bond and the spontaneous magnetization, then extremely interesting properties can be predicted for ferromagnetic solid solutions. Let us consider a solid solution with two components A and B. In the absence of ferromagnetism, the distribution of bonds is isotropic, i.e. among all the bonds having a well defined orientation, the relative proportions of bonds A–A, A–B and B–B are always the same and are independent of the given orientation. This is no longer true in the presence of a spontaneous magnetization J_s with a given direction, since the bond energies depend on the angle they make with J_s. It is quite possible, for example, that among the bonds whose orientation is along J_s the proportion of A–A bonds is a little bigger than among the bonds lying perpendicularly to J_s. In other words, if the temperature is high enough so that the atoms can change position, a form of uniaxial anisotropy will arise in the solid solution, whose axis is that of the spontaneous magnetization. This anisotropy is certainly very weak, since the energies involved are much smaller than kT, but it should nevertheless be detectable magnetically, on account of the extraordinary sensitivity of the methods of measurement of magnetic anisotropy. In particular, if a certain uniaxial anisotropy is generated at high temperature by giving the magnetization a well defined orientation by means of an external magnetic field H_1, and the sample is then rapidly quenched to a temperature low enough to arrest diffusion, the alloy retains a permanent anisotropy which is subsequently insensitive to the orientation of J_s and is convenient for magnetic measurements. In this state the A and B atoms are arranged in a special way: there is no longer complete short range disorder, but a beginning of anisotropic order. We shall call this an *orientational superlattice*.

Such orientational superlattices are probably the cause of many of the anisotropic phenomena described in the literature as resulting from heat treatment in a magnetic field.

In what follows, we shall examine this question in greater detail. We shall restrict ourselves firstly to the case of ideal cubic solid solutions with two components (clearly this phenomenon does not occur in a pure substance). By ideal are meant solutions where the atoms A and B are distributed completely at random, with no correlation between the types of neighbouring atoms. The magnetic energy of the three kinds of bond is defined by the three corresponding values l_{aa}, l_{ab}, l_{bb} of the coefficient l in expression (31). It should be shown first of all that of these quantities only the combination

$$l' = l_{aa} + l_{bb} - 2 l_{ab} \tag{32}$$

enters into the statistics of the phenomena: furthermore l' should not be confused with the quantity l defined by equation (18).

In fact, it is possible to go from one arbitrary configuration of the atoms A and B to another arbitrary configuration by a series of elementary operations each one consisting of an exchange of two atoms A and B. Now let us consider the bond with orientation $\alpha_1, \alpha_2, \alpha_3$ between these two atoms. If the two atoms lying at the extremities of these two bonds are identical, the permutation of A and of B does not change the fraction of the total energy corresponding to these bonds with orientation $\alpha_1, \alpha_2, \alpha_3$. But if the atoms are different, with the initial state corresponding for example to two A–B bonds and the final state to one A–A bond and one B–B bond, the change in energy produced by the permutation, as far as the orientation considered is concerned, is

$$\pm (l_{aa} + l_{bb} - 2l_{ab})(\cos^2 \varphi - 1/3).$$

(10) The case of dilute solid solutions: The problem becomes particularly simple when the atomic concentration c_b of one of the constituents B, for example, is small. Among the B atoms, those which are surrounded only by A atoms cannot participate in the superlattice since their environment remains isotropic. The only ones that count are the B atoms having a B neighbour, i.e. the B–B bonds.

Let there be *one* of these bonds; it can occupy n different orientations, where n is the number of nearest neighbour atoms. Its energy is equal to $l_1(\cos^2 \varphi - 1/3)$. The quantity l_1 is the value of l' given by relation (32), corresponding to the temperature T_1 of the heat treatment which was carried out in a magnetic field of orientation $\beta_1, \beta_2, \beta_3$; φ is defined by

$$\cos \varphi = \Sigma \alpha_i \beta_i \qquad (33)$$

The probability $\bar{\omega}(\alpha)$ that the given bond has the orientation $\alpha_1, \alpha_2, \alpha_3$ is given by the expression

$$\bar{\omega}(\alpha) = \frac{\exp\left\{ -l_1 \dfrac{\left(\cos^2 \varphi - \dfrac{1}{3}\right)}{kT_1} \right\}}{\displaystyle\sum_n \exp\left\{ \dfrac{-l_1\left(\cos^2 \varphi - \dfrac{1}{3}\right)}{kT_1} \right\}}. \qquad (34)$$

Since l_1 is much smaller than kT_1, this probability simplifies to

$$\bar{\omega}(\alpha) = \frac{1}{n}\left[1 - \frac{l_1}{kT_1}\left(\cos^2 \varphi - \frac{1}{3}\right) \right]. \qquad (35)$$

Let us now assume that after being kept for a sufficiently long time at the temperature T_1 for equilibrium to be established, the solid solution is cooled to the temperature T_0 sufficiently quickly for the equilibrium not to be modified. We also suppose that T_0 is low enough to make configurational changes impossible. We define the direction of the magnetisation by $\gamma_1, \gamma_2, \gamma_3$ and set

$$\cos \varphi' = \Sigma \alpha_i \gamma_i. \qquad (36)$$

Moreover, let l_0 be the value of l' corresponding to the new value of the temperature, T_0. The anisotropy energy w for one bond can then be written

$$w = l_0 \sum_n \bar{\omega}(\alpha)\left(\cos^2 \varphi' - \frac{1}{3}\right). \qquad (37)$$

Lastly, recalling that the number of B–B bonds in 1 cm^3 is equal to $nc_b^2 N_a/2V$ and denoting by $\overline{\cos^2 \varphi \cos^2 \varphi'}$ the mean value of $\cos^2 \varphi \cos^2 \varphi'$ over all possible values of $\alpha_1, \alpha_2, \alpha_3$, one gets for the anisotropy energy density D_u connected with the orientational superlattice

$$D_u = \frac{nc_b^2 l_0 l_1 N_a}{2VkT_1}\left(\frac{1}{9} - \overline{\cos^2 \varphi \cos^2 \varphi'}\right). \qquad (38)$$

Besides, it is easy to show that

$$\overline{\cos^2 \varphi \cos^2 \varphi'} = (s_4 - s_{22})\Sigma \beta_i^2 \gamma_i^2 \\ + 4s_{22}\Sigma' \beta_i \beta_j \gamma_i \gamma_j + s_{22}; \qquad (39)$$

The values of this quantity are listed in Table III for different lattice types.

The result is particularly simple in the case of isotropic distribution of the bonds i.e. for a polycrystalline material made up of an assembly of

TABLE III.

Lattice type	Values of $1/9 - \overline{\cos^2 \varphi \cos^2 \varphi'}$
Isotropic bond distribution	$\dfrac{2}{45} - \dfrac{2}{15}\Sigma^2 \beta_i \gamma_i$
Face centred cube	$\dfrac{1}{36} - \dfrac{1}{12}\Sigma \beta_i^2 \gamma_i^2 - \dfrac{4}{12}\Sigma' \beta_i \beta_j \gamma_i \gamma_j$
Body centred cube	$-\dfrac{4}{9}\Sigma' \beta_i \beta_j \gamma_i \gamma_j$
Simple cube	$\dfrac{1}{9} - \dfrac{1}{3}\Sigma \beta_i^2 \gamma_i^2$

randomly oriented crystallites: the anisotropy energy is proportional to the cosine squared of the angle between the existing magnetization direction and the direction of the magnetic field applied during the heat treatment. For single crystals the anisotropy is of a more complicated type.

(11) Effect of concentration: To calculate the effect of concentration when c_b is no longer very small, one can use a result from the beautiful paper of Yvon [6], which is a generalization of Bethe's method concerning the probability n_{BB} of finding two atoms B in ith neighbour position. Restricting the fundamental equations to terms with two nodes, one obtains the approximation

$$n_{BB} = \frac{1 + 2c_b(w_i - 1) - [(1 - 2c_b)^2 + 4c_b(1 - c_b)w_i]^{\frac{1}{2}}}{2(w_i - 1)}.$$

(40)

in which the quantity w_i takes the value

$$w_i = \exp\left\{-\frac{v + l_1\left(\cos^2\varphi - \frac{1}{3}\right)}{kT_1}\right\}, \quad (41)$$

where v is equal to $v_{aa} + v_{bb} - 2v_{ab}$, and v_{aa}, v_{bb} and v_{ab} are the ordinary binding energies of the nearest neighbour atoms in the solid solution, taken pairwise.

In an ideal solution v is zero and, since l_1/kT_1 is small, equation (40) reduces to the following form noted by Fournet [7]

$$n_{BB} = c_b^2 + c_b^2 c_a^2 (w_i - 1). \quad (42)$$

Comparison of this relation with (35), (37) and (38) shows that to account for the influence of the concentration, c_b^2 in equation (38) should be replaced by $c_a^2 c_b^2$.

In a non-ideal solid solution, v is non zero. In this case the term c_b^2 in equation (38) must be replaced by a quantity S equal to the derivative of n_{BB} with respect to $-l_1(\cos^2\varphi - 1/3)/kT_1$ and l_1 then set equal to zero. This gives

$$S = \frac{4c_a^2 c_b^2 \exp\left(-\frac{v}{kT_1}\right)}{\sqrt{1+x}\,(1 + \sqrt{1+x})^2}, \quad (43)$$

with

$$x = 4c_a c_b \left[\exp\left(-\frac{v}{kT_1}\right) - 1\right]. \quad (44)$$

Finally, in a substance with isotropically distributed bonds (by compensation) and where D_u has the form $C_s \cos^2\theta$, the anisotropy constant C_s is given by

$$C_s = \frac{n\,SL_0L_1}{15\,VRT_1} \quad (45)$$

where $L_0 = N_a l_0$, $L_1 = N_a l_1$ and where R is the gas constant. For $v = 0$, S is simply equal to $c_a^2 c_b^2$.

In γ ferronickels, specific heat measurements made on crossing the order–disorder transition indicate than $N_a v$ is positive and approximately equal to 6×10^{10} ergs: v/kT_1 is therefore about 1 and w_i approximately 0.37. Formula (43) then shows that to take into account the non-ideality of the solution, the ideal solution value of S must be multiplied by 0.37 for $c_a = 0$ or $c_b = 0$, by 0.68 for $c_a = 1/4$ or $c_b = 1/4$, and by 0.94 for $c_a = c_b = 1/2$.

(12) Order of magnitude of the anisotropy caused by the orientational superlattice: In the first part of this paper it was shown that the values of $N_a l$ for a pure substance were of the order of 10^8 or 10^9 ergs. The combination $L_{aa} + L_{bb} - 2L_{ab}$ which occurs in the expression for C_s must have a similar magnitude, on the basis of its values L_0 and L_1 at the temperatures T_0 and T_1. Using the values calculated for the ferronickels (c.f. § 5 and § 6) one gets

$$L_{NiNi} + L_{FeFe} - 2L_{FeNi} = 2.0 \times 10^9 \text{ ergs},$$

in confirmation of the above argument. Lastly, for $c_a = 1/2$, one obtains values of C_s of the order of 10^3 to 10^5 ergs/cm^3, with $T_1 = 800$ K.

In ferrocobalts, the magnetostriction goes through a very marked maximum in the neighbourhood of $c_a = 1/2$, which appears to suggest that the value of L_{CoFe} is rather big. Neglecting the values of L_{FeFe} and L_{CoCo} one finds that L_{CoFe} is of the order of $2\lambda VC_3$, and that therefore L_0 and L_1 are of the order of $4\lambda VC_3$. As C_3 is approximately 10^{12}, V about 7 cm^3 and λ about 70×10^{-6}, the values of L_0 are still approximately equal to 2×10^9 ergs.

(13) Orientational superlattices and the problem of Permalloy: It is known that when ferronickels of the Permalloy class are cooled in a magnetic field, they acquire a magnetic anisotropy which can be as high as 2×10^3 ergs/cm^3. Several interpretations of this phenomenon have been proposed. One of these, defended by Kaya [8] postulates the presence of two phases, one disordered, and the other ordered, the latter being the FeNi$_3$ superlattice, with a difference in spontaneous magnetizations ΔJ_s assumed to be of the order of 40 e.m.u. If one of these phases takes on a needle structure oriented in the direction of the magnetic field applied during the heat treatment, then anisotropy of the order of $\pi \Delta J_s^2 V(1-V)$ occurs, where V is the fraction of the total volume occupied by one of the phases: the corresponding maximum uniaxial anisotropy is approximately 10^3 ergs/cm^3. This explanation is open to objection partly because the order of magnitude is rather small, partly because ferronickels also display anisotropy for the composition FeNi, close to which FeNi$_3$ does not form, and also because it is not clear why the FeNi$_3$ phase should grow in the form of needles given that the difference between the parameters of the ordered phase and the disordered phase is very small.

Another interpretation proposed some time ago by Bozorth [9] is based on the idea that at high temperatures plastic deformation relaxes the magnetostrictive constraints corresponding to the imposed direction of magnetization, so that at low temperatures, where such relaxation is no longer possible, energy must be dissipated in order to deflect the magnetization from the initially imposed direction. One can object to this explanation that on the one hand the anisotropy generated by magnetostriction can hardly be as much as 600 ergs/cm^3, and on the other hand, it is difficult to imagine the appearance of constraints in a single crystal which is free to deform in all directions: such monocrystals nevertheless display uniaxial anisotropy. By direct experiment, moreover, Chikazumi [10] has shown that in order to obtain a magnetic anisotropy equal to that observed in fact, the internal strains must be at least 3.4 kg/mm^2, while magnetostriction cannot generate strains of greater than 0.08 kg/mm^2.

A much more fruitful idea is that of *directional order* due to Chikazumi [11] who points out that the Fe–Ni bonds must be shorter than the others, because the formation of the FeNi$_3$ superlattice, which obtains a relatively larger number of such bonds, is associated with a relative contraction of 4×10^{-4}. If the concentration of these bonds is greater in one direction than in another, elastic constraints are generated which can give rise to magnetic anisotropy through strain relaxation in the hot state, as described in the Bozorth mechanism. Unfortunately the author gives no clear indication as to the origin of the forces liable to give rise to directional order and their relation to the direction of the magnetic field present during the cooling. Also, he seems to suggest that directional order acts on the magnetization direction only through the magnetostriction. We shall return to this point in paragraph 21.

We propose an explanation of the uniaxial anisotropy of Permalloys in terms of the formation of an orientational superlattice according to the process described above. The numbers quoted above indicate that the required order of magnitude is amply attained.

It should be mentioned that the orientational superlattice obtained by heat treatment in a magnetic field is quite unrelated to the FeNi$_3$ superlattice of the order–disorder type: these two superlattices have completely different properties and symmetries and are caused by different forces. Quite on the contrary, the presence of the FeNi$_3$ phase hinders the formation of the orientational superlattice, since in this phase all the atoms occupy completely defined positions. This consideration explains the fact, pointed out by Rathenau [12] from the work of Snoek and Smit, that an alloy of composition FeNi$_3$ displays no uniaxial anisotropy if it has been ordered before cooling in a magnetic field. From a kinetic standpoint, the orientational superlattice must appear first, since it is only a short range anisotropic order. The ordinary superlattice, which involves mean- and long-distance order, appears afterwards and reduces the orientational superlattice.

Interesting experiments by Chikazumi, quoted by Kaya [8], confirm our interpretation. The former author investigated a Permalloy crystal disc (with 75% Ni) cut in the (110) plane, and heat treated it in magnetic fields H_1 parallel to different directions contained in the plane of the disc. He found uniaxial anisotropies in the plane of the disc, superimposed on the ordinary magnetocrystalline anisotropy, of the form $-C_0 \cos^2 \theta$, where θ is the angle between the existing magnetization and the field H_1. On applying H_1 along the directions [001], [110], [111], the constant C_0 respectively takes the value 0.46×10^3, 0.69×10^3 and 1.25×10^3 ergs/cm^3. Now the theory described above, and the equations (38) and (45), give for the density of the anisotropy energy D_u in this case

$$D_u = \frac{5}{8} C_s \left(\Sigma \beta_i^2 \gamma_i^2 + 4 \Sigma' \beta_i \beta_j \gamma_i \gamma_j \right). \quad (46)$$

It follows that the theoretical values of C_0 for the three field directions specified above, are respectively proportional to 2, 3 and 4. The agreement with the experimental data is satisfactory, considering the difficulties of the measurement and the approximations of the theory, in which, for example, the effect of second neighbours has been neglected.

(14) Treated Alnicos: It seems reasonable to attribute to the same mechanism the phenomena of orientation by cooling in a magnetic field observed in many series of alloys: Fe–Co, Co–Ni, Fe–Co–Ni, Fe–Si, etc.

It is also possible that *part* of the magnetic anisotropy in Alnicos V caused by an orientational superlattice forming in one of the two phases that occur in this alloy, or in both phases at once. The Fe–Co bonds, whose anisotropy energy appears to be particularly big, are indeed present here. According to the current interpretation [13], the anisotropy is attributed to the precipitation of a second phase, in the form of rods whose axes grow parallel to one of the fourfold axes of the lattice: those rods whose orientation lies closest to the treatment field H_1 grow most. In this interpretation, it is not clear why the treatment with the field H_1 parallel to a three fold axis should generate anisotropy, unless one assumes that the axes of the rods line up with the field H_1. This hypothesis may be satisfactory for Alnicos treated at 800°C in which the precipitated grains have become very big and have certainly lost all coherence with the matrix, as can be seen from the small value of the coercive field: it does not seem to be valid, however, for very small grains, of the order of 40 Å, which must stay rigidly oriented with respect to the matrix. Now, it is just these very small grains which are characteristic of well treated Alnicos. The existence of an additional anisotropy mechanism is therefore probable.

(15) The case of cobalt ferrite: Treatment in a magnetic field in the region of 300 to 400°C imparts to cobalt ferrite, and to ferrites of similar composition, a magnetic anisotropy which can be greater than 10^6 ergs/cm³. On account of its various particularities, this phenomenon is very similar to the preceding ones and differs only in its order of magnitude. It should be pointed out that, from a theoretical standpoint, even pure ferrites can give rise to uniaxial anisotropy and to orientational superlattices, since on the B sites there exist Fe^{+++} and M^{++} ions distributed randomly. Furthermore, in pure ferrites these two kinds of ions occur in equal quantities and, given the lattice type (f.c.c.), do not favour the formation of a superlattice of the ordinary kind, but favour rather the creation of an orientational superlattice, either within the sublattice B, or through interactions between ions located on the sublattice B. We wish to know whether the required order of magnitude can be obtained by the proposed mechanism.

Let us assume for the sake of argument that the orientational phenomena are due only to the sublattice B. The atomic volume which enters the formulae is that occupied by N_a atoms on the B sites, i.e. 21.5 cm³. We assume, as before, that l_{FeFe}, m_{FeFe}, l_{CoCo}, m_{CoCo} are negligible compared with l_{FeCo} and m_{FeCo}. In these conditions, we adopt the values given by Bozorth [14]

$$\lambda_{100} = -512 \times 10^{-6} \quad \text{and} \quad \lambda_{111} = -45 \times 10^{-6},$$

and for the elastic constants $C_2 = 0.5 \times 10^{12}$ and $C_3 = 1 \times 10^{12}$. Substitution of these values into equation (17) gives

$$L_{FeFe} = N_a/FeCo = 2 \times 10^{10} \text{ ergs}.$$

At 700 K, the ratio L/RT will thus be approximately 1/3; moreover, in formula (45) one must now set $L = -2L_{FeCo}$, which gives for the anisotropy constant

$$C_s = 3.2 \times 10^7 \text{ ergs/cm}^3.$$

This is more than is necessary, even if it is supposed that L_1 is three or four times smaller than L_0, as is likely on account of the rise in temperature.

(16) The changes in magnetostriction caused by orientational superlattices: The creation of an orientational superlattice is accompanied by a change in magnetostriction which we shall evaluate. To simplify the nomenclature, we shall restrict ourselves to the case of a cubic substance possessing an orientational superlattice around the fourfold axis Ox, and we shall take the two other rectangular axes Oy and Oz along the two remaining fourfold axes. It should be added that these calculations are also valid for the case of substances which are isotropic by compensation. The superlattice is defined by the probability

$$\frac{1}{n}\left[1 - \frac{l_1}{kT_1}\left(\alpha_1^2 - \frac{1}{3}\right)\right]$$

of finding a given B–B bond with the orientation $\alpha_1, \alpha_2, \alpha_3$. The contribution of this bond to the total magnetoelastic energy, whose deformation is A_{ij} and spontaneous magnetization orientation is $\beta_1, \beta_2, \beta_3$, is obtained by averaging over the n possible positions of the bond the quantity (c.f. § 3)

$$w = -\frac{l_1}{kT_1}\left(\alpha_1^2 - \frac{1}{3}\right)\{2l_0 \Sigma A_{ij}\beta_i\alpha_j \Sigma \alpha_i\beta_i \\ + (m_0 r_0 - 2l_0)\Sigma^2 \alpha_i\beta_i \Sigma A_{ij}\alpha_i\alpha_j\}. \quad (47)$$

The energy density associated with this deformation A_{ij} is obtained by multiplying this mean value of w by the number $c_b^2 nN_a/2V$ of B–B bonds contained in 1 cm³: this is a linear function of the A_{ij}. This must be added to the quadratic form (2) which gives the elastic energy associated with this deformation, and then the sum must be minimized with respect to each of the A_{ij}. In this way the values of the A_{ij} and of the magnetostriction $\delta\lambda/\lambda$ along a direction $\gamma_1, \gamma_2, \gamma_3$ are deduced. The final calculation gives, neglecting terms independent of the orientation, because they are not experimentally observable,

$$\frac{\delta\lambda}{\lambda} = \frac{c_b^2 n L_0 L_1}{2VRT_1}\left[\frac{1}{C_2}(s_4\beta_1^2\gamma_1^2 + s_{22}\beta_2^2\gamma_2^2 + s_{22}\beta_3^2\gamma_3^2) + \frac{1}{C_3}(s_4 + s_{22})\Sigma'\beta_i\beta_j\gamma_i\gamma_j + \frac{1}{C_3}(s_{22} - s_4)\beta_2\beta_3\gamma_2\gamma_3\right]$$

$$+ \frac{nc_b^2 L_1(M_0 r_0 - 2L_0)}{2VRT_1}\left[\frac{1}{2C_2}(s_6 - s_{42})\beta_1^2\gamma_1^2 + \frac{1}{2C_2}(s_{222} - s_{42})(\beta_3^2\gamma_2^2 + \beta_2^2\gamma_3^2)\right.$$

$$\left. + \frac{2}{C_3}s_{42}(\beta_1\beta_2\gamma_1\gamma_2 + \beta_1\beta_3\gamma_1\gamma_3) + \frac{2}{C_3}s_{222}\beta_2\beta_3\gamma_2\gamma_3\right]. \tag{48}$$

In the absence of experimental data, it is superfluous to discuss this general formula. We restrict ourselves to the simple case where the magnetostriction is measured along Oy ($\gamma_1=0$, $\gamma_2=1$, $\gamma_3=0$) as a function of the position of the magnetization in the xOy plane ($\beta_3=0$). One finds simply

$$\frac{\delta\lambda}{\lambda} = \frac{nL_0 L_1 c_b^2 s_{22}}{2VRT_1 C_2}\beta_2^2. \tag{49}$$

In the case of an isotropic substance, s_{22} is equal to 1/15.

This magnetostriction can be expressed in a remarkable way as a function of the anisotropy constant C_s defined above (equation 45). One has

$$\frac{\delta\lambda}{\lambda} = -\frac{C_s}{2C_2}\beta_2^2. \tag{50}$$

Taking for the uniaxial anisotropy energy the large value of 10^6 ergs/cm^3, and taking $2C_2 = 10^{12}$, the maximum magnetostriction is equal only to 10^{-6}. This is completely negligible in comparison with the large magnetostrictions (or the order of 100×10^{-6}) of the solid solutions which acquire significant uniaxial anisotropies.

Thus it seems that the changes in magnetostriction, observed by different authors following thermal treatment in a magnetic field, do not correspond to any deep underlying phenomenon. They are more probably due to the plain fact that in the presence of a uniaxial anisotropy, the initial demagnetized state no longer corresponds to an isotropic distribution of the spontaneous magnetization in the different elementary domains, but rather to a distribution along or opposed to one direction: that of easy magnetization. In these conditions, in the limiting case, the saturation longitudinal magnetostriction vanishes in the easy magnetization direction, and in the perpendicular direction takes a value equal to 1.5 times the initial isotropic state value.

III
ORIENTATIONAL SUPERLATTICES CAUSED BY MECHANICAL DEFORMATIONS

(17) Production of orientational superlattices by deformation: In the second part of this article we showed how in a ferromagnetic solid solution the spontaneous magnetization could create an orientational superlattice, characterized by the anisotropy of the short-range disorder. But such superlattices can be generated by other means, through a homogeneous deformation of the lattice, produced, for example, by traction or compression: the energy of the bonds will then depend on their orientation. This phenomenon must occur to a greater or lesser extent in all solid solutions whether they are ferromagnetic or not, but it is probably too weak to be detected by other means than by the uniaxial anisotropy of the magnetic properties. It is therefore appropriate to investigate it in ferromagnetic substances. If the thermal treatment with deformation is performed in a temperature region where the substance is still ferromagnetic, two orientational superlattices appear simultaneously, one due to deformation, and the other due to spontaneous magnetization, but it is possible to separate them by changing their relative orientation. It is also possible by deformation to create a pure orientational superlattice by carrying out the treatment above the Curie point. However, in order to maintain this superlattice at low temperature by quenching without it being modified by the appearance of the spontaneous magnetization, the Curie point must be lower than the temperature at which diffusion becomes important.

To determine the order of magnitude of this phenomenon, it is essential to know, at least roughly, how the energy of a pair of nearest neighbour atoms in the solid solution depends on the type and the separation of these two atoms.

(18) Brief theory of the elastic properties of a solid solution: Just as in the investigation of the magnetic properties, we shall take the total elastic energy to be the sum of energies of the different bonds between nearest neighbour atoms. We assume that the energy of a single bond w_{ij} depends only on the type of the two atoms and on the distance r separating them. In the narrow range in which r can vary, one can take for a pair composed of one atom of species i and the other of species j:

$$w_{ij} = g_{ij}(r - r_{ij})^2 \qquad (51)$$

If the components A and B of the solid solution have similar properties, the characteristic constants g_{ij} and r_{ij} should not be too different and one can set

$$\left. \begin{array}{l} g_{ij} = g_0 + \delta g_{ij}, \\ r_{ij} = r_0 + \delta r_{ij}, \end{array} \right\} \qquad (52)$$

where δg_{ij} and δr_{ij} are much smaller than g_0 and r_0.

Let us now determine as a function of the atomic concentrations c_i and c_j the variation of the distance r_m between two nearest neighbour atoms. We assume this distance to be the same whatever the type of the two atoms of the pair. We assume, moreover, that the solid solution is ideal, i.e. the atoms are distributed at random. The average energy of a bond can then be written

$$E' = \Sigma c_i c_j g_{ij}(r - r_{ij})^2 \qquad (53)$$

where the summation sign is extended to the values A and B of i and j.

Replacing the r_{ij} and the g_{ij} by their values and minimizing E' gives for the value of r_m desired

$$r_m = r_0 + \Sigma c_i c_j \delta r_{ij} \qquad (54)$$

where we have assumed that the numbers of A–A, A–B and B–B bonds remain proportional to c_A^2, $2c_A c_B$ and c_B^2, i.e. the alloy has been quenched from a very high temperature.

The energy of a bond ij is thus $w_{ij}(r_m)$. When the crystal undergoes a deformation, the distance between two near neighbour atoms becomes $r_m + \delta r$, and the bond energy becomes

$$w_{ij}(r_m) + \frac{\partial w_{ij}}{\partial r} \delta r.$$

Neglecting terms of second order, one gets

$$\frac{\partial w_{ij}}{\partial r} = 2g_{ij}(r - r_{ij}) = 2g_0(\Sigma c_i c_j r_{ij} - \delta r_{ij}). \qquad (55)$$

However, for the rearrangements induced by the orientational superlattice, it is not the quantity w_{ij} alone which is of interest, but rather the combination

$$w = w_{aa} + w_{bb} - 2w_{ab},$$

whose derivative with respect to r, Q is

$$Q = \frac{\partial w}{\partial r} = -2g_0(\delta r_{aa} + \delta r_{bb} - 2\delta r_{ab}). \qquad (56)$$

For the sake of brevity we set

$$\mathbf{D_0} = \delta r_{aa} + \delta r_{bb} - 2\delta r_{ab}. \qquad (57)$$

The quantity D_0 is very simple to interpret. The variation of r_m as a function of concentration is represented in Fig. 1 by the parabola ACB. If the alloy obeyed Vegard's Law, the straight line AB would be obtained. For the equiatomic alloy ($c_b = 1/2$), the difference CC″ between the straight line and the curve is equal to $D_0/4$. The quantity D_0 is thus a measure of the deviation from Vegard's law.

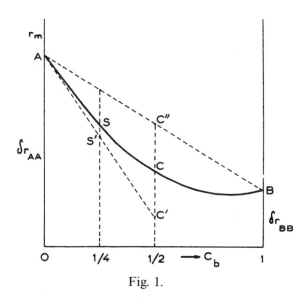

Fig. 1.

In certain alloys such as the γ ferronickels, only part of the curve AB is experimentally observable: the tangent at the origin AC′ can then be drawn. The deviation CC′ of the curve from the tangent at the concentration $c_b = 1/2$ is also equal to $D_0/4$. Application to the ferronickels, on the nickel-rich side, gives $D_0 = r_0/150$, according to the data collected by Bozorth [15].

This value is confirmed by a completely independent method by the observations of Chikazumi [10], who found by dilatometry that the transition of FeNi$_3$ ($c_b = 1/4$) from the disordered to the ordered state was accompanied by a relative contraction of 4×10^{-4}.

Since the FeNi$_3$ superlattice contains no Fe–Fe bonds, this contraction is represented by S'S in Fig. 1, where S' and S are the intersections of the ordinate $c_b = 1/4$ with the tangent AC' and the curve AB. Theoretically, S'S = D$_0$/16, which gives here D$_0$ = r_0/156, in complete agreement with the above value.

As for the factor g_0, it is directly related to the compressibility coefficient χ_0. The elastic energy stored in 1 cm^3 in an isotropic expansion δr is equal, on the one hand to $(3\delta r/r_0)^2/2\chi_0$, and on the other to $nN_a g_0 \delta r^2/2V$, since the number of bonds per cubic centimetre is equal to $nN_a/2V$. This gives

$$\frac{nN_a g_0}{2V} = \frac{9}{2\chi_0 r_0^2}. \quad (58)$$

For iron, nickel, cobalt and their alloys, χ_0, which is equal to $3/(3C_1 + 2C_2)$ is approximately 0.5×10^{-12}; $nN_a g_0/2V$ is therefore about 1.5×10^{28} ergs/cm^3.

(19) Determination of the orientational superlattice: Consider a single crystal of a solid solution with two constituents A and B such that the concentration c_b of the B component is small, and having a certain B–B bond with orientation $\alpha_1, \alpha_2, \alpha_3$. In the absence of deformation, this bond is oriented at random and the average value of α_i^2 is equal to 1/3.

Now let us apply to the crystal a simple traction P, oriented in the direction $\beta_1, \beta_2, \beta_3$. Starting with relation (2), a straightforward elasticity calculation shows that components of the deformation tensor take the values

$$A_{ii} = \frac{P\beta_i^2}{2C_2} - \frac{C_1 P}{2C_2(3C_1 + 2C_2)}, \quad A_{ij} = \frac{P\beta_i \beta_j}{2C_3}. \quad (59)$$

The variation δr in the interatomic distance corresponding to the bond of orientation α is given by the following expression, in which a constant independent of α has been added so as to make the average value of δr_α equal to zero (this simplifies the subsequent calculations without sacrificing rigour):

$$\delta r_\alpha = \frac{Pr_0}{2C_2}\left(-\frac{1}{3} + \Sigma \alpha_i^2 \beta_i^2\right) + \frac{Pr_0}{C_3} \Sigma' \alpha_i \alpha_j \beta_i \beta_j. \quad (60)$$

The elastic energy associated with this bond therefore increases by an amount $Q\delta r_\alpha$, according to equations (55) and (56), and the probability of observing this bond in the orientation $\alpha_1, \alpha_2, \alpha_3$ at temperature T$_1$ is proportional to $\exp(-Q\delta r_\alpha/kT_1)$ or to $1 + 2g_0 D_0 \delta r_\alpha/kT_1$, since $Q\delta r_\alpha/kT_1$ is small.

The equilirbium distribution of the atoms in the alloy at the temperature T$_1$ is thus defined by the above probability. By means of a rapid quench to a temperature T$_0$ sufficiently low to stop diffusion, this state can be kept in pseudo equilibrium, when the traction P is removed.

(20) Magnetic anisotropy connected with the orientational superlattice produced by deformation: Let $\gamma_1, \gamma_2, \gamma_3$ be the existing direction of the spontaneous magnetization at temperature T$_0$: the magnetic energy of a bond whose orientation is $\alpha_1, \alpha_2, \alpha_3$ is given by $l_0(\Sigma^2 \alpha_i \gamma_i - 1/3)$. The mean energy w_m of a bond for all possible orientations, taking into account the probability factor calculated above, can be written:

$$w_m = \frac{2Pg_0 D_0 r_0 l_0}{kT_1}\left[\frac{s_4 - s_{22}}{2C_2} \Sigma \beta_i^2 \gamma_i^2 \right.$$
$$\left. + \frac{2s_{22}}{C_3} \Sigma' \beta_i \beta_j \gamma_i \gamma_j\right]. \quad (61)$$

The corresponding energy density is obtained by multiplying w_m by the number $c_b^2 nN_a/2V$ of B–B bonds contained in 1 cm^3.

Applying this formula to the face centred cube ($n = 12$) in the [100] direction, the anisotropy energy density obtained has the form

$$D_u = C'_s \left(\cos^2 \theta - \frac{1}{3}\right), \quad (62)$$

where θ is the angle between the magnetization and the direction of the traction P. The coefficient C'_s is given by

$$C'_s = \frac{3SL_0 PD_0}{8C_2 \chi_0 RT_1 r_0}. \quad (63)$$

For the [111] direction, the result is obtained by replacing 2C$_2$ by C$_3$ in equation (63). In this equation, following paragraph 11, we have replaced c_b^2 by S to make the formula valid in the whole concentration range, S being defined by equation (43).

We are now ready to calculate the order of magnitude of the expected phenomenon. Let us take

$$c_a = c_b = 0.5; \quad v = 0; \quad C_2 = 0.45 \times 10^{12}$$

$$\chi_0 = 0.5 \times 10^{-12}; \quad R = 8.32 \times 10^7;$$

$$T_1 = 800 \text{ K};$$

$$P = 10^9, \text{ i.e. } 10 \text{ kg/mm}$$

$$L_0 = 10^9 \text{ ergs/cm}^3; \quad D_0/r_0 = 1/150.$$

This gives $C'_s = 10^4$ ergs/cm^3.

This effect should therefore easily be detectable.

(21) The Chikazumi effect: In an article which we referred to earlier (§ 13), Chikazumi assumes that directional order is produced by magnetostriction [11]. Now the difference in length between a bond parallel to the spontaneous magnetization direction and a perpendicular bond is equal to $3r_0\lambda_{100}/2$. From the results of paragraph 18 it follows, therefore, that the permutation of two atoms A and B, which causes the switch of an A–B bond from the direction parallel to J_s to a perpendicular direction, is accompanied by a change in energy w'' given by

$$w'' = \pm 3g_0 D_0 \lambda_{100}. \tag{64}$$

In the case of Permalloy at 25% Ni, taking

$$N_a g_0 = 1.5 \times 10^{28}, \quad r_0 = 2.5 \times 10^{-8}$$

$$D_0/r_0 = 1/150 \quad \text{and} \quad \lambda_{100} = 10 \times 10^{-6}$$

one finds that $N_a w''$ is approximately 2×10^{-6} ergs. This energy is small. The magnetic anisotropy energy densities to which it gives rise are given by equation (45), in which L_1 is replaced by $N_a w''$. Since L_1 is equal at least to 10^8 ergs and $N_a w''$ about 2×10^6 ergs, it follows that the Chikazumi effect is negligible compared to the effects that we have investigated in the second part of this article.

(22) Magnetic anisotropy in cold rolled ferronickels: The above theory appears to hold the key to the very remarkable properties of cold-rolled ferronickels, and in particular the results of Rathenau and Snoek [16]. Plates, which are heavily cold-rolled and then recrystallized at high temperature, achieve a quasi single crystal state: the orientations of all the crystallites are practically the same, with the (100) plane in the plane of the plate and the [010] direction in the direction of the rolling. The two directions [010] and [001] are then magnetically equivalent. If now the plate is rolled again, but moderately this time, either in the [010] or in the [001] direction, so that the thickness is reduced by a third, the crystallite orientation does not change but the plate nevertheless acquires a very strong uniaxial magnetic anisotropy such that the direction of easy magnetization lies perpendicular to the direction of rolling.

The magnetic energy density has the form $-K_0 \cos^2\theta$, where θ still denotes the angle between the magnetization and the direction of rolling. The constant $-K_0$, which is zero for pure nickel, increases steadily with the iron concentration, going through a maximum of about 2×10^5 ergs/cm^3 at 50 per cent iron, then drops quickly thereafter. The hypothesis of residual internal stresses does not explain the facts: firstly these internal stresses should be of different signs and orientations and should cancel out on average to give a practically zero total effect; secondly, the effect vanishes for nickel in which the magnetostriction is large, while it is large ($K_0 = -0.4 \times 10^5$ ergs/cm^3) at about 20% iron, just where the magnetostriction vanishes.

We propose an interpretation of these phenomena by assuming an *orientational superlattice* formed by the pressure exerted during the rolling: this pressure must be of the order of 50 to 100 kg/mm^2 — the required order of magnitude is achieved, as shown earlier (§ 20). Plastic deformation plays an essential part here, since it allows the atoms to take up their equilibrium distribution, even though the ordinary temperature of operation is well below the recrystallization temperature. It has been known for a long time that cold working and plastic deformation tend to accelerate the onset of short-range order, as was recalled by Averbach [17] in a recent article.

The assumption of an orientational superlattice allows an interpretation of the essential features of the phenomenon, such as its absence in pure nickel, the maximum at about 50 Ni–50 Fe, and the insensitivity to changes in the magnetostriction with concentration. As far as the form of the K_0 dependence on composition is concerned, it appears that near pure nickel $-K_0$ initially increases more slowly with c_b than is indicated by the formula in $c_a^2 c_b^2$. This may be due to the fact that the pressure required to produce the same rate of plastic deformation is much smaller near the pure metal than near the composition FeNi. It may also be caused by the non-ideality of the solid solution, as was pointed out at the end of paragraph 11.

When a large single crystal is cold-rolled the results are similar to the previous ones and can be interpreted in the same way. The same is also true of cold-drawing an almost single crystal strip: the direction of easy magnetization is always perpendicular to the drawing direction. On the other hand, the phenomena are quite different when polycrystalline strips made of randomly oriented crystallites are used. A second rolling or drawing of such strips produces a uniaxial magnetic anisotropy whose sign is opposite to that obtained with quasi single crystal strips. The direction of easy magnetization is now in the direction of rolling or of drawing.

These very strange results may be explained by a mechanism similar to one proposed, following the ideas of Masing [18], to interpret the process of restoration after creep; this interpretation, incidentally, is identical with that of the Bauschinger effect, and was given previously by Heyn [19]. In a polycrystal, the deformability depends enormously on its orientation and on the orientation of the neighbouring crystallites. Certain crystallites A, which have a particularly high elastic limit, undergo only an elastic deformation during the overall plastic deformation. On the other hand, other crystallites B have a low elastic limit and are subjected to a large plastic deformation which gives rise to an orientational superstructure. When the system of external stresses is relaxed, the A crystallites tend to take up their original shape again and subject the neighbouring B crystallites to a plastic deformation of opposite sign to that produced by the system of external stresses applied previously. This second plastic deformation undergoes the original orientational superlattice, and produces a second, complementary as it were to the first, which is accompanied by a magnetic anisotropy of opposite sign to the initial anisotropy.

The main features of the phenomena observed by Rathenau and Snoek thus seem to be explained. As far as the detailed experimental results are concerned, it is unjustified to try to analyze them in the absence of data on the mechanical forces applied during the rolling or drawing, and until experiments of this kind are performed again under better defined conditions, in particular concerning the orientation of the magnetization during the plastic deformation: it was in fact shown above (§ 17) that the formation of orientational superlattices is determined not only by the orientation of the spontaneous magnetization but also by the system of mechanical stresses applied.

REFERENCES

[1] C. Kittel, *Rev. Mod. Physics* 1949, **21**, 541.
[2] L. Neel, *C.R. Acad. Sci.* 1953, **237**, 1613.
[3] R. Becker, *Z. Physik*, 1930, **62**, 253.
[4] R. M. Bozorth, *Rev. Mod. Physics*, 1953, **25**, 42.
[5] K. Kronenberg, Private communication.
[6] J. Yvon, *Cahiers de Physique*, 1945, **28**, 1.
[7] G. Fournet, *J. Physique Rad.* 1953, **14**, 374.
[8] S. Kaya, *Rev. Mod. Physics*, 1953, **25**, 49.
[9] R. M. Bozorth and J. F. Dillinger, *Physics*, 1935, **6**, 285.
[10] S. Chikazumi, *J. Phys. Soc. Japan*, 1950, **5**, 327.
[11] S. Chikazumi, *J. Phys. Soc. Japan*, 1950, **5**, 333.
[[2] G. W. Rathenau, *Rev. Mod. Physics*, 1953, **25**, 55.
[13] R. D. Heidenreich and E. A. Nesbitt, *J. Appl. Phys.*, 1952, **23**, 352; E. A. Nesbitt and R. D. Heidenreich, *J. Appl. Phys.*, 1952, **23**, 366.
[14] R. M. Bozorth and J. G. Walker, *Phys. Rev.*, 1952, **88**, 1209.
[15] R. M. Bozorth, Ferromagnetism, D. Van Nostrant Co, New York, 1951.
[16] G. W. Rathenau and J. L. Snoek, *Physica*, 1941, **8**, 555.
[17] B. L. Averbach, A Seminar on the cold working of Metals, *Amer. Soc. Metals*, Cleveland, Ohio, 1949, p. 262.
[18] G. Masing, *Wiss. Siemens Konz.*, 1924, **3**, 231; 1925, **4**, 244; 1925, **5**, 135.
[19] E. Heyn, *Festband Kaiser Wilhelm Gesell.*, 1921, p. 134; *J. Inst. Metals*, 1914, **12**, 3.

A119 (1962)

FORMATION OF AN ORDERED FeNi STRUCTURE BY IRRADIATION WITH NEUTRONS
(with J. PAULEVE, D. DAUTREPPE and J. LAUGIER)

Note (1977): The formation of the FeNi superlattice, described here and unknown until that time, is due to the increase in the number of vacancies in the crystal lattice and to the accompanying increase in diffusion and in ordering rate of the atoms. Irradiation with electrons produces similar effects (c.f. article A-123 and patent B-4, not reproduced here).
 Irradiation of an initially cubic crystal subdivides it into quadratic crystallites. The effects of this subdivision are investigated in article A127 (Ch. VI).

Abstract: Neutron irradiation of 50% iron–nickel in a magnetic field gives rise to very strong magnetic anisotropy due to the formation of an FeNi structure of AuCu type. The critical temperature of the corresponding order–disorder transformation is approximately 330°C. The ordering is sensitive to the direction of the magnetic field.

The first experiments on irradiation of polycrystalline alloys of iron–nickel by neutrons in a magnetic field at room temperature [1] revealed a uniaxial magnetic anisotropy whose appearance we interpreted as the effect of the formation of an orientational superlattice [2].

Further irradiation experiments in a large range of temperature show that, for the case of the 50% iron–nickel alloy, the phenomenon is more complex. Figures 1, 2 and 3 show the observed variation of the anisotropy energy E at various temperatures as a function of the total flux, with polycrystalline samples. The irradiation was carried out in the swimming pool reactor *Mélusine*. The flux shown in the total fast neutron flux of energy greater that 1 MeV. It amounted to about 15% of the total neutron flux.

The anisotropy produced is always uniaxial and the direction of easy magnetization coincides with that of the magnetic field applied during the irradiation. At 300°C following an irradiation of 1.7 nvt (fast) it rises to 2×10^6 ergs/cm^3; this value is incompatible with an orientational superlattice.

The X-ray analysis of this sample, irradiated at 300°C revealed the presence in the crystallites of a superlattice of the AuCu (L 10) type. No quadratic deformation or parameter variation is observed with respect to the initial lattice, to within 5×10^{-4}.

Formation of long range order in the iron and nickel atoms has thus occurred. Examination of the superlattice lines of the crystallites reveals the presence of (100) and (110) planes of iron and nickel which are roughly perpendicular to the magnetic field of the treatment. In view of the observed macroscopic

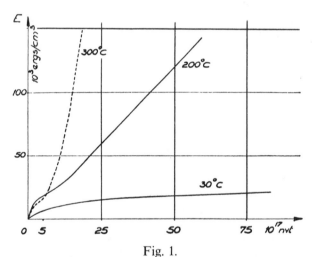

Fig. 1.
Anisotropy energy at 30 and 200°C as a function of the integrated fast neutron flux.

magnetic anisotropy, one must admit that as a result of the magnetic interactions between the bonds and the magnetic field, the ordered domains do not form randomly. In particular, for crystallites whose [100] or [110] directions are roughly parallel to the magnetization imposed by the treating field, the order preferentially forms alternate planes of iron and nickel perpendicular to these axes. It may be concluded that in an ordered single crystal the directions of easy magnetization will be [100], [0$\bar{1}$1], and [011]. These three directions, which, in the ordered structure, are perpendicular to alternate planes of iron and nickel,

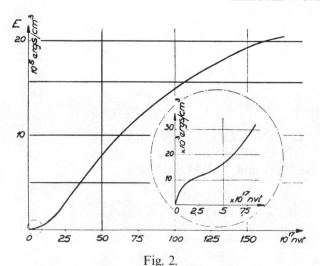

Fig. 2.
Anisotropy energy at 300°C as a function of the integrated fast neutron flux.

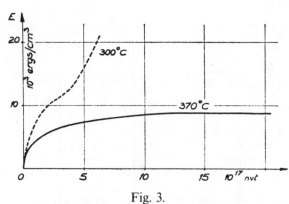

Fig. 3.
Anisotropy energy at 370°C as a function of the integrated fast neutron flux.

are highly differentiated from the other directions. The [0$\bar{1}$1] and [011] directions are of course equivalent.

In a polycrystal, in spite of the presence of these three directions of easy magnetization, of which two are oriented randomly in a plane roughly perpendicular to the treatment field, the field direction remains macroscopically the only easy magnetization direction. Investigations on single crystals are currently in progress to determine the anisotropy energies associated with the different axes.

The appearance of a strong anisotropy in the crystallites is of course accompanied by a strong increase in the coercive field, which rises above 40 Oe, instead of 0.5 Oe in the unirradiated sample, and a large drop in permeability. The values of the saturation magnetization, measured [3] in a field of 30,000 Oe, are not noticeably affected. In contrast, on account of the variation in the directions of easy magnetization from one crystallite to another, saturation is reached only at fields of 6,000 Oe parallel to the treatment field, and 15,000 Oe in the perpendicular direction. The appearance of an ordered state at 300°C is confirmed by measurement of the electrical resistance inside the reactor which shows a sharp decrease characteristic of the onset of order. These investigations of kinetics are still under way, but the results are already quite clear.

Examination of the curves of the onset of the magnetic anisotropy E for different temperatures shows that the strong anisotropy due to ordering no longer occurs above 330°C, which would correspond to the critical temperature of this order–disorder transformation. At the lowest temperatures, the irradiation times were insufficient to reach very high values of E, but a sample having been formed with a strong anisotropy at 300°C, then irradiated at a lower temperature, always gives an increase in E; this shows that the degree of ordering permissible at low temperature has not yet been reached, which is in agreement with the theories of ordering.

Above 330°C, a limiting value of E is quickly reached (Fig. 3). This anisotropy may be due to short range order or to an orientational superstructure; the values extrapolated from the experiments of Ferguson [1] are in fact compatible with the observed values.

The rapid increase in E apparent at the start of the irradiation and below the critical temperature might be explained in the same way, with the long range order setting in later.

We point out that Dekhtyar and Kazantseva [4] have already shown for a 50% iron–nickel alloy around 300–360°C the presence of anomalies in permeability, coercive field and saturation magnetization which they attributed to short range order.

In conclusion, activation of the diffusion process by irradiation, already used by several authors for the investigation of order–disorder transformations [5], [6], has revealed a new order–disorder transformation which has been hitherto unknown, and which cannot be induced by heat treatment on account of its low critical temperature.

In addition, during the ordering process, the controlling influence of the magnetic field is observed to discriminate between different planes which are crystallographically identical.

REFERENCES

[1] J. Pauleve and D. Dautreppe, *Comptes rendus*, 1960, **250**, 3804.
[2] L. Neel, *J. Phys. Rad.*, 1954, **15**, 225.
[3] R. Pauthenet, Private communication.
[4] M. V. Dekhtyar and N. M. Kazantseva, *Fiz. Metal. Metalloved*, 1959, **8**, 412.
[5] Dienes and Damask, *J. Appl. Phys.*, 1958, **29**, 1713.
[6] R. H. Kernohan and M. S. Wechsler, *J. Phys. Chem. Solids*, 1961, **18**, 175.

Chapter XIII

ROCKS AND BAKED CLAY
A69, C27

Articles A101, A102 and A104, which deal with the reversal of the thermoremanent magnetization of rocks due to intrinsic magnetic properties of the material, are not included here. In C27 some references are made to this mechanism which only applies in quite exceptional circumstances for the observed reversals. We now know that magnetic reversals in rocks are almost always due to true reversals of the Earth's magnetic field.

A69 (1949)

THEORY OF THE MAGNETIC AFTER-EFFECT IN FERROMAGNETICS IN THE FORM OF SMALL PARTICLES, WITH APPLICATIONS TO BAKED CLAYS

Summary: In the introduction (§ 1 and § 2) are recalled the remarkable magnetic properties of bricks, pottery and lavas in weak fields, of the same order of magnitude as the Earth's field, which have been highlighted by the experiments of F. Thellier. Notable are the additivity of the partial thermoremanent magnetizations (§ 4) and the very great stability with respect to demagnetizing forces of the thermoremanent magnetization compared with the isothermal remanent magnetization (§ 3). It is proposed to attribute these properties to the subdivision of the constituents into independent particles, sufficiently small for each one to constitute a single elementary domain.

In the first part, the magnetic properties of an isolated small particle as given by the classical theory are briefly exposed. Then the possibility and the consequences of spontaneous fluctuations of the magnetic moment from one easy direction of magnetization to another are considered. These arise from the influence of thermal fluctuations in magnetic fields well below the critical field (§§7–10).

In the second part the mechanism for these fluctuations is studied (§§11–13) and it is shown that they are due to parturbing couples of magnetoelastic (§§ 14 and 15) or purely magnetic origin (§ 16) produced by the thermal elastic deformations of the particles. The relaxation time is thence calculated (§§ 17 and 19) as a function of the particle volume, the temperature and the magnetic field. That is the time after which thermal equilibrium is established. The calculations are applied to fine particles of iron (§ 18).

Next, in the third part, are examined the magnetic properties of an ensemble of independent particles of different volumes and critical fields, with relaxation times spread over a broad time scale (§§ 20–24, 29, 30). In particular, the acquisition of isothermal remanent magnetization (§§ 26, 28), thermoremanent magnetization (§ 38) and their disappearance under the influence of time (§ 27), temperature (§§ 31, 32, 42) and magnetic field (§ 34) are examined. The existence of a magnetic after-effect proportional to the logarithm of time and losses in alternating fields given by a Jordan term (§§ 35–37) are demonstrated. All the results of Thellier are qualitatively interpreted, in particular the differences in stability of the remanent magnetizations (§§ 33, 39), the saturation of the thermoremanence (§ 40) the laws of additivity (§ 41) etc. Finally, a theoretical relation between the decrease in remanent magnetization under the influence of time and under the influence of annealing is verified experimentally, and the order of magnitude of the volume and critical field of particles which play a role in the above phenomena is given (§§ 43 and 44).

INTRODUCTION

(1) Magnetic properties of baked clay and lavas in weak magnetic fields: It has long been known that bricks and certain rocks such as lavas are likely to acquire a relatively strong and extremely stable permanent magnetization on cooling in a weak magnetic field, of the same order of magnitude as the Earth's field. The study of this permanent magnetization permits at any time after the event the determination and intensity of the magnetic field that produced it. Hence there follow numerous possible geophysical or historical applications, seeking for example the values of the Earth's field in the past or, conversely, dating baked clay. We are concerned with a sort of high temperature magnetization like the one we have referred to in an earlier paper [1] in connection with stable substances with a large coercive field. A detailed study of these phenomena in baked clay has allowed E. Thellier [2] to work out a very curious series of laws which seem to be fairly general. These results have been extended to synthetic rhombohedral iron sesquioxide by J. Roquet [3] and to lavas by T. Nagata [4].

Before analyzing these properties, it is useful briefly to summarize the chemical nature of the substances in which they have been observed. In the first place, there is a sesquioxide αFe_2O_3 obtained by precipitation from a solution of iron chloride by ammonia. Then there are lavas, the magnetic constituents of which are iron sesquioxide, magnetite and titanomagnetites in various proportions depending on the volcano of origin. Finally there are baked clays: bricks, pottery etc. Using clay from Nontron which contains about 7% Fe_2O_3 as raw material, the products fired in an oxidizing atmosphere (air) at any temperature or in an inert atmosphere (nitrogen) at temperatures below 750°C show a Curie point at 675°C which shows that at least a part of their magnetic properties arises from αFe_2O_3. However, the study of the variations of the remanent magnetization produced by successive refiring at increasing temperatures also reveals the presence of magnetite.

Finally, the clay baked in an inert atmosphere above 800°C or baked in a reducing atmosphere (town gas) only shows the Curie point of magnetite, and therefore contains no rhombohedral iron sesquioxide.

(2) Remanent magnetization of baked clays: There exist two processes for importing a remanent magnetization in zero field to a ferromagnetic substance at a temperature T.

(A) It may be cooled down from its Curie point θ to the temperature T_0 in a magnetic field h, and then the field cut off. In this way, a *thermoremanent magnetization* (TRM) is obtained, following the terminology of Thellier.

(B) It is also possible to apply a field h and then to cut if off while maintaining the temperature at T_0. In this way an *isothermal remanent* magnetization (IRM) is obtained.

These two remanent magnetizations are functions of h with very different dependences, as shown in Fig. 1, drawn following Thellier for a clay baked in an inert atmosphere at 870°C. In particular, the TRM tends to a limit in a field of about one hundred Oersteds. For the same field, the TRM is much larger than the IRM. Miss Roquet has obtained similar results for αFe_2O_3.

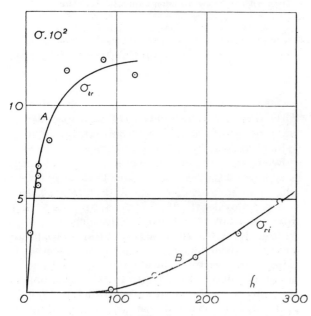

Fig. 1.
Variation of the thermoremanent magnetization σ_{tr} and the isothermal remanent magnetization σ_{ri} of a clay baked in an inert atmosphere at 870°C as a function of the field h which has produced them. (From the data of Thellier.)

(3) Stability of the TRM and IRM: The stability of the two types of magnetization is very different. The TRM is much more stable than the IRM.

Stability in time — Consider first the stability in time. Experiment shows that the TRM imparted to baked clay remains invariant, even after several months. It is stable. On the contrary, when an initially unmagnetized baked clay is given a certain IRM, this magnetization decays with time. The analysis of Thellier's results shows that over a very broad interval of time it decreases in proportion to the logarithm of time (Fig. 2). The time at the end of magnetization is taken as the origin.

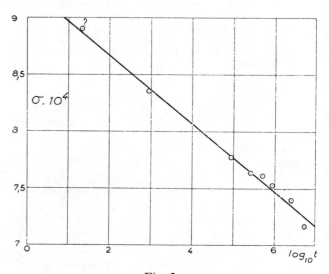

Fig. 2.
Time decay of the isothermal remanent magnetization of a clay baked in an inert atmosphere at 870°C. (From the data of Thellier.)

Thermal stability — Let us now examine the influence of raising the temperature. For that, starting from T_0 with a sample provided with a certain remanent magnetization σ_r, let us successively refire it at increasing temperatures $T_1, T_2, \ldots, T_i$ bring it back each time between firings to the temperature T_0 and measure the remanent magnetization. Plotting the remanent magnetization measured at T_0 as a function of T_i the two curves in Fig. 3 are obtained, one for the TRM, the other for the IRM. In relative terms, the temperature dependence of the TRM is seen to be much less than that of the IRM. So long as the refiring temperature is not too high, the former remains roughly constant whereas the latter decreases linearly with T_i. The curve is drawn for a clay baked in an inert atmosphere (N_2 at 870°C), but Miss Roquet has obtained analogous results for synthetic αFe_2O_3.

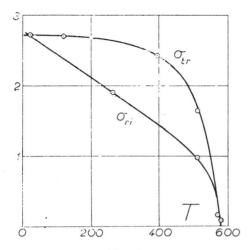

Fig. 3.
Comparison of the variations of thermoremanent magnetization σ_{tr} and isothermal remanent magnetization σ_{ri} of a clay baked in an inert atmosphere at 870°C as a function of refiring temperature. (From the data of Thellier.)

(4) Properties of partial thermoremanent magnetizations: The TRM which we were discussing in the preceding paragraphs is the thermoremanent magnetization $\sigma(T_0 h\theta)$ acquired by the substance when the field h is acting *throughout* the cooling. But *partial thermoremanent magnetization*, abbreviated as PTRM may also be considered. It is acquired at the temperature T_0 when the field h acts from the Curie point θ down to the temperature T_1 and is then suppressed between the temperatures T_1 and T_0. We denote it by $\sigma(T_1 h\theta)$. We can also consider anoterh PTRM denoted as $\sigma(T_0 h T_1)$ which is acquired during cooling when the field is zero in the interval θ, T_1 and equal to h in the interval T_1, T_0. For the substances mentioned in paragraph 1, these two PTRMs obey the Thellier relation

$$\sigma(T_0 h\theta) = \sigma(T_0 h T_1) + \sigma(T_1 h\theta). \qquad (1)$$

Furthermore, E. Thellier has shown that on refiring a sample possessing the complete TRM to T_1 in zero field, the sample only possesses a magnetization equal to $\sigma(T_1, h, \theta)$ once it is brought back to the temperature T_0.

In an even more general way, E. Thellier has established that thermoremanent magnetizations display a series of remarkable properties which are best described in this authors own words [2]. "For a given sample and a constant magnetic field h, in any interval of temperature

$$T_1, T_2 (T_0 < T_1 < T_2 < \theta)$$

there is associated a certain magnetic moment which is acquired by the body when, on cooling in the field h, it passes through the interval T_1, T_2. This moment is oriented in the same direction as h with respect to the body. It is insensitive to any reheating at temperatures less than T_1 and it disappears completely on reheating to the temperature T_2. Furthermore, it is independent of other thermoremanent moments acquired in intervals of temperature outside T_1, T_2, moments which may moreover be due to fields different from h in magnitude and direction. The effects of all the moments add geometrically but, paradoxical as it may seem, each one of them preserves a real independence and, in some sense, an exact memory of the temperatures and the field which have defined it."

Generality of the properties — These remarkable properties belong to a whole series of substances which differ profoundly one from another from a chemical point of view. They are therefore not specific to any one substance, but are rather related to a particular physical state, probably the finely-divided state of the magnetic components of the substances which has already been demonstrated by the large value of the coercive field [1].

We assume that these ferromagnetic components are formed of particles sufficiently small to behave like single elementary domains in the sense of P. Weiss.

It is fitting to begin by setting out the magnetic properties of such an isolated particle.

FIRST PART
MAGNETIC PROPERTIES OF AN ISOLATED SMALL PARTICLE

(5) Spontaneous magnetization and free energy: At a given temperature, the magnitude m_0 of the magnetic moment $\vec{m}$ of a particle G depends only on its volume v, since we suppose that it behave as a single elementary domain, then

$$m_0 = v \mathcal{J} \qquad (2)$$

where $\mathcal{J}$ denotes the *spontaneous magnetization*. It is a function of temperature which depends only on the nature of the material making up the particle and which goes to zero at the Curie point θ. So long as one is not too close to θ, the variation of $\mathcal{J}$ remains fairly insignificant. In a general way, we will denote by $\mathcal{J}_i$ the value of $\mathcal{J}$ at the temperature T_i.

Because of the existence of the shape dependent demagnetizing field and magnetocrystalline coupling, the energy E_a of a particle in the absence of an external field depends on the orientation of the magnetic moment $\vec{m}$ with respect to fixed axes associated with the particle. To simplify, we suppose that the particle possesses rotation symmetry with respect to an axis $\bar{A}$, and we write the energy in the form

$$E_a = -\frac{v\mathcal{J}H}{2}\cos^2\theta, \qquad (3)$$

where H is a positive constant and θ is the angle between $\vec{m}$ and $\bar{A}$. The constant H depends on the shape of the particle and on the anisotropy constant K. For our purposes, it is enough to know that it is a function of the temperature which goes to zero at the Curie point. We denote its value at temperature T_i by H_i.

In the absence of an external field, the moment $\vec{m}$ takes up a position which corresponds to a minimum of E_a. It orients itself along the axis $\bar{A}$, either in one sense or the opposite sense.

(6) Influence of a magnetic field: Let us now apply to the grain G a magnetic field h of variable intensity but constant direction. The component of the magnetic moment of the particle m in the direction of h varies in a way which depends essentially on the direction of h with respect to the rotation axis $\bar{A}$.

(1) *The applied field is parallel to $\bar{A}$* — The observed moment varies with h following a rectangular hysteresis loop shown in Fig. 4, composed of two horizontal arms with ordinates $+m_0$ and $-m_0$ with discontinuities at the abscissae $h=H$ and $h=-H$. The constant H is therefore the *critical field* which produces the rotation of the magnetization.

(2) *The applied field is perpendicular to $\bar{A}$* — The observed moment then varies linearly with field according to the formula

$$m = \frac{v\mathcal{J}h}{H}, \qquad (4)$$

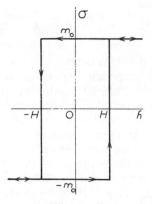

Fig. 4.
Hysteresis loop of an isolated particle whose axis of revolution is parallel to the applied field.

provided the field h lies between $-H$ and $+H$. Outside these limits, m is equal to the saturation value m_0.

Hence, so long as the modulus of the applied field is less than H, the magnetic moment remains invariant in the first case, and varies in proportion to the field in the second case.

(7) Influence of thermal fluctuations: The properties which we have just described schematically apply to particles for which the magnetic moment $\vec{m}$, once oriented in a certain sense along a special direction, conserves the orientation indefinitely so long as it is undisturbed by an external magnetic field.

In reality, for very small particles, the situation cannot remain like this because thermal fluctuations will disturb the initial distribution. When kT is of the same order of magnitude as the energy $VH\mathcal{J}/2$ needed to reach the top of the potential barrier which separates the two equilibrium positions of the magnetic moment, the moment can flip spontaneously so that thermal equilibrium is rapidly established. If, for example, the external field is zero, the probabilities of finding the moment $\vec{m}$ directed in one sense or the other along $\bar{A}$ become equal after a certain time, independently of the initial conditions. If the particle only contains a few atoms, the interval of time which separates two flips should be of order of magnitude of the period of oscillation of atoms in solids, but it must increase very rapidly with the particle dimensions.

(8) The relaxation time: Following these spontaneous flips, the expectation value $\bar{m}$ of the projection of the magnetic moment of a particle on the rotation axis $\bar{A}$ in zero field at time t is written

$$\bar{m} = m_0 e^{-\frac{t}{\tau_0}} \qquad (5)$$

where m_0 is the initial value of m at time $t=0$, the saturation value for example. τ_0 is a constant which we will call the *relaxation time*.

For the observer, the properties of the grain G depend essentially on the value of the relaxation time τ_0 in the experimental conditions compared with the *duration* of the experiment. It is useful to distinguish several important cases.

(1) *The relaxation time is very long compared with the duration of possible experiments* — We are then in the conditions of the classical theory, and we have described the properties of the grain G in paragraph 6. In particular, after application of a magnetic field h parallel to $\bar{A}$ and smaller in magnitude than the value of the critical field H, the initial state of magnetization of the grain G is unchanged.

(2) *The relaxation time is very short compared with the duration of possible experiments* — In this case, the

particle achieves the state of thermodynamic equilibrium completely. Let us suppose for example that the applied field h is parallel to the direction of the particle's axis. The magnetic moment of the particle possesses two possible equilibrium positions, one along the direction of the field, the other in the opposite direction. Applying Boltzmann statistics, the average magnetic moment $\bar{m}$ of the particle is given by

$$\bar{m} = m_0 \, \text{th} \, \frac{m_0 h}{kT} \qquad (6)$$

To write this equation, we have neglected the small thermal oscillations of the magnetic moment about its two equilibrium positions. This approximation is legitimate in the case which interests us[1]. Forumla (6) shows that the average magnetization of the particle depends on the field h in a reversible manner. The hysteresis has disappeared.

(3) *In the case where the applied field is perpendicular to the particle axis* the phenomena, already reversible when the relaxation time was long compared with the duration of the experiment, are not modified when the relaxation time becomes shorter than the duration of the experiment.

(9) The rôle of the temperature: Let us consider a particle G with its axis $\bar{A}$ parallel to the field h whose relaxation time at room temperature is much longer than possible experimental times. Theory shows what is already fairly obvious, that the relaxation time decreases continually as the temperature is increased from zero to the Curie point. There exists therefore a temperature T_r, necessarily less than the Curie temperature, at which the relaxation time of the particle under consideration becomes of the same order of magnitude as the experimental time. We call T_r the *relaxation temperature*.

Strictly speaking, it is really a finite interval of temperature rather than one particular temperature since the duration of possible experiments may vary from a few seconds to a few hours, but the relaxation time varies very rapidly with temperature so this interval is very narrow. Hence, it is not just convenient, but quite legitimate to speak of a relaxation temperature without specifying the length of the experiment to which it is related.

The relaxation temperature T_r plays an extremely important rôle because, above this temperature the particle is always in magnetic equilibrium in the field h, whereas below it the influence of the same field is unable to modify the initial state[2].

(10) The mechanism of magnetization of a particle: Suppose for example that we cool a particle in the magnetic field h, possibly varying with temperature, from the Curie point to room temperature. So long as the temperature T remains greater than T_r, there is magnetic equilibrium in the field h, and the magnetic moment of the particle is given by formula (6) where m_0 and h must be considered as functions of temperature. As soon as the temperature decreases below T_r, flips of the magnetization becomes impossible. In each particle the magnetic moment keeps the orientation it possessed at the temperature T_r. The hyperbolic tangent which appears in formula (6) and which measures the probability of finding the moment oriented in one sense or the other, therefore retains the value it has taken at the temperature T_r. From that instant, the only variation of moment that may be observed comes from the temperature dependence of the spontaneous magnetization. Hence, at a temperature T, less than T_r, the average magnetic moment is a reversible function of temperature given by the formula

$$\bar{m} = v \mathcal{J} \, \text{th} \, \frac{v \mathcal{J}_r h_r}{kT_r}, \qquad (7)$$

in which $\mathcal{J}$ and $\mathcal{J}_r$ represent the values of the spontaneous magnetization at temperature T and T_r respectively, and where h_r is equal to the value of the applied field at the instant when the particle crosses the relaxation temperature T_r.

To produce irreversible variations in the magnetic moment of the particle, it is therefore essential to heat it to a temperature equal to or greater than the relaxation temperature.

This proposal is in apparent contradiction with the classical theory of the magnetic properties of the particle G, set out above, according to which irreversible changes are obtained at a given temperature T by applying a field h greater than the critical field H, without there being any need to raise the temperature of the particle. This contradiction is easily resolved, because, as we will see below, the relaxation temperature depends on the value of applied field and tends to absolute zero as the field h approaches the critical field H. It follows that when h approaches H, the relaxation temperature T_r decreases below T, the temperature of the experiment.

What has just been said about the properties of a

---

[1] An elementary statistical mechanical calculation based on the formulae of paragraph 11 shows that the true moment is obtained by multiplying the expression for m given in (6) by $(1 - kT/2v \mathcal{J} H)$. Now, according to formula (48), $kT/2v \mathcal{J} H$ is equal to $1/4(Q + \log \tau)$. Since this quantity is of order one percent according to the numerical estimates in paragraph 44, it is permissible to neglect the term.

[2] It should be noted here that the relaxation temperature is a function of the field h (cf. § 10, at the end).

particle G shows that it should be possible to explain the remarkable properties of baked clays by considering them as ensembles of particles with different relaxation temperatures. Nevertheless, before going any further, it seems essential to develop the calculation of the relaxation time as a function of the various parameters which define the particle. This will be the aim of the second part of this dissertation.

Readers who are happy to accept the formulae given at the end of the second part may go straight onto the third part, where we specify the properties of an ensemble of particles like G.

SECOND PART
DETERMINATION OF THE RELAXATION TIME OF A PARTICLE

(11) Equilibrium distribution of the magnetic moment: Let a particle of volume v be sufficiently small to constitute a single elementary domain. All the spins are rigidly coupled together and give a resultant of constant magnitude equal to $v\,\mathcal{J}$, where $\mathcal{J}$ denotes the value of the spontaneous magnetization at the temperature under consideration. As a result of the existence of the shape dependent demagnetizing field and the coupling forces between the magnetic moment and the crystal lattice, the energy of the particle depends on the orientation of the moment with respect to the axes associated with the particle. In the simplest case, of rotation symmetry about an axis Ox, the term E_a in the energy of the system which depends on the orientation of the moment is written, for 1 cm^3, in the form

$$E_a = B \sin^2 \theta \qquad (8)$$

where B is a positive constant and θ the angle of moment with Ox.

We consider the particle in the rectangular coordinate system $Oxyz$; the direction Ox or the opposite direction Ox' corresponding to the minimum energy is called the easy magnetization direction. It is the direction adopted by the magnetic moment at absolute zero. At temperature T, when thermodynamic equilibrium is attained in the absence of external forces, the probability $g(\theta)d\theta$ of finding θ lying between θ and $\theta + d\theta$ is given by the Boltzmann formula

$$g(\theta)\,d\theta = C \exp \left\{ -\frac{Bv \sin^2 \theta}{kT} \right\} \sin\theta\, d\theta \qquad (9)$$

in which the coefficient C is determined by writing that the probability of finding θ lying between 0 and π is 1.

We limit ourselves to the cases in which Bv/kT is sufficiently large for the directions occupied by the magnetic moment to be concentrated in small solid angles centred on Ox and Ox', so that in the calculation of C, $\sin\theta$ may be replaced by θ. We then find

$$C = \frac{Bv}{kT} \qquad (10)$$

which finally gives

$$g(\theta) = \frac{Bv}{kT} \exp \left\{ -\frac{Bv \sin^2 \theta}{kT} \right\} \sin\theta, \quad 0 < \theta < \pi. \qquad (11)$$

Naturally, the probability of finding the magnetic moment in the neighbourhood of Ox or Ox' is 1/2.

Let us now suppose that the moment is initially directed along Ox, by the application for example, of a magnetic field of sufficient intensity along Ox, and then switching it off. A first state of temporary equilibrium is very quickly established in which the probability of finding the moment lying between θ and $\theta + d\theta$ is $g'(\theta)d\theta$, given by

$$g'(\theta) = 2g(\theta), \quad \theta < \theta < \frac{\pi}{2} \qquad (12)$$

where θ lies between 0 and $\pi/2$. The probability of finding θ lying between $\pi/2$ and π is negligible since the moment must surmount the potential barrier at $\theta = \pi/2$ to reach this region. We intend to look for the rate of establishment of the final equilibrium state, characterized by equation (11) with lying between 0 and π.

(12) The establishment of equilibrium: In a more general way, let us consider a system of N identical particles in which, in time t, P particles are magnetized in the neighbourhood of Ox and $N-P$ particles in the neighbourhood of Ox'. An initial temporary equilibrium state is established in which the number dN of particles whose directions of spontaneous magnetization lie between θ and $\theta + d\theta$ is given by one or the other of the two formulae

$$dN = 2P g(\theta)\,d\theta ; \qquad (13)$$

$$dN = 2(N - P) g(\theta)\,d\theta , \qquad (14)$$

according to whether θ lies between 0 and $\pi/2$ or $\pi/2$ and π.

The system then tends to a symmetric distribution by the following mechanism. During the interval of time dt, the magnetic moment of a certain number of particles passes from the region of Ox to the region of Ox' by crossing the equatorial plane yOz. This number is obviously proportional to the interval of time dt and to the number P of particles situated in the region of Ox. Similarly, the moment of certain other particles passes from the region of Ox' to the region of Ox. The balance of particles gained and lost in the region of Ox can therefore be written in the form

$$dP = \frac{N-P}{2\tau_0} dt - \frac{P}{2\tau_0} dt, \qquad (15)$$

where τ_0 is a constant having the dimension of time.

Macroscopically, the resultant magnetization of the system is proportional to the quantity $S = 2P-N$, the solution of a differential equation which is obtained directly from (15). One finds

$$S = S_0 e^{-\frac{t}{\tau_0}}, \qquad (16)$$

where S_0 denotes the initial value of S at time $t=0$. Thus, after a time τ_0, the initial magnetization of the system is divided by e. We have already given τ_0 the name *relaxation time*. It is a basic property of the system which has to be calculated.

(13) The rôle of perturbing forces: The constant τ_0 depends on the number of magnetic moments which cross the equatorial plane. These passages would be impossible were it not for the influence of perturbing forces capable of transferring to the system of coupled spins the thermal agitation energy of the crystalline lattice. In effect, this system of spins is equivalent to a gyroscope of angular momentum $\vec{\sigma}$ equal to $2m\mathcal{J}/eg$ per cm^3, g is the Landé factor, here equal to 2, while e and m represent the charge and mass of the electron. Furthermore, the forces of coupling with the lattice are equivalent to a couple $\vec{\Gamma}$ acting in the equatorial plane, perpendicular to $\vec{\mathcal{J}}$, of strength

$$\Gamma = -\frac{\partial E_a}{\partial \theta} = -2B \sin\theta \cos\theta. \qquad (17)$$

The general theorem for angular momentum

$$\frac{d\vec{\sigma}}{dt} = \vec{\Gamma} \qquad (18)$$

shows that the magnetic moment of the particle precesses around Ox keeping the initial value of the angle θ and that the angular velocity ω about Ox (the angular velocity of the meridian plane which contains the magnetization) is given by

$$\omega = \frac{2eB}{m\mathcal{J}} \cos\theta. \qquad (19)$$

In particular, when the moment is directed in the equatorial plane, its rate of precession is zero and it retains its original orientation indefinitely in the absence of perturbations.

Let us now suppose that there exists some perturbing couples associated with thermal deformations of the crystal lattice, and let us denote by $|\overline{\Gamma}|$ the average absolute value of the component of these couples along an arbitrary direction in the equatorial plane. It follows that the magnetic moments, which are oriented in the neighbourhood of the equatorial plane, are then endowed with an angular velocity $d\theta/dt$ in the meridian plane which contains them. The absolute value $|d\theta/dt|$ is given by

$$\left|\overline{\frac{d\theta}{dt}}\right| = \frac{e}{m\mathcal{J}} |\overline{\Gamma}|. \qquad (20)$$

The number $-dP$ of particles whose magnetic moment crosses the equatorial plane in the sense of increasing θ during the time interval dt is therefore equal to the number of particles whose moments at time t are included between the equatorial plane and the cone of angle $\pi/2 = |\overline{d\theta/dt}|dt$ and whose velocity is positive. This number is also equal, by definition, to $(P/2\tau_0)dt$. It then follows from equation (13) that

$$\frac{P}{2\tau_0} dt = P g\left(\frac{\pi}{2}\right) \left|\overline{\frac{d\theta}{dt}}\right| dt, \qquad (21)$$

and finally

$$\frac{1}{\tau_0} = \frac{2Bve|\overline{\Gamma}|}{mkT\mathcal{J}} \exp\left\{-\frac{Bv}{kT}\right\}. \qquad (22)$$

The problem is thus reduced to the determination of the average value of the perturbing couples.

(14) Nature of the perturbing couples: Where do the couples come from? It is clear that they should not be sought in the variations of the Heisenberg exchange interactions related to thermal variations of the interatomic distances because the interactions are invariant with respect to rotations of the ensemble of spins. It remains therefore to consider the changes of the demagnetizing field and magnetocrystalline and magnetoelastic forces of coupling with the crystal lattice coming from deformations produced by the influence of thermal agitation.

Strictly speaking, it would be necessary to know the different vibrational modes of the particle to solve the problem. This is not generally possible, but an approximate solution can be obtained by limiting the consideration to homogeneous deformations characterized by the six components of a symmetric tensor A_{ij}. As we are concerned with couples lying in the plane yOz, and acting on a magnetic moment situated in the same plane, only the three components A_{22}, A_{23} and A_{33} of the deformation tensor play a rôle in this respect, or, what is equivalent, the three combinations $A_{22} + A_{33}$, $A_{22} - A_{33}$ and A_{23}. Finally, it is only necessary to take into account the two latter combinations because the first one, which corresponds to a uniform expansion, cannot give rise to any couple.

Furthermore, the elastic deformation energy E_d of a particle of volume v, supposed to be elastically isotropic, is expressed in the form of a sum of six squares of which the only two that interest us are [5, p. 132]

$$E_d = \frac{vG}{2}(A_{22} - A_{33})^2 + 2GvA_{23}^2 + \cdots, \quad (23)$$

where G is the shear modulus. The average value of each of these terms is equal to $kT/2$, which gives

$$\overline{(A_{22} - A_{33})^2} = \frac{kT}{vG}; \quad (24)$$

$$\overline{A_{23}^2} = \frac{kT}{4vG}. \quad (25)$$

(15) Couples due to magnetoelastic energy: In a substance with isotropic magnetostriction for which the direction of the spontaneous magnetization is situated in the plane yOz making an angle φ with Oy, the magnetoelastic energy E_e is written [5, p. 132]

$$E_e = -3G\lambda[(A_{22} - A_{33})\cos^2\varphi + \\ + 2A_{23}\cos\varphi\sin\varphi], \quad (26)$$

where λ is the longitudinal saturation magnetostriction. An expression is deduced for the component Γ of the couple acting on the spontaneous magnetization on this account.

$$\Gamma = 6G\lambda[(A_{22} - A_{33})\cos\varphi\sin\varphi - \\ - A_{23}(\cos^2\varphi - \sin^2\varphi)]. \quad (27)$$

Taking into consideration equations (24) and (25) and the fact that the fluctuations of A_{23} are independent of those of $A_{22} - A_{33}$, $\overline{\Gamma^2}$ the mean square value of this couple is written

$$\overline{\Gamma^2} = \frac{9G\lambda^2 kT}{v}; \quad (28)$$

It is independent of φ.

As the values of Γ are probably distributed according to a gaussian law, the average value sought, $|\overline{\Gamma}|$ is finally deduced from the positive values of Γ.

$$|\overline{\Gamma}| = 3\lambda\sqrt{\frac{2kT}{\pi vG}}. \quad (29)$$

(16) Couples due to the demagnetizing field: To obtain the order of magnitude of these couples, let us suppose the particle has the shape of an ellipsoid of revolution stretched out along Ox with a magnetic moment directed in the equatorial plane making an angle φ with Oy. The energy E_d per cm³ arising from variations of the demagnetizing field associated with the deformation is written

$$E_d = -D\mathcal{J}^2[(A_{22} - A_{33})\cos^2\varphi + \\ + 2A_{23}\cos\varphi\sin\varphi], \quad (30)$$

where D is numerical coefficient which varies from $4\pi/5$ for a sphere to π for a cylinder. On average we may take $D = 3$.

The expression for E_d has the same analytical form as that for E_e. The calculation of the average couple is therefore identical. Taking into account the two effects which we have just analyzed, a definitive value for the average couple is obtained

$$|\overline{\Gamma}| = |3G\lambda + D\mathcal{J}^2|\sqrt{\frac{2kT}{\pi vG}}. \quad (31)$$

In general, the term in $D\mathcal{J}^2$ is small compared with $3G\lambda$. Unless the magnetostriction is zero, the magnetoelastic effect is much greater than the effect of the shape demagnetizing field.

(17) Calculation of the relaxation time: Finally, bringing together the formulae (22) and (31) and remembering that according to formulae (3) and (8) one must put

$$B = \frac{H\mathcal{J}}{2}, \quad (32)$$

where H is the magnetic field which is necessary to turn the magnetization of the particle in the absence of thermal fluctuations, one obtains

$$\frac{1}{\tau_0} = \frac{eH}{m} |3G\lambda + D\mathcal{J}^2| \sqrt{\frac{2v}{\pi G k T}} \exp\left\{-\frac{vH\mathcal{J}}{2kT}\right\}. \tag{33}$$

(18) Application to small iron particles: Let us apply formula (33) to small iron particles obtained by reducing an organic salt at low temperature. These particles possess a high critical field H, close to 1000 Oe [6]. Taking the following numerical values, $\mathcal{J} = 1700$, $H = 1000$, $e/m = 1.76 \times 10^7$, $\lambda = 20 \times 10^{-6}$, $G = 0.77 \times 10^{12}$, and $k = 1.38 \times 10^{-16}$, the formula (33) may be written, neglecting the term in $D\mathcal{J}^2$ as

$$\log_{10} \tau_0 = 2{,}68 \cdot 10^{21} \frac{v}{T} - 19{,}80 - 0{,}5 \cdot \log_{10} \frac{v}{T}. \tag{34}$$

Table I gives the values of v/T corresponding to several values of τ_0.

τ_0	10^{-1}	10	10^3	10^5	10^7	10^9	sec
v/T	3,2	3,9	4,7	5,4	6,2	7,0	10^{-21}

For the particles to retain the magnetization that was initially imparted to them after several years have passed, it is necessary that v/T should be at least equal to 7×10^{-21}. At room temperature, $T = 300$ K, their volume must therefore exceed 2×10^{-18} cm^3 which is the volume of a sphere about 160 Å in diameter.

Furthermore, we have shown previously [7] that particles must have a diameter less than 320 Å if they are to constitute the single elementary domains which are necessary to obtain a high coercive field. It therefore appears that to be suited for making good permanent magnets, iron powder should be made up of particles of well defined diameters lying between 160 and 320 Å. In fact, F. Bertaut [8] has shown that the diameters of particles of the best powders lie between 200 and 300 Å.

(19) Influence of an external field h on the value of the relaxation time: By applying a constant magnetic field h to the particle in the direction Ox, the energy becomes

$$E'_a = -\frac{v\mathcal{J}H}{2} \cos^2\theta - v\mathcal{J}h\cos\theta. \tag{35}$$

Provided that the absolute value of h is less than H, the positions $\theta = 0$ and $\theta = \pi$ remain the equilibrium positions, but the barrier to be overcome to pass from one to the other depends on the sense of the passage. An elementary calculation shows that the barrier to be surmounted to pass from $\theta = 0$ to $\theta = \pi$ is given by

$$E(0, \pi) = \frac{v\mathcal{J}(H+h)^2}{2H}. \tag{36}$$

Similarly,

$$E(\pi, 0) = \frac{v\mathcal{J}(H-h)^2}{2H}. \tag{37}$$

As a result of the asymmetry it is necessary to replace formula (15) by the more general formula

$$dP = \frac{(N-P)}{\tau(\pi, 0)} dt - \frac{P}{\tau(0, \pi)} dt \tag{38}$$

with the different time constants $\tau(0, \pi)$ and $\tau(\pi, 0)$ which correspond respectively to the passage from $\theta = 0$ to $\theta = \pi$ and the opposite passage. Arguments and calculations quite similar to those of §§ 11, 12 and 13 allow these constants to be calculated. One thereby obtains

$$\frac{1}{\tau(0, \pi)} = C\left(1 + \frac{h}{H}\right)\left(1 - \frac{h^2}{H^2}\right)^{1/2} \exp\left\{-\frac{v\mathcal{J}(H+h)^2}{2HkT}\right\}, \tag{39}$$

$$\frac{1}{\tau(\pi, 0)} = C\left(1 - \frac{h}{H}\right)\left(1 - \frac{h^2}{H^2}\right)^{1/2} \exp\left\{-\frac{v\mathcal{J}(H-h)^2}{2HkT}\right\}, \tag{40}$$

and

$$C = \frac{eH}{2m} |3G\lambda + D\mathcal{J}^2| \sqrt{\frac{2v}{\pi G k T}}. \tag{41}$$

Variation as a function of time of the resultant magnetic moment of N grains in a constant field. If $v\mathcal{J}$ is the magnitude of the moment of a grain G, the resultant moment of N such grains is equal to $Sv\mathcal{J}$ with $S = 2P - N$. Integrating equation (38) we find

$$S = S_1 + (S_0 - S_1) e^{-\frac{t}{\tau_h}} \tag{42}$$

where S_0 and S_1 are the initial ($t=0$) and final ($t=\infty$) values of S. The relaxation time τ_h is expressed in terms of the two time constants calculated above, in the form

$$\frac{1}{\tau_h} = \frac{1}{\tau(\pi,0)} + \frac{1}{\tau(0,\pi)}. \qquad (43)$$

Furthermore, we must have

$$S_1 = N \frac{\tau(0,\pi) - \tau(\pi,0)}{\tau(0,\pi) + \tau(\pi,0)}. \qquad (44)$$

It is not far wrong to take the two coefficients of the exponentials in formulae (39) and (40) as being equal since the variations of the exponentials themselves play a dominant role. With this assumption, we find

$$S_1 = N \, \text{th} \, \frac{v \mathcal{J} h}{kT} \qquad (45)$$

in agreement with the equation (6).

When the field h tends to zero, or, more precisely, when $v \mathcal{J} h/kT$ is much less than one, τ_h tends to the value τ_0 calculated above. On the contrary, when h is positive and $v \mathcal{J} h/kT$ is somewhat greater than one, τ_h becomes practically equal to the value $\tau(\pi, 0)$ given by equation (40).

THIRD PART
MAGNETIC PROPERTIES OF AN ENSEMBLE OF PARTICLES WITH DIFFERENT RELAXATION TIMES

(20) The ensemble of particles of type G: Let us now study the properties of a substance composed of an ensemble E of these *independent* particles, with the same spontaneous magnetization but with different volumes v and critical fields H. The ensemble E is defined by the number $n(H_0, v)dH_0 dv$ of particles whose volume is included between v and $v+dv$ and whose critical field at room temperature H_0 is included between H_0 and $H_0 + dH_0$. To each particle there corresponds a point in the (H_0, v) plane.

To simplify the question of the orientation of the particles, we suppose that the rotation axes of one third of the particles are parallel to the magnetizing field h and that the axes of the other two thirds are perpendicular to this field.

(21) Magnetic properties of particles whose rotation axes are perpendicular to the field direction: We have indicated above (§ 6) that each particle contributes to the total magnetization through its magnetic moment m proportional to the applied field h.

$$m = \frac{v \mathcal{J} h}{H}, \qquad |h| < H. \qquad (46)$$

There lies one of the origins of the large magnetic susceptibility that baked clay and lavas show in *weak fields*. In strong fields, greater than the critical field H, saturation of the grains G is reached and the susceptibility tends to zero. It is not a question of paramagnetism in the proper sense of the term.

Naturally, if the substance contains αFe_2O_3, the susceptibility due to the antiferromagnetism of this material [1], which is of order 20×10^{-6} per gram, must also appear. Nevertheless, in the clay from Nontron which contains 7% of Fe_2O_3, the contribution of this constituent to the susceptibility is only 1.5×10^{-6}. It is negligible compared with the susceptibility of baked clays which is at least 18×10^{-6}.

We will pay no further attention to the relatively uninteresting properties of the particles whose axes of revolution are perpendicular to the field, and we will now examine the particles whose axis of revolution is parallel to the field h.

(22) The particles whose axes of revolution are parallel to the direction of the field: The properties of the particles G depend essentially on their relaxation time τ_h that we have just calculated (equations (33), (39), (40) and (43)) as a function of their volume v, their critical field H, the temperature T and the applied field h. When this time is known, it is possible in principle to calculate exactly the expectation value of the magnetic moment of a particle, knowing the experimental conditions to which it has been subject, and then to calculate the properties of the ensemble E knowing the distribution of the particles in the (H_0, v) plane. In practice, such calculations can only be developed with particularly simple distributions and they remain tedious even when the temperature and the applied field vary in a very simple way.

It is useful, nevertheless, to get a general idea of the properties of small ensembles. For that we will classify the particles in the plane (H_0, v) into the categories according to the value of their relaxation time τ_h compared with the duration t of an experiment consisting for example of applying a field h at temperature T for this time.

We will suppose that the particles whose relaxation time is less than t attain perfect thermodynamic equilibrium in the field h at the end of the time t, whereas the particles whose relaxation time is greater than t retain the magnetic state which they possessed before the beginning of the experiment under consideration.

This division of the (H_0, v) plane into two well-separated regions is naturally very schematic. In reality, the region containing particles which attain their equilibrium at the end of a time t is separated from the region containing particles whose magnetization is unchanged during this period of time by a transition region with a certain width, and not simply by a line. Nonetheless, the approximation is valuable because of its simplicity. Besides, an involved discussion shows that it gives quite precise results except in the case of a field h which is *variable* and *weak*. We will therefore use it to sort out the main features of the principal magnetic properties of the ensemble E.

Note: We are led here to study the properties of an ensemble of particles with relaxation times spread over a scale of several tens of octaves. F. Preisach [9; 5, p. 244], inspired by the theory of dielectrics, has already used such an ensemble, assumed to exist *a priori*, to interpret formally the magnetic after-effect and its relation to the loss angle. We will therefore find two or three results here which have been already established by that author. In fact, the theory proposed here goes much further because it specifies the origin of the phenomenon and the value of the relaxation time. What is more, it differs profoundly from the theory of Preisach in that the relaxation time is not a constant, but depends on the *temperature* and on the *magnetic field*. Numerous interesting relationships are therefore established between very varied phenomena, and difficulties are raised concerning the possible inapplicability of the principle of superposition [5].

(23) The relaxation time and the $\mathscr{H}$-curves: As the particles of the ensemble E have been classed according to the values of H_0, their critical field at room temperature T_0, it is helpful to set

$$H = FH_0 \qquad (47)$$

where F is a function of T, independent of v, equal to 1 for $T = T_0$, slowly varying in the vicinity of T_0 and tending to zero when T tends to the Curie point.

The expressions which give the relaxation time τ_h can now be transformed and written according to the value of $v\, h/kT$ as

$$vH_0 = \frac{Q + \log \tau_h}{RF}, \qquad \frac{v\mathcal{J}h}{kT} \ll 1 \; ; \quad (48)$$

$$\frac{v\left(H_0 - \dfrac{h}{F}\right)^2}{H_0} = \frac{Q' + \log \tau_h}{RF}, \qquad \frac{v\mathcal{J}h}{kT} > 4 \; ; \quad (49)$$

where R, Q and Q' are given by the equations

$$R = \frac{\mathcal{J}}{2kT}; \qquad (50)$$

$$Q = \log\left\{\frac{eH}{m}|3G\lambda + D\mathcal{J}^2|\sqrt{\frac{2v}{\pi GkT}}\right\}; \qquad (51)$$

$$Q' = Q + \log\left\{\frac{1}{2}\left(1 - \frac{h}{H}\right)^{3/2}\left(1 + \frac{h}{H}\right)^{1/2}\right\}; \qquad (52)$$

The coefficients Q and Q' are functions of v, T, H and h, but have a logarithmic dependence so that their variations are negligible compared with the other terms in equations (48) and (49). As a first approximation Q and Q' may be supposed to be equal and constant. We will see later (§ 43) that this common value is close to 22 when we are concerned with baked clay and the particles which arise in the phenomena studied here.

In the region of intermediate values of $v\, h/kT$, the expression for the function $f(H_0, v)$ is much more complicated.

We denote by the name $\mathscr{H}$-curve the locus of points in the (H_0, v) plane corresponding to particles with a relaxation time equal to t, for a given field h and temperature T. When it is necessary to specify the parameters, we will use the notation $\mathscr{H}(T, h, t)$.

(29) General appearance of the $\mathscr{H}$-curves: It is useful to distinguish two cases.

(a) *The applied field* h *is zero* — In this case, the equation of the locus $\mathscr{H}$ is equation (48). The $\mathscr{H}$-curves are rectangular hyperbolae whose axes are the coordinate axes. The right hand side of equation (48) is a complicated function of temperature which begins by increasing slowly with temperature in the region of room temperature, then more and more rapidly until it finishes by becoming infinite at the Curie point, as is shown schematically in Fig. 5 in which curves are drawn for different temperatures. The figures on the curves represent the reduced temperatures T/θ where θ is the Curie temperature.

All the points to the left of any $\mathscr{H}$-curve, that is in the region between the $\mathscr{H}$-curve and the coordinate axes, correspond to particles whose relaxation time is less than t. At the end of a time t, the particles belonging to the region have taken up the magnetic equilibrium corresponding to the field h so that their average individual magnetic moment is given by the formula

$$\overline{m} = v\mathcal{J}\,\text{th}\,\frac{v\mathcal{J}h}{kT}. \qquad (53)$$

On the other hand, the points to the right of the -curve correspond to particles whose relaxation time is

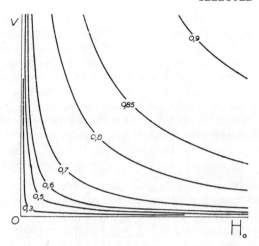

Fig. 5.
Schematic $\mathscr{H}$-curves as a function of reduced temperature T/θ.

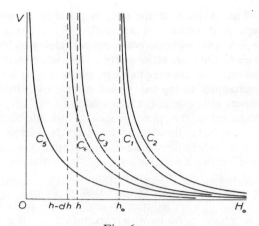

Fig. 6.
An illustration of the way in which isothermal remanent magnetization is acquired.

greater than t and consequently to particles which retain at the end of the time t their original state that they had before the field h was applied.

(b) *The field h is not zero* — The general equation of the curve $\mathscr{H}(T, H, t)$ is now very complicated. When $v\mathfrak{J}h/kT$ is somewhat greater than 1, it reduces to equation (49) and represents a cruve with a hyperbolic appearance (Fig. 6, curves C_1, C_2, C_3, C_4) whose asymptote is the line

$$H_0 = \frac{h}{F}. \qquad (54)$$

When $v\mathfrak{J}h/kT$ becomes small, that is to say when H_0 becomes Q times greater than h/F (Q is close to 20), the curve $\mathscr{H}(T, h, t)$ approaches the curve $\mathscr{H}(T, 0, t)$ and becomes indistinguishable from it.

In the same way as in (a) the points situated to the right of the curve $\mathscr{H}(T, h, t)$ correspond to particles whose relaxation time is greater than t which consequently retain their initial state for a time t after the field is applied. On the other hand, the points situated to the left of the curve possess a relaxation time shorter than t and attain thermodynamic equilibrium in the field h by the end of the time t.

(25) Magnetization of the ensemble E in a field h_0 at temperature T_0: Let us start initially from a virgin ensemble E with no magnetization. That is to say that the expectation value of the magnetic moment of each one of the particles is zero. At constant temperature T_0 we apply a field h_0 at time $t=0$. By time t_0, all the particles to the left of the curve $\mathscr{H}(T_0, h_0, t_0)$ drawn as C_1 on Fig. 6, are magnetized. The average magnetic moment of one of the particles is given by the formula

$$\bar{m} = v\mathfrak{J}_0 \,\text{th}\, \frac{v\mathfrak{J}_0 h_0}{kT_0}. \qquad (55)$$

The average magnetic moment of the particles situated to the right of the curve $\mathscr{H}(T_0, h_0, t_0)$ remains zero.

At a later time t, all the particles situated to the left of the curve $c(T_0, h_0, t)$, drawn as C_2 in Fig. 6, are magnetized. The magnetization of E is therefore a function of time. If the difference $\log t - \log t_0$ is small compared with Q, the examination of the equation defining the $\mathscr{H}$-curves shows that the area between the curves $\mathscr{H}(T_0, h_0, t)$ and $\mathscr{H}(T_0, h_0, t_0)$ is proportioned to $\log t - \log t_0$. In these conditions, the magnetization σ of E is written

$$\sigma = A + B(\log t - \log t_0), \qquad (56)$$

where B is small compared with A. It depends on time and there is a *magnetic after-effect*. The range of validity of this formula depends on the way in which the density $n'(H_0, v)$ of particles varies near the $\mathscr{H}$-curves in question. If this density varies sufficiently slowly as in certain baked clays studied by Thellier, $\log_{10} t$ may vary from 0 to 10 or 12 without formula (56) ceasing to be valid.

(26) Acquisition of the isothermal remanent magnetization: After having magnetized the ensemble E in a field h_0 for a time t, we keep the temperature constant at T_0 and decrease the field from h_0 to zero. The curve $\mathscr{H}(T_0, h, \tau)$ corresponding to a value h of this decreasing field is represented by C_3 on Fig. 6. The time τ associated with this $\mathscr{H}$-curve is related to the rate of decrease of the field. It becomes smaller as the rate is inzcreased, but it is unnecessary here to specify its value.

When the field decreases from the value h to value $h - dh$, the magnetic moment of particles situated between the curves $\mathcal{H}(T_0, h, \tau)$ and $(T_0, h - dh, \tau)$ represented by C_3 and C_4 on Fig. 6 are stabilized at the value

$$m = v\,\mathcal{J}_0\,\text{th}\,\frac{v\,\mathcal{J}_0\,h}{kT_0} \qquad (57)$$

which does not vary any further thereafter, whereas the particles situated to the left of $\mathcal{H}(T_0, h - dh, \tau)$ come to equilibrium in the new field $h - dh$. Finally, when h becomes zero, all the particles between the curves $\mathcal{H}(T_0, h_0, \tau)$ and $\mathcal{H}(T_0, 0, \tau)$ represented by C_2 and C_5 on Fig. 6, remain magnetized with a moment given by equation (57) where h is the field corresponding to the curve (T_0, h, τ) which passes through the point representing the particle under consideration. In this way the isothermal remanent magnetization (IRM) is obtained.

For a sufficiently strong field h_0, the great majority of the particles lying between $\mathcal{H}(T_0, h_0, t)$ and $\mathcal{H}(T_0, 0, \tau)$ remain magnetized at the saturation $v\,\mathcal{J}_0$ because $v\,\mathcal{J}_0 h/kT$ is almost everywhere much greater than unity.

(27) Influence of time on the isothermal remanent magnetization: It follows from the end of the last paragraph that if h_0 is large enough, most of the particles which remain magnetized retain their saturation magnetization $v\,\mathcal{J}_0$ which was imparted by the influence of the field h_0 after that field is cut off. They do not change their state, and consequently the rate of decrease of the field should have no appreciable influence on the final magnetization of the ensemble.

We therefore do not take account the time needed to decrease the field, and we suppose that we measure the moment of the ensemble E at an instant t' after cutting off the field h_0. The first of the curves $\mathcal{H}(T_0, h_0, t)$ and $\mathcal{H}(T_0, 0, t')$ which delimit the region of magnetized particles depends on t, the second on t'. An argument analogous to that of paragraph 25 shows that the remanent magnetization σ_{ri} of the ensemble E takes the form

$$\sigma_{ri} = A + B(\log t - \log t_0) - C(\log t' - \log t'_0) \quad (58)$$

in which A, B, and C are positive quantities independent of t and t', and of t_0 and t'_0, times whose order of magnitude must be t and t' if the formula is to remain valid.

Certain experiments on baked clay reported in Thellier's thesis [2] lead to a formula of the type (58), so far as the variation of σ_{ri} with t' is concerned. Such a formula represents the situation well when t' varies in the ratio from 1 to 10^4, and maybe even from 1 to 10^6.

Notice that the term B in formula (58) comes from variations of the magnetization of particles situated close to the curve $\mathcal{H}(T_0, h_0, t)$ whereas the term in C is due to particles situated near the curve $\mathcal{H}(T_0, 0, t')$. These curves diverge most in the regions of the (H_0, v) plane where the densities $n(H_0, v)$ differ more and more as h_0 increases. It follows that B and C must have similar values to begin with when h_0 is small, and then diverge when h_0 increases.

(28) Influence of the field h_0 on the value of the IRM: If the density of particles in the (H_0, v) plane varies slowly, if h_0 is small, and if $\log(t/t_0)$ and $\log(t'/t'_0)$ do not greatly exceed unity, the number n of particles included between the curves $\mathcal{H}(T_0, 0, t')$ and $\mathcal{H}(T_0, h_0, t)$ varies linearly with h_0 and with $\log(t/t_0) - \log(t'/t'_0)$.

$$n = ah_0 + b\left(\log\frac{t}{t_0} - \log\frac{t'}{t'_0}\right). \qquad (59)$$

The numbers a and b are independent of h_0, t and t'. It is then necessary to distinguish two cases according to the values of h_0.

(a) h_0 *is small enough for the product* $v\,\mathcal{J}_0 h_0$ *to be small compared with* kT_0 *in almost all the particles* — Denoting the mean square of volume of the particles by $\bar{v}^2$, one finds

$$\sigma_{ri} = \frac{\bar{v}^2\,\mathcal{J}_0^2}{kT_0}\left[ah_0^2 + bh_0\left(\log\frac{t}{t_0} - \log\frac{t'}{t'_0}\right)\right]. \qquad (60)$$

(b) h_0 *is large enough for the product* $v\,\mathcal{J}_0 h_0$ *to be large compared with* kT_0 *in almost all the particles* — In these conditions, each particle possesses a moment $v\,\mathcal{J}_0$ directed in the sense of the field h_0 and one finds

$$\sigma_{ri} = \bar{v}\,\mathcal{J}_0\left[ah_0 + b\left(\log\frac{t}{t_0} - \log\frac{t'}{t'_0}\right)\right] \qquad (61)$$

where $\bar{v}$ is the average particle volume. The terms which depend on time are now independent of h_0.

When h_0 increases from zero and when the terms describing the magnetic after-effect are separated, the IRM σ_{ri} first increases as h_0^2 and then with h_0, in agreement with Thellier's observations [2]. Naturally, if the values of h_0 become large enough for the conditions set out at the beginning of the section to be no longer valid, then it is necessary to take into account the variation of the particle density in the (H_0, v) plane and the formulae for σ_{ri} become complicated.

(29) Description of a hysteresis loop: After having brought the magnetizing field down to zero, let us now apply a negative field $-h_0$ for a time t''. All the particles situated to the left of $\mathscr{H}(T_0, h_0, t'')$ must be negatively magnetized. Following the cut off of the field $-h_0$, and t''' seconds after this cut off, the particles lying between $\mathscr{H}(T_0, 0, t''')$ and $\mathscr{H}(T_0, h_0, t'')$ remain negatively magnetized and those lying to the right of $\mathscr{H}(T_0, h_0, t'')$ keep their earlier state. If we neglect the effect of time by putting $t = t' = t'' = t'''$, it may be seen that the magnetization acquired in the field $-h_0$ is equal and opposite to that acquired in the field h_0.

It is interesting to plot the *hysteresis loop of the remanent magnetization*. This loop differs from an ordinary hysteresis loop in that for each value of the field, it is not the magnetization in the field that is measured, but the magnetization remaining after the field is cut off. Supposing that condition (b) of § 28 is satisfied, a parallelogram loop is obtained composed of two horizontal branches and two sides with slope equal to $2a\bar{v}_0$ (Fig. 7), very similar to the loops determined by Thellier [2] and Miss Roquet [3].

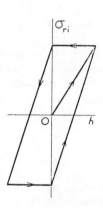

Fig. 7.
Hysteresis loop of the isothermal remanent magnetization.

(30) Demagnetization by a decreasing a.c. field: Arguments analogous to the preceding ones allow the determination of the effect of a decreasing a.c. field on the magnetization of the ensemble E. Let us denote by $+h_0, -h_0, +h_2, -h_3$ etc. the successive maxima and minima of the field and by $\mathscr{H}_i$ the curve (T_0, h_i, τ) corresponding[3] to the maximum (or the minimum) h_i. The plane (H_0, v) is then decomposed into successive bands $H_0 - H_1, H_1 - H_2, H_2 - H_3$ etc. which are oppositely magnetized, the first positively, the

[3] The constant τ in $\mathscr{H}_i$ is the time for which the maximum value of the field h_i acts. It is therefore a time which is of order a quarter of the period of the applied alternating field.

second negatively, the third positively, etc. If the amount of decrease of the field in a period is small, the bands will be very narrow so that an arbitrary element $\delta H_0 \delta v$ of the (H_0, v) plane chosen to lie between the extreme curves $\mathscr{H}(T_0, h_0, \tau)$ and $\mathscr{H}(T_0, 0, \tau)$ will be crossed by several bands and consequently has zero magnetic moment. The substance will therefore be demagnetized on condition that the initial value of the alternating field was at least as large as the magnetic fields which had previously acted on the substance.

(31) Effect of a refiring on the isothermal remanent magnetization: Consider an ensemble having acquired at temperature T_0 a certain remanent magnetization under the influence of a field h_0. Let us keep the field at zero, and take the ensemble up to a temperature T_1, greater than T_0, for a time t, and then bring it down to the initial temperature T_0. This reheating brings about the demagnetization of the magnetized particles lying to the right of the curve $(T, 0, t)$ represented by C_2 on Fig. 8.

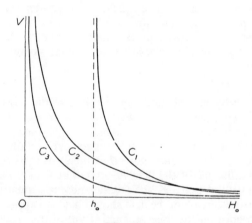

Fig. 8.
The effect of refiring on the isothermal remanent magnetization is to demagnetize the particles lying between C_3 and C_4.

If we neglect the effect of time, it is easily verified that the decrease of remanent magnetization is proportional to $T_1 - T_0$, at least to begin with, so long as this quantity is not too large. It may be seen also that all subsequent refirings carried out at temperatures less than or equal to T_1, will not alter the remanent magnetization any further. There has been a *stabilization*.

(32) Comparison of the effects of time and of refiring on the remanent magnetization: In zero field, the equation of the $\mathscr{H}$-curves is written

$$vH_0 = K \tag{62}$$

with

$$K = \frac{2kT}{\mathcal{J}F}(Q + \log t). \tag{63}$$

On increasing K by dK we demagnetize the ndK particles lying between the curves $vH_0 = K$ and $vH_0 = K + dK$. Denoting by $\bar{m}$ the average magnetic moment of the particles, the change $d\sigma_r$ in the remanence is written

$$d\sigma_r = -n\bar{m}dK. \tag{64}$$

The factor K depends on time and on temperature. Therefore,

$$dK = \frac{\partial K}{\partial \log t} d\log t + \frac{\partial K}{\partial T} dT. \tag{65}$$

Bringing together equations (63) and (64), one finds

$$\frac{\partial \sigma_r}{\partial \log t} = -n\bar{m} \frac{\partial K}{\partial \log t}; \tag{66}$$

$$\frac{\partial \sigma_r}{\partial t} = -n\bar{m} \frac{\partial K}{\partial T}. \tag{67}$$

There also exists a close relation between the decrease of the remanent magnetization produced by refiring and the decrease produced by time. From the preceding equations one deduces

$$\frac{\dfrac{\partial \sigma_r}{\partial \log t}}{\dfrac{\partial \sigma_r}{\partial T}} = \frac{\dfrac{\partial K}{\partial \log t}}{\dfrac{\partial K}{\partial T}}. \tag{68}$$

This relation is independent of the initial value of the remanent moment. It may be verified numerically, hence its significance (cf. § 44).

(33) Stabilization of the remanent magnetization by refiring: The same decrease in remanence *arising from the same particles* can be produced either by a refiring of duration τ at the temperature T_1 or by keeping the temperature at its initial value T_0 and by waiting for a long enough time t. The values of t must be the same in the two cases, and neglecting the temperature dependences of $\mathcal{J}$ and F, this equality is represented by the relation

$$T_1(Q + \log \tau) = T_0(Q + \log t) \tag{69}$$

whence

$$\log \frac{t}{\tau} = \frac{(Q + \log \tau)(T_1 - T_0)}{T_0}. \tag{70}$$

In an example studied below (§ 44), we will see that $(Q + \log t)/T_0$ at room temperature for $\tau = 5$ mins is close to 1/9. With $T_1 - T_0 = 90°$, $\log t/\tau = 10$, whence $t = \tau e^{10} = 22,000\tau$, or approximately 76 days. A refiring of five minutes at the given temperature is therefore equivalent to an ageing of 76 days. Instead of waiting for the stabilization of a remanent magnetization by natural ageing, it is much better to induce an artificial ageing by refiring.

Conversely, a prolonged ageing produces the same effects as raising the temperature of an unaged substance. One hundred million years corresponds to $\log_e t = 35.6$. If formula (70) were still valid, such an ageing at room temperature (300 K) would produce the same effects as heating for five minutes at 650 K, supposing that $Q = 22$. Refiring at such a high temperature would destroy a large part of the remanent magnetization. It is therefore understandable that certain very ancient rocks now only possess according to Koenigsberger [10], a permanent magnetization that is much smaller than the one they should normally show as a result of their cooling in the Earth's field. It is not necessary to suppose that at the time of cooling the Earth's field was much less than it is at present.

(34) Stabilization of the isothermal remanent magnetization by the influence of a magnetic field: The particles which give rise to the magnetic after-effect at the temperature T_0 are located near the curve $\mathcal{H}(T_0, 0, t)$ (cf. Fig. 9). To stabilize the remanent magnetization it is necessary to demagnetize these particles by some method or other, at least on average, so that as t varies the curve $\mathcal{H}(T_0, 0, t)$ sweeps out regions of the (H_0, v) plane that are already demagnetized. For this, as we showed previously, (cf. § 30) it is sufficient to apply a slowly decreasing alternating field, decreasing from a maximum value h_0. The particles lying to the left of $\mathcal{H}(T_0, h_0, \tau)$ where τ is of order a quarter of the period of the applied field, are thereby effectively demagnetized. Naturally, it is necessary to give h_0 a value sufficient for the band swept out to be broad enough, but not too broad so as to avoid demagnetizing all the particles and leave a certain remanent magnetization.

(35) Susceptibility of the ensemble E in an alternating magnetic field: It follows from the analysis of the

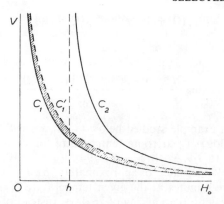

Fig. 9.
The isothermal remanent magnetization acquired in the field h corresponds to particles lying between the curves C_1 and C_2. Particles liable to be demagnetized as a function of time or of a small increase in temperature are those lying in the shaded region.

phenomena which arise in the course of the description of the hysteresis loop (cf. § 29) that the particles lying to the left of the curve $\mathscr{H}(t_0, 0, \tau)$ are reversibly magnetized in the variable applied field h. Moreover, these particles do not constitute the only source of reversible magnetization for the ensemble E. We have shown previously (cf. § 21) that all the particles whose rotation axes were perpendicular to the direction of the applied field also show a magnetization varying reversibly with h.

Let us now return to the particles lying to the left of $\mathscr{H}(T_0, 0, \tau)$. We will suppose that h is small enough for none of the particles to be saturated. In these conditions, denoting their total number by N and the mean square volume by $\bar{v}^2$, the corresponding induced magnetization σ_i is written

$$\sigma_i = \frac{N\bar{v}^2 \mathcal{J}^2 h}{kT} \qquad (71)$$

The susceptibility $\chi_i = \sigma_i/h$ depends on temperature in a complicated way, not only through the term in T but also through the number N which varies with T since the $\mathscr{H}$ curve is displaced with T. Furthermore, the susceptibility χ_p coming from the particles with their rotation axes perpendicular to the field must be superposed on χ_i, and this susceptibility χ_p is also a function of T through the critical field H (cf. equation (46) in § 21). The total effect is therefore quite complicated.

More interesting is the variation of χ_i, with the frequency ω of the alternating applied field $h = h_0 \sin \omega t$, because the susceptibility χ_p does not depend on ω. The interpretation is therefore more straightforward. Roughly speaking, the particles following the variation of the field are those whose relaxation time is less than $1/\omega$. When the frequency changes from ω to $\omega + d\omega$, the magnetization undergoes a change $d\sigma_i$ due to the fact that the particles whose relaxation time lies between $1/\omega$ and $1/(\omega + d\omega)$ no longer contribute to the magnetization. Looking back to paragraph 32, it is observed that the number of these particles is equal to $-n(\partial K/\partial \log t)d \log \tau$ with $-d \log \tau = d \log \omega$. Denoting the average magnetic moment taken by these particles in a constant field h_0 by $\bar{m}$ one obtains

$$-d\sigma_i = n\bar{m} \frac{\partial K}{\partial \log t} d \log \omega \cdot \sin \omega t . \qquad (72)$$

Introducing the variation $d\chi_i$ of the susceptibility corresponding to $d\sigma_i$ and comparing with formula (66) for the loss of magnetization by the magnetic after-effect, the following interesting relation is found which seems to be open to numerical confirmation

$$\frac{\partial \chi_i}{\partial \log \omega} = \frac{1}{h_0} \frac{\partial \sigma_r}{\partial \log t} . \qquad (73)$$

In this formula, σ_r represents the remanent magnetization acquired in the field h_0. For this formula to be rigorously true it is necessary that

(1) the values of the derivatives of χ_i and of σ_r apply to the same value of K. This is to say that the time t, after which the measurement of the derivative of σ_r is made at the end of the magnetization, should be of the same order of magnitude as $1/\omega$.

(2) χ_i is measured in a field of the same order of magnitude as h_0. Luckily it emerges from the experiments of Thellier that $\partial \chi/\partial \log t$ only varies slowly with $\log t$. It must be the same for the variation of $\partial \chi/\partial \log \omega$ with $\log \omega$, which should facilitate the confirmation of formula (73).

In a substance studied by Thellier[4]

$$\frac{1}{h_0} \frac{\partial \sigma_r}{\partial \log t},$$

when h_0 is small, is approximately equal to 0.5×10^{-6}. It follows that the susceptibility should decrease by 1.15×10^{-6} each time the frequency is multiplied by 10. As the total susceptibility is 18×10^{-6}, the effect should be quite easy to demonstrate.

(36) Losses in the ensemble E in an alternating field: A particle G with time constant τ_0 subjected to an alternating field $h = h_0 \sin \omega t$ of very small amplitude

---
[4] Sample C25.

adopts a moment m given by the solution of the differential equation

$$\tau_0 \frac{dm}{dt} + m = m_0 \sin \omega t, \quad (74)$$

where m_0 denotes the moment taken by the particle at the end of an infinite time in the constant field h_0. One therefore finds

$$m = \frac{m_0 \sin \omega t}{1 + \omega^2 \tau_0^2} - \frac{m_0 \omega \tau_0 \cos \omega t}{1 + \omega^2 \tau_0^2}. \quad (75)$$

The magnetizing system must therefore supply to the particle G a certain energy w which appears in the form of heat given in ergs per second by the formula

$$w = -\int_{t=0}^{t=1} m\,dh = \frac{\frac{1}{2} m_0 h_0 \omega^2 \tau_0}{1 + \omega^2 \tau_0^2} \quad (76)$$

For the ensemble E, the total energy W thus transformed into heat, that is to say the losses, is written in ergs per second, using the notation of paragraph 22, as

$$W = \int_0^\infty nw\,dK \quad (77)$$

but

$$dK = \frac{\partial K}{\partial \log \tau_0} d\log \tau_0 \quad (78)$$

Now it follows from the expression (76) that w only has an appreciable value in the case where τ is of the same order of magnitude as ω. In such a region, n is nearly independent of τ. Equation (77) can then be integrated to give

$$W = \frac{\pi}{4} n m_0 \omega h_0 \frac{\partial K}{\partial \log t}. \quad (79)$$

The losses per cycle only depend on frequency through n. If the variation of n with τ is negligible[5], the losses per cycle are independent of the frequency. This result is to be expected because it is a standard one for any system, such as the ensemble E, involving an extended

---

[5] It is now a matter of variations of τ extending over several octaves, whereas for the integration in (77) to be valid, it is enough that n remains roughly constant when τ varies in an octave centred around $1/\omega$.

spectrum of time constants. From an experimental point of view, one recognizes here what is known as the Jordan term in magnetic losses.

It is interesting to relate these losses to the magnetic after-effect. Taking into account equation (66), it follows that

$$W = -\frac{\pi \omega h_0}{4} \frac{\partial \sigma_r}{\partial \log t}, \quad (80)$$

where σ_r is the remanent magnetization acquired in the field h_0. For this formula to be strictly correct, it is necessary that the frequency ω at which W is measured should be of the order of magnitude of $1/t$, t being the time involved in measuring the derivative of σ_r. Otherwise this formula is only approximate.

(37) Formulae involving the loss angle: It is often helpful in expressing losses to bring in the phase angle δ, known as the loss angle, which the magnetic induction makes the field h. Let us call the conservative permeability μ, that is the ratio with the magnetic field as denominator and the component of the induction which is in phase with the field as numerator. The losses W_1 in ergs per second per cubic centimeter are then given by

$$W = \frac{\omega \mu h_0^2 \, \text{tg}\, \delta}{8\pi} \quad (81)$$

Comparison with the formula (80) gives

$$\text{tg}\, \delta = -\frac{2\pi^2}{\mu h_0} \frac{\partial \sigma_r}{\partial \log t}. \quad (82)$$

This expression takes on a more elegant form when h is very weak. In this case, denoting the change in induction on applying a constant field h_0 as B and the corresponding remanent induction as B_r, according to the results in § 28 one has approximately

$$\frac{\partial B}{\partial \log t} = -\frac{\partial B_r}{\partial \log t}. \quad (83)$$

One can therefore write

$$\text{tg}\, \delta = \frac{\pi}{2} \cdot \frac{1}{B} \cdot \frac{\partial B}{\partial \log t}. \quad (84)$$

This formula is only rigorously true in the case where the frequency ω for which δ is measured is of the same order of magnitude as $1/t$, t being the time elapsed since the moment of application of the field h corresponding to the measurement of the derivative of B.

Moreover, combining equations (73) and (82), we find

$$\text{tg}\,\delta = -\frac{2\pi^2}{\mu}\frac{\partial \chi_i}{\partial \log \omega}. \quad (85)$$

But according to paragraph 52 we have

$$\mu = 1 + 4\pi(\chi_i + \chi_p), \quad (86)$$

where χ_p is independent of the frequency. Equation (85) may therefore be written in the form

$$\text{tg}\,\delta = -\frac{\pi}{2}\cdot\frac{1}{\mu}\cdot\frac{\partial \mu}{\partial \log \omega}. \quad (87)$$

The two formulae (84) and (87) lend themselves readily to numerical confirmation. They apply to any system of particles whose relaxation times are spread out over a broad range, no matter what the relaxation mechanism is. They are generalizations of two of Preisach's results.

(38 Thermoremanent magnetization. *Its acquisition:* Let us choose, between room temperature and the Curie point θ, two temperatures T_1 and T_2, where T_1 is less than T_2. After first bringing the ensemble E to a temperature greater than θ, let us then cool it in a magnetic field which is zero outside the interval of temperature T_1, T_2 and equal to h within the interval.

To simplify matters we suppose that h is weak. We plot as C_1 and C_2 on Fig. 10 the two curves $\mathscr{H}(T_1, h, \tau)$ and $\mathscr{H}(T_2, h, \tau)$ which could be confused with $\mathscr{H}(T_1, 0, \tau)$ and $\mathscr{H}(T_2, 0, \tau)$ since h is small.

When the temperature is decreased from the value T_2 the application of the field h magnetizes all the particles lying to the left of C_2, then when T_1 is reached, bringing the field back to zero demagnetizes all the particles lying to the left of C_1. In the end, only those remain magnetized that lie in the shaded zone between C_1 and C_2 shown in Fig. 10. When the system is brought back to the temperature T_0, it then possesses a certain remanent magnetization which we denote as $\sigma(T_1, h, T_2)$ to recall the circumstances of its creation.

The order of magnitude of the time τ — The order of magnitude of τ as it appears for example in the equation of curve C_2, can be estimated in the following way. It is the time needed for the relaxation time of a particle situated near the curve $\mathscr{H}(T_2, 0, \tau)$ to be multiplied by e by the effect of cooling. Attributing to $Q + \log \tau$ the value 27, it is seen that τ represents the time needed for the substance to cool by $T/27$ degrees, for instance by 25° when $T = 675$ K. In fact a particle whose relaxation time is much longer than τ can only

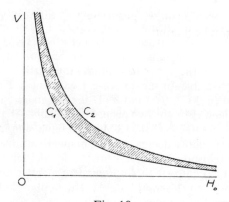

Fig. 10.
An illustration of the way in which thermoremanent magnetization is acquired.

be weakly magnetized because τ seconds after switching on the field h, the relaxation time of the particle is already multiplied by e. Conversely, if the relaxation time of the particle under consideration is significantly less than τ, the particle will have taken on its equilibrium magnetization well before its relaxation time is multiplied by e.

Effect of refiring — It is important to note that refirings in zero field carried out at temperatures less than or equal to T_1 do not alter the value of the remanent magnetization at room temperature since the corresponding $\mathscr{H}$ curves are always to the left of C_1. To being about a decrease in the remanence, the refiring temperature T' must exceed the temperature T_1, which will then demagnetize particles lying between $\mathscr{H}(T', 0, \tau)$ and $\mathscr{H}(T_1, 0, \tau)$. A refiring at a temperature above T_2 causes the remanence to disappear entirely.

Definitions — One important case is that where the temperatures T_1 and T_2 coincide with T_0 and θ. The magnetic field h acts throughout the cooling, and is only switched off when the temperature becomes equal to T_0. All the particles lying to the right of $\mathscr{H}(T_0, 0, \tau)$ are involved in this phenomenon since the curve $\mathscr{H}(\theta, 0, \tau)$ is at infinity. The remanent magnetization $\sigma(T_0, h, \theta)$ thus acquired is the *complete thermoremanent magnetization*, abbreviated TRM. When T_1 is greater than T_0, or when T_2 is less than θ, we will speak of the partial thermoremanent magnetization, abbreviated PTRM.

(39) Comparison of the properties of the TRM and IRM: The TRM acquired in a moderate field h is much greater than the IRM acquired in the same field. The first corresponds to all the particles lying to the right of $\mathscr{H}(T_0, 0, \tau)$ whereas the second corresponds only to the particles lying between $\mathscr{H}(T_0, 0, \tau)$ and $\mathscr{H}(T_0, h, \tau)$.

The decrease in the remanent magnetization produced by a refiring at temperature T' is practically the same in absolute value whether it is a TRM or an IRM provided that these magnetizations were acquired in the same field. In fact the particles which are responsible for the decrease, located near $\mathcal{H}(T_0, 0, \tau)$ between $\mathcal{H}(T_0, 0, \tau)$ and $\mathcal{H}(T', 0, \tau)$ are magnetized in the same way by the two processes. But since the TRM is much greater than the IRM in the same magnetizing field, the *relative* decrease of the magnetization caused by the refiring is much less for the TRM than for the IRM. The complete thermoremanent magnetization is therefore much more stable thermally than the isothermal remanent magnetization. Analogous arguments show that the stability as a function of time or in negative magnetic fields is likewise much better for the TRM than for the IRM.

(40) The effect of saturation: The average magnetic moment $\bar{m}$ taken on in the experiment described in paragraph 38 by a particle G lying between the curves C_1 and C_2 is given, at room temperature T_0 by the formula

$$\bar{m} = v\mathcal{J}_0 \, \text{th} \, \frac{v\mathcal{J}h}{kT}, \quad (88)$$

where T is the temperature corresponding to the curve $\mathcal{H}(T, 0, \tau)$ which passes through the coordinates H_0, v of the particle G, and where $\mathcal{J}$ is the value of the spontaneous magnetization, at this temperature T. The moment remains proportional to h so long as $v\mathcal{J}h/kT$ is small compared with unity, and it tends to a limit equal to $v\mathcal{J}$ when $v\mathcal{J}h/kT$ exceeds unity. It follows that the TRM begins by increasing proportionally with h and then tends to a limit, in accord with Thellier's observations. Naturally, it is possible to attain the same limit by isothermal magnetization, but the fields needed are very much greater.

(41) Additivity of partial thermoremanent magnetizations: All the strange properties of additivity of the PTRMs acquired in intervals of temperature that do not overlap found by Thellier in baked clays (§ 4) and confirmed by Nagata on lavas, can be interpreted immediately by invoking the mechanism of magnetization described in paragraph 38. In particular, the PTRM $\sigma(T, h \theta)$ acquired by cooling in a field h acting from the Curie point down to the temperature T_1 is due to all the particles lying to the right of $\mathcal{H}(T, h, \tau)$ whereas the PTRM $\sigma(T_0 h T)$ acquired on applying the field only from T_1 down to room temperature T_0 is due to the particles lying between $\mathcal{H}(T_0, 0, \tau)$ and $\mathcal{H}(T_1, h, \tau)$. The sum of these two magnetizations is therefore equal to the complete TRM $\sigma(T_0 h \theta)$ due to all the particles situated to the right of $\mathcal{H}(T_0, 0, \tau)$.

$$\sigma(T_0 h \theta) = \sigma(T_0 h T_1) + \sigma(T_1 h \theta). \quad (89)$$

Similarly, it is shown that the TRM acquired on cooling from T_1 to T_0 is exactly equal to the decrease in TRM produced by reheating to T_1, etc.

(42) Reversible temperature dependence of the PTRMs: For example, let $\sigma(T h \theta)$ be the PTRM acquired during cooling in a field acting from the Curie point down to the temperature T_1. Below T_1, the PTRM is a reversible function of temperature. In fact at the temperature $T' < T_1$, each particle lying to the right of the curve $\mathcal{H}(T_1, h, \tau)$ possesses an average magnetic moment

$$m = v\mathcal{J}' \, \text{th} \, \frac{v\mathcal{J}h}{kT} \quad (90)$$

where $\mathcal{J}$ and T are completely determined for each particle and characterize the conditions in which the particle in equation has acquired its remanent magnetization. On the other hand, $\mathcal{J}'$ is the value of the spontaneous magnetization at the temperature T' of observation of the remanent magnetization of the system. It is a reversible function of temperature. Thus, below T_1, the remanent magnetization of the system varies proportionally to the spontaneous magnetization. Nagata has observed this variation in some special cases.

(43) Calculation of the value of Q: We have shown previously that at the temperature T the particles responsible for the effects of time and temperature on the remanent magnetization were located near the curve $\mathcal{H}(T, 0, \tau)$ with equation

$$vH = \frac{2kT}{\mathcal{J}} (Q + \log \tau) \quad (91)$$

where τ is a time of the same order of magnitude as the length of the experiment. Furthermore, Q is really a slowly varying function of v and H given by the formula (51). Eliminating v from these two equations, a relation between H and Q is found having the form

$$\log H = aQ - b \log (Q + c) \quad (92)$$

so that for considerable changes of H, from 100 to 5000 Oersteds, covering the range of likely values, the corresponding changes of Q do not exceed 5% on either side of the average. It is therefore possible to determine the value of Q to within this accuracy.

We will suppose that the particles are made of a substance with properties similar to magnetite. Therefore, we will take $\mathcal{J} = 500$, $G = 10^{12}$, $\lambda = 10 \times 10^{-6}$, $T = 300$ K and we will neglect the terms in $D\mathcal{J}^2$ in formula (51). The numerical values chosen could be in error by 50% yet only produce an error of a few percent in Q. For τ, we will adopt the value of 100 seconds, the approximate length of a measurement on the magnetometer.

After substituting in the numerical values, equation (92) becomes

$$\log_{10} H = 0{,}868\, Q - 14{,}86 - \log_{10}(Q + 4{,}6)\,. \quad (93)$$

The values of Q corresponding to some values of H are shown in Table II. It seems reasonable to adopt the value $Q = 22$.

H	10	10^2	10^3	10^4
Q	19,9	21,0	22,2	23,4

(44) Numerical confirmation: We are concerned with the relation between the time variation of the remanent magnetization and its temperature dependence. It we set

$$\rho = \frac{\partial \sigma_r}{\partial \log t} : \frac{\partial \sigma_r}{\partial T}\,, \quad (94)$$

we have shown (§ 32) that

$$\rho = \frac{\partial K}{\partial \log t} : \frac{\partial K}{\partial T}\,. \quad (95)$$

In taking the derivative of K with respect to T, it must be realized that $\mathcal{J}$ and F are functions of temperature. For want of any more accurate data, we will suppose that F, which characterizes that temperature dependence of the coercive field varies proportionally to the spontaneous magnetization. Finally, we find

$$\frac{1}{\rho} = \frac{Q + \log \tau}{T}\left(1 - \frac{2T}{\mathcal{J}}\frac{\partial \mathcal{J}}{\partial T}\right). \quad (96)$$

From the study of the temperature dependence of the spontaneous magnetization of magnetite [11], it follows that close to room temperature $(2T/\mathcal{J})(\partial \mathcal{J}/\partial T) = 0.29$. As the ferromagnetic substance which constitutes the particles has a Curie point at 675°C, slightly higher than that of magnetite which is at 590°C, we will multiply this value of 0.29 by the inverse of the ratio of the absolute Curie temperatures to obtain 0.26. We will take τ, which is a sort of residence time at the temperature T during the measurement of $\partial \sigma_r/\partial T$, as roughly 5 minutes, giving finally

$$\rho = \frac{300}{(22 + 5{,}7)\,1{,}26} = 8{,}6\,. \quad (97)$$

Unfortunately there is only one case where $\partial \sigma_r/\partial \log t$ and $\partial \sigma_r/\partial t$ have both been measured on the same sample, allowing an experimental determination of the ratio ρ. It is the sample C25 of Thellier, consisting of Nontron clay heated in nitrogen to 665°C and magnetized at room temperature in a field of 470 Oersteds. Figs. 2 and 3, from the work of the author, indicate the variations of the specific remanent magnetization as a function of time and of refiring temperature respectively.

From there we deduce

$$\frac{\partial \sigma_r}{\partial \log t} = -12{,}8 \cdot 10^{-6}\quad;\quad \frac{\partial \sigma_r}{\partial T} = -1{,}43 \cdot 10^{-6} \quad (98)$$

hence a ratio

$$\rho = 8{,}95 \quad (99)$$

in excellent agreement with the theoretical value given above.

Average volume of the particles — The same sample allows us to obtain some indication of the average volume of the particles responsible for the magnetic after-effect and the variation of the IRM on reheating. For this it suffices to compare, after the same refiring, the decrease of the TRM and that of the IRM. In both cases the curve sweeps out the same part of the (H_0, v) plane. The only difference arises because in the second case the particles in this region have been magnetized to saturation by a large field of 470 Oersteds, whereas in the first case the TRM is acquired in a field h too weak to produce this saturation. The ratio of these two magnetizations is equal to $\tanh(v\, h/kT)$. Hence

$$\frac{\partial \sigma_{tr}}{\partial T} : \frac{\partial \sigma_{ri}}{\partial T} = \text{th}\,\frac{v\mathcal{J}h}{kT}\,. \quad (100)$$

We have already given the value of $(\partial \sigma_{ri}/\partial T)$ above. As for $(\partial \sigma_{tr}/\partial T)$, it may be taken from the experiments of Thellier on the sample C21, if not identical to C25 at least very similar to it.

$$-\frac{\partial \sigma_{tr}}{\partial T} = 0{,}84 \cdot 10^{-6} \quad \text{with} \quad h = 13 \text{ oe}. \quad (101)$$

We deduce

$$\text{th}\,\frac{v\mathcal{J}h}{kT} = \frac{0{,}84}{1{,}43} = 0{,}587. \quad (102)$$

Finally, with $\mathcal{J} = 500$ we obtain

$$v = 3{,}4 \cdot 10^{-18} \text{ cm}^3$$

Critical field — Substituting this value into equation (91) which gives vH as a function of known quantities, we finally deduce

$$H = 1\,300 \text{ œrsteds} \quad (103)$$

which represents the average critical field of particles involved in the phenomena of magnetic after-effect and demagnetization by refiring. Substituting this value into equation (93), we obtain

$$Q = 22{,}3 \quad (104)$$

a value very close to the one we have adopted.

(45) Conclusions: In summary, the magnetic properties of the ensemble E set out in particular in paragraphs 25 to 32 and 38 to 42 are in qualitative agreement down to the last detail with the remarkable properties of baked clays and lavas found by Thellier, Roquet and Nagata. It therefore seems correct to seek there the general explanation of these phenomena. Furthermore, paragraphs 43 and 44 give an example of a quantitative confirmation of the theroetical predictions by experiment.

The phenomena of stabilization and the properties in an alternating magnetic field treated in paragraphs 33 to 47 have not so far been the subject of any particular study so far as the baked clays and lavas are concerned. There lie some interesting possibilities for quantitative verification of the theory.

To be in a position to put the theory to the test for a given sample, it is necessary to determine the function $n(H_0, v)$ which governs the distribution of the particles and carefully compare the theoretical predictions with the experimental results. Unfortunately there are too few experimental data and too many variable parameters for this problem to be of much interest. There will be no possible confirmation. Furthermore some serious difficulties present themselves. (1) We do not know the exact temperature dependence of the spontaneous magnetization and of the critical field H of the substance which makes up the particles. If the errors on the account are small near room temperature, this is not the case near the Curie point. (2) Baked clays and particularly lavas are *mixtures* of substances with different Curie points [12, 4]. Nevertheless it seems that in any case there should be no difficulty in finding for each sample at least one distribution $n(H_0, v)$ which can explain the observed experimental properties.

REFERENCES

[1] L. Néel, *Ann. de Phys.*, 1949, **4**, 249.
[2] E. Thellier, Thèse, Paris, 1948; *Ann. Inst. Phys. Globe*, 1938, **16**, 157; *C.R. Acad. Sci.*, 1941, **212**, 281; 1941, **213**, 59; 1941, **213**, 1019; 1946, **223**, 319; E. Thellier and O. Thellier, *C.R. Acad. Sci.*, 1942, **214**, 382.
[3] J. Roquet and E. Thellier, *C.R. Acad. Sci.*, 1946, **222**, 1288; J. Roquet, *C.R. Acad. Sci.*, 1946, **222**, 727; 1947, **224**, 1418.
[4] T. Nagata, *Bull. Earthquake Res. Inst.*, 1943, **21**, March, 1.
[5] R. Becker and W. Doring, Ferromagnetismus, Springer (Berlin, 1939).
[6] L. Néel, L. Weil and J. Aubry, *Brev. Fr. dép.* Champbéry, 7 Avril 1943, no. PV, 323; L. Néel, *C.R. Acad. Sci.*, 1947, **224**, 1488; L. Weil, *C.R. Acad. Sci.*, 1947, **225**, 229.
[7] L. Néel, *C.R. Acad. Sci.*, 1947, **224**, 1488.
[8] F. Bertaut, Thèse, Grenoble, 1949.
[9] F. Preisach, *Z.f. Physik*, 1945, **94**, 277.
[10] Koenigsberger, *Beitr. angew. Geophys.*, 1935, **5**, 193; *Terr. Magn. Atm. Elect.*, 1935, **43**, 119 and 299.
[11] P. Weiss and R. Forrer, *Ann. de Phys.*, 1929, **12**, 279.
[12] R. Chevalier and J. Pierre, *Ann. de Phys.*, 1932, **18**, 383.

C27 (1955)
SOME THEORETICAL ASPECTS OF ROCK-MAGNETISM

Summary: The memoir is devoted to a brief theoretical study of the most typical magnetic properties of rocks. In particular §§ 3–16 are on ferrimagnetism, §§ 17–35 on single domain particles and §§ 36–57 on large multi-domain particles.

Theoretical studies are made of the following aspects of the subject and compared with the experimental results: remanent magnetization (§ 38), initial susceptibility (§ 39), variation with applied field of thermo-remanent magnetization (abbreviated to T.R.M.) (§§ 40, 41, 57), the ratio Q_k of T.R.M. acquired in a given field to the induced magnetization in the same field (§ 42), the additivity of partial T.R.M.'s in the case both the small grains (§ 28) and large grains (§ 57).

Considerable space is devoted to the magnetic 'viscosity' due to thermal agitation in small grains (§§ 24–27) and in larger one (§§ 49–56). Expressions are given for magnetic 'viscosity' in the range of Rayleigh's relations (§ 51) particularly with a demagnetizing field present (§ 54). The theoretical and experimental results on the irreversible decrease in isothermal remanent magnetization are briefly quoted both for small (§ 30) and laege (§ 55) grains.

Different reversing mechanisms are reviewed which could cause a negative T.R.M., that is one directed in the opposite sense to the field applied during cooling. Some are related to negative Weiss-Heisenberg exchange forces: reversal by diffusion involving ionic exchange between the two sub-lattices in a ferrimagnetic (§ 7), reversal by anomalous thermal variation in spontaneous magnetization (§§ 11, 12), reversal by diffusion with complete change of composition (§ 16). The others are effects of the demagnetizing field: reversal in mixtures of two constituents with different Curie Points (§§ 31–34), reversal by segregation, allotropy and chemical alteration (§ 35). The actual examples so far known are recalled.

§ 1. Introduction

In the general scheme of magnetic theory, the problems of the magnetic properties of rocks are distinguished by certain peculiarities which it is convenient to make clear at the outset.

In the first place, the carriers of magnetic properties are the various more or less pure oxides of iron, magnetite, titanomagnetites, hematite, maghemite, etc., that is, substances which are ferrimagnetic or antiferromagnetic rather than classical ferromagnetics. A notable result is that the thermal variation of saturation magnetization of rocks can be of very different types from that of iron or nickel.

A second point to be emphasized is that the magnetic constituents are a small proportion, perhaps a few per cent, distributed amongst the practically non-magnetic bulk of the rock. The problem is that of magnetic grains more less far from one another so that the demagnetizing field due to shape is very important.

A third point is that, since the geophysicist is especially interested in the magnetic properties of rocks relative to the earth's field, the properties in fields small compared with the coercive force are of particular interest.

Finally, magnetic viscosity and the effects of time on magnetic phenomena in general are of great importance, because the time scale concerned is the geological one.

§ 2. The three processes of change in magnetization

A detailed study and investigation of the magnetic properties of rocks could be the basis of a complete treatise on magnetism, and naturally there is no place for that here. We shall consider only the most important particular points, referring for the rest to the classical works on ferromagnetism [1].

According to Weiss' already classic theory, a ferromagnetic substance is divided into *elementary domains* of varying size within which the magnetization is uniform. The direction of this magnetization varies from one domain to another but its magnitude J_s remains constant and is called the *spontaneous magnetization*. In practice, J_s depends only on the temperature T and the temperature at which it falls to zero is called the Curie Point. On the other hand the shape and size of the elementary domains as well as the orientation of their spontaneous magnetizations depend on many factors: the applied field, the internal demagnetizing field and that due to shape, magneto-crystalline and magneto-elastic couplings, the presence of dislocations or impurities, etc.

The magnetization of ferromagnetic bodies, considered as a function of magnetic field and of temperature, thus depends on three distinct processes: (a) the reversible change in the *magnitude* of the spontaneous magnetization with temperature: (b) the changes of direction or rotations of the spontaneous magnetization within the domains, whose boundaries remain fixed; (c) the displacement of the walls separating elementary domains whose direction of magnetization remains fixed.

We have to study the effects of these different processes on the properties of rocks and baked clays.

THE FERRIMAGNETISM OF ROCKS

§ 3. Definition of ferrimagnetism

The simplest kind of ferromagnetic substance is typified by iron: all the atoms are identical and have the same permanent magnetic moment. A coupling exists between the moments of neighbouring atoms which favours the parallel orientation of these moments so that, at temperatures low enough for the energy of thermal agitation kT to be small compared with the energy of coupling, all the atomic moments are parallel to each other: this is the strict definition of *ferromagnetism*.

But more complicated kinds can be imagined, such as where there are two types of atoms with couplings of such magnitude and sign that at low temperatures the atomic moments of one type all point in one direction and those of the other type all in exactly the opposite direction. One is then dealing with *ferrimagnetism* [2].

In both cases a small element of volume dv of the substance, large compared with atomic dimensions, has a permanent spontaneous magnetic moment $J_s dv$, where J_s is the spontaneous magnetization defined above.

§ 4. The simple inverse ferrites

Many of the ferromagnetic constituents of rocks, in particular the titanomagnetites, are of the spinel type with properties more or less analogous to those of the ferrites MFe_2O_4, where M is a divalent metal. In these ferrites the metallic ions can occupy two crystallographically diffefent types of sites: the lattice has, per unit molecule, one tetrahedral site A, surrounded by 4 oxygen atoms and two octahedral sites B surrounded by 6 oxygen atoms.

It has been shown experimentally [3] that several types of spinel ferrites exist: the *normal ferrites* have the M ion on the A site and the two Fe^{+++} ions on the B sites, as in zinc ferrite. These ferrites are not magnetic. There are also the *inverse ferrites* in which the A site is occupied by an Fe^{+++} ion and one of the B sites is occupied by the M ion and the other by the second Fe^{+++} ion; these ferrites are magnetic.

The interatomic couplings are, from a magnetic point of view, most powerful when they act between an ion on an A site and one on a B site; these AB interactions are *negative* and tend to align the atomic moments of the two interacting atoms antiparallel. The interactions between atoms occupying the same types of site, the AA or BB interactions, are generally much weker and play only a secondary role. Owing to the strong negative AB interactions the magnetic lattice is, at low temperatures, divided into two sub-lattices A and B, corresponding to the A sites and B sites, whose spontaneous magnetizations J_A and J_B are oriented in opposite directions: this is a typical case of *ferrimagnetism* [2].

These conclusions were reached from a profound study of the magnetic properties of these substances and have been very directly confirmed by neutron diffraction. Shull and his co-workers have shown [4], for example in magnetite (where $M = Fe^{++}$), that the magnetic moments of atoms on A sites are in fact antiparallel to those of atoms on B sites. Thus in the inverse ferrites the molecular moment, close to absolute zero, is simply given by the atomic moment of the M ion, since the moments of the Fe^{+++} ions on A sites exactly cancel those of the Fe^{+++} ions on B sites. In fact it is found that with $M = Ni^{++}, Co^{++}, Fe^{++}, Mn^{++}$ the molecular moments of the corresponding ferrites are close to 2, 3, 4 and 5 μ_B which are the values of the spin magnetic moments of these ions. The slight remaining differences are probably due to residual orbital moments.

§ 5. The distribution of ions on the different sites in spinels

Assuming that in the ferrites the elementary atomic moments of sub-lattice are aligned antiparallel to those of sub-lattice B, then the magnetic properties will be closely related to the distribution of cations between the two types of sites. This distribution depends on factors which are not yet fully understood. According to Verwey and Heilmann [3] and Romeijn [5], Zn^{++}, Cd^{++}, Ga^{+++}, In^{+++} and Ge^{+++} ions have a preference for A sites, Ni^{++}, Cr^{+++}, Ti^{++++} and Sn^{++++} ions for B sites while Mg^{++}, Al^{+++}, Fe^{++}, Co^{++}, Mn^{++}, Fe^{+++} and Cu^{++} ions can occupy either A or B sites according to circumstances.

In certain cases, the tendency of the ions to be placed in a definite ordered distribution is feeble enough for the distribution to change considerably with temperature. This effect is observable in $ZnFe_2O_4$ and $NiFe_2O_4$ and is important in $CdFe_2O_4$ while in $MgFe_2O_4$ and $CuFe_2O_4$ a detailed quantitative study has been made of it [6].

Letting w be the energy necessary to transfer a bivalent ion M from a B site to an A site, this transfer being of course accompanied by that of a ferric ion in the opposite direction, and letting y and $1-y$ be the proportions of M ions, on the A and B sites at temperature T, then Boltzmann's law shows that [2]:

$$\frac{y(1+y)}{(1-y)^2} = \exp\left(-\frac{w}{kT}\right).$$

Thus y is a function of T. The state of thermodynamic equilibrium at any temperature T can in principle be preserved in false equilibrium at room temperature by quenching. If the atomic moments of Fe^{+++} and M^{++} ions are respectively 5 and m Bohr magnetons, the saturation molecular moment M_s observed at low temperatures will be:

$$M_s = [m + 2y(5-m)]\mu_B.$$

These ideas have been confirmed by Pauthenet's work [6] on Mg and Cu ferrites in which he followed the variation of saturation moment with the temperature of quenching and found values of w/k equivalent to 1220°K and 1540°K respectively.

If the quenching temperature T is below 500–600°K thermodynamic equilibrium is no longer established at that temperature even after several hours. This shows the possibility of preserving a false equilibrium at room temperature for a time which is long compared with the length of experiments. But this would no longer be true after some tens of millions of years; then the moment M_s of the ferrite considered would correspond to y_0, the value of the equilibrium parameter appropriate for the prevailing temperature T_0. If this true equilibrium state happens to be destroyed, for instance by heating, no laboratory method can ever restore it. This is an example of the important effect that the enormous scale of geological time can have on the physical state of rocks as well as on their magnetic properties.

§ 6. Substituted ferrites: the case of aluminium

Important results are given by a study of substituted ferrites such as $NiFe_{2-m}Al_mO_4$, which can be described as $NiFe_2O_4$ in which some Al^{+++} ions have replaced Fe^{+++} ions. If all the Al ions replaced ferric ions on octahedral B sites, a decrease of saturation molecular moment at absolute zero would be observed proportional to the amount m of substituted Al ions. In particular for $m=0.4$, the saturation moment ought to be zero and the negative values given by the formula for $m>0.4$ would simply correspond to the spontaneous magnetization of sub-lattice A becoming greater than that of sub-lattice B.

In fact as E. W. Gorter has shown [7] this is roughly what happens for well-annealed specimens except that the moment is zero for $m=0.62$ showing that a small fraction of the Al^{+++} ions occupy A sites. On the contrary for quenched specimens the moment does not fall to zero and thus never becomes negative even for the composition $NiFeAlO_4$: the necessary conclusion is that in the quenched state a much larger proportion of Al^{+++} ions occupy A sites. Thus annealing causes some Al ions to move from A sites to B sites, giving a decrease of spontaneous magmetization since the number of Fe^{+++} ions, carrying $5\mu_B$, on A sites increases. This effect is most marked for $m=1$ when the molecular moment of the quenched specimen is $+0.42\mu_B$ while that of the annealed one is $-0.64\mu_B$.

§ 7. Change of sign of spontaneous magnetization by annealing

Since the spontaneous magnetization of $NiFeAlO_4$ changes sign on annealing the remanent magnetic moment of a specimen of this substance, magnetized at high temperature and then quenched, must change sign spontaneously on reheating: there is an *inversion of spontaneous magnetization*. This inversion is produced by diffusion of Al ions from A to B. To produce the inversion in the laboratory several hours at 400°C are needed, but the same effect would probably occur at ordinary temperaatures in some millions of years. Naturally this inversion of the spontaneous magnetization involves reversal of any remanent magnetization. From Gorter's data [7] the same inversion effect should occur over the whole range $0.62<m<1$. Also it seems probable that the compounds in the range $0.50<m<0.62$ which in the laboratory always have a positive moment, whether they are quenched or annealed, could, after some millions of years at ordinary temperatures acquire a negative moment: it is only necessary that, at the rather high temperature at which annealing in the laboratory still has a noticeable effect, the equilibrium concentration of the Al^{+++} ions be but slightly positive. Then the lower annealing temperature, made possible by the enormous length of the process, will displace the equilibrium towards negative moments. Such a substance would in all possible laboratory experiments always have a positive thermoremanence but in the course of geological tome would acquire a negative spontaneous magnetization and a remanent magnetization opposite to its original one.

This curious behaviour of the ferrites of the type $NiFe_{2-m}Al_mO_4$ is important although similar effects could probably not be observed in the natural substituted magnetites which form many of the ferromagnetic constituents of rocks. The reason is that in the nickel ferrites the magnetic moments of the two sub-lattices A and B are very close 5 μ_B and 7 μ_B so that it is much easier in this material for the A sub-lattice to have a greater moment than the B sub-lattice than in magnetite where the moments of the two sub-lattices, at 5 and 9 μ_B are much less close.

§ 8. Substituted ferrites: the case of titanium

Many of the ferromagnetic constituents of rocks are titanomagnetites, derived from magnetite Fe_3O_4 by substitution of Ti^{++++} ions for Fe^{+++} ions, together with a change of ionization of an Fe^{+++} ion to an Fe^{++} ion. This gives solid solutions of spinel type with the general formula $Ti_mFe_{3-m}O_4$ which were first studied by Michel and Pouillard [8]. If it is assumed that the Ti^{++++} ions always occupy B sites and that the remaining B sites are filled by Fe^{++} ions, the saturation molecular moment ought to be given by $(4-6m)\mu_B$ in the interval $0 < m < 0.5$ and by $(2-2m)\mu_B$ when $0.5 < m < 1$. The experimental facts are not enough to decide if these suggestions, especially on the distribution of Fe^{++} ions, are correct. All that is known is that the saturation moment and the Curie Point fall regular as m incfeases.

Recently Kawai, Kuma and Sasajima [9] have observed that solid solutions of this type, which immediately after quenching have a single clearly defined Curie Point, change gradually so that after several months at room temperature they have two Curie Points, one at about 100°C and the other close to that of magnetite. It is tempting to interpret this change as a segregation into two phases, one rather weakly magnetic and the other near to magnetite.

Finally mention must be made of maghemite which is similar to the cubic sesquioxyde $Fe_2O_3\gamma$ and has a spinel structure rather like that of magnetite but with some B sites empty: it is also a ferrimagnetic.

§ 9. The thermal variation of the magnetization of ferrites

The thermal change of spontaneous magnetization J_s in rocks is particularly interesting as they have generally acquired their natural magnetization in a temperature range close to their Curie Point where J_s is much smaller than at ordinary temperatures. In an assemblage of elementary domains, in which the boundaries and direction of spontaneous magnetization of each domain remain fixed, the mean magnetization of the aggregate will be proportional to, and vary revefsible with, J_s. Thus a small magnetization acquired near the Curie Point can increase reversibly during cooling of the specimen. As will be shown later (cf. §§ 32–35) this can have curious results in mixtures of several ferromagnetic constituents with different Curie Points. For most ordinary ferromagnetic substances such as iron, nickel, cobalt and their alloys, the curves of J_s against temperature have very similar shapes, typically a slow variation of J_s at low temperature (less than 10% up to $T = 0.6\theta$) and a rapid one near the Curie Point θ. In the case of rocks, although the $J_s - T$ curve of pure magnetite is very similar to that of metals, very different types are also found, such as a practically linear change of J_s with T from absolute zero right up to the Curie Point. This variability is typical of the ferrimagnetics.

§ 10. The molecular field theory applied to ferrimagnetism

As a first approximation to a theory of the thermal variation of spontaneous magnetization the Weiss molecular field theory can be used. This, being purely phenomenological, replaces the interactions between the elementary magnetic moments by an imaginary magnetic field called the molecular field which is proportional to the magnetization **J** and is given by:

$$\mathbf{H}_m = N\mathbf{J}$$

where N is the molecular field coefficient.

The magnetization law for a paramagnetic in which the atomic moments are quite independent of each other and can be freely oriented is of the form:

$$J = B(H/T).$$

If then a ferromagnetic can be considered as a paramagnetic, which is magnetized and given its spontaneous magnetization J_s by its own molecular field H_m, J_s will be the folution of the implicit equation:

$$J_s = B(NJ_s/T).$$

The classical discussion of this equation need not be given here except to recall that the shape of the $J_s(T)$ curve depends on the form of the function $B(H/T)$ which itself depends on the value of j, the total angular momentum quantum number of the magnetic atom concerned. The $J_s(T)$ curves of ferromagnetic metals correspond roughly to the value $j = 1/2$, indicating that the magnetic carriers are isolated spins.

Néel [10] has slightly modified the molecular field theory by supposing H_m to be proportional to the mean magnetic moment of closely neighbouring atoms. For substances like the ferrimagnetics having two sub-lattices A and B with magnetizations $\mathbf{J}_a$ and $\mathbf{J}_b$ two molecular fields must be considered, $\mathbf{H}_a$ acting on atoms on A sites and $\mathbf{H}_b$ on those on B sites. They are given by:

$$\mathbf{H}_a = n_{aa}\mathbf{J}_a + n_{ab}\mathbf{J}_b ,$$
$$\mathbf{H}_b = n_{ab}\mathbf{J}_a + n_{bb}\mathbf{J}_b ,$$

involving three molecular field coefficients n_{aa}, n_{ab}, n_{bb} representing the AA, AB and BB interactions. If $B_a(H/T)$ and $B_b(H/T)$ are the paramagnetic magnetizations of ions on the A and B sub-lattices respectively, it can be shown that, in the basence of an applied field, the system has a spontaneous magnetization $\mathbf{J}_s$:

$$\mathbf{J}_s = \mathbf{J}_a + \mathbf{J}_b .$$

It is the vector sum of the two partial spontaneous magnetizations of the two sub-lattices A and B, $\mathbf{J}_a$ and $\mathbf{J}_b$ which are solutions of the complementary equations:

$$\mathbf{J}_a = \mathbf{B}_a(\mathbf{H}_a/T) ,$$
$$\mathbf{J}_b = \mathbf{B}_b(\mathbf{H}_b/T) .$$

In the important case of the ferrites, n_{ab} *is negative and large compared with the other two coefficients* n_{aa} and n_{bb}. The partial spontaneous magnetizations are then antiparallel and J_s is simply the arithmetic difference between J_a and J_b. If these have different absolute values the system will have a finite resultant spontaneous magnetization, that is to say an apparent ferromagnetism, created by negative interactions.

The two partial spontaneous magnetizations J_a and J_b disappear together at the Curie Point θ above which simple paramagnetism is observed. However the reciprocal of the susceptibility χ does not vary linearly with temperature, following the Curie-Weiss law of ordinary ferromagnetics, but according to a hyperbolic law of the form

$$\frac{1}{\chi} = \frac{T}{C} + \frac{1}{\chi_0} - \frac{\sigma}{T-\theta} ,$$

where C is the theoretical Curie constant of the ions present and χ_0, σ and θ depend on n_{aa}, n_{bb} and n_{ab}. This hyperbolic variation is characteristic of ferrimagnetic substances.

The three coefficients of the molecular field can be found from a study of the susceptibility above the Curie Point [11] and hence the thermal variation of spontaneous magnetization below the Curie Point can be deduced without recourse to any other data. Good agreement between theory and experiment has been found by Pauthenet [6] for nickel, cobalt an iron ferrites and more recently by Clark and Sucksmith [12] for manganese ferrite.

Despite these good agreements it must not be forgotten that theory based on the molecular field approximation sometimes gives very inaccurate results as in the case of linear chains of atoms.

§ 11. Inversion of the spontaneous magnetization

Although the ferrites mentioned above already have a very different thermal change of spontaneous magnetization from the classical ferromagnetics, much more extraordinary curves can be conveived. That this is likely can be seen *a priori*, given the number of different factors on which the variation of $J_s(T)$ depends: values of j for the A and B ions, of the ratios n_{aa}/n_{ab} and n_{bb}/n_{ab}, and of the spontaneous magnetizations at absolute zero J_{as} and J_{bs} of the sub-lattices A and B. Curious forms are found especially when J_{ab} and J_{bs} have closely similar *magnitudes* as in observing their slight difference their own irregularities are magnified.

Thus curves of type P (Fig. 1) are possible, showing an *increase* of spontaneous magnetization from absolute zero to a maximum value. Such curves have actually been observed by E. W. Gorter [7] in ferrites of general formula $Ni_{1.5-a}Mn_aFeTi_{0.5}O_4$ when a lies in the range 0.4–0.675.

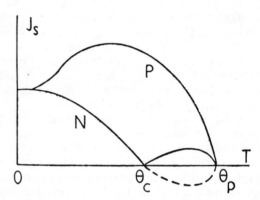

Fig. 1.
Two unusual types of thermal change of spontaneous magnetization characteristic of ferrimagnetic substances.

N is an even more remarkable type of curve in which the spontaneous magnetization decreases normally and disappears for the first time at temperature θ_c but then begins to increase again, reaches a maximum and finally disappears at a Curie Point θ_p. θ_c is not a Curie Point but only the temperature at which the spontaneous magnetization changes sign. Below θ_c the spontaneous magnetization J_a of sub-lattice A, say, is greater than that of B while above θ_c the reverse is true. θ_c can be called the *compensation temperature* at which the absolute values of the spontaneous magnetizations of the two sub-lattices are equal. Two conditions would appear to favour the appearance of this curious phenomenon: the two sub-lattices must have rather similar saturation magnetizations and at the same time rather different values of j.

This type of curve whose existence was forecast theoretically in 1948 was found some years later by E. W. Gorter [7] in the mixed ferrites $Li_{0.5}Fe_{2.5-a}Cr_aO_4$ for values of a between 1.00 and 1.70 (cf. Fig. 2).

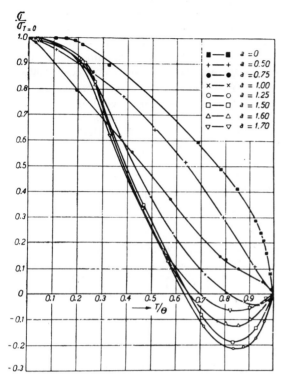

Fig. 2.
Curves of thermal change of spontaneous magnetization in the ternary ferrites $Li_{0.5}Fe_{2.5-a}Cr_aO_4$ after E. W. Gorter.

The interpretation of θ_c as a temperature at which the spontaneous magnetization changes sign can be confirmed by an elegant experiment. A remanent magnetization is given to the specimen at a temperarure below θ_c. It is then heated in a zero magnetic field past θ_c and the remanent magnetization is found to change sign spontaneously when the temperature of the specimen is θ_c. This is a very neat and direct proof of the existence of two magnetic sub-lattices in ferrites.

More recently other instances of the N type have been found in a new group of ferrites discovered by Forestier. These are the rare earth ferrites of general formula $Fe_2M_2O_6$ when M is a trivalent rare earth element [13]. Guiot-Guillain, Pauthenet and Forestier [14] have shown if M is Gd, Er or Dy these ferrites have N-type curves. They have also shown that the first point where the spontaneous magnetization disappears is in fact a compensation temperature by means of the spontaneous reversal of remanent magnetization at θ_c. Néel [15] has shown that as there are probably three sub-lattices the explanation is still more complicated[1]. In other respects the magnetic properties of these are earth ferrites are somewhat similar to those of rhombohedral ferric oxide $Fe_2O_3\alpha$ which will be considered later.

§ 12. Distinction between compensation temperature and Curie point

Although the spontaneous magnetization disappears at the compensation temperature just as at the Curie Point, the possibility of a thermo-remanent magnetization in a small field H which distinguishes the Curie Point, does not occur at the compensation temperature. In fact at this temperature while the couple HJ_s exerted by the external field on the resultant spontaneous magnetization J_s falls to zero because J_s falls to zero, the perturbing magnetocrystalline and magnetoelastic forces remain finite as they are of the order of the squares J_a^2, J_b^2 of the partial spontaneous magnetizations; thus there is no possibility of magnetization by a small field.

The result is that if a substance with a compensation point θ_c is acted on by a small field H when cooled from above its Curie Point, it will acquire near the latter a thermo-remanent magnetization (or T.R.M.) which from a few tens of degrees lower, will change practically reversibly remaining proportional to the spontaneous magnetization, independent of any variations in the applied field provided this remains of the order of H. In particular the T.R.M. will change sign with the spontaneous magnetization at θ_c. At room temperature such a substance will then have a T.R.M. opposite to the sense of the field to which it is due.

It should be noticed that this reversed T.R.M. is fundamentally due to a difference between the thermal changes of the two sub-lattices. It is a different mechanism from that described before (II 7) involving an exchange of ions between the two sub-lattices below the Curie Point.

§ 13. Triangular systems

Finally mention should be made in the study of ferrimagnetism of more complicated phenomena which appear if within say the B sub-lattice, negative BB interactions occur which are comparable with the AB interaction between the sub-lattices, the B sub-lattice can then be divided in its turn into two secondary sub-lattices. Such as the *triangular* systems

---

[1] It was realized a year after the writing of this article that these properties should in fact be attributed to a new type of compounds, the ferrite-garnets of the formula $Fe_5M_{13}O_{12}$.

studied by Yafet and Kittel [16]; no certain examples of them are yet known.

§ 14. Rhombohedral ferric oxide

Rhombohedral ferric oxide $Fe_2O_3\alpha$, the mineral *hematite*, is an important ferromagnetic constituent of rocks and baked clays. Unfortunately its magnetic properties are not yet fully understood despite much study. The effects [17] are approximately those of a paramagnetism only slightly dependent on temperature, with a specific susceptibility of about 20×10^{-6}, superimposed on a very feeble ferromagnetism with a molecular saturation magnetization of about one-hundredth of a Bohr magneton and a Curie Point at 675°C, definitely higher than that of magnetite (595°C).

The paramagnetism derives from a *fundamental antiferromagnetism* due to the distribution of ferric ions on two equal sub-lattices with equal and opposite spontaneous magnetizations. This distribution was suggested by detailed analysis of the magnetic properties and has subsequently been directly confirmed by neutron diffraction (Shull *et al.* [4]). At temperature below 250°K, the direction of antiferromagnetism, that is the direction to which the spontaneous magnetizations of the two sub-lattices are parallel or antiparallel, lies in the basal plane of the crystal [18] and can turn freely in that plane. The result is that the paramagnetic susceptibility, which is related to distortion of the antiparallel arrangement of spontaneous magnetizations under the action of the applied field, is practically independant of the crystallographic orientation of this field. On the other hand, above 250°K, the direction of antiferromagnetism changes and becomes parallel to the ternary axis of the crystal with the result that the susceptibility becomes much smaller in this direction than in the basal plane, as is found experimentally.

The temperature at which ordering in the two antiferromagnetic sub-lattices disappears (the Néel temperature) is probably 675°C because there is, at this temperature, a large specific heat anomaly which, having a magnitude comparable to that of magnetite, cannot be attributed to the feeble ferromagnetism.

The problem of the superimposed ferromagnetism is much more difficult. It has sometimes been thought, owing to the low value of the saturation magnetization, that it was due to ferrimagnetic impurities such as magnetite or the cubic sesquioxide $Fe_2O_3\gamma$, but this hypothesis runs into difficulties over the change of magnetic properties according to thermal treatment. Also it has been shown that the ferromagnetism is oriented with respect to the lattice; for instance, above 250°K, the ferromagnetic part of the magnetization is much greater in directions in the basal plane than along the ternary axis. To explain this it has been suggested that the ferromagnetism is due to small crystals of magnetite, deformed and oriented by intergrowth in the larger crystals of the sesquioxide. This explanation may be partly true in some cases, but it seems more likely that the source is rather in lattice defects, such as errors in the regular alternation of layers of ion atoms magnetized antiparallel, which defects might be related to dislocations of the lattice; there would then be a sort of ferromagnetic plane surrounded by an antiferromagnetic matrix. Or it could be described as an imperfection in the antiferromagnetism of the rhombohedral sesquioxide which caused a slight asymmetry between the two sub-lattices and thus produced a small resultant spontaneous magnetization: in general there would be a slight ferrimagnetism. The strongest argument for this hypothesis is the coincidence of the ferromagnetic Curie Point with the Néel temperature of the fundamental antiferromagnetism.

Whether they be due to impurities or to lattice defects, it seems that the magnetic properties of $Fe_2O_3\alpha$ can be represented as those of very small feroomagnetic domains, having the properties of hysteresis, buried in a paramagnetic matrix. The magnetization of the matrix is always small and proportional to the applied field so that it is of little interest in rock-magnetism.

§ 15. The ilmenite-hematite series

Another series of compounds occurs in rocks, with the general formula $Fe_{2-m}Ti_mO_3$ which can be considered [19] as a combination in varying proportions of ilmenite $FeTiO_3$ and hematite Fe_2O_3; these compounds form two series of solid solutions, one of ilmenite type and one of hematite type separated by a solubility gap which probably increases in size as the temperature falls [19].

Ilmenite is antiferromagnetic. Crystallographically its structure is derived from that of $Fe_2O_3\alpha$ by feplacing every other layer of iron atoms by a layer of Ti atoms. These, in the ionic state Ti^{++++}, are not magnetic so that the alternations of the layers of cations can very probably be represented as: $(+Fe)(Ti)$ $(-Fe)(Ti)(+Fe)\dots$ where each layer is represented by brackets and the + and − signs represent the antiparallel orientations of the spontaneous magnetizations of the iron layers. In $Fe_2O_3\alpha$ the alternation is $(+Fe)(-Fe)(+Fe)\dots$

Near the composition $Fe_{4/3}Ti_{2/3}O_3$ which is close to the limit of solid solution the compounds become definitely ferromagnetic with a molecular saturation moment of about $2\,\mu_B$. Such a large magnetic moment cannot be attributed to impurities or lattice defects but rather to a ferrimagnetic arrangement such as $(+A)$ $(-B)(+A)(-B)\dots$ in which the positive spontaneous magnetization of the A layers is a little greater than the negative one of the B layers. For instance, Nagata [20] suggests the compositions $(+Fe_{1/6}^{++} Fe_{2/3}^{+++} Ti_{1/6}^{++++})$ for the A layers and $(-Fe_{1/2}^{++} Ti_{1/2}^{++++})$ for the B layers giving a resultant molecular moment of $2\,\mu_B$.

§ 16. A possible reversal of spontaneous magnetization by change of composition

As stated above, the solubility gap which exists around $m=0.5$ probably becomes wider when the temperature becomes lower. Suppose that this process extends to the composition $Fe_{4/3}Ti_{2/3}O_3$: a new phase close to Fe_2O_3 and very rich in iron will be formed at the expense of the original phase, which will then lose iron and gradually tend towares ilmenite in which the alternation of layers is $(Fe^{++})\,(Ti^{++++})\,(Fe^{++})\ldots$ This leads to an increase in the spontaneous magnetization of the A layers and a decrease in that of the B layers. At the beginning the magnetic alternation of layers is $(+A)\,(-B)\,(+A)\,(-B)\ldots$ and remains thus under the influence of the negative interactions between consecutive layers. But from the antiferromagnetic structure of ilmenite given above, there must also be negative interactions between the A layers. These negative interactions of AA type may become greater and greater as the spontaneous magnetization of A increases while on the other hand the AB interactions decrease with the decrease in spontaneous magnetization of B. A point would then be reached when the magnetic structure $(+A)\,(-B)\,(+A)\,(-B)\,(+A)\ldots$ with a positive *resultant* spontaneous magnetization, would become unstable and would invert to the more stable structure $(+A)\,(-B)\,(-A)\,(-B)\,(+A)$ by reversal of the spontaneous magnetization of alternate A layers. This structure has *negative* resultant spontaneous magnetization. This reasoning presupposes that the spontaneous magnetization of the B layers maintains its original orientation despite both the reversal of some of the A layers and the thermal agitation. For this to be true it seems that there must be large enough positive interactions between B atoms in the same layer.

Whether this is true or not it is *a priori* not absurd that segregation of a solid solution into two phases during cooling could cause a reversal of T.R.M. by an atomic process involving negative molecular fields. It is evidently necessary to study the magnetic properties of compounds of the $Fe_{2-m}Ti_mO_3$ type carefully and to determine, preferably by neutron diffraction, the details of the antiferromagnetic structure and its modifications according to thermal treatment and change of concentration.

It will be seen later that unmixing can also give rise to a reversed T.R.M. by means of a demagnetizing field due to shape (cf. §35).

SINGLE DOMAIN PARTICLES

§ 17. Wall thickness and grain size

In the range of magnetic fields in which hysteresis occurs, change of magnetization in ferromagnetic bodies is due to *rotation* of the spontaneous magnetization of the elementary domains and no *movement of the walls* separating them. In general the second process is more important in fields small compared with the coercive force.

The separating walls must not be regarded as infinitely thin surfaces but rather as zones of transition of finite thickness in which the magnetization gradually changes from the direction on one side to that on the other. The theory of these walls was put forward by F. Bloch [21] and has since been developed by others [22]: the *thickness* of the zone is given in order of magnitude by:

$$p = a(NJ_s^2/K)^{1/2}$$

where a is the distance between neighbouring magnetic atoms, N the Weiss molecular field coefficient and K the magnetocrystalline anisotropy energy. This gives wall thicknesses of the order of a few hundred to a few thousand Angströms.

These results show immediately that very small grains of dimensions of the order of and less than p, can only contain a single domain since there is no space for a wall. In these very fine *single domain* grains, the magnetization is uniform and equal to the spontaneous magnetization J_s. The upper limit of grain size has been determined elsewhere (Néel [23]).

With such single domain grains changes of magnetization can only occur by the rotation process so that the phenomena are particularly simple.

As the ferromagnetic constituents of rocks are dispersed in very small grains it is natural to see how far their magnetic properties can be explained on the model of single-domain grains.

§ 18. Elementary magnetization cycles [24]

In a single domain grain the internal magnetization energy depends only on the orientation of the magnetic moment with respect to certain axes in the grain. Considering, for simplification, magnetically uniaxial grains, the energy E is given by:

$$E = Kv \sin^2 \theta$$

where v is the volume of the grain and θ the angle between the magnetic moment and the axis.

The anisotropy constant K can arise in various ways. It can be due to a magnetocrystalline coupling so that K is equal to the anisotropy constant of a large crystal of the same material. It can arise from an anisotropy of shape: if the grain is an ellipsoid of revolution with demagnetizing field coefficients n along the axis and m in the equatorial plane, then:

$$K = \frac{1}{2}(m-n)J_s^2.$$

Finally an anisotropy due to mechanical stress gives a value of K:

$$K = \frac{3}{2}\lambda\sigma$$

where σ is a tension parallel to the axis and λ is the longitudinal saturation magnetostriction of the material of the grain. The three mechanisms can of course act simultaneously.

In an actual system these would be a very large number of grains, oriented at random. But the properties of a grain depend greatly on its orientation with respect to the applied field.

When the axis of the ellipsoid is parallel to H, the hysteresis cycle is rectangular, with the overall height of $2J_s$ and overall width of $2H_c$. The coercive force H_c being given by $2K/J_s$. At $H = +H_c$ and $H = -H_c$ there are discontinuities in the magnetization, elsewhere the susceptibility is zero.

Conversely there is no hysteresis at all when the axis of the grain is perpendicular to the magnetic field. The component of the magnetic moment along the field is constant and equal to $-vJ_s$ when $H < -H_c$, varies linearly from $-vJ_s$ to $+vJ_s$ as H changes from $-H_c$ to $+H_c$ and is constant at $+vJ_s$ when $H > +H_c$.

In a randomly assemblage of independent grains, the average limiting cycle is [25] similar to Fig. 3 with a remanent magnetization J_r equal to $1/2J_s$ and a coercive force of $0.96K/J_s$ about half the maximum coercive force of the individual grains [26]. The initial susceptibility is $J_s^2/3K$. The limiting hysteresis cycle is analogous to that of "hard" magnetic materials (Alnico, Riconal, etc.) which are now used for good permanent magnets.

§ 19. Objections to the interpretation of the magnetic properties of rocks by means of the theory of single domain grains

It is now possible to interpret the magnetic properties of a rock as those of an assemblage of randomly oriented domain grains.

First there is a rather theoretical fundamental objection. The grains of ferromagnetic minerals in rocks vary from one to several hundred microns in size. Now for magnetite the critical diameter should be about 0.02–0.03 μ and certainly less than 0.1 μ. This value should be reasonably accurate as an order of magnitude since the formula used to get it [23] gives

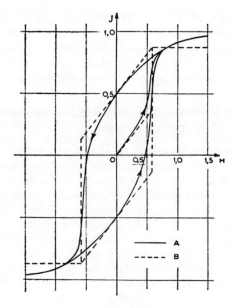

Fig. 3.
Limiting hysteresis cycle and initial magnetization curve (full lines) for an assemblage of independent randomly oriented grains.

results for iron which are confirmed by direct measurement of the grain size with x-rays or an electron microscope.

The second objection is more experimental and concerns the ratio of the remanent magnetization of a rock specimen to its saturation magnetization: it can be shown experimentally ([II] 38) that this ratio is proportional to the coercive force, whereas the theory of single-domain grains predicts a constant value of 1/2. Also, Nagata [27] has shown that the coercive force of rocks increases as the grain-size decreases while theoretically it should not depend on size after the single domain size is reached.

However, all these facts will later be found explicable in terms of multi-domain grains.

The above arguments are based on the experimental data for rocks with fairly low coercive forces of less than 1000 oe in which the magnetic constituents are magnetites or titanomagnetites. They do not apply to rocks with minerals of the ilmenite-hematite type which have far higher coercive forces sometimes of several thousand oersteds.

In these, as already mentioned, it is possible that the model of single domain grains does correspond to the actual structure of the rock. To prove this definitely their magnetic properties must be more closely studied. Valuable information can be got in this way particularly from application of Rayleigh's relations.

§ 20. Lord Raylieh's relations and single-domain particles

The Rayleigh relations concerned here are those applying to magnetization in fields very weak compared to the coercive force. A statement of these relations will be given later with an interpretation of them obtained by considering the macroscopic hysteresis cycle as the sum of a large number of unsymmetrical elementary cycles. Each unsymmetrical cycle will correspond with a point (a, b) in a plane where a and b are the lower and higher coercive forces (cf. § 44).

An assemblage of single domain grains does not obey the two Rayleigh relations. Each grain in fact has a *symmetrical* cycle, so that $a = -b$ and all the points representing elementary cycles lie on the second bisector 0δ of the plane (a, b). Thus although it is possible to satisfy the first of Rayleigh's relations by an appropriate density of representative points on 0δ, it is impossible to satisfy the second relation because on the returning branch of the hysteresis cycle when the field H decreases from its maximum H_m to zero, the magnetization J maintains its maximum value J_m and only begins to decrease when H becomes negative. See § 44 and Figs. 13–16.

§ 21. The role of magnetic interaction between grains

In principle this provides a method of deciding whether the independant grains are multi-domains or single domains by finding whether they do or do not obey Rayleigh's second relation.

But when single-domain grains are sufficiently close to each other interactions become important and the problem is somewhat altered. We have shown elsewhere [28] that it is then convenient to distinguish different effects.

First, the magnetic shape anisotropy is decreased by proximity and the mean coercive force is thus decreased from H_c to $H_c(1-v)$ where v is the volume of ferromagnetic grains in unit volume.

Then there is a sort of Lorentz field to be considered equal to $4/3\pi J_m$ where J_m is the mean magnetization per unit volume; it is *magnetizing* field which makes the whole hysteresis cycle more upright.

Last but most important from the present point of view are the variations of the Lorentz field from its mean value from poace to place. The interacting magnetic field h applied to a grain by its neighbours is different for each grain but remains constant in direction and intensity provided the external field is small. The effect is as if the representative point of the primitive cycle $(-a, a)$ became instead $(-a-h, a-h)$ thus representing an unsymmetrical cycle. The two Rayleigh relations are then automatically obeyed without the need for any special distribution of anisotropy of the elementary grains.

It is interesting [28] that when the grains are very close ($v=0, 5$) the ratio BH_c/A at 0.70 is similar to the value 0.58 for the same ratio given by the domain wall theory [29] when the walls are very close to each other[1]. Similarly the values of coercive force given by the wall theory and by the theory of single-domain grains are about the same when the volume occupied by magnetic substance and the unoccupied volume are of the same order. Generally speaking the properties of very imperfect substances are treated equally well by the wall theory or by the theory of single domains very close to each other.

Usually the volume of ferromagnetic constieutnts in a rock is very small, of the order of a few per cent so that interactions are weak but not negligible compared with the earth's field. In fact it has been shown [28] that the root-mean-square of the fluctuating field is of the order of vJ_s. If $J_s = 500$ and $v = 0.02$ this gives 10 oe. 2% of single-domain grains dispersed in a non-magnetic matrix should thus obey the two Rayleigh relations between -10 and $+10$ oe. If the coercive force were large, say several hundred or several thousand oersteds, the discrepancies from Rayleigh's relations outside this interval, -10 to $+10$ oe should be easy to detect and should give a conclusion as to whether or not single-domain grains were concerned.

The range of validity of Rayleigh's relations is much greater in more complicated structures such as in aggregates of grains where the total relative volume of the aggregates is small but within each aggregate the grains are very close to each other. The theory of interacting single-domain grains is then very useful as will be seen later (cf. § 32).

§ 22. Thermo-remanent and isothermal remanent magnetization

Although the dimensions of ferromagnetic grains in igneous rocks are generally too large for the theory of single-domain grains to be applied and although it leads to some conclusions which do not agree with certain experimental facts it is still of great interest because of its simplicity and the fact that in some cases (such as $Fe_2O_3\alpha$) it is probably correct.

In particular it allows of a very simple interpretation of the *magnetization by heating* of rocks in general. This is the property by which a rock, heated above its Curie Point and cooled to room temperature in a magnetic field H small compared with the coercive force, acquires a *permanent* magnetization at ordinary temperatures called the *thermo-remanent magnetization* (T.R.M.) which has the following

---

[1] A and B are defined in paragraph 43; H_c is the coercive force.

properties [31]. It is much greater and more stable than the *isothermal remanent magnetization* (I.R.M.) obtained by the classical process of applying the same field H and then removing it, at constant temperature. Sometimes the T.R.M. acquired in a given field can be hundreds of times greater than the I.R.M. produced by the same field. The negative field required to destroy the magnetization is much greater for a T.R.M. than for an I.R.M. when only a field about the same as or even less than the original magnetizing field is needed. Finally the I.R.M. due to a small field can be destroyed by heating to a quite moderate temperature while the corresponding T.R.M. is unaffected.

§ 23. Outline interpretation of T.R.M. [32]

To obtain an approximate explanation of these properties consider an assemblage of single domain grains whose individual coercive forces are distributed between zero and an upper limit which is a little higher than the overall coercive force H_c; H_c being large compared to the magnetizing field H.

The I.R.M. produced by the field H is exclusively due to the grains having a coercive force less than H and therefore magnetized irreversibly by this field. This I.R.M. is small cince only a small proportion of the order of H/H_c of the grains are affected and evidently it can be destroyed by an opposing field of about the magnitude of H.

To understand the thermo-remanent case it must be remembered that the individual coercive forces of the grains are of the order of K/J_s and tend to zero at the Curie Point. This is an experimental fact explained theoretically by the variation of K as J_s^2 whether K is due to shape anisotropy or a magneto-crystalline anisotropy arising from interactions similar to the magnetic dipolar coupling: thus H_c varies as J_s and tends to zero at the Curie Point (cf. § 41 again). Therefore even a very small magnetic field will suffice, near the Curie Point, to magnetize irreversibly all the grains of the assembly. As the temperature falls the intensity of the magnetization of the grains will increase proportionally to the spontaneous magnetization while the direction in each grain remains fixed. At room temperature a T.R.M. is thus produced which is equal to the maximum possible I.R.M. J_r. To produce J_r isothermally a nagnetic field of the order of the maximum coercive force H_c would be necessary. Evidently such a T.R.M. will be much higher and much more stable than the I.R.M. acquired in the same field.

This reasoning succeeds in explaining some of the curious properties of the T.R.M. but unfortunately leads to the conclusion that it is independent of the field H applied during cooling and always equal to the maximum remanence J_r of the limiting hysteresis cycle. This does not agree with the experimental facts that in small fields the T.R.M. varies as H and in slightly higher fields as $H^{1/2}$ (cf. § 43). Some factor, possibly the thermal agitation, must have been neglected.

§ 24. Thermal agitation in a single-domain grain [32]

The magnetic moment vJ_s of a single-domain grain such as those considered earlier ($\Pi\Pi$ 17 and 18) can, in the absence of any applied magnetic field, take up two orientations both of equal minimum energy: $\theta = 0$ and $\theta = \pi$. Obviously if the height of the potential barrier between these two positions is very large compared with kT, thermal agitation cannot move the magnetic moment from one position to the other so that it remains always in the direction to which it was originally brought by a magnetic field. However since the height of this potential barrier is $vH_cJ_s/2$ a value of v can always be found, so small that its height is of the order of kT, in which case thermal agitation can cause the moment to change spontaneously from one position to the other.

In these conditions if a remanent moment M_0 is given to each grain of an assembly of identical grains, it will tend to zero exponentially, thus:

$$M_r = M_0 \exp(-t/\tau_0)$$

where τ_0 is the relaxation time of the grain.

§ 25. Determination of the relaxation time

This relaxation time depends on perturbing couples acting on the magnetic moment of the grain. When there are no such couples the only force acting on the moment is the classical couple due to the energy term $Kv \sin^2 \theta$ which, since the atomic moments are analogous to gyroscopes, produces a continuous precession of the moment around the axis of symmetry of the grain, θ remaining constant. This is no longer true when there are perturbing couples in the equatorial poane capable of changing the polar angle θ and allowing the magnetic moment to cross the equatorial plane.

It seems that the most active of these perturbing couples arise from elastic deformations of the grain by thermal agitation. Because of these deformations the grain loses its spheroidal symmetry and takes up the shape of, for instance, a triaxial ellipsoid. This causes the appearance of a magneto-elastic couple and a transverse demagnetizing field both of which tend to turn the magnetic moment in a meridian plane. Boltzmann's principle gives the distribution in energy levels close to the equatorial plane. Using this and knowing the speed of precession due to the perturbing couples it is easy to calculate the number of moment vJ_s which cross the equatorial plane per second and hence find the relaxation time. Thus if we put:

$$C = \frac{eH}{2m} |3G\lambda + DJ_s^2| \left[\frac{2v}{\pi GkT}\right]^{1/2}$$

then:

$$\frac{1}{\tau_0} = C \exp\left\{-\frac{vH_c J_s}{2kT}\right\}, \quad (1)$$

where e, m are the charge and mass of the electron, G the shear modulus and D a numerical constant depending on the shape of the grain equal to about 3.

Similar calculations can be applied to the case where there is an external magnetic field h, parallel to the axis of the grain. There are two relaxation times $\tau(0, \pi)$ and $\tau(\pi, 0)$ for movement from the position $\theta = 0$ to $\theta = \pi$ and conversely. These are:

$$\frac{1}{\tau(0,\pi)} = C\left(1 + \frac{h}{H_c}\right)\left(1 - \frac{h^2}{H_c^2}\right)^{1/2} \exp\left\{-\frac{vJ_s(H_c + h)^2}{2H_c kT}\right\},$$

$$\frac{1}{\tau(\pi,0)} = C\left(1 - \frac{h}{H_c}\right)\left(1 - \frac{h^2}{H_c^2}\right)^{1/2} \exp\left\{-\frac{vJ_s(H_c - h)^2}{2H_c kT}\right\}.$$

§ 26. Critical diameter: blocking temperature

It is interesting to apply formula (1) to grains of iron in which $H_c = 1000$, $J_s = 1700$, $e/m = 1.76 \times 10^7$, $\lambda = 20 \times 10^{-6}$, $G = 0.77 \times 10^{12}$ and $k = 1.38 \times 10^{-16}$. This gives relaxation times τ_0 varying from 10^{-1} sec to 10^9 sec as v/T varies from 3.2×10^{-21} to 7.0×10^{-21}. Thus at room temperature spheres of iron 160 Å in diameter can keep the initial magnetization indefinitely while spheres only a little smaller, of 120 Å diameter are degmagnetized almost instantaneously, in less than 0.1 sec.

These figures show that at a given temperature there is a fairly well defined critical diameter D_T separating the grains into two groups.

Grains in the first group with diameters greater than D_T keep the initial orientation of the magnetic moments indefinitely in false equilibrium despite the action of any magnetic field h which is small compared with the normal coercive force H_c.

In the second group of grains with diameters less than D_T, thermodynamic equilibrium is reached almost instantaneously. In a small field h, parallel to the axis of the grain the only two possible positions for the moment are parallel or antiparallel to the field; then the mean magnetic moment of the grain is given by:

$$M(T) = vJ_s(T) \operatorname{th} \frac{vhJ_s(T)}{kT}. \quad (2)$$

The expression $J_s(T)$ is used to emphasize that J_s is a function of temperature. This moment $M(T)$ follows exactly any variations occurring in the applied field h.

The *blocking temperature* T_B is the temperature at which the diameter of the grain is the critical one or in other words the temperature at which the relaxation time becomes of the order of the duration of the experiments made on a rock. Thus in an *assemblage* of grains of very different volumes the blocking temperature can vary from the Curie Point for large grains to nearly absolute zero for very small grains.

§ 27. Process of acquiring T.R.M. [32]

The process by which such an assemblage of grains acquires a T.R.M. can now be analysed. Consider the system cooled from above its Curie Point to room temperature in a weak magnetic field $h(T)$ which is a function of the temperature. At any temperature above T_B the grains of diameter D_T acquire a mean moment given by equation (2) with $h = h(T)$ and then keep this same moment as the temperature falls below T_B because the relaxation times become too long for the moment to change from one equilibrium position to the other. Strictly, taking account of the thermal variation of J_s the mean magnetic moment $M(0)$ at room temperature T_0 is:

$$M(0) = vJ_s(T_0) \operatorname{th} \frac{vJ_s(T_B)h(T_B)}{kT_B} \quad (3)$$

Thus at room temperature each grain has a mean magnetic moment which depends on the field $h(T)$ which was acting at the temperature T when the critical diameter became equal to the actual diameter of the grain. Hence the name blocking temperature.

The nmean moment acquired in this way is not affected by chance variations in the field $h(T_0)$. On heating the system, thus magnetized, progressively in zero field the grains with blocking temperature T only lose their moment at T, that is at exactly the same temperature as they acquired this moment. As Thellier has put it, these grains have a magnetic memory and preserve the history of the magnetic field in qhich they cooled.

Equation (3) shows that the T.R.M. is proportional to the applied field h, when this is small enough.

Except for the small proportion acquired close to room temperature, the T.R.M. is particularly stable. In fact a field of the order of the coercive force of the grains, which may be very high, is needed to change it; similarly the relaxation time for spontaneous demagnetization is enormously long.

In contrast to this peculiar stability of T.R.M., an I.R.M. can be removed by a opposing field only about as great as that which produced it; in the same way the relaxation times are very much smaller than for a T.R.M. since the height of the potential barrier between the two stable positions of the magnetic moment is only of the order of $vJ_s h/2$ instead of $vJ_s H_c/2$.

§ 28. Partial thermo-remanent magnetizations (P.T.R.M.'s)

The properties ascribed by this theory to the T.R.M. of an assemblage of single-domain grains correspond remarkably well to the thermo-remanent properties of bricks and baked clays first described by Thellier [31] and to those of the rocks studied by Nagata [33]. These properties have been well summarized by Thellier as follows [34]: "To any temperature interval T_1, T_2 ($T_0 < T_1 < T_2 < \theta$) there correspond for a given specimen and a given magnetic field h, a particular magnetic moment which is acquired by the specimen when cooled from T_2 to T_1 in this field h. This moment is directed parallel to h and is unaffected by any heating to temperatures less than T_1 but disappears completely by heating to T_2. Further it is quite independent of other thermoremanent moments acquired in temperature intervals outside T_1 and T_2 even though they be due to fields h that are different in magnetide and direction. All these moments are added geometrically but, paradoxical though it may seem, each is quite independent and preserves a sort of exact memory of the temperature and field which produced it".

Can it be concluded from this agreement of theory and experiment that the ferromagnetic constituents of these specimens are single-domain grains? This is quite likely if the constituent is rhombohedral ferric oxide in which the ferromagnetic elements must be very small. But magnetites and titanomagnetites generally occur in grains of diameter much greater than a micron and the theory of single-domain grains cannot be applied to them. In addition, as will be shown later, the theory of large grains accounts for their properties much better than that of small grains. Thus it ought to be possible to explain the properties described by Thellier on other models than that of single domain grains (cf. § 57).

Thellier's [31] and Nagata's [33] experiments show that the blocking temperatures of magnetic rocks range from room temperature up to the Curie Point. But when there is only one ferromagnetic constituent, most of the T.R.M. is acquired in a fairly small temperature range about 50° below the Curie Point, so that for simplification each ferromagnetic constituent can be associated with a single blocking temperature.

§ 29. Time-variation of isothermal remanent magnetization (I.R.M.)

Suppose that an assemblage of single-domain grains has been magnetized by a field H_m and that the remanent magnetization at time t_0 is σ_{r_0}, the time origin being the instant at which the field H_m was removed. Roughly speaking, at time t_0 those grains with time constants τ less than t_0 will have lost their magnetization while those with $\tau > t_0$ will have kept theirs unchanged. Later at time t the remanent magnetization will have fallen to a lower value σr since grains with time constants between t_0 and t will have been demagnetized in their turn. Now the relation between the constants of a grain (v, H_c, ..., etc.) and its relaxation time τ is:

$$\frac{vH_c J_s}{2k} = T_0(Q + \log \tau), \quad (4)$$

where $Q = \log C$. Hence the decrease in remanent magnetization $\sigma_{r_0} - \sigma_r$ must be proportional to the change in $T(Q + \log t)$ from t to t_0 provided this change is small, thus:

$$\sigma_r = \sigma_{r_0} - AT_0(\log t - \log t_0'). \quad (5)$$

The constant A depends on the distribution of the grains as a function of v and H_c.

Thus the decrease in remanent magnetization is proportional to the logarithm of the time. Some interesting results of Thellier's can be explained in this way; in particular, for some specimens he found a logarithmic law for values of t from 20 sec to 5×10^6 sec. This decrease in remanent magnetization with time is only one of many aspects of the phenomenon of 'magnetic viscosity' shown by an assemblage of fine single-domain grains; fuller details will be found in an earlier paper [32].

§ 30. The effect of rise of temperature on remanent magnetization

As above, let σ_{r_0} be the remanent magnetization remaining at time t_0 at room temperature T_0, and suppose that the temperature is raised very quickly to T_1 and maintained constant at that level for a period of the same order as t_0. During this process $T(Q + \log t)$ increases by about $Q(T_1 - T_0)$, producing demagnetization of a certain number of grains which remain demagnetized when the temperature is lowered again. A decrease of thermal origin has occurred in the spontaneous magnetization, given by:

$$\sigma_r = \sigma_{r_0} - AQ(T_1 - T_0), \quad (6)$$

where A has the same value as above. The decrease in remanent magnetization is proportional to rise in temperature.

This phenomenon has been observed by Thellier [31]. From (5) and (6) one can obtain:

$$\frac{\partial \sigma_r}{\partial \log t} \bigg/ \frac{\partial \sigma_r}{\partial T} = \frac{T_0}{Q} \quad (7)$$

The ratio of the two derivatives is equal to T_0/Q. Assuming reasonable values for the different parameters in the expression for Q one gets that Q varies from 19.9 to 23.4 as H_c varies from 10 to 10^4 oe.

Taking the values $Q=22$ and $T=300°K$, T_0/Q is about 13.5. Thellier found for one specimen (No. C 25) the value 8.95.

Actually this argument is a little too simple because in the calculation of $\partial \sigma_r/\partial T$ the variations of H_c and J_s in equation (4) have been ignored. Both these decrease with rising temperature though exactly how is still poorly known. Therefore the decrease in σ_r with T is a little more than that calculated from equation (4) and the ratio of the two derivatives, equal to T_0/Q according to equation (7), is in fact a little less as in the experimental results.

§ 31. T.R.M. of a mixture of two constituents [35]

When there is a mixture of two ferromagnetic constituents A and B with clearly different Curie Points, the T.R.M. of the mixture sometimes has curious properties. As explained above (cf. § 26) the T.R.M. of constituent A is fixed at a blocking temperature T_a which is assumed here to be above the Curie Point of constituent B. Below T_a the T.R.M. of A increases reversibly with the spontaneous magnetization of A as the temperature falls, independently of the applied field. But the T.R.M. of B is entirely dependent on the field acting on B at its own blocking temperature T_b. This field is the geometrical sum of the external applied field H and the demagnetizing field created by the previously magnetized grains of A. It can happen that this resultant field is opposite in direction to H and strong enough for the T.R.M. of B at room temperature to be greater than that of A. In these conditions, since the T.R.M. of B is negative, the total T.R.M. of the mixture is in the opposite sense to the field which caused it.

This phenomenon only occurs with a highly concentration of ferromagnetic constituents. It certainly cannot happen if the grains of A and B are independently dispersed in a non-magnetic medium in the low proportions of a few per cent characteristic of igneous and sedimentary rocks. But it is the local proportions of A and B that matter, not the overall proportion in the rock. Negative T.R.M. can appear if the grains of A and B are *gathered in concentrated aggregates*. As the total concentration is small these aggregates must be far apart but that does not affect the issue.

The existence of such aggregates is not an artificial hypothesis: it amounts to saying that the two constituents have a common origon; they could for instance have formed by exsolution into two phases, during slow cooling, of a homogeneous solid solution that was stable at higher temperatures, as must often occur in the system Fe_2O_3–TiO_2–FeO (cf. § 35). Furthermore such exsolution often gives rise to crystallites of very unsymmetrical shapes such as plates or rods which are, as will be explained later, favourable to the appearance of negative T.R.M.'s.

§ 32. Reversed T.R.M.: single-domain grains

The problem for single-domain grains has been discussed elsewhere by the author [35]. The result is as follows for the simple case in which the concentrations of the two ferromagnetic constituents in the aggregate are both equal to 3c. Suppose that the volume susceptibilities of the two constieuents at their respective blocking temperatures are both equal to s, and that the magnetic grains are of elongated shape with a demagnetizing field coefficient to $4/3\pi - n$. Let R be the relative increase of spontaneous magnetization in each constituent from the blocking temperature to room temperature (the increase in the spontaneous magnetization of A from the blocking temperature T_b of B to room temperature is assumed negligible).

In these conditions it can be shown that the aggregates acquire a *negative* T.R.M. provided that:

$$cnRs > \frac{2 - 8c + 16c^2/3}{1 - 2c - 2(1 - 3c)/R}. \qquad (8)$$

Adopting the values $R=4$, $s=7$, $n=1.7$ which the magnetic properties of rocks show to be reasonable, the inequality is satisfied if the concentration 3c of each of the constituents is greater than 0.27.

As 3c can vary from 0 to 0.5 it appears that negative T.R.M. is certainly possible in grains with properties similar to those of actual grains in ordinary lavas.

The inequality (8) also shows that negative thermoremanence is favoured by a large anisotropy of the grains (n), by a large irreversible susceptibility (s) and by a high concentration of the two constituent (c). The Curie Points of A and B must also be far enough apart for the relative increase in spontaneous magnetization of A from T_a to T_b to be close to its final value R.

§ 33. Reversed T.R.M.: large grains

The theory just summarized applies to single-domain grains but it is easy to show that aggregates of large grains could also develop negative T.R.M.'s. The argument depends only on the assumption that large grains have a blocking temperature; this assumption will be proved later (cf. § 57). Consider a dense spherical aggregate composed of alternate layers of A and B, the thicknesses of the layers being equal and small compared to the radius of the aggregate; each layer is equivalent to a large grain. At its blocking temperature T_a constituent A takes on a magnetization J_a such that the mean internal field of the aggregate is zero, hence:

$$J_a = \frac{3H}{4\pi},$$

where H is the applied field. At the blocking temperature T_b of the second constituent the magnetization of A has increased to RJ_a, where R is the ratio of spontaneous magnetizations at T_b and T_a. The field within the aggregate which is the resultant of the applied field and the demagnetizing field of A is therefore $H(1-R)$. This is negative and must be compensated by the positive demagnetizing field due to a negative magnetization J_b of B given by:

$$J_b = \frac{3H(1-R)}{4\pi}.$$

In this argument the reversible part of the susceptibility is neglected and the irreversible susceptibility of both constituents is assumed to be large compared to $3/4\pi$.

At room temperature the magnetizations of the two constituents become $R'RJ_a$ and $R''J_b$, R' and R'' being the relative increases of spontaneous magnetization for A and B respectively from T_b to room temperature. The mean T.R.M. J_{tr} of the whole aggregate at room temperature is therefore:

$$J_{tr} = \frac{3H}{4\pi}(R'R + R'' - RR'').$$

If the blocking temperatures T_a and T_b are sufficiently far apart R' will be close to 1, while if the magnetic properties of A and B are similar (except for the blocking temperatures) one can assume $R = R''$. The T.R.M. then has the sign of $(2-R)$ and is negative when the relative increase in spontaneous magnetization between the blocking temperature and room temperature is greater than 2.

This result is very similar to that for single domains. The inequality (8) shows that in fact that for $3c = 0.5$, R must be greater than 1.5 for there to be negative T.R.M. even if s is infinite.

§ 34. Experimental results on the reversed T.R.M. of mixtures of two constituents

Such phenomena as described above appear to have been experimentally observed by Nagata and his coworkers [38] in the dacitic lavas of Mt. Haruna and Mt. Asio, Japan. These rocks acquire a reversed thermoremanence in the laboratory and contain two distinct ferromagnetic constituents A and B separable by a magnetic extraction at a temperature between the two Curie Points. The constituent A with a Curie Point at about 500°C is a titanomagnetite spinel and has a magnetization curve very similar to that of magnetite. The B constituent having a Curie Point of about 200°C is a rhombohedral ferrimagnetic solid solution of ilmenitehematite type and has a spontaneous magnetization which changes practically linearly with temperature all the way from absolute zero to the Curie Point. The magnetic properties of A and B are obviously very dissimilar so that the simplified theories given above are scarcely applicable. The concentration of B is extraordinarily small, only a few per cent of that of A. The fact that despite this the mixture acquires a reversed T.R.M. must be due to some peculiar capacity of B for acquiring a large T.R.M. This could be due to the high coercive force of rhombohedral lattices and also to the high values of R which are probably associated with the linear variation of spontaneous magnetization.

§ 35. Reversed T.R.M. due to unmixing

It is shown in § 32 that under certain conditions the shape demagnetizing field can produce a negative T.R.M. in a mixture of two constituents. It is there supposed that the two constituents exist before the thermal treatment in a magnetic field. But this condition is not absolutely necessary; as the following example shows a negative T.R.M. can also develop during a slow exsolution at room temperature.

Consider a spherical single-domain grain of a constituent A with spontaneous magnetization J_s and suppose that an unmixing occurs at room temperature precipitating a second magnetic phase B, accompanied of course by a change in composition of A. The first crystallites of the new phase B grow in the demagnetizing field of A, given by $4/3\pi J_s$. If the magnetocrystalline field of B is less than this, which it usually is, and if the exchange coupling between A and B is negligible the spontaneous magnetization of the crystallites of B will necessarily be oriented antiparallel to that of A. Further, once the process has started it will continue automatically during formation of phase B from phase A. This is true even if A disappears completely as in a allotropic transformation, when the final phase B would have a spontaneous magnetization opposite to that of the original A phase.

In an assemblage of grains of A having originally amean T.R.M. in a particular direction, this process necessarily implies that as the B phase is precipitated the T.R.M. will decrease in intensity and, when the concentration of B is high enough, will change sign.

A detailed but more complicated analysis shows that similar, but perhaps less definite, phenomena can also occur in large multi-domain grains.

It should be noticed that this process of development of reversed T.R.M. is of great generality since it can accompany any allotripic transformation and any exsolution into two phases and does not depend on the relative positions of the Curie Points of the two phases. The ternary system TiO_2–Fe_2O_3–FeO which is of enormous geomagnetic importance, is particularly rich in the kind of transformation concerned. Not only are there solubility gaps at room temperature in the systems $Fe_{2-m}Ti_mO_3$ (Pouillard) and $Fe_{3-m}Ti_mO_4$ (Kawai, Kume and Sasajima) but also, according to Pouillard [8], intermediate compounds $Fe_{3-m-n}Ti_mO_{4-n}$ can exsolve into two phases each belonging to one of the first two series.

Besides these another process has been suggested by Graham [37] consisting of an alteration of constituent A to another constieutnt B, particularly oxidation of magnetite to maghemite; magnetically the process is very much as that described above.

Finally it is important to realize that the demagnetizing fields which cause these reversals are *local* demagnetizing fields, often of great intensity, related to the topography of the elementary domains. The exchange coupling between the crystallites of A and B has been neglected; this is reasonable when there is no crystallographic continuity between A and B but is doubtful if B is oriented in A by intergrowth. This exchange coupling, arising from the Weiss-Heisenberg forces can sometimes favour the development of reverse T.R.M. but more probably inhibits it. To say exactly what it does do, much more information is needed than is usually to hand, on the development and detailed structure of the precipatated phase. Thus negative T.R.M. by exsolution must be considered possible but not certain.

LARGE GRAINS

§ 36. Variation of coercive force with grain size

Generally speaking substances have a much smaller coercive force in the massive state than they have when finely divided in single-domain grains. The reason for this is that in the massive state the changes of magnetization in the hysteresis cycle are due to wall displacements, which in an unstrained and perfect substance require no expenditure of energy.

It can also be shown by experiment that as a massive substance is powdered into finef and finer grains, by physical or chemical means, the coercive force increases regularly and gradually tends towards the limiting value for single-domain grains [38]. The coercive force is roughly speaking inversely proportional to the diameter p of the grains.

Although this result is extremely simple it has not yet been given any satisfactory explanation. The basic reason for this failure is that we have no precise idea at all of the mechanism of reversal of magnetization in medium-sized grains. Although Kittel proposed a mechanism some year ago [39] involving "nucleation" of a phase magnetized antiparallel to the principal phase, quantitative application of his ideas gives results in complete disagreement with experiment.

Perhaps it is possible to suggest an approach which will give at the very most a lower limit to the coercive force of a large grain. During a complete hysteresis cycle the grain must twice pass through a state of zero total magnetic moment, corresponding to a particular subdivision of the grain into elementary domains. This state, with internal energy W (say), must be reached irreversibly so that the energy expended in single cycle, which is of the order of $2H_cJ_s$, must be at least 2 W.

Consider a cube of side p of a magnetically uniaxial substance which has a magnetocrystalline energy, E_c, given by:

$$E_c = K \sin^2 \theta,$$

where θ is the angle between the spontaneous magnetization J_s and the axis. In the demagnetized state the simplest domain system consists of elementary lamellar domains of thickness e completed, when K is small compared to J_s^2, by triangular closure domains (Fig. 4). The corresponding energy is:

$$W = \frac{\gamma p^3}{e} + \frac{1}{2} Kep^2,$$

where γ is the wall surface energy. The thickness e takes the value at which this energy W is a minimum. Then equation H_cJ_s to this minimum value of W gives:

$$H_c = \left(\frac{2K}{pJ_s^2}\right)^{1/2}$$

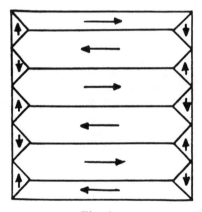

Fig. 4.
Subdivision of a cube into elementary domains giving zero total magnetic moment.

This formula has been applied to magnetite, assuming $K = 10^5$ erg/cm^3, $\gamma = 1$ erg/cm^2 and $J_2 = 450$, and the results compared with Gottschalk's [38] experimental results for powdered magnetite. Figure 5 shows that for p greater than 10^{-3} cm, equation (9) is in good agreement but that for smaller grains it gives too low a value for the coercive force.

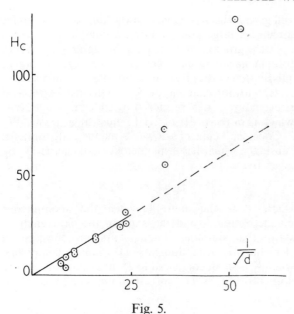

Fig. 5.
Coercive force of powdered magnetite as a function of $1/\sqrt{d}$, when d is the mean diameter of the grains. (Gottschalk's experimental results compared with the theoretical straight line given by formula (9).)

This discrepancy must be due to the difficulty of forming nuclei magnetized in the opposite direction to the principal phase. Since a demagnetizing field and a multiplicity of directions of easy magnetization favour the formation of such nuclei, the discrepancy from equation (9) should be very large in substances with only one direction of easy magnetization and with a demagnetizing field which is small compared to the magnetocrystalline fields. In fact for Mn–Bi which is such a substance, eqnation (9) gives coercive forces ten times smaller than those determined by Guillaud [40].

To summarize, it can be said that a satisfactory theory of the coercive force of large grains has still to be proposed.

§37. The demagnetizing field of large grains

The variation of coercive force H_c with the diameter p of the grain has just been described. Fundamentally this increase of coercive force derives from the fact that the size of the elementary domains becomes smaller as the grain becomes smaller. A similar increase occurs in a massive substance if a decrease in the size of the domains is produced by other means such as the formation of internal stresses varying irregularly on a small scale of distance. It is this decrease in domain size which causes an increase in the coercive force and therefore a general expansion of the hysteresis cycle along the H axis although the remanent magnetization is not changed and remains about half the saturation magnetization J_s.

In large grains another factor appears, the magnitude of the demagnetizing field. The demagnetizing field coefficient n is of the order of that for a sphere $4\pi/3$ and gives for an intensity of magnetization of 500 e.m.u. a considerable demagnetizing field of the order of 2000 oe. The effect of this can be simulated by "inclining" the hysteresis cycle. This is done by using the same cycle and the same axis of magnetic fields in the (J, H) plane but a new sloping axis of intensity of magnetization with a gradient of $-1/n$.

The hysteresis cycle then appears flatter than the original cycle; in particular the remanent magnetization decreases and can be much less than half the coercive force. This characteristic property allows the presence of a demagnetizing field to be detected purely experimentally.

§38. Remanent magnetization of an assembly of large grains

The remanent magnetization J_r of a large grain with a demagnetizing field coefficient n, after saturation in a field much higher than the coercive froce, is given by the intersection of the descending branch of the limiting hysteresis cycle with a line of slope $1/n$ passing through the origin. If n is large the ordinate of the point of intersection is approximately equal to H_c/n. Then the ratio of the remanent to the saturation magnetization j_r/j_s is given by:

$$\frac{j_r}{j_s} = \frac{H_c}{nJ_s}.$$

And, other things being equal, this ratio is proportional to the coercive force H_c.

Figure 6 constructed from data gathered by Nagata [27] for different rock specimens shows that this is a approximately true. The experiment value of the constant of proportionality $1/nJ_s$ is about $1/600$. Assuming the average value of the saturation magnetization of the rocks given by Nagata [27] as 230 e.m.u. a mean value of $1/2.6$ is got for $1/n$ which is very reasonable.

The approximation used above become inaccurate when j_r/j_s exceeds 0.3 since for high coercive forces it tends to a limiting value of 0.5. Amongst Nagata's data there can in fact be found two specimens with coercive forces of 465 oe and 345 oe and values of the j_r/j_s ratios J_r/J_s equal to 0.40 and 0.35 respectively. As expected they give points well below the straight line of Fig. 6.

Thus the study of the remanent magnetization of rocks of low coercive force shows that the ferromagnetic constituents are in large grains and not in fine single-domain grains.

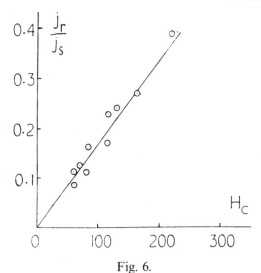

Fig. 6.
The ratio of remanent magnetization j_r to saturation magnetization j_s plotted against coercive force for some eruptive rocks. (Nagata's experimental results compared with the straight line of formula (10).)

§ 39. Initial susceptibility of an assemblage of large grains

The apparent initial susceptibility of a ferromagnetic substance is strongly affected by the presence of a demagnetizing field (with coefficient n, say). Its apparent and real values, A' and A are related by:

$$A' = \frac{A}{1+nA}.$$

When nA is large compared to 1 the apparent susceptibility A' is nearly equal to $1/n$.

In most magnetic substances A varies inversely as the coercive force and is about 0.2 when $H_c = 500$ oe. For a sphere $1/n = 0.24$. Hence the apparent susceptibility of rocks with coercive forces of less than 500 oe, such as the eruptives studied by Nagata, must depend more on the demagnetizing field effects than on the intrinsic susceptibility of the material.

However ferromagnetic grains with very irregular shapes are better represented by triaxial ellipsoids than by spheres. The three demagnetizing coefficients along the three principal axes n_1, n_2 and n_3 are all different and:

$$n_1 + n_2 + n_3 = 4\pi.$$

The apparent susceptibility of an aggregate of grains, oriented at random is then given by:

$$A' = \frac{A}{3}\left(\frac{1}{1+An_1} + \frac{1}{1+An_2} + \frac{1}{1+An_3}\right).$$

It is easily shown that this is always greater than that of a sphere of the same material.

Obviously it will appear as if the demagnetizing coefficient n of the average grain is less than $\frac{4}{3}\pi$. Nagata, from an experimental study of the susceptibility of some igneous rocks, estimates that the average value of n is about 3.5 with $A' = 0.23$ and $A = 0.12$. With artificial mixtures of sand and large grains of powdered magnetite, n is found to be 3.15 when the susceptibility A of the same magnetite in the massive state is 0.43. The smaller value of A in natural specimens is probably connected with the higher value of their coercive force, which would imply a smaller value of A.

Thus the experimental results on the initial susceptibility of rocks are in excellent agreement with the large grain hypothesis. Though much less naturally explained by the theory of single-domain grains, they are not actually inconsistent with it.

§ 40. Movement of a single wall

To interpret the magnetic properties of large grains more fully it is necessary to consider the mechanism of the magnetization process. Wall displacements are, as stated above, the mechanism concerned. Consider the substance as composed of elementary domains which can be crossed from end to end by a wall.

The fundamental process in the movement of a single wall P (supposed plane) of surface-area S, its position being represented by the abscissa, z, of its intersection with an axis OZ perpendicular to the wall. This wall P separates two elementary domains with spontaneous magnetizations $+J_s$ and $-J_s$ and can move from $z=0$ to $z=L$, that is through a distance L which is called the *free path of the wall*. For $z=0$ the magnetic moment of the domain is $-LSJ_s$ and for $z=L$, $+LSJ_s$.

In a perfectly pure substance, the energy of the system W has a constant value W_0 in zero magnetic field, which is independent of the position z of the wall. In a field H, parallel to J, a pressure of $2HJ_s$ acts on the wall. An infinitely small positive field is enough to move the wall to its limit at $z=L$ and in infinitely small negative field brings it back to $z=0$. There is no hysteresis.

But in a real substance, which is always somewhat imperfect, W is a complicated function of z which varies haphazardly around the mean value W_0. With no applied field the wall is in equilibrium at points when W is a maximum. Application of a field H moves the wall until the equilibrium condition:

$$2J_s H = \frac{1}{S}\frac{dW}{dz} \qquad (10)$$

is satisfied. If the equilibrium is to be stable the second derivative d^2W/dz^2 must be positive. Thus to saturate the domain in the positive direction a field of at least the maximum value of $(1/2SJ_s)(dW/dz)$ must be applied. Therefore the coercive force depends on the maximum gradient of W.

Liboutry [41] gives a very suggestive graphical representation of these phenomena. He plots the quantity $(1/2)dW/dz$ which he calls the *opposition* as abscissa against the position z of the wall as ordinate. As stated above, z is proportional to the magnetization. This gives a curve C whose intersections, where its slope is positive, with the lines given by $x = 2J_s H$ give points M corresponding to possible equilibrium states of the system. Except for certain volume changes, movements of M correspond to variations in magnetization with H.

The simplest model which has hysteresis, is obtained by representing the opposition as a sine curve $A \sin pz$ where p is large compared with $1/L$. This is the full line of Fig. 7. The limiting hysteresis cycle ABCD (dotted line) approached a rectangle as p becomes greater. If at

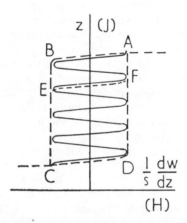

Fig. 7.
Hysteresis cycle given by a free path obstructed by equally spaced barriers all of the same height; the full line represents the opposition, the broken one the hysteresis cycle.

the point E on the vertical part of the return cycle the movement is stopped and the magnetic field increased again, the point representing the magnetization will follow the path EF which is nearly horizontal when p is large. For completeness it must be supposed that the coefficient A decreases as the temperature rises and tends to zero at the Curie Point.

§41. Thermo-remanent magnetization of the model with a single wall

This model can be used to determine the thermo-remanent magnetization J_{tr} produced by cooling in a constant field H. It depends essentially on the mode of variation of the spontaneous magnetization J_s and the coercive force H_c with temperature, especially close to the Curie Point. Unfortunately experimental data on this are extremely scarce, while Forrer and Baffie [42] have shown that these phenomena are often very complicated.

Generally speaking theory and experiment agree that the true remanent magnetization J_r and the spontaneous magnetization J_s vary as $(\theta - T)^{1/2}$ in the neighbourhood of θ. But the form of the thermal variation of coercive force H_c is very different from one substance to another. This is not surprising since the coercive force is a complicated function of many independent factors. Forrer [43] found a variation of H_c with $(\theta - T)^{1/2}$ in large crystals of magnetite which had the low coercive force of 4 oe at room temperature, while for the very fine magnetite grains of a ferromagnetic rock with a very high coercive force Akimoto [44] found that H_c varied nearly as $(\theta - T)$.

Being concerned here with rocks of high coercive force such as these latter it seems better to use this last result. Then one has:

$$\frac{H_c}{H_{c0}} = \left[\frac{J_s}{J_{s0}}\right]^2, \qquad (11)$$

H_{c_0} and J_{s_0} being the values of H_c and J_s at room temperature. The simplest form of hysteresis cycle is considered, as before; a rectangular limiting cycle of height $2J_s$ and width $2H_c$ with all the partial cycles of the same width $2H_c$.

The demagnetizing field of the grain with a mean coefficient n plays an important part. Very close to the Curie Point the hysteresis cycle has the form of Fig. 8(a) and in an applied field $H = OP$ the magnetization is represented by the point of intersection Q, of a line of

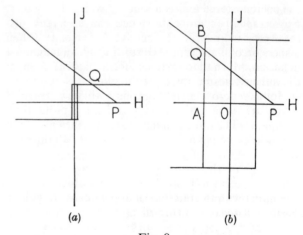

Fig. 8.
The process of acquiring T.R.M. (see text).

slope $-1/n$ through P, with the upper branch of the hysteresis cycle. As the temperature falls the cycle widens and takes the form of 8(b) while the line PQ remains fixed, so that the point Q moves back along the upper branch of the cycle. It continues to move back as long as the ratio $r = AQ/AB$ continues to decrease. Now:

$$r = \frac{H + H_c}{nJ_s}$$

and the derivative dr/dT vanishes when:

$$\frac{dH_c}{dT} = \frac{H + H_c}{J_s} \frac{dJ_s}{dT}$$

or from equation (11) when $H = H_c$ and r has the value $n_0 = 2H/nJ_s$. From this temperature r increases as the temperature falls but Q the point representing the magnetization cannot move back up the descending part of the clcle BA. It describes (in reduced coordinates) a horizontal branch of the cycle of reduced height r_0.

At room temperature the specimen thus has a T.R.M. given by:

$$J_{tr} = r_0 J_{s_0}.$$

It is not altered when H is removed and hence from equation (11):

$$J_{tr} = \frac{2H^{1/2} H_{c_0}^{1/2}}{n} \qquad (12)$$

The T.R.M. is thus proportional to the square root of the applied field and the square root of the coercive force.

§ 42. Comparison with experimental results; value of Q_k

This theoretical formula is compared in Fig. 9 with Nagata's [27] experimental results on a rock specimen (No. 60), with Mlle. Roquet's [45] results for a dispersion of fine grains $(1/10 \mu)$ of artificial magnetite in kaolin (M'_2) and for artificial ferric oxide (F_5). In all cases the parabolic law represents the facts quite well, except in very low fields. This point will be considered later (^{II} 57).

Equation (12) shows that if J^1_{tr} is the T.R.M. acquired in a field of 1 oe (Nagata's saturated T.R.M.) the ratio J^1_{tr}/J_{s_0} should be equal to $2H^{1/2}_{c_0}/nJ_{s_0}$. Figure 10 shows the value of this ratio determined by Nagata for different specimens plotted against the value of H_{c_0} and compared with the theoretical curve obtained with an assumed value of 1800 for nJ_{s_0} (given by the theoretical values of $J_{s_0} = 450$ and $n = 4.19$).

Q_k is the ratio of T.R.M. in the earth's field Z to the magnetization induced by the same field. Since the

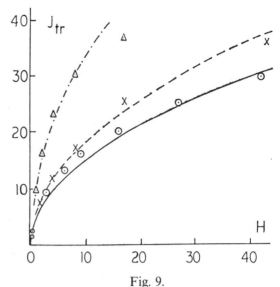

Fig. 9.
Variation of T.R.M. J_{tr} with the applied

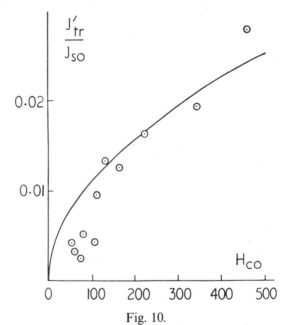

Fig. 10.
The ratio of T.R.M. in a field of 1 oe. J^1_{tr} to the saturation magnetization J_{s_0} plotted against the coercive force H_{c_0}. (Comparison of Nagata's results with the theoretical parabola of formula (12).)

induced magnetization is about Z/n:

$$Q_k = 2\left(\frac{H_c}{Z}\right)^{1/2}. \qquad (13)$$

For $Z = 0.5$, Q_k should vary from 4 to 400 as H_c varies from 2 to 20 000 oersteds. Experiments do in fact show that Q_k is greater for rocks with a higher coercive force.

§ 43. Statement of Lord Rayleigh's relations

The expressions for the magnetization of ferromagnetic bodies in *weak fields*, called Rayleigh's relations are particularly interesting in the case of rocks because the earth's magnetic field is always very weak compared to the coercive force of a rock.

Only a brief statement of them is necessary. The magnetization of a body, which has been previously demagnetized by an a.c. field gradually decreasing to zero, is a quadratic function of the applied field H:

$$J = AH + BH^2.$$

This is *Rayleigh's first relation*.

If, when the magnetic field has reached a value H_m, producing a magnetization J_m, it is reduced by an amount ΔH, the magnetization changes by ΔJ which has the same sign as ΔH and is given numerically by:

$$|\Delta J| = A|\Delta H| + \tfrac{1}{2}B|\Delta H|^2.$$

This formula is only valid provided H remains inside the interval $-H_m$ the $+H_m$. Brown [47] calls this *Rayleigh's second relation*. The term of first order in H always has the form AH hence the coefficient A is called the reversible susceptibility.

From these formulae one can deduce that a magnetic field H_m leaves, when removed, a remanent magnetization J_r, given by:

$$J_r = \tfrac{1}{2}BH_m^2.$$

§ 44. Theory of Rayleigh's relations

L. Néel [29] has shown that Rayleigh's relations can be explained on the single wall model. For this purpose a slightly more refined expression than that of § 40 must be used to represent the opposition $(1/S)dW/dz$. The curve of $(1/S)dW/dz$ is simulated by a series of straight lines $A_1A_2, A_2A_3, \ldots$ The projection of each section on the z-axis is the same and equal to $2l$ and the ordinates of the ends have a Gaussian random distribution about the mean value zero (Fig. 11).

In a field H_1 the wall is in equilibrium at the point M on the section A_1A_2 which has a positive slope (therefore stable equilibrium). The abscissa of M is

$$V_1 = 2H_1J_s = (1/S)dW/dz.$$

When the field is increased from H_1 to H_2 the point representing the position of the wall follows the path MA_2NA_4PQ with two discontinuities A_2-N and A_4-P; the return path is $QPA_7A_6A_5RM$ and hysteresis therefore occurs.

In the range of weak fields these cycles reduce to the two imaginary elementary cycles of Fig. 12, the first (a) representing the the irreversible part of the phenomena and the second (b) the reversible part. If a and b are the critical fields corresponding to the abscisse of the

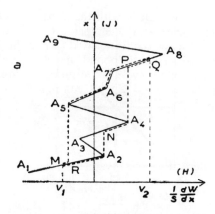

Fig. 11.
Diagram showing the course of a hysteresis cycle according to Lliboutry. The full line is the opposition and the broken one the hysteresis cycle.

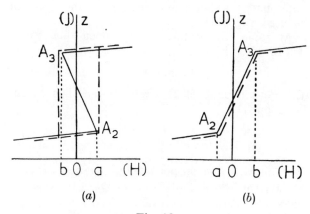

Fig. 12.
(a) Irreversible and (b) reversible imaginary cycles corresponding to small wall displacements against an opposition represented by straight lines as in Fig. 11.

points A_2 and A_3 each of the imaginary cycles can be represented by a point (a, b) in the plane Q of Fig. 13. When there is a large number of walls, there will therefore be a large number of representative points whose density will be uniform close to O, the origin of coordinates in the plane Q.

The upper half of the plane (a, b) in which $a < b$ corresponds to AH the reversible term in the magnetization expression, the other half-plane where $a > b$ to the irreversible term in H^2. It was shown long ago by Weiss and Freudenreich [48] and later by Preisach [49] that an assembly of such unsymmetrical cycles corresponding to this second half-plane allows of an immediate explanation of Rayleigh's laws. According to these authors the cycles concerned were real ones such as those of isolated grains whereas in

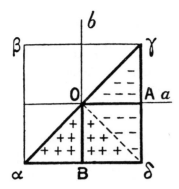

Fig. 13.
The magnetic state of the imaginary irreversible cycles after demagnetization in a decreasing alternating field. After demagnetization by heating above the Curie Point, cycles in the quadrant BOA are magnetized at random, some positively and some negatively.

Néel's theory [29], here described briefly in a form due to Lliboutry [41], they are imaginary cycles corresponding to partial displacement of walls.

In zero applied field, the domains of the sector αOB (Fig. 13) are always positively magnetized while those of sector AOγ are always negatively magnetized. But the domains of the quadrant BOA can have either sign according to the initial conditions. It is however easily shown that after demagnetization in a decreasing alternating field, the situation is as in Fig. 13 when the domains are positively or negatively magnetized according as their representative points are below or above the second bisector Oδ. Application of an increasing field H causes reversal of the domains in the triangle OEC, the process being represented in Fig. 14 where CD is a line with abscissa $a = H$. Similarly the return curve corresponds to the situation in Fig. 15 where the abscissa of the line C'D' is equal to the decreasing field. It is obvious without further argument that Rayleigh's two laws can be interpreted in this way.

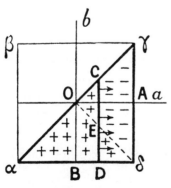

Fig. 14.
Magnetic state of the imaginary cicles during magnetization, after demagnetizing in decreasing c.c.

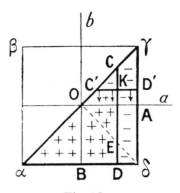

Fig. 15.
Magnetic state of the imaginary cycles when the descending branch of the real hysteresis cycle is being followed.

§45. Demagnetization by heating to above the Curie Point

This demagnetizing process leads to a very different initial state from that produced in a decreasing alternating field. The two critical fields a and b of the imaginary cycles increase continuously from zero at the Curie Point to room temperature. There is no reason why the magnetization of these cycles should be of one sign rather than another so that in the limit the magnetizations of those in the quadrant BOA (Fig. 13) are, after cooling, distributed at random. The result of this is that the initial magnetization curve of a thermally demagnetized body differs from the usual curve. The BH^2 term is smaller (since there are fewer domains to be reversed in the lower half of triangle OEC in Fig. 14) and the AH term is larger since it must include terms due to reversal of a proportion of the domains of the trapezium BOED.

In the simple way Mlle. Roquet's interesting results, on the I.R.M. of artificial rocks, can be interpreted. She found [45] that the I.R.M. obrained after the first application of a field H is less than that produced after several cycles from $+H$ to $-H$. The difference between the two is roughly proportional to H for small fields while the I.R.M. obtained after several cycles ($\pm H$) is proportional to H^2 so that the phenomenon is more obvious when H is small. In the model outlined above the I.R.M. obtained after cyclical treatment corresponds as usual to the domains of the lower half of triangle OEC which remain positively magnetized. On the other hand the I.R.M. acquired after a single application of the field involves also the irreversible positive remagnetization of those domains in the area OBDE which have been left negatively magnetized by thermal demagnetization. During the next half-cycle this extra I.R.M. is compensated by one in the opposite direction due to another trapezium similar to OBDE but lying symmetrically on the other side of $O\gamma$; after completion of the cycle everything follows as before (see Fig. 14).

§ 46. Anhysteretic magnetization

The higher values of anhysteretic magnetization as compared with the ordinary magnetization are also easily explained on this model. The anhysteretic magnetization is obtained by applying a field H and at the same time an alternating field which decreases to zero. It is easily shown that this will correspond to the positive magnetization of the domains of the area δEOCF whereas the ordinary magnetization produced by simple application of the same field H would be due only to those domains in the region COE (Fig. 16).

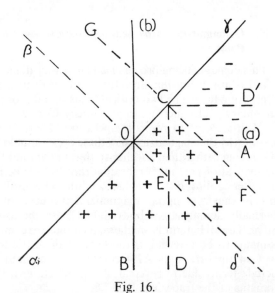

Fig. 16.
Magnetic state of the imaginary cycles after the action of a constant magnetic field together with a decreasing alternating one (the anhysteretic curve).

§ 47. The reversible term and the ratio BH_c/A

Points in the plane (a, b) lying above the first bisector $\alpha\gamma$ correspond to reversible cycles such as in Fig. 12(b). In zero field the magnetic state is represented by all the domains below the second bisectrix 0β being positively magnetized and those above negatively. When a field H is applied the boundary between these two types domains moves from 0β to CG, when C is the point with abscissa $a = H$. Thus the change in magnetization of the domains in this upper half of the plane represents the reversible AH term in Rayleigh's relations.

Theory shows [29] that the ratio BH_c/A should vary from 0.58 for substances in which $2l$ is of the order of L to about 1 when $2l$ is very small compared to L. It is found experimentally that it is about 0.5 in "hard" materials (such as used for permanent magnets) and up to several units in "soft" materials, in qualitative agreement with the theory. In rocks a value of about 0.5 should be expected since their ferromagnetic constituents tend to be "hard".

§ 48. The effect of the shape demagnetizing field on Rayleigh's relations

The form of Rayleigh's relations is not altered by the presence of a demagnetizing field. The initial magnetization relation is still expressed by an equation of the type:

$$J = A'H + B'H^2$$

but the values of the coefficients are modified. If n is the demagnetizing coefficient it is easily shown that:

$$A' = \frac{A}{1+nA}, \qquad B' = \frac{B}{(1+nA)^3}.$$

It is also known that the presence of a demagnetizing field does not affect the apparent coercive force H'_c which remains equal to the real value H_c. Hence the relation between the apparent values A', B', H'_c and the real values A, B, H_c is:

$$\frac{B'H'_c}{A'} = \frac{BH_c}{A} \frac{1}{(1+nA)^2}.$$

In magnetite A should be of the order 0.5–1 and for the grains found in rocks n is about 3 (cf. § 39). This means that the ratio $B'H'_c/A'$ is probably 6–16 times smaller than the value of 0.5 found for BH_c/A in magnetically hare materials. This is a considerable decrease and provides a very good method for determining whether a material is composed of *large ferromagnetic grains* dispersed in a non-magnetic matrix. It does not seem to have been used yet.

MAGNETIC VISCOSITY IN LARGE GRAINS[1]

§ 49. The different types of magnetic viscosity [50]

Time can play an important part in the phenomena of magnetization through a variety of effects.

The fundamental magnetic process, that is the alignment of an elementary atomic moment along a magnetic field, is itself not instantaneous. The magnetic moment undergoes a precession while its direction is moving into that of the applied field. However the duration of this process of alignment is very small, of the order of 10^{-10} sec; it is only of importance in high

---

[1] Another name for "magnetic viscosity" is "magnetic after-effect".

frequency magnetic fields and so is of no interest in rock magnetism.

But *diffusion viscosity* and *thermal agitation viscosity* are important. The first is connected with diffusion of elementary material particles, atoms or electrons, within the crystal lattice. Obviously if the distribution of these particles depends on the orientation of the spontaneous magnetization relative to the crystallographic axes any change of magnetization must involve a non-instantaneous rearrangement of the particles. The example of carbon in the lattice of α-iron is the best known. The carbon atoms can occupy 3 different types of interstitial sites since each C atom has two diametrically opposite Fe neighbours, the line of whose centres can make different angles with the spontaneous magnetization and hence have different energies. At room temperature a time of about 1 min is required for a carbon atom to jump from one side to the next. The effect is easily observable since to reach a new equilibrium after a change in the direction of magnetization about the same amount of time is required.

Magnetite, with which we are particularly concerned, shows [52] a similar diffusion viscosity due to movement of electrons from ferrous to ferric ions thus changing their valency. At 100°K these movements require times of about 1 min but at ordinary temperatures they are practically instantaneous and probably do not affect the normal process of magnetization in rocks. Possibly other diffusion phenomena occur in the magnetic constituents of rocks at high temperatures but as there is no information whatever about them, the subject will not be discussed further, in spite of its great interest.

Generally speaking thermal agitation viscosity is due to irregular fluctuations in the forces acting on the spontaneous magnetization which enable it to cross barrier which it otherwise could not. In this way thermal agitation makes irreversible changes of magnetization possible and is an effect common to all ferromagnetic substances.

The effects of this viscosity on rotation processes have already been dealt with in a special application to single-domain particles. Its effects on wall movements must now be considered.

§50. Ths viscosity field [50]

These effects can be described very simply by supposing that a fluctuating viscosity field $\pm H_f(t)$ is always added to the applied field H and that:

$$H_f(t) = S(Q + \log t) \quad (14)$$

where Q is a numerical constant of the order of 40 or 50, t is the time interval since the application of the field H and S, which has the dimensions of a magnetic field, is a constant characteristic of the specimen considered and dependent on the temperature.

In the absence of the viscosity field, we know that after a magnetic field H has given to a ferromagnetic substance a magnetization J_0, a small increase in the applied field of h produces a change in magnetization of j given by $j = (A + C)h$ while a small decrease gives a change:

$$j' = -Ah.$$

A and C are the reversible and irreversible differential susceptibilities at the point (J_0, H) of the hysteresis cycle. With a fluctuating field the magnetization J after a time t is thus:

$$J = J_0 + CS(Q + \log t). \quad (15)$$

Barbier's experiments [53] show that S has about the same value all round the hysteresis cycle.

§51. The fluctuation viscosity in the range of Rayleigh's relations

Formula (15) can be applied to some particular cases. In the Rayleigh region where the initial magnetization is given by:

$$J = AH + BH^2$$

the irreversible differential susceptibility at the point (J, H) is C = 2BH as that including the viscosity term:

$$J = AH + BH^2 + 2BHS(Q + \log t). \quad (16)$$

After application of a field H and then a return to zero field the remanent magnetization is $1/2 BH^2$ and the irreversible susceptibility is $C = -BH$ so that the remanent magnetization after a lapse of time is:

$$J_r = \frac{1}{2} BH^2 - BHS(Q + \log t). \quad (17)$$

This expression shows that the visocsity of isothermal remanent magnetization increases in relative importance with the size of the field which caused the magnetization.

At different points of the limiting cycle, obtained after saturation in a very high field the magnitudes of the viscosity are greater in absolute value but less in relative value than in the preceding case. They can be calculated from formula (15) with $C = -kJ_s/H_c$ where k is a numerical coefficient of about 1/10 for H = 0 (remanent magnetization) and several units for $H = H_c$, at which point the absolute value of the viscosity is a maximum.

§52. The values of S [53]

Barbier has made a thorough experimental study of these theoretical conclusions and shows them to be well founded.

He has also measured values of S for very diverse specimens and has found that S increases with the coercive force. Figure 17 summarizes his results and shows that S varies from about one thousandth of the coercive force for soft materials to about four thousandths for hard materials. These results are particularly interesting for they show that in substances with a coercive force of about 200 oersteds the viscosity field S log t is of the order of the earth's magnetic field.

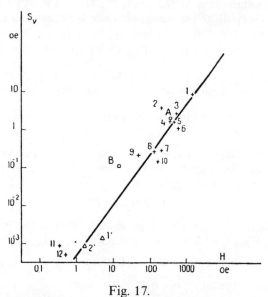

Fig. 17.
Values of the viscosity constant S for various specimens plotted against their coercive force (after J. C. Barbier).

§ 53. Theoretical calculation [50] of S

These thermal agitations may be considered to act through the mechanism of the internal dispersion fields produced by thermal oscillations of the spontaneous magnetization about its mean value. Suppose that the fluctuation field which helps a wall to cross an obstacle, that is to pass through a volume v (vJ_s corresponds to a Barkhausen discontinuity), is equal to the mean component H_m in a given direction of the dispersion field, over a similar volume v within an elementary domain. This mean value is a random function of time t whose mean square is given by equating the mean energy to kT. More exactly H_ρ the root mean square of H_m is:

$$H_\rho = \sqrt{\left(\frac{4\pi kT}{3v}\right)}.$$

Consider the possible values of H_m as having a Gaussian distribution and let θ be the reorganization time, that is the minimum time interval which must separate two observations in order that the corresponding values of H_m should be practically independent. In these conditions if ε is the probability that H_m is greater than a certain value H_f, a time $t = \theta/\varepsilon$ must elapse for H_m to reach the value H_f with a probability of about 1. Then it is found that t is related to H_f by an expression of the type of (14), with the following values of Q and S:

$$Q \sim -2 \log \theta; \quad S = H_\rho Q^{-1/2}.$$

The reorganization time θ is estimated to be of order of 10^{-10} sec which gives values of Q of 40 to 50.

As Street and Woolley [54] have proposed, the thermal agitations can also be considered as affecting the heights of the barriers which oppose wall movements. If the fluctuations in the energy required to cross a barrier are of the order of kT the effects are the same as if the wall were acted on by a fluctuating field whose root mean square H_ρ was:

$$H_\rho = \frac{kT}{2J_s v}.$$

The argument then proceeds as before.

These considerations show that, if v is constant, S varies as $T^{1/2}$ if the thermal agitations are those of an internal dispersion field but as T is they are fluctuations in the heights of barriers. Barbier has studied the variations of S in a very wide temperature range and has found that in fact S varies approximately as $T^{3/4}$. Thus it seems probable that both the above mechanisms occur simultaneously but possible changes in v with temperature must also be considered.

§ 54. The effect of the demagnetizing field on viscosity of thermal agitation

To obtain expressions for magnetic viscosity in large grains the demagnetizing field must now be taken into account. The method used before (cf. § 37) can be employed for these calculations. In the Rayleigh region, where the term in H^2 is small compared to that in H, the initial magnetizaation relation, for instance, becomes:

$$J = RAH + R^3BH^2 + 2R^3BHS(Q + \log t) \quad (18)$$

where the coefficient R is given by:

$$R = \frac{1}{1 + nA},$$

n being the demagnetizing coefficient.

Similarly the remanent magnetization left after application and removal of a field H_m is:

$$J_r = \frac{1}{2} R^3 B H_m^2 - R^3 B H_m S(Q + \log t) \quad (19)$$

Thus in large grains as well as small grains the viscosity is proportional to $\log t$ agreeing with Thellier's experimental results [31]. It is impossible to distinguish between the two cases from this point of view. The experimental facts on magnetic viscosity in rocks are also much too meagre to decide whether it varies with H_m as in (19). It would be equally interesting to know the values of S and whether they are relat3d to H_c in Fig. 17.

Outside the Rayleigh region the effect of the demagnetizing field on the magnetic viscosity is more complicated. In particular when the irreversible susceptibility C becomes appreciable compared to the reversible susceptibility A the change of magnetization with time, J_t, is given by:

$$J_t = \frac{CS(Q + \log t)}{(1 + nA)(1 + nA + nC)}$$

which reduces, when C is large compared to A, to Barbier's [53] simple formula:

$$J_t = \frac{CS(Q + \log t)}{(1 + nC)}.$$

§ 55. The effect of a rise of temperature on the remanent magnetization [50]

The arguments of ⁿ 53 show that the viscosity field:

$$H_f = S(Q + \log t)$$

not only increases with time but also with temperature since S varies as $(T/v)^{1/2}$ or as T/vJ_s according to the mechanism involved.

Thus as well as an irreversible linear decrease in remanent magnetization with $\log t$ there should also be an irreversible decrease due to the rise in the viscosity field with temperature.

To observe this phenomenon it is necessary to exclude the reversible thermal variation in the principal term, $\frac{1}{2}R^3 BH^2$, of J_r. To achieve this one takes advantage of the fact, predicted by Rayleigh's relations and confirmed by experiment, that a first order *decrease* in the viscosity field only produces a second order effect on the magnetization. Hence the fraction of the thermal variation of J_r due to the thermal increase of H_f can be obtained by determining the decrease in J_r produced by heating the specimen form T_0 to T_1 and then cooling to T_0 again.

However, it is not the increase in the viscosity field itself but the increase in the ratio of the fiscosity field to the coercive force H_f/H_c that is important. In fact the walls have to cross barriers whose height decfeases as the temperature rises and the coercive force is the measure of these heights. We are therefore concerned with the increase in the quantity:

$$H'_j = H_{c0} S(Q + \log t)/H_c.$$

The phenomenon is found experimentally with about the predicted order of magnitude. Barbier [53] has studied it in ferromagnetic metal alloys and some artificial oxides with high coercive force. It has been observed by Thellier [31] and Mlle. Roquet [46] in baked clays and in dispersions of magnetite.

The ratio of the time variation to the thermal variation of remanent magnetization can be expressed quite simply. We have:

$$\frac{\partial J_r}{\partial \log t} \bigg/ \frac{\partial J_r}{\partial T} = \frac{\partial H'_j}{\partial \log t} \bigg/ \frac{\partial H'_j}{\partial T}.$$

Simplifying the right-hand side by neglecting $\log t$ compared to Q and assuming the empirical variation of S with $T^{3/4}$ found by Barbier which lies between the two theoretical predictions, we get finally:

$$\frac{\partial J_r}{\partial \log t} \bigg/ \frac{\partial J_r}{\partial T} = \frac{T}{Q}\left(\frac{3}{4} - \frac{\partial H_c}{\partial T}\right)^{-1}$$

in which all terms except Q can be determined experimentally. This gives a direct method of finding Q.

Assuming that Q is about 40 to 50 and neglecting the thermal change of coercive force, the formula predicts values of the ratio:

$$\frac{\partial J_r}{\partial \log t} \bigg/ \frac{\partial J_r}{\partial T}$$

of the order of 8 to 10 at room temperature. Barbier has in fact found values of between 10 and 13 for magnetically hard materials which are exactly of the expected order of magnitude. Thellier [31] has found a ratio very close to 9 for a specimen of clay from Noron baked in nitrogen at 665°C. On the other hand Mlle. Roquet [46] in specimens containing mainly magnetite obtained ratios of 3.5 to 2.9. Possibly these low values are at least partly due to thermal variation of the coercive force. In fact from Forrer and Baffie's [42] experiments the value of $(T/H_c) \partial H_c/\partial T$ should be about -0.5 which gives at $T = 300°K$ and $Q = 50$ a theoretical value equal to 4.8 of the ratio:

$$\frac{\partial J_r}{\partial \log t} \bigg/ \frac{\partial J_r}{\partial T}.$$

The phenomenon dealt with in the next section may also be responsible.

§ 56. Irreversible decrease of remanent magnetization due to any change of temperature, positive or negative

It is found experimentally that in certain relatively soft materials irreversible decreases of remanent magnetization may occur quite independently of those considered above, due to either *positive or negative* temperature changes. A mechanism entirely different in principle from the preceding ones must be concerned. It is probably a thermal change in the functions relating the energy of the system of walls to their position in the crystal lattice. The result is an additional decrease of remanent magnetization with temperature and a decfease of the ratio $(\partial J_r/\partial \log t)/(\partial J_r/\partial T)$. In this way Barbier has found a ratio of 2.5 for a ferro-cobalt with a coercive force of 155 oe. The existence of this parasitic phenomenon is also revealed by the irreversible decrease of remanent magnetization J_r produced by cooling below room temperature and then allowing the temperature to rise again. This does not occur in the normal substances with which the original theory dealt.

Finally it should be noticed that thermal agitation produces analogous effects in both single domain and large grains; for instance in both, the same order of magnitude is found for the ratio of the decfeases in I.R.M. due to time and due to rise in temperature.

§ 57. Thermo-remanent magnetization in very weak fields

The brief theory of T.R.M. given earlier shows that it is proportional to $H^{1/2}$ (§ 41). In fact it is found experimentally that in very weak fields of about 1 oe it is more proportional to H than to $H^{1/2}$. The theory should not actually be valid for such small fields for no account has been taken of the thermal agitation field. In particular the blocking of walls probably cannot take place until the coercive force has reached a value at least equal to H_f, the thermal agitation field at temperature near the Curie Point. In this region assuming the validity of (11), the spontaneous magnetization J_s is given by:

$$\frac{H_f}{H_{c_0}} = \left(\frac{J_s}{J_{s_0}}\right)^2.$$

The magnetization acquired in a field H is then equal to H/n and assuming that the blocking occurs at this moment the T.R.M. at room temperature J''_{tr} is given by multiplying again by J_{s_0}/J_s:

$$J''_{tr} = \frac{H}{n}\left(\frac{H_{c_0}}{H_f}\right)^{1/2}. \qquad (20)$$

This gives a proportionality to H in very small fields. From (12) and (20) J_{tr} should be equal to J''_{tr} for $H = 4H_f$, showing that the transition from one law to the other occurs for a field of 4 times the value of the fluctuation field at the Curie Point.

There are no experimental data to verify formula (20).

§ 58. The additivity of partial T.R.M.s in the large grain model

To what extent can the model of an assembly of large grains account for Thellier's laws of the additivity of partial T.R.M.? It has already been shown (cf. §§ 27, 28) that all Thellier's laws are simply and naturally explained on the single domain model by the action of thermal agitation. Similar considerations will probably provide a start for attacking the problem ih the large grain model.

Consider the free path L of a wall. A first approximation (§§ 40, 41) assumes the free path to be obstructed by a number of equally spaced barriers all of the same height; the second approximation, used in considering Rayleigh's relations (cf. ПП 44–46) supposes that the barriers are of different heights. Suppose now that the substance is acted on by a field H, small compared with the fluctuation field but large enough for the ratio vJ_sH/kT to be much greater than 1 (v is the volume through the wall passes in moving from one barrier to the next). In these conditions the possible crossing of the barrier depends on the magnitude of the fluctuation field compared to the height of the barrier.

Consider a temperature T_1 less than the Curie Point θ and compare the total T.R.M. due to field H with the partial T.R.M.'s acquired from θ to T_1 and from T_1 to room temperature in the same field. For this purpose the barriers are classed into two types: the first, α, including all barriers which can be crossed with the help of the thermal fluctuation field at temperatures below T; the second type, β, includes all the others, numbering q.

After demagnetization the wall lies, on the average, at the centre M of its free path AB (AB = L). If the field H acts continuously from θ to room temperature the wall crosses all the barriers and moves to the end A of its free path AB where it is found at room temperature definitely blocked. The path AM corresponds to the complete T.R.M.

If the field H acts only between θ and T_1, the wall again crosses all the barriers and reaches A, but after the field is removed at T_1 it can cross back over barriers of type α and is only stopped by those of type β. Thus on the average the wall is finally at a point N at a distance $AN = L/2q$ from A. The partial T.R.M. acquired between θ and T_1 is then proportional to the distance MN.

If H is applied only from T_1 to room temperature the wall remains at M on the average until T_1 is reached. After the field is applied at this point, it crosses all α-type barriers and stops when it meets the first of β type at N' at a mean distance from M of $L/2q$. The partial T.R.M. from T to room temperature is then proportional to the distance MN'.

Since $NM + N'M = AM$ the total T.R.M. is equal to

Fig. 18.

the sum of the two partial ones, and an important law of Thelliers is explained. The others can be derived similarly.

At first sight this argument seems to show that the different T.R.M.'s are independent of the field H. This is not true since in reality the free path AB must itself be considered as depending on H.

The whole argument is based on the assumption that the two types of barrier, high and low, are independent and that this distribution is not affected by temperature. It is no doubt a rough and ready mode of reasoning and could be improved, but it has the advantage of showing that it is possible to interpret Thellier's laws in the scheme of a theory of wall displacements[1].

---

[1] Note (1977). The above reasoning may also be used with the Preisach–Néel diagram; it leads to the same conclusions.

BIBLIOGRAPHY

[1] R. Becker and W. Doring, *Ferromagnetismus*, Springer (Berlin, 1949); R. M. Bozorth, *Ferromagnetism*, Van Nostrand (New York, 1951); L. F. Bates, *Modern Magnetism*, University Press (Cambridge, 1951; F. Pawlek, *Magnetische Werkstoffe* Springer (Berlin, 1952), etc.
[2] L. Néel, *Ann. Phys.*, 1948a, **3**, 137; *Ann. Inst. Fourier, Grenoble*, 1948b, **1**, 163.
[3] T. F. W. Barth and E. Posnjak, *Z. Kristallogr.*, 1932, **82**, 325; E. J. W. Vervey and E. L. Heilmann, *J. Chem. Phys.*, 1947, **15**, 174.
[4] C. G. Shull et al., *Phys. Res.*, 1951a, **83**, 208, 333; *Ibid.*, 1951b, **84**, 912.
[5] F. C. Romeijn, *Philips Res. Rep.*, 1953, **8**, 304, 321.
[6] R. Pauthenet, *Thèse*, Grenoble, 1952a; *Ann. Phys.*, 1952b, **7**, 710.
[7] E. W. Gorter, *Thesis*, Leiden, 1954a; *Philips Res. Rep.*, 1954b, **9**, 295, 321, 403.
[8] A. Michel, *Ann. Chimie*, 1937, **8**, 317.
[9] N. Kawai, S. Kume and S. Sasajima, *Proc. Japan Acad.*, 1954, **7**, 588.
[10] L. Néel, *Ann. Phys.*, 1932, **17**, 5; *C.R. Acad. Sci., Paris*, 1934, **198**, 1311.
[11] M. Fallot and P. Maroni, *J. Phys. Rad.*, 1951, **12**, 253.
[12] C. C.ark and W. Sucksmith, *Proc. Roy. Soc.* A, 1954, **225**, 147.
[13] H. Forestier and G. Guiot-Guillain, *C.R. Acad. Sci., Paris*, 1950, **230**, 1844; *Ibid.*, 1951, **232**, 1832; *Ibid.*, 1952, **235**, 48; *Ibid.*, 1953, **237**, 1554.
[14] R. Pauthenet and P. Blum, *C.R. Acad. Sci., Paris*, 1954, **239**, 33; G. Guiot-Guillain, R. Pauthenet and H. Forestier, *C.R. Acad. Sci., Paris*, 1954, **239**, 155.
[15] L. Néel, *C.R. Acad. Sci., Paris*, 1954, **239**, 8.
[16] Y. Yafet and C. Kittel, *Phys. Rev.*, 1952, **87**, 290.
[17] L. Néel, *Ann. Phys.*, 1949, **4**, 249.
[18] L. Néel and R. Pauthenet, *C.R. Acad. Sci., Paris*, 1952, **234**, 2172.
[19] R. Chevallier and J. Pierre, *Ann. Phys.*, 1932, **18**, 383; cf. ref. (8).
[20] T. Nagata, S.. Akimoto and S. Uyeda, *J. Geomagn. Geoelectr.*, 1953, **5**, 168.
[21] F. Bloch, *Z. Phys.*, 1932, **74**, 295.
[22] L. Néel, *Cahiers de Phys.*, 1944, No. 25, p. 1.
[23] L. Néel, *C.R. Acad. Sci., Paris*, 1947, **224**, 1488.
[24] L. Néel, *Ann. Geophys.*, 1949, **5**, 99.
[25] E. C. Stoner and E. P. Wohlfarth, *Phil. Trans. Roy. Soc.* A, 1948, **240**, 599.
[26] L. Néel, *C.R. Acad. Sci., Paris*, 1947, **224**, 1550.
[27] T. Nagata, *Rock-Magnetism*, 1953, Maruzen (Tokyo).
[28] L. Néel, *Appl. Sci. Res., Hague*, 1954, B, **4**, 13.
[29] L. Néel, *Cahiers de Phys.*, 1942, No. 12, p. 1; *Ibid.*, 1943, No. 13, p. 18.
[30] L. Néel, *Ann. Univ. Grenoble*, 1946, **12**, 299.
[31] E. Thellier, *Ann. Inst. Phys. Globe*, 1938, **16**, 157; *J. Phys. Rad.*, 1951, **12**, 205.
[32] L. Néel, *Ann. Géophys.*, 1949, **5**, 99; cf. also ref. (17).
[33] T. Nagata, *Bull. Earthquake Res. Inst.*, 1940, **18**, 281; *Ibid.*, 1941a, **19**, 49; *Ibid.*, 1941b, **19**, 304; *Ibid.*, 1942, **20**, 192; *Ibid.*, 1943, **921**, 1.
[34] E. Thellier, *C.R. Acad. Sci., Paris*, 1946, **223**, 319.
[35] L. Néel, *Ann. Géophys.*, 1951, **7**, 90.
[36] T. Nagata, S. Uyeda and S. Akimoto, *J. Geomagn. Geoelectr.*, 1952, **4**, 22, 102; *Ibid.*, 1963, **5**, 168.
[37] J. W. Graham, *J. Geophys. Res.*, 1953, **58**, 243.
[38] V. Gottschalk, *Physics*, 1935, **6**, 127; V. Gottschalk and F. Wartman, *U.S. Bur. Mines, Rep. Investig.*, 1935, No. 3, 268, 67 and 83.
[39] C. Kittel, *Phys. Rev.*, 1948, **73**, 810.
[40] C. Guillaud, *Thèse*, Strasbourg, 1943.
[41] L. Lliboutry, *Ann. Phys.*, 1951, **6**, 731.
[42] R. Forrer and R. Baffie, *J. Phys. Rad.*, 1944, **5**, 97.
[43] R. Forrer, *J. Phys. Rad.*, 1931, **2**, 312.
[44] S. Akimoto, *J. Geomagn. Geoelectr.*, 1951, **3**, 47.
[45] J. Roquet, *Thèse*, Paris, 1953.
[46] Lord Rayleigh, *Phil. Mag.*, 1887, **23**, 225.
[47] W. F. Brown, *Phys. Rev.*, 1949, **75**, 147.
[48] P. Weiss and J. de Freudenreich, *Arch. Sc. Phys. Nat., Genève*, 1916, **42**, 449.
[49] F. Preisach, *Z. Phys.*, 1935, **94**, 227.
[50] L. Néel, *J. Phys. Rad.*, 1950, **11**, 49; *Ibid.*, 1951, **12**, 339; *Ibid.*, 1952, **13**, 249.
[51] J. L. Snoek, *Physica*, 1939, **6**, 161; *Ibid.*, 1941, **8**, 711; *New Developments in Ferromagnetic Materials*, Elsevier (Amsterdam, 1947).
[52] H. P. J. Wijn and H. P. J. van der Heide, *Rev. Mod. Phys.*, 1953, **25**, 98; H. P. J. Wijn, *Thesis*, 1953, Leiden.
[53] J. C. Barbier, *Ann. Phys.*, 1954, **9**, 84.
[54] R. Street and J. C. Woolley, *Proc. Phys. Soc.*, 1949, A, **62**, 562; *Ibid.*, 1950, B, **63**, 509; *Ibid.*, 1952, B. **65**, 461, 679.

Chapter XIV
SURFACE PROBLEMS
A121, A122

A121 (1962)
ON A MAGNETOSTATIC PROBLEM RELATING TO THIN FERROMAGNETIC LAYERS

Summary: A study of the magnetostatic energy and of the distribution of the magnetization in two ferromagnetic layers separated by a thin nonmagnetic layer, with plane parallel faces, and covered by a distribution of magnetic charges.

Introduction: The study of the distribution of the spontaneous magnetization in the interior of a magnetic material, under the influence of magnetic charges arbitrarily distributed on a given surface, may appear to be of purely academic interest: it is, however, capable of some interesting applications. Consider for example a ferromagnetic material bounded by a plane surface parallel to which it is magnetized. If, as in an orange peel, the surface has small irregularities, each hump or depression constitutes a small magnetic dipole, with a positive or negative moment, under the influence of which the magnetization, initially supposed to be uniform, takes on a new equilibrium state. Such a problem is difficult to solve rigorously since it brings into play an analytically complicated boundary condition. It is happily almost equivalent to the much simpler problem of the magnetization of a ferromagnetic material bounded by a plane on which one imposes a certain distribution of magnetic charges, of density equal to the density, projected on the plane, of magnetic charges created on the real uneven surface by a uniform distribution of magnetization.

This method allows, for example, the study of the energy of magnetic coupling between two thin ferromagnetic layers separated by a third nonmagnetic layer, with uneven interfaces showing correlations.

Statement of the problem: We thus propose to study the perturbations, assumed to be small, brought about in the distribution of spontaneous magnetization of two ferromagnetic layers A and A' by an arbitrarily imposed distribution of magnetic charges on the two interface planes A–B and B–A', separating A and A' by a third nonmagnetic intermediate layer B of thickness b. We refer this system to three rectangular axes $Oxyz$, with the plane Oxy parallel to the plane of the layers.

We limit ourselves here to the study of the effects of distributions of magnetic charges dependent on the single variable x. These distributions may be described by an expansion as a Fourier series and, given the linear character of the problem, it is sufficient to study separately each of the pairs of homologous terms. Thus let

$$\mathbf{M} = m \sin px \quad \text{and} \quad \mathbf{M}' = m' \sin px \qquad (1)$$

be the densities of charges imposed at the two interfaces.

The most important quantity to determine is the energy E_p associated with the distribution (1): it is equal to the work done in bringing up step by step, from infinity, the charges constituting the distribution. If we denote by

$$\mathbf{V}_0 = v_0 \sin px \quad \text{and} \quad \mathbf{V}'_0 = v'_0 \sin px \qquad (2)$$

the values of the magnetic potentials at the interfaces and if we assume, as we will ultimately prove, that v_0 and v'_0 are linear and homogeneous functions of m and m', the energy E_p is written

$$E_p = \int \frac{1}{2} \mathbf{M} \mathbf{V}_0 \, ds + \int \frac{1}{2} \mathbf{M}' \mathbf{V}'_0 \, ds' = \frac{1}{4}(mv_0 + m'v'_0), \qquad (3)$$

the integrals being carried out over 1 cm² of the interfaces A–B and B–A'.

The total energy is the sum of the E_p corresponding to the different terms in the Fourier expansion.

Magnetization and potentials of the different materials: We denote by J, J' the spontaneous magnetizations of the two materials A, A', by N, N' their molecular field coefficients, by a, a' the separation of the magnetically active nearest neighbour atoms. We further suppose that A and A′ possess a uniaxial magnetic anisotropy about Ox defined by the anisotropy constants C and C′.

When the densities are zero, $m = m' = 0$, the magnetizations of A and A′ are uniform and parallel to Ox: we assume them also to be in the same sense. When m and m' differ from zero, the magnetization of material A is deflected from Ox, in the neighbourhood of the inrerface A–B, by an angle φ which we call the deviation and we assume always to be small. We assume, in addition, that the magnetization always remains parallel to the plane Oxz.

The components I_x, I_y, I_z of the magnetization of material A are written, neglecting second order corrections:

$$I_x = J \quad ; \quad I_y = 0 \quad ; \quad I_z = \Phi J, \qquad (4)$$

with an analogous notation for material A′, with the letters primed.

We denote by V, V′, V_b the potentials, functions of x and z which arise from the magnetic fields in the materials A, A′ and B.

The potential V_b in B obeys the Laplace equation $\Delta V_b = 0$, and taking account of the boundary conditions which we have written, and in which $\sin px$ will be a factor, takes the form:

$$V_b = (v_1 e^{-pz} + v_2 e^{pz}) \sin px, \qquad (5)$$

in which v_1 and v_2 are constants to be determined.

Considering V and V′, they obey the Poisson equation $\Delta V + 4\pi\omega = 0$, in which the density of charge ω is equal to $-\text{div}\, I$. Using equation (4), this becomes

$$\frac{\partial^2 V}{\partial x^2} + \frac{\partial^2 V}{\partial z^2} - 4\pi J \frac{\partial \Phi}{\partial z} = 0 \qquad (6)$$

with an analogous equation for V′.

Equilibrium magnetization: At each point, the spontaneous magnetization is in equilibrium under the action of a certain number of couples: (1) the couple due to the applied magnetic field, namely $-J(\partial V/\partial z)$; (2) the restoring couple of the magnetocrystalline forces, equal to $-C\Phi$; (3) the couple arising from the exchange forces between the nearest neighbour atoms and which may be replaced by an effective magnetic field equal to $(1/6)Na^2\nabla^2 J$.

On writing that the sum of these three couples is zero, one obtains, for the part relating to material A, the following partial differential equation

$$\frac{Na^2 J^2}{6}\left(\frac{\partial^2 \Phi}{\partial x^2} + \frac{\partial^2 \Phi}{\partial z^2}\right) - J\frac{\partial V}{\partial z} - 2C\Phi = 0, \qquad (7)$$

with an analogous equation for the material A′.

Solution of the system of equations (6) and (7): We look for a solution of the form:

$$V = v e^{-qz} \sin px \quad ; \quad \Phi = \varphi e^{-qz} \sin px, \qquad (8)$$

where v and φ are constants. On substituting these expressions into equations (6) and (7), we obtain the system

$$\left.\begin{array}{l}(q^2 - p^2) v + 4\pi J q \varphi = 0, \\[4pt] Jqv + \dfrac{Na^2 J^2}{6}(q^2 - p^2)\varphi - 2C\varphi = 0.\end{array}\right\} \qquad (9)$$

In order that this system, linear and homogeneous in v and φ, should have a solution q must be a root of the biquadratic equation:

$$q^4 - q^2\left(2p^2 + \frac{12C}{Na^2 J^2} + \frac{24\pi}{Na^2}\right) +$$
$$+ p^4 + \frac{12C}{Na^2 J^2} p^2 = 0, \qquad (10)$$

the roots of which we denote by $\pm q_1$, $\pm q_2$: q_1 and q_2 being positive.

The deviation is given in general by

$$\Phi = (\varphi_1 e^{-q_1 z} + \varphi_2 e^{-q_2 z} + \varphi_3 e^{+q_1 z} +$$
$$+ \varphi_4 e^{+q_2 z}) \sin px. \qquad (11)$$

It is necessary to keep all four terms as long as the thickness of layer A is finite, the two additional constants φ_3, φ_4 being determined by suitably written boundary conditions for the exterior face of layer A. For simplicity we assume here that layer A extends to infinity on the side of positive z. Since the potential cannot become infinite at infinity we only use the first two terms of equation (11).

On a boundary condition particular to ferromagnetic materials: We are dealing with materials for which, as in the example which we are considering, one takes account of exchange forces: the derivative of the spontaneous magnetization with respect to the normal to the surface, must be zero.

In fact, in the interior of the ferromagnetic material,

the effective magnetic field equivalent to the exchange forces, is proportional to $\nabla^2 J$: it thus depends only on the second derivatives of the spontaneous magnetization. The terms containing the first derivatives and which are an order of magnitude larger cancel out in pairs, since the action of an atom situated at a distance $\vec{r}$ from the central atom being considered, cancels the action of another atom situated at a distance $-\vec{r}$. It is not the same for the atoms situated at the surface, which only possess neighbours on one side. They would thus be in equilibrium conditions extremely different from thos for atoms belonging to the neighbouring interior layers, if the first derivatives of the magnetization perpendicular to the surface were not zero. Since representing the deviations of the magnetization by a function Φ, without a singularity at the surface, implies that there is no difference between the direction of magnetization of the surface atoms and that of their nearest interior neighbours, we conclude that the first derivatives must in fact be zero.

In the case which interests us, since $\partial I_z/\partial z$ is zero at the surface $z=0$, we may write

$$\Phi = \varphi_0 (q_2 e^{-q_1 z} - q_1 e^{-q_2 z}) \sin px, \quad (12)$$

with the single arbitrary constant φ_0.

Expression of the potential, continuity: Thus, in equation (9), φ may take the two values $\varphi_0 q_2$ and $-\varphi_0 q_1$: one may therefore deduce the two corresponding values of v, which give finally the expression for the potential

$$V = 4\pi J q_1 q_2 \varphi_0 \left(\frac{e^{-q_1 z}}{p^2 - q_1^2} - \frac{e^{-q_2 z}}{p^2 - q_2^2} \right) \sin px. \quad (13)$$

One thus deduces the value $V_0 - v_0 \sin px$ of the potential at the interface A–B, where

$$v_0 = 4\pi J q_1 q_2 \varphi_0 \left(\frac{1}{p^2 - q_1^2} - \frac{1}{p^2 - q_2^2} \right). \quad (14)$$

As the potential must be continuous, V_0 must be equal to the value of V_b at the interface A–B, which gives, on taking, in the expression for V_b, the origin of the coordinates in the middle of layer B:

$$v_0 = v_1 e^{-\frac{1}{2}pb} + v_2 e^{\frac{1}{2}pb}. \quad (15)$$

Charges on the inferfaces: Gauss's law: In the immediate neighbourhood of the interface A–B, Φ takes the value:

$$\Phi_0 = \varphi_0 (q_2 - q_1) \sin px. \quad (16)$$

The total density of the magnetic charges carried by the interface A–B is thus the sum of the imposed density $m \sin px$ and the induced density $-\varphi_0 J(q_2 - q_1) \sin px$.

Gauss's law in the case of this interface is written:

$$\left(\frac{\partial V_b}{\partial z} \right)_{z=\frac{b}{2}} - \left(\frac{\partial V}{\partial z} \right)_{z=0} =$$
$$= 4\pi [m - \varphi_0 J(q_2 - q_1)] \sin px. \quad (17)$$

But, using equations (13) and (14), one may write:

$$-\left(\frac{\partial V}{\partial z} \right)_{z=0} + 4\pi \varphi_0 J(q_2 - q_1) \sin px =$$
$$= p \rho v_0 \sin px, \quad (18)$$

where ρ is a dimensionless coefficient given by the relation:

$$\rho = \frac{p(q_1^2 + q_2^2 + q_1 q_2 - p^2)}{q_1 q_2 (q_1 + q_2)}. \quad (19)$$

With these notations, equation (17) is written:

$$\rho v_0 = \frac{4\pi m}{p} + v_1 e^{-\frac{1}{2}pb} - v_2 e^{\frac{1}{2}pb}. \quad (20)$$

For the second material we have to write the analogous equations to (15) and (20), so that

$$v'_0 = v_1 e^{\frac{1}{2}pb} + v_2 e^{-\frac{1}{2}pb}; \quad (21)$$

$$\rho' v'_0 = \frac{4\pi m'}{p} - v_1 e^{\frac{1}{2}pb} + v_2 e^{-\frac{1}{2}pb}, \quad (22)$$

where ρ' is given by an analogous equation to (19) in which q'_1 and q'_2 are the positive roots of an analogous equation to (10) obtained by replacing C, N, a and J by C', N', a' and J'. The four equations (15), (20), (21) and (22) allow the calculation as a function of m and m' of the four unknowns v_0, v'_0, v_1 and v_2. On eliminating v_1 and v_2, one finds:

$$v_0 = \frac{4\pi}{p} \frac{m \operatorname{ch} pb + m\rho' \operatorname{sh} pb + m'}{(1 + \rho\rho') \operatorname{ch} pb + (\rho + \rho') \operatorname{sh} pb} \quad (23)$$

and a symmetric equation for v'_0.

Finally, on applying equation (3), the energy E_p is written

$$E_p = \frac{\pi}{p} \frac{m^2(\operatorname{ch} pb + \rho' \operatorname{sh} pb) + 2mm' + m'^2(\operatorname{ch} pb + \rho \operatorname{sh} pb)}{(1 + \rho\rho') \operatorname{sh} pb + (\rho + \rho') \operatorname{ch} pb} . \quad (24)$$

Study of the values of p and p' — If we put:

$$\lambda = \frac{Na^2 p^2}{24\pi} ; \quad \mu = \frac{C}{2\pi J^2} , \quad (25)$$

where λ and μ are two dimensionless constants, the biquadratic equation (10) takes the form:

$$q^4 - q^2 p^2 \left(2 + \frac{\mu}{\lambda} + \frac{1}{\lambda}\right) + p^4 \left(1 + \frac{\mu}{\lambda}\right) = 0 . \quad (26)$$

One may then express ρ, a symmetric function of the roots of (26), as a rational function of the coefficients and one obtains:

$$\rho = \frac{1 + \lambda + \mu + \lambda^{\frac{1}{2}}(\lambda + \mu)^{\frac{1}{2}}}{(\lambda + \mu)^{\frac{1}{2}}[1 + \mu + 2\lambda + 2\lambda^{\frac{1}{2}}(\lambda + \mu)^{\frac{1}{2}}]^{\frac{1}{2}}} . \quad (27)$$

When μ and λ become infinite, that is to say when the material A loses its magnetism or when its magnetization becomes fixed, the parameter ρ tends towards unity, and the energy E_p takes, for $\rho = \rho' = 1$, the simple form:

$$E_p = \frac{\pi}{2p}(m^2 + m'^2 + 2mm' e^{-pb}) . \quad (28)$$

This is also the magnetostatic energy in vacuum of a system of two planes, a distance b apart, carrying respectively the densities $m \sin px$ and $m' \sin px$.

When μ is very small with respect to λ, that is to say when one may neglect the magnetocrystalline anisotropy, C being very small, or p being large, one may write:

$$\rho = \frac{1 + 2\lambda}{\lambda^{\frac{1}{2}}(1 + 4\lambda)^{\frac{1}{2}}} . \quad (29)$$

Generally, in ferromagnetic materials, μ is very small with respect to unity; if p is so small that λ itself is small with respect to unity, one may write:

$$\rho = \frac{1}{(\lambda + \mu)^{\frac{1}{2}}} . \quad (30)$$

We note finally that the equality $\lambda = \mu$ allows the definition of a certain value p_c of p, given by the equation:

$$\frac{1}{p_c} = \frac{L_e}{2\pi} = \left(\frac{Na^2 J^2}{12C}\right)^{\frac{1}{2}} . \quad (31)$$

However, one may remark that L_c, the wavelength corresponding to the distribution of magnetic charges, is also, from the same formula (31), of the same order of magnitude as the thickness of the 180° Bloch walls separating two single domains in the material in question. Thus when the wavelength of the charge distribution is small compared with the thickness of the wall, we may neglect the magnetocrystalline anisotropy and put $\mu = 0$; conversely when the wavelength is large compared with the wall thickness one may neglect the exchange energy. This is a result which one would expect intuitively by noticing that the wall thickness corresponds to the spread of a spatial region such that the magnetocrystalline energy and the exchange energy are of the same order of magnitude.

A122 (1963)
ON A NEW MODE OF COUPLING BETWEEN THE MAGNETIZATIONS OF TWO THIN FERROMAGNETIC LAYERS

Summary: A theoretical study of the magnetostatic coupling capable of being produced between two thin ferromagnetic layers separated by a third nonmagnetic layer, and due to correlations between the topographical irregularities of the two interfaces under consideration.

The experimental study of the magnetic properties of stratified structures, composed of two thin ferromagnetic layers A and A' separated by a third nonmagnetic layer B, shows that energies of coupling between the magnetizations of the two outside layers may exist. Thus, Bruyère has shown [1] that, the magnetization of the layer A' stays constant and close to the saturation magnetization, the layer A shows in certain cases an asymmetric hysteresis cycle as if it were subjected, by layer A, to a type of magnetic field of coupling H_m, changing sign with the magnetization of layer A' and capable of attaining several oersteds. In a manner of speaking, perhaps preferably, one may also say that there exists a surface energy of coupling which is close to $H_m Jd$, where J and d designate the magnetization and the thickness of layer A, that is of the order of 0.01 erg/cm^2 in the cases observed. With regard to the origin of these couplings, various explanations may be invoked: we interest ourselves here solely in the magnetostatic couplings.

There exist several types of this: the most simple is the geometric demagnetizing field. Each of the ferromagnetic layers is placed in the demagnetizing field of the other, so that their magnetizations tend to orient themselves antiparallel. This coupling is particularly weak when the thickness of the layers is small and their diameter is large.

A second type of magnetostatic coupling arises from the interactions between the walls separating the elementary domains: one knows [2] in fact, that in thin layers these walls are always associated with distributions of magnetic charges and with magnetic fields of dispersion. We will not return to these phenomena which have already been well studied theoretically [5] and experimentally [4].

But there also exist magnetostatic couplings related to the topography of the interfaces A–B and B–A', separating respectively layer A from layer B and layer B from layer A'. In fact thanks to the method of fabrication of the layers, in general evaporation under vacuum, and to their polycrystalline nature, their surface is certainly not plane and possibly possesses even acute angles. Furthermore, if the interface A–B possesses a hump, the deposition of B is not able to remove it if B is thin, so that the interface B–A' tends to reproduce this hump, but in a form becoming more attenuated as the thickness of B becomes greater. We suppose now that the layer A and A' possess fixed magnetizations J and J', parallel to a direction Ox in the plane Oxy of the layer, and in the same sense. The walls of the hump in A–B are then inclined to the magnetization J: this hump therefore constitutes a small magnetic dipole with a moment parallel to Ox. But the homologous hump in the interface B–A', forming in practice a depression in layer A, also constitutes a dipole with a moment antiparallel to the first. These two dipoles are put into Gauss's second relation and their moments possess the most stable relative orientation. Each pair of homologous humps thus introduces into the energy of the system a term tending to give A and A' magnetizations parallel and in the same sense.

Naturally this phenomenon is opposed by the effect of the dispersion field, which tends to orient parallel to each other the spontaneous magnetizations of the regions close to the surface and to diminish the surface dipole moments, in spite of the exchange forces and the forces of uniaxial anisotropy which tend, on the contrary, to maintain the uniformity of magnetization in the whole layer. The true coupling is, therefore, less strong than that which one calculates with the hypothesis of a fixed magnetization.

In order to quantify these phenomena, we expand as a Fourier series the functions z and z' of x and y

representing the interfaces A–B and B–A', the origin of the coordinates being chosen in both cases on the same axis zz' but at different points to cancel out the mean values $\bar{z}$ and $\bar{z}'$ of z and z'. A similarity of form between the two interfaces translates itself mathematically into a correlation between z and z'; the mean value $\overline{zz'}$ of the product zz' is not zero.

The most simple case is that of undulations parallel to Oy. Two homologous terms in the expansion of z and z' are:

$$z = h \sin px \; ; \quad z' = h' \sin px \qquad (1)$$

corresponding to a correlation $\overline{zz'} = hh'/2$. Magnetically the layers A and A' are defined by their spontaneous magnetizations J and J', their molecular field coefficients N and N' and by the distances a and a' of the nearest magnetically active neighbour atoms. We assume in addition that a magnetothermal treatment has given to the layers A and A' a uniaxial anisotropy about Ox defined by the anisotropy coefficients C and C': the quantities, J, J', N, N', C, C' are defined for unit volume. Finally, let b be the mean thickness of layer B.

When the magnetization of the layers A and A' is fixed, uniform and parallel to Ox, the interfaces A–B and B–A' carry magnetic charges of which the densities M and M', projected on the plane Oxy, are written:

$$M = ph \, J \cos px \quad \text{and} \quad M' = -ph' \, J' \cos px. \qquad (2)$$

The negative sign in the expression for M' arises because layer A' is below the interface B–A' whereas A is above A–B.

If we remove the rigidity of the magnetization a new equilibrium state is established, but in the general case it is difficult to calculate rigorously. The question is only to study the perturbations produced to a simple uniform distribution, parallel to Ox, by the magnetic charges created by this same distribution on the actual undulating interfaces, and of a density given by equation (2). At the very least if ph is small with respect to unity, one may replace this problem by the analogous problem of the perturbations produced to a simple uniform distribution parallel to Ox, by the densities of magnetic charges *imposed on a plane surface boundary* and precisely equal to M and M'. One thus gains by being able to write the boundary conditions on the two planes $z=0$ and $z'=0$.

This problem is easy to solve by assuming that the direction of spontaneous magnetization always stays close to Ox. The calculation shows [3] that the increase in the energy of the system, per unit of surface area of the layered structure, is the sum of two terms E_p and E'_p which, in the case where the thicknesses of the layers A and A' are large with respect to the thickness b of the intermediate layer B, are given by the formulae:

$$E_p = -\frac{2 \pi phh' \, JJ'}{(1 + \rho\rho') \operatorname{sh} pb + (\rho + \rho') \operatorname{ch} pb} ; \qquad (3)$$

$$E'_p = \frac{\pi ph^2 J^2 (\operatorname{ch} pb + \rho' \operatorname{sh} pb) + \pi ph'^2 J'^2 (\operatorname{ch} pb + \rho \operatorname{sh} pb)}{(1 + \rho\rho') \operatorname{sh} pb + (\rho + \rho') \operatorname{ch} pb}$$

$$(4)$$

in which ρ is given by:

$$\rho = \frac{1 + \lambda + \mu + \lambda^{1/2} (\lambda + \mu)^{1/2}}{(\lambda + \mu)^{1/2} [1 + \mu + 2\lambda + 2\lambda^{1/2} (\lambda + \mu)^{1/2}]^{1/2}}, \qquad (5)$$

with the abbreviations:

$$\lambda = \frac{N \, a^2 p^2}{24\pi} ; \quad \mu = \frac{C}{2\pi J^2}. \qquad (6)$$

The coefficient ρ' is given by the analogous formulae in which λ, μ, N, a, C, J are respectively replaced by $\lambda', \mu', N', a', C', J'$.

We discuss first the values of ρ and ρ'. For this we quantify the characteristics of the ferromagnetic materials forming the layers A and A' by choosing the following numerical values: $N = N' = 6000$; $J = J' = 900$ e.m.u.; $a = a' = 2.9$ Å; $C = C' = 5000$ ergs/cm³ which correspond approximately to permalloy aligned by thermal treatment in a magnetic field, a material currently being used for the fabrication of thin magnetic layers.

With these data one obtains $\mu = 0.001$. The parameter μ thus possesses a value which is very small with respect to unity, and constant as long as it relates to the same ferromagnetic material. On the contrary, the parameter λ depends not only on the properties of the material but also on the wavelength $L = 2\pi/p$ of the undulations of the interface.

In particular λ is equal to μ for a certain value L_e of L given by:

$$L_e = \pi \sqrt{\frac{N a^2 J^2}{3C}}, \qquad (7)$$

where $L_e = 5200$ Å with the numerical data adopted. One notices that L_e is also of the same order of magnitude as a 180° Bloch wall. In fact, in a uniaxial substance, the thickness of a particular wall, taken from a point where the magnetization makes an angle θ_1 with the axis to a corresponding point where it makes an angle $\pi - \theta_1$, is given [6] by

$$L_b = \sqrt{\frac{Na^2 J^2}{3C}} \text{ Log cotg } \frac{\theta_1}{2}. \tag{8}$$

One sees, therefore, that for $\theta_1 = 5°$, L_b is precisely equal to L_e.

On the other hand, examinations by electron microscopy seem to show that the scale L of the surface irregularities is less than 200 Å, which is considerably smaller than L_e. The result is that λ is much larger than μ (650 times larger for L = 200 Å) so that μ is negligible with respect to λ and equation (5) reduces to:

$$\rho_e = \frac{1 + 2\lambda}{\lambda^{1/2} (1 + 4\lambda)^{1/2}}. \tag{9}$$

The parameter ρ_e is a continually decreasing function of λ (itself proportional to p). Thus when the scale of the irregularities of the interfaces is small with respect to the thickness of the Bloch walls, one may neglect the magnetocrystalline anisotropy.

Given this, we return to the meaning of E_p and E'_p. Concerning the energy E'_p, it is the sum of the two terms corresponding to the uniaxial anisotropy of the films A and A' related to the particular form of the model chosen for the interface: undulations parallel to Oy. The magnetization tends to align itself parallel to the undulations. These terms are not particularly interesting: they disappear moreover with an isotropic model of the interface, consisting for example of two spatial components juxtaposed one on the other, one undulating parallel to Oy, the other parallel to Ox.

Concerning E_p, it is a negative quantity when J and J' are of the same sign: the magnetizations of the two layers tend to arrange themselves parallel to each other in the same sense, the coupling is positive.

This energy is a function of p, and is inversely proportional to a certain dimensionless factor G given by $G = [(1 + \rho^2)\text{sh } pb + 2\rho \text{ ch } pb]/pb$. We have calculated G for various values of the product pb, which is also dimensionless, on the assumption that the thickness b of B is equal to 100 Å:

pb ...	12	6,0	3,0	2,4	2,1	1,8	1,4	0,8	0,4	0,2
G ...	28	200	157	22,1	18,5	17,9	18,5	21,9	45,8	158	608

One sees that G passes through a pronounced minimum, equal to 17.9 for $pb = 2.1$, that is for a wavelength L of 300 Å. The corresponding maximum coupling energy E_{pm}, with the numerical data adopted, is $2.8\ hh' \times 10^{11}$ ergs/cm². To obtain 0.01 ergs/cm², the order of magnitude of the coupling energies observed, hh' must be equal to 360 Å², and thus a correlation $\overline{zz'}$ of 180 Å². This value corresponds to highly chaotic interfaces, composed for example of monocrystalline blocks 150 Å in diameter and 30 to 40 Å high.

If instead of choosing values of the parameters applicable to permalloy, we assume the magnetization of the layers A and A' to be fixed, N and N' become infinite while ρ tends to unity. The energy E_p tends to the value:

$$E_{pr} = -\pi phh' JJ' e^{-pb}, \tag{10}$$

for which the maximum value is equal to $9.4\ hh' \times 10^{11}$ ergs/cm², for $pb = 1$. Thus the act of passing from ferromagnetic to fixed magnetization in permalloy divides the maximum coupling energy by 3.4 and the corresponding optimum wavelength by 2.1.

Let us say a few words about the case where the irregularities of the interfaces are on average isotropic, along the directions of the plane Oxy. To avoid complicated calculations, we assume that there is the same ratio between the coupling energy of a body with a fixed magnetization and that of permalloy, and we consider isotropic irregularities or irregularities with cylindrical undulations. It suffices therefore to study the very simple case of fixed magnetization.

We treat the problem for the fundamental term in the Fourier expansion of z and z', where

$$z = h \sin px \sin py\ ;\quad z' = h' \sin px \sin py, \tag{11}$$

corresponding to a correlation $\overline{zz'} = hh'/4$, two times smaller than in the previous case, for the same values of h and h'. We assume that the fixed magnetizations J and J' of layers A and A' are parallel to the plane Oxy and make the angles φ and φ' with Ox. The density M of magnetic charges created on the real interface A–B and projected on to the plane Oxy is written:

$$M = hp\ J\ (\cos \varphi \cos px \sin py + \sin \varphi \sin px \cos py), \tag{12}$$

with an analogous expression for the density M' of charges created on the interface B–A', obtained on replacing M, h, φ, and J by $-M'$, h', φ' and J' in the previous equation.

The magnetostatic energy associated with the distribution of densities M and M' on the planes $z = 0$ and $z' = 0$ is the sum of the two terms E_p and E'_p:

$$E_p = -\frac{\pi p}{2\sqrt{2}}\ hh' JJ' \cos(\varphi - \varphi')\ e^{-pb\sqrt{2}} \tag{13}$$

and

$$E'_p = \frac{\pi p}{4\sqrt{2}} (h^2 J^2 + h'^2 J'^2).\quad (14)$$

The term E'_p, constant and independent of the orientation of the magnetizations, is of no interest. The term E_p corresponds to a cosinusoidal type of coupling between J and J'. It is a function of p which goes through a maximum for $pb=\sqrt{2}/2$. The maximum energy of coupling, for $b=100$ Å, $J=-J'=900$ e.m.u., is equal to 2.34 $hh' \times 10^{11}$ ergs/cm².

By the method of reasoning given above, one estimates therefore that the maximum coupling energy for permalloy is close to 0.69 $hh' \times 10^{11}$ ergs/cm², for isotropic surface irregularities of scale L close to 420 Å. At the same values of the correlation $\overline{zz'}$, the maximum coupling energy due to isotropic surface irregularities is approximately a factor of two smaller than that of cylindrical undulations.

We conclude that a correlation between the topographical irregularities of the two interfaces may effectively produce a coupling energy of the order of 0.01 ergs/cm² between the magnetizations of two thin ferromagnetic layers separated by a third nonmagnetic layer of thickness 100 Å. It is necessary however for this that the "wavelength" and the corresponding amplitudes of the irregularities should be of the order of 400 and 40 Å respectively.

REFERENCES
[1] Private communication.
[2] L. Néel, *Comptes rendus*, 1955, **241**, 533.
[3] H. W. Fuller and D. L. Sullivan, *J. Appl. Phys.*, Suppl., 1962, **33**, 1063.
[4] I. B. Puchalska and R. J. Spain, *Comptes rendus*, 1962, **254**, 2937.
[5] L. Néel, *Comptes rendus*, 1962, **255**, 1545.
[6] L. Néel, *Cah. Phys.*, 1944, **25**, 1.

Chapter XV

ANTIFERROMAGNETIC HYSTERESIS
A142

A142 (1967)
A THEORETICAL STUDY OF THE FERRO–ANTIFERROMAGNETIC COUPLING BETWEEN THIN FILMS

Summary: In the first part, the author studies the magnetic properties of a system formed from very fine antiferromagnetic grains which are epitaxially formed on a thin ferromagnetic film. From a statistical point of view, he shows that the superficial magnetic moments of these grains, corresponding to the interface of contact with the thin film behave under the action of the ferromagnetic film as volumetric moments of fine ferromagnetic grains under the action of a magnetic field. Upon this basis, he interprets the unsymmetric hysteresis cycles presented by the system, the variation of the critical fields with the number of order of the cycle, and the hysteresis of rotation. Systems of this sort thus permit the experimental study of antiferromagnetic hysteresis.

In the second part, the author studies the magnetic behaviour of a ferromagnetic thin film possessing parallel faces, in which the magnetic moments of the superficial atomic planes are submitted to arbitrary couples. He deduces the behaviour, in a rotating magnetic field of two thin films, one ferromagnetic and the other antiferromagnetic, coupled by their epitaxy. He introduces the notion of oscillating hysteresis, which is to be distinguished from the alternating hysteresis and from the hysteresis of rotation. He shows that the ratio of the oscillating hysteresis (for a total amplitude equal to π) to the hysteresis of rotation can be zero in certain cases. Upon these bases, he suggests the interpretation of the small values of this ratio, for a system of two coupled films.

PART ONE
ANTIFERROMAGNETIC HYSTERESIS AND ITS ANALOGY WITH FERROMAGNETIC HYSTERESIS

(1) Introduction: By the partial oxidation of a thin ferromagnetic film, for example of nickel or cobalt, it is possible to cover it with a coating of antiferromagnetic crystallites NiO or CoO. These oxide crystallites are crystallographically more or less coherent with the crystallite of metal on which they are formed. This then results in the appearance of intense magnetic exchange interactions between the metal and the oxide, of the same order of magnitude as the interactions which exist between the atoms of the metal or between the cations of the oxide. These interactions tend to ensure, for either side of the interface, the parallelism of the spontaneous magnetization of the ferromagnetic film with the direction of antiferromagnetism of the antiferromagnetic film. One has thus produced a multiple film with *ferro-antiferromagnetic oupling*.

The magnetic properties of multiple thin films have been studied in detail recently by several authors [1]–[4] in particular by Mme Schlenker, who has discovered some very curious hysteresis phenomena. We propose to interpret these.

The thickness of the initial ferromagnetic film is of the order of 1000 Å. In such a film, the spontaneous magnetization retains approximately the same direction relative to the interior of a domain so that its size, in the plane of the film, is much larger than both the thickness of the film and the dimensions of the crystallites which form the film. This magnetic coherence is ensured principally by the large value of the exchange interactions, by the thinness of the film which, together with the condition div $\vec{J} = 0$ for the vanishing of the internal dispersion field, requires that the magnetization of an atomic plane parallel to the plane of the film to be uniform and to depend only on the distance from the surface. The *direction* of this *coherent magnetization* depends on the induced uniaxial anisotropy and on the direction of the applied magnetic field.

As a result of these coherent domains being large with respect to the crystallites and on condition that the latter are randomly oriented, the individual anisotropy of the crystallites disappears through compensation: there only remains the induced uniaxial anisotropy, created intentionally during the deposition of the film by the application of a magnetic field.

Because of the very process of their formation, the dimensions of the antiferromagnetic crystallites should be at most equal to the dimensions of the ferromagnetic crystallites. The result is that a single ferromagnetic domain corresponds to a large number of antiferromagnetic crystallites.

(2) Surface antiferromagnetic magnetization: To an element $d\sigma$ of the surface of an antiferromagnetic crystallite in contact with a supporting ferromagnetic film corresponds a spontaneous magnetic moment $\vec{dm}$ and a *surface magnetization*

$$\vec{s} = \frac{\vec{dm}}{d\sigma}. \qquad (1)$$

This moment $\vec{dm}$ is that of the atoms of the atomic plane which constitutes the surface of the antiferromagnetic crystallite. The vector $\vec{s}$ rotates at the same time as the direction Δ of antiferromagnetism, changing sense whenever one goes from a given antiferromagnetic domain to a domain in antiphase, and changing sense also at the crossing of an atomic "step" in the surface of the crystallite. For certain orientations of the surface with respect to the crystal lattice, $\vec{s}$ is zero. *A priori* any orientation of $\vec{s}$ with respect to $d\sigma$ is possible. In the absence of physico-chemical alterations of the multiple film, $\vec{s}$ may change according to two principal processes: by rotation of the direction of antiferromagnetism Δ in the small volume adjacent to $d\sigma$; by displacement by the passage across $d\sigma$ of a wall separating two domains in antiphase or, more generally, two domains with different directions of Δ. To be complete, one should add to these two mechanisms the variation of the intensity and direction of the magnetizations of each of the magnetic sub-lattices produced by very high magnetic fields. These three mechanisms correspond, in ferromagnetic bodies, to processes of rotation, of displacement of a wall and of a variation of the magnitude of the spontaneous magnetization.

As in ferromagnetic bodies, it must be the displacements of the walls which are the easiest to produce in large antiferromagnetic crystallites, since in principle there is no energy to provide. However, when the dimensions of the crystallites are smaller than the thickness of the wall the only remaining process possible is the rotation of Δ.

Contrarily to what happens for ferromagnetic bodies, it is in general impossible in large antiferromagnetic bodies to move the walls or to change the orientation of Δ, by means of a magnetic field H. In fact, in ferromagnetic materials, the pressures on the walls or the couples acting on the direction of $\vec{J}$ are of the order of JH, whereas for antiferromagnetic materials they are of the order of χH^2, χ and J being the susceptibility and the magnetization per cubic centimetre. With $H = 10^3$, $J = 10^3$, $\chi = 10 \times 10^{-6}$, in e.m.u., the pressures and the couples are thus of the order of 10^6 in ferromagnets and only of 10 in antiferromagnets: the difference is enormous.

(3) Ferro–antiferromagnetic coupling: If $\vec{J}$ denotes the magnetization of the ferromagnetic film in the neighbourhood of the element $d\sigma$, the coupling energy relating to this element is written:

$$dw = - a\vec{s}\,\vec{j}\, d\sigma, \qquad (2)$$

where a characterizes the exchange coupling. The order of magnitude of asj is twenty ergs per square centimetre: this is in fact of the order of the exchange energy per cubic centimetre (10^9 ergs) divided by the number of atomic planes per centimetre (5×10^7).

One sees, therefore, that it is much easier to act on the magnetism of an antiferromagnet by the intermediary of ferro–antiferromagnetic coupling than by the direct action of a magnetic field. In the second case one has available a couple of 10 c.g.s. for 1 cm³ of material. In the first case a couple of 20 c.g.s. is available, of the same order of magnitude as in the previous case, but which this time produces its effects in a thin film of oxide the thickness of which may for example be reduced to 1000 Å: it is a million times more effective than in the first case.

Equation (2) shows that everything behaves as if the magnetic moment $\vec{s}\,d\sigma$ were subjected to a magnetic field $\vec{a}\,j$. On assuming that the quantities a, $\vec{s}$, $\vec{j}$ vary independently of each other, the energy W of coupling per square centimetre is written

$$W = - A\vec{S}\vec{J}, \qquad (3)$$

where A is an average coupling constant, where the magnetic moment $\vec{S}$ is equal to the sum of $\vec{s}\,d\sigma$ scaled to 1 cm² and where $\vec{J}$ is the mean value of $\vec{j}$, which we associate with the magnetization of the ferromagnetic film: everything behaves, therefore, as if the moment $\vec{S}$ were subjected to an effective magnetic field equal to A J: its order of magnitude is a million oersteds.

Clearly the approximations which allow one to write equation (3) are somewhat gross, but this is not of great importance since we are trying to obtain a general idea of the phenomena, rather than to construct a quantitative theory.

(4) The principle of analogy: Just as the magnetization $\vec{M}$ of a ferromagnet is the resultant of the magnetizations of a multitude of small elementary volumes, so the moment $\vec{S}$ is the resultant of the moments of a multitude of small elementary surface domains. On the other hand the mechanisms which control the orientation of the moments of these elementary surface domains, are, as we have established above, very similar to the mechanisms of control of the orientation of the magnetization of the elementary ferromagnetic volumes. Finally, just as there exist various types of magnetic interactions between the elementary ferromagnetic volumes, there exist also various interactions between the elementary surface domains.

It thus follows that the laws which describe the variations of $\vec{S}$ as a function of the variations of the effective field $A\vec{J}$, must be analogous to those which describe the variations of the magnetization $\vec{M}$, of a ferromagnetic body, as a function of the variations of magnetic field $\vec{H}$: in particular the forms of the hysteresis cycles must be similar.

(5) Thermal treatment of multiple films: its role and its effects: Experiments show [1] that a multiple film with ferro–antiferromagnetic coupling, when cooled in a magnetic field so that it crosses the Néel temperature θ_N of the antiferromagnetic oxide, subsequently shows asymmetric hysteresis cycles, as long as the temperature is maintained below θ_N. It is necessary of course, as is almost always the case, that the Curie point θ_F of the film of ferromagnetic metal should be considerably higher than θ_N.

To interpret these facts, we use the principle of analogy stated above.

The cooling of the system S and the crossing of the point θ_N take place under the influence of a constant magnetic field of direction $\varphi = 0$, which gives the magnetization $\vec{J}$ of the ferromagnetic film, and consequently the effective field $A\vec{J}$, a palpably constant magnitude, as well as an fixed orientation $\varphi = 0$. Under this influence the system S acquires an anhysteretic moment $\vec{S}_a$, parallel to $A\vec{J}$, whose properties must be analogous to those of the *anhysteretic thermal magnetization* of a ferromagnetic body.

It is necessary, therefore, firstly to remind ourselves briefly of the latter.

(6) Properties of a ferromagnetic body in the neighbourhood of the anhysteretic state: A ferromagnetic body under the influence of a constant field $\vec{H}_a$, acquires during its cooling from a temperature above the Curie point θ_F down to room temperature, a magnetization $\vec{M}_a$ parallel to $\vec{H}_a$. This is called the anhysteretic thermal magnetization. It is shown as a function of field by the dashed curve in Fig. 1: where $M_a = OA'$.

We examine the room temperature properties of a body so treated. The state $A(H_a, M_a)$ exhibits similar properties to those of the state obtained after demagnetization in zero field: i.e. disregarding the reversible changes of magnetization, the irreversible changes of magnetization are proportional to the square of the changes of field provided that these changes are small. The result is that in the neighbourhood of A, the irreversible differential susceptibility is zero whatever the direction. One may

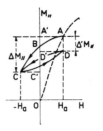

Fig. 1.

also state the laws, valid in the neighbourhood of A, similar to those of Rayleigh given for the states close to the origin O.

If, the body being in the state A, one removes the field $\vec{H}_a$, the magnetization takes the value OB (Fig. 1): called the *thermoremanent magnetization*. If one applies a field $-\vec{H}_a$, antiparallel to $\vec{H}$, the magnetization takes the value $OC' = M_a - \Delta M_\parallel$. On reapplying the field $\vec{H}_a$, the magnetization takes the value $OD' = M_a - \Delta' M_\parallel$. In further cycles the representative points return alternatively to C and D, at least if one neglects the phenomena of "tilting" and "creep" which we will discuss further (§ 13). With regard to the reversible changes of $\vec{M}$, the initial tangents to the branches of the reverse cycles at A and D, as well as the tangent at C for the ascending branch, are parallel to the axis OH of the abscissae. If H_a is small with respect to the coercive field of the body, the changes $\Delta M_\parallel$ and $\Delta' M_\parallel$ are small with respect to M_a and the Rayleigh laws are applicable: the curves ABC and CD are thus parabolae and $\Delta M_\parallel = 2\Delta' M_\parallel$.

If in the initial state $A(H_a, M_a)$, one applies an additional field H of magnitude H_a perpendicular to the initial field H_a and then one removes this field H, the point representing the variation $\Delta M_\perp$ of the magnetization in a direction perpendicular to H_a follows the path OAB (Fig. 2) and the body has a final magnetization $\Delta M_\perp = OB$ perpendicular to M_a. If M_a is small and if Rayleigh's laws are applicable, $\Delta M_\perp = \tfrac{1}{2}\Delta M_\parallel$. If on the contrary M_a is close to the saturation magnetization, $\Delta M_\perp$ decreases and tends to zero: one sees in fact that if H_a is very large in the direction $\varphi = 0$, a perpendicular magnetization cannot exist in the absence of a perpendicular driving field.

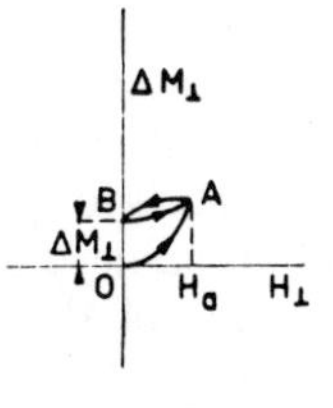

Fig. 2.

Further cycles O, H_a, O, identical to the first, give, in the first approximation, the same value ΔM of the perpendicular magnetization.

The initial tangent at O to the curve, for the first perpendicular magnetization, the tangent at A to the reverse branch and the tangent to the rising branch of the cycle BA at B are horizontal, with regard to reversible changes in M.

(7) Properties of the ferromagnetic film in the absence of ferro–antiferromagnetic coupling: The properties of the ferromagnetic film serving as a support may be explained by the existence of a uniaxial anisotropy energy, so that the expression relating to a cubic centimetre may be written in the form

$$W_a = -\frac{1}{4} H_a J \cos 2\varphi, \qquad (4)$$

where H_a is a certain constant and φ is the angle between the magnetization $\vec{J}$ and the axis. We recall that this axis is parallel to the direction of the magnetic field applied at the moment of deposition of the film. The equilibrium positions of $\vec{J}$, in the absence of any external perturbation, correspond to $\varphi = 0$ and $\varphi = \pi$.

If one applies a magnetic field $\vec{H}$ in the direction $\varphi = 0$, the couple Γ applied to $\vec{J}$ is equal to

$$\Gamma = -J(H_a \cos \varphi + H) \sin \varphi. \qquad (5)$$

The magnetization $\vec{J}$ remains stable in the position $\varphi = 0$, as long as Γ remains negative, i.e. when H is greater than $-H_a$. When H decreases below the critical value $-H_a$, the magnetization rotates in an irreversible way from $\varphi = 0$ to $\varphi = \pi$. If $H'_1, H'_2, H'_3, \ldots$ are the critical fields corresponding to decreasing values of the field, in the course of successive field cycles starting from an initial cycle, and if $H_1, H_2, H_3, \ldots$ are the critical fields corresponding to increasing values of the field, we have $H_n = -H_a$ and $H'_n = H_a$ for all n.

(8) Determination of the first descending critical field H_1, taking account of coupling: The system S becomes magnetized under the infleunce of the effective field $A\vec{J}$. This magnetization is the sum of reversible and irreversible terms. If the system S is isotropic, the reversible moment S_r acquired under the influence of the effective field $A\vec{J}$ is parallel to $\vec{J}$ and its magnitude is independent of the orientation of $\vec{J}$: one may thus put it equal to $C \cdot \vec{J}$. The corresponding coupling energy $-ACJ^2$ is constant, as long as $\vec{J}$ has a constant magnitude. There is thus no need to concern ourselves with the reversible terms of the system S. We will disregard them henceforth.

In the course of the thermal treatment the system S has acquired the anhysteretic moment S_a parallel to the axis.

On the other hand, if we denote by e the thickness of the ferromagnetic metal film, the coupling energy may be written in the form $-(A/e)\vec{S} \cdot \vec{J} e$ showing that everything acts as if the magnetization $\vec{J}$ of the whole ferromagnetic film were subject to a magnetic field equal to AS/e.

If we assume for a moment that S_a is constant, the couple Γ applied to $\vec{J}$ becomes, on neglecting terms of higher order than φ

$$\Gamma = -J\left(H_a + \frac{A}{e} S_a + H\right)\varphi, \qquad (6)$$

which shows that the first descending critical field is equal to

$$H_1 = -H_a - \frac{A}{e} S_a. \qquad (7)$$

In practice S_a does not remain rigorously constant since, when $\vec{J}$ rotates by a small angle φ, the equivalent effective magnetic field decreases by $\frac{1}{2}AJ\varphi^2$ in the direction $\varphi = 0$ and increases by $AJ\varphi$ in the direction $\varphi = \pi/2$. But following the principle of analogy and the remarks of paragraph 6, since the changes in $\vec{S}$ are proportional to the square of the changes in $A\vec{J}$, they are negligible to the order of approximation used above.

(9) Determination of the first ascending critical field H: When H becomes smaller than $-H_1$, $\vec{J}$ rotates from the direction $\varphi = 0$ to a direction close to $\varphi = \pi$. Naturally, this operation changes the magnetic moment of the system S.

In the direction $\varphi = 0$, the effective field goes from the value $+A\vec{J}$ to the value $-A\vec{J}$. Following the principle of analogy and paragraph 6, the component of S changes from the value S_a to the value $S_a - \Delta S_\parallel$.

In the direction $\varphi = \pi/2$, the effective field begins by increasing from 0 to AJ as $\vec{J}$ rotates from $\varphi = 0$ to $\varphi = \pi/2$ and then decreases from AJ to O as $\vec{J}$ rotates from $\varphi = \pi/2$ to $\varphi = \pi$. The corresponding component of $\vec{S}$ goes from the value O to the value $\Delta S_\perp$.

Following the principle of analogy, when S_a is small with respect to saturation, $\Delta S_\perp$ should just be an eighth of $\Delta S_\parallel$, but as S_a increases and tends towards saturation $\Delta S_\perp$ becomes very small with respect to $\Delta S_\parallel$ and finally becomes zero.

Under the influence of the equivalent field $A\Delta S_\perp/e$ created in the ferromagnetic film by this perpendicular component, the magnetization does not align itself exactly along the direction $\varphi = \pi$ but follows a nearby direction $\varphi = \pi - \varepsilon$ where

$$\epsilon = \frac{A\Delta S_\perp}{(H_a - H)e - A(S_a - \Delta S_\parallel)} \qquad (8)$$

For $H = H_1$, ϵ is equal to $\epsilon_1 = A\Delta S_\perp/(2H_a e + A\Delta S_\parallel)$. When H rises above H_1, ϵ increases and becomes infinite for $H = H_a - A(S_a - \Delta S_\parallel)/e$. The "ascending critical field" H'_1 at which a discontinuous change in the orientation of $\vec{J}$ occurs, of amplitude less than π, must therefore correspond to a value a little smaller than the above, which we put equal to:

$$H'_1 = H_a - \frac{A}{e}(S_a - \Delta S_\parallel) - h_\perp . \qquad (9)$$

If we assume that the discontinuity occurs in the neighbourhood of $\epsilon = 1$, one obtains the order of magnitude for H

$$h_\perp \sim \frac{A}{e}\Delta S_\perp . \qquad (10)$$

We calculate further on the exact value of h for a particular case. But it is extremely important to note here that the discontinuity is preceded by a region where ϵ varies as $(H'_1 - H)^{-1}$, and where the magnetization varies as $\cos \epsilon$, as is shown schematically in Fig. 3. This behaviour is quite different from that in the neighbourhood of H_1, where the magnetization remains constant and equal to J until the field reaches exactly the value H_1.

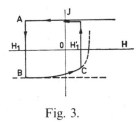

Fig. 3.

These theoretical predictions are completely in agreement with the experimental results of Mme Schlenker [1]: the difference in shape between the neighbouring branches of the cycle, on the one hand of H_1, and, on the other hand, of H'_1 is striking, as much for the films of Co–CoO for those of NiFe–NiFeMn.

(10) Proof of the existence of the discontinuity and determination of its position: When ϵ varies from 0 to π, the perpendicular moment varies between $\Delta S_\perp$ and $2\Delta S_\perp$, being put on average equal to $\tfrac{3}{2}\Delta S_\perp$. The corresponding couple acting on J is equal to

$$\frac{3AJ}{2e}\Delta S_\perp \cos \epsilon .$$

On the other hand, when δ varies from 0 to π, the moment $S_\parallel$ parallel to $\varphi = 0$ varies from $S_a - \Delta S_\parallel - \Delta' S_\parallel$, a dependence which we represent approximately by

$$S_\parallel = S_a - \frac{1}{2}(\Delta S_\parallel + \Delta' S_\parallel) - \frac{1}{2}(\Delta S_\parallel - \Delta' S_\parallel)\cos \epsilon . \qquad (11)$$

The corresponding couple acting on J is equal to

$$\frac{AJ}{e} S_\parallel \sin \epsilon .$$

Finally the moment J is subject to a couple $J(H - H_a \cos \epsilon) \sin \epsilon$ resulting from the applied field and the uniaxial anisotropy.

Ultimately the couple Γ replied to J may be written in the form

$$\Gamma = C(a \sin \epsilon - \sin 2\epsilon + \lambda \cos \epsilon) \qquad (12)$$

where C is a constant and where one puts

$$a = \frac{H + \dfrac{A}{e}\left(S_a - \dfrac{1}{2}\Delta S_\parallel - \dfrac{1}{2}\Delta' S_\parallel\right)}{\dfrac{1}{2}H_a + \dfrac{H}{4e}(\Delta S_\parallel - \Delta' S_\parallel)} \qquad (13)$$

and

$$\lambda = \frac{\dfrac{3A}{2e}\Delta S_\perp}{\dfrac{1}{2}H_a + \dfrac{A}{4e}(\Delta S_\parallel - \Delta' S_\parallel)} . \qquad (14)$$

When H increases above the value H_1, the moment J takes up an equilibrium position corresponding to the value of ϵ given by $\Gamma = 0$. This equilibrium position becomes unstable for $\dfrac{d\Gamma}{d\epsilon} = 0$. The discontinuity occurs at the value $\epsilon = \epsilon_1$ as long as these two relations are satisfied.

One easily finds:

$$\sin \epsilon_i = \left(\frac{\lambda}{2}\right)^{1/2}, \text{ with } \epsilon_i < \frac{\pi}{2}, \qquad (15)$$

$$a = 2\cos^3 \epsilon \qquad (16)$$

The critical field is given by

$$H'_1 = H_a - \frac{A}{e}(S_a - \Delta S_\parallel) - h_\perp \qquad (17)$$

with

$$h_\perp = \left[H_a + \frac{A}{2e}(\Delta S_\parallel - \Delta' S_\parallel)\right](1 - \cos^3 \epsilon_i). \qquad (18)$$

(11) The position of subsequent discontinuities: When, after the application of a field H'_1, the magnetization returns to the neighbourhood of the direction $\varphi = 0$, the system describes a reverse cycle in the direction $\varphi = 0$ and the component of S in this direction takes the value $S_a - \Delta' S_\parallel$, as we have already indicated above. As for the perpendicular component, it returns to the same value ΔS since the perpendicular component of the effective field has described a new cycle O, AJ, O identical to the preceding one.

The angle ϵ' of equilibrium for J with the direction $\varphi = 0$ is given for small ϵ' by:

$$\epsilon' = \frac{A \Delta S_\perp}{H_a + H + A(S_a - \Delta' S_\parallel)} \qquad (19)$$

Its value ϵ'_1 for $H = H'_1$ is close to ϵ_1.

A similar reasoning to that of the previous paragraph based on the same hypotheses shows that the second descending critical field is equal to

$$H_2 = -H_a - \frac{A}{e}(S_a - \Delta' S_\parallel) + h_\perp \qquad (20)$$

where h has been defined above, ϵ_i designates this time the angle of the magnetization $\vec{J}$ with the axis $\varphi = 0$, corresponding to the discontinuity.

For the subsequent cycles, we find the same values:

$$H_2 = H_3 = H_4 = \ldots$$

and

$$H'_1 = H'_2 = H'_3 = \ldots$$

(12) Order of magnitude of the quantities used: We have attempted to determine approximately the values of the different quantities which occur in the formulae above: in particular the five quantities H_a, h, $\Delta S_\parallel$, $\Delta' S_\parallel$, $\Delta S_\perp$, for the films Co–CoO–I and Co–CoO–II studied by Paccard and Mme Schlenker. We set out initially the experimental values of H_1, H'_1 and H_2, at $-196°$C. We have assumed that H_a was equal to the coercive field of the same film at a temperature of 20°C, below θ_N: there is thus no further coupling. Finally we have estimated from the experimental curves [1] the values of ϵ_i for $H = H'_1$. These initial data are summarized in Table I.

With the aid of the data in Table I, using equations (7), (9), (18), and (20), we have calculated the values of Table II.

TABLE I.
Initial data.

	H_1	H'_1	H_2	H_a	$\cos \epsilon_i$
Co-CoO-I ...	−288	77	−172	19	0,70
Co-CoO-II ...	−187	−13	−157	29	0,77

TABLE II.
Values of the coupling fields in e.m.u.

	$h_\perp$	$\frac{A}{e}\Delta S_\parallel$	$\frac{A}{e}\Delta' S_\parallel$	$\frac{A}{e}\Delta S_\perp$	$\frac{A}{e}S_a$
Co-CoO-I	96	357	99	52	269
Co-CoO-II	39	131	13	18	157

Table II prompts the following remarks. In the first place the uniaxial anisotropy field H_a is weak with respect to the coupling fields. In the second place, for the film Co–CoO–I, $\Delta S_\parallel$ is larger than S_a; the sum $S_a - \Delta S_\parallel$ is negative, thus the effective field $A\vec{J}$ is larger than the coercive field of the system S.

In the third place, the values of $\Delta S_\perp$ are in the region of seven times smaller than $\Delta S_\parallel$, which is evidently very close to the value eight predicted by Rayleigh's laws. This remark is related to the fact that $A\vec{J}$ is of the order of magnitude of the coercive field, showing that in the system S, the region of applicability of Rayleigh's laws extends up to the neighbourhood of the coercive field: the system S thus possesses similar properties to those of a substance which is magnetically very hard.

(13) The narrowing of subsequent cycles: "tilting" and "creep": The above theory shows that the ascending critical fields (H_1, H_2, ...) are all equal and that the same is true for the descending critical fields starting from H_2. From the experimental point of view, this conclusion is not confirmed: further small variations occur in the course of subsequent cycles. Mme Schlenker has shown that these variations obey approximately the following empirical laws, except for large values of n

$$H_\infty - H_n = -An^{-\frac{1}{2}}, \quad n \geq 2;$$
$$H'_\infty - H'_n = -Bn^{-\frac{1}{2}}, \quad n \geq 1;\quad (21)$$

the two coefficients A and B are positive; A is generally larger than B: 2 to 2.6 times more, for the films Co–CoO for example.

On invoking again the analogy between the properties of the system S and those of ferromagnets, it is possible to explain this behaviour by the intervention of the phenomena of *tilting*, related to the existence of very strong negative interactions between ferromagnetic domains.

We have shown [5] that starting from a demagnetized initial state the successive cycles of magnetization described between O and H were as indicated schematically in Fig. 4. In the course of the

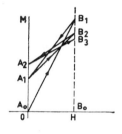

Fig. 4.

successive cycles, the magnetizations B_0B_1, B_0B_2, B_0B_3, ..., taken in the field H, decrease, while the remanent magnetizations A_0A_1, A_0A_2, ..., increase: in one word, the small asymmetric partial cycles described between O and H undergo a tilting movement which tends to render the line joining the two points of the cycle parallel to the axis OH.

By analogy with these results, starting from an initial anhysteretic state (AJ, S_a), $\Delta S_\parallel$ is the analogue of B_0B_n and $\Delta' S_\parallel$ the analogue of A_0A_n. As the order number n increases, $\Delta S_\parallel$ must decrease while $\Delta' S_\parallel$ must increase: therefore H_n must increase and H'_n decrease with n.

The same phenomenon must occur for βS which corresponds to a complete cycle; at each reversal of $\vec{J}$, ΔS is thus the analogue of A_0A_n and should increase, as should h, *as a function of n*. This effect should be added to that above.

Finally, the principle of analogy allows us to predict the existence of a *creep*. The laws of this phenomena have been described in a series of works by Ngyuyen Van Dang [6]. Fig. 5 shows schematically their behaviour. As n increases the partial cycles undergo a progressive downward translation corresponding to

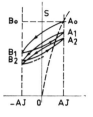

Fig. 5.

progressive increases and equal to $\Delta S_\parallel$, represented by B_0B_n, and $\Delta' S_\parallel$, represented by A_0A_n, with n: H_n and H'_n should increase with n. The combination of the effects of tilting and creep means that H_n should increase with n more rapidly than H'_n decreases.

All these conclusions are in agreement with the experimental results.

Tilting and creep constitute manifestations of the interactions between the ferromagnetic domains. Creep is correctly interpreted by a statistical theory taking into account the interactions between a large number of domains, and consequently the interactions between domains are relatively far apart. On the contraty tilting requires very strong interactions, but between a much more restricted number of domains: it thus relates to interactions between domains fairly close together.

In ferromagnetic bodies, the effects of creep are much more pronounced than those of tilting. This is perhaps because there are no very strong interactions between neighbouring domains and because magnetic dipole interactions are capable of acting at a long range.

In the antiferromagnetic films examined above the effects of tilting prevail, on the contrary, over those of creep. The reason perhaps lies in the almost complete absence of magnetic dipole interactions in antiferromagnetic bodies, since there is no spontaneous magnetization, and in the presence, in the case of epitaxial films, of an extremely efficient mechanism giving rise to powerful negative interactions. Assuming in effect that the ferromagnetic metal film presents a "step": the film of oxide which forms epitaxially on top probably also presents a step. The atomic planes of the oxide, in contact with the ferromagnet, on either side of the step, belong to different magnetic sub-lattices: the corresponding moments $\vec{S}$ are thus normally antiparallel, in the absence of ferro–antiferromagnetic coupling. Everything thus takes place as if we had a system of two domains very strongly negatively coupled: their decoupling requires in fact the rotation of the direction of antiferromagnetism by π, in a small region centred around the step.

(14) Interpretation of the hysteresis of rotation: Paccard and Mme Schlenker have established that thin films with ferro–antiferromagnetic coupling show hysteresis of rotation: the losses in a rotating magnetic field tend towards a nonzero limit when the magnetic field increases indefinitely, while in the same conditions they tend to zero for ferromagnetic bodies.

This behaviour is immediately explained in the proposed schemes. The rotating field $\vec{H}$ drives the magnetization $\vec{J}$ of the ferromagnetic film, with a certain angular delay. When one changes the magnitude of H the angular delay increases or decreases, but nothing more: $\vec{J}$ always rotates in the same manner, and retains the same magnitude. The system S is thus always subject to a rotating field $A\vec{J}$ of constant amplitude independent of H. The hysteretic losses of rotation should thus remain constant as H increases: we denote these by W_r, per cycle and per cm^3 of the ferromagnetic film.

To estimate W_r, we note that in the direction $\varphi = 0$, the amplitude of the variation of S is equal to $\Delta S_\parallel - \Delta' S_\parallel$, for a variation of effective field of amplitude $2AJ$. One may write

$$W_r = \frac{2kk'\,AJ}{e}(\Delta S_\parallel - \Delta' S_\parallel), \qquad (22)$$

where k and k' are two numerical coefficients which take account of the two principle effects:

(1) The shape of the hysteresis cycle. If we are in the Rayleigh domain, one must put $k = 1/3$. The comparison of the values of $\Delta S_\perp$ and $\Delta S_\parallel$ seems to justify this hypothesis.

(2) The ratio of the hysteresis of rotation to the unidirectional AC hysteresis. In the Rayleigh domain, $k' = \frac{3\pi}{4} = 2.36$. We therefore obtain $kk' = 0.75$. On taking $kk' = 1$, we are almost certain to be adopting an overestimated value.

The values of $\Delta S_\parallel$ and $\Delta' S_\parallel$ are deduced respectively from the values of H'_1 and H_2. One finds finally

$$W_r = 2kk'\,J(H'_1 - H_2 - 2H_a + 2h_\perp). \qquad (23)$$

If we neglect H_a and h with respect to $H'_1 - H_2$, we find for the films Co–CoO, with $J = 1400$

$$W_r\,(\text{calc.}) \sim 2\,800\,(H'_1 - H_2). \qquad (24)$$

One can compare this result with the empirical law which Mme Schlenker has deduced from her experiments

$$W_r\,(\text{obs.}) \sim 26\,000\,(H_\infty - H_1). \qquad (25)$$

Experiments show however that $H'_1 - H_2$ is two to four times larger than $H_\infty - H_1$. Thus the calculated values of W_r are at least three times smaller than the observed values.

The proposed theory thus allows one to show the existence of a hysteresis of rotation, but the calculated values are appreciably smaller than the real values. It is unlikely that the various approximations cause this and are responsible for an error as large as a factor of three. It seems rather that it is a question of a fundamental imperfection of the model, or rather of the absence of a model, since we have always argued by analogy. To clear up this discrepancy it thus seems indispensible to have recourse to a precise model and to study the ratios of the hysteresis of rotation to the AC hysteresis with it. It is this which we propose to do in the second part of this work.

(15) Magnetic after-effects: The dimensions of the antiferromagnetic crystallites are probably extremely small, possibly of the order of 50 Å. One should thus expect from them that the rotations of $\vec{s}$ and Δ, under the influence of the changes of effective field $A\vec{J}$, show magnetic after-effects similar to those which exist in a system of very fine ferromagnetic grains. Mme Schlenker has actually observed that the couples felt by the thin film, after an angular variation of the applied external magnetic field, vary as a function of time.

A careful study of these effects is very interesting because the knowledge of the time constants involved should allow one to obtain a supplementary piece of information: the mean value of the product of the volume of the antiferromagnetic grains and the magnetocrystalline anisotropy constant.

PART TWO
THE EQUILIBRIUM OF MAGNETIC MOMENTS IN A MIXED FILM: HYSTERESIS OF ROTATION AND OSCILLATORY HYSTERESIS

(16) Introduction: We propose to examine quantitatively, with a simple model, several aspects of the phenomena of magnetic coupling between two thin films, one ferromagnetic, the other antiferromagnetic, grown directly on top of each other. In particular, if there is an epitaxy, that is to say coherence between the crystal lattices of the two films, the coupling is very strong since it brings into play exchange or superexchange energies, of a similar nature to those which give rise to ferromagnetism or antiferromagnetism. This results, notably, in a continuity between the direction of the spontaneous magnetization of the ferromagnetic film and the direction of antiferromagnetism of the antiferromagnetic film.

Since it is possible, by means of an external magnetic field, to orient at will the direction of the spontaneous magnetization of the ferromagnetic film, one may thus, by an indirect process, also influence the direction of antiferromagnetism of the antiferromagnetic film, something which is otherwise extremely difficult to do. In such a process, the spontaneous magnetization does not remain in the same direction throughout the thickness of the ferromagnetic film, similarly for the direction of antiferromagnetism. It is advisable thus to study first the equilibrium state of a single ferromagnetic, or antiferromagnetic, film, taking account of the boundary conditions.

(17) Equilibrium of the magnetic moments of the atomic planes: We consider a plane ferromagnetic thin film possessing a uniaxial anisotropy direction in its plane and uniformly magnetized in this direction, which is also called the soft direction of magnetization.

We imagine that by exercising a suitable couple, of axis perpendicular to the plane of the film, on the magnetic moments of the atoms of the surface, one changes the direction of magnetization of one of the surface planes. The question is then to determine in what way this perturbation penetrates to the interior of the film, in the absence of any applied external magnetic field.

We assume that the thin film is made up of n atomic planes, parallel to the plane of the film, labelled by an index i running from 1 to n. We assume that the magnetization of each plane i is uniform and parallel to the plane of the film, and we denote by $\vec{m}_i$ the corresponding magnetic moment per unit of surface. We call $\theta_i/2$ the angle of $\vec{m}_i$ with the direction of soft magnetization.

We assume that the magnetocrystalline energy of each atomic plane is simply equal, per unit surface, to $-K \cos \theta_i$, where K is a positive constant. For brevity we designate henceforth the direction $\theta_i = 0$ by the soft direction, and the direction $\theta_i = \pi$ by the hard direction. The corresponding magnetocrystalline couple acting on $\vec{m}_i$ is equal to $2K \sin \theta_i$.

We assume that the magnetic exchange forces exist only between adjacent planes and that the corresponding energy, per surface unit, is equal to $-I \cos \frac{\theta_j - \theta_i}{2}$ with $j - i = \pm 1$. This causes $\vec{m}_i$ to be subject to a couple equal to $I \sin \frac{\theta_j - \theta_i}{2}$.

In order for $\vec{m}_i$ to be in equilibrium, the sum of the three couples applied to it should be zero:

$$-2K \sin \theta_i + I \sin \frac{\theta_{i+1} - \theta_i}{2} + I \sin \frac{\theta_{i-1} - \theta_i}{2} = 0. \quad (26)$$

If θ_i is a function of i varying fairly slowly with i, one may write:

$$\theta_{i+1} = \theta_i + \theta'_i + \frac{1}{2} \theta''_i,$$
$$\theta_{i-1} = \theta'_i - \theta'_i + \frac{1}{2} \theta''_i, \quad (27)$$

θ'_i and θ''_i denoting the first and second derivatives of θ with respect to i. The equilibrium condition is then written, omitting the index i

$$-4K \sin \theta + I \theta'' = 0. \quad (28)$$

This second order differential equation may be integrated once after having been multiplied by $2\theta'$. One thus finds

$$8K \cos \theta + I \theta'^2 \doteq 8KC \quad (29)$$

the constant of integration being $8KC$. On separating the variables this equation may be written

$$\frac{d\theta}{\sqrt{C - \cos \theta}} = \pm \sqrt{\frac{8K}{I}} di. \quad (30)$$

This equation (30) shows us that in order to satisfy the condition of continuity invoked above, I must be large with respect to K, which we have already assumed to be the case.

Since the film consists of n planes, the values θ_1 and θ_n of θ corresponding to the surface planes must satisfy the equation:

$$\int_{\theta_1}^{\theta_n} \frac{d\theta}{\sqrt{C - \cos \theta}} = n \sqrt{\frac{8K}{I}} = \rho, \quad (31)$$

where ρ is a dimensionless quantity characterizing the film.

(18) Boundary conditions: The two extreme atomic planes $i = 1$ and $i = n$ are subject to special conditions, because they only possess one nearest neighbour atomic plane, and the equilibrium condition (26) is not satisfied. It is thus necessary that external couples Γ_1 and Γ_n should be applied respectively to the moments $\vec{m}_i$ and m_n, to replace the exchange couples due to the missing planes of indices 0 and $n+1$. In the first approximation, these exchange couples are respectively equal to $-\frac{1}{2}\theta'_1$ and $\frac{1}{2}\theta'_n$, giving the boundary conditions

$$\Gamma_1 = -\frac{1}{2} I\theta'_1 \, ; \quad \Gamma_n = \frac{1}{2} I\theta'_n, \qquad (32)$$

where on taking the value of θ' determined from equation (30) and using the positive solution

$$\begin{aligned}\Gamma_1 &= -\sqrt{2KI}\,\sqrt{C-\cos\theta_1}\,;\\ \Gamma_n &= \sqrt{2KI}\,\sqrt{C-\cos\theta_n}.\end{aligned} \qquad (33)$$

These couples Γ_1 and Γ_n are the sum of the applied external couples and the couples due to the magnetocrystalline anisotropy of the surface.

This surface anisotropy is not well known; it is, however, of the order of magnitude of K rather than of the order of I. The resulting couples are thus small vis-à-vis the actual applied external couples which, as we shall see further on, are due to the influence of exchange and are therefore of the order of magnitude of I. We thus neglect, henceforth, the surface anisotropy.

(19) The constant C: The phenomena are very different depending on the value of C.

(1) $C>1$. In this case θ' never becomes zero: θ increases continuously in the same direction and its variation between θ_1 and θ_n may reach any number of times 2π. But for that it is necessary to apply to the two faces of the film nonzero external couples of opposite signs.

(2) $C<1$. One may put $C=\cos\theta_0$: θ may thus vary between θ_0 and $2\pi-\theta_0$. When θ reaches one of these extremal values, for a value of i different from 1 or n, the solution for θ' changes and one must change in equation (30) the sign $+$ to $-$, or conversely. Thus θ oscillates about $\theta=\pi$: it is a *periodic function* of i, returning to the same value when i increases by i_T. The periodicity i_T is defined by:

$$i_T\sqrt{\frac{8K}{I}} = 2\int_{\theta_0}^{2\pi-\theta_0} \frac{d\theta}{\sqrt{\cos\theta_0-\cos\theta}}. \qquad (34)$$

This periodicity tends to infinity as θ_0 tends to zero.

When the two couples at the extremities are zero, θ_1 and θ_n are equal to θ_0 or $2\pi-\theta_0$: n is thus a multiple of $i_T/2$.

It is easy to imagine an ideal experiment illustrating the two cases where C is larger or smaller than unity. Let us start with a ferromagnetic film uniformly magnetized along the soft axis $\theta=0$. The two moments at the extremities m_1 and m_n are parallel. If we apply to the two moments equal and opposite couples, the solution is non-periodic, of type $C>1$; if the couples are equal and in the same sense the solution is of the periodic type $C<1$.

We are particularly interested here in the case where one of the extremities is free ($\Gamma_1=0$) and a couple Γ_n is applied to the other extremity. In the interior of the film the angle θ should thus vary from θ_0 to a certain value θ_n, smaller than $2\pi-\theta_0$.

(3) $C=1$. In this interesting case, the elliptic integral (31) is soluble. The integration gives the equation:

$$\sqrt{2}\,\text{Log tg}\,\frac{\theta}{4} = \sqrt{\frac{8K}{I}} + \text{Cte} \qquad (35)$$

which defines the variation of θ as a function of i. In particular as θ varies from 0 to 2π, i varies from $-\infty$ to $+\infty$: one has something like a Bloch wall of 180°. In principle, its thickness is thus infinite. We may, however, define a half-thickness corresponding to half the variation of θ, from $\pi/2$ to $3\pi/2$; $\tan\theta/4$ thus changes from $\sqrt{2}-1$ to the inverse value and the left hand side of (31) changes by 2.494. The number of planes contained in this half-thickness corresponds to $\rho=2.494$. If we agree to take twice this value as the practical thickness of a 180° wall, we obtain

$$\rho = 5, \quad \text{i.e.} \quad n = 5\sqrt{\frac{I}{8K}}.$$

To assess these ideas we recall that in ordinary ferromagnets the thickness of a 180° wall is of the order of 1000 Å, n reaches several hundred.

(20) The two types of wall: When C is close to unity, whether above or below, and when the ferromagnetic film is thick, the film clearly subdivides into *walls* and *domains*. The variation of θ with i is shown schematically in the curves of Fig. 6.

In both cases the domains D and D′ appear relatively broad, with the magnetization of the interior fairly uniform, and are separated by thin regions P and P′ in the interior of which the magnetization changes rapidly and which are the walls. In the case of the periodic solution $C=1-\varepsilon$, one should note that the direction of the magnetization does not reach the positions $\theta=0$ and $\theta=2\pi$ but reaches an angle equal to approximately $\sqrt{2\varepsilon}$.

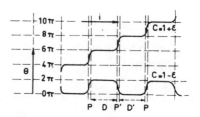

Fig. 6.

There exist two types of domains D and D': those with their magnetization in one direction, and those with their magnetization in the opposite direction. In the case of the periodic solution, there also exist two types of wall P and P', differing by the positive or negative sense of the variation of the direction of magnetization of the interior. In the case of the non-periodic solution $C + 1 = \varepsilon$, all the walls are of the same sign.

Let us note that the thickness of the domains tends towards infinity as ε tends to zero, while the thickness and the structure of the wall remains about the same as for the wall studied at the end of the previous paragraph which corresponds to $C = 1$.

(21) A detailed study of the case where C is smaller than unity: For this it is necessary to know the values of the integral

$$\rho = \int_{\theta_1}^{\theta_n} \frac{d\theta}{\sqrt{\cos\theta_1 - \cos\theta}} \qquad (36)$$

It reduces to an elliptic integral by the change of variable

$$\sin\varphi = \cos(\theta/2)/\cos(\theta_1/2).$$

On putting

$$\psi_n = \arcsin\left(\frac{\cos(\theta_n/2)}{\cos(\theta_1/2)}\right),$$

one obtains

$$\rho = \int_{\psi_n}^{\frac{\pi}{2}} \frac{d\varphi}{\sqrt{1 - \cos^2(\theta_1/2)\sin^2\varphi}}, \qquad (37)$$

which is expressed as a function of the elliptical integrals K and F, complete and incomplete, of second degree, of modulus $k = \cos(\theta_1/2)$:

$$\rho = \sqrt{2}\,\{K[\cos(\theta_1/2)] - F[\cos(\theta_1/2), \psi_n]\}. \qquad (38)$$

With the aid of the Institute of Applied Mathematics of Grenoble and thanks to the obliging help of M. Sabonnadière, I have assembled tables of the function ρ and the function $\gamma_n = \sqrt{\cos\theta_1 - \cos\theta_n}$, to 7 decimal places, for values of θ_1 and θ_n varying in steps of 2 degrees starting from 0.

Starting from these tables, one may determine the

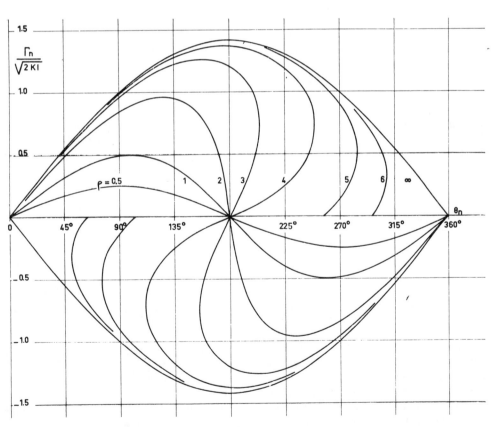

Fig. 7.

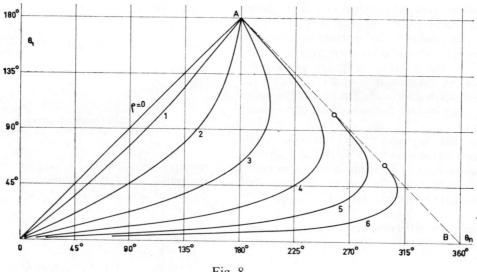

Fig. 8.

variation of the reduced couple $\gamma_n = \Gamma_n/\sqrt{2KI}$ as a function of θ_n, for different values of the parameter ρ, i.e. for different values of the reduced thickness of the film. These curves have been shown in Fig. 7 for $\rho = 0.5$, 1, 2, 3, 4, 5, 6 and $+\infty$.

We have similarly shown in Fig. 8 the variations of θ_1 as a function of θ_n for different values of ρ.

(22) A study of the curves γ These curves belong to three different types, U, V and W depending on the values of ρ: type U when ρ is less than 2.221, type V when ρ is between 2.221 and 4.442 and finally type W when ρ is greater than 4.442.

These are of course periodic curves, with a period equal to 2π. The three types are shown schematically in Fig. 9, in the interval 0 to 2π. While for types U and V,

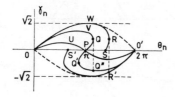

Fig. 9.

there is only a single continuous curve in the whole region of variation of θ, on the contrary for type W the curve γ_n is composed of distinct segments centred about the points $O, O', \ldots$ of the abscissae $0, 2\pi, 4\pi, \ldots$

In all cases, in the absence of any external influence, all the moments $\vec{m}_i$ are initially oriented in the soft direction $\theta = 0$. One thus rotates the moment $\vec{m}_n$ by an angle $\theta_n/2$ on applying a reduced couple γ_n.

(1) *Type U* — When one increases $\theta_n/2$ from 0 to $\pi/2$, γ_n begins by increasing, passes through a maximum γ_{nm} for a value $\theta_{nm}/2$ a little larger than $\pi/4$, and then decreases to zero. Correspondingly the gap $\Delta\theta = \theta_n - \theta_1$ in the spread of directions of the moments m_i, initially zero, spread of directions increases up to a maximum value and then decreases again to zero for $\theta_n/2 = \pi/2$. When one continues to increase $\theta_n/2$ from $\pi/2$ to π, the same phenomena are reproduced symmetrically but this time γ_n is negative while θ_1 becomes larger than θ_n.

The whole process takes place in a reversible manner: the orientation of each moment $\vec{m}_i$ changes in a continuous and reversible way.

When ρ is small, all the moments $\vec{m}_i$ stay almost parallel, the system rotates as a whole, and the couple becomes simply:

$$\gamma_n = \frac{\rho}{2} \sin \theta_n. \qquad (39)$$

Fig. 7 shows that by $\rho = 0.5$, this situation is already almost realized.

(2) *Type V* — The couple γ_n begins by increasing up to a maximum γ_{nm} for $\theta = \theta_{nm}$ and then decreases to a point Q of abscissa θ_{ne} and ordinate γ_{ne}, with a tangent parallel to the ordinate axis. Subsequently γ_n continues to decrease, while θ_n decreases to θ_{ne} at π. Correspondingly $\Delta\theta = \theta_n - \theta_1$ increases, passes through a maximum, and then decreases to zero at the point P.

In the interval $2\pi - \theta_{ne} < \theta_n < \theta_{ne}$ three values of γ_n exist for each value of θ_n.

(3) *Type W* — The corresponding curves greatly

resemble those of type V, with a point R, of abscissa θ_{ne} and ordinate γ_{ne}, with a tangent parallel to the ordinate axis, but differ in that for $\gamma_n = 0$ they do not meet the axis at the point P, of abscissate π, but at a point S of abscissate θ_{n0}. Similarly the negative branch of γ_n meets the axis at S', symmetric to S about P. Each value of θ_n gives 2 corresponding values of γ_n in the interval $(2\pi - \theta_{n0}, \theta_{n0})$ and 3 values in the intervals $(2\pi - \theta_{ne}, 2\pi - \theta_{n0})$ and $(\theta_{n0}, \theta_{ne})$.

Concerning the gap $\theta_n - \theta_1$ of the spread of directions, contrary to what happens at P for type V, this gap remains finite at S and S'. In the latter case the spread has a finite gap and takes a symmetric configuration with respect to the hard direction, whereas at P all the moments $\vec{m}_i$ had been parallel to the hard direction. At S and S', the applied external couples being zero, it is only the magnetocrystalline forces which maintain the gap in the spread of directions, in spite of the opposing exchange forces.

When ρ tends to infinity, θ_1 tends to zero almost independently of the value of θ_n, as is shown in Fig. 8, and γ_n tends to $\sqrt{1-\cos\theta_n}$, i.e. to $\sqrt{2}\sin(\theta_n/2)$. Fig. 7 shows that by $\rho = 6$, i.e. for a thin film of thickness hardly greater than that of a 180° Bloch wall, γ_n is given by $\sqrt{2}\sin(\theta_n/2)$ as θ_n varies from 0 to 270°.

Thus, when ρ varies from 0 to $= +\infty$, the family of curves γ_n gives the impression of a continuous change from the sinusoid $\sin\theta_n$ to the sinusoid $\sin(\theta_n/2)$, of double the periodicity: certainly an interesting curiosity.

It is equally interesting to note that the dependence of θ_n as $\sin(\theta_n/2)$ is precisely equal to that which occurs for two thin layers with *rigid* and *uniform* magnetization, linked by a coupling energy proportional to the cosine of the angle between the magnetization of the two films.

(23) The variation of θ_{ne} with ρ: As we will show below, the abscissae θ_{ne} of the vertical tangent to the curves γ_n, plays an important role in the comparison of the AC hysteresis with the hysteresis of rotation. It is thus necessary to know the dependence of θ_{ne} as a function of ρ. Table III gives several values of θ_{ne}, and

TABLE III.

ρ	θ_{1e}	θ_{ne}
2,221	180°	180°
2,62	127°	190°
3	110°	205°
4	80°	249°
5	60°	285°
6	41°	309°
7,25	27°	328°
7,81	22°	334°
8,54	17°	340°
9,52	12°	346°
∞	0°	360°

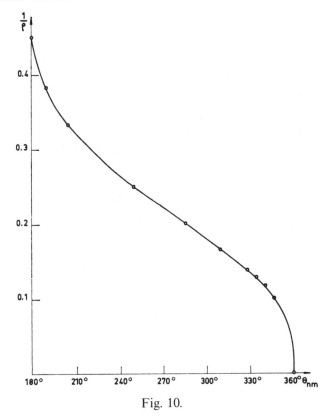

Fig. 10.

the corresponding values of θ_{1e}, for a series of values of ρ. We have in addition shown in Fig. 10 the dependence of θ_{ne} as a function of $1/\rho$.

(24) The case of an antiferromagnetic film: The theory that we have just developed relating to a thin ferromagnetic film is applicable also to a thin antiferromagnetic film, consisting of atomic planes parallel to the plane of the film, the planes of even index i being for example magnetized in one sense and the planes of odd index i magnetized in the opposite sense. In fact, in the two systems, the expressions for the energy are identical, since there is no external magnetic field applied, since the magnetocrystalline energy is unchanged when the magnetization changes sense, and finally because the expression for the exchange energy is also unchanged since if the exchange constant changes sign, one of the interacting magnetizations also changes sign.

Instead of observing a progressive rotation of the vector $\vec{m}_i$ we observe a progressive rotation of the vector $(-1)^i \vec{m}_i$: this is the only difference. In the following we assume that the index n of the final plane is even, which is not a restriction since n is implicitly assumed large with respect to unity.

We note, moreover, that an antiferromagnetic film is practically unaffected by the influence of an external magnetic field, since the sum of the two couples acting on two consecutive atomic planes is almost zero. It is thus only necessary to take account of the effects undergone by the two semi-moments at the extremities $\frac{1}{2}\vec{m}_i$ and $\frac{1}{2}\vec{m}_n$ which are uncompensated. We neglect here these phenomena, to which we have already devoted a study [7].

It is however on making the contact between the two thin films, the one ferromagnetic, the other antiferromagnetic, that it is possible to put the theory developed above to the test. In fact if the two films in contact are ferromagnetic, it is not possible by applying a magnetic field to rotate the magnetizations of the films with respect to each other, since the magnetic field acts simultaneously on both. On the contrary, if one of the two films is antiferromagnetic, it is almost unaffected by the application of a magnetic field, so that it is possible to rotate, with respect to it, the magnetization of the ferromagnetic film.

The only problem is to produce a magnetic coupling between the two films. The best way to realize this is to produce an epitaxy between the two films: there is thus a coupling by exchange.

If this epitaxy is perfect there is almost continuity between the direction of antiferromagnetism and the direction of the spontaneous magnetization of the ferromagnetic film. The angular divergence between $\vec{m}_n$ and $\vec{m}'_n$ (the primed letters corresponding to the ferromagnetic film, the non-primed letters to the antiferromagnetic film) is in fact of the same order of magnitude as the angular divergence between $\vec{m}_j$ and $-\vec{m}_{j+1}$, or between $\vec{m}'_j$ and $\vec{m}'_{j+1}$, since the coupling energies between nearest neighbour planes are in all cases of the same order of magnitude: so these latter angular divergences are small by hypothesis.

If this epitaxy is imperfect, the coupling may be much weaker and the angle between $\vec{m}_n$ and $\vec{m}'_n$ is no longer negligible.

(25) The coupling of an antiferromagnetic film with a ferromagnetic film of fixed magnetization: Let us take a thin antiferromagnetic film to which the theory developed in paragraphs 17 to 23 is applicable. We assume now that the couple Γ_n acts on the magnetic moment $\vec{m}_n$ of the last atomic plane, so that it should make an angle $\theta_n/2$ with the soft direction, being produced by the action of a thin ferromagnetic film. We assume at first that the magnetization $\vec{J}$ of this film always stays *uniform, rigid* and parallel to the plane of the films: it is thus completely defined by the angle φ of $\vec{J}$ with the soft direction of the antiferromagnetic film.

The energy W of coupling between the two films is of the form

$$W = -A \cos \frac{\psi}{2} \qquad (40)$$

where A is a positive constant and $\psi/2$ is the angle of $\vec{m}_n$ with $\vec{J}$. With these definitions

$$\varphi = \frac{1}{2}(\theta_n - \psi) \qquad (41)$$

while the couple Γ_e exercised on $\vec{m}_n$ by the ferromagnetic film is written:

$$\Gamma_e = -A \sin \frac{\psi}{2} \qquad (42)$$

Equilibrium is given by the condition $\Gamma_e = \Gamma_n$. To determine this, one first traces the curve Γ_n as a function of θ_n, calculated in paragraph 21, with 0 as the origin, then one traces the curve Γ_e as a function of ψ, with the point M of abscissate $OM = 2\varphi$ as origin. The point N of intersection of the two curves corresponds to equilibrium.

(26) Stability of the equilibrium: If, starting from the position N of equilibrium, one gives θ_n a small positive increase, the reverse couple undergoes an increase $\delta\Gamma_n$ and the driving couple an increase $\delta\Gamma_e$. The equilibrium is stable if $\vec{m}_n$ tends to return to its initial position i.e. in $\delta\Gamma_n - \delta\Gamma_e$ is positive: this requires that the curve Γ_n passes above the curve Γ_e when θ_n increases.

(27) The variation of θ_n as a function of φ: We study now the evolution of the phenomena when the magnetization $\vec{J}$ of the driving ferromagnetic film rotates with respect to the soft direction of the first film, i.e. when φ increases from 0. It is necessary to distinguish the two cases where the coupling constant A is larger or smaller than the maximum value Γ_{nm} of Γ_n.

(1) $A < \Gamma_{nm}$ — Let P and P' be the points of the curve Γ_n of ordinates $+A$ and $-A$ (Fig. 11). When starting from 0 φ increases indefinitely, the point N leaves O, reaches first the point P, goes back to P', returns to P, and then oscillates indefinitely between P' and P, with a period equal to 4π. The process is completely reversible.

If the curve Γ_n is of type V, it appears that the arc of the curve Q'PQ (Q and Q' are the points of Γ_n where the tangent is vertical) may also be traversed in a

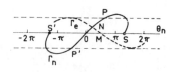

Fig. 11.

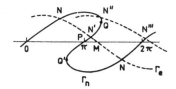

Fig. 12.

reversible manner, if A is smaller than the ordinate Γ_{ne} of Q (Fig. 12). The point N' shows a corresponding effect to the points N, at stable equilibrium: but this is rather a case of metastable equilibrium, since the potential energy of the state N' is larger than that of state N.

If A is contained between Γ_{ne} and Γ_{nm}, the equilibrium becomes unstable when N' passes above Q, at N'', while the two other points of intersection N''' correspond to stable equilibria: an irreversible rotation of magnetization is thus triggered at the moment when N' passes above Q.

When Γ_n is of type W, an irreversible rotation is similarly triggered when φ increases and the point N, originally on the arc R'S', reaches the point S', the representative point goes to N' (Fig. 13).

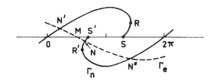

Fig. 13.

(2) $A > \Gamma_{nm}$ — When φ increases from zero, the point N of intersection moves along the arc of the curve OR, beginning from the point O, until it reaches the point N_t corresponding to the moment where at M, on reaching the point M_t of abscissa $2\varphi_t$, the curve Γ_e becomes tangential to Γ_n (Fig. 14). When φ increases a little beyond φ_t, the point of equilibrium moves in a discontinuous and irreversible manner from N_t to N'', on a negative branch of the curve Γ_n.

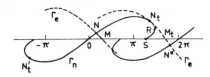

Fig. 14.

On the contrary, if φ begins to decrease before having reached the value φ_t, the point N goes back along the arc $N_t'ON_t$ where N_t' is the symmetric equivalent of N_t with respect to O. The range of the spread of reversible change of φ is thus equal to $2\varphi_t$.

One should, however, note that these irreversible rotations do not necessarily occur only with the types V and W of curve Γ_n. In fact, with the curve of type U, the curve Γ_e in order to become tangential to the curve Γ_n, must have a slope at M less in absolute value than the slope $-\Gamma_p$ of the curve Γ_n at P. In particular, when ρ is small ($\rho < 0.5$), Γ_n being almost identical to the sinusoid

$$\Gamma_n = \Gamma_{nm} \sin \theta_n,$$

different processes take place depending on the values of A:

(a) $O < A < \Gamma_{nm}$ — The point N oscillates reversibly about O. The vectors $\vec{m}_i$ oscillate reversibly about the soft direction.

(b) $\Gamma_{nm} < A < \Gamma_p$ — The point N accompanies M in its motion, with a discontinuity each time θ_n increases by 2π: the vectors $\vec{m}_i$ accompany $\vec{J}$ in its rotation, with an angular discontinuity each half-turn.

(c) $\Gamma_p < A$ — The point N accompanies M in its motion, always in a reversible manner: the vectors $\vec{m}_i$ rotate reversibly at the same time as $\vec{J}$, sometimes a little in advance, sometimes a little behind.

When the coupling constant A becomes very large with respect to Γ_{nm}, the portion of the sinusoid Γ_e capable of intercepting the curve Γ_n reduces to a vertical line passing through M and parallel to the ordinate axis (Fig. 15). The abscissate $2\varphi_e$ of the point

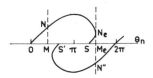

Fig. 15.

M_e corresponding to the discontinuity, for types U and V, is thus just equal to θ_{ne}. The limits between which 2φ can change reversibly, extend from $-\theta_{ne}$ to $+\theta_{ne}$ and tends towards 4π when ρ tend towards infinity.

(28) **On the physical significance of the magnitude of A:** If the epitaxial contact between the ferromagnetic film and the antiferromagnetic film is made with a unique common inter-reticular plane, the exchange constant A must be of the same order of magnitude as I: A is thus large with respect to Γ_{nm} which is just of the order of magnitude of $\sqrt{KI}$, knowing already that $\sqrt{I/K}$ is large.

In reality the interface must present "steps". From

one step to the next, the constant A changes sign since from one atomic plane to the next the magnetization in an antiferromagnet changes sign. There results a more or less complete compensation of the ferro–antiferro coupling, in the interior of a domain with coherence of magnetization $\vec{J}$. In addition, the interface may be polluted by impurity atoms, being deposited in the interval of time separating the fabrication of the two films, and which also contribute to decreasing the exchange interactions between the two films. For all these reasons, A may take any value, from zero up to a large value, of the order of I.

(29) Coupling with a ferromagnetic film of deformable magnetization: A ferromagnetic film of fixed and uniform magnetization, the existence of which we have postulated in order to align the moment of the surface film of the antiferromagnetic body, does not exist. We have in practice a ferromagnetic film in which the orientation of the magnetic moments of the successive planes varies with the distance from the interface. Nevertheless, the theory developed in paragraphs 17 and 18 is not applicable here, since the anisotropy of the film in which we are not interested is zero, or at least very small. In addition, one now applies a magnetic field to it in order to modify the orientation of its magnetization. It is thus only the exchange forces and the external magnetic field which tend to ensure the uniformity of its magnetization. It is not necessary to take account of the possibility of internal dispersion fields, because with the atomic planes, like their magnetization, being parallel to the plane of the film, div $\vec{J}$ is zero.

If m is the magnitude of the magnetic moment of an atomic plane and φ_i is the angle of the magnetic moment of the plane of index i with the direction of the magnetic field H, the corresponding energy is equal to $-m\mathrm{H}\cos\varphi_i$ and the restoring couple is equal to $-m\mathrm{H}\sin\varphi_i$. On the other hand the exchange energy between two consecutive planes is $-\mathrm{I}\cos(\varphi_{i+1}-\varphi_i)$. By a reasoning identical to that used above, one deduces that the φ_i should satisfy the second order differential equation

$$-m\mathrm{H}\sin\varphi + \mathrm{I}\varphi'' = 0 \qquad (43)$$

identical to equation (28) when one replaces θ by φ and 4K by mH.

In particular, the direction of the magnetic field H being fixed, in order to rotate the moment $\vec{m}_n$ of the last plane of index n to an angle φ_n it is necessary to apply a couple Γ_n which, following (33), is written:

$$\Gamma_n = \sqrt{\frac{m\mathrm{HI}}{2}}\sqrt{C-\cos\varphi_n}. \qquad (44)$$

The curves giving Γ_n as a function of φ_n are thus the same as the curves giving Γ_n as a function of θ_n (Fig. 7), but if one chooses the same unit angle *they correspond to double the period*. We are now in a position to describe what happens when this ferromagnetic film is coupled with the antiferromagnetic film studied at the start of this work, and when one rotates the field $\vec{H}$. In order to shorten the discussion, we will limit ourselves to the case where the two coupling curves are of type V or type W. Concerning type U however, where ρ is small, the results are not very different from those for fixed magnetization ($\rho = 0$). To avoid confusion, we shall henceforth prime all the quantities relating to the ferromagnetic film.

We shall first assume that the coupling constant A between the two films is of the order of magnitude of I or I′ so that there is a continuity between the orientations of $\vec{m}_i$ and $\vec{m}'_i$. In particular, the vector $\vec{m}_n$ must be almost parallel to the vector $\vec{m}'_n$, and, since there is an equilibrium, the two couples Γ_n and Γ'_n must be equal and opposite.

Starting from an origin O, corresponding to the soft direction of the antiferromagnetic film, and as a function of $\varphi = \theta_n/2$, we plot the curve Γ_n, of period π, and then starting from the point M of abscissate φ_0, corresponding to the direction of the magnetic field, we plot the curve $-\Gamma'_n$, of period 2π. The intersection N gives the state of equilibrium (Fig. 16).

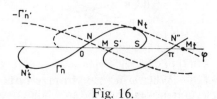

Fig. 16.

When φ_0 increases from zero, the point M moves from O towards the right, and the point N moves from O along the curve Γ_n until it reaches the point N_t where the curve $-\Gamma'_n$ is tangential to the curve Γ_n; M has thus arrived at M_t of abscissa φ_t. If φ_0 passes φ_t, an irreversible rotation occurs and N moves in a discontinuous manner from N_t to N''.

If M reverses before having reached M_t, N can then move in a reversible manner along the arc of the curve Γ_n between N_t and its symmetric equivalent about O, N'_t. Thus, φ_0 may move, without setting off any irreversible rotations, through a range of values $2\varphi_t$ between $-\varphi_t$ and $+\varphi_t$.

The value of φ_t depends essentially on the ratio $r = \Gamma'_{n'm'}/\Gamma_{nm}$ of the maximum value of Γ'_n to the maximum value of Γ_n. We recall that this ratio is of the order of magnitude of $\sqrt{\dfrac{4\mathrm{KI}}{m\mathrm{HI}'}}$. If this ratio is large, the segment of the curve Γ'_n, in which we are interested, reduces to a straight line parallel to the ordinate axis: φ_t reduces to φ_{ne}, the abscissate of the point R with a tangent to the curve Γ_n which is vertical. If this ratio is equal to one, the two curves become tangential at the poisition of their maximum. If ρ and ρ' are fairly large, being above the value 5 for example, these maxima are situated at $\varphi_m = \pi/2$ for Γ_n and at $(\varphi - \varphi_0)_m = \pi$ for $\Gamma'_{n'}$, so that φ is then equal to $3\pi/2$.

If this ratio is smaller than one, the two curves become tangential in the position shown in Fig. 17. The abscissa φ_t of M_t becomes close to 2π especially when ρ' is large: the range of reversibility tends therefore to 4π when ρ' tends towards infinity.

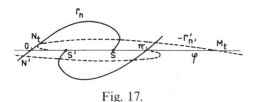

Fig. 17.

One should note an interesting difference in the appearance of these phenomena, depending on whether r is larger or smaller than one, when φ increases indefinitely in one direction.

When r is larger than one, at the moment of the discontinuity the point N jumps from one arc of the curve Γ_n to the next: in other words, by a series of successive jumps, the direction of antiferromagnetism accompanies the rotation of the magnetic field H: the vectors $\vec{m}_n$ and $\vec{m}'_n$ are dragged along by the field H as it rotates. On the contrary, if r is smaller than one, the point N jumps from N_t to N' on the same arc of the curve Γ_n. The direction of antiferromagnetism just oscillates about the soft direction: the vectors $\vec{m}_n$ and $\vec{m}'_{n'}$ do not follow the field H and remain in the neighbourhood of the soft direction.

To return to this limited discussion of the case where the two curves Γ are of types V and W, i.e. to the case where ρ and ρ' are larger than 2.22, we find that the limits of the range within which φ can change reversibly extend from $2\varphi_e = \theta_{ne}$ up to 4π.

(30) The case where the coupling energy A becomes small: When the coupling constant between the planes n and n' of the two films becomes weak, i.e. an order of magnitude less than I or I', it is no longer possible to neglect the angle between the magnetic directions of these two planes. For example if Γ_t is the ordinate of the point N_t corresponding to the discontinuity, in order to cancel this couple it is necessary that the magnetic directions of the planes n and n' should subtend an angle Ψ_t defined by

$$\Gamma_t = A \sin \Psi_t. \quad (45)$$

In the first approximation, the discontinuity does not occur for $\varphi = \varphi_t$ but for a larger angle φ_a equal to the sum of φ_t and Ψ. The limits of the reversible range of φ may thus exceed 4π.

In addition, particular discontinuities related to the small size of A may also appear. We will not discuss this problem further.

(31) The energy of the states of equilibrium: The internal energy W_i of any atomic plane is the sum of the magnetocrystalline energy $-K \cos \theta_i$ and the exchange energy

$$-\frac{I}{2}\left[\cos\frac{\theta_i - \theta_{i-1}}{2} + \cos\frac{\theta_i - \theta_{i+1}}{2}\right],$$

where the coefficient $\frac{1}{2}$ has been introduced to take account of the fact that this energy is shared with the two nearest neighbour atomic planes. Since θ_i varies slowly with i one may write

$$W_i = -K \cos \theta_i + \frac{I}{8}\theta_i'^2. \quad (46)$$

Taking account of equation (29), one may also write

$$W_i = \frac{I}{4}\theta_i'^2 - KC. \quad (47)$$

The internal energy $W(\theta_1, \theta_n)$ of the film, per square centimetre, becomes finally, using equation (5)

$$W = \frac{I}{4}\int_1^n \theta'^2 \, di - nKC$$

$$= \sqrt{\frac{KI}{2}}\int_{\theta_1}^{\theta_n}\sqrt{C - \cos\theta}\, d\theta - nKC. \quad (48)$$

By means of the change of variable indicated in paragraph 21, one reduces the integral of the second side of (48) to the difference of two first order elliptic integrals, one complete and the other incomplete.

The total energy W_T of the system of two films in equilibrium is, on assuming Ψ_t is negligible

$$W_T = W_1(\theta_1, \theta_n) + W_2(\varphi_1, \varphi_{n'}) \quad (49)$$

with

$$\theta_n = 2\varphi_{n'}.$$

When an irreversible process is triggered, the system goes from a state $\theta_1, \theta_n, \varphi_1, \varphi_{n'}$ of energy W_T to another state $\theta'_1, \theta'_n, \varphi'_1, \varphi'_{n'}$ of lower energy W'_T. The difference $\Delta W = W_T - W'_T$ is dissipated in the form of heat in the two films: the energy corresponding to the terms W_1 is dissipated in the antiferromagnetic film, the energy corresponding to the terms W_2 in the ferromagnetic film.

(32) The different processes of hysteretic loss: The study of the hysteretic losses of coupled films provides us with some particularly interesting results. Certain particular types of loss seem in fact to be specific to coupled films, and should not occur in other ferromagnetic systems.

It is possible to distinguish three loss processes for which we will now define the nomenclature: *AC hysteretic* losses, referring to a magnetic field of fixed direction varying between the two limits $+H$ and $-H$; losses by *hysteresis of rotation*, referring to a magnetic field H of constant magnitude, rotating continuously in the same sense; finally, losses by *oscillatory hysteresis* referring to a magnetic field H of constant magnitude showing angular oscillations between two fixed directions.

While for the first two types the losses depend principally on the sole parameter H, for oscillatory hysteresis, they depend principally on the two parameters of a complete oscillation, namely the field H and the angular amplitude ψ. The study of the losses as a function of these two parameters should thus be rich in information.

Concerning the coupled films, the losses by oscillatory hysteresis or hysteresis of rotation appear *a priori* to be easier to interpret than the losses by AC hysteresis: in fact in this latter case the magnitude of H varies, which brings in the variation of the parameter ρ' characterizing the rigidity of the film and thus complicates the phenomena. We note, however, that there should exist a certain similarity between AC hysteresis and oscillatory hysteresis of amplitude $\psi = \pi$.

From the experimental point of view, while the AC hysteresis is very well known, our knowledge of the hysteresis of rotation is rather less well developed, and as to oscillatory hysteresis it has been hardly ever studied.

(33) Hysteretic losses in a system of thin coupled films: The theories which we have developed at the start of this work, together with the remarks of paragraph 31 on the calculation of the energy, allow us to calculate the losses in an element composed of two coupled films, characterized by a given orientation to the soft direction, i.e. of the direction of the film along which the direction of antiferromagnetism of the antiferromagnetic film spontaneously aligns itself. In practice, it is only very exceptionally that we have structures that consist of isolated elements of the above type: more generally, one has systems made up of a large number N of such elements.

We propose to compare the losses by hysteresis of rotation with those by oscillatory hysteresis in such a system. The first, per turn, are equal to $2Nw$, where w denotes the energy dissipated in the form of heat in each element, at each discontinuity of orientation of the magnetization.

For oscillatory hysteresis, we occupy ourselves first with the case where the amplitude of the complete oscillation is equal to π. Let φ be the direction of the field $\vec{H}$, with respect to any reference direction. Let $A_1, A_2, \ldots, A_j$, be the points on the axis of φ, for a given element, corresponding to the discontinuities for rotation in the positive sense, and $B_1, B_2, \ldots, B_j$, be the points corresponding to the discontinuities for rotation in the negative sense. Each segment $B_j A_j$ corresponds to a range of reversible change of φ, with a spread equal to $2\varphi_e$.

Concerning the irreversible effects, everything takes place as if the field H oscillates between a point M and a point M' such that $MM' = \pi$, for a given element. For the ensemble of N elements of the system, everything occurs as if the N points M were uniformly distributed between B_j and B_{j+1}. We have therefore two cases to consider, depending on the value of φ_e.

(a) $2\varphi_e < 2\pi$ — The points A and B are located on the φ axis in the order $B_j A_{j-1} B_{j+1} A_j, \ldots$ Firstly the point M may be situated at M_1, between B_j and A_{j-1}, and the point M' at M'_1 between B_{j+1} and A_j. There is therefore no discontinuity in the course of the oscillation of $\vec{H}$ since $M_1 M'_1$ is completely contained in the reversible segment $B_j A_j$. On the contrary if the point M is situated at M_2 between A_{j-1} and B_{j+1}, and the point M' is situated at M'_2 between A_j and B_{j+2}, one observes a discontinuity at A_j in the course of the positive oscillation and a discontinuity at B_{j+1} in the course of the negative oscillation. The ratio R of the losses for a complete oscillation, to the rotation losses per turn is given by:

$$R = \frac{A_{j-1} B_{j+1}}{B_j B_{j+1}} = 2 - \frac{2\varphi_e}{\pi} \qquad (50)$$

This ratio decreases from 1 to 0 when φ_e increases from $\pi/2$ to π.

(b) $2\varphi_e > 2\pi$ — One may easily show that there are no losses due to oscillation: $R = 0$.

These results may be easily generalized to the case where the amplitude ψ of the oscillation is different from π. One finds:

$$R = 1 - \frac{2\varphi_e - \psi}{\pi} \qquad (51)$$

The ratio R decreases from 1 to 0 when $2\varphi_e$ increases from ψ to $\psi + \pi$.

In ferro–antiferromagnetically coupled films, the reversible range may reach 4π, or even greater. For such films, it is thus necessary to give the oscillations an amplitude of 4π in order to make the oscillatory losses equal to the losses by rotation.

This very large value contrasts with the larger value, $3\pi/2$, which one may obtain with monodomain anisotropic domains.

One sees, therefore, the interest in pursuing systematic measurements of the oscillatory losses both as a function of ψ and H.

It seems, however, that the apparent difficulty in the interpretation of the experiments of Mme Schlenker, explained in the first part of this work, to understand the abnormally strong hysteresis of rotation, compared with the oscillatory hysteresis, may be resolved in the outline of the ideas which we are going to discuss.

(34) AC hysteresis and oscillatory hysteresis of amplitude π: One knows that in very thin ferromagnetic films, the changes of magnetization are due to a coherent rotation of magnetization in a more or less extended domain. A change in magnetic field from +H to −H, in the direction $\varphi = 0$, rotates the magnetization from $\varphi = 0$ to $\varphi = \pi$, or more precisely, since the magnetization must remain a little behind the field, from $\varphi = 0$ to $\varphi = \pi - \varepsilon$. In reverse, when the field returns from −H to +H, the direction of the magnetization changes from $\varphi = \pi - \varepsilon$ to $\varphi = \varepsilon'$, where ε and ε' are small positive angles. Because of the existence of ε and ε', the rotations, in the course of subsequent cycles, always take place in the same half-plane, i.e. in the initial half-plane. There is thus a similarity between AC hysteresis and oscillatory hysteresis of amplitude π. This similarity is particularly strong when the magnetization of the ferromagnetic film is more rigid, i.e. when ρ' is small.

REFERENCES

[1] D. Paccard, C. Schlenker, O. Massenet, R. Montmory and A. Yelon, *Phys. Stat. Sol.*, 1966, **16**, 301.
[2] A. A. Glazer, A. P. Potapov, R. I. Tagirov and Ya. S. Shur, *Phys. Stat. Sol.*, 1966, **16**, 745.
[3] C. Schlenker and D. Paccard, *Bull. Sci. Franc., Phys.*, 1966, **6**, 8.
[4] Paccard, D. and Schlenker, C. *J. de Phys.*, 1967, **28**, 611.
[5] L. Néel, *Comptes Rendus*, 1958, **246**, 1963.
[6] Nguyen Van Dang, Thesis, Grenoble, 1961, *Acta Electronica*, 1968, **11**, 181.
[7] L. Néel, *Low Temperature Physics*, p. 413 (Cours de l'Ecole des Houches, 1961; Gordon and Breach, Ed., New York, 1962).

CHAPTER XVI

APPARATUS AND TECHNIQUES

A11, A31, A39

Articles A-11 and A-31 describe briefly two apparatus, a potentiometer and joulemeter, which have novel arrangements. Also attached is an article by B. Persoz (*Bull. Soc. Fr. Phys.*, **429**, 1939, p. 39) giving a description of a wattmeter constructed according to the suggestions of L. Néel. These apparatus disappeared at Strasbourg during the 1939–1945 war.

In article A-39, an original method for measuring the true specific heats of metals, in the form of wires, is described which uses two ballistic wattmeters. One will also find in article A-43 (chap. XVII) another novel method for measuring the true specific heats of bulk samples of metals.

This is also the place to note an original method for the construction of *model ships* intended for the simulation and study of their properties in the earth's magnetic field. This method, called '*netting*', was the subject of a note to the Navy in 1950.

The construction of models on a reduced scale, 1/100th for example, runs in fact into the difficulty of providing plates of a suitable thickness and initial permeability. There are also the problems of soldering.

These difficulties have been overcome by replacing each plate by a rectangular mesh of iron wires, in such a way that one has $\mu e = \mu' s/d$, where μ and e are the permeability and the thickness of the plate, μ' and s are the permeability and cross section of the wires and d their spacing. In particular d is very easily adjusted to the desired value. The wires representing girders may be easily introduced into the mesh. It is equally simple to solder the assembly electrically at points. Since it is only an approximation, one thus obtains a precision of a few per cent, which is usually sufficient.

A11 (1933)
A HIGH PRECISION POTENTIOMETER FOR MEASURING THERMOELECTRIC ELECTROMOTIVE FORCES

The measurement of changes of temperature with a 1/100th of a degree precision, up to 1100°, by means of a thermocouple, requires a high precision potentiometer for which the parasitic electromotive forces should be less than 10^{-7} volts. To satisfy this condition I have replaced the sliding contacts with mercury filled contacts by means of which one opposes the electromotive force to be measured with a variable portion of the potentiometer circuit. I have removed the equilibriating wire and the cursor: one straddles the electromotive force to be measured and interpolates with the aid of a galvanometer. The precision of this interpolation is limited, since on two successive contacts the external resistance of the galvanometer circuit is different, thus the corresponding sensitivities are different; it is also advantageous, for ease of use, to reduce as much as possible the final interpolation. For these two reasons the divisions of the scales should be small. With a normal potentiometer with a single circuit, one can hardly divide the total scale into more than two hundred parts. I have thus adopted a potentiometer *with two independent circuits* each supplied by an accumulator (this arrangement had already been proposed by 1907 by White).

A first circuit A (Fig. 1) contains at *ef* ten divisions each of 1 millivolt, related to 11 contacts numbered from 0 to 10, and, at *gh*, ten divisions of 0.01 millivolts. The second circuit B contains, at *ab*, ten divisions of 10 millivolts and, at *cd*, ten of 0.1 millivolt. These two circuits remain completely unaltered: the normal state once reached is not changed in use. A system of movable bridges, such as *p*, and a commutator C allow one to connect in series: a fraction of circuit A, a fraction of circuit B, the electromotive force to be measured E and a galvanometer G. One may therefore oppose E by a variable electromotive force between 0 and 111.11 millivolts in divisions of 0.01 millivolt. One determines the remaining fraction by interpolation by means of a sensitive galvanometer. I have used an excellent Hartmann and Braun instrument with a sensitivity of 10^{-10} ampère.

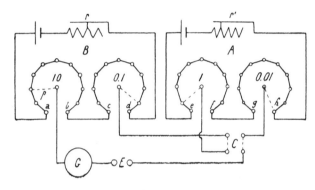

Fig. 1.

Circuits A and B, as well as the regulating resistances *r* and *r'*, contain coils, not shown in the figure, opposable to a calibrating battery. A three-directional commutator allows one to go instantaneously from the measuring configuration to the regulating configuration, using the calibrating battery and, naturally, the same galvanometer. False operation is impossible. One notes that the commutator C, allows one *to oppose* two fractions of circuits A and B. One may thus calibrate the potentiometer without any auxiliary apparatus: by a suitable arrangement of bridges *p*, one opposes, for example, ten millivolts taken between any two consecutive contacts in the portion *ab* of circuit B, to the fraction *ef* of A which also gives ten millivolts. The complete calibration of the apparatus, with the maximum possible precision, can be done in relative units.

All the circuits are in manganin. The terminals are fixed in a sheet of bakelite. The upper part of the potentiometer is protected against air currents and dust by a thick sheet of glass. All the handling is done from the outside. The current regulation boxes, *r* and *r'*, are placed outside the potentiometer itself. They provide regulation to an accuracy of one millionth.

This potentiometer has given excellent results. The

parasitic electromotive forces are less than 1/10th of a microvolt. One sets therefore an electromotive force of 100 millivolts to an accuracy of one millionth, if the electromotive force of the calibrating battery remains defined to this precision. In practice, one reaches a hundred thousandth on condition that the temperature of the calibrating cell is held constant to a 1/10th of a degree, since one knows that the temperature coefficient of the international Weston cell is four one hundred thousandths per degree.

A31 (1937)
JOULEMETER
(with BERNARD PERSOZ)

The apparatus, presented to the French Physical Society and entirely constructed at the Institut de Physique at Strasbourg, is a meter intended to measure, with a precision of a few thousandths, energies of a few tens of joules, dissipated between two points in an electrical circuit, during an interval of time which may be very short or on the contrary, may reach several minutes.

One may describe this as an electrodynamic wattmeter which will be damped proportionately to the speed and thus one will replace the suspension wire by a wire of negligible torsion constant. The damping is produced by a heavy piece of solid copper (cross-section = 1 cm^2), suspended from the lower moving part of the equipment and placed in the magnetic field of a large permanent magnet. The voltage circuit consists of a very light bobbin made of ivory with a winding of about a hundred ohms. The inlet and outlet for the current were made with long ribbons of very thin copper. This coil is placed at the centre of symmetry of a pair of Helmholz coils which constitute the magnetic circuit. An auxiliary magnet compensates for terrestial magnetic field if necessary, and particularly the stray field from the damping magnet at the voltage coil. The additional electric damping, variable moreover, produced by the voltage coil is completely negligible.

One measures the rotation of the moving part by Poggendorff's method. The apparatus functions in the same way as a fluxmeter, i.e. in the absence of current or voltage the spot is practically in equilibrium at any point on the scale. It is only when the duration of the experiment exceeds thirty seconds that one has to correct for drift due to the torsion constant of the suspension wire not being completely negligible.

This apparatus was conceived in order to perform accurate measurements of specific heats, but it lends itself to numerous applications amongst which are the measurement of the area of the whole or a part of a hysteresis cycle. Experiments of this type are in progress. One has also measured the energy required to melt lead wires of different lengths. The extrapolation to zero length gives the energy dissipated in the fusing arc.

The simplicity with which this apparatus allows one to approach certain problems suggests that it will be highly useful in electrical engineering. It is also an excellent apparatus for experiments.

Appendix to A31
BALLISTIC WATTMETER
Bernard Persoz (Bull. Soc. Fr. Phys. 429, 1939, p. 395)

M. Néel and myself have been led to envisage the use of two wattmeters in a ballistic fashion, measuring instantaneously dissipated energies, for the determination of the true specific heat of a thin metallic wire by a novel method. The wire is heated by the passage of a very short pulse of electric current. The energy supplied to it is given directly by the deviation of one of the wattmeters. The rise in temperature may be determined through the increase in resistance of the wire, thanks to the other wattmeter.

This apparatus, entirely constructed in the workshop of the Institut de Physique, consists of two fixed, parallel coils, traversed by the same intensity as the sample wire, and of a small movable coil, of thin wire, connected to the terminals of the fraction of the circuit studied (voltage u). The axes of these coils should clearly be perpendicular so that the couple acting should be a maximum and proportional to the product ui.

The moving part, suspended by a torsion wire (tungsten) and carrying a small mirror to display its displacements, has in addition, attached to its lower section, an electromagnetic damping system, consisting of a simple vertical cylinder of copper, positioned in the rectangular gap of a fixed permanent magnet, and a system for altering the moment of inertia of the unit. Two symmetrical runners can slide along two horizontal rods. It is in fact useful that the apparatus should be damped, if one does not want to take too long when one makes several successive measurements of energy (ballistic), or when one measures a power (ordinary wattmeter). The current connections consist of strips of berrylium copper.

The stray fields have not been compensated. It has been found easier to make two additional correction measurements: one with only the voltage and the other with only the current; this second measurement is to take account of the inevitable ferromagnetism of the moving part.

Each apparatus is calibrated as a normal wattmeter with the aid of a current and a voltage measured by a potentiometer. It is then simple to get the sensitivity for instantaneous energy, knowing the period and the damping.

Several experiments have been done: a measurement of the energy required for melting a wire, a measurement of the areas of different hysteresis cycles.

A39 (1939)

A NEW METHOD FOR MEASURING THE TRUE SPECIFIC HEATS OF METALS AT HIGH TEMPERATURES
(with M. Bernard Persoz)

The electrical method for the measurement of true specific heats may be applied in a particularly simple way [1] to metallic wires. However the corrections due to heat losses reach 20% by 400°C. It is possible to reduce these corrections to about 1% at 1000°C, by reducing the heating time to 0.1 s and by using the wire itself as a resistance thermometer.

The wire AB to be studied (diameter 0.55 mm, length 10 cm) is placed in a furnace at a temperature θ_0 at which one wishes to measure the specific heat. The energy sent into the wire by means of a short pulse of high current, turned on and off by a suitable switch, is measured with the specially constructed wattmeter W_1, used ballistically. The rise in temperature (10° to 80°) is determined in the following manner. The wire constitutes the branch AB of a Thomson double

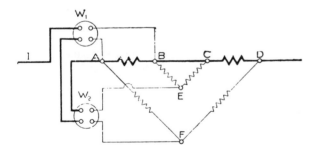

bridge, in which the branch CD consists of a *constant* reference resistor. The bridge is in equilibrium before the passage of the heating current, so that the potential difference EF is proportional to the product of the intensity I of the main current and the increase in resistance of the sample. The voltage circuit of a second ballistic wattmeter is put on this diagonal, while the current circuit is crossed by the main current I. Let W_1 and W_2 be the energies measured by the first and second wattmeters, and p a function of the resistance R of the sample, close to $1/R \, dR/d\theta$ when the rise in temperature $\theta_1 - \theta_0$ is small; a simple calculation shows that the mean calorific capacity M of the sample between θ_1 and θ_0 is

$$M = \frac{1}{2} \frac{K}{J} p \frac{W_1^2}{W_2},$$

where J is the mechanical equivalent of the calorie.

The constant K, which depends on the bridge, is determined on calibrating the second wattmeter, by the introduction of a known potential difference in one of the arms of the bridge. Before each measurement of specific heat, one measures the resistance R of the wire, and one thus obtains the thermal dependence of R necessary for calculating p.

The wire is placed in a solid iron container intended to define its initial temperature, measured with a Pt, Pt–Rh thermocouple, the junction of which is clamped to the container. The voltage contacts consist of knife-edge clips. It is not necessary to evacuate the container, it is sufficient to eliminate the oxygen with a little powdered manganese placed inside the furnace, which should be airtight.

This method gives a precision of 1 to 2% up to 1000°C. It has been applied to copper, nickel and platinum. If the wire is ferromagnetic and below its Curie point, the magnetic field produced by the current I aligns the magnetization of the elementary domains transversely and appreciably modifies the resistance of the wire. The method is no longer applicable, unless one fixes the magnetization by an auxiliary magnetic field produced by an external coil.

[1] Lapp, *Ann. de Physique*, 1929, **12**, 442.

Chapter XVII
EXPERIMENTAL WORK
A7, A9, A20, A36

A7 (1932)

A STUDY OF THE PARAMAGNETIC PROPERTIES OF IRON–TIN AND IRON–SILICON ALLOYS

Note (1977): This is a second extract from the thesis A7. The essential parts of the thesis have been given in chapter I. The study of the solid solutions Fe–Si and Fe–Sn allows one to determine the Curie constant of α iron with a fairly high precision, by extrapolation to pure iron. One thus obtains $C_M = 1.26$.

The $\beta \rightarrow \gamma$ transition, in the region of 900°, limits the study of β iron at high temperatures.

One knows that at this temperature the iron transforms from a body centred to a face centred cubic lattice. At 1350° the $\gamma \rightarrow \delta$ transition occurs, which is exactly the inverse of the previous transition, so that the β iron and the δ iron possess the same crystal structure [1].

But the addition to iron of certain metals, such as tin [2] and silicon [3], raises the $\beta \rightarrow \gamma$ transition point and lowers the $\gamma \rightarrow \delta$ transition point, so that 2% of tin, or of silicon, is sufficient to cause the complete disappearance of the γ phase: the iron keeps the same crystal structure, body centred cubic, until it melts. Fig. 17 shows schematically the corresponding phase diagram, for iron–tin or iron–silicon alloys.

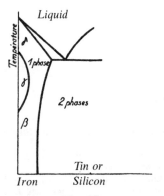

Fig. 17.

M. Weiss has proposed the use of this fact to extend the study of the paramagnetic properties of iron to above 900°C.

I have studied four alloys of iron with tin and three alloys of iron with silicon. All are solid solutions in which the γ phase has disappeared.

In all cases the dependence of the coefficient of magnetization as a function of temperature, has the same appearance: from 1100°, the limit of the measurements, to 950°, $1/\chi$ varies linearly with temperature: the extrapolation of this line provides a paramagnetic Curie point θ_p. Below 950°, the experimental points fall on a curve, situated above the line, for which the curvature increases up to the moment where one reaches the ferromagnetic Curie point θ_f, about forty degrees below θ_p.

I have assembled in Table I some of the series of measurements, keeping only the high temperature points. I have also given the products $\chi(T - \theta_p)$, which allow one to appreciate the precision with which the points follow a straight line.

In Table II, I give, for the seven alloys studied, the values of the Curie constant C, of θ_p and θ_f. I have also given the values of the atomic composition and of the atomic Curie constant.

I have shown these values in Figs. 1 and 2. The extrapolation to pure iron, i.e. the prime object of the work, gives the following values:

$$C_M = 1.26 \qquad \theta_f = 772° \qquad \theta_p = 815°$$

One sees also that the two sets of alloys provide consistent results.

TABLE I.

T°	$\chi \times 10^6$	$\chi(T-\theta_p) \times 10^4$	T°	$\chi \times 10^6$	$\chi(T-\theta_p) \times 10^4$
Fe Sn 10			Fe Si 5		
1033,0	100,32	2232	1122,5	65,64	2130
1022,8	104,40	2216	1109,3	68,23	2124
1011,7	110,50	2223	1089,1	73,0	2125
1007,1	113,68	2235	1068,4	78,69	2128
985,1	127,67	2229	1052,4	83,59	2127
962,1	147,14	2231	1030,7	91,67	2133
933,7	180,97	2229	1010,2	100,63	2130
			986,4	113,16	2132
Fe Sn 13			956,0	134,59	2127
1047,1	93,81	2215	936,8	152,46	2116
1034,3	99,07	2212			
1023,7	104,14	2215	Fe Si 4		
1015,8	108,32	2218	1111,8	57,57	2037
998,6	118,24	2218	1094,8	60,37	2033
942,2	168,83	2215	1076,4	63,73	2029
929,4	186,09	2203	1066,1	66,01	2034
			1050,6	69,48	2033
Fe Sn 22			1033,6	73,75	2033
1012,3	106,03	2187	1020,1	77,46	2030
1011,3	106,40	2184	1012,8	79,85	2035
993,5	116,94	2187	1002,3	83,37	2037
987,5	120,89	2194	987,3	88,66	2033
964,1	138,64	2192	969,5	95,96	2030
960,2	142,67	2199	946,9	107,52	2031
936,4	168,19	2193			
924,0	185,57	2190	Fe Si 7		
			1066,1	51,85	1883
Fe Sn 16			1027,3	57,93	1879
1022,1	101,19	2182	991,0	65,28	1880
998,7	113,47	2181	954,3	74,76	1879
985,9	121,94	2188	916,4	88,22	1883
956,6	145,7	2187	907,6	91,92	1881
947,2	155,17	2183			
935,4	169,21	2181			

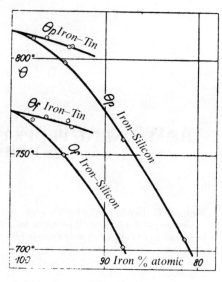

Fig. 1.

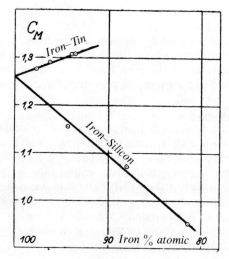

Fig. 2.

TABLE II.

N°	Fer % by weight	Fer % atomic	$C \times 10^5$	C_M	θ_p	θ_f
					degrees	degrees
Sn 10	95,23	97,70	2228	1,276	810,5	768,1
Sn 13	92,35	96,25	2216	1,290	811	769,4
Sn 12	87,95	93,95	2196	1,307	806	767,5
Sn 16	87,21	93,55	2183	1,308	806,5	764,2
Si 5	97,04	94,32	2129	1,155	798	749,8
Si 4	93,59	88,03	2033	1,072	758	701,6
Si 7	89,72	81,41	1881	0,953	703	,,

REFERENCES

[1] Westgren and Phragmen, *J. of Iron and Steel Inst.*, 1922, **105**, 241.
[2] Wever and Reinecken, *Mitteilungen aus dem Kaiser Wilhelm Institut für Eisenforschung*, 1925, **7**, 69.
[3] Wever and Giani, *Mitteilungen aus dem Kaiser Wilhelm Institut für Eisenforschung*, 1925, **7**, 59; Phragmen, *J. of Iron and Steel Inst.*, 1926, **109**, 397.

A9 (1932)
THE MAGNETIC PROPERTIES OF MANGANESE AND CHROMIUM IN DILUTE SOLID SOLUTIONS

Abstract: In the first part of the work, the author shows that manganese in dilute solid solution in copper or silver, and chromium in solution in gold, possess a paramagnetism which obeys Weiss's law: $\chi = C/(T - \Theta)$. The Curie constants are proportional to the quantity of chromium or of manganese: the moment of chromium is 23.47 Weiss magnetons in these conditions.

In the second part, the author goes on to survey the directions in which one might look for the interpretation of a temperature independent paramagnetism. Pauli's theory does not provide the solution for manganese and chromium. The low temperature anomalies of a substance with a negative molecular field, using the scheme proposed by the author, allow one to interpret the experimental results in a satisfactory way. Chromium and manganese thus possess negative magnetic moments and molecular fields which, expressed in degrees, are for manganese: $\Theta = -1720$ K and for chromium $\Theta = -4150$ K.

I
EXPERIMENTAL RESULTS

(1) Introduction: A large number of metals, more than twenty, possess a paramagnetism independent of temperature [1]. However, there must be many more, since diamagnetism, which is a general atomic property, may mask a weak paramagnetism. In fact, copper, weakly diamagnetic under normal conditions, becomes paramagnetic following a permanent mechanical deformation [2].

The interpretation of constant paramagnetism is difficult, particularly since it is impossible to have precise data on its magnitude: in fact, the superimposed diamagnetism is in most cases of the same order of magnitude and is not calculable, *a priori*, exactly.

Amongst the metals with constant paramagnetism, chromium and above all manganese are particularly interesting. They have, respectively, specific susceptibilities of [19] 4.3×10^{-6} and 9.9×10^{-6}, large values in comparison with their likely diamagnetic susceptibilities.

(2) Goal of the experiments: It is important to study the evolution of the magnetic properties of chromium and manganese diluted in a metal not possessing any constant paramagnetism. From this point of view, the interesting solid solutions are unfortunately of a restricted number.

Three series have been considered: (a) Cu–Mn; (b) Ag–Mn; (c) Au–Cr.

(a) Manganese and copper give solid solutions [3]. From 0 to 30% manganese, they crystallize in a face centred cubic lattice [4].

(b) Manganese and silver give solid solutions from 0 to 20% manganese [5].

(c) Chromium and gold give solid solutions from 0 to 7% chromium [6]. The specific susceptibilities of copper, silver and gold are -0.086×10^{-6}, -0.20×10^{-6}, and -0.15×10^{-6} respectively.

Their constant paramagnetism, if it exists, is thus weak: the conditions are thus favourable for studying the dilution of the magnetism of chromium and manganese.

(3) Experimental technique: The alloys have been prepared from pure metals, by melting in a vacuum, by means of an induction furnace installed by M. Ribaud of the Institut de Physique at Strasbourg. I have taken the greatest precautions in order to ensure the homogeneity of the alloys: the consistency of the analyses is moreover a guarantee.

The copper–manganese has been analyzed by M. Rosentein at the Institut de Chimie at Strasbourg. The silver–manganese and the gold–chrome have been analysed by Hoepfner at Hamburg. In addition, I have repeated, myself, a large number of analyses: the agreement was excellent. The analyses of silver–

manganese and gold–chromium were the most precise.

The magnetic measurement technique is identical to that which I described in a previous paper [7]. The accuracy is a thousandth in the measurements of susceptibility and a few tenths of a degree in the temperatures.

At high temperatures, the susceptibility measurements are rather difficult because a slow migration towards the exterior occurs of the diluted metal: the alloy is depleted of chromium or manganese. This effect is particularly important for small samples. I have been able, by a prolonged anneal at 700° in vacuum, to decrease the concentration of manganese by $\frac{2}{3}$ in a manganese–silver. For the measurements of susceptibility, it is thus necessary to operate quickly and reduce the interval between the measurements to that which is strictly necessary for the equalization of the temperatures of the sample and the thermocouple.

(4) Results: Fourteen alloys were studied for which the compositions were:

Cu – Mn 1	1,4 p. 100 Mn	Ag – Mn 1	2,5 p. 100 Mn	Au – Cr 1	1,59 p. 100 Cr
– 2	2,0 –	– 2	3,45 –	– 2	2,93 –
– 3	3,2 –	– 3	5,02 –	– 3	5,71 –
– 4	5,0 –	– 4	7,02 –		
– 5	10,0 –	– 5	7,03 –		
		– 6	12.25 –		

I give in Tables I and V the results of the measurements of the specific susceptibility for the manganese–silvers and the chromium–golds.

The susceptibility has been corrected for the specific diamagnetism of the diluent, assumed, from our point of view, to constitute the complete mass of the alloy. This procedure is legitimate since the correction is small and the diluant is 9/10 of the solid solution.

A single glance shows that the susceptibility of these alloys, instead of being independent of temperature, decreases rapidly with it and suggests the idea of a paramagnetism obeying Weiss's law. The plots of $1/\chi$ confirms this analogy and Figs. 1, 2 and 3 show the approximately linear dependence of $1/\chi$ on temperature; the lines which one thus plots extrapolate to temperature different from 0 K. I will examine to what extent these experiments allow one to determine a Curie constant and a Curie point.

(5) Manganese–silver (Fig. 1.: All the curves plotted in Fig. 1 show a weak convexity towards the temperature axis. For reasons which I will explain in the second part of this paper, it is logical to think that the substance will only obey Weiss's law at high temperatures. Experience shows that if a straight

TABLE I.
Silver–manganese.

T	$\chi \times 10^6$	T	$\chi \times 10^6$	T	$\chi \times 10^6$	T	$\chi \times 10^6$	T	$\chi \times 10^6$	T	$\chi \times 10^6$
Ag – Mn 1		Ag – Mn 2		Ag – Mn 3		Ag – Mn 4		Ag – Mn 5		Ag – Mn 6	
662°8	1,452	12°1	6,731	637°7	2,834	661°7	4,237	12°2	14,874	670°3	7,176
619,7	1,539	62,6	6,871	637,3	2,951	624,3	4,389	107,0	11,228	632,5	7,449
581,4	1,617	108,0	5,268	597,7	3,090	582,5	4,600	147,8	10,072	593,1	7,768
538,8	1,705	142,6	4,848	558,6	3,253	541,6	4,825	194,5	9,031	559,6	8,038
487,0	1,843	179,8	4,459	505,2	3,497	494,2	5,142	258,9	7,954	518,2	8,456
442,7	1,971	11,8	6,917	450,9	3,775	448,8	5,478	297,6	7,324	475,1	8,926
392,6	2,133	179,0	4,471	401,1	4,077	393,7	5,974	368,0	6,441	420,0	9,670
333,6	2,355	216,7	4,130	314,7	4,731	340,2	6,542	416,4	5,951	370,9	10,523
280,0	2,603	250,4	3,848	260,9	5,264	320,0	6,794	472,2	5,483	315,7	11,650
226,7	2,895	294,4	3,543	205,9	5,960	268,2	7,550	517,6	5,163	259,2	13,259
174,9	3,249	347,5	3,244	157,1	6,672	209,7	8,479	11,7	14,740	205,0	14,889
117,3	3,763	386,9	3,036	112,4	7,508	161,4	9,598	516,5	5,161	154,8	16,604
61,6	4,372	425,3	2,855	62,6	8,655	117,4	10,789	554,4	4,929	110,6	18,738
11,1	5,104	467,4	2,684	11,0	10,184	61,8	12,664	594,2	4,669	59,2	21,562
		521,2	2,457			10,9	14,961	642,6	4,426	11,1	25,389
		567,2	2,312					677,5	4,258		
		620,1	2,156								
		678,8	2,010								
		11,5	6,542								

EXPERIMENTAL WORK

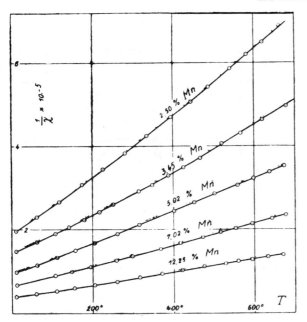

Fig. 1.

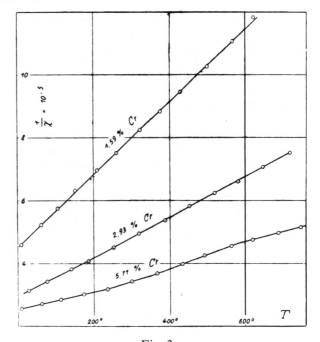

Fig. 3.

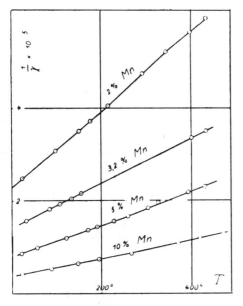

Fig. 2.

TABLE II.
Silver–manganese.

Alloy	$C \times 10^3$	θ (*)		$C_{Mn} \times 10^2$
Ag-Mn 1	1,23	−180°	480°-620°	4,92
2	1,722	−178°	420°-680°	4,99
3	2,539	−222°	150°-650°	5,06
4	3,686	−223°	160°-542°	5,25
5	3,886	−236°	370°-680°	5,53
6	7,140	−325°	470°-670°	5,83

(*) In degrees centigrade

section exists in this region, it is short. The specific Curie constants C and the Curie points Θ so calculated are thus dubious. I give these in Table II just as an indication, together with the temperature interval in which Weiss's law was verified.

In Fig. 4, I have plotted the specific Curie constants as a function of the manganese concentration. In Table II, I give the values C_{Mn} of the Curie constant per gramme of manganese.

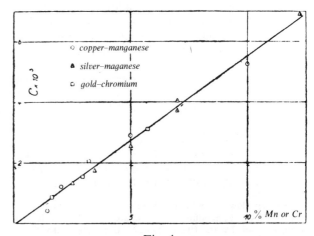

Fig. 4.

They vary systematically with the concentration; but, if one considers that when the concentration changes from 2.5% to 12.5%, C_{Mn} only changes from 4.92×10^{-2} to 5.83×10^{-2}, it is legitimate to think that the fundamental phenomenon is the proportionality of C to the concentration.

A secondary problem also affects this law.

The Curie points show only a weak systematic variation with the concentration. Poorly defined factors, traces of impurities, tempering, have as much influence on the Curie point as the proportion of manganese.

(6) Manganese–copper (Fig. 2): The experimental results on the alloy Mn–Cu with 1 to 1.4% of manganese are well described by the formula

$$\chi = 4.167/(T + 312°) \times 10^{-4}$$

as demonstrated in Table III. Weiss's law is exactly verified.

TABLE III.
Cu–Mn 1.

T	$\chi \times 10^6$ obs.	$\chi \times 10^6$ calc.
24°9	1,238	1,237
76,6	1,072	1,072
165,1	0,875	0,873
188,4	0,832	0,833
218,3	0,784	0,786
259,4	0,729	0,729
45,6	1,166	1,166
320,4	0,659	0,659
345,8	0,634	0,633

For the more concentrated alloys, the agreement between the observed values and the calculated values is a little less good. I have assembled in Table IV the values of specific C and θ which one may draw from the study of these alloys.

I have indicated the temperature interval in which Weiss's law is obeyed.

The Curie points increase slightly with the concentration.

In the column C_{Mn} I give the values of the Curie constant per gramme of manganese. With the exception of the value corresponding to Cu–Mn 1, clearly in disagreement with the other results[1], C_{Mn}

[1] This very low concentration alloy (1.4%) was alloyed for a long time at high temperature. This probably produced a loss of manganese so that the sample studied was not identical with the sample analyzed.

TABLE IV.
Copper–manganese.

Alloy	$C \times 10^3$	θ (*)		$C_{Mn} \times 10^2$
Cu-Mn 1	0,417	−312°	25°-346°	2.98
2	1,22	−278°	0°-215°	6,11
3	2,06	−292°	30°-430°	6,31
4	2,89	−226°	20°-260°	5,78
5	5,26	−196°	20°-200°	5,26
	5,31	−180°	20°-250°	5,31

(*) In degrees centigrade

decreases slightly with the concentration. When it goes from 2 to 10%, C_{Mn} only decreases from 6.11×10^{-2} to 5.31×10^{-2}, so that my conclusion will be the same as for the manganese–silvers: to a first approximation, the Curie constant is proportional to the concentration. But our uncertainty in the interpretation of possible stray impurity phenomena, only allows us to give a very mediocre confidence to the average value

$$C_{Mn} = 5,28 \times 10^{-2}$$

deduced from all the measurements on the solid solutions of manganese.

(7) Chromium–gold (Fig. 3): Alloys 1 and 2 follow Weiss's law with a high precision over a large temperature interval, as demonstrated by the comparison of the values of the observed specific susceptibility with the calculated specific susceptibility in Table V.

The specific Curie constants and the Curie points used in the calculations are given in Table VI.

Alloy 3 possesses a curious thermal variation with three consecutive straight sections. These sections are well characterized: their Curie constants are clearly different.

In Fig. 4, I have shown the specific Curie constants of alloys 1 and 2, and the Curie constant of the high temperature straight line for alloy 3. These points are on a straight line pointing to the origin, so confirming the idea of the great simplicity of the high temperature phenomena. The Curie constants of the other two sections for alloy 3 seem to bear no relation to the above.

The Curie points decrease very rapidly when the proportion of manganese increases. In the last column of Table VI, I have given the values of the Curie constants C_{Cr}, per gramme of chromium. Given the slight uncertainties in the analyses, one may consider these values to be identical and take the specific Curie

EXPERIMENTAL WORK

TABLE V.
Gold–chromium.

Au-Cr 1			Au-Cr 2			Au-Cr 3		
T	χ × 10⁶ obs.	χ × 10⁶ calc.	T	χ × 10⁶ obs.	χ × 10⁶ calc.	T	χ × 10⁶ obs.	χ × 10⁶ calc.
10°2	2,155	2,162	20°4	3,175		10°7	3,891	3,889
61,7	1,912	1,916	78,6	2,910		63,0	3,671	3,680
104,8	1,739	1,749	140,2	2,622		113,4	3,495	3,500
151,0	1,586	1,600	187,0	2,450	2,449	173,8	3,309	3,305
210,2	1,436	1,443	252,9	2,223	2,218	236,4	3,132	3,126
261,0	1,331	1,330	320,2	2,020	2,023	301,4	2,917	2,918
323,7	1,215	1,214	386,7	1,860	1,861	367,1	2,706	1,697
375,6	1,133	1,132	453,2	1,716	1,723	433,7	2,508	2,505
10,7	2,163	2,159	520,3	1,605	1,604	496,5	2,353	2,348
374,7	1,131	1,133	581,2	1,511	1,509	10,8	3,801	
427,7	1,059	1,059	647,3	1,416	1,417	494,3	2,351	2,353
499,7	0,974	0,974	718,8	1,332	1,330	565,0	2,195	2,197
566,1	0,904	0,906	25,3	3,165		619,6	2,103	2,104
621,6	0,845		90,2	2,937		687,7	2,010	2,011
688,8	0,788					745,7	1,937	1,938
10,1	2,135							

TABLE VI.
Gold–chromium.

Alloy	C × 10³	θ (*)		$C_{Cr} × 10^2$
Au-Cr 1	0,8674	− 391°	10°-630°	5,45
2	1,691	− 503°	30°-190°	5,77
—	1,548	− 445°	190°-720°	5,28
Au-Cr 3	3,596	− 914°	10°-270°	6,29
—	2,344	502°	270°-600°	4,11
—	3,100	− 854°	600°-750°	5,42

(*) In degrees centigrade

constant of chromium to be the mean of the three observed values at 5.38×10^{-2}.

The corresponding number of Weiss magnetons is 23.47. One should consider this as a paramagnetic moment. I note that the ferromagnetic moment of chromium, determined by Sadron [8] is 19 magnetons. One finds therefore a difference of 4.47 magnetons between the two moments. This difference is close to the difference of 5 magnetons found for nickel [9] and 4.79 magnetons found for iron [16].

(8) Conclusions of the experimental work: The fundamental facts which emerge from this study are the following:

Chromium and manganese, in a state of very dilute solid solution (0 to 5%) possess a paramagnetism obeying Weiss's law; the Curie constants vary proportionately with the concentration. These magnetic properties relate therefore to manganese and chromium. These metals have a magnetic moment and a molecular field.

II
THEORETICAL INTERPRETATIONS

(9) Variable paramagnetism: In a small external field H, when there are no interactions between the carriers of the magnetic moment, classical or quantum theories lead to the following formula which gives the susceptibility of a group of N carriers

$$\chi = \frac{N\mu^2}{3\,KT}. \qquad (1)$$

In the classical theory of Langevin [11] μ represents the moment of the carrier, while in the quantum theories μ is related to the Bohr magneton by the formula [12]:

$$\mu = g\,\sqrt{j(j+1)}\,\mu_B$$

j being the number of quanta of rotation of the carrier and g the Landé factor. Formula 1 is valid for high temperatures, provided however that kT is small with respect to the difference of the internal energies of the carrier between the stable level at normal temperatures and the higher levels[2], a change of moment obeying such a mechanism could not explain a constant paramagnetism at low temperatures.

(10) The quantum degeneracy of paramagnetism: Pauli has based his theory [14] of constant paramagnetism on the degeneracy of a gas of electrons at low temperatures.

Assuming that in a metal there exist λ free electrons per atom; per gramme-atom, at very high temperatures, the susceptibility of the electrons will be[3]

$$\chi = \frac{\lambda N \mu^2}{3\,KT}$$

with $\mu^2 = 3\mu_B^2$.

At low temperatures there is degeneracy, the susceptibility becomes independent of the temperature and takes the value χ_0. On designating by V the atomic volume and m the mass of the electron one has

$$\chi_0 = \frac{\lambda N \mu^2 m}{h^2}\left(\frac{8\pi V}{3N\lambda}\right)^{\frac{2}{3}} = 1{,}87 \times 10^{-6} \times \lambda^{\frac{1}{3}} V^{\frac{2}{3}}. \quad (2)$$

As T increases from 0 to $+\infty$, $1/\chi$ is represented by a curve similar to Γ in Fig. 5. Defining the temperature T_0 by the relation

$$\frac{\lambda N \mu^2}{3\,kT_0} = \chi_0,$$

which gives

$$T_0 = 2{,}01 \times 10^5 \lambda^{-\frac{1}{3}} V^{-\frac{2}{3}} \quad (3)$$

T_0 occurs in a mixed region which separates constant paramagnetism from variable paramagnetism.

To apply the theory to sodium one takes $\lambda = 1$. The agreement with theory is good. In particular one finds

$$T_0 = 24{,}000\ \text{K}.$$

At normal temperatures the degeneracy is complete. If, instead of taking the electrons as the moment carriers, we had taken the atoms, it would have been necessary to divide T_0 by the ratio of the mass of the atom to the mass of the electron, which is of the order of 10^5. The degeneracy temperatures obtained would have been of the order of 1 K, and thus inaccessible to our measurements.

Although Pauli's theory provides satisfactory results for the alkali metals, it does not seem however that one may invoke it to explain the paramagnetism of manganese and some other metals with a strong constant paramagnetism.

For manganese, the atomic susceptibility is 545×10^{-6}, the atomic volume 7.64 cm^3, formula (2) gives

$$\lambda = 420{,}000$$

and formula (3)

$$T_0 = 290{,}000{,}000\ \text{K}$$

420,000 free electrons per atom are quite impossible. In addition, even for a solid solution of manganese diluted by 1/1000, the degeneracy temperature T_0 would still be 2,900,000 K while the experiments show that a solution of 1 in a 100 is no longer degenerate at room temperature.

The Pauli mechanism provides only a small part of the paramagnetism of manganese and chromium.

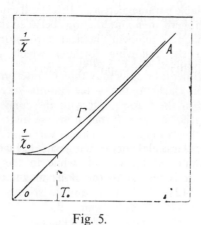

Fig. 5.

[2] In nitric oxide this condition is not realized. The theory predicts thus a variation of moment verified exactly by experiment [13].
[3] Cf. the very clear exposition of Fermi statistics and Pauli's theory, by L. Brillouin in his work: Quantum statistics [1], Presses Universitaires de France, Paris.

(11) Cryomagnetic anomalies: As G. Foëx has shown [15], crystal field effects may give rise, at low temperatures, to deviations from Curie's law. However, it is unreasonable to attribute the constant paramagnetism to such effects, since on the one hand experiments show that in crystals these anomalies only manifest themselves at temperatures below 100 K and on the other hand they do not lead generally to a temperature independent paramagnetism.

(12) Paramagnetism with a molecular field: It only remains to examine the properties of paramagnets with a molecular field. With this hypothesis one obtains Weiss's law at high temperatures [16]

$$\chi = \frac{N\mu^2}{3K(T-\Theta)}. \qquad (4)$$

Θ may be positive or negative. If Θ is positive, for $T < \Theta$ one has ferromagnetism.

Θ is negative in several metals such as platinum and palladium.

At low temperatures, the deviations from the law (4) are due to:

(a) quantum degeneracy.
(b) crystal field effects.
(c) fluctuations in molecular field.

In previous paragraphs, I have shown that the first two causes can only play a secondary role in the interpretation of constant paramagnetism.

We examine the third.

(13) The theory of constant paramagnetism: In a recent paper [7] I have shown that a negative molecular field gives, at low temperatures, a temperature independent paramagnetism.

Taking the schematically simple case of a substance, crystallized in the body centred cubic system, in which the mutual interactions reduce to interactions between immediate neighbours, such that $\omega \cos\alpha$ [4] represents, at a constant separation, the relative potential energy of two immediate neighbours for which the magnetic moments make the angle α.

At high temperatures, the law is of the same type as that of Weiss:

$$\chi = \frac{C}{T-\Theta} \qquad \Theta < 0.$$

On the contrary, at low temperatures χ takes the value χ_0 independent of T:

$$\chi_0 = -\frac{C}{3\Theta}.$$

For $T = -\Theta$ the difference $\chi - \chi_0$ is no larger than $0.014\,\chi_0$.

The dependence of $1/\chi$ upon temperature is shown in Fig. 6.

The remarkable magnetic properties of dilute solid solutions of manganese and chromium, demonstrated in the first part of this work, are thus immediately interpretable: when one dilutes the manganese or the

[4] ω = a positive constant.

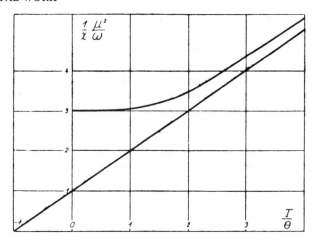

Fig. 6.

chromium the Curie point of the alloy approaches absolute zero as the dilution becomes greater and, at room temperature but with a variable concentration, one passes from the region of constant paramagnetism to that of variable paramagnetism. The experiments on the dilute solid solutions provide the value of C. On applying formula (5) one finds:

for manganese: $\Theta = -1720°$

for chromium: $\Theta = -4150°$.

The order of magnitude is satisfactory.

Using this estimate, the susceptibility of manganese at $1720°$ differs from the susceptibility at absolute zero by only 1.4%, while in an alloy of 1% manganese, the corresponding point will be 17.2 K so that at room temperature one will be clearly in the region where the susceptibility obeys Weiss's law.

(14) Variation of the Curie point: The essential part of the experimental results may thus be interpreted in a satisfactory fashion; but a few points such as the variation of the Curie point with the concentration remain rather unclear. If the atoms are distributed at random, the theory predicts a linear variation from 0 to Θ, when the atomic fraction of the paramagnetic metal varies from 0 to 1.

In the chromium–golds one observes a rapid fall in the Curie point which agrees quite well with the large negative value of Θ for pure chromium, but the variation is not linear. In the manganese–coppers and the manganese–silvers the Curie points vary in opposite ways but in a less pronounced way.

One may attribute these differences to the undue simplicity of the hypotheses on which the theory is based, and to the fact that the atoms are not randomly distributed in the crystals.

(15) Imperfections of the theory: It only takes account of the interactions between nearest neighbour atoms. One should also take account of the more distant atoms which produce a molecular field with relatively weaker fluctuations. In the pure metal, the short range molecular field is the true source of the constant paramagnetism, the long range molecular field then intervening by a different process. In the dilute solid solutions, on the contrary, the two molecular fields intervene by the same mechanism, so that the experiments on such substances only provide us with the total molecular field, without allowing us to determine the short range part of the molecular field, which would be necessary to calculate, *a priori*, the susceptibility of the pure metal.

(16) The role of superlattices: The atoms of the two constituents of a solid solution do not arrange themselves at random on the lattice points: there are often regular arrangements. The existence of these structures has been established on a sound experimental basis [17].

In solid solutions of gold and copper, for which X-ray studies are facilitated by the large difference in the atomic numbers of these two metals, the existence of two structures corresponding to AuCu and $AuCu^3$ has been established [18]. We note that in the latter, all the neighbours of an atom of gold are copper atoms, whereas a statistical distribution will give 3 neighbours of gold and 9 neighbours of copper.

The Curie point of an alloy with superlattices should differ considerably from the Curie point of the same alloy without the superlattices, since the neighbours play a dominant role in the molecular field. In fact in the solid solutions of Cu–Mn, containing 10 to 20% manganese, several experimental facts depend on these phenomena.

Starting from the same alloy heated to 500° for a few hours, one obtains substances which, between 0° and 100°, possess the same Curie constants but different Curie points, depending on the speed of cooling. For a slow cooling, there is the onset of a superlattice: the Curie point is higher by about twenty degrees.

The same phenomenon occurs in cobalt–platinum with 10% cobalt [7].

One understands therefore why the Curie points vary rather irregularly: they are extremely sensitive to the least change of structure.

To interpret the magnetic properties of an alloy completely one has to take account of the possible structures.

Finally, it is not certain that in the Cu–Mn and the Ag–Mn, the experiments having been done at fairly high temperatures, one is really in the region of Weiss paramagnetism. One understands thus the weak systematic variation of the constant C_{Mn}; all the observed values will be too large.

(17) Conclusion: The schéma which I am proposing for interpreting the constant paramagnetism of metals allows one also to explain the magnetic properties of these metals when diluted, as demonstrated in the first part of this work.

It is necessary now, from the point of view of quantum mechanics, to determine the nature and the importance of forces between atoms and to know why these are forces of disorientation in manganese and chromium, and on the contrary are forces of orientation in iron, nickel and cobalt.

REFERENCES

[1] International Critical Tables, New York, t. VI, p. 354.
[2] F. Bitter, *Phys. Rev.*, 1930, **36**, 978; K. Honda and Shimizu, *Sc. Rep. Töh, Imp. Univ.*, 1931, **20**, 460.
[3] Sahmen, *Z.f. Anorg. Chemie*, 1908, **57**, 1.
[4] R.-A. Patterson, *Phys. Rev.*, 1924, **23**, 552.
[5] Siebe, *Z.f. Anorg. Chemie*, 1919, **108**, 161.
[6] Vogel and Trilling, *Z.f. Anorg. Chemie*, 1923, **129**, 276.
[7] Néel, Thesis, Strasbourg, 1932.
[8] Sadron, *C.R. Acad. Sci.*, 1931, **193**, 1070.
[9] Weiss and Foex, *Le Magnétisme*, Paris, A. Colin, p. 200.
[10] Neel, *C.R. Acad. Sci.*, 1931, **193**, 1325.
[11] Langevin, *Ann. de Phys.*, 1905, **5**, 70.
[12] Pauli, *Physik, Z.*, 1920, **21**, 615; Sommerfeld, *Z.f. Phys.*, 1923, **19**, 221.
[13] Van Vleck, *Phys. Rev.*, 1928, 585; Aharoni and Scherrer, *Z.f. Physik.*, 1929, **58**, 749; Wiersma, de Haas and Capel, *Comm. of Leiden*, no. 212b.
[14] Pauli, *Z.f. Physik.*, 1927, **41**, 97.
[15] Foex, *Ann. de Phys.*, 1921, **16**, 174.
[16] Weiss, *J. Phys.*, 1907, **6**, 661.
[17] Ewald U. Hermann, *Strukturbericht* (Leipzig), 1931, p. 484.
[18] Kurnatov and Zemczusny, *Z.f. Anorg. Chemie*, 1907, **54**, 149; C.-H. Johansson and J.-O. Linde, *Ann. de Phys.*, 1925, **78**, 439; C.-H. Johansson and J.-O. Linde, *Ann. der Physik*, 1927, **82**, 449; Borelius, Johansson and Linde, *Ann der Physik.*, 1928, **86**, 291; Le Blanc, Richter and Schiebold, *Ann. der Physik*, 1928, **86**, 929.
[19] J. Safranek, *Revue de métallurgie*, february 1924.

A20 (1935)

THE MAGNETIC PROPERTIES OF PURE NICKEL IN THE NEIGHBOURHOOD OF THE CURIE POINT

Note (1977): In the neighbourhood of the Curie point, the variation with temperature of the inverse $1/\chi$ of the paramagnetic susceptibility no longer obeys the Curie–Weiss law. The experiments reported here were undertaken to determine whether there was a continuous curve or, as previously suggested by several authors, whether there was a juxtaposition of straight lines each corresponding to a different 'magnetic state'.

A continuous curve was found. On using the current representation of 'scaling laws', the results are given by the relation

$$\chi = (T - \Theta)^{-\gamma}$$

with $\gamma = 1.18 + 0.05$, valid from 2°5 to 7° from the Curie point.

Abstract: A precision apparatus is described for measuring magnetic susceptibilities, in fields between 50 and 1000 Gauss, with an arrangement allowing one to adjust the temperature to a precision of better than 0°005. Nickel has thus been studied from the Curie point (358°) up to 368°. In this interval, the dependence of the susceptibility upon temperature and magnetic field has been determined.

(1) Goal of the experiments: The ferromagnetic metals only obey the Curie–Weiss law at temperatures of at least 100° above the Curie point. Close to the Curie point, the curve $\frac{1}{\chi} = f(T)$ is convex towards the temperature axis (Fig. 1). In addition, a few tens of degrees above the Curie point, the curvature becomes very weak and measurements made in a small temperature interval suggest the existence of a straight line. Very close to the Curie point, on the contrary, the curvature becomes strong and in particular the question of determining the initial tangent comes up.

My preliminary measurements on iron have shown that this curve is made up of juxtaposed straight sections. But the experiments were difficult and I preferred to start again with nickel, with a much greater precision: the Curie point of nickel, 358°, is much more accessible than that of iron: 770°.

I have thus set out to study the magnetic properties of nickel in an interval of a few degrees above the Curie point, by the classic method of attraction in a non-uniform field, by means of a moving balance [1]. Two points require special attention: (1), the production and the measurement of temperature; (2) the production of a non-uniform magnetic field of a suitable intensity.

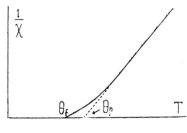

Fig. 1.

(2) The production and measurement of temperature: In the interval to be studied, the susceptibility changes by 50%, when the temperature changes by 1°. To make measurements of the susceptibility to a relative precision of 0.1%, it is necessary to define and to measure the temperatures to a precision of 0°002. At a temperature of 400°, it is possible to maintain constant, to 0°03, the temperature of a poorly insulated externally water-cooled furnace, for at least about ten minutes. If the substance studied is well insulated at the centre of the furnace, its temperature oscillations are strongly damped. The thermocouple being placed in the substance itself, its temperature at each instant can differ only very little from the average temperature of the substance. In fact, experiment shows that this difference is less than 0°005. I used a thermocouple of 'Imphy' (B.T.E.–C.T.E.), practically stable up to 400° and giving an electromotive force of 50 μV per degree.

For the measurement of the electromotive forces I have constructed a high precision potentiometer which has been previously described [2]. This is a deflection potentiometer, with two independent circuits which enable a simple autocalibration. It allows the measurement of an electromotive force of 20 millivolts to an accuracy of 0.1 μV. The calibrating Weston cells were placed in a thermostat which was maintained constant to $\pm 0°,1$.

The couple had been calibrated, by means of a certain number of fixed points, with a precision estimated at 0°,2. Its characteristic: $e = f(T)$ is linear to 0°,002 accuracy in the 10° interval of temperature where the measurements were made. The corresponding slope $\frac{de}{dT}$ is known to an accuracy of 1/200. In this work, the temperatures will be given to 0°,001: this merely signifies that a temperature *interval* greater than 1° is known to 0.5%. The absolute values of the temperatures may contain an error of $\pm 0°,3$.

(3) The production of a non-uniform magnetic field: In the region studied, the susceptibility is a function of magnetic field so that one wants to find the limiting value in a field of zero. One must, therefore, work in a series of exactly known weak magnetic fields, which is only possible in an iron free coil.

I have calculated approximately a form of coil such that the attraction $KH\frac{\partial H}{\partial x}$ has a very flattened maximum along the axis, outside the coil, to allow a simple positioning of the furnace. I constructed a model which reproduced, with $\frac{1}{4}$ of the linear dimensions, the definitive coil and by successive adjustments to this model I obtained the cross-section given below (Fig. 2). The winding is made of two sections each consisting of 95 metres of copper tube of 8 mm diameter. The two sections are arranged in parallel for the water flow and in series for the electric current. In the continuous regime, one can use an intensity of 150 amperes.

Along the axis, the maximum of $H\frac{\partial H}{\partial x}$ is 10 mm from the front surface of the coil. In addition the corresponding cheek of the coil, 4 mm thick, had a rectangular opening 2 × 15 cm cut into it; one may thus put a furnace of 28 mm diameter around the sample. The investigation of the field of attraction was made by means of a small sphere of nickel of 2 mm diameter. Along the axis, the results are shown in Fig. 3.

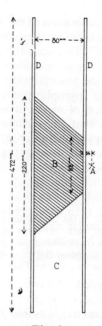

Fig. 2.
A, the position of the maximum attraction; B, core in oak; C, space occupied by the windings; D, cheeks in brass.

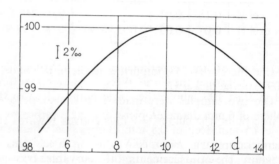

Fig. 3.
On the ordinate, the attraction in arbitrary units; on the abscissa, the distances in mm from the plane of the coil.

Perpendicular to the axis, the relative variations of the attraction are less than 1 in 1000, up to 4.5 mm from the axis.

The intensity of the electric current in the coil is measured with a potentiometer with a relative precision always greater than 0.05%.

For a given value of current, the magnetic field, at the position of maximum attraction, was compared by means of a fluxmeter with the field of a calibrating magnet, itself measured with a Cotton balance. One thus finds: 7.02 gauss per ampere, to 1% precision.

(4) The magnetic calibration of the apparatus: In the study of nickel, the use of small magnetic fields does not give rise to any difficulty, since the susceptibilities are very large, but the magnetic calibration of the balance is delicate, since the susceptibility of the calibrating body, manganese pyrophosphate, is very small compared with that of nickel. It is necessary to use large ampules so that the outside parts are clearly outside the region of maximum attraction. Knowing the topology of the field of attraction and the geometric shape of the ampule, assumed to be homogeneous, I have calculated that one should add 0.3% to the measured attraction. Two ampules of pyrophosphate, one of which had been compared with water by M. Lallemand, gave similar results. The absolute values of the susceptibilities have thus been determined to 0.6% and the relative values of 0.2%.

(5) Substances studied: I have studied two samples of nickel, the one prepared by the Mond Nickel Co., the other by W. C. Heraeus [3]. Subsequently, they will be denoted respectively by the letters M and H. An analysis gave the following results in per cent:

	Fe	Cu	Co	Mn	Others	Ni
Nickel M	0,18	0,01			0,05	99,76
Nickel H	0,01	0,03	0,005	0,005	0,04	99,91

so that there are 0.24% impurities in the nickel M and 0.09% in the nickel H.

The two samples were made into the shape of a sphere of 6 mm diameter, pierced along a radius by a hole 4.5 mm deep of 0.8 mm diameter in which the thermocouple was mounted.

Before the measurements the two spheres were annealed for 36 h at 800° and cooled in 24 h to room temperature.

(6) Order of the measurements: I have established that a few series of changes of temperature are sufficient to stabilize the samples, i.e. in the course of an experimental session of 4 h, during which the substance was always kept at a temperature above the Curie point, the measurements were reversible and the substance did not change. But after having left the substance to cool and remain at room temperature for a long time, or at a higher temperature which is still less than the Curie point, one finds *sometimes* a displacement of the Curie point, very weak however, and only exceptionally reaching 0.1°.

The possibility of such a displacement, in the interval between two series of measurements, affects the mode of operation in the following way:

(1) For a series of fixed temperatures, one studies the variations of the susceptibility with the magnetic field, taking care to make all the measurements relating to one temperature in the same session.

(2) For a fixed magnetic field, one studies the variation of the susceptibility with temperature, in a few sessions as close as possible, and one uses the coefficients of the variation of the susceptibility with the magnetic field, previously determined, to correct, if necessary, the observations to zero field.

(7) The determination of the coefficients of the demagnetizing field: The true susceptibility χ is related to the apparent susceptibility $\chi_{app.}$, given directly by the experiments, by the formula[1]

$$\frac{1}{\chi} = \frac{1}{\chi_{app.}} - n,$$

where n is the demagnetizing factor. The measurement of the apparent susceptibility, in conditions where the true susceptibility is practically infinite, below the Curie point for example, allows one to calculate n.

Table I, relating to nickel H, shows that n thus

TABLE I.

T	H	$\frac{1}{\chi_{app.}} = n$
352°24	31,7 60,1	35,53 35,64
353°35	59,2 115,0 201,6	35,56 35,48 35,56
353°68	31,6 59,8	35,57 35,60
354°36	60,2 117,0	35,75 35,50
355°74	31,3 58,9	35,57 35,53

determined is effectively independent of the temperature and the magnetic field.

---

[1] The values of χ, $\chi_{app.}$ etc., are always expressed in C.G.S. and given for unit mass.

TABLE II.

T	h	$1/\chi$
	Nickel M	
359°140	49,6	162,2
	63,3	165,4
	123,5	171,0
	169,4	175,4
	281,5	182,6
	335,9	185,7
	423,1	190,4
	558,6	197,9
	669,8	203,9
	781,9	210,6
	Nickel H	
358°100	14,3	29,78
	30,3	38,05
	65,5	47,10
	123,0	57,17
	216,1	70,35
	350,2	85,02
	482,7	99,06
358°400	16,1	36,92
	33,3	44,32
	69,3	53,19
	127,7	62,53
	220,7	73,95
	353,8	88,92
	487,5	102,35
360°240	41,3	171,0
	80,0	171,0
	139,9	174,4
	228,9	178,7
	351,7	186,4
	472,7	193,4
	635,3	202,9

I have adopted the following values

Nickel M: $n = 34.70$,

Nickel H: $n = 35.58$.

The internal magnetic field really acting h, is related to the external field H by the formula:

$$h = H \frac{\chi_{app.}}{\chi} .$$

In the following I give only the values of the true susceptibility and the field really acting, unless otherwise indicated.

(8) The susceptibility in an interval of 2.5° above the Curie point θ_f: Table II and Figs. 4 and 4A show a few values of χ as a function of h.

Theory [4] indicates that for a sufficiently small magnetic field, $1/\chi$ is of the form

$$\frac{1}{\chi} = \frac{1}{\chi_0} + ch^2 . \quad (1)$$

Such a variation is incompatible with the experimental results, since even with a plot against h, the curves are concave towards the field axis. It is, therefore, impossible to define an initial susceptibility χ_0.

Except for very low fields, the curve for nickel M, at 359.14°, is identical to the curve for nickel H at 360.24°, which indicates that the difference between the Curie points of these two substances is approximately 1.10°.

(9) The variation of the susceptibility with magnetic field for temperatures more than 2.5° above the Curie point: The problem becomes clearer. This time I have plotted $1/\chi$ as a function of h^2. One sees that for nickel H, in fields greater than 300 gauss, a relation of the form (1) is verified; for nickel M it is only verified for fields greater than 600 gauss. For fields less than these limits, the susceptibility takes abnormally large values, particularly large as the field becomes smaller. All things being equal however, the effect is more pronounced in nickel M.

This phenomenon is new.

The normal, regular phenomenon is certainly the dependence given by equation (1), on the one hand, because of the theoretical considerations which support this and on the other hand, because the coefficients c, considered as functions of χ_0, are the same for the two samples of nickel, while the initial curve, different in the two cases, looks like a parasitic phenomenon.

One should undoubtedly assume that, in spite of the prolonged annealings, the substances are not homogeneous. A few thousandths of the mass of the sample, with a Curie point 5° higher than the rest, would be sufficient to produce similar effects. I thus attribute the initial curvature to a sort of *spread* of the Curie point. The origin is without doubt complex, nevertheless the impurities must play a role, since it is the nickel M, the more impure, which shows the largest anomalies.

Very close to the Curie point, i.e. within the spread,

EXPERIMENTAL WORK

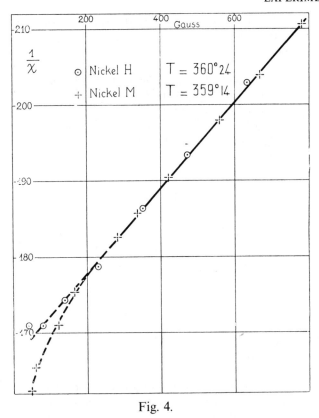

Fig. 4.

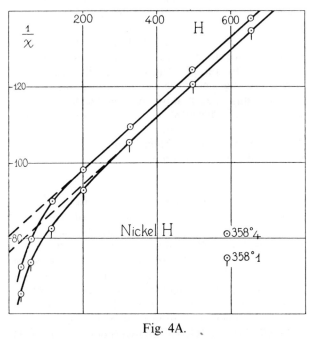

Fig. 4A.

TABLE III.
Nickel M.

T	h	$1/\chi$	$h^2 \times 10^{-3}$
360°540	293,1	276,4	86
	444,6	282,1	198
	588,2	285,9	346
	708,0	290,3	501
	832,4	294,1	693
360°640	53,8	264,9	3
	68,1	270,3	5
	103,7	275,0	11
	132,0	277,4	17
	181,9	279,2	33
	224,3	282,9	50
	303,3	286,8	92
	361,7	289,7	131
	437,5	291,2	209
	599,7	296,2	360
	786,8	301,8	619
361°340	132,9	323,6	18
	223,9	328,4	50
	363,3	338,0	132
	522,6	343,2	273
	663,3	347,1	440
	866,4	350,4	751

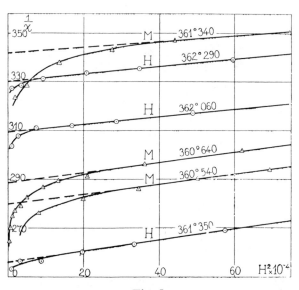

Fig. 5.

the effects of the mixture completely mask the ideal phenomena relating to a pure substance. This is why it is impossible, from this point of view, to take advantage of the results of the preceding paragraph.

Table V gives the determinations of c, and of $1/\chi_0$, in the temperature interval where these determinations are possible. This interval is however restricted, in the

TABLE IV.
Nickel H.

T	h	1/χ	$h^2 \times 10^{-3}$
361°350	102,8	253,3	11
	180,0	256,5	32
	294,0	256,7	86
	443,3	260,3	196
	578,2	263,5	334
	758,5	269,2	575
362°290	106,0	327,4	11
	184,3	328,6	34
	303,0	331,3	92
	453,7	333,7	206
	590,0	335,4	348
	772,0	339,2	596
362°060	95,1	303,8	9
	166,2	307,8	28
	271,7	311,2	74
	407,5	312,0	166
	535,3	313,7	287
	700,5	317,1	491

TABLE V.

Nickel	T	$c \times 10^6$	$\dfrac{1}{\chi_0}$	σ_0
M	360°640	21,4	288,4	68,6
	540	20,7	279,8	71,7
	361°340	11,8	341,8	78.6
	359°733	35,8	222,0	68,0 ?
	940	25,4	238,8	75,3 ?
	360°314	26,4	262,0	67,6 ?
	362°350	7,0	422,2	83,8 ?
	520	8,8	431,8	73,3 ?
	950	5,4	466,8	86,9 ?
H	362°060	15,4	309,8	75,5
	290	14,5	330,5	73,1
	361°350	23,4	255,8	73,5

limit of the magnetic fields employed, at low temperatures for the reason given above, and at high temperatures, because as c varies as the inverse of the square of the distance from the Curie point [4], the relative changes of the susceptibility, for the same change of field, vary as the inverse of the cube of the distance from the Curie point and thus decrease extremely quickly.

(10) The interpretation of the coefficient c: If one attributes the paramagnetism of nickel to carriers of angular momentum $S = \frac{1}{2}$, uniquely due to spin, I have shown [4] that c is related to σ_0, the total moment obtained on aligning the carriers parallel, by the equation:

$$c = \frac{\lambda}{\sigma_0^2} \frac{\chi_0 + n}{\chi_0^2} \qquad (2)$$

n and λ being determined by the study of the dependence of the initial susceptibility upon temperature.

One finds $n = 2400$ and $\lambda = 3.11$ between 360° and 370°.

Equation (2) allows the calculation of σ_0 (the last column of Table V). For nickel M, owing to the importance of the perturbing term indicated in the previous paragraph, the values of σ_0 are less well grouped than for nickel H. Nevertheless, the mean value of σ_0 for nickel M: 74.8 is very close to the mean value of 74.0 found from the study of nickel H. The good agreement of these two determinations justifies the level of discussion adopted. The mean maximum magnetization used in these determinations of c was $\sigma = 2.5$.

These results, together with those provided by the study of nickel in strong fields, from the experiments of Weiss and Forrer [7], show that σ_0 is, in reality, a slowly varying function of the magnetization σ, which tends to $\sigma_0 = 77$, for $\sigma = 0$. This value leads to the attribution of a moment for the elementary carrier of $\mu = 5782$, very close to the Bohr magneton: $\mu_B = 5564$. The initial hypothesis put forward at the start of this paragraph is thus confirmed.

(11) The initial susceptibility as a function of the temperature: This study having already been accurately made over a large temperature range, I will not repeat it. I set out simply to determine in a temperature interval close to the Curie point, the shape of the variation.

I therefore give only a few experimental results obtained between 361° and 370° for nickel H, in an external magnetic field of 470 gauss. The values of $1/\chi$ have not been corrected for the demagnetizing field. These values were corrected to zero field using the values of c given by equation (2), with $\sigma_0 = 74.0$. At the lowest temperatures the correction is uncertain since the plot $1/\chi = f(h^2)$, only possesses a vanishing linear section. The values of $1/\chi_0$, thus determined, corrected

EXPERIMENTAL WORK

TABLE VI.

Nickel $H - \frac{1}{\chi}$ measured in $H = 370$ G, then corrected to 0 field.

T	$\frac{1}{\chi_{app.}}$	$\frac{1}{\chi_0}$	T	$\frac{1}{\chi_{app.}}$	$\frac{1}{\chi_0}$	T	$\frac{1}{\chi_{app.}}$	$\frac{1}{\chi_0}$
365°070	604,8	568,2	353°758	495,1	458,0	360°549	239,6	198,1
364°904	590,4	553,7	684	488,8	451,7	720	251,6	210,8
892	592,0	555,3	582	479,7	442,5	962	267,1	227,5
365°019	603,1	566,5	525	475,5	438,3	361°122	280,5	240,7
046	605,9	569,3	414	466,3	429,0	072	277,0	237,1
364°970	599,7	563,1	238	451,8	414,4	115	280,2	240,4
932	597,7	561,0	170	446,5	409,1	426	303,1	263,9
854	591,5	554,8	107	441,3	403,9	625	319,1	280,2
781	585,5	548,8	362°966	429,2	391,7	625	319,9	281,0
903	594,3	557,6	888	423,7	386,1	741	328,7	289,9
859	592,6	555,9	808	415,7	378,1	676	323,9	285,1
719	578,1	541,4	636	401,7	363,9	951	345,2	306,7
573	565,9	529,1	617	400,5	362,7	362°084	356,1	317,7
635	571,2	534,4	523	391,7	353,8	171	362,2	323,9
616	571,6	534,8	415	384,5	346,5	199	365,9	327,6
574	568,6	531,8	282	373,5	335,4	371	379,4	341,4
457	556,3	519,5				514	392,3	354,4
322	543,0	506,2	361°951	346,8	308,3	594	398,0	360,2
232	536,1	499,2	899	342,2	303,6	577	397,0	359,2
233	535,4	498,5	761	331,2	292,5	994	430,8	393,3
349	542,8	506,0	762	330,6	291,9	363°018	433,2	395,7
290	538,6	501,7	712	327,3	288,6			
230	533,1	496,2	606	321,0	282,1	366°658	751,5	715,9
173	530,6	493,7	616	320,0	281,1	888	776,2	740,6
090	525,7	488,8	523	313,0	274,0	367°777	861,8	826,2
			418	305,1	265,9	368°282	915,1	879,5
364°125	527,1	490,2				303	915,4	879,8
179	532,3	495,4	359°993	202,8	159,5			
078	523,5	486,6	360°190	214,8	171,8			
363°934	510,6	473,6	350	227,3	185,1			
867	503,5	466,5	554	239,2	197,7			

for the demagnetizing field, are given in Table VI. They correspond to five distinct series of measurements.

Displayed graphically, these values lie on a regular curve. To elaborate this point, I have divided the experimental points, in order of increasing temperature, into eight groups, and I have calculated the mean of T and the mean of $1/\chi_0$ for each group[2] (Table VII). The fictitious point, with coordinates thus calculated, should be very close to a true point since in the interior of each group the curvature is almost negligible.

To represent the results graphically on a large scale, I have shown as a function of T, the difference:

$$\Delta = \frac{1}{\chi_0} - \frac{1}{\chi_1} \quad \text{with} \quad \frac{1}{\chi_1} = 81 \, (T - 358°140).$$

Fig. 6 shows that one thus obtains a completely regular curve, without any trace of linear sections. Therefore, even in weak fields, there is no straightening;

[2] In Table VII, I have indicated by n the number of points in each group.

TABLE VII.

n	T	$\dfrac{1}{\chi_0}$	Δ
6	360°393	187,16	+ 4,63
7	361°234	249,97	− 0,68
13	770	292,97	− 1,10
11	362°445	348,09	− 0,65
15	363°330	422,12	+ 1,69
13	364°256	501,06	+ 5,62
15	851	552,93	+ 9,30
5	367°581	808,42	+ 43,70

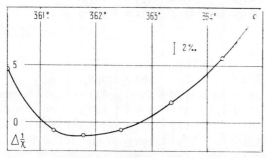

Fig. 6.

the *principle of uniformization* [6] does not apply. The continuous curvature which one observes is just that which allows one to predict the influence of fluctuations of the molecular field [7].

The following parabolic equation may be used to represent the experimental results between 360.4° and 367.8°, over an interval where the susceptibility changes by a factor of 4

$$\frac{1}{\chi_0} = 295{,}4 + 81\,(T - 361°8) + 1{,}25\,(T - 361°8)^2 \ . \quad (3)$$

The discrepancy between the experimental values and the calculated values from this formula, is of the order of a few thousandths, except for the lowest point (T = 360.393°) where the discrepancy reaches exceptionally 1.8%.

From equation (3) one calculates the temperature of the Curie point $\Theta = 357.92°$ and the tangent at the Curie point

$$\left(\frac{\partial T}{\partial \dfrac{1}{\chi}}\right)_\theta = 71{,}3 \ .$$

These values are not too certain since the extrapolation is made over an interval of more than two degrees, in which it is possible that the shape of the curve could change.

(12) A remark on the determination of the Curie point: The method for the determination of the Curie point, given above, leads, in the case of nickel H for example, to the use only of observations in fields greater than 300 gauss. It provides correctly a mean Curie point which is the centre of gravity of the Curie points of the different elements which constitute the substance. On the contrary, when one operates only in a weak magnetic field, the study of the curves gives a higher Curie point than the mean Curie point and if it were possible to experiment in an infinitely weak field, one would thus obtain the Curie point of the portions of the substance which had the highest Curie point.

In short, a correct and precise determination of the Curie point necessitates a complete knowledge of the magnetic equation of state in its neighbourhood. Only under this condition is it possible to define it to a precision of 0°1.

(13) Conclusions: (1) The study of nickel in weak fields, in the neighbourhood of the Curie point, shows that $1/\chi$ varies in a regular manner with temperature, as does its first derivative. The principle of uniformization does not apply: there are no different magnetic states.

(2) The study of the magnetic properties very close to the Curie point, shows the probable existence of a spread of Curie points, of the order of a degree, the larger less pure the nickel.

(3) The values of $c = \dfrac{\partial 1/\chi}{\partial H^2}$, as a function of T agree with the determinations made from the experiments of Weiss and Forrer and confirm the attribution to the elementary carrier of the magnetism of nickel, of a moment equal to a Bohr magneton.

(4) The tangent to the curve $1/\chi = f(T)$ has, at the Curie point, the value 71.3.

REFERENCES
[1] G. Foex and R. Forrer, *J. Phys.*, 1926, **7**, 180.
[2] L. Néel, *Bull. Soc. Fr. Phys.*, 1933, **334**, 20.
[3] P. Weiss and R. Forrer, *Ann. de Phys.*, 1929, **12**, 324.
[4] L. Néel, *J. Phys.*, **1934**, **5**, 104.
[5] P. Weiss and R. Forrer, *Ann. de Phys.*, 1926, **5**, 153.
[6] P. Weiss, *J. Phys.*, 1911, **1**, 900; Congrès Int. Electr., 1932, **1**, rapp. 15; G. Foex, *J. Phys.*, 1931, **2**, 353.
[7] L. Néel, *Ann. de Phys.*, 1932, **17**, 5.

A36 (1938)

THE APPLICATION TO NICKEL OF A NEW METHOD FOR THE MEASUREMENT OF TRUE SPECIFIC HEATS

Note (1977): It was a matter of knowing whether the anomaly of the true specific heat of ferromagnets disappears suddenly above the Curie point, corresponding to the theory of P. Weiss, or whether, as predicted by the theory of molecular field fluctuations, one observes a spread.

The method of measurement used here comprises an original arrangement: the presence, between the sample and the calorimetric container, of an intermediate container which serves to collect and measure the heat losses from the sample.

The recent determinations of the true specific heat of nickel agree up to the Curie point, but, beyond they diverge notably. To resolve this disagreement, I have devised an electrical method which allows the study of a very pure (99.91%) *bulk sample*.

The nickel to be studied, a cylinder of approximately 100 g., is placed, in vacuum, inside a thermostat at the temperature at which one wishes to measure the specific heat. One measures, with a thermocouple, the rise in temperature which corresponds to the quantity of heat given to the cylinder by an electrical resistor situated in its interior and through which one passes a current for a short time. The main difficulty is the precise determination of the heat losses between the heating period and the instant when, the temperature having become uniform, one measures its rise. The originality of the method consists of surrounding all parts of the cylinder by an isolated container, of known calorific capacity, of which one measures the rise in temperature; one has thus directly the heat lost by the cylinder, while the heat lost by the container only intervenes to the second order.

With a homogeneous couple, the precision is 2 or 3 thousandths, up to 500°C. Here, it was a little less, since I used very sensitive couples which were moderately homogeneous which allowed small rises in temperature (two degrees). The table and the figure give the measured specific heats, with constant accuracy, compared with the results of Lapp [1] and of Ahrens [2]. The sudden decrease of the specific heat begins at $357.5° \pm 1°$, while the Curie point of the same nickel has been found [3] to be $357.9°$. There is a *spread* of the specific heat above the Curie point, the results of Ahrens are confirmed. The lower Curie point which he found ($351°$) comes from the impurities in his sample (0.6%).

t	c	t	c	t	c
390,2	0,1313	91,7	0,1129	18,2	0,1060
376,1	0,1332	156,4	0,1202	410,6	0,1305
368,6	0,1341	225,4	0,1289	425,8	0,1305
363,6	0,1375	303,2	0,1385	14,1	0,1052
358,0	0,1556	361,3	0,1404	19,7	0,1057
354,3	0,1539	365,2	0,1383	35,0	0,1074
346,3	0,1487	371,1	0,1342	456,8	0,1297
334,7	0,1449	378,1	0,1337	457,2	0,1301
262,3	0,1314	211,6	0,1268	487,7	0,1303
192,2	0,1238	368,5	0,1322	540,1	0,1307
120,8	0,1154	395,2	0,1315		
60,9	0,1096	13,9	0,1057		

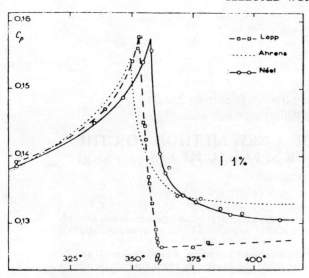

REFERENCES
[1] Lapp, *Ann. de Phys.*, 1929, **12**, 442.
[2] Ahrens, *Ann. der Physik*, 1934, **21**, 169.
[3] Néel, *J. Phys. Rad.*, 1935, **6**, 27.

INDEX

after-effects
  diffusion 343, 369-84
    irreversible 356, 365-8
    in Rayleigh region 343-55
    reversible 356, 358-65
    in small particles 407-27
alloys
  magnetic properties 125-6, 128-9, 133-5
  molecular field fluctuations 19-22
  neutron irradiation of iron-nickel 403-4
  paramagnetic properties 499-500
  saturation magnetic moments 52
  superposed paramagnetism of nickel 132-3
Alnicos, treated 396-7
alternating fields
  apparent susceptibility and 267-9
  diffusion after-effects 379-84
aluminium 430
anhysteretic magnetization 260-5
anisotropy
  coercive force and 304
  magnet steels 306-7
  neutron irradiation of iron-nickel alloys 403-4
  superparamagnetism 109-10
  surface magnetic 387-93
    and orientational superlattices 393-402
anisotropy constant 170-6
anomalous after-effect 379
antiferromagnetism 105
  coupled with ferromagnetism 469-87
  ferrites 72-5, 86-7
  hysteresis 469-76
  local molecular field 58
  superparamagnetism and 107-13
  see also constant paramagnetism;
    superantiferromagnetism
apparent Curie constant 22
apparent susceptibility 267-9
approach law 161-78
  anisotropy constant and 170-6
  magnetostriction 177-8
  materials with cavities 164-9
  porous iron 163-4
asymptotic Curie point 66-7

baked clays 407-27
ballistic wattmeter 494
band shapes and magnetic properties 46-7

Bloch wall energies 234-7
blocking temperature 439

cadmium ferrite 77, 81-2
carbon, diffusion after-effect and 369-84
cavities
  approach law 163, 164-9
  coercive force 274-85, 293, 302-3
  see also inclusions
Chikazumi effect 401
chromium 51, 501-8
clays, baked 407-27
closure domains 195-6
cobalt 50
  true Curie constant 156-7
cobalt ferrite 397
coercive field
  magnetic hardness and 169
  multiple domains and 252-8
coercive force 286-303
  cavities, inclusions and 274-85
  ferromagnetic powders 304
constant paramagnetism
  interaction energies 131-5
  manganese 105-6
  molecular field fluctuations 22-5
  see also antiferromagnetism
constant size anisotropy 390-1
copper 49
copper ferrite 79-80, 81, 89-90
creep
  coupled domains 333-5
  in hysteresis loops 318-21
  random with random thermal fluctuations 336-9
critical diameters 206-7
crystallites, quadratic 208-10
crystals
  interaction energies 130-2
  magnetocrystalline energy 208-10
  molecular field theory 54-5
cubic iron sesquioxide 82-4
Curie point(s)
  ferrites 66-7, 76, 77, 78, 97-8
  molecular field fluctuations 3-4, 15-16, 21-2, 66-7
  nickel's magnetic properties near 509-16
  pressure and 150
  true 155-8

INDEX

see also ferromagnetic Curie point; paramagnetic Curie point

demagnetizing field and anhysteretic magnetization 260-5

diffusion after-effect 343, 369-84
   see also reversible after-effect
domains 54, 181-2
   antiferromagnets 74-5
   displacements of isolated walls 241-51
   grains of single 435-43
      critical diameter 206-7
   hard materials 308-12
   high frequency permeability and 211-18
   interactions between 326-35
   monocrystalline iron 182-94
      secondary surface structure 194-202
   multiple and coercive field 252-8
   walls of ferromagnetic 221-30
   see also Bloch wall energy; coercive force

elastic energy 139-40
elementary domains see domains
electrons
   paramagnetism of nickel 44-5
   paramagnetism in a rectangular band 46-7
energy, elastic isotropic deformation and 139-41
energy molecular field 56
exchange energy 223-4
expansion
   anomalies of ferromagnets 138-45
      interaction energy curve 145-50
   theory of 141-2

ferric oxide, rhombohedral 434
ferrimagnetism
   ferrites 58, 66-72, 93-6
   rocks 429-35
ferrites 63-91
   antiferromagnetism 72-6
   ferrimagnetism 58, 66-72
   interactions 85-90
   rare earth 99-100
   saturation magnetization 93-6
   thermal expansion and Curie constant 97-8
ferromagnetic Curie point 15-16, 21-2, 76, 77, 78
   see also Curie point(s)
ferromagnetic materials
   fine-grained cubic 206-7
   powders 304, 415
   thin layers 459-62
      coupled 463-6
      coupled with antiferromagnetic films 469-87
ferromagnetism
   domain walls 221-30
   ferrites 75
   interacting domains 326-35
   interacting grains 322-5
   molecular field fluctuations 14-18
   moment and molecular field 138-58
      expansion anomalies 138-45
      interaction energy curve 145-50
      temperature variation of molecular field and paramagnetism 151-8
   quantum mechanics treatment 38-9
   random magnetic fields and 315-17
ferronickels 401-2, 403-4
fluctuation after-effect 343-55, 384
   see also irreversible after-effect
fluctuations in the molecular field
   alloys 19-22
   constant paramagnetism 22-5
   ferromagnetism 14-18
   initial susceptibility 11-14
   internal energy and specific heat 18-19
   linear lattices 5-11
   nickel 41-2
   semi-classical model and 27-38
free anisotropy 390, 391

gadolinium 157-8
garnets, rare earth 101-2
gases 43

hardness, magnetic 161-9
heat treatments 80-1
hematite 434-5
hysteresis
   antiferromagnetic 469-76
   creep 318-21
   interacting ferromagnetic grains 322-5
   loss by 486-7
   tilting effect 326-8

ilmenite-hematite series 434-5
inclusions 274-85, 287, 293, 302-3
   see also cavities
indirect bonds 88-90
initial susceptibility 11-14
   nickel 29-31
interaction energies 125-35, 145-58
interionic equilibrium 48-51
internal energy and specific heat 18-19
internal stresses 287, 293-302
inverse ferrites 64, 82
iron 50
   approach to saturation of porous 163-4
   Curie constant 157
   inclusions in 283, 302-3

INDEX

interaction energy and magnetic shell diameter 133-4, 149-50
magnetization of monocrystalline 181-202
   elementary domains 191-4
   magnetization modes 182-91
   secondary surface structure 194-202
particle diameter for powders 415
iron-nickel alloys 401-2, 403-4
iron sesquioxide, cubic 82-4
iron-silicon alloys 499-500
iron-tin alloys 499-500
irradiation of iron-nickel with neutrons 403-4
irreversible after-effect 356, 365-8
see also fluctuation after-effect
isothermal remanent magnetization (IRM) 408-9, 418-24, 424-5

joulemeter 493

lattices, linear 5-11
lead ferrite 81
local molecular field 57-9

magnesium ferrite 81
magnetic energy 140-1
magnetic hardness 161-9
magnetic orbitals 133-5
magnetic viscosity 450-4
magnetite 82-4, 86
magnetization process 242
   anhysteretic 260-5
   monocrystalline iron 181-202
   pyrrhotite 203-5
   see also Rayleigh's laws; saturation
magnetocrystalline anisotropy 390-1
magnetocrystalline energy 208-10, 222-3
magnetons 54
magnetostatic coupling 463-6
magnetostriction
   approach to saturation 177-8
   orientational superlattices 397-8
   surface magnetic anisotropy 388-93
magnets
   manufacture of permanent 273
   properties of hard materials 308-12
   steels for 259, 306-7
manganese 50
   constant paramagnetism 105-6
   interaction energy and magnetic orbitals 134
   magnetic properties 501-8
manganese antimonide 84-5
molecular field
   approximation 65-90
   energy 56
   fluctuations see fluctuations

local 57-9
   temperature variation and paramagnetism 151-8
   to correct the equation of state 56
   Weiss 53-7, 58-9
molybdenum 135
monocrystals 182-91

neutron irradiation of iron-nickel 403-4
nickel
   alloys' superposed paramagnetism 132-3
   approach law 172-3
   electrons and paramagnetism of 44-5
   elementary carriers 42
   expansion anomaly 143-5
   inclusions in 283
   interionic equilibrium 49-51
   internal stresses 302
   magnetic equation of state 26-42
   magnetic properties near Curie point 509-16
   true Curie constant 155-6
   true specific heat 517-18
   see also iron-nickel alloys
nickel-zinc ferrites 95-6
normal after-effect 378-9

orbitals, magnetic 133-5
ordering Curie point see paramagnetic Curie point
orientational superlattices
   mechanical deformations 398-402
   spontaneous magnetization 393-8

palladium 49-50
paramagnetic Curie point 15-16, 67, 76, 77, 78
   rare earth elements 136-7
   see also Curie point(s)
paramagnetism 3, 75, 76
   chromium 501-8
   constant see constant paramagnetism; antiferromagnetism
   electrons in a rectangular band 46-7
   ferrites 76-9, 88
   manganese 105-6, 501-8
   nickel 44-5
   temperature variation of molecular field 151-8
   see also superparamagnetism
particles
   magnetic after-effect
      ensembles of 416-27
      isolated 409-12
      relaxation times 412-16
   single domain 206-7, 435-43
   superantiferromagnetism 114-21
   superparamagnetism and antiferromagnetism 107-13
Permalloys 395-6
permeability, high frequency 211-18

INDEX

platinum 50
polycrystalline materials 170-6
potentiometer, high precision 491-2
powder patterns 191-2, 197-8, 200-1
powders, ferromagnetic 304, 415
pyrrhotite 54-5, 203-5

quadratic crystallites 208-10
quantum mechanics 38-41

random fields
    creep and thermal fluctuations 336-9
    magnetic and ferromagnetism 315-17
rare earth elements 136-7
rare earth ferrites 99-100
rare earth garnets 101-2
Rayleigh's laws 241-2
    isolated wall displacements 241-51
    magnet steels 259
    magnetic after-effects and 344-5
    multiple domains and coercive field 252-8
relaxation time 410-11
    of a particle 412-16
remanence patterns 191-3
reversible after-effect 356, 358-65
    see also diffusion after-effect
rhombohedral ferric oxide 434
rocks 428-55
    ferrimagnetism 429-35
    laarge grains 443-55
    single domain particles 435-43

saturation magnetization
    approach to see approach law
    ferrites 93-6
silicon see iron-silicon alloys
single domain particles 435-43
    critical diameter 206-7
specific heat
    internal energy and 18-19
    nickel 37-8, 517-18

    true 495, 517-18
spinels 64, 429-30
steels, magnet
    anisotropy 306-7
    Rayleigh's laws 259
stresses, internal 287, 293-302
superantiferromagnetism 114-16
    temperature dependence 118-21
superexchange 88
superparamagnetism 107-13
surface magnetic anisotropy 387-93
    orientational superlattices 393-402
surface tension 216-18
susceptibility, apparent 267-9

temperature
    superantiferromagnetism and 118-21
    variation of molecular field and paramagnetism 151-8
thermoremanent magnetization (TRM)
    baked clays 408-9, 424-6
    large grains 454-5
    single domain particles 437-43
thin films
    Bloch wall energy 232-3
    ferro-antiferromagnetic coupling 469-87
    ferromagnetic 459-66
tilting 326-8, 331-3
tin see iron-tin alloys
titanium 51, 431
transition elements 50-1, 52
tungsten 51

vanadium 51
viscosity, magnetic 450-4

wattmeter, ballistic 494
Weiss molecular field 53-7, 58-9

zinc ferrite 81-2
    see also nickel-zinc ferrites